General Equilibrium and Welfare Economics

James C. Moore

General Equilibrium and Welfare Economics

An Introduction

With 40 Figures and 11 Tables

Springer

Professor James C. Moore
Purdue University
Department of Economics
100 S. Grant Street
West Lafayette, IN 47907-2076
USA
moorejc@purdue.edu

ISBN-10 3-540-31407-5 Springer Berlin Heidelberg New York
ISBN-13 978-3-540-31407-3 Springer Berlin Heidelberg New York

Cataloging-in-Publication Data
Library of Congress Control Number: 2006932262

Springer is a part of Springer Science+Business Media

springeronline.com

Hardcover-Design: WMXDesign GmbH, Heidelberg

SPIN 11660033 42/3153-5 4 3 2 1 0 – Printed on acid-free paper

To Donna, Donovan, Brian, Jerry, Linda, Ted, Julie, and Bradley,

...and to the University of Minnesota Faculty who taught me General Equilibrium and Welfare Economics:

John Chipman, Leo Hurwicz, Ket Richter, and Hugo Sonnenschein,

this book is most affectionately, and respectfully, dedicated.

Preface

This book is intended as a graduate- (or perhaps, advanced undergraduate-) level textbook in general equilibrium and welfare economics. General equilibrium theory is, of course, at the very heart of our fledgling science of economics, and welfare economics provides the normative basis for all professional policy recommendations, as well as most applied work. In developing this text, I hope that I have not slighted the needs of the aspiring economic theorist, but at the same time, I have tried to take account of the fact that most of the students who have studied or will study this text will not go on to specialize in advanced theory. Consequently, I have attempted to include and concentrate upon that material which I believe would be most useful to students who will go on to specialize in, for example, international economics, public economics, or economic development. How well I have succeeded in this endeavor only time will tell.

This book has been developed from lecture notes and hand-outs which I have used over the past several years in the course, 'General Equilibrium and Welfare Economics,' (Economics 609) which I have taught at Purdue University. Before going further, however, let me quickly confess that I have never covered all of the material in this book in one semester. On the other hand, I have taught all of it at one time or another, so the whole book has been classroom-tested to some extent.

The course for which the book was written is the second semester of microtheory required of students in the first year of our PhD program. Consequently, I have written the book assuming that the reader is familiar with, say, the partial equilibrium portion of Mas-Colell, Whinston, and Green [1985], which is used as the text in the first semester of our microtheory sequence. I also assume that the reader has the usual mathematical background required of a first-year graduate student in economics: competence in calculus, and some background in Linear Algebra, as well as familiarity with the elementary concepts of set theory: membership, union, intersection, and set-theoretic difference. I do not often use game theory in any very essential way in this work, but the reader should be familiar with the definitions of Nash equilibrium and the core. I have included a glossary of the basic mathematical notation which is used in this book at the end of this preface.

I have included a number of exercises at the end of each chapter, and I would strongly recommend that a student who is encountering this material for the first time work through as many of these problems as her or his schedule permits. In Chapter 19 I have also included solutions for a number of these problems, but I hope that it goes without saying that a student should make every effort to work through a problem on her or his own before consulting Chapter 19 for its solution!

A number of people have contributed to this project in various ways, and I very much want to express my gratitude for their help. In particular, Dan Kovenock, John Ledyard and Bill Novshek have read various parts of the manuscript, and have made a number of helpful comments thereon. Several research assistants have done yeoman work in trying to rid this manuscript of all the 'typo's' and other errors which I always manage to accumulate. I particularly want to thank Dan Nguyen, Jennifer Pate Offenberg, Daniela Puzzello, and Brian Roberson, who have gone 'above and beyond' the usual requirements of a research assistant in helping to clean up this manuscript. Thanks are also due Paola Boel and Curtis Price for their help in this regard, as well as to my secretary, Karen Angstadt, who has handled the various organizational chores which I have inflicted upon her with her usual efficiency and dispatch. In addition, of course, several 'generations' of graduate students in our economics program have endured assignments in, and lectures oriented toward this material with no (or little) complaint.

I would also like to thank my colleagues in the economics group of the Krannert School here at Purdue, who have been remarkably tolerant of the death grip I have maintained on Economics 609 over the past several years. I would also like to thank Deans Rick Cosier and Bob Plante, who maintained an atmosphere which encourages scholarly work in a variety of dimensions and directions. Finally, of course, I must thank my wife, Donna, without whose tolerance and encouragement this book could not possibly have been written.

Mathematical Notation

I will use '$\mathbb{R}^n$' to denote n-dimensional Euclidean space, and I will use bold letters to denote elements therein (vectors). Thus, if $\boldsymbol{x} \in \mathbb{R}^n$, $\boldsymbol{x}$ is of the form:

$$\boldsymbol{x} = (x_1, \ldots, x_j, \ldots, x_n),$$

with 'x_j' denoting its jth coordinate. It will only very rarely make any difference whether we consider elements of $\mathbb{R}^n$ to be row or column vectors, but on those few occasions in which it does, I will take them to be column vectors, despite the fact that I will almost always write them as in the above equation (it does, after all, save a lot of space).

I use what seems to be the standard notation for vector inequalities on $\mathbb{R}^n$:

$$\begin{aligned} \boldsymbol{x} \geq \boldsymbol{y} &\iff x_i \geq y_i, \text{ for } i = 1, \ldots, n, \\ \boldsymbol{x} > \boldsymbol{y} &\iff \boldsymbol{x} \geq \boldsymbol{y} \;\&\; \boldsymbol{y} \ngeq \boldsymbol{x}, \text{ and} \\ \boldsymbol{x} \gg \boldsymbol{y} &\iff x_i > y_i, \text{ for } i = 1, \ldots, n. \end{aligned}$$

Making use of these inequalities, we define the:

$$\begin{aligned} \textbf{nonnegative orthant}\colon \mathbb{R}^n_+ &= \{\boldsymbol{x} \in \mathbb{R}^n \mid \boldsymbol{x} \geq \boldsymbol{0}\} \\ \textbf{semipositive orthant}\colon \mathbb{R}^n_+ \setminus \{\boldsymbol{0}\} &= \{\boldsymbol{x} \in \mathbb{R}^n \mid \boldsymbol{x} > \boldsymbol{0}\}, \text{and} \\ \textbf{strictly positive orthant}\colon \mathbb{R}^n_{++} &= \{\boldsymbol{x} \in \mathbb{R}^n \mid \boldsymbol{x} \gg \boldsymbol{0}\}, \end{aligned}$$

where '$\mathbf{0}$' denotes the origin in $\mathbb{R}^n$, and we use the symbol '$\setminus$' to denote set-theoretic difference; that is:

$$A \setminus B = \{x \in A \mid x \notin B\}.$$

Since we will often be considering ordered pairs, for example, $(\boldsymbol{p}, w) \in \mathbb{R}^n_{++} \times \mathbb{R}_+$, where $\boldsymbol{p} \in \mathbb{R}^n_{++}$ and $w \in \mathbb{R}_+$, and in general need to distinguish between the ordered pair $(x, y) \in \mathbb{R}^2$ and the open interval in $\mathbb{R}$ bounded by x and y; we will use a somewhat unorthodox notation for intervals of real numbers, thus:

$$\begin{aligned}
[x, y] &= \{z \in \mathbb{R} \mid x \leq z \leq y\}, \\
[x, y[&= \{z \in \mathbb{R} \mid x \leq z < y\} \\
]x, y] &= \{z \in \mathbb{R} \mid x < z \leq y\}, \text{ and} \\
]x, y[&= \{z \in \mathbb{R} \mid x < z < y\}.
\end{aligned}$$

Incidentally, in the above material I have made use of the notation '$\boldsymbol{x} \not\geq \boldsymbol{y}$,' to indicate that it is not the case that $\boldsymbol{x} \geq \boldsymbol{y}$, and whenever possible I will use a similar notation, a diagonal line through a symbol, to denote the negation of the relation indicated. Unfortunately, the limitations on the symbols available to me in the typesetting program will mean that I can't always do this. Thus, for example, we will often use the notation '$\boldsymbol{x}G\boldsymbol{y}$' to mean that a consumer considers the commodity bundle $\boldsymbol{x}$ to be at least as good as $\boldsymbol{y}$. However, we will have to use the notation '$\neg\boldsymbol{x}G\boldsymbol{y}$' to indicate the opposite situation (the negation); that is, to indicate that the consumer does *not* consider $\boldsymbol{x}$ to be at least as good as $\boldsymbol{y}$.

I will make fairly extensive use of universal and existential quantifiers. Thus we might write, assuming that A and B are sets of real numbers:

$$(\forall x \in A)(\exists y \in B): y \geq x;$$

which is read verbally as, "for every x in the set A, there exists an element, y, in the set B such that y is at least as great as x." In general, the end of a string of quantifiers will be indicated by a colon (:), and you should be careful to take note of the order in which the quantifiers occur. Thus, for example, the statement:

$$(\forall x \in \mathbb{R})(\exists y \in \mathbb{R}): y > x,$$

is true, whereas the statement:

$$(\exists y \in \mathbb{R})(\forall x \in \mathbb{R}): y > x,$$

is not! If you have not been introduced to this notation previously, it may be quite intimidating at first; but I think that you will quickly find that its use is very advantageous in stating complicated conditions. In fact, you might begin to convince yourself of this by comparing the equation in which I introduced this notation with the verbal interpretation which follows it.

W. Lafayette, IN J. C. M.
June, 2006

Contents

Preface vii

1 An Introduction to Preference Theory 1
1.1 Introduction 1
1.2 Binary Relations and Orderings 2
1.3 Preference Relations and Utility Functions 11

2 Algebraic Choice Theory 21
2.1 Introduction 21
2.2 The General Algebraic Theory of Choice 22
2.3 Some Criticisms of the Model 25
2.4 Stated Preferences versus Actual Choices 28
2.5 The Specification of the Primitive Terms 30
2.6 Weak Separability of Preferences 33
2.7 Additive Separability 39
2.8 Sequential Consumption Plans 41
2.9 The BPL Experiment Reconsidered 44
2.10 Probabilistic Theories of Choice 45
2.11 Are Preferences Total? 47
2.12 Are Preferences Transitive? 50
2.12.1 'Just Noticeable Difference,' or 'Threshold Effects' 50
2.12.2 Decision Rules Based On Qualitative Information 51
2.12.3 Priorities and Measurement Errors 52
2.12.4 Group Decisions: The Dr. Jekyll and Ms. Jekyll Problem 53
2.13 Asymmetric Orders 54

3 Revealed Preference Theory 59
3.1 Introduction 59
3.2 Choice Correspondences and Binary Relations 59
3.3 Regular-Rational Choice Correspondences 64
3.4 Representable Choice Correspondences 67
3.5 Preferences and Observed Demand Behavior 70
3.6 The Implications of Asymmetric Orders* 77

4 Consumer Demand Theory **85**
4.1 Introduction 85
4.2 The Consumption Set 85
4.3 Demand Correspondences 88
4.4 The Budget Balance Condition 90
4.5 Some Convexity Conditions 94
4.6 Wold's Theorem 96
4.7 Indirect Preferences and Indirect Utility 97
4.8 Homothetic Preferences 104
4.9 Cost-of-Living Indices 108
4.10 Consumer's Surplus 111
4.11 Appendix 125

5 Pure Exchange Economies **131**
5.1 Introduction 131
5.2 The Basic Framework 131
5.3 The Edgeworth Box Diagram 133
5.4 Demand and Excess Demand Correspondences 138
5.5 Pareto Efficiency 142
5.6 Pareto Efficiency and 'Non-Wastefulness' 150

6 Production Theory **155**
6.1 Introduction 155
6.2 Basic Concepts of Production Theory 155
6.3 Linear Production Sets 161
6.4 Input-Output Analysis 166
6.5 Profit Maximization 171
6.6 Profit Maximizing with Constant Returns to Scale* 176
6.7 Production in General Equilibrium Theory 178
6.8 Activity Analysis* 182

7 Fundamental Welfare Theorems **191**
7.1 Introduction 191
7.2 Competitive Equilibrium with Production 191
7.3 Some Diagrammatic Techniques 196
7.4 Walras' Law with Production 201
7.5 The 'First Fundamental Theorem' 205
7.6 'Unbiasedness' of the Competitive Mechanism 211
7.7 A Stronger Version of 'The Second Theorem' 219

8 The Existence of Competitive Equilibrium **227**
8.1 Introduction 227
8.2 Examples, Part 1 229
8.3 Assumption (c) and the Attainable Set 235
8.4 The Gale and Mas-Colell Theorem 239
8.5 An (Especially) Simple Existence Theorem 241

8.6 Appendix 244

9 Examples of General Equilibrium Analyses **249**
9.1 Introduction 249
9.2 Optimal Commodity Taxation: Initial Formulation 249
9.3 A Reconsideration of the Problem 252
9.4 The Simplest Model of Optimal Commodity Taxation 255
9.5 Some Results 256
9.6 Optimal Income Taxation 259
9.7 Monopoly in a General Equilibrium Model 269
9.8 Money in a General Equilibrium Model 272
9.9 Indivisible Commodities 276

10 Comparative Statics and Stability **281**
10.1 Introduction 281
10.2 Aggregate Excess Demand 282
10.3 The 'Law of Demand' 286
10.4 Gross Substitutes 291
10.5 Qualitative Economics 294
10.6 Stability in a Single Market 298
10.7 Multi-Market Stability 302
10.8 A Note on Non-Tâtonnement Processes 307

11 The Core of an Economy **311**
11.1 Introduction 311
11.2 Convexity and the Attainable Consumption Set 314
11.3 The Core of a Production Economy 317
11.4 The Core in Replicated Economies 320
11.5 Equal Treatment 329
11.6 Appendix 330

12 General Equilibrium with Uncertainty **333**
12.1 Introduction 333
12.2 Arrow-Debreu Contingent Commodities 333
12.3 Radner Equilibrium 339
12.4 Complete Markets 345
12.5 Complete Markets and Efficiency 350
12.6 Concluding Notes 355

13 Further Topics **359**
13.1 Introduction 359
13.2 Time in the Basic Model 359
13.3 An Infinite Time Horizon 366
13.4 Overlapping Generations 369
13.5 A Continuum of Traders 372
13.6 Suggestions for Further Reading 379

14 Social Choice and Voting Rules **383**
14.1 Introduction . . . 383
14.2 The Basic Setting . . . 384
14.3 Voting Rules . . . 387
14.4 Arrow's General Possibility Theorem . . . 392
14.5 Appendix. A More Sophisticated Borda Count . . . 402

15 Some Tools of Applied Welfare Analysis **407**
15.1 Introduction . . . 407
15.2 The Framework . . . 408
15.3 Measurement Functions . . . 409
15.4 Social Preference Functions . . . 411
15.5 The Compensation Principle . . . 416
15.6 Indirect Preferences: Individual and Social . . . 420
15.7 Measures of Real National Income . . . 422
15.8 Consumers' Surplus . . . 428

16 Public Goods **437**
16.1 Introduction . . . 437
16.2 A Simple Model . . . 437
16.3 Public Goods . . . 441
16.4 A Simple Public Goods Model . . . 442
16.5 Lindahl and Ratio Equilibria . . . 446
16.6 The 'Fundamental Theorems' for Lindahl Equilibria . . . 455
16.6.1 The 'First Fundamental Theorem' . . . 456
16.6.2 The 'Second Fundamental Theorem' . . . 458
16.6.3 The 'Metatheorem' . . . 462

17 Externalities **467**
17.1 Introduction . . . 467
17.2 Externalities: A First Look . . . 468
17.3 Extending the Model . . . 475
17.4 The 'Coase Theorem' . . . 480
17.5 Lindahl and Externalities . . . 484
17.6 Postscript . . . 487

18 Incentives and Implementation Theory **489**
18.1 Introduction . . . 489
18.2 Game Forms and Mechanisms . . . 490
18.3 The Gibbard-Satterthwaite Theorem . . . 495
18.4 Implementation Theory . . . 502
18.5 Single-Peaked Preferences and Dominant Strategies . . . 504
18.5.1 Single-Peaked Preferences . . . 504
18.5.2 The Bowen Model . . . 509
18.6 Quasi-Linearity and Dominant Strategies . . . 511
18.7 Implementation in Nash Equilibria . . . 520

18.8 Nash Implementation with Public Goods . . . 522
18.9 The Revelation Principle Reconsidered . . . 525
18.10 Notes and Suggestions for Further Reading . . . 527

19 Appendix. Solutions for Selected Exercises **531**
19.1 Chapter 1 . . . 531
19.2 Chapter 2 . . . 533
19.3 Chapter 3 . . . 533
19.4 Chapter 4 . . . 536
19.5 Chapter 5 . . . 537
19.6 Chapter 6 . . . 540
19.7 Chapter 7 . . . 542
19.8 Chapter 8 . . . 543
19.9 Chapter 9 . . . 548
19.10 Chapter 10 . . . 548
19.11 Chapter 11 . . . 548
19.12 Chapter 12 . . . 550
19.13 Chapter 13 . . . 551
19.14 Chapter 14 . . . 551
19.15 Chapter 15 . . . 551
19.16 Chapter 16 . . . 551
19.17 Chapter 17 . . . 552
19.18 Chapter 18 . . . 553

References **555**

Author Index **569**

Subject Index **573**

Chapter 1

An Introduction to Preference Theory

1.1 Introduction

Choice, or more precisely, choice under constraint, is central to economic theory. Choice theory is the foundation of the economic theory of demand, of welfare and public economics, and is crucially important in decision and game theory. Since so many of these topics are critical parts of this course, it is only appropriate that we begin our study by investigating the foundations of choice theory itself; namely abstract preference theory.

For us, of course, the most important single application of choice theory, is to consumer demand theory. In its most basic general equilibrium form, this theory postulates that we can think of consumers as making choices of '**commodity bundles**,' which for us will be vectors $\boldsymbol{x} = (x_1, \ldots, x_j, \ldots, x_n) \in \mathbb{R}^n$, where the jth coordinate of $\boldsymbol{x}$, x_j, denotes the quantity of the jth commodity available for consumption. We suppose further that, irrespective of prices and income or wealth, the consumer's choice is constrained (presumably by physiological and/or technological requirements) to some subset, X of $\mathbb{R}^n$. It is also usual to suppose that the consumer's choice of commodity bundles in X is consistent with the consumer's preferences over the set X; which preferences are modeled as a binary relation over X. This, of course, leads very naturally into the next section, which is concerned with beginning our investigation of binary relations in the abstract.

You are probably already familiar with the fact that binary relations are used as an abstract representation of consumers' preference relations in economic theory. What you may not be aware of is that the theory of binary relations is also central to welfare economics, and to index number theory, as well as to a number of other applications in economic theory. Consequently, we will devote a considerable amount of time to the study of binary relations in the abstract. This will represent a bit of 'overkill,' insofar as consumer demand theory is concerned, but we will be developing the theoretical foundations for much of our work in welfare economics as well as for the theory of consumer behavior.

1.2 Binary Relations and Orderings

Whether or not you have encountered a formal definition of a 'binary relation,' you certainly have encountered examples of such before this. The weak and strong inequalities for the real number system, for instance, are both examples of binary relations. Informally, a **binary relation**, R, on a set X, is simply a rule such that for each x and y in X, we can determine whether xRy, yRx, or neither, or both. Thus, for example, for any non-empty set, X, we can define the relation E (for equality) by:

$$xEy \iff x = y.$$

Another example is the relation G defined on $\mathbb{R}$ by:

$$yGx \iff y \geq x^2.$$

Notice that this last example is a special case of the following. Suppose $f\colon \mathbb{R} \to \mathbb{R}$, and define the relation G on $\mathbb{R}$ by:

$$yGx \iff y \geq f(x).$$

In this section, we will consider the following properties of binary relations. *In the definition to follow, and throughout the remainder of this chapter, we shall suppose that the set on which the binary relation is defined is non-empty.*

1.1 Definition. Let G be a binary relation on a nonempty set X. We shall say that G is:

1. **total** iff:
$$(\forall x, y \in X)\colon xGy \text{ or } yGx \text{ or } x = y.$$
2. **reflexive** iff:
$$(\forall x \in X)\colon xGx.$$
3. **irreflexive** iff:
$$(\forall x \in X)\colon \neg xGx.$$
4. **symmetric** iff:
$$(\forall x, y \in X)\colon xGy \Rightarrow yGx.$$
5. **asymmetric** iff:
$$(\forall x, y \in X)\colon xGy \Rightarrow \neg yGx.$$
6. **antisymmetric** iff:
$$(\forall x, y \in X)\colon [xGy \;\&\; yGx] \Rightarrow x = y.$$
7. **transitive** iff:
$$(\forall x, y, z \in X)\colon [xGy \;\&\; yGz] \Rightarrow xGz.$$

The key here is the ∀ quantifiers

Notice that a number of the relations which appear to be negations of one another actually are not. For example, irreflexivity is *not* the negation of reflexivity; that is, if a relation is not reflexive, it is nonetheless not necessarily irreflexive, and conversely. Similarly, a relation which is not symmetric is not necessarily asymmetric; conversely, a relation may fail to satisfy asymmetry, yet not be symmetric.

$X=\{x,y,z\}$ $x \succsim x$, $y \not\succsim y$, $z \succsim z$ $\Rightarrow$ $\succsim$ is neither reflexive nor irreflexive

1.2 Examples/Exercises.

1. Let X be the set of all persons alive on earth at the present date, and define the relation R on X by:

$$xRy \iff x \text{ is the brother of } y;$$

that is, xRy if, and only if: (a) x is a male person, and (b) x and y have the same (pair of) parents.

Insofar as the normal English definition of the phrase 'is the brother of' is concerned, the relation is irreflexive; whereas, in the way we have defined it here, the relation is *not* irreflexive. How could you modify the definition in order to make it correspond more closely to normal English usage?

2. Let X be the set of all physical objects on the earth at the present time, and define the relation R on X by:

$$xRy \iff x \text{ has at least as much mass as } y.$$

Show that R is total, reflexive, and transitive. (This is something of a trick question, since it is really an empirical, and not a mathematical issue. In order to arrive at something which you can prove, assume that mass can be measured to any degree of accuracy that we choose.)

3. Consider the usual weak inequality relation, $\geq$, on the real numbers. Show that $\geq$ is total, reflexive, antisymmetric, and transitive. Incidentally, here is an example of a binary relation which is neither symmetric nor asymmetric.

4. Show that the usual strict inequality relation, $>$, on the real numbers is total, irreflexive, asymmetric (and thus antisymmetric, since asymmetry implies antisymmetry), and transitive.

5. Let $f\colon X \to \mathbb{R}$, where X is any nonempty set, and define E on X by:

$$xEy \iff f(x) = f(y).$$

Show that E is reflexive, symmetric, and transitive. □

Incidentally, before proceeding further with our discussion of binary relations, I should mention that my insistence on having the set X be nonempty in Definition 1.1 is, essentially, for one reason; namely, a binary relation on the empty set satisfies all of the conditions, 1–7, in Definition 1.1. Consequently, if we include binary relations on the empty set in our definitions, the relationships among the conditions defined in 1.1 become somewhat confused!

Most of the binary relations which we encounter in economic theory are orderings of one type or another, where we use the term **ordering** to mean any transitive binary relation. Before considering the types of orderings which we will study in connection with consumer preference relations, however, let's take a look at some orderings from mathematics which we will find particularly useful.

1.3 Definitions. For $\boldsymbol{x}, \boldsymbol{y} \in \mathbb{R}^n$, we define:

1. $\boldsymbol{x} \geq \boldsymbol{y}$ [read '$\boldsymbol{x}$ is greater than or equal to $\boldsymbol{y}$'] iff:

$$x_i \geq y_i \quad \text{for } i = 1, \ldots, n.$$

2. $\boldsymbol{x} > \boldsymbol{y}$ [read '$\boldsymbol{x}$ is semi-greater than $\boldsymbol{y}$'] iff $\boldsymbol{x} \geq \boldsymbol{y}$, but $\boldsymbol{y} \not\geq \boldsymbol{x}$.
3. $\boldsymbol{x} \gg \boldsymbol{y}$ [read '$\boldsymbol{x}$ is strictly greater than $\boldsymbol{y}$'] iff:

$$x_i > y_i \quad \text{for } i = 1, \ldots, n.$$

Notice that if $n = 1$, the distinction between $>$ and $\gg$ disappears. On the other hand, for $n \geq 2$, there is a real difference between the two; for example, in the case of $\mathbb{R}^3$, if we take:

$$\boldsymbol{x} = (1,1,1), \boldsymbol{y} = (1,2,0), \text{and } \boldsymbol{z} = (0,0,0),$$

we have:

$$\boldsymbol{x} \gg \boldsymbol{z}, \boldsymbol{y} > \boldsymbol{z}, \text{but } \neg(\boldsymbol{y} \gg \boldsymbol{z}).$$

The weak inequality relation for $\mathbb{R}^n$ is an example of a **partial order**; that is, it is reflexive, antisymmetric, and transitive. This is stated formally in Theorem 1.4, which follows. The proof of 1.4 is fairly easy, and will be left as an exercise (those of you who have not been through a proof of this result before, however, should be sure to try to work out a proof now).

1.4 Theorem. *The weak inequality ($\geq$) for $\mathbb{R}^n$ is a partial order (that is, it is reflexive, antisymmetric, and transitive). However, $\geq$ is not total for $n \geq 2$.*

1.5 Definitions. We shall say that $\boldsymbol{x} \in \mathbb{R}^n$ is:
1. **nonnegative** iff $\boldsymbol{x} \geq \boldsymbol{0}$,
2. **semi-positive** iff $\boldsymbol{x} > \boldsymbol{0}$, and
3. **(strictly) positive** iff $\boldsymbol{x} \gg \boldsymbol{0}$,

where '$\boldsymbol{0}$' denotes the origin in $\mathbb{R}^n$ in each of the above statements.

1.6 Definitions. We define $\mathbb{R}^n_+$, the **nonnegative orthant** in $\mathbb{R}^n$, as the set of all nonnegative vectors in $\mathbb{R}^n$; that is:

$$\mathbb{R}^n_+ = \{\boldsymbol{x} \in \mathbb{R}^n \mid \boldsymbol{x} \geq \boldsymbol{0}\};$$

and $\mathbb{R}^n_{++}$, the **strictly positive orthant** in $\mathbb{R}^n$, by:

$$\mathbb{R}^n_{++} = \{\boldsymbol{x} \in \mathbb{R}^n \mid \boldsymbol{x} \gg \boldsymbol{0}\}.$$

Be careful to note the distinction between the strictly positive orthant and the 'semi-positive orthant:'

$$\mathbb{R}^n_+ \setminus \{\boldsymbol{0}\} = \{\boldsymbol{x} \in \mathbb{R}^n \mid \boldsymbol{x} > \boldsymbol{0}\};$$

although in $\mathbb{R}$ (that is, in the case where $n = 1$), the two sets coincide.

In economic theory, it is quite usual to base consumer demand theory on the assumption that an individual consumer's (weak) preference relation over the set of commodity bundles, X, is a weak order; which we formally define as follows.

1.7 Definition. Let G be a binary relation on a set X. We shall say that G is a **weak order** (or that **G is a weak ordering of X**) iff G is total, reflexive, and transitive.

1.8 Examples/Exercises.

1. It follows immediately from Example 1.2.3–4 that the usual weak inequality, $\geq$, on the real numbers is a weak order, but that the strict inequality for the real numbers, $>$, is not. Since $\geq$ is antisymmetric, it is an example of a more restrictive type of order than a weak order, called a **linear order** (which, by definition, is a relation which is total, reflexive, antisymmetric, and transitive).

2. Let X be any non-empty set, and let f be a real-valued function defined on X. If we define the relation G on X by:

$$xGy \iff f(x) \geq f(y),$$

show that G is a weak order on X.

3. Is the weak inequality, $\geq$, a weak order of $\mathbb{R}^n$, for $n \geq 2$? □

1.9 Proposition. *Let G be a binary relation on a set X, and define P and I on X by:*

Any relation!

$$xPy \iff [xGy \ \& \ \neg yGx],$$

and

$$xIy \iff [xGy \ \& \ yGx],$$

respectively. Then P is asymmetric and irreflexive, and I is symmetric.

Proof. We will only prove that P is asymmetric; the proof that P is irreflexive is immediate, and the proof that I is symmetric will be left as an exercise. In all three cases the argument is almost so simple as to not need doing, but when one first encounters this sort of material, it is difficult to know exactly where to begin in constructing a proof of these facts. Consequently, we will illustrate.

Let x and y be arbitrary elements of X, and suppose that xPy. Then by definition of P, we have:

$$xGy \text{ and } \neg yGx. \tag{1.1}$$

But then we see that we cannot have yPx; because by definition of P, this would require that yGx and $\neg xGy$; and by (1.1), neither of these conditions holds. Thus, if xPy, we cannot have yPx as well, and it follows that P is asymmetric. □

From the proposition just established, we see that the terminology in the following is indeed justified.

1.10 Definitions. If G is a binary relation on X, we define:

1. the **asymmetric part of G**, P, by:

$$xPy \iff [xGy \text{ and } \neg yGx].$$

2. the **symmetric part of G**, I, by:

$$xIy \iff [xGy \ \& \ yGx].$$

If 'G' denotes a consumer's weak preference relation over the set of commodity bundles, X, then the asymmetric part of G, P, would clearly be interpretable as the consumer's strict preference relation, and the symmetric part, I, is the consumer's indifference relation. Under these assumptions, the indifference relation is an example of an equivalence relation, defined as follows.

1.11 Definition. If X is a non-empty set, and R is a binary relation on X, we shall say that R is an **equivalence relation on X** iff R is reflexive, symmetric, and transitive.

1.12 Examples/Exercises.

1. In the terminology just introduced, you were asked in Example 1.2.5 to show that the relation, E, defined there is an equivalence relation.

2. Let $X = \mathbb{R}_+$, and define the relation R on X by:

$$xRy \iff |x - y| < 1.$$

Is R an equivalence relation? Explain. Is R an equivalence relation if, instead of $X = \mathbb{R}_+$, we take X to be the set of nonnegative integers; that is, $X = \{0, 1, 2, \dots\}$?

3. Suppose we take X to be the set of people in this room, and define $f\colon X \to \mathbb{R}$ by:

$$f(x) = \text{the height of } x \text{ to the nearest inch}$$

(rounding up to $n+1$ if the exact height is $n.500\dots0\dots$). If we now define E on X by:

$$xEy \iff |f(x) - f(y)| < 1,$$

is E an equivalence relation? □

1.13 Theorem. *If G is a transitive binary relation, then:*

1. the asymmetric part of G, P, is irreflexive, asymmetric, and transitive.

2. the symmetric part of G, I, is symmetric and transitive;

3. for any w, x, y and z in X:

$$[wGx, xPy, \& \ yGz] \Rightarrow wPz.$$

and, if G is reflexive (as well as being transitive), then:

4. I is an equivalence relation.

We need Reflexivity to get an equivalence Relation idea.

Proof. We will only prove part 1 of the conclusion; leaving parts 2–4 as exercises.

It follows at once from Proposition 1.9 that P is irreflexive and asymmetric. To prove that P is transitive, let x, y and z be elements of X such that:

$$xPy \ \& \ yPz.$$

Then, from the definition of P, we have:

$$xGy \ \& \ \neg yGx, \tag{1.2}$$

and:

$$yGz \ \& \ \neg zGy. \tag{1.3}$$

From the first parts of (1.2) and (1.3), and the transitivity of G, we then have:

$$xGz. \tag{1.4}$$

I equivalence:

Reflex: $a \succsim a \ \forall a$

Suppose that we were to have zGx as well. Then from the first part of (1.3) and the transitivity of G, we would have yGx; which contradicts the second part of (1.2). Therefore we must have $\neg zGx$, and combining this with (1.4), we see that xPz. □

Notice that the assumptions of the above result do not require G to be a weak order; if G is a weak order, the asymmetric part satisfies somewhat stronger properties. In particular, in this case, P will satisfy the following, as we will prove shortly.

1.14 Definition. We shall say that a relation, P, on a set X is **negatively transitive** iff, for all $x, y, z \in X$, we have:
if xPz, then either xPy or yPz.

While the condition defining negative transitivity undoubtedly appears odd at first reading, notice that if P is a strict preference relation, what it says is the following. If x is preferred to z, and y is any other alternative, then if x is not preferred to y (so that, in the usual interpretation of preference, it must be true that y is at least as good as x), it must be the case that y is preferred to z.

1.15 Theorem. *If G is a weak order on X, and P and I are the asymmetric and symmetric parts of G, respectively, then for all $x, y \in X$:*
1. we have:

$$\neg xPy \iff yGx, \tag{1.5}$$

or, equivalently:

$$\neg yGx \iff xPy. \tag{1.6}$$

2. exactly one of the following conditions holds:

$$xPy,\ yPx,\ \text{or } xIy.$$

3. P is negatively transitive.

Proof. I will leave the proof of parts 1 and 2 as exercises. To prove part 3, suppose that xPz, but that $\neg xPy$. Then by (1.5), yGx; and, since xPz, it then follows from part 3 of Theorem 1.13 that yPz □

In our work thus far, we have generally been considering the properties which will be satisfied by the asymmetric part of a (usually reflexive) binary relation. Suppose we turn things around, and begin with an asymmetric binary relation which we use to define a reflexive relation, as follows.

1.16 Definition. Suppose P is a binary relation on X. We define the **negation of *P***, which we will denote by 'G,' by:

$$xGy \iff \neg yPx. \tag{1.7}$$

Why are we interested in the negation of a binary relation? Well, suppose we begin with the idea of a strict preference relation for a consumer; instead of first introducing the idea of a weak preference (or 'at least as good as') relation, and using it to define the strict preference relation. If we once again denote this strict preference relation by 'P,' then we can *define* the weak preference relation as the negation of P. What is the point of this? The basic reason is that many scholars have expressed doubts concerning the transitivity of the weak preference relation; and in any event, most economists and psychologists are much more comfortable in assuming that strict preference relations are transitive than they are assuming that the weak preference relation is transitive. We will discuss some of the reasons for this in the next chapter; in the meantime, let's investigate some of the properties of the negation of an asymmetric binary relation.[1]

1.17 Proposition. *If P is an asymmetric binary relation on X, then its negation, G, is total and reflexive. Moreover, P is the asymmetric part of G, and G is the* only *total and reflexive binary relation having P as its asymmetric part.*

Proof. To prove that G is total, suppose x and y are elements of X such that $\neg xGy$. Then it follows from (1.7) that we must have yPx; and, since P is asymmetric, it then follows that $\neg xPy$. Therefore yGx.

That G is reflexive follows immediately from the fact that an asymmetric relation is also irreflexive.[2]

To prove that P is the asymmetric part of G, let $x,\ x' \in X$. If we then have xGx' and $\neg x'Gx$, it is obvious from the definition of G that we have xPx'. Conversely, suppose xPx'. Then by definition of G, we must have $\neg x'Gx$. Moreover, since P is asymmetric, we also have $\neg x'Px$. Therefore, xGx', and we see that we have:

$$xGx' \ \&\ \neg x'Gx.$$

It now follows that P is the asymmetric part of G.

In order to establish uniqueness, suppose now that G^* is a total and reflexive relation having P as its asymmetric part, and let $x, x' \in X$ be arbitrary. If xG^*x', then, since P is the asymmetric part of G^*, it follows that $\neg x'Px$, and thus by definition of the negation that xGx'.

Conversely, suppose we have $\neg xG^*x'$. Then, since G^* is total and reflexive, we must also have $x'G^*x$. But then, since P is the asymmetric part of G^*, it follows that $x'Px$. Therefore, using the definition of the negation, we see that $\neg xGx'$; and we conclude that $G \equiv G^*$. □

While this last proposition establishes that there can be only one total and reflexive binary relation of which P is the asymmetric part, there may be other reflexive and possibly transitive binary relations of which P is the asymmetric part, *even if P is transitive, as well as asymmetric*. This is shown by the following example.

1.18 Example. Consider the 'semi-greater-than' relation on $\mathbb{R}^2$, defined by:

$$\boldsymbol{x} > \boldsymbol{x}' \iff [\boldsymbol{x} \geq \boldsymbol{x}' \ \&\ \boldsymbol{x}' \not\geq \boldsymbol{x}].$$

Obviously $>$ is the asymmetric part of $\geq$. On the other hand, we will show that G, the negation of $>$, is given by:

$$\boldsymbol{x}G\boldsymbol{x}' \iff \boldsymbol{x}' \not> \boldsymbol{x} \iff \big[\max\{x_1 - x_1', x_2 - x_2'\} > 0 \text{ or } \boldsymbol{x} = \boldsymbol{x}'\big].$$

[1] Strict preferences are asymmetric, by the very definition of the word 'prefer.'

[2] Therefore it also follows that G will be reflexive if P is simply irreflexive, and not necessarily asymmetric.

To prove the above statement, suppose first that:

$$\neg\big[\max\{x_1 - x_1', x_2 - x_2'\} > 0 \text{ or } \boldsymbol{x} = \boldsymbol{x}'\big].$$

Then:

$$\max\{x_1 - x_1', x_2 - x_2'\} \leq 0 \text{ and } \boldsymbol{x} \neq \boldsymbol{x}',$$

from which it follows that $\boldsymbol{x}' > \boldsymbol{x}$; and we conclude that:

$$\boldsymbol{x}' \not> \boldsymbol{x} \Rightarrow \big[\max\{x_1 - x_1', x_2 - x_2'\} > 0 \text{ or } \boldsymbol{x} = \boldsymbol{x}'\big].$$

Conversely, if:

$$\max\{x_1 - x_1', x_2 - x_2'\} > 0 \text{ or } \boldsymbol{x} = \boldsymbol{x}',$$

then it is apparent that $\boldsymbol{x}' \not> \boldsymbol{x}$.

Consequently, since G is different from the weak inequality on $\mathbb{R}^2$, and is obviously reflexive, we see that there is in this case more than one reflexive binary relation of which $\geq$ is the asymmetric part. □

Our next result shows the reason that the property defined in 1.14 is called 'negative transitivity.'

1.19 Proposition. *If P is a binary relation on X, and G is its negation, then G is transitive if, and only if, P is negatively transitive.*

Proof. To prove that the negation of P is transitive, let x, y, and z be elements of X such that:

$$xGy \;\&\; yGz, \tag{1.8}$$

and suppose, by way of obtaining a contradiction, that $\neg xGz$. Then it follows from the definition of G that zPx. But this is impossible; for it would then follow from negative transitivity that either zPy or yPx, and either of these conditions contradicts (1.8). Consequently, we see that G is transitive.

To prove the converse, suppose that G, the negation of P, is transitive; let x and y be elements of X such that xPy, and let $z \in X$. If $\neg xPz$, then we have, by definition, zGx. If we also have $\neg zPy$, then it would necessarily be the case that yGz, and the transitivity of G would imply yGx; which contradicts the assumption that xPy. □

The last couple of results we have established have several interesting implications, which we will state as corollaries; the proof of which I will leave as exercises.

1.20 Corollary. *If P is an asymmetric and negatively transitive binary relation on a non-empty set X, then its negation, G, is a weak order on X, and P is its asymmetric part.*

Our next corollary is an immediate consequence of 1.20 and 1.13.

1.21 Corollary. *If P is a binary relation which is asymmetric and negatively transitive, then P is also transitive.*

Supp. xPy and yPz. Assume $\neg xPz$. By asymmetry $yPz \Rightarrow \neg zPy$
Now $\neg xPz$ and $\neg zPy \overset{(NT)}{\Rightarrow} \neg xPy$, contradicting xPy.

While Corollary 1.21 may suggest that negatively transitive binary relations are also transitive, there exist binary relations which are negatively transitive and irreflexive, but which are *not* transitive. A very simple example of such a relation is the usual inequality relation, $\neq$. It is easy to show that this relation is irreflexive and negatively transitive, but it is *not* transitive.

It is important to notice that a binary relation may be asymmetric and transitive without being negatively transitive, as is shown by the following (generic) example.

1.22 Examples.

1. Let X be any non-empty set, let $f\colon X \to \mathbb{R}$ be any real-valued function defined on X, and let δ be a strictly positive real number. If we then define the relation P on X by:

$$xPy \iff f(x) > f(y) + \delta, \tag{1.9}$$

you should have no difficulty in proving that P is irreflexive, asymmetric, and transitive. On the other hand, P will generally not be negatively transitive. For example, let $X = \mathbb{R}_+$, let f be the identity function, and let $\delta = 1$. Then, letting $x = 3/2$, $y = 3/4$, and $z = 0$, we have xPz, but neither xPy nor yPz.

2. Let $X = \mathbb{R}^2_+$, let $\boldsymbol{z} = (a, b)$ be a fixed vector, where $a > 0$ and $b > 0$, and define:

$$Z = \{\boldsymbol{x} \in \mathbb{R}^2_+ \mid (\exists\lambda \geq 0)\colon \boldsymbol{x} = \lambda\boldsymbol{z}\},$$

Imagine now an extremely cautious consumer who has been maximizing preferences at the bundle $\boldsymbol{z}$, and who is only willing to compare bundles whose proportions are the same as those at $\boldsymbol{z}$; in other words, his effective strict preference relation is defined by:

$$\boldsymbol{x}P\boldsymbol{x}' \iff \boldsymbol{x}, \boldsymbol{x}' \in Z \;\&\; \boldsymbol{x} \gg \boldsymbol{x}'.$$

In this case, is P asymmetric? Is it transitive?

3. Suppose our cautious consumer of the previous example now decides that, given a bundle, $\boldsymbol{x} \in Z$, any bundle $\boldsymbol{x}^*$ which is not on the ray Z should be preferred to $\boldsymbol{x}$ if some of one or both commodities could be taken away from $\boldsymbol{x}^*$ to yield a point $\boldsymbol{x}'$ on Z which is such that $\boldsymbol{x}' \gg \boldsymbol{x}$. Suppose further that our consumer now notices that, for a given $\boldsymbol{x}^*$ not on Z, the best (largest) bundle he can obtain on Z by giving up one of the commodities is the bundle $\boldsymbol{x}'$ defined by:

$$\boldsymbol{x}' = \left[\min\left\{\frac{x_1^*}{a}, \frac{x_2^*}{b}\right\}\right]\boldsymbol{z}.$$

Because of this, our consumer now decides that for $\boldsymbol{x} \in Z$, any bundle $\boldsymbol{x}^* \in \mathbb{R}^2_+$ satisfying:

$$\left[\min\left\{\frac{x_1^*}{a}, \frac{x_2^*}{b}\right\}\right]\boldsymbol{z} \gg \boldsymbol{x},$$

should be preferred to $\boldsymbol{x}$. Is the relation P so defined asymmetric? Is it transitive?

4. Continuing with our cautious friend of the previous two examples, suppose he now realizes that any bundle in $\mathbb{R}^2_+$ can be converted to one having the right proportions via the formula in the previous example. Because of this, our consumer

now decides that a bundle $\boldsymbol{x}$ should be preferred to a bundle $\boldsymbol{x}^* \in \mathbb{R}^2_+$ if, and only if:

$$\left[\min\left\{\frac{x_1}{a}, \frac{x_2}{b}\right\}\right]\boldsymbol{z} \gg \left[\min\left\{\frac{x_1^*}{a}, \frac{x_2^*}{b}\right\}\right]\boldsymbol{z}.$$

Is the relation P so defined asymmetric? Is it transitive? Can you think of a simpler way of representing it? □

1.3 Preference Relations and Utility Functions

In general equilibrium theory it is usual to suppose that a consumer's choice of a 'commodity bundle' is limited to some non-empty subset, X, of $\mathbb{R}^n$; which subset we will refer to as the **consumption set**. If $\boldsymbol{x} = (x_1, \ldots, x_n)$ is an element of X (a '**commodity bundle**'), then 'x_j' will denote the quantity of the jth commodity available to the consumer per unit of time, if $x_j \geq 0$. If, on the other hand, $x_j < 0$, then we will take this to mean that the consumer is offering to supply the jth commodity in the amount $-x_j = |x_j|$ per unit of time.

In this context, the consumer is supposed to choose according to his or her[3] 'preference relation,' G, defined over the consumption set. Typically one assumes that this preference relation is a weak order (although we will introduce a weaker assumption in the next chapter). It is important to note that, while we refer to G as a 'preference relation,' it would be more appropriate to call it something like the 'at-least-as-good-as relation;' since if $\boldsymbol{x}$ and $\boldsymbol{y}$ are elements of X, we would say that the consumer considers $\boldsymbol{x}$ at least as good as $\boldsymbol{y}$ (or that $\boldsymbol{y}$ is no better than $\boldsymbol{x}$) if, and only if $\boldsymbol{x}G\boldsymbol{y}$ (notice that this is consistent with the assumption that G is reflexive). Where it is important to make this distinction, we will refer to G as the consumer's **weak preference relation**. In any event, the asymmetric part of G, P, is called the consumer's **strict preference relation**, and the symmetric part of G, I, is called the consumer's **indifference relation**. It follows from 1.13 and 1.20 that if G is a weak order, then P is irreflexive, asymmetric and negatively transitive (and transitive as well); while I is an equivalence relation.

1.23 Definition. If G is a binary relation on X, we define the **upper contour set for x**, Gx, and the **lower contour set for x**, xG, by:

$$Gx = \{y \in X \mid yGx\} \text{ and } xG = \{z \in X \mid xGz\},$$

respectively.

In the case where G is a consumer's (weak) preference relation, we will often wish to consider the sets Px and xP, where P is the asymmetric part of G. We will refer to these two sets as the **strictly preferred** and **strictly inferior to x sets**, respectively.

[3] Hereafter we will use the word 'its' in place of this awkward circumlocution. The word is more appropriate in any case, since the term 'consumer,' as generally used in economics, should not necessarily be interpreted to be an individual. A safer, more correct, general identification is to interpret 'consumer' to mean 'household.'

Notice that if G is a binary relation on a set X, that the upper- and lower contour set ideas define very natural correspondences from X into itself; specifically, we can define the correspondences $\Gamma\colon X \mapsto X$ and $\Phi\colon X \mapsto X$ by:

$$\Gamma(x) = xG \text{ and } \Phi(x) = Gx,$$

respectively. I have used symbols different from G to denote these correspondences, because (a) there are two such correspondences, and (b) in principle, the correspondences should be distinguished, to some extent, from the binary relation. However, notice that either of these two correspondences completely defines the binary relation. Conversely, if Γ is *any* correspondence such that $\Gamma\colon X \mapsto X$, then Γ defines a binary relation, G, on X by:

$$xGy \iff y \in \Gamma(x); \tag{1.10}$$

or, by:

$$xG'y \iff x \in \Gamma(y), \tag{1.11}$$

for that matter. In the first instance, we are identifying $\Gamma(x)$ with the lower contour set for x (xG), while in the second definition, we are identifying $\Gamma(y)$ with the upper contour set for y (that is, with $G'y$).

In practice, we will often find it very convenient to use a correspondence to define a binary relation; although when we do, we will generally identify the values of the correspondence with the upper contour sets, rather than the lower; that is, we will generally define the binary relation as in (1.11), above, rather than as in (1.10).

$$xG'y \iff y \in \Gamma(x).$$

However, things are greatly simplified, and confusion minimized, by using the upper- or lower-contour set notation in the first place; and this is what we will do hereafter. As an example of this method of defining a binary relation, notice that the preference relation of Example 1.22.3 of the previous section can be defined by the upper contour set correspondence, given by:

$$G\boldsymbol{x} = \begin{cases} \left\{ \boldsymbol{x}^* \in \mathbb{R}^2_+ \mid \min\left\{\frac{x_1^*}{a}, \frac{x_2^*}{b}\right\} > \frac{x_1}{a} \right\}, & \text{for } \boldsymbol{x} \in Z, \\ \emptyset & \text{for } \boldsymbol{x} \in \mathbb{R}^2_+ \setminus Z. \end{cases}$$

Hopefully, you will agree that this represents a much simpler method of defining the consumer's (strict) preference relation than was used in our original development of the example. (For other examples of defining preference relations by this method, see Examples 1.31, below.)

1.24 Definitions. If G is a reflexive binary relation on X, we shall say that a function $f\colon X \to \mathbb{R}$ **represents G on X** iff, for all x and y in X, we have:

$$xGy \iff f(x) \geq f(y). \tag{1.12}$$

If a function exists which represents G on X, we shall say that **there exists a representation for G**, or that **G admits of a real-valued representation**. In the special case in which G is a consumer's (weak) preference relation, we shall say that a function f which represents G on X is an (**ordinal**) **utility function for G**.

In Example 1.8, we showed that if a binary relation, G admits of a real-valued representation, then G is a weak order. Thus, a *necessary* condition for a binary relation to be representable by a real-valued function is that G be a weak order. Shortly, we will consider *sufficient* conditions for a binary relation to be representable, but before we do, let's consider a further aspect of the definition of representability itself.

Notice that (1.12) can be written as the compound statement:

$$xGy \Rightarrow f(x) \geq f(y), \tag{1.13}$$

and:

$$f(x) \geq f(y) \Rightarrow xGy. \tag{1.14}$$

Once again letting 'P denote the asymmetric part of G, we see that the contrapositive of (1.13) is:

$$f(y) > f(x) \Rightarrow yPx$$

(recall Theorem 1.15), while the contrapositive of (1.14) is:

$$yPx \Rightarrow f(y) > f(x).$$

Therefore condition (1.12) is equivalent to:

$$xPy \iff f(x) > f(y); \tag{1.15}$$

which leads us to the following.

1.25 Definition. We shall say that an asymmetric relation, P, on X is **representable** iff there exists a function $f\colon X \to \mathbb{R}$ satisfying (1.15), above.

Thus, if G is a weak order, then G is representable by Definition 1.24 if, and only if, its asymmetric part, P, is representable according to Definition 1.25. Accordingly, the two definitions are equivalent in the sense just stated, and where we find it more convenient to use (1.15) rather than (1.12) as our definition of representability, we shall not hesitate to do so.

1.26 Proposition. *Suppose that X is a finite set, and that G is a reflexive binary relation on X. Then there exists a function, $f\colon X \to \mathbb{R}$ which represents G on X if, and only if, G is a weak order.*

Proof. It follows at once from 1.8.2 that if there exists a function representing G, then G must be a weak order. To prove the converse, suppose G is a weak order, and define the real-valued function f on X by:

$$f(x) = \#\{y \in X \mid xGy\} = \#xG; \tag{1.16}$$

that is, $f(x)$ is the number of elements, y, of X such that xGy. To prove that f represents G, suppose first that xGy. Then if $z \in X$ is such that yGz, it follows from the transitivity of G that xGz as well. Therefore:

$$yG = \{z' \in X \mid yGz'\} \subseteq \{z \in X \mid xGz\} = xG,$$

and it follows that:

$$f(x) = \#xG \geq \#yG = f(y).$$

Conversely, suppose that x' and y' are such that $x'Py'$. Then:

$$x'Gy', \tag{1.17}$$

and:

$$\neg y'Gx' \tag{1.18}$$

From (1.17) and the transitivity of G, it is easy to see that:

$$y'G \subseteq x'G; \tag{1.19}$$

while from the reflexivity of G and (1.18), we see that:

$$x' \in x'G \text{ and } x' \notin y'G. \tag{1.20}$$

From (1.19) and (1.20) it follows that:

$$f(x') = \#x'G > \#y'G = f(y').$$

Thus we have shown that;

$$x'Py' \Rightarrow f(x') > f(y'),$$

which, since G is total, is equivalent to:

$$f(y') \geq f(x') \Rightarrow y'Gx'. \quad \square$$

1.27 Corollary. *Suppose that X is a finite set, and that P is an asymmetric binary relation on X. Then there exists a function, $f\colon X \to \mathbb{R}$ satisfying:*

$$xPy \iff f(x) > f(y),$$

for all $x, y \in X$ if, and only if, P is negatively transitive. In other words, an asymmetric binary relation on a finite set, X, is representable if, and only if, it is negatively transitive.

While the results just presented provide very simple and straightforward necessary and sufficient conditions for a binary relation to be representable for the case in which X is a finite set, things get more complicated if X is an infinite set, as is demonstrated by the following example.

1.28 Example. (**The lexicographic order**.) Let $X = \mathbb{R}^2_+$, and define $>_L$, the **lexicographic order**, on X, by:

$$(x_1, x_2) >_L (y_1, y_2) \iff \begin{cases} x_1 > y_1 & \text{or:} \\ x_1 = y_1 & \text{and } x_2 > y_2. \end{cases} \tag{1.21}$$

It is easy to show that $>_L$ is total and asymmetric (and thus is antisymmetric). We will prove that it is negatively transitive, from which it will follow that it is also transitive.

To prove negative transitivity, suppose $\boldsymbol{x} >_L \boldsymbol{z}$. Then either:

$$x_1 > z_1, \tag{1.22}$$

or:

$$x_1 = z_1 \text{ and } x_2 > z_2. \tag{1.23}$$

Now let $\boldsymbol{y} \in \mathbb{R}^2_+$, and suppose $\boldsymbol{y} \not>_L \boldsymbol{z}$. Then either:

$$y_1 < z_1, \tag{1.24}$$

or:

$$y_1 = z_1 \text{ and } y_2 \leq z_2. \tag{1.25}$$

However, if (1.22) holds, then either (1.24) or (1.25) implies $x_1 > y_1$, and thus $\boldsymbol{x} >_L \boldsymbol{y}$. Similarly, if (1.23) and (1.24) hold, then $\boldsymbol{x} >_L \boldsymbol{y}$. On the other hand, if (1.23) and (1.25) hold, then we have $x_1 = y_1$ and $x_2 > y_2$; so that $\boldsymbol{x} >_L \boldsymbol{y}$ in this case as well.

One can show, however, that $>_L$ does *not* admit of a real-valued representation. We will present only an outline of a proof of this here. For details, see Debreu [1959, pp. 72–3]. Hopefully, the basic idea of the argument will be clear enough, despite the fact that it formally depends upon some cardinal number concepts which you may not have previously encountered.

Suppose, by way of obtaining a contradiction, that $>_L$ admits of a real-valued representation, so that there exists a function $f\colon \mathbb{R}^2_+ \to \mathbb{R}$ satisfying:

$$(\forall \boldsymbol{x}, \boldsymbol{y} \in \mathbb{R}^2_+)\colon \boldsymbol{x} >_L \boldsymbol{y} \iff f(\boldsymbol{x}) > f(\boldsymbol{y}); \tag{1.26}$$

and for the sake of convenience in the remainder of our argument, let us use the generic notation '(x, y)' to denote elements of $\mathbb{R}^2_+$. Then, for each $x \in \mathbb{R}_+$, we can define real numbers a_x and b_x by:

$$a_x = f(x, 0) \text{ and } b_x = \sup_{y \in \mathbb{R}_+} f(x, y). \tag{1.27}$$

Moreover, from (1.26) we see that, since for each $x \in \mathbb{R}_+$, $(x, 1) >_L (x, 0)$, we must have:

$$a_x = f(x, 0) < f(x, 1) < b_x;$$

while for $x, x^* \in \mathbb{R}_+$ such that $x^* > x$, similar considerations establish that $b_x \leq a_{x^*}$. Thus we see that the family, $\mathcal{I}$ given by:

$$\mathcal{I} = \{[a_x, b_x[\mid x \in \mathbb{R}_+\},$$

is a family of disjoint, non-degenerate intervals of real numbers; a distinct such interval for each nonnegative real number, x. But this is impossible, because there are only a countable number of such intervals; whereas there are an uncountable number of nonnegative real numbers. More crudely put, there are simply too many nonnegative real numbers to obtain a non-degenerate interval for each, such that no two (distinct) intervals have any points in common! □

In order to pursue the question of when a weak order on an *infinite* set will be representable by a real-valued function, we will need to begin by considering the following generalization of the notion of a closed set in $\mathbb{R}^n$.

1.29 Definition. If X is a non-empty subset of $\mathbb{R}^n$, and A is a non-empty subset of X, we shall say that **A is closed relative to X**, or that **A is closed in X**, iff, whenever $\langle \boldsymbol{x}_q \rangle$ is a sequence of points from A which converges to a point $\boldsymbol{x}$ which is an element of X, then we must have $\boldsymbol{x} \in A$.

1.30 Definitions. Let X be a non-empty subset of $\mathbb{R}^n$, and let G be a weak order on X. We shall say that G is:

1. **upper semi-continuous on X** iff, for each $\boldsymbol{x} \in X$, the set $G\boldsymbol{x}$ is closed in X.
2. **lower semi-continuous on X** iff, for each $\boldsymbol{x} \in X$, the set $\boldsymbol{x}G$ is closed in X.
3. **continuous on X** iff G is both upper- and lower semi-continuous on X.

1.31 Examples.

1. Let $X = \mathbb{R}^2_+$, and define the correspondence Γ on X by:

$$\Gamma(\boldsymbol{x}) = \begin{cases} \mathbb{R}^2_+ & \text{if } \boldsymbol{x} = \mathbf{0}, \\ \mathbb{R}^2_+ \setminus \{\mathbf{0}\} & \text{if } \boldsymbol{x} \neq \mathbf{0}. \end{cases}$$

We then let G be the binary relation defined by Γ; that is:

$$\boldsymbol{x}G\boldsymbol{x}^* \iff \boldsymbol{x} \in \Gamma(\boldsymbol{x}^*).$$

Then G is lower semi-continuous on X, but is *not* upper semi-continuous on X. Notice, however, that G is representable; for example by the function $f\colon X \to \mathbb{R}$ defined by:

$$f(\boldsymbol{x}) = \begin{cases} 0 & \text{for } \boldsymbol{x} = \mathbf{0}, \text{ and:} \\ 1 & \text{for } \boldsymbol{x} > \mathbf{0}, \end{cases}$$

is a function which represents G on X.

Notice that in defining the preceding example, the correspondence Γ is simply the upper contour correspondence. *Hereafter, where there appears to be no danger of confusion, we will use the same symbol to refer to both the correspondence and the relation; as is done in the next example.*

2. Let $X = \mathbb{R}^2_+$, and define G on X by:

$$G(\boldsymbol{x}) = \begin{cases} \{\boldsymbol{x}\} & \text{for } \boldsymbol{x} = (1,1), \text{ and} \\ \mathbb{R}^2_+ & \text{for } \boldsymbol{x} \neq (1,1). \end{cases}$$

In this case, it is easy to see that G is upper semi-continuous on X, but is *not* lower semi-continuous on X. Once again G is representable on X, however; for example, by the function f defined on X by:

$$f(\boldsymbol{x}) = \begin{cases} 0 & \text{for } \boldsymbol{x} \neq (1,1), \text{ and:} \\ 1 & \text{for } \boldsymbol{x} = (1,1). \end{cases}$$

3. The lexicographic ordering, $>_L$, defined in 1.28, above, is *neither* upper-, nor lower semi-continuous on X.

4. Let $X = \mathbb{R}^2_+$, and define the subset, D, by:

$$D = \{\boldsymbol{x} \in \mathbb{R}^2_+ \mid x_1 = x_2\}.$$

We then define G by:

$$G(\boldsymbol{x}) = \begin{cases} X & \text{if } \boldsymbol{x} \notin D, \\ \{\boldsymbol{x}' \in D \mid \boldsymbol{x}' \geq \boldsymbol{x}\} & \text{if } \boldsymbol{x} \in D. \end{cases}$$

Is G representable on X? by a continuous real-valued function? □

One can make use of the above definitions to prove the following result; although here we will simply state the result without providing a proof.

1.32 Proposition. *If G is a binary relation on a non-empty set, X, and if there exists a continuous function, $f\colon X \to \mathbb{R}$, which represents G on X, then G is a continuous weak order on X.*

The next definition is not one which I will expect you to remember, and is stated only for completeness.

1.33 Definition. Let X be a subset of $\mathbb{R}^n$. A pair of subsets of X, A and B, will be said to be a **separation for $\boldsymbol{X}$** iff A and B satisfy:

1. A and B are both closed in X,
2. A and B are both non-empty, and
3. $A \cap B = \emptyset$ and $X = A \cup B$.

The set X will be said to be **connected** iff there exists no separation for X.

Intuitively, a subset of $\mathbb{R}^n$ is connected if it is 'of one piece.' The space $\mathbb{R}^n$ itself is connected (even if $n = 1$), and $\mathbb{R}^n_+$ is connected; in fact, any convex subset of $\mathbb{R}^n$ is connected. Debreu has proved the following theorem, which we will state without proof.[4]

1.34 Theorem. *If X is a connected subset of $\mathbb{R}^n$, and G is a continuous weak order on X, then G is representable on X. In fact, there exists a continuous real-valued function which represents G on X.*

Be sure to note that the above result establishes the fact that, if X is a connected subset of $\mathbb{R}^n$, then the continuity of G is a *sufficient condition* for G to be representable. The fact that it is not necessary is shown by the following example.

1.35 Example. Let $X = \mathbb{R}^2_+$, and define the upper contour correspondence $G\colon X \mapsto X$ by:

$$G(\boldsymbol{x}) = \begin{cases} \mathbb{R}^2_+ & \text{if } \boldsymbol{x} = \boldsymbol{0}, \\ \mathbb{R}^2_+ \setminus \{\boldsymbol{0}\} & \text{if } \boldsymbol{x} \in \mathbb{R}^2_+ \setminus \{\boldsymbol{0}, (1,1)\}, \\ \{(1,1)\} & \text{if } \boldsymbol{x} = (1,1). \end{cases}$$

[4] For a proof, see Debreu [1959, pp. 56–9].

Here it can be shown that the binary relation, G is *neither* upper-, nor lower semi-continuous. However, it is clear that the function $f\colon \mathbb{R}^2_+ \to \mathbb{R}_+$ defined by:

$$f(\boldsymbol{x}) = \begin{cases} 0 & \text{for } \boldsymbol{x} = \mathbf{0}, \\ 1 & \text{for } \boldsymbol{x} > \mathbf{0} \;\&\; \boldsymbol{x} \neq (1,1), \\ 2 & \text{for } \boldsymbol{x} = (1,1). \end{cases}$$

represents G on X. □

Exercises.

1. Prove Theorem 1.4.
2. Show that the inequality, $>$, on $\mathbb{R}^2$ is *not* negatively transitive.
3. Prove directly (that is, without using Proposition 1.17) that the 'semi-greater-than' relation on $\mathbb{R}^2$, $>$, is the asymmetric part of the relation G defined in Example 1.18.
4. Let X be any nonempty set, let $f\colon X \to \mathbb{R}$, and define the relation G on X by:

$$xGy \iff f(x) \geq f(y).$$

Show that G is a weak order.

5. Show that $\geq$, the usual weak inequality on $\mathbb{R}^n$ is reflexive, antisymmetric, and transitive.
6. Show that if $f\colon X \to \mathbb{R}$ represents the weak order G, and $F\colon f(X) \to \mathbb{R}$ is any strictly increasing function, then the composition of F and f, the function g defined by:

$$g(x) = F[f(x)] \quad \text{for } x \in X,$$

also represents G. It is because of this consideration that such representations are called **ordinal utility functions** in consumer demand theory.

7. Suppose X is a nonempty set, that P is a binary relation on X, and that $f\colon X \to \mathbb{R}$ is a function satisfying:

$$(\forall x, x' \in X)\colon xPx' \iff f(x) > f(x').$$

Show that P is asymmetric and negatively transitive.

8. Suppose X is a nonempty set, that P is a binary relation on X, and that $f\colon X \to \mathbb{R}$ is a function satisfying:

$$(\forall x, y \in X)\colon xPy \Rightarrow f(x) > f(y).$$

Prove the following statements, or provide a counterexample:

a. P is irreflexive.
b. P is asymmetric.
c. P is transitive.
d. P is negatively transitive.

9. Assume the same conditions as in Exercise 8, above, except this time assume:

$$(\forall x, y \in X)\colon f(x) > f(y) \Rightarrow xPy.$$

Answer the same questions as in Exercise 8.

10. Let G be the relation defined on $\mathbb{R}$ by:

$$xGy \iff x \geq f(y),$$

where $f\colon \mathbb{R} \to \mathbb{R}$. Can you provide sufficient conditions for G to be:

a. reflexive?

b. total?

c. transitive?

d. asymmetric?

e. antisymmetric? (If you have found sufficient conditions, don't worry at this point about whether they're necessary as well.)

11. Show that if G is a weak order on a finite set, X, then the following function represents G on X:

$$u(x) = \#X - \#Px,$$

where P is the asymmetric part of G. How does this function compare with that used in the proof of Proposition 1.26?

12. Show that if G is a weak order on a finite set, X, then the following function represents G on X:

$$u^*(x) = 1 - \left(\frac{\#Px}{\#X}\right),$$

where P is the asymmetric part of G.

13. Show that if G is a weak order on a finite set, X, then the following function represents G on X:

$$\bar{u}(x) = \frac{1}{[(\#Px) + 1]},$$

where P is the asymmetric part of G.

14. Define the set $E \subseteq \mathbb{R}^2_+$ by:

$$E = \{\boldsymbol{x} \in \mathbb{R}^2_+ \mid x_1 = x_2\},$$

and define the relation P on $\mathbb{R}^2_+$ by:

$$P\boldsymbol{x} = \begin{cases} \{\boldsymbol{x}' \in E \mid \min\{x'_1, x'_2\} > \min\{x_1, x_2\}\} & \text{if } \boldsymbol{x} \in \mathbb{R}^2_+ \setminus E, \\ \{\boldsymbol{x}' \in \mathbb{R}^2_+ \mid \min\{x'_1, x'_2\} > \min\{x_1, x_2\}\} & \text{if } \boldsymbol{x} \in E. \end{cases}$$

Is the relation P asymmetric? Is the relation P transitive? In each case, provide a justification for your answer, either a brief proof or a counterexample.

15. Suppose $\succsim$ is a linear order on a nonempty set, X, that $\{Y, Z\}$ is a **partition** of X,[5] and that Z is a binary relation on Y. Define the relation R on X by:

$$xRy \iff \begin{cases} xQy & \text{if } x, y \in Y, \\ x \succsim y & \text{if } x, y \in Z, \text{ or} \\ x \in Y \ \&\ y \in Z. & . \end{cases}$$

a. Show that if Q is a weak order on Y, then R is a weak order on X.

b. Show that if Q is a linear order on Y, then R is a linear order on X.

[5] So that Y and Z are both nonempty, $Y \cap Z = \emptyset$, and $X = Y \cup Z$.

Chapter 2

Algebraic Choice Theory

2.1 Introduction

In this chapter we will examine the foundations of the economic theory of consumer demand. In particular, we will begin a critical examination of the appropriate interpretation of two of the primitives (undefined, basic terms) in general equilibrium theory; namely 'consumer' and 'commodity.'

One's initial tendency is to identify 'consumers' in the theory with individual 'consumers,' as we define the term in the popular press. That is, a 'consumer' would be an individual adult human being who is not in the care of others. As it turns out, however, in economic application, we are on firmer ground (for reasons to be explained shortly) if an individual consumer in the theory is identified with a household in the 'real world.' This creates some potential problems in our theoretical development, and is one of the things which we will discuss at some length in this chapter.

Insofar as the primitive 'commodity' is concerned, in most of abstract general equilibrium theory, commodities are differentiated by four characteristics:

- physical characteristics,
- time of availability,
- location at which the commodity is available, and
- state of the world in which the commodity is available.

Thus, suppose that in a given economy, we have only two physically distinct commodities; say No. 1 paper clips, and sheets of 8 1/2 by 11 one hundred per cent rag content bond paper. However, suppose that we also are considering two distinct locations, two time periods (today and tomorrow), and two possible states of the world (with each a possibility both today and tomorrow). Then in our general equilibrium model we would distinguish 2^4 commodities; so that n, the number of commodities in our analysis (and the dimension of the commodity space), is equal to 16. Thus, in most of our basic theory we can be considered to be taking into account location, time, and undertainty. There is, however, a problem with this; if, for example, we are going to analyze the effects of location, we need to put more structure in the model than we will be doing. In particular, for example, we need to take into account the fact that No. 1 paper clips available today at location one and given state of the world one, as opposed to paper clips available today at location two, given state of the world one, are differentiated from one another in a different way than are paper clips available today at location one and given state of the world one versus paper clips available tomorrow at location one, given state of the world one. In other words, we need to add more structure to the model than we will generally be doing in this book in order to analyze the effect of location, time, or uncertainty. In later chapters, we will devote some attention to the analysis of the effects of time and of undertainty, but we will do very little in the way of analyzing the effects of location. In this omission I am not alone, basic economic theory and application is rather remiss in analyzing the effect of location, and I will have to leave a more thorough analysis of this topic to a specialized course in location theory.

In any case, the notions of 'commodity' and quantities thereof can be interpreted in many different ways in the context of the bulk of the general equilibrium theory which we will be studying, and in this chapter we will be taking what can reasonably be termed the 'applied microeconomics' interpretation of these notions. Specifically, in this chapter we will generally suppose that commodities are differentiated solely by physical characteristics, that there is no uncertainty as to availability; and, correspondingly, that the jth coordinate of a commodity bundle represents the quantity of the jth commodity available for consumption 'now,' per unit of time. In particular, then, we will suppose that quantities of commodities represent 'flows' per unit of time. In the next section we will introduce what we will call the general algebraic choice model, which can be regarded as setting forth the 'bare bones' of the theory. The remaining sections of the chapter can essentially be regarded as a critique of this basic model within the context of the applied microeconomics interpretation of the commodity space; we will be concerned with a critical appraisal of some standard interpretations of the basic model, and with a number of criticisms which have been leveled at this type of choice theory.

2.2 The General Algebraic Theory of Choice

The term 'algebraic' is used to distinguish the theory to be studied here from probabilistic choice theories, which will be considered briefly later in the chapter. Roughly speaking, an algebraic theory is deterministic in nature; in the sense that the basic assumption of the theory is that, if a decision-maker were repeatedly offered a choice between a given pair of alternatives, he or she would make the same choice from the pair each time it was offered. In a probabilistic theory, the basic assumption is that there is a probability that one of the alternatives, call it 'x,' would be chosen over the other (denoted by 'y'); and that if the choice set $\{x, y\}$ were offered a large number of times, the proportion of times that x would be chosen from this set would be approximately equal to this probability. We will discuss the distinction between these two types of choice theory in more detail in Section 10 of this chapter.

In the basic algebraic theory of choice, it is assumed that the decision-maker has well-defined preferences, $\succsim$, over a nonempty set of alternatives, X; and that, if her choice is constrained to a nonempty subset, B, of X, the alternative chosen will be at least as good (in terms of her preferences, $\succsim$) as any other alternative in B. More formally, the theory deals with:

X a nonempty set, the set of alternatives,

$\succsim$ the 'preference relation,' assumed to be a weak order on X, and

$\mathcal{B}$ a nonempty family of nonempty subsets of X.

Sets contained in (elements of) $\mathcal{B}$ will be called **budget sets**; and the pair $\langle X, \mathcal{B}\rangle$ will be called a **budget space**. The fundamental assumption of the theory is that, if the decision-maker's choice is confined to the set $B \in \mathcal{B}$, the element, or alternative actually chosen from B will be an element of the set $h(B)$ defined as:

$$h(B) = \{x \in B \mid (\forall y \in B)\colon x \succsim y\}. \tag{2.1}$$

The elements of X should be considered to be distinct and mutually exclusive alternatives. In the economic theory of consumer demand, X is usually taken to be a subset of $\mathbb{R}^n$; with the ordered n-tuples:

$$\boldsymbol{x} = (x_1, x_2, \ldots, x_n) \in X,$$

taken to be **commodity bundles**, a list of quantities (per unit of time) of the n commodities. In this context, the budget sets, B, are usually taken to be of the form:

$$B = b(\boldsymbol{p}, w) = \{\boldsymbol{x} \in X \mid \boldsymbol{p} \cdot \boldsymbol{x} \leq w\},$$

where $\boldsymbol{p}$ is an element of $\mathbb{R}^n$, and represents a vector of prices, and w represents the consumer's wealth (or income, per time period, depending upon the context). However, the framework being presented here is, in principle, applicable to many other situations; for example the elements of X might be interpreted as cash flows, military strategies, inventory policies, legislative programs, potential marriage partners, sound energies (at fixed frequency, but varied decibel levels; or at a fixed decibel level and varying frequencies), and so on. More detailed examples (specific interpretations) representing typical applications of the general algebraic choice model, are presented in the following.

2.1 Examples. 1. Let '$Z = \{z_1, z_2, z_3, z_4\}$' denote a set of objects. We might formulate the problem of analyzing a consumer's choice of the objects in Z in one of two different ways, depending upon the choice context.

a. Define the entities a_i $(i = 0, 1, \ldots, 15)$ in the following way:

$$\begin{aligned} &a_0 = \emptyset, a_i = \{z_i\} \text{ for } i = 1, \ldots, 4,\\ &a_5 = \{z_1, z_2\}, a_6 = \{z_1, z_3\}, \ldots, a_9 = \{z_2, z_4\}, a_{10} = \{z_3, z_4\},\\ &a_{11} = \{z_1, z_2, z_3\}, a_{12} = \{z_1, z_2, z_4\}, a_{13} = \{z_1, z_3, z_4\}, a_{14} = \{z_2, z_3, z_4\},\\ &a_{15} = \{z_1, z_2, z_3, z_4\}; \end{aligned}$$

and define the set of alternatives, X, by:

$$X = \{a_0, a_1, \ldots, a_{15}\}. \tag{2.2}$$

If the consumer has well-defined preferences over this set of alternatives (in particular, if the consumer's preference relation is a weak order), and if she chooses an element from X in accordance with these preferences, then the general algebraic choice model is applicable here. We would expect that in a typical case, the actual choice would be constrained to some proper subset of X; for example:

$$B = \{a_0, a_1, \ldots, a_{10}\}$$

('you can have at most any two of the four elements...'). The prediction of the theory would then be that the consumer's actual choice would be an element of the set $h(B)$ given by:

$$h(B) = \{a_i \in B \mid a_i \succsim a_j, \text{ for } j = 0, 1, \ldots, 10\}.$$

Moreover, if the objects are all desirable, then in this case we would expect to find that:

$$h(B) \subseteq \{a_5, \ldots, a_{10}\}.$$

b. Suppose now that the four objects are four different kinds of (new) refrigerators. In this case, it would appear that the general algebraic choice model is applicable to the consumer's choice of a refrigerator. However, the structure of the problem is greatly simplified by taking what we might call the 'marketing approach' to the problem. The general idea here is to restrict the claimed applicability of the theory to the case wherein the consumer has already decided to buy (choose) exactly one refrigerator. The set of alternatives then becomes:

$$Z = \{z_1, \ldots, z_4\}.$$

The relationship between the sort of preference relation considered in part (a) of this example and that on Z is of particular interest at this point.

Notice first that a weak order, $\succsim$, on the set X defined in (2.2) induces a weak order, G, on the set Z by the definition:

$$z_i G z_j \iff a_i = \{z_i\} \succsim a_j = \{z_j\};$$

however, the converse is not true. More specifically, if G is the consumer's preference relation on Z, then G will generally provide very little information about the consumer's preference relation, $\succsim$, on X. In fact, while it might at first glance appear that if:

$$z_1 P z_2 P z_3 P z_4$$

(where P is the asymmetric part of G, the consumer's 'strict preference relation'), then we would surely have:

$$a_5 = \{z_1, z_2\} \succ a_8 = \{z_2, z_3\},$$

a little thought should convince you that even this is not the case. Even in the special case of different kinds of refrigerators, we may well have $a_8 \succ a_5$ in this case; for suppose z_1 and z_2 are full-size refrigerators, and z_3 is a smaller ('apartment-size' (?)) refrigerator. Abstracting from questions of price, the consumer might very well prefer to have the combination of a full-size and smaller refrigerator (which could, perhaps, fit in a rec room in the basement) to having two full-size refrigerators.

Returning to the 'marketing approach,' however, notice that the consumer may have well-defined preferences over the set Z (over the four refrigerators), and make a choice consistent with these preferences, without having given a thought to the question of whether, for example, the combination $\{z_1, z_2\}$ is preferred to $\{z_2, z_3\}$.

2. Let X be a set of lottery tickets of the form (more correctly, which can be denoted by):

$$x = (\pi; a, b),$$

where 'π' denotes the probability of winning the prize a; the probability of winning the alternative prize, b, being equal to $1 - \pi$. If a decision-maker can be regarded as having a preference relation over these alternatives which is total, reflexive, and transitive (and thus is a weak order), then the general algebraic choice model is applicable to the analysis of this situation.

3. Let X be the collection of all legislation of a particular type which has been proposed on the floor of the U. S. House of Representatives as of a certain date. Would you expect to be able to apply the general algebraic choice model to the actions of the House sub-committee having jurisdiction over this type of legislation? How about to the President's choice of legislative policy in the area?

4. Let 'x_1' denote the quantity of food available for a specific period of time (say a month), and let 'x_2' denote the quantity of clothing available during the same period of time. What kinds of difficulties might we encounter in trying to analyze a particular consumer's choice of food and clothing within the context of the general algebraic choice model, taking $X = \mathbb{R}^2_+$? Before jumping to any conclusions here, carefully consider the problem of comparing a pair $\boldsymbol{x} = (x_1, x_2)$ with a second pair, $\boldsymbol{x}^* = (x_1^*, x_2^*)$.

5. Suppose there are n types of soft drinks (excluding coffee and tea) available in a particular locality as of the beginning of a given month, and that we label these soft drinks with the numbers $1, \ldots, n$ (for example, '1' might be Coca-Cola, '2' Pepsi-Cola, '3' Royal Crown Cola, '4' Diet coke, etc.). Letting $X = \mathbb{R}^n_+$, the general algebraic choice model *might* be applicable to a particular consumer's choice of soft drinks for the month, if said consumer has well-defined preferences over X. This simple example can be used to illustrate a number of difficulties in the applicability of the model, however, and we will return to a consideration of various aspects of this example in the following sections. □

2.3 Some Criticisms of the Model

Somewhat paradoxically, the general algebraic choice model suffers from two seemingly contradictory flaws: it is so general as to have very little predictive power, yet at the same time, there is a very real question as to whether the assumptions of the model are satisfied in very common individual choice situations. We will return to the issue of the predictive power of the model in the next chapter; in this section and the remainder of this chapter, we will briefly consider a number of criticisms which have been levied at the model regarding its applicability.

Among many objections which have been raised concerning the realism or applicability of the assumptions of the general algebraic choice model, there are seven types of criticisms which we will consider in this chapter.

1. The model may be inapplicable to certain kinds of choices under uncertainty; specifically, to situations in which the decision-maker's choice does not uniquely determine the outcome.
2. The model, as set forth here, does not admit of a genuinely dynamic analysis.
3. Individuals' preferences over alternatives may depend upon the way the alternatives are presented; that is, a given alternative may be describable or representable in two different ways, and an individual's preference for it as compared to a second alternative may depend upon which of these representations is chosen. This is the issue of 'framing,' and will be discussed in the next section.
4. Individuals' stated preferences may be inconsistent with their actual choices.
5. In actual choice situations (particularly in experiments), individuals often exhibit inconsistencies. Thus a probabilistic (as opposed to an algebraic) theory of choice may be needed.
6. It may be unreasonable to suppose that preferences are total.
7. It may be unreasonable to suppose that preferences are transitive.

We will provide only an extremely cursory consideration of the first three of these objections here. The remaining four difficulties will be given a more extensive consideration in the remaining sections of this chapter.

As an illustration of the kind of difficulties presented by choice under uncertainty, consider the situation in which a business manager has a choice among three inventory policies, a_1, a_2 and a_3; with policy a_i best if event E_i occurs, for $i = 1, 2, 3$. Without further assumptions about (or better, knowledge of) the policies and the probabilities of the events E_i, we could hardly assert with any confidence that our manager has well-defined deterministic preferences over the three policies. The difficulty here, notice, is that we would suppose that the decision-maker's preferences are not defined directly over the objects of choice (a_1, a_2 and a_3), but rather over *outcomes*, which are determined jointly by the policy variables a_i and the random events, E_j.

It is usual in economics to analyze choice under uncertainty (more correctly, under risk) via the expected utility (EU) model, which we can develop here as follows. Denote the outcome associated with action a_i, given state j, E_j by 'x_{ij},' and let p_j denote the probability of the occurence of state E_j, where:

$$p_1 + p_2 + p_3 = 1.$$

If the von Neumann-Morgenstern utility of the the outcomes is given by a function $u\colon X \to \mathbb{R}$, then action a_i is preferred to action a_k if, and only if:

$$p_1 u(x_{i1}) + p_2 u(x_{i2}) + p_3 u(x_{i3}) > p_1 u(x_{k1}) + p_2 u(x_{k2}) + p_3 u(x_{k3}).$$

While this model creates an elegant theory of decision-making under risk, and has been widely-used in economics, there is a great deal of empirical evidence which casts doubt on some of the assumptions of this model. We will discuss the model, and some of this empirical evidence later in the course; in the meantime, let me mention that Loomes, Starmer, and Sugden [1991], Machina [1987], Starmer [1996, 2000], and Tversky and Thaler [1990] all provide quite readable and useful discussions of the empirical findings which have been at odds with this theory, as well as some alternative theories which have been proposed in response to the empirical findings.

Insofar as the difficulties with a dynamic analysis are concerned, suppose we consider the typical formulation in economics, where X is taken to be a non-empty subset of $\mathbb{R}^n$, and 'x_i' denotes the quantity of the ith commodity (per unit of time) for $i = 1, \ldots, n$. In this context, we can distinguish between the physically identical commodity available now, as opposed to T periods from now.[1] Thus 'x_1' might denote the quantity of #1 paper clips available 'now' (at the beginning of period 1), 'x_2' the quantity of #1 paper clips available at the beginning of the next period ($t = 2$), and so on. The trouble with this is that in order to apply this interpretation, we need to assume that the decision-maker knows his or her budget set now (at $t = 1$) even though it involves commodities available only at later dates. This would appear to be reasonable only (if at all) in the presence of much more pervasive and efficient futures markets than appear to exist currently.[2] Furthermore, actual preferences may change over time (if a learning process takes place as commodities are consumed, for example), and there is no adequate allowance for this effect in the present formulation. We will return to a more complete discussion of some difficulties connected with the dynamic analysis of consumer choice in Section 8 of this chapter.

As to the 'framing' problem, consider an example/experiment which was reported by Tversky and Kahneman [1988]. A group of subjects was presented with the following material (their Problem 3).

> Problem 3 [N = 150] Imagine that you face the following pair of concurrent decisions. First examine both decisions, then indicate the options you prefer.
>
> Decision (i) Choose between:
>
> A. a sure gain of $240
>
> B. 25% chance to gain $1000, and a 75% chance to gain nothing.
>
> Decision (ii) Choose between:
>
> C. a sure loss of $750,
>
> D. 75% chance to lose $1000, and 25% chance to lose nothing.

When this problem was presented to 150 subjects, 84% chose (A) as the first decision and 87% chose (D) as Decision (ii). However, consider the following, which is, notice, exactly the same as the concurrent choice in the previous problem.

[1] We might equally well want to distinguish on the basis of location and on the basis of the state of the world in which the commodity will be available, but we will postpone a consideration of these complications for the moment.

[2] For excellent discussions of this kind of interpretation of the model, and some of the pitfalls involved therein, see Debreu [1959, pp. 28–36 & pp. 50–5], or Chapter 20 of Mas-Colell, Whinston, and Green [1995].

Problem 4. Choose between:

A & D. 25% chance to win \$ 240, and 75% chance to lose \$ 760.
B & C. 25% chance to win \$ 250, and 75% chance to lose \$750.

When the problem was presented in this way to 86 subjects, all of them chose B & C; but notice that the problem is exactly the same as Problem 3, it is simply presented ('framed') in a different way.

This is only one type of 'framing' difficulty which has been investigated in the literature. A much more complete discussion of the problem, as well as some other anomalies, is provided in the references mentioned earlier: Loomes, Starmer, and Sugden [1991], Machina [1987], Starmer [1996, 2000], and Tversky and Thaler [1990], as well as Tversky and Kahneman [1988] and Tversky *et al* [1990].

2.4 Stated Preferences versus Actual Choices

The general algebraic choice model is applicable to a specific choice situation only if the decision-maker has well-defined preferences (a weak order) over the underlying set of alternatives, X; *and* if, when presented with a subset of X from which a choice must be made, said decision-maker always makes a choice consistent with these preferences. In terms of the notation of Section 2, if B represents the available or feasible set, then the actual choice must be an element of $h(B)$, where:

$$h(B) = \{x \in B \mid (\forall y \in B) \colon x \succsim y\}. \tag{2.3}$$

The difficulty to be considered in this section is the claim that, even in cases where the weak order condition is satisfied, the actual choice from some subset, B, of X may not be an element of $h(B)$.

We are all familiar with the fact that stated preferences may be very different from actual choices in situations where there are 'face-saving' motives, or group pressure present. Thus, for example, we have all heard of situations in which an individual who was actually watching a particular (low-brow) television show would claim, when asked by a pollster, that he was watching some other show which he perceived to be more socially respectable. Similarly, we have probably all, at one time or another when out with a group, yielded to group preferences, and attended a movie that was our second choice or lower. Neither of these phenomena really represents a fundamental theoretical difficulty with the general algebraic choice model, however. The first situation is simply a matter of not stating true preferences, while the second situation may be reconciled with the model by noting that there is nothing inconsistent about the fact that an individual may prefer the alternative of attending movie x in the company of friends to the alternative of attending movie y alone, even though he prefers attending movie y alone to attending movie x alone.[3]

In contrast, the objection to be considered in this section is that, even when these face-saving, or group pressure difficulties are absent, there may nonetheless

[3]However, to dismiss this sort of difficulty this glibly is to ignore the possibility that the model may be extremely sensitive to the correct specification of the alternative set, X. We will return to this question in the next section.

be inconsistencies between stated preferences and actual choices. For example, in an experiment conducted at Purdue University by Professors F. M. Bass, E. A Pessemier, and D. R. Lehmann [1972], 280 subjects were required to select a 12-ounce can of soft drink four days a week for three weeks from the set of alternatives shown in Table 1.

	Non-Diet	**Diet**
Cola	Coca-Cola	Tab
	Pepsi-Cola	Diet Pepsi
Lemon-Lime	7-Up	Like
	Sprite	Fresca

Table 2.1: Soft Drink Choices.

For participating in the experiment, the subjects each received $3 in addition to 12 cans of soft drink. To quote from the study (p. 533):

> In order to keep the selection as natural as possible and to control the effect of specific purchase and use contexts, subjects were allowed to make their choice any time between 9:00 a.m. and 12:30 p.m. in a room which adjoined the student lounge where soft drinks, candy, etc., are available in vending machines. All the subjects had a reason to be in the building daily between those times.[4]

In addition to making the choices, participants were required to fill out three different questionnaires at various times; in one of which (Questionnaire 2), they were asked to rank-order the eight brands in terms of their own preferences. This questionnaire was filled out at the beginning of the experiment, and at the end of each of the three weeks. The accuracy of the choice predictions based on the preference rankings, as well as on the basis of the last-period choice, are summarized in Table 2, below (Bass, Pessemier, and Lehmann [1972], Table 4, p. 537).[5]

Model	**Percentage of Correct Choice Predictions**
Stated first choice (Post)	52.1
Stated first choice (Pre)	50.8
Last period choice	37.1
Random	12.5

Table 2.2: Choice Probabilities.

There are a number of facets of this experiment which deserve further consideration, and in fact in the next four sections we will be discussing various aspects of this

[4] The subjects were all Purdue students and/or secretaries.

[5] In Table 2, 'Post' means that the first choice is based upon Questionnaire 2 the first time it was asked *after* the choice, while 'Pre' means the Questionnaire 2 response obtained most recently *before* the actual choices.

experiment and their relationship to the conventional economic theory of demand. For the sake of convenience, and with apologies to Professors Bass, Pessemier, and Lehmann, we will hereafter refer to their experiment as the 'BPL experiment.'

2.5 The Specification of the Primitive Terms

Suppose we begin by considering the basic framework of the economic theory of demand, as developed in a typical textbook. From a formal point of view, we can say that this theory is the special case of the general algebraic choice model which is obtained by taking X to be a non-empty subset of $\mathbb{R}^n$ (for the sake of convenience at this point, we will take $X = \mathbb{R}^n_+$), and $\mathcal{B}$ to be the family of all subsets of X having the form:

$$B = b(\boldsymbol{p}, w) = \{\boldsymbol{x} \in \mathbb{R}^n_+ \mid \boldsymbol{p} \cdot \boldsymbol{x} \leq w\} \quad \text{for } \boldsymbol{p} \in \mathbb{R}^n_{++}, w \in \mathbb{R}_+.$$

If $\boldsymbol{x} = (x_1, x_2, \ldots, x_n)$ is an element of $\mathbb{R}^n_+$, the ith coordinate of $\boldsymbol{x}$, x_i, denotes the quantity of the ith commodity available per unit of time. Similarly, 'p_i' denotes the price per unit of the ith commodity. In terms of the formal development of the theory, however, 'commodity,' 'price,' 'wealth,' 'unit of time,' and 'consumer' are all **primitives** of the theory; that is, they are undefined terms, just as 'point' and 'line' are undefined terms, or primitives, in Euclidean geometry. To be sure, since economics is (or at least is partially) an empirical science, rather than a branch of mathematics, there is an implicit claim that there are empirical counterparts, or specific interpretations, of these primitives such that the assumptions of the theory are satisfied in actual choice situations (given these interpretations).[6] However, most textbooks are conveniently vague as to the claimed applicability of the theory; that is, most textbooks never state explicitly under which interpretations of the primitives it is being claimed that the assumptions of the model will hold in actual choice situations. It would appear, however, that most members of the economics profession would feel safest with something like the following specifications of these primitive terms.

S.1 'Individual consumer' is (for economists in the U. S.) taken to be an individual household, as defined by the U. S. Bureau of the Census.[7]

S.2 The list of consumer commodities should be exhaustive in terms of the immediate locality involved (including everything available within, say, a two-hour drive, or by mail order, or the web, in the locality in question).

S.3 The 'commodities' should be specified precisely enough so that different units of what we are calling the 'same' commodity should be essentially indistinguishable, with regard to physical characteristics, location, time of availability,

[6] Hopefully it is obvious that this distinction between a primitive term in a theory and an empirical counterpart, or possible interpretation of the term, has nothing whatsoever to do with whether we are undertaking a 'mathematical' or a 'non-mathematical' development of the theory.

[7] Because of this, to remind ourselves that the term 'consumer' is a primitive, and to avoid sexist connotations, in theoretical discussions we will generally use the pronoun 'it' in referring to an individual consumer.

and state of the world in which the commodity is available. This last aspect of the definition of a commodity brings uncertainty into consideration, and will be ignored for the remainder of this chapter; although we will return to this problem in a later chapter.

S.4 The time involved will be taken to be one month; thus we would interpret x_i to be the quantity of the ith commodity available per month.

S.5 The 'unit' in which a given commodity is measured can be any convenient unit in terms of which the commodity is actually sold, implicitly or explicitly (thus we might use 'fluid ounces' as the unit of measurement for milk or soft drinks, weight in ounces [or grams] for bread, fruits, vegetables, etc.).

S.6 'Price' will be interpreted as price per unit at the beginning of the month.

S.7 'Wealth' will be interpreted as 'planned total consumption expenditure' for the month.

The above list of interpretations (or specifications) is not claimed to be definitive, and in fact, the definitions set out are not really sufficiently precise for econometric work; although they should be sufficiently precise and detailed for the purposes of our present discussion. Furthermore, while (for what it's worth) I feel most confident about the empirical validity of the theory under the above list of specifications, I am certainly *not* claiming that the economic theory of demand is only valid under the above list of specifications of the primitives. I have set out this list here primarily to make one basic point: even if we are confident that the economic theory of consumer demand is empirically valid under one set of specifications of the primitives of the theory, there is no reason to suppose, on *a priori* grounds, that it is empirically valid under some alternative specification of the primitives, *unless we can demonstrate (deductively) that its validity under the first specification implies its validity under the second specification as well.*

We will refer to the theory with the above list of specifications as the **standard economic theory of demand**. This terminology is simply a convenient label, and should not be taken to mean that all economists feel most confident about the validity of the model with these specifications; however, as the reader will probably agree, this specification has some claim to a consensus status. Since the reader has no doubt already encountered many discussions presenting *a priori* and/or introspective reasons for believing the theory to be empirically valid under something like the above specifications, we will confine our discussion here to one or two remarks about these specifications.

First of all, insofar as item S.1 is concerned, the basic reason for this specification is that most economists' confidence as to the validity of the theory is inversely proportional to the size of the decision-making unit; in fact, the theory seems to have been developed with the idea that 'individual consumer' should be specified to mean an 'individual human being.' On the other hand, many individuals (most notably, dependent children) do not make consumption decisions for themselves, and adult members of a multiple-person household presumably make joint decisions on many consumption items. Consequently, it would appear that the household is the smallest entity that can be treated as an independent consumption unit.

The reason for specifying the unit of time to be a month hinges around two basic considerations. First, the time period should be long enough to allow for *expected* variety-seeking behavior (most people would make a different selection from a given restaurant menu at breakfast than they would at lunch even if the expenditure involved was exactly the same), but not so long as to allow for *expected* changes in taste (a given individual's preferences over commodities are likely to be very different at 60 than they were at 20 years of age). Secondly, it appears that most households actually do some formal budget planning for each month (or so I hear).

Insofar as the specification of price is concerned, p_i would probably be best interpreted as the 'expected average (per unit) price of the ith commodity for the forthcoming month.' However, because of the difficulty in predicting a household's expected average price, most economists would probably generally settle for interpreting p_i to be the price per unit at the beginning of the month (since this is when most formal budget planning seems to be done) and hope for the best.

Finally, it should be mentioned that the 'wealth' specification is a rather tautological definition that I have used here only for the sake of convenience. A more meaningful specification of the wealth variable can only be made, however, after we have gone into some aspects of the specification of the consumption set, X, which we will not take up until a later chapter.

Suppose we now re-consider the BPL experiment in light of the above discussion. As we shall see, this single experiment could be regarded as a test of many different theories; and what is more important, from the standpoint of our present discussion, it can be regarded as a test of the empirical validity of the algebraic choice model under a number of different specifications of the primitives. However, it is probable that the simplest way of viewing the experiment is as a test of the theory obtained when items S.1 and S.4 in the specification of the 'standard theory' are changed to:

S.1′ 'Individual consumer' is taken to be an 'individual subject' of the experiment.

S.4′ The time unit involved is taken to be one day: thus we will interpret x_i to be the quantity of the ith soft drink (in number of 12-ounce cans) available per day.

We will also simplify things drastically by assuming that there are only two brands of soft drinks available. While this is obviously unrealistic, the presence of two distinct brands in our theoretical model will suffice to illustrate the points to be made in our discussion. We will also suppose that the consumers' consumption set, X, can be taken to be $\mathbb{R}^n_+$, and that the first two coordinates measure the quantities available for consumption of these two different brands of soft drink; with

$$x_j = \text{the quantity of brand } j \text{ in number of 12-ounce cans, for } j = 1, 2.$$

We will hereafter refer to the special case of the economic theory of demand in which $S.1'$ and $S.4'$ are substituted for $S.1$ and $S.4$, respectively, and with the convention indicated for the first two coordinates as the **soft drink model**.

Before proceeding further with our discussion, we should take note of the fact that a consumer could, in principle, satisfy all of the assumptions of the standard economic theory of demand, yet not have well-defined preferences over daily consumption. We will ignore this possibility for the moment (more correctly, we will simply assume that assumption $S.4'$ holds), but we will return to this issue in Section 8 of this chapter.

There remains a further bit of difficulty concerning the nature of the specification tested in the BPL experiment, however, stemming from the fact that, in the context of the model, the notion of brand preference is not necessarily well-defined. In the next several sections of this chapter, we will consider several different ways of defining what is meant by the statement that a consumer prefers one brand of soft drink over all others. The issue which is our initial concern is this: if we ask a given subject to rank-order Brands 1 and 2 in order of preference, and he or she responds that Brand 1 is preferred to Brand 2, what should we take this to mean? The first problem we face in trying to define a straightforward interpretation of brand preference is that a given subject's preference for soft drinks might depend upon his or her other consumption for the day. If this possibility sounds slightly far-fetched to you, consider the preferences commonly expressed by wine afficiandos: white wine, rather than red, with fish or fowl; red wine with red meat. In particular, in the BPL experiment, any given subject's preferences over brands of soft drinks might depend upon what he or she was having for lunch. In any event, in the next section we will tackle a formal consideration of this problem.

2.6 Weak Separability of Preferences

As noted in the previous section, in the 'usual' case, preferences over, for example, soft drinks will depend upon the consumption of other items. The case in which such dependence does not occur is, by definition, that in which preferences are separable. Our discussion of separability will be confined to a very simple situation, as compared to other such discussions in the literature; in that we will limit our consideration to the case in which the (exhaustive) list of commodities available can be divided into two groups in such a way that preferences over one of the commodity groups is weakly separable. Other authors deal with many commodity groups, and with other forms of separability.[8]

Throughout the remainder of this section, we will suppose that X is a subset of $\mathbb{R}^n$, where $n \geq 2$, and that X can be written in the form:

$$X = Y \times Z,$$

where:

$$Y \subseteq \mathbb{R}^{k_1} \text{ and } Z \subseteq \mathbb{R}^{k_2}, \quad k_i \geq 1, \text{ for } i = 1, 2, \text{ and } k_1 + k_2 = n;$$

[8]For more comprehensive treatments, see Blackorby and Davidson [1991], Blackorby and Russell [1994], and Mak [1986]. For earlier surveys, see Katzner [1970], pp. 27–32 and 78–90; and, for the definitive treatment of the mathematics of separability, see Blackorby, Primont, and Russell [1978].

and that $\succsim$ is a weak order on X. We will then follow the convention of denoting points (commodity bundles) in X by:

$$\boldsymbol{x} = (\boldsymbol{y}, \boldsymbol{z}) \quad \text{where } \boldsymbol{y} \in Y \text{ and } \boldsymbol{z} \in Z.$$

2.2 Definition. For each $\boldsymbol{z}^* \in Z$, we define the conditional preference relation, $\succsim_{\boldsymbol{z}^*}$, on Y by:

$$\boldsymbol{y} \succsim_{\boldsymbol{z}^*} \boldsymbol{y}' \iff (\boldsymbol{y}, \boldsymbol{z}^*) \succsim (\boldsymbol{y}', \boldsymbol{z}^*).$$

Similarly, given any $\boldsymbol{y}^* \in Y$, we define $\succsim_{\boldsymbol{y}^*}$ on Z by:

$$\boldsymbol{z} \succsim_{\boldsymbol{y}^*} \boldsymbol{z}' \iff (\boldsymbol{y}^*, \boldsymbol{z}) \succsim (\boldsymbol{y}^*, \boldsymbol{z}').$$

The proof of the following result is more or less immediate, and will be left as an exercise.

2.3 Proposition. *Given any $(\boldsymbol{y}^*, \boldsymbol{z}^*) \in Y \times Z$, the conditional preference relations, $\succsim_{\boldsymbol{z}^*}$ and $\succsim_{\boldsymbol{y}^*}$, are weak orders on Y and Z, respectively.*

2.4 Example. Let $n = 3, X = \mathbb{R}^3_+$, and consider the weak order, $\succsim$ defined on X by:

$$\boldsymbol{x} \succsim \boldsymbol{x}^* \iff u(\boldsymbol{x}) \geq u(\boldsymbol{x}^*),$$

where the utility function $u(\cdot)$ is defined on X by:

$$u(\boldsymbol{x}) = (x_1 x_2)^{1/2} + (x_2 x_3)^{1/2} \quad \text{for } \boldsymbol{x} \in X.$$

Notice that X can be written in the form:

$$X = Y \times Z,$$

where:

$$Y = \mathbb{R}^2_+ \text{ and } Z = \mathbb{R}_+$$

(and thus $k_1 = 2$ and $k_2 = 1$). Clearly, if $z^* \in Z = \mathbb{R}_+$ (since $Z = \mathbb{R}_+$, we will use simply 'z,' rather than '$\boldsymbol{z}$,' to denote the second sub-vector of $\boldsymbol{x}$), the conditional preference relation, $\succsim_{z^*}$, is given by:

$$\boldsymbol{y} \succsim_{z^*} \boldsymbol{y}' \iff (y_1 y_2)^{1/2} + (\sqrt{z^*})\sqrt{y_2} \geq (y_1' y_2')^{1/2} + (\sqrt{z^*})\sqrt{y_2'}.$$

Thus, for example, if $z^* = 1$, $\succsim_{z^*}$ is representable by the conditional utility function:

$$u(\boldsymbol{y}; z = 1) = (y_1 y_2)^{1/2} + \sqrt{y_2};$$

while if $z^* = 9$, $\succsim_{z^*}$ is representable by the conditional utility function:

$$u(\boldsymbol{y}; z = 9) = (y_1 y_2)^{1/2} + 3\sqrt{y_2}.$$

Since $u(\cdot; z = 1)$ is *not* an increasing transformation of $u(\cdot; z = 9)$, it is clear that the corresponding conditional preference relations are not the same. We can verify this by considering the points:

$$\boldsymbol{y} = (16, 16) \text{ and } \boldsymbol{y}^* = (1, 64).$$

We have:

$$u_1(\boldsymbol{y}; z = 1) = 16 + 4 = 20 > u(\boldsymbol{y}^*; z = 1) = 8 + 8 = 16,$$

so that for $z = 1$, we have $\boldsymbol{y} \succ_z \boldsymbol{y}^*$. On the other hand, for $z = 9$:

$$u(\boldsymbol{y}; z = 9) = 28 < u(\boldsymbol{y}^*; z = 9) = 32;$$

and thus, with $z^* = 9$:

$$\boldsymbol{y}^* \succ_{z^*} \boldsymbol{y}.$$

We see then, that if $\succsim$ is the preference relation of a consumer, then said consumer prefers having $\boldsymbol{y} = (16, 16)$ to having $\boldsymbol{y}^* = (1, 64)$ if there is only one unit of the third commodity available; but the consumer prefers having $\boldsymbol{y}^*$ to having $\boldsymbol{y}$ if it has 9 units of the third commodity. □

The above example is illustrative of the usual case; we will usually find that if, say $\boldsymbol{z} \neq \boldsymbol{z}^*$, then the conditional preference relations, $\succsim_{\boldsymbol{z}}$ and $\succsim_{\boldsymbol{z}^*}$, will not be the same. Thus, if 'hours of automotive use' and 'shoes' are included in the first group of commodities, and 'gallons of gasoline per month' in the second group, we would likely find that the marginal rate of substitution between the former two commodities would depend upon the quantity of gasoline available, so that $\succsim_{\boldsymbol{z}}$ would be different from $\succsim_{\boldsymbol{z}^*}$, for $\boldsymbol{z} \neq \boldsymbol{z}^*$. On the other hand, if, for example, the first commmodity group contained all foodstuffs, while the second group contained all other commodities, then we might find the assumption that all marginal rates of substitution between commodities in the first group are independent of the quantities in the second group to be a little more plausible. More formally, we might in this latter case expect the following condition to hold.

2.5 Definition. If $X = Y \times Z$, and Z^* is a non-empty subset of Z, we shall say that $\succsim$ is **weakly separable in $\boldsymbol{y}$ over Z^*** iff, for each $\boldsymbol{z}$ and $\boldsymbol{z}^*$ in Z^*, we have:

$$\succsim_{\boldsymbol{z}} \equiv \succsim_{\boldsymbol{z}^*} .$$

If $\succsim$ is weakly separable in $\boldsymbol{y}$ over Z, we will simply say that $\succsim$ **is weakly separable on $\boldsymbol{Y}$**.

Similarly, if Y^* is a non-empty subset of Y, we shall say that $\succsim$ is **weakly separable in $\boldsymbol{z}$ over $\mathbf{Y}^*$** iff, for each $\boldsymbol{y}$ and $\boldsymbol{y}^*$ in Y^*, we have:

$$\succsim_{\boldsymbol{y}} \equiv \succsim_{\boldsymbol{y}^*};$$

and if $Y^* = Y$, we will say that $\succ$ **is weakly separable on $\boldsymbol{Z}$**.

In other words, for example, $\succsim$ is weakly separable in $\boldsymbol{y}$ if the consumer's preferences over sub-bundles from Y are independent of how much $\boldsymbol{z}$ is available to it. We have already looked at a case in which preferences were not weakly separable; the concept will probably be a great deal more clear, however, if we also take a look at a case in which preferences *are* weakly separable.

2.6 Example. Let $X = \mathbb{R}^n_+$, where $n \geq 2$, let k_1 and k_2 be positive integers such that $k_1 + k_2 = n$, and let $\succsim$ be representable on X by the Cobb-Douglas function:

$$u(\boldsymbol{x}) = \prod_{j=1}^{n} x_j^{a_j}, \tag{2.4}$$

where:

$$a_j \geq 0, \text{ for } j = 1, \ldots, n; \text{ and } \alpha \stackrel{\text{def}}{=} \sum_{j=1}^{n} a_j = 1. \tag{2.5}$$

We also define two subsets of $\mathbb{R}^{k_2}_+$, Z_1 and Z_2, by:

$$Z_1 = \Big\{\boldsymbol{z} \in \mathbb{R}^{k_2}_+ \mid \prod_{j=k_1+1}^{n} z_j > 0\Big\},$$

and:

$$Z_2 = \Big\{\boldsymbol{z} \in \mathbb{R}^{k_2}_+ \mid \prod_{j=k_1+1}^{n} z_j = 0\Big\},$$

respectively.

If $\boldsymbol{z}^* \in Z_1$, we have:

$$\boldsymbol{y} \succsim_{\boldsymbol{z}^*} \boldsymbol{y}' \iff \Big(\prod_{j=1}^{k_1} (y_j)^{a_j}\Big) \cdot \Big(\prod_{j=k_1+1}^{n} (z_j^*)^{a_j}\Big) \geq \Big(\prod_{j=1}^{k_1} (y_j')^{a_j}\Big) \cdot \Big(\prod_{j=k_1+1}^{n} (z_j^*)^{a_j}\Big);$$

or equivalently, since $\prod_{j=k_1+1}^{n} (z_j^*)^{a_j} > 0$:

$$\boldsymbol{y} \succsim_{\boldsymbol{z}^*} \boldsymbol{y}' \iff \prod_{j=1}^{k_1} (y_j)^{a_j} \geq \prod_{j=1}^{k_1} (y_j')^{a_j}$$

Thus, defining the function $u_1 \colon \mathbb{R}^{k_1}_+ \to \mathbb{R}_+$ by:

$$u_1(\boldsymbol{y}) = \prod_{j=1}^{k_1} (y_j)^{a_j}, \tag{2.6}$$

we see that if we let '$\succsim_1$' denote the weak order induced on Y by $u_1(\cdot)$, we have, for each $\boldsymbol{z}$ from Z_1:

$$\succsim_{\boldsymbol{z}} \equiv \succsim_1 .$$

I will leave as an exercise the task of showing that $\succsim$ is also weakly separable in $\boldsymbol{y}$ over Z_2, but that, for each $\boldsymbol{z} \in Z_2$, $\succsim_{\boldsymbol{z}}$ is the trivial ordering of $Y \equiv \mathbb{R}^{k_1}_+$. □

It is obvious that any weak order will be weakly separable in $\boldsymbol{y}$ over Z^* if Z^* is a singleton; and, for less obvious reasons, the concept of weak separability in $\boldsymbol{y}$ is also not very interesting[9] in the case where $k_1 = 1$. Consequently, our interest in this definition centers around the situation in which $k_1 > 1$, and Z^* contains more than one element. However, the known and interesting results concerning weak separability do not generally require these restrictions (except that they may

[9] 'Most' weak orders of interest in connection with demand theory are weakly separable in a single variable. In particular, you should have no difficulty in showing that if $\succsim$ is strictly increasing in the first variable, or first commodity, then it is weakly separable in that variable.

require that $Z^* = Z$). Consequently, there is no reason to complicate our definition of weak separability by excluding these two cases.

Another fact of which we should take notice in connection with weak separability is that the condition is *not* symmetric. That is, for example, we may find that $\succsim$ is weakly separable in $\boldsymbol{y}$ over Z, but that it is *not* weakly separable in $\boldsymbol{z}$ over Y. Our next example illustrates such a case.

2.7 Example. Let $X = \mathbb{R}^4_+$, $k_1 = k_2 = 2$, $Y = Z = \mathbb{R}^2_+$, define the function $u\colon X \to \mathbb{R}_+$ by:

$$u(\boldsymbol{x}) = \big[\min\{x_1, x_2\}\big] \cdot \big[x_3 + 1\big] + x_4;$$

and let $\succsim$ be the weak order induced on X by $u(\cdot)$. It is then very easy to show that, while $\succsim$ is weakly separable in $\boldsymbol{y}$ over Z, $\succsim$ is *not* weakly separable in $\boldsymbol{z}$ over Y. □

Now, one might very well argue that there is no *a priori* reason to suppose that a given individual's preferences would be weakly separable in any commodity sub-group; but notice that the theory would be much more useful in application if weak separability were the rule, rather than the exception! In fact, the data requirements in dealing with an exhaustive list of finely-differentiated commoties are so enormous that I know of no empirical study which has been undertaken in such a context. Moreover, the notion of separability of preferences has a number of interesting implications; for example to provide sufficient conditions for 'two-stage budgeting' (Blackorby and Russell [1997]), and to some issues of Social Choice (LeBreton and Sen [1999]).

Now, the question of immediate concern is, what does this notion of weak separability have to do with defining brand preference? Well, even before attempting a formal definition of brand preference, we can already note that it is apparent that unless consumer preferences are weakly separable in the soft drink component, the notion of brand preference is going to be more than a bit ambiguous. In other words, if the consumer's conditional preferences over soft drinks is *not* independent of other consumption, then it is not immediately apparent how one could unambiguously define what is meant by the statement that one brand of soft drink is preferred to the others.

In order to formally define a connection between stated brand preference and preferences over commodity bundles, let's begin by supposing that $X = \mathbb{R}^n_+$. Given that this is the case, we can write:

$$X = \mathbb{R}^2_+ \times \mathbb{R}^m_+,$$

where $m = n - 2$. We will then use the generic notation:

$$\boldsymbol{x} = (\boldsymbol{y}, \boldsymbol{z}),$$

to denote commodity bundles, $\boldsymbol{x} \in X$, where $\boldsymbol{y} \in \mathbb{R}^2_+$ and $\boldsymbol{z} \in \mathbb{R}^m_+$. As already suggested, we will suppose throughout the remainder of this section, that each subject's preference ordering is weakly separable in $\boldsymbol{y}$ over $\mathbb{R}^m_+$.

Now, suppose a subject tells us that Brand 1 (of soft drink) is preferred to Brand 2.[10] The question is, what can we interpret this to mean? In developing a formal definition, let's introduce the notation: $\langle 1, 2\rangle$ as shorthand for the statement (by the subject), "Brand 1 is preferred to Brand 2;" with $\langle 2, 1\rangle$ indicating that the subject's response is: "Brand 2 is preferred to Brand 1." Furthermore, making use of the assumption of weak separability, we define the relation G on $\mathbb{R}^2_+$ by:

$$\boldsymbol{y}G\boldsymbol{y}' \iff (\boldsymbol{y}, \boldsymbol{z}^*) \succsim (\boldsymbol{y}', \boldsymbol{z}^*), \tag{2.7}$$

for some 'reference bundle,' $\boldsymbol{z}^* \in \mathbb{R}^m_+$;[11] using '$P$' and '$I$' to denote the asymmetric and symmetric parts of G, respectively. Using this notation, we can make the following assumption.

A1. The response $\langle 1, 2\rangle$ by a given subject implies that, in terms of this subject's preference ordering, we must have:

$$\big((1,0), \boldsymbol{z}^*\big) \succ \big((0,1), \boldsymbol{z}^*\big),$$

for some 'reference bundle,' $\boldsymbol{z}^* \in \mathbb{R}^m_+$; or, more compactly:

$$(1,0)P(0,1).$$

Similarly, the response $\langle 2, 1\rangle$ implies that:[12]

$$(0,1)P(1,0).$$

Now the question is, if the other assumptions of the soft drink model are satisfied, and if A1 is correct, will the subjects' statements about brand preferences correlate perfectly with the choices actually made? A moment's thought will undoubtedly suffice to raise some doubt in your mind about this. The basic difficulty (given separability of preferences) is that the can of soft drink chosen as part of the experiment may or may not be the *only* can of soft drink consumed by the subject in a given day; and it may well be, for example, that for a given subject:

$$(1,1)P(2,0),$$

even though:

$$(1,0)P(0,1).$$

It would appear that what is happening in the experiment is that something (namely, a can of soft drink) is being added to the subject's budget for the day. Consequently, the difficulty alluded to in the above paragraph may well cause a discrepancy between stated preferences and actual choice *unless* preferences are additive in the soft drink component; a condition which we will consider in the next section.

[10] Remember that for the sake of simplicity we are now supposing that there are only two brands of soft drink available.

[11] Notice that, given weak separability of preferences, it makes no difference what reference bundle is chosen.

[12] Recall that each subject was allowed to choose exactly one can of soft drink per day, as a part of the experiment.

2.7 Additive Separability

The formal difficulty with the assumption concerning brand preferences which was discussed in the previous section is that, in the notation of that section, we may find that, for some $\boldsymbol{y}, \boldsymbol{y}' \in \mathbb{R}^2_+$, we have $\boldsymbol{y}P\boldsymbol{y}'$, yet, for some $\boldsymbol{x}^* \in \mathbb{R}^n_+$:

$$\big((\boldsymbol{y}', \mathbf{0}) + \boldsymbol{x}^*\big) \succ \big((\boldsymbol{y}, \mathbf{0}) + \boldsymbol{x}^*\big).$$

In other words, even if a consumer's preferences are weakly separable in $\boldsymbol{y}$ on all of Z, and we find that commodity bundle one is preferred to commodity bundle two, it does not follow that if we add bundle one to her other consumption that she will consider herself better off than if bundle two had been added to her other consumption. This assertion is verified by the following.

2.8 Example. Let $X = \mathbb{R}^3_+$, and let $\succsim$ be the weak order induced on X by the utility function:

$$u(\boldsymbol{x}) = (x_1 + 1)^2 \cdot (x_2 + 1) \cdot (x_3 + 1).$$

In this case, if we define $Y = \mathbb{R}^2_+$ and $Z = \mathbb{R}_+$, it is easily verified that $\succsim$ is weakly separable in $\boldsymbol{y}$ over all of $Z = \mathbb{R}_+$, and that the conditional preference ordering on Y is representable by the function $v\colon \mathbb{R}^2_+ \to \mathbb{R}_+$ given by:

$$v(\boldsymbol{y}) = (y_1 + 1)^2 \cdot (y_2 + 1).$$

We also have:

$$v(1, 0) = 4 > v(0, 1) = 2;$$

so that, if the first two coordinates of each commodity bundle represent quantities of Brands 1 and 2 of soft drinks, respectively, then according to our assumption A1, this consumer prefers Brand 1 to Brand 2. However, letting:

$$\boldsymbol{x}^* = (2, 0, 1),$$

we have:

$$\big((1, 0), 0\big) + \boldsymbol{x}^* = (3, 0, 1),$$

and thus $u(3, 0, 1) = 32$. On the other hand:

$$\big((0, 1), 0\big) + \boldsymbol{x}^* = (2, 1, 1),$$

and $u(2, 1, 1) = 36$, so that:

$$\big[\big((0, 1), 0\big) + \boldsymbol{x}^*\big] \succ \big[\big((1, 0), 0\big) + \boldsymbol{x}^*\big]. \quad \square$$

So, the above example indicates a problem with our definition/assumption A1 even if preferences are weakly separable in soft drinks. However, consider the following definition.

2.9 Definition. Suppose $\succsim$ is a weak order on a non-empty subset of $\mathbb{R}^n$, X, which is of the form:

$$X = Y \times Z,$$

where $Y \subseteq \mathbb{R}^{k_1}$, $Z \subseteq \mathbb{R}^{k_2}$, with $k_i \geq 1$, for $i = 1, 2$, $k_1 + k_2 = n$; and suppose Y is a convex cone. We shall say that $\succsim$ is **additively separable in $\boldsymbol{Y}$** iff:

1. $\succsim$ is weakly separable in $\boldsymbol{y}$ over Z, and:
2. for every $\boldsymbol{y}_1, \boldsymbol{y}_2 \in Y$, and every $\boldsymbol{x}^* \in X$, we have:[13]

$$\boldsymbol{y}_1 G \boldsymbol{y}_2 \iff (\boldsymbol{y}_1, \mathbf{0}) + \boldsymbol{x}^* \succsim (\boldsymbol{y}_2, \mathbf{0}) + \boldsymbol{x}^*.$$

where G is the conditional weak preference order on Y derived from $\succsim$.

2.10 Example. Let $X = \mathbb{R}^3_+ = \mathbb{R}^2_+ \times \mathbb{R}_+$, and let $\succsim$ be the weak order induced on X by the function:

$$u(\boldsymbol{x}) = x_1 + x_2 + 2\sqrt{x_3},$$

and define $Y = \mathbb{R}^2_+$ and $Z = \mathbb{R}_+$. It is then easy to see that, in terms of the notation introduced in the above definition, if $\boldsymbol{y}_i \in Y$, for $i = 1, 2$, and $\boldsymbol{x}^*$ is any element of $X = \mathbb{R}^3_+$, then:

$$\boldsymbol{y}_1 G \boldsymbol{y}_2 \iff (\boldsymbol{y}_1, \mathbf{0}) + \boldsymbol{x}^* \succsim (\boldsymbol{y}_2, \mathbf{0}) + \boldsymbol{x}^*.$$

Consequently, $\succsim$ is additively separable in Y in this case. □

Our assumption about the meaning of stated brand preference says that if a subject states that brand 1 is preferred to brand 2 (retaining the assumption that there are only two brands to be concerned with), then this means that, in terms of conditional *strict* preference:

$$(1, 0)P(0, 1). \tag{2.8}$$

This brings us back to the problem mentioned earlier, namely, if (8) holds, does this also mean that:

$$(2, 0)P(1, 1)?$$

Moreover, up to this point we have ignored a possible stronger definition of preference: we might interpret the statement, "Brand 1 is preferred to Brand 2," to mean that if $\boldsymbol{y}$ and $\boldsymbol{y}'$ are elements of $\mathbb{R}^2_+$ for which:

$$y_1 + y_2 = y_1' + y_2'$$

then:

$$y_1 > y_1' \Rightarrow \boldsymbol{y} P \boldsymbol{y}'$$

[13]Notice that, since $\boldsymbol{x}^*$ is of the form:

$$\boldsymbol{x}^* = (\boldsymbol{y}^*, \boldsymbol{z}^*),$$

the sum $((\boldsymbol{y}_i, \mathbf{0}) + \boldsymbol{x}^*$ is of the form:

$$((\boldsymbol{y}_i, \mathbf{0}) + \boldsymbol{x}^*) = (\boldsymbol{y}_i + \boldsymbol{y}^*, \boldsymbol{z}^*).$$

Therefore, since Y is a convex cone, $((\boldsymbol{y}_i, \mathbf{0}) + \boldsymbol{x}^*)$ is an element of X. See Proposition 6.6, in Chapter 6.

So, how does this stronger definition of brand preference compare with the definition originally set out?

Well, as it turns out, there is no conflict between these two definitions of brand preference if preferences are additively separable in Y, and are representable; for, given some technical qualifications which needn't concern us here, under these conditions, there must be a utility function, u, which takes the form:

$$u(\boldsymbol{y}, \boldsymbol{z}) = \phi(\boldsymbol{a} \cdot \boldsymbol{y}) + \psi(\boldsymbol{z}), \tag{2.9}$$

where $\boldsymbol{a}$ is a (fixed) semi-positive vector in $\mathbb{R}^{k_1}$, $\phi\colon \mathbb{R}_+ \to \mathbb{R}_+$, and $\psi\colon Z \to \mathbb{R}_+$. I will leave it to you to show that, given that preferences can be represented by a utility function of the form (2.9), if we find that (maintaining our assumption that there are only two brands of soft drinks, and thus that $k_1 = 2$):

$$(1, 0)P(0, 1), \tag{2.10}$$

then, for any two vectors, $\boldsymbol{y}, \boldsymbol{y}' \in Y$, we have that:

$$[y_1 + y_2 = y_1' + y_2' \ \&\ y_1 > y_1'] \Rightarrow \boldsymbol{y}P\boldsymbol{y}'. \tag{2.11}$$

Thus, if preferences over daily consumption are additively separable in the soft drink component, and *are constant from day to day,* it would appear that either interpretation of the meaning of brand preference which we have considered would imply perfect agreement between stated brand preference, and actual choice in the soft drink experiment. However, this brings us back to the issue of whether it is reasonable to suppose that preferences over daily consumption remain constant from day to day. We will consider this question in the next section.

2.8 Sequential Consumption Plans

It was noted in Section 5 that a consumer might satisfy all the assumptions of the standard economic theory of demand, and yet not have well-defined preferences over daily consumption. Thus, a given consumer may satisfy all the assumptions of the standard economic theory of demand, and yet not satisfy the Soft Drink Model; at least not in the sense of having invariant preferences from one day to the next. In order to establish this fact, and to explore the reasons for it, we will develop a model in this section which *explains* the standard economic theory of demand. Because this model *explains* the standard theory (that is, its assumptions imply those of the standard theory, with the appropriate specifications of the primitives) it is, in effect, a special case of the standard theory; and, it should be emphasized, there are other special cases of the standard economic theory of demand in which the consumer's daily preferences would be well-defined. On the other hand, the model to be developed here [which we will call the **sequential consumption plan (SCP) model**] seems sufficiently plausible and interesting as to merit the time which we will spend on its development.

In order to motivate our discussion, let's begin by considering a consumer that satisfies all of the assumptions of the standard theory. Let us further assume, for the sake of simplicity, that when said consumer chooses the bundle $\boldsymbol{x}^*$ from the budget set, B, this means that it purchases the bundle $\boldsymbol{x}^*$ at the beginning of the month. In terms of the context in which the standard theory was formulated, this means that the consumer makes $\boldsymbol{x}^*$ available for its consumption[14] over the month to come. The next question which arises, however, is how, more specifically, at what rate, will the consumer choose to consume the bundle $\boldsymbol{x}^*$? We would probably be quite surprised if the consumer proceeded to consume $\boldsymbol{x}^*$ at the rate:

$$\boldsymbol{z}^* \equiv (1/30)\boldsymbol{x}^*,$$

per day (assuming, for the sake of convenience, that there are 30 days in a month); but isn't this exactly what would happen if the consumer's preferences over daily consumption were exactly the same from day to day?

Let's take a look at this question from a bit different point of view. Suppose that at the beginning of each month, our consumer considers alternative sequences of consumption of the form:

$$\boldsymbol{z} = (\boldsymbol{z}_1, \boldsymbol{z}_2, \ldots, \boldsymbol{z}_{30}),$$

where $\boldsymbol{z}_t \in \mathbb{R}^n_+$ denotes planned consumption of the tth day ($t = 1, \ldots, 30$). Suppose further that our consumer's preferences over such sequences of consumption plans constitutes a weak order, G, on the set $\mathfrak{Z}$ of admissible sequences of this type, where $\mathfrak{Z}$ is a subset of $\mathbb{R}^{30n}_+$. In fact, will suppose that $\mathfrak{Z}$ is of the form:

$$\mathfrak{Z} = \prod\nolimits_{t=1}^{30} Z_t \tag{2.12}$$

where:

$$Z_t = \mathbb{R}^n_+ \quad \text{for } t = 1, \ldots, 30. \tag{2.13}$$

In particular, we will suppose that Z_t, the feasible consumption set for day t, is invariant from day to day. For the sake of convenience, we will also suppose that G is representable by a continuous utility function, $U(\cdot)$; that is, we will suppose that $U \colon \mathfrak{Z} \to \mathbb{R}$, and satisfies, for all $\boldsymbol{z}, \boldsymbol{z}' \in \mathfrak{Z}$:

$$\boldsymbol{z}G\boldsymbol{z}' \iff U(\boldsymbol{z}) \geq U(\boldsymbol{z}').$$

We can then relate this situation to the standard theory in the following way. Define the function $u \colon \mathbb{R}^n_+ \to \mathbb{R}$ by:

$$u(\boldsymbol{x}) = \max\left\{U(\boldsymbol{z}) \mid \boldsymbol{z} \in \mathfrak{Z} \;\&\; \sum\nolimits_{t=1}^{30} \boldsymbol{z}_t \leq \boldsymbol{x}\right\}. \tag{2.14}$$

In other words, $u(\boldsymbol{x})$ is the maximum utility which could be obtained from an admissible sequence, $\boldsymbol{z}$, whose daily components add up to a bundle less than or

[14]There is a bit of confusion between stock and flow going on here, which, it is hoped, will not cause you undue confusion. Strictly speaking, we should distinguish between $\boldsymbol{x}^*$, which is a flow, and the consumer's inventory of commodities at the beginning of the month, which is a stock. However, since x_j is measured in terms of quantity per month, the real number x_j will here also be the quantity of the jth commodity on hand at the beginning of the month.

equal to $\boldsymbol{x}$.[15] The function, $u(\cdot)$ defined in (2.14) obviously induces a weak order, $\succsim$, on $\mathbb{R}^n_+$. Consequently, this model is actually a special case of the standard economic theory of demand. The next question, however is this: under what conditions will the corresponding daily preferences be invariant from day to day?

After our discussion in Section 6, it should be clear to you that it is not apparent that we can speak of 'daily preferences' in an unambiguous fashion here unless the weak order, G, is separable. To be more specific, consider an arbitrary $t \in \{1, \ldots, 30\}$, and denote values of $\boldsymbol{z} = (\boldsymbol{z}_1, \ldots, \boldsymbol{z}_{30})$ by:

$$\boldsymbol{z} = (\boldsymbol{z}_{-t}, \boldsymbol{z}_t),$$

where:

$$\boldsymbol{z}_{-t} = (\boldsymbol{z}_1, \ldots, \boldsymbol{z}_{t-1}, \boldsymbol{z}_{t+1}, \ldots, \boldsymbol{z}_{30}).$$

Notice that, for a fixed value of $\boldsymbol{z}_{-t}$, G induces a weak order on $Z_t = \mathbb{R}^n_+$, $G_t(\boldsymbol{z}_{-t})$, defined by:

$$\boldsymbol{z}_t G_t(\boldsymbol{z}_{-t}) \boldsymbol{z}'_t \iff (\boldsymbol{z}_{-t}, \boldsymbol{z}_t) G (\boldsymbol{z}_{-t}, \boldsymbol{z}'_t). \tag{2.15}$$

However, if $\boldsymbol{z}^*_{-t} \neq \boldsymbol{z}_{-t}$, there is no reason to suppose that we will necessarily have:

$$G_t(\boldsymbol{z}_{-t}) \equiv G_t(\boldsymbol{z}^*_{-t}).$$

To see the point of this statement, simply ask yourself whether or not your preferences for pizza versus meatloaf for dinner tonight might not be different if you had had pizza each night for the preceding twenty-nine days than would be the case if you had had meatloaf for each of the preceding twenty-nine dinners.

In the special case in which $G_t(\boldsymbol{z}_{-t})$ is independent of the value of $\boldsymbol{z}_{-t}$, for $t = 1, \ldots, 30$, we shall say that G is **weakly separable in daily consumption**. However, notice that, even if G is weakly separable in daily consumption, it will not necessarily be the case that:

$$G_t = G_1 \quad \text{for } t = 2, \ldots, 30. \tag{2.16}$$

In other words, daily preferences may be different even if G is weakly separable in daily consumption. If, on the other hand, G is weakly separable in daily consumption, and, in addition satisfies (2.16), we shall say that **G is stationary**. It is no doubt abundantly clear to you that there are a number of subtle and difficult problems connected with the analysis and interpretation of this sort of model, and that we have no more than begun to analyze these problems here. However, our goal was simply to introduce the general idea of the SCP model, and to point out some of the reasons why the appropriate specification of the 'unit of time' is so important in the empirical testing of the general algebraic choice model.

Rabin [1998] provides a very interesting little example which is of particular interest in connection with the SCP model. I quote his introduction as follows.

> Say you eat at one of two restaurants every night, either *Blondie's* or *Fat Slice*. You enjoy *Fat Slice* more, but because you also enjoy variety, your utility each evening is as follows.

[15] It can also be shown that, under the assumptions being employed here, the function $u(\cdot)$ will be (well -defined and) continuous on $\mathbb{R}^n_+$.

Utility from *Fat Slice* = 7 if you ate at *Blondie's* last night,
Utility from *Fat Slice* = 5 if you ate at *Fat Slice* last night,
Utility from *Blondie's* = 4 if you ate at *Fat Slice* last night,
Utility from *Blondie's* = 3 if you ate at *Blondie's* last night.

Suppose now that you have eaten at *Blondie's* last night, and let's just consider consumption decisions over a five-day horizon. Obviously, your utility-maximizing choice for tonight is to dine at *Fat Slice*; however, what about tomorrow? If, in fact, your preferences satisfy the assumptions of the SCP model, you will dine at *Blondie's* tomorrow, at *Fat Slice* the next night, and so on; since this provides a total utility of 29 for the five nights together. On the other hand, if you only consider tomorrow's preferences tomorrow, and so on, you will dine at *Fat Slice* each of the following four nights, since this provides a marginal utility of 5 for each night. However, your total utility for the five nights is then only 27! In general, if you alternate consumption between the two restaurants from night to night, you obtain an average utility of 5.5; whereas if you always eat at *Fat Slice*, your average utility will be only 5.[16]

The case discussed here is, I'm afraid, a situation similar to what a lot of us face in real life; a similar sort of anomaly arises in connection with procrastination, for example. For the classic development of this idea, see Phelps and Pollak [1968]. In general this sort of example highlights the possibility of an inconsistency between planning for the future and carrying out those plans; a theoretical possibility which seems to have first been pointed out by Strotz [1955].[17] We will not discuss the theory of intertemporal choice further here, but let me recommend Koopmans [1972a, 1972b], Goldman [1979, 1980], and Loewenstein and Prelec [1992].

2.9 The BPL Experiment Reconsidered

In Section 5, we noted that a subject might satisfy all of the assumptions of the standard economic theory of demand, and Assumption A1 as well, yet nonetheless display inconsistencies between stated brand preference and actual choice in the BPL experiment. In the last three sections, we have discussed a number of factors which could lead to such inconsistencies in this context; moreover, in the process, we have implicitly developed a special case of the standard economic theory of demand which would predict perfect agreement between stated brand preference and actual choice. This special theory occurs when the subject satisfies the assumptions of the standard economic theory of demand[18] (call this Assumption A0), Assumption A1 from Section 6, and Assumptions A2 and A3, defined as follows.

A2. The subject satisfies the stationary SCP model;, and, denoting the common value of the daily preferences, G_t, by 'G,' the weak order, G:

A3. is additively separable in the soft drink component.

[16]This is an example of 'melioration.' See Herrnstein and Prelec [1992] and Rabin [2002].

[17]For more recent discussions of this sort of inconsistency, see Goldman [1979, 1980]; and for an excellent general discussion of anomalies connected with intertemporal choice, see Loewenstein and Prelec [1992].

[18]Assuming that each of the subjects is a 1-person household.

If the subject satisfies Assumptions A0–A3, then, as you can readily verify for yourself, there should be perfect agreement between stated brand preferences and actual choice. Since the results of the BPL experiment do not substantiate this agreement, we can regard the BPL experiment as having rejected the joint hypothesis:

$$H_0 \equiv A0 \;\&\; A1 \;\&\; A2 \;\&\; A3.$$

Thus we can conclude that at least one of the Assumptions A0–A3 is not satisfied, at least for the population sampled in the experiment.

The experiment does not, of course, tell us which of the Assumptions A0–A3 is false; although from other experiments and/or statistical studies, we might be inclined to believe that A0 is true, and therefore that the culprit must be one or more of Assumptions A1–A3. Further experiments may yet shed some light on which of these latter assumptions, if any, is empirically tenable. However, just now the broader lesson to be gained from our study of the BPL experiment is that, even if we consider the standard economic theory of demand to be empirically correct, we must guard against the presumption that this implies that the theory is empirically correct under alternative specifications of the primitives of the theory.

2.10 Probabilistic Theories of Choice

Returning to the general algebraic choice model of Section 2, and recalling our convention of denoting the asymmetric and symmetric parts of $\succsim$ by '$\succ$' and '$\sim$,' respectively; we can give an operational definition of $\succ$ (strict preference) as follows:

'$x \succ y$' means that if the decision-maker were presented with repeated choices between x and y, he/she would always choose x. (2.17)

If we introduce the notation '$p(x, y)$' to denote the probability that x would be chosen over y, given that only x and y are available (so that the budget set is $\{x, y\}$), (2.17) is equivalent to the statement:

$$x \succ y \iff p(x, y) = 1. \tag{2.18}$$

While it is a bit harder to come up with a satisfactory operational definition of indifference along these lines, the following is certainly a possibility:

$$x \sim y \iff p(x, y) = 1/2. \tag{2.19}$$

If we accept (2.18) and (2.19) as operational definitions of $\succ$ and $\sim$, then the assumption made implicitly in the algebraic choice model is that for each $x, y \in X$, we have:

$$p(x, y) \in \{0, 1/2, 1\}.$$

In contrast, a probabilistic theory of choice assumes only that $p(x, y) \in \{0, 1\}$.

2.11 Definition. Let X be a non-empty set. We shall say that a function, $p\colon X \times X \to [0, 1]$ is a **binary preference probability** (on X) iff $p(\cdot)$ satisfies:

$$\big(\forall (x, y) \in X \times X\big)\colon p(x, y) + p(y, x) = 1.$$

Probabilistic, as opposed to algebraic theories of choice are the rule, rather than the exception, in psychology. As to the reasons for this, we can probably do no better than to quote from a classic text in the field (Coombs, Dawes, and Tversky [1970, p. 148]):

> Inconsistency is one of the basic characteristics of individual choice behavior. When faced with the same alternatives, under seemingly identical conditions, people do not always make the same choice. Although the lack of consistent preferences may be attributable to factors such as learning, saturation, or changes in taste over time, inconsistencies exist even when the effects of such factors appear negligible. One is led, therefore, to the hypothesis that the obeserved inconsistency is a consequence of an underlying random process.
>
> The randomness may reflect uncontrolled momentary fluctuations such as attention shifts, or it may correspond to a choice mechanism that is inherently probabilistic. Be that as it may, the most natural way of coping with inconsistent preferences is by replacing the deterministic notion of preference by a probabilistic one....

Certainly a great many choice experiments have revealed an inconsistency in stated binary preferences. To quote Starmer [2000, p. 374]:

> ...a common finding is that individuals confronted with the same pairwise choice problem twice within a given experiment frequently give different responses on the two occasions. Stochastic choice is more convincing than indifference as an account for such intrinsic variability...

Starmer goes on to cite a number of such recent experimental studies, among them Hey and Orme [1994] and Ballinger and Wilcox [1997], in which between one-quarter and one-third of subjects 'switch' preferences on repeated questions.

Despite all of this experimental evidence, it cannot fairly be said that the algebraic choice model has been obviously refuted, for several reasons. For us, the most important of these reasons is that the objects over which choice has been made in the experiments cited have not been those appearing in the economic theory of demand; namely commodity bundles. It is entirely possible that preferences over gambles, for example, are much less consistent than are preferences over commodity bundles; and all of the studies cited here have involved choices over uncertain prospects.[19] In any event, in the remainder of the course, our focus will be upon algebraic choice theory, since the concensus of the profession seems to be that this theory is appropriate for the issues to be examined in our remaining discussion.

Things are a bit different when it comes to applied work and/or forecasting, however. As noted by McFadden [2001], before the 1960's empirical applications of demand theory generally proceeded by positing the existence of a '... *representative agent* [consumer], with market-level behavior given by the representative agent's behavior writ large.' (McFadden [2001, p. 351]). Deviations from preference-maximizing behavior in the market as a whole were then attributed to an error term, generally assumed to be additive, with zero mean.

All of this is consistent, to a certain extent, with the fact that the focus of interest in economics is upon market behavior, rather than individual choice behavior; at least

[19] It should be noted, however, that choice over commodity bundles often involves risk; for example, in assessing the desirability of a new product.

in the theory of demand.[20] However, the conditions under which aggregate demand will exhibit the same qualitative properties as are implied for individual consumer behavior by the economic theory of demand are extremely restrictive; as we will demonstrate in a later chapter. In contrast, McFadden has pioneered the theoretical and statistical development of methods for extending estimates of individual behavior to arrive at estimates of market demand. An integral part of such methods is the assumption that consumer preferences can be represented by a well-behaved function of the (p. 357) '...characteristics of the consumer, and consumption levels and attributes of goods.' Differences in choice (randomness) are then attributed to unobserved characteristics. This technique has been particularly effective in predicting choice over discrete alternatives. For details, see McFadden [2001]

2.11 Are Preferences Total?

Suppose there are exactly n commodities available, and that the consumers' budget sets are subsets of $\mathbb{R}^n$, but that consumers' rankings of these commodity bundles may depend upon a vector of 'environmental variables,' $\boldsymbol{y}$; which might include such variables as season of the year, the expected mean temperature over the planning period, the number of interesting concerts scheduled in the area during the planning period, and so on. More precisely, suppose the environmental variables, $\boldsymbol{y}$, can take on any value in some set Y. It then seems natural to suppose that consumers have preference orderings, $\succsim$, over the set:

$$Z \stackrel{\text{def}}{=} X \times Y. \tag{2.20}$$

We then have a situation analogous to that discussed in Section 6. In particular, if a consumer's preference relation, $\succsim$, is a weak order, then for each $\boldsymbol{y} \in Y$, $\succsim$ induces a weak order, $\succsim_{\boldsymbol{y}}$, on X, defined by:

$$\boldsymbol{x} \succsim_{\boldsymbol{y}} \boldsymbol{x}' \iff (\boldsymbol{x}, \boldsymbol{y}) \succsim (\boldsymbol{x}', \boldsymbol{y}); \tag{2.21}$$

however, for $\boldsymbol{y} \neq \boldsymbol{y}^*$, we may have $\succsim_{\boldsymbol{y}} \neq \succsim_{\boldsymbol{y}^*}$.

Thus in such a context we may find that in one period, the consumer chose $\boldsymbol{x}_1 \in X$ at prices $\boldsymbol{p}_1$, while in the next period it chose $\boldsymbol{x}_2 \in X$ at prices $\boldsymbol{p}_2$, where:

$$\boldsymbol{p}_1 \cdot \boldsymbol{x}_1 > \boldsymbol{p}_1 \cdot \boldsymbol{x}_2 \tag{2.22}$$

while:

$$\boldsymbol{p}_2 \cdot \boldsymbol{x}_2 > \boldsymbol{p}_2 \cdot \boldsymbol{x}_1. \tag{2.23}$$

This is, of course, apparently inconsistent behavior, in that (2.22) indicates that (since $\boldsymbol{x}_1$ was chosen when $\boldsymbol{x}_2$ would have cost less);

$$\boldsymbol{x}_1 \succ \boldsymbol{x}_2;$$

while (2.23) would appear to indicate that $\boldsymbol{x}_2 \succ \boldsymbol{x}_1$. In the context of the present discussion, however, there is nothing inconsistent about such a situation; for the

[20] As we shall find later, however, the theory of individual choice which we have been developing here, and will continue to develop in the next chapter, will be of considerable value in our analysis of social, or group choice.

environmental variable prevailing in period one (which we will denote by '$\boldsymbol{y}_1$') may be different from that in period two (call in '$\boldsymbol{y}_2$'), and it may be that:

$$(\boldsymbol{x}_1, \boldsymbol{y}_1) \succ (\boldsymbol{x}_2, \boldsymbol{y}_1),$$

and yet:

$$(\boldsymbol{x}_2, \boldsymbol{y}_2) \succ (\boldsymbol{x}_1, \boldsymbol{y}_2).$$

Of course, this difficulty disappears if $\succsim$ is weakly separable in $\boldsymbol{x}$ on Y, but this is an assumption that is very difficult to justify as a universal rule; most consumers' preferences for 'cut-offs' versus coats, or for swim suits versus ski pants, are likely to be quite different in June than in January. However, if the consumer's preference relation varies from month to month, the standard economic theory of demand loses virtually all of its predictive power; since, as was pointed out by Samuelson some time ago (Samuelson [1938], [1947], [1948]), the entire operational content of the standard economic theory of demand is bound up in revealed-preference conditions like:

$$\boldsymbol{p}_1 \cdot \boldsymbol{x}_1 \geq \boldsymbol{p}_1 \cdot \boldsymbol{x}_2 \Rightarrow \boldsymbol{p}_2 \cdot \boldsymbol{x}_1 \geq \boldsymbol{p}_2 \cdot \boldsymbol{x}_2 \tag{2.24}$$

If $\boldsymbol{x}_1$ and $\boldsymbol{x}_2$ are the commodity bundles chosen by the consumer in two successive periods, however, such a statement will hold[21] (that is, the implication will be logically correct) only if the consumer's preference relation is the same in the two periods.[22]

There are several ways in which we might attempt to circumvent the difficulty under discussion here. The simplest way out of it, which constitutes the reason for the title of this section, is to drop the assumption that the consumer's preference relation is total. Why is this? Well, if $\succsim$ is a weak order on $X \times Y$, we can use it to define a relation, G, on X by:

$$\boldsymbol{x} G \boldsymbol{x}^* \iff [(\forall \boldsymbol{y} \in Y)\colon (\boldsymbol{x}, \boldsymbol{y}) \succsim (\boldsymbol{x}^*, \boldsymbol{y})]. \tag{2.25}$$

It is easy to prove that if $\succsim$ is a weak order on $X \times Y$, then G will be reflexive and transitive, but, in general, will not be total on X. If fact, G will be total if, and only if, $\succsim$ is weakly separable in $\boldsymbol{x}$ on Y.

On the other hand, there is a problem with this approach; specifically, in relating this relation, G, to demand behavior. If we return to the framework set out in Section 2, with a family of budget sets, $\mathcal{B}$, we see that we cannot characterize the consumer's demand correspondence, h, by:

$$h(B) = \{\boldsymbol{x} \in B \mid (\forall \boldsymbol{x}' \in B)\colon \boldsymbol{x} G \boldsymbol{x}'\} \quad \text{for } B \in \mathcal{B}. \tag{2.26}$$

[21] Given certain additional assumptions (for example, strict quasi-concavity) on the preference relation, we can strengthen (2.24) to:

$$[\boldsymbol{p}_1 \cdot \boldsymbol{x}_1 \geq \boldsymbol{p}_1 \cdot \boldsymbol{x}_2 \;\&\; \boldsymbol{x}_1 \neq \boldsymbol{x}_2] \Rightarrow \boldsymbol{p}_2 \cdot \boldsymbol{x}_1 > \boldsymbol{p}_2 \cdot \boldsymbol{x}_2.$$

[22] Even if we assume that the consumer's preferences change from month to month, however, it might be well worth testing the hypothesis that these changes cycle with the seasons of the year. Thus, for example, we might test the hypothesis that a consumer's preferences were the same in June, 2000, as in June, 2001; compare July, 2000, with July, 2001, and so on.

In fact, the set defined in (2.26) will be empty, for most budgets $B \in \mathcal{B}$; and thus the consumer's actual choice will not generally be an element of the set. Suppose we instead define a correspondence, h, by

$$h(B) = \{\boldsymbol{x} \in B \mid (\forall \boldsymbol{x}' \in X)\colon \boldsymbol{x}'P\boldsymbol{x} \Rightarrow \boldsymbol{x}' \notin B\} \quad \text{for } B \in \mathcal{B}; \tag{2.27}$$

where P is the asymmetric part of G. With this definition we have almost the same problem; in general, the set defined in (2.27) will not be empty,[23] but for a given budget, B, the consumer may make a choice which is not an element of the set $h(B)$ defined in (2.27). I will leave the task of explaining why this may happen under the present assumptions as an exercise. Suppose, however, that instead of defining the relation P as the asymmetric part of G, we define a new relation P by:

$$\boldsymbol{x}P\boldsymbol{x}' \iff (\forall \boldsymbol{y} \in Y)\colon (\boldsymbol{x}, \boldsymbol{y}) \succ (\boldsymbol{x}', \boldsymbol{y}); \tag{2.28}$$

and then take (2.27) to be our definition of the correspondence h, using this new definition of P. A moment's reflection should then suffice to convince you that the consumer's choice will now always be an element of $h(B)$, for each $B \in \mathcal{B}$. The natural question to ask about this, however, is whether a theory based on these assumptions and this definition has any real predictive power. We will consider this question further in the next chapter. In the meantime, let's take a look at some alternative approaches to the solution of this environmental variable problem which are also of interest.

The first alternative, and perhaps that most consistent with the literature on demand theory, is to re-interpret the notion of a commodity. Thus, we might think of 'food,' 'clothing,' 'housing,' 'recreational goods,' and so on, as individual commodities; rather than using the specification $S.3$ (of Section 5) of what we have been calling the 'standard economic theory of demand.' The point of this change is that it is a very intuitively appealing notion that preferences over broadly-defined commodity groups like 'food' versus 'clothing,' might exhibit much less variability (over seasons of the year, in particular) than preferences for more narrowly-defined items, like 'cut-offs' versus 'chinos,' 'cold-cuts' versus clam chowder, or beer versus (hot-buttered) rum. Unfortunately, this leaves us with the very messy scientific problem of determining exactly which such specification of the primitive 'commodity' will work; and with the concomitant problem of how to measure such conglomerates. In all fairness, it would appear that we would have to admit that the economics profession has not succeeded in fully solving this problem.

Another interesting way of handling the difficulty under consideration here is to treat the variable $\boldsymbol{y} \in Y$ as random. If the weak order $\succsim$ on $X \times Y$ is representable by a utility function, $U(\boldsymbol{x}, \boldsymbol{y})$, one is then led to the notion of random utility (over X). This is one of two alternative assumptions underlying probabilistic theories of choice. We will not be able to pursue this topic further in this course, but let me recommend the book by Train [1986] as a very readable introduction to both the theory and estimation techniques which have been developed in this area.[24]

[23] We will provide a justification for this statement in Chapter 4.

[24] See also McFadden [2001]. It is he who developed most of the theory and many of the estimation techniques which are used in this area.

Yet another way of reacting to this difficulty is to attempt to systematically exploit this variability in preferences. After all, as economists we are fundamentally more concerned with market demand than with individual consumer demand, and to the extent that individual preferences change from month to month (or from quarter to quarter) in a systematic manner which is much the same for different individuals (and casual observation suggests that there is at least some basis for believing this to be the case), we may be able to develop a much more useful and powerful theory of market demand because of this variability than would be possible without it.

2.12 Are Preferences Transitive?

Over the years, a number of writers have questioned the correctness of assuming that preferences are transitive. In this section, we will consider four kinds of objections, or difficulties, which have been raised regarding the transitivity assumption.

2.12.1 'Just Noticeable Difference,' or 'Threshold Effects'

More than sixty years ago, W. E. Armstrong [1939] objected to the transitivity assumption on the grounds that an alternative, x, may be indifferent to y, y may be indifferent to z; and yet x may be preferred to z. The reasoning behind this sort of contention is that the difference between x and y may be too small to notice, and such may also be the case as regards y and z; yet the difference between x and z may nonetheless be sufficiently great as to result in the preference of x over z. As a theoretical device to deal with this and similar phenomena in the area of psychophysics, R. Duncan Luce [1956] developed the notion of a semi-order. An example of this sort of binary relation was presented in Chapter 1, but we will briefly review the example here.[25]

Let $f\colon X \to \mathbb{R}$, and suppose that $X = \mathbb{R}^n_+$, and that a consumer's preferences on X satisfy:

$$\boldsymbol{x} \succsim \boldsymbol{x}' \iff f(\boldsymbol{x}) \geq f(\boldsymbol{x}') - \delta, \tag{2.29}$$

where δ is a positive constant. I will leave it as an exercise to show that the asymmetric part of $\succsim$ (the strict preference relation) is given by:

$$\boldsymbol{x} \succ \boldsymbol{y} \iff f(\boldsymbol{x}) > f(\boldsymbol{y}) + \delta; \tag{2.30}$$

while the symmetric part (the indifference relation) is given by:

$$\boldsymbol{x} \sim \boldsymbol{y} \iff |f(\boldsymbol{y}) - f(\boldsymbol{x})| \leq \delta. \tag{2.31}$$

The constant, δ, is therefore identified with the threshold level of perception, or 'just noticeable difference.' Thus $\boldsymbol{x}$ and $\boldsymbol{y}$ are indifferent if the absolute value of the difference between the value of f at $\boldsymbol{x}$ and its value at $\boldsymbol{y}$ is not greater than δ; whereas, if, say, $f(\boldsymbol{x}) > f(\boldsymbol{y}) + \delta$, then $\boldsymbol{x}$ is preferred to $\boldsymbol{y}$. We showed in Chapter 1 that, while strict preference is transitive in this case, indifference is not.

[25]The reader interested in learning more about semi-orders should be warned that later authors have used a set of axioms different (logically equivalent, but somewhat more transparent) from those originally formulated by Luce. Luce's and Suppes' [1965] survey article, or Fishburn's [1970] book both contain good, and extensive introductions to the concept.

2.12.2 Decision Rules Based On Qualitative Information

Suppose a prospective home buyer has a choice of three houses, labeled 'A,' 'B,' and 'C;' all of which are selling at the same price. Suppose further that our home buyer is interested in three attributes of a home, other than price: square feet of floor space, location (convenience with respect to school, shopping centers, and so on), and appearance; and ranks these three houses with respect to these characteristics as follows.

	Floor Space	**Location**	**Appearance**
A	Best	Middle	Poorest
B	Middle	Poorest	Best
C	Poorest	Best	Middle

Table 2.3: Housing Attributes.

Consider the following decision rule:

$$x \succ y \iff x \text{ is better than } y \text{ in at least two characteristics;}$$

and show, on the basis of this decision rule, that we have:[26]

$$A \succ B,\ B \succ C,\ \text{and } C \succ A.$$

In an experiment involving 62 college students, K. O. May [1954] examined a similar problem. We quote from his description of the experiment (p. 6).

> ...The alternatives were three hypothetical marriage partners, x, y and z. In intelligence they ranked xyz, in looks yzx, in wealth zxy. The structure of the experiment was not explained, but subjects were confronted at different times with pairs labeled with randomly chosen letters. On each occasion, x was described as very intelligent, plain looking, and well off; y as intelligent, very good looking, and poor; z as fairly intelligent, good looking, and rich. All prospects were described as acceptable in every way, none being so poor, plain, or stupid as to be automatically eliminated. ...Parts of the experiment were repeated to test for consistency and possible capriciousness. The results, as well as the behavior of the subjects, indicated practically no random element in the choices. In terms of the probability definition of preference given in the first section, it was evident that 0 and 1 were the only possible probabilities, and that repeated trials were not necessary.
>
> Since indifference is ruled out, there are six possible orderings and two circular patterns designated by $xyzx$ and $xzyx$. If group preferences be defined by majority vote, the results indicate a circular pattern, since x beat y by 39 to 23, y beat z by 57 to 5, and z beat x by 33 to 29. The number of individuals having each of the possible patterns was xyz: 21; $xyzx$: 17; yzx: 12; yxz: 7;

[26]Before leaving this discussion, you may wish to consider the following question. Suppose that, in a given local housing market, all prospective buyers are interested in the attributes, or characteristics, discussed here, and only those (other than price). Would you expect to find any configuration very different from that shown in Table 3 for any three houses which were all selling at the same price?

zyx: 4; *xzy*: 1; *zxy*: 0; *xzyx*: 0. The intransitive pattern is easily explained as the result of choosing the alternative that is superior in two out of three criteria. The orders *xyz* and *yzx* seem to have resulted from giving heavier weight to intelligence and looks, respectively. The four who chose inversely with respect to intelligence (*zyx*) were men, and may indicate the extent of male fear of intelligent women....

What is the significance of this experiment? Of course it does not prove that individual patterns are always intransitive. It does, however, suggest that where choice depends on conflicting criteria, preference patterns *may* be intransitive unless one criteria dominates....

2.12.3 Priorities and Measurement Errors

The following is an example of what Tversky [1969] has dubbed a '**lexicographic semi-order**.' Suppose a decision-maker is attempting to rank-order a group of college applicants, having available only three pieces of information: their examination scores on tests of intelligence, emotional stability, and social facility. Our decision-maker decides that for him the order of importance of these scores is the order in which they are listed above. On the other hand, he also recognizes the fact that all of these tests are subject to a great deal of measurement error; thus he arrives at the following rule (perhaps based upon his perception of the standard errors of the testing techniques). If individual 1's intelligence score is more than 3 points higher than individual 2's score, he will rank 1 above 2, whatever their remaining two scores. If their intelligence scores differ by no more than 3 points, he will look at their emotional stability examination scores. If this difference is more than 6 points (this examination being somewhat less reliable than the first), he will rank the individual having the higher score above the other, and ignore the third score. Finally, if the emotional stability scores for the two individuals differ by no more than 6 points, he will look at the third score; ranking 1 above 2 if 1's social facility score is 9 or more points higher than 2's. Formally, if we denote the intelligence, emotional stability, and social facility scores of applicant x by 'I_x,' 'E_x,' and 'S_x,' respectively, the decision rule we have just described verbally supposes that there exist positive constants, δ_1, δ_2, and δ_3 such that the decision-maker will order pairs of candidates, x and y, in the following way:

$$x \succ y \iff \begin{cases} I_x > I_y + \delta_1 & \text{or} \\ |I_x - I_y| \leq \delta_1 \text{ and } E_x > E_y + \delta_2, & \text{or} \\ |I_x - I_y| \leq \delta_1, |E_x - E_y| \leq \delta_2 \text{ and } S_x > S_y + \delta_3. & \end{cases} \tag{2.32}$$

Of course, in the special case which we have described verbally, the decision rule is of the form (2.32), with:

$$\delta_1 = 3, \ \delta_2 = 6, \ \text{and } \delta_3 = 9.$$

Now suppose that the candidates have the following test score profiles set out on the next page.

If our decision-maker only makes adjacent pair comparisons of the candidates (that is, compares a with b, b with c, and so on) he will end up ranking the candidates inversely with respect to their intelligence scores; despite the fact that the

Applicant	I	E	S
a	69	84	75
b	72	78	65
c	75	72	55
d	78	66	45
e	81	60	35

Table 2.4: Test Scores.

decision rule ostensibly gives first priority to intelligence! Furthermore, as in the case of the semi-order considered earlier, the decision-maker will exhibit intransitive indifference. However, in contrast to the simple semi-order considered earlier, this decision-maker will also display intransitive strict preference.[27]

2.12.4 Group Decisions: The Dr. Jekyll and Ms. Jekyll Problem

Yet a further problem in our economic theory of choice arises from the fact that we usually specify our individual decision-making unit ('individual consumer') as being, or corresponding to 'individual household' in the census data. The difficulty with this is that most households contain more than one individual, and the collective choices of a group may not be transitive even though all the individuals in the group have a transitive ordering over the alternatives available, as the following example demonstrates.[28]

Suppose three individuals, A, B, and C, rank three alternatives (for example, political candidates, proposed budgets, etc.) in the following way, and that majority voting is to be used to rank-order the alternatives. If we denote group preference

A	B	C
x	y	z
y	z	x
z	x	y

Table 2.5: A Preference Profile.

by 'P,' it is easy to show that in this case we have:

$$xPy,\ yPz,\ \text{and}\ zPx.$$

An interesting aspect of this sort of situation (although it has no necessary connection with the question of whether a household's preferences can reasonably be assumed to be transitive) is that if a group choice of one of the three alternatives is to be made by pair-wise elimination:

[27] A different sort of systematic violation of transitivity has been observed in experiments involving choice over risky prospects. For an excellent summary, as well as some particularly interesting experimental results, see Loomes, Starmer, and Sugden [1991].

[28] Notice the formal similarity between this example and that used in the May [1954] experiment discussed earlier.

> one pair of alternatives is compared first, and the alternative which wins the majority vote is then voted upon vis-a-vis the third alternative,

then the alternative actually chosen depends upon which pair was compared first. For example, given the preference profile in Table 5, if x and y are compared first, we have xPy; so that x would then be compared with z. Since zPx, the final choice (that is, the alternative actually chosen) would be z. On the other hand, if we first compared y and z, the final choice would be x, and so on.

Returning to the issue of whether a multiple-person household will display transitive preferences, consider, for a moment, the simplest sort of non-single-person household; namely one containing just two persons, whom we will suppose are husband and wife. Suppose further that each of them has a well-defined preference relation, G_i ($i = 1$ for the wife, and $i = 2$ for the husband), over the household consumption set, X, and denote the asymmetric part of G_i by 'P_i,' for $i = 1, 2$. Then if $\boldsymbol{x}^*$ is the commodity bundle chosen from $b(\boldsymbol{p}, w)$ for some period, it seems quite unlikely that this choice will satisfy:

$$\big(\forall \boldsymbol{x} \in b(\boldsymbol{p}, w)\big) \colon \boldsymbol{x}^* G_i \boldsymbol{x}, \tag{2.33}$$

for either $i = 1$ or $i = 2$ [although a lot of Ms. 1's and Mr. 2's might claim that (2.33) is satisfied with $i = 2$ or $i = 1$, respectively]; rather, it seems that some sort of compromise solution is likely to be reached. However, regardless of how Ms. 1 and Mr. 2 go about reconciling their diverse preferences, it seems likely that the final choice, $\boldsymbol{x}^*$, will be such that $\boldsymbol{x}^* \in b(\boldsymbol{p}, w)$ and will be such that there exists no $\widehat{\boldsymbol{x}} \in b(\boldsymbol{p}, w)$ such that:

$$\widehat{\boldsymbol{x}} P_1 \boldsymbol{x}^* \;\&\; \widehat{\boldsymbol{x}} P_2 \boldsymbol{x}^*. \tag{2.34}$$

The considerations of the above paragraph suggest that it may be worthwhile to pursue the following approach. We first define the binary relation, P, on X by:

$$\boldsymbol{x} P \boldsymbol{x}' \iff [\boldsymbol{x} P_1 \boldsymbol{x}' \;\&\; \boldsymbol{x} P_2 \boldsymbol{x}']. \tag{2.35}$$

The relation P can be thought of (and we will often refer to it as) the household's **unanimity ordering**. Furthermore, it makes sense to think of P as the household's strict preference relation, in that, if we accept the argument of the preceding paragraph, then the household will behave as if it attempted to maximize the binary relation, P; that is, the household will, given a price-wealth pair $(\boldsymbol{p}, w)$, choose a commodity bundle $\boldsymbol{x}^* \in b(\boldsymbol{p}, w)$ satisfying:

$$(\forall \boldsymbol{x} \in X) \colon \boldsymbol{x} P \boldsymbol{x}^* \Rightarrow \boldsymbol{p} \cdot \boldsymbol{x} > w.$$

2.13 Asymmetric Orders

Because of the difficulties discussed in the previous two sections, we will wish whenever possible to consider a more general form of ordering than a weak order as our 'model' of consumer preferences. Specifically, we will whenever possible (especially in developing the theory of welfare economics), suppose only that a consumer's strict preference relation is an asymmetric order; where we define this as follows.[29]

[29] Recall that the asymmetric part of a weak order is also negatively transitive.

2.12 Definition. Let P be a binary relation on a non-empty set, X. We shall say that P is an **asymmetric order** iff P is asymmetric and transitive.

Making use of this definition, it is easy to prove the following.

2.13 Proposition. *Suppose $\succ_y$ is an asymmetric order on the non-empty set X, for each $y \in Y$, and define the binary relation, P on X by:*

$$xPx' \iff [(\forall y \in Y) \colon x \succ_y x'].$$

Then P is also an asymmetric order on X.

Thus it follows from this proposition that the binary relations defined in equation (2.28) of Section 11, equation (2.30) of Section 12, and equation (2.35) of Section 12, are all asymmetric orders (given the assumptions of the respective sections). Consequently, we can see that there is a real gain in generality in assuming, wherever possible, that a consumer's (strict) preference relation, P, is an asymmetric order; and, correspondingly, that, given a budget space, $\langle X, \mathcal{B}\rangle$, that the consumer's demand correspondence takes the form:

$$h(B) = \{x \in B \mid (\forall x' \in X) \colon x'Px \Rightarrow x' \notin B\} \quad \text{for } B \in \mathcal{B}. \tag{2.36}$$

In the next chapter, we will investigate the implications of these assumptions. In the meantime, let's take a look at the way in which our continuity assumptions need to be reformulated in order to apply to asymmetric orders.

2.14 Definitions. Let X be a non-empty subset of $\mathbb{R}^n$, and let P be an asymmetric binary relation on X. We shall say that P is:

1. **upper semi-continuous on X** iff, for each $\boldsymbol{x}, \boldsymbol{y} \in X$, if $\boldsymbol{x}P\boldsymbol{y}$, then there exists a neighborhood, $N(\boldsymbol{y})$, such that, for all $\boldsymbol{y}' \in N(\boldsymbol{y}) \cap X$, $\boldsymbol{x}P\boldsymbol{y}'$.
2. **lower semi-continuous on X** iff, for each $\boldsymbol{x}, \boldsymbol{y} \in X$, if $\boldsymbol{x}P\boldsymbol{y}$, then there exists a neighborhood, $N(\boldsymbol{x})$, such that, for all $\boldsymbol{x}' \in N(\boldsymbol{x}) \cap X$, $\boldsymbol{x}'P\boldsymbol{y}$.
3. **continuous on X** iff it is both upper and lower semi-continuous on X.
4. **strongly continuous on X**, iff, for each $\boldsymbol{x}, \boldsymbol{y} \in X$, if $\boldsymbol{x}P\boldsymbol{y}$, then there exist neighborhoods, $M(\boldsymbol{x})$ and $N(\boldsymbol{y})$, respectively, such that, for all $\boldsymbol{x}' \in M(\boldsymbol{x}) \cap X$, and for all $\boldsymbol{y}' \in N(\boldsymbol{y}) \cap X$:
$$\boldsymbol{x}'P\boldsymbol{y}'.$$

One can prove the following relationships.

2.15 Theorem. *Let X be a non-empty subset of $\mathbb{R}^n$, and let P be an asymmetric binary relation on X. If P is also negatively transitive, and if we let 'G' denote the negation of P, then:*

1. *P is upper semi-continuous on X if, and only if, G is upper semi-continuous on X.*
2. *P is lower semi-continuous on X if, and only if, G is lower semi-continuous on X.*
3. *P is strongly continuous on X if, and only if, G is continuous on X.*

The last part of Theorem 2.15 probably looks a bit strange, but may be cleared up by noting the following facts: (a) there exist asymmetric orders (which are not negatively transitive) which are continuous, but which are not strongly continuous, and (b) if an asymmetric order is negatively transitive and continuous, then it is strongly continuous. An example of an asymmetric order, and one of which we will make a great deal of use, is defined in the following.

2.16 Definition. We will say that a relation P on a non-empty set X is a **semi-order** iff there exists a function $f\colon X \to \mathbb{R}$ amd a postive constant, $\delta \in \mathbb{R}_{++}$ such that, for all $x, y \in X$:

$$xPy \iff f(x) > f(y) + \delta.$$

We will say that P is a **continuous semi-order** iff the function f is continuous.

It is fairly easy to prove that a continuous semi-order is strongly continuous. (See Exercise 2, at the end of this chapter.)

A useful generic example of an asymmetric order is provided in the following example.

2.17 Example. Let X be any non-empty subset of $\mathbb{R}^n$, let $u\colon X \to \mathbb{R}$ be any real-valued function defined on X, and let α and β be any nonnegative constants. If we then define P on X by:

$$\boldsymbol{x}P\boldsymbol{y} \iff u(\boldsymbol{x}) > u(\boldsymbol{y}) + \alpha\|\boldsymbol{x} - \boldsymbol{y}\| + \beta,$$

then P is an asymmetric order. □

Given that we have an interest in asymmetric orders, it is obvious that the following type of binary relation is of interest.

2.18 Definition. We will say that a binary relation, G, on a nonempty set, X, is a **quasi order** iff G is total, reflexive, and its asymmetric part, P, is transitive.[30]

Notice that it follows from Proposition 1.17 that a binary relation, P, is an asymmetric order if, and only if, its negation is a quasi order. Obviously a weak order is a special case of a quasi order.

Exercises.

1. Show that the relation $>$ is an asymmetric order on $\mathbb{R}^n$.

2. Show that a continuous semi-order is strongly continuous. (Where we say that **a semi-order is continuous** iff the function f by which it is defined is a continuous function.)

3. Show that the binary relation, $\succ$, defined in equation (2.30) of Section 12 is irreflexive, asymmetric, and transitive; and that $\sim$ [the symmetric part of the

[30] There does not seem to be an established term to denote a binary relation satisfying these properties. However, Sen [1986] refers to binary relations whose asymmetric part is transitive as being 'quasi-transitive.' Consequently, it seems reasonable to apply the term 'quasi order' in the present case.

relation $\succsim$ defined in equation (2.29) of Section 12] is reflexive and symmetric, but not transitive.

4. Prove that the binary relation, P, defined in Example 2.17 is an asymmetric order.

5. Show that if a consumer's preferences are representable by a utility function of the form (2.9) of Section 7 (with $\boldsymbol{a} \gg \mathbf{0}$), then the condition of equation (2.10) $[(1,0)P(0,1)]$ insures the implication given in equation (2.11) of Section 7.

6. As in Section 12.4, consider a two-person household, with G_i being the (weak) preference relation of the ith person ($i = 1, 2$). Define the relation G by:

$$\boldsymbol{x}G\boldsymbol{y} \iff [\boldsymbol{x}G_1\boldsymbol{y} \ \& \ \boldsymbol{x}G_2\boldsymbol{y}],$$

and let P be the asymmetric part of G. Do you think that it will be the case that the household consumption choice will always be an element of:

$$h(\boldsymbol{p}, w) = \{\boldsymbol{x} \in b(\boldsymbol{p}, w) \mid (\forall \boldsymbol{y} \in X) \colon \boldsymbol{y}P\boldsymbol{x} \Rightarrow \boldsymbol{p} \cdot \boldsymbol{y} > w\}?$$

Why or why not? How does the correspondence defined here compare with that defined in Section 12.4?

Chapter 3

Revealed Preference Theory

3.1 Introduction

We noted earlier that if a decision-maker has (a) a well-defined preference relation which is, in mathematical terms, a weak order, and (b) always makes a choice consistent with said preferences, then the general algebraic choice model is applicable as a description of the choice situation. In effect, then, these two conditions together constitute sufficient conditions for the application of the general algebraic choice model. They are not necessary conditions, but nonetheless the model can be applied only if the decision-maker behaves *as if* (a) and (b) hold. In this chapter, we will investigate the implications of this last statement; that is, the implications of the assumption that a decision-maker behaves as if (a) and (b) hold. In Sections 2 through 4 we will look at the implications of conventional demand theory; that is, the implications of the assumption that consumer preferences can be modeled as a weak order. In Section 5 we will consider the testable implications of the model when only a finite number of observations of quantities demanded can be made. Finally, in Section 6, we will take a brief look at the implications of the assumption that consumer strict preferences are assumed only to be an asymmetric order. The approach to be followed in the first four sections, as well as most of the definitions and results, are due to Richter [1966, 1971].[1]

3.2 Choice Correspondences and Binary Relations

In traditional demand theory, we suppose that a consumer makes a unique choice from a budget set of the form:

$$b(\boldsymbol{p}, w) = \{\boldsymbol{x} \in \mathbb{R}^n_+ \mid \boldsymbol{p} \cdot \boldsymbol{x} \leq w\},$$

where 'w' denotes the consumer's wealth, or income. These choices result in a demand function, $\boldsymbol{h}\colon \mathbb{R}^n_{++} \times \mathbb{R}_+ \to \mathbb{R}^n_+$. The following definition extracts the key elements of these concepts, and generalizes the idea to a very broad concept of choice.

[1] Although I have taken the liberty of changing Richter's terminology slightly.

3.1 Definitions. A **budget space**, $\langle X, \mathcal{B}\rangle$, is a nonempty set, X, and a family, $\mathcal{B}$, of nonempty subsets, B, of X. A **choice correspondence** on a budget space $\langle X, \mathcal{B}\rangle$ is a correspondence, h, which to each $B \in \mathcal{B}$, assigns a non-empty subset, $h(B)$, satisfying:

$$(\forall B \in \mathcal{B})\colon h(B) \subseteq B.$$

We will sometimes, particularly in Sections 4 and 6 of this chapter (and in later chapters), be interested in a special kind of choice correspondence; those satisfying the condition:

$$(\forall B \in \mathcal{B})(\exists x \in B)\colon h(B) = \{x\}. \tag{3.1}$$

In this case, we will refer to h as a **choice function**, and (at the expense of strictly proper mathematics useage), for $B \in \mathcal{B}$ we will think of $h(B)$ as being an element of B rather than a subset of B. That is, in the case where h is a choice function, if $B \in \mathcal{B}$ and $x \in B$ satisfy (3.1), above, we will write:

$$h(B) = x,$$

rather than '$h(B) = \{x\}$.'

3.2 Definition. Let h be a choice correspondence on a budget space, $\langle X, \mathcal{B}\rangle$. We shall say that a binary relation, G, on X, **rationalizes h on** $\langle X, \mathcal{B}\rangle$ iff:

$$(\forall B \in \mathcal{B})\colon h(B) = \{x \in B \mid (\forall y \in B)\colon xGy\}. \tag{3.2}$$

In the economic theory of consumer choice, it is generally assumed that a decision-maker has a preference relation, R, which is a weak order, and that given a budget set, $B \in \mathcal{B}$, chooses an element from the set $h(B)$ defined by:

$$h(B) = \{x \in B \mid (\forall y \in B)\colon xRy\}.$$

Obviously in this case h will be a choice correspondence. Furthermore, the weak order R will in this case rationalize h on $\langle X, \mathcal{B}\rangle$. We will now begin examining the converse question of whether a given choice correspondence can be rationalized by *some* binary relation.

3.3 Definition. Let h be a choice correspondence on a budget space $\langle X, \mathcal{B}\rangle$. We shall say that h is:

1. **rational** iff there exists a binary relation, G, which rationalizes h on $\langle X, \mathcal{B}\rangle$.
2. **reflexive-rational** iff there exists a reflexive binary relation, G, on X which rationalizes h on $\langle X, \mathcal{B}\rangle$.
3. **transitive-rational** iff there exists a transitive binary relation, G, on X which rationalizes h on $\langle X, \mathcal{B}\rangle$.
4. **regular-rational** iff there exists a weak order, G, on X which rationalizes h on $\langle X, \mathcal{B}\rangle$.
5. **irrational** iff it is not rational; that is, iff there exists no binary relation, G, which rationalizes h on $\langle X, \mathcal{B}\rangle$.

3.4 Proposition. *There exist irrational choice correspondences; that is, there exist choice correspondences which cannot be rationalized by any binary relation.*

In order to prove this proposition, it obviously suffices to exhibit an irrational choice correspondence. This is done in the following example.

3.5 Example. Let $X = \{a, b, c\}$, and $\mathcal{B} = \{B_1, B_2\}$, where:

$$B_1 = X = \{a, b, c\} \qquad h(B_1) = \{b\},$$
$$B_2 = \{a, b\} \qquad h(B_2) = \{a\}.$$

Suppose, by way of obtaining a contradiction, that there exists a binary relation, G, which rationalizes h on $\langle X, \mathcal{B}\rangle$. Then by (3.2) and the definition of $h(B_1)$, we see that:

$$bGa, bGb, \text{ and } bGc.$$

However, it then follows that:

$$(\forall x \in B_2)\colon bGx,$$

which implies, if G rationalizes h, that $b \in h(B_2)$; contrary to the definition of $h(B_2)$. □

3.6 Proposition. *There exist choice correspondences which can be rationalized by a reflexive binary relation, but not by any* total *binary relation.*

In order to prove this, it again suffices to produce an example, as in the following.

3.7 Example. Let $X = \{a, b, c\}, \mathcal{B} = \{B_1, B_2, B_3\}$, and:

$$B_1 = \{a, c\} \qquad h(B_1) = \{a, c\},$$
$$B_2 = \{b, c\} \qquad h(B_2) = \{b, c\},$$
$$B_3 = \{a, b, c\} \qquad h(B_3) = \{c\}.$$

Suppose that G is a binary relation which rationalizes h on $\langle X, \mathcal{B}\rangle$. Then, from the definition of h, it follows that G must satisfy:

$$\begin{array}{cccc} & \boldsymbol{a} & \boldsymbol{b} & \mathbf{c} \\ \boldsymbol{a} & aGa & \ldots & aGc \\ \boldsymbol{b} & \ldots & bGb & bGc \\ \mathbf{c} & cGa & cGb & cGc. \end{array} \tag{3.3}$$

The entries in the first row of the above matrix follow from the definition of $h(B_1)$, those in the second row from $h(B_2)$, and so on. Notice that the particular binary relation defined in (3.3) [that is, if we take (3.3) to be the definition of G] rationalizes h on $\langle X, \mathcal{B}\rangle$; which establishes the fact that h is rational. On the other hand, if G is defined as in (3.3), then it is not total, since we have neither bGa, nor aGb.

Now suppose we try to extend G in such a way as to make it total. If we have aGb, then it follows that

$$(\forall x \in B_3)\colon aGx,$$

so that G no longer rationalizes h [since $a \notin h(B_3)$]. On the other hand, if we let bGa, then we have:

$$(\forall x \in B_3)\colon bGx;$$

and, since $b \notin h(B_3)$, G no longer rationalizes h. Since any binary relation which rationalizes h must satisfy (3.3), it then follows that there exists no binary relation which is both total and rationalizes h in this case. □

The logic of the argument developed in the last paragraph of the preceding example may not be all that clear at this point. In any case, consideration of the following material may make said logic clearer, as well as improving our understanding of the theory of choice correspondences in general.

3.8 Definitions. Let h be a choice correspondence on a budget space $\langle X, \mathcal{B}\rangle$. We then define the relations V and W on X by:

$$xVy \iff (\exists B \in \mathcal{B})\colon x \in h(B) \;\&\; y \in B, \tag{3.4}$$

[read 'x is directly revealed preferred to y'], and xWy iff there is a finite sequence, $\langle u_i\rangle_{i=1}^m$, satisfying:

$$xVu_1V\ldots Vu_mVy. \tag{3.5}$$

[read 'x is revealed preferred to y'].

Our immediate concern at the moment is with the V relation (we will return to a discussion of the W relation later on). The first, and most important, thing to notice about the V relation is that if h is a choice correspondence, and G rationalizes h on $\langle X, \mathcal{B}\rangle$, then G must extend V on X, defined as follows.

3.9 Definition. If R and S are binary relations of a nonempty set X, we shall say that ***S* extends *R* on *X*** iff we have:

$$(\forall x, y \in X)\colon xRy \Rightarrow xSy.$$

In the case at hand, then, if G rationalizes h on $\langle X, \mathcal{B}\rangle$, we must have:

$$(\forall x, y \in X)\colon xVy \Rightarrow xGy. \tag{3.6}$$

[I will leave the verification of (3.6) as an exercise; it follows at once from the definitions.]

Returning now to Example 3.7, notice that the relation defined in (3.3) is actually the V relation corresponding to the given h; and what we established in the last paragraph of the example is that any binary relation which extends V and is also total cannot rationalize h on $\langle X, \mathcal{B}\rangle$.

If we think about the Examples 3.5 and 3.7 in connection with the V relation, it quickly becomes apparent that if h can be rationalized by the V relation, then h is a rational choice correspondence [let $G = V$ in Definition 3.2]. We might also suspect that h is rational only if h can be rationalized by V, and it turns out that this is indeed the case, as we shall now establish.

3.10 Definition. Let h be a choice correspondence on the budget space $\langle X, \mathcal{B}\rangle$. We shall say that ***h* satisfies the V-axiom** (Richter [1971, p. 33]) iff:

$$(\forall x \in X)(\forall B \in \mathcal{B})\colon [x \in B \;\&\; (\forall y \in B)\colon xVy] \Rightarrow x \in h(B). \tag{3.7}$$

3.11 Theorem. *A choice correspondence, h, on a budget space $\langle X, \mathcal{B}\rangle$, satisfies the V-axiom if, and only if, it is rational.*

Proof.
1. Suppose h satisfies the V-axiom, and let $B \in \mathcal{B}$ be arbitrary. If $x \in h(B)$, then we obviously have (by definition of the V relation):

$$(\forall y \in B)\colon xVy.$$

Conversely, if $x' \in B$ satisfies:

$$(\forall y \in B)\colon x'Vy,$$

then it follows from the assumption that h satisfies the V-axiom that $x' \in h(B)$. Consequently, since $B \in \mathcal{B}$ was arbitrary, it follows that:

$$(\forall B \in \mathcal{B})\colon h(B) = \{x \in B \mid (\forall y \in B)\colon xVy\};$$

and thus the relation V rationalizes h. Therefore h is rational.

2. Suppose h is rational, and that G is a binary relation on X which rationalizes h (so that G satisfies 3.2 [equation (3.2)]). If $B \in \mathcal{B}$ and $x \in B$ satisfy:

$$(\forall y \in B)\colon xVy.$$

then, since G must extend V, we have:

$$(\forall y \in B)\colon xGy.$$

Since G rationalizes h, it then follows that $x \in h(B)$. Therefore, h satisfies the V-axiom. □

Notice that in the first part of the above proof we have established that a choice correspondence, h, is rational if, and only if, it can be rationalized by the V relation which it defines. However, let me hasten to add that a rational choice correspondence can generally be rationalized by many different binary relations, as is illustrated by the following example.[2]

3.12 Example. Let $X = \{a, b, c\}$, and $\mathcal{B} = \{B_1, B_2\}$, where:

$$B_1 = \{a, b\} \qquad h(B_1) = \{a\},$$
$$B_2 = \{a, c\} \qquad h(B_2) = \{a\}.$$

In this case the V relation determined by h, which does rationalize h, is given by:

	a	***b***	***c***
a	aVa	aVb	aVc
b	...	...	...
c	...	...	....

[2]For conditions implying that the binary relation rationalizing a given choice correspondence is unique, see Arrow [1959] (for the case in which X is finite) and Chipman and Moore [1977] (for the case in which X is infinite).

However, each of the following two relations also rationalize h (and there are many other relations which rationalize h as well):

$$\begin{array}{cccc} & \boldsymbol{a} & \boldsymbol{b} & \boldsymbol{c} \\ \boldsymbol{a} & aGa & aGb & aGc \\ \boldsymbol{b} & \ldots & bGb & \ldots \\ \boldsymbol{c} & \ldots & \ldots & cGc \end{array} \qquad \begin{array}{ccc} \boldsymbol{a} & \boldsymbol{b} & \boldsymbol{c} \\ aG'a & aG'b & aG'c \\ \ldots & bG'b & bG'c \\ \ldots & \ldots & cG'c. \end{array}$$

Probably a few moments' thought will suffice to convince you that if a choice correspondence is rational, then it is reflexive-rational. However, we will nonetheless take the time to prove this.

3.13 Proposition. *If a choice correspondence, h is rational, then it is reflexive-rational.*

Proof. Suppose h is rational. Then by (the proof of) Theorem 3.11, h can be rationalized by the direct preference relation, V. Define the binary relation G on X by:

$$Gx = Vx \cup \{x\} \quad \text{for each } x \in X.$$

Then G is reflexive, and we can show that it rationalizes h, as follows.

First, let $B \in \mathcal{B}$ be arbitrary, and let $x \in h(B)$. Then, by definition we have:

$$(\forall y \in B)\colon xVy;$$

and thus, since G extends V, we also have $(\forall y \in B)\colon xGy$.

Conversely, suppose $x' \in B$ is such that $x' \notin h(B)$. Then it follows from the V-Axiom that there exists $x^* \in B$ such that $x^* \neq x'$ and:

$$\neg x'Vx^*.$$

But then, since:

$$Gx^* = Vx^* \cup \{x^*\},$$

we see that $\neg x'Gx^*$ as well. Consequently, we conclude that:

$$x' \notin h(B) \Rightarrow (\exists x^* \in B)\colon \neg x'Gx^*;$$

or, equivalently: if $x' \in B$ satisfies:

$$(\forall x \in B)\colon x'Gx,$$

then $x' \in h(B)$. □

3.3 Regular-Rational Choice Correspondences

In order to study regular rational choice correspondences, we begin by establishing the following.

3.14 Lemma. *Suppose h is a choice correspondence on $\langle X, \mathcal{B}\rangle$. If G is a transitive binary relation on X which rationalizes h, then G must extend W.*

Proof. Suppose G is a transitive binary relation which rationalizes h, and let $x, y \in X$ be such that xWy. Then, by definition of the W relation, there exist $u_1, \ldots, u_s \in X$ such that:

$$xVu_1V \ldots Vu_sVy.$$

Thus, since G must extend V, we have:

$$xGu_1G \ldots Gu_sGy;$$

and, since G is transitive, it then follows that xGy. □

3.15 Proposition. *There exist choice correspondences which can be rationalized by a total and reflexive binary relation, but not by any transitive binary relation.*

3.16 Example. Let $X = \{a, b, c\}, \mathcal{B} = \{B_1, B_2, B_3\}$, and:

$$\begin{aligned} B_1 &= \{a, b\} & h(B_1) &= \{a\}, \\ B_2 &= \{b, c\} & h(B_2) &= \{b\}, \\ B_3 &= \{a, c\} & h(B_3) &= \{c\}. \end{aligned}$$

It is easy to show that the following relation, which is the V relation defined from h, is total and reflexive, and rationalizes h on $\langle X, \mathcal{B} \rangle$:

$$\begin{array}{cccc} & \boldsymbol{a} & \boldsymbol{b} & \boldsymbol{c} \\ \boldsymbol{a} & aVa & aVb & \ldots \\ \boldsymbol{b} & \ldots & bVb & bVc \\ \boldsymbol{c} & cVa & \ldots & cVc. \end{array} \tag{3.8}$$

(notice that G is identical to the V relation in this case). The fact that h cannot be rationalized by any transitive binary relation follows easily from 3.18, below. □

3.17 Definition. (Richter [1966]). We shall say that a choice correspondence, h, on $\langle X, \mathcal{B} \rangle$, satisfies the **Congruence Axiom** iff we have:

$$(\forall x, y \in X)(\forall B \in \mathcal{B})\colon [x \in h(B) \;\&\; y \in B \;\&\; yWx] \Rightarrow y \in h(B).$$

3.18 Theorem. (Richter) *Let h be a choice correspondence on a budget space, $\langle X, \mathcal{B} \rangle$. Then there exists a transitive binary relation rationalizing h on $\langle X, \mathcal{B} \rangle$ if, and only if, h satisfies the Congruence Axiom.*

Proof.

1. Suppose h can be rationalized by a transitive binary relation, G, and suppose $B \in \mathcal{B}$ and $x, y \in B$ satisfy:

$$x \in h(B) \;\&\; yWx. \tag{3.9}$$

Then, by Lemma 3.13, we have:

$$yGx. \tag{3.10}$$

Furthermore, since $x \in h(B)$, it follows that:

$$(\forall u \in B)\colon xWu;$$

and, again using Lemma 3.13, we then have:

$$(\forall u \in B)\colon xGu. \tag{3.11}$$

Combining (3.10) and (3.11) with the fact that G is transitive, we then have:

$$(\forall u \in B)\colon yGu;$$

and, since G rationalizes h, it then follows that $y \in h(B)$. Therefore, h satisfies the Congruence Axiom.

2. Suppose h satisfies the Congruence Axiom, and let $B \in \mathcal{B}$. From the definition of the W relation it is obvious that:

$$\big(\forall x \in h(B)\big)\big(\forall u \in B\big)\colon xWu.$$

If, on the other hand, $y \in B$ satisfies:

$$(\forall u \in B)\colon yWu,$$

then it follows at once from the Congruence Axiom, and the fact that $h(B) \neq \emptyset$, that $y \in h(B)$. Since $B \in \mathcal{B}$ was arbitrary, we have shown that h satisfies:

$$(\forall B \in \mathcal{B})\colon h(B) = \{x \in B \mid (\forall u \in B)\colon xWu\}; \tag{3.12}$$

that is, W rationalizes h. Since W is obviously transitive, our result follows. □

The relation W is the *transitive closure* of V, for a given choice correspondence. For our purposes, the transitive closure of a relation, R, is defined as follows.

3.19 Definition. Let R be a binary relation on a nonempty set, X. We will say that a binary relation, G, on X is the **transitive closure of *R*** iff:

1. G is transitive,
2. G extends R on X, and:
3. given any *transitive* binary relation, $\succsim$, which extends R on X, $\succsim$ must also extend G on X.

Suppose now that R is a binary relation on a nonempty set, X, and define the relation G on X by xGy iff there exists a finite sequence, $\langle u_i\rangle_{i=1}^m \subseteq X$ satisfying:

$$xRu_1 \;\&\; u_1Ru_2 \;\&\; \ldots \;\&\; u_mRy,$$

for $x, y \in X$. It is easy to show that G is then transitive, and obviously G extends R on X. Furthermore, one can establish, by an argument similar to the proof of Lemma 3.13 (details are left as an exercise), that if $\succsim$ is a transitive binary relation which extends R on X, then $\succsim$ must also extend G. Consequently, it follows that G is the transitive closure of R,[3] and, as a special case of this result, it follows that for a given choice correspondence, h, the revealed preference relation, W, determined by h is the transitive closure of the relation V determined by h. From these considerations and a careful study of the proof of Theorem 3.18, one can easily prove the following (again the details will be left as an exercise).

[3]Notice also that R is its own transitive closure if it is itself transitive.

3.20 Proposition. *If h is a choice correspondence on a budget space $\langle X, \mathcal{B}\rangle$, then h is transitive-rational if, and only if, it can be rationalized by W.*

In light of the above proposition, and the discussion which preceded it, let's consider another example of a choice correspondence which can be rationalized by a total and reflexive binary relation, but not by any transitive binary relation. This example will also be particularly useful to us in our consideration of social choice functions in Chapter 14.

3.21 Example. Let $X = \{a, b, c\}, \mathcal{B} = \{B_1, B_2, B_3, B_4\}$, and:

$$\begin{aligned} B_1 &= \{a, b\} & h(B_1) &= \{a, b\}, \\ B_2 &= \{b, c\} & h(B_2) &= \{b, c\}, \\ B_3 &= \{a, c\} & h(B_3) &= \{a\}. \\ B_4 &= X & h(B_4) &= \{a, b\}. \end{aligned} \tag{3.13}$$

In this case, the V relation determined by h is given by the following table:

$$\begin{array}{cccc} & \boldsymbol{a} & \boldsymbol{b} & \boldsymbol{c} \\ \boldsymbol{a} & aVa & aVb & aVc \\ \boldsymbol{b} & bVa & bVb & bVc \\ \boldsymbol{c} & \dots & cVb & cVc. \end{array} \tag{3.14}$$

It is then easily seen that V is total, reflexive, and rationalizes h. However, it is also more or less immediate that the W relation is in this case the *trivial relation* defined by:

$$xWy \iff x, y \in X.$$

in particular, we have cWa, and from this fact you can easily show that (a) W does not rationalize h, or (b) h does not satisfy the Congruence Axiom (take your pick). In any case it follows that h cannot be rationalized by any transitive binary relation.

□

Theorem 3.18 has been extended (in one direction) by Richter [1966] to the form presented in Theorem 3.21, below. Since the proof of this extended result involves a considerably more sophisticated argument than that used in the proof of 3.18, however, we will not provide a proof here. On the other hand, notice that Theorem 3.18 is *not* a special case of 3.21. In fact, while the sufficiency part of 3.21 generalizes the sufficiency part of 3.18, the necessity part of 3.21 is a special case of the necessity part of 3.18.

3.22 Theorem. (Richter [1966, p. 639]). *Let h be a choice correspondence on a budget space $\langle X, \mathcal{B}\rangle$. Then h is regular rational if, and only if, h satisfies the Congruence Axiom.*

3.4 Representable Choice Correspondences

In modern discussions of demand theory, authors often make the statement that the economic theory of consumer behavior assumes that consumers behave *as if* they were maximizing a real-valued utility function. The following definition provides a precise definition of this statement.

3.23 Definition. A choice correspondence, h, on a budget space $\langle X, \mathcal{B}\rangle$ will be said to be **representable** iff there exists a function, $g\colon X \to \mathbb{R}$ satisfying:

$$(\forall B \in \mathcal{B})\colon h(B) = \{x \in B \mid (\forall y \in B)\colon g(x) \geq g(y)\}. \tag{3.15}$$

It follows at once from Theorem 3.17 that if h is representable, then h must satisfy the Congruence Axiom. On the other hand, it is possible for a choice function to satisfy the Congruence Axiom, and yet not be representable (see Richter [1971, pp. 46-7]). In order to state sufficient conditions for representability, we consider a special class of choice functions, defined as follows.

3.24 Definition. A choice correspondence, h, will be said to be **competitive** iff h is a choice correspondence on the budget space $\langle \mathbb{R}^n_+, \mathcal{B}^*\rangle$, where:

$$\mathcal{B}^* = \big\{B \subseteq \mathbb{R}^n_+ \mid \big(\exists(\boldsymbol{p}, w) \in \Omega\big)\colon B = b(\boldsymbol{p}, w)\big\};$$

where we define:

$$\Omega = \{(\boldsymbol{p}, w) \in \mathbb{R}^{n+1} \mid \boldsymbol{p} \in \mathbb{R}^n_{++} \;\&\; w \in \mathbb{R}_+\},$$

and where, for $(\boldsymbol{p}, w) \in \Omega$, we define:

$$b(\boldsymbol{p}, w) = \{\boldsymbol{x} \in \mathbb{R}^n_+ \mid \boldsymbol{p} \cdot \boldsymbol{x} \leq w\}.$$

[Notation: For competitive choice correspondences, we will write $B = b(\boldsymbol{p}, w)$ and $h(B) = h(\boldsymbol{p}, w)$.]

For a competitive choice correspondence, h, define:

$$h(\Omega) = \bigcup_{(\boldsymbol{p},w)\in\Omega} h(\boldsymbol{p}, w).$$

3.25 Examples/Exercises.

1. Consider the Cobb-Douglas utility function, $g\colon \mathbb{R}^n_+ \to \mathbb{R}_+$, given by:

$$g(\boldsymbol{x}) = \prod\nolimits_{i=1}^n (x_i)^{a_i},$$

where:

$$a_i > 0 \quad \text{for } i = 1, \ldots, n; \quad \text{and} \quad \sum\nolimits_{i=1}^n a_i = 1. \tag{3.16}$$

In this case, as is well known, the corresponding demand functions are given by:

$$h_i(\boldsymbol{p}, w) = \frac{a_i w}{p_i} \quad \text{for } i = 1, \ldots, n. \tag{3.17}$$

It should then be clear that here we have:

$$h(\Omega) \subseteq \mathbb{R}^n_{++} \cup \{\boldsymbol{0}\}; \tag{3.18}$$

that is:

$$\big(\forall(\boldsymbol{p}, w) \in \Omega\big) \colon h(\boldsymbol{p}, w) \in \mathbb{R}^n_{++} \cup \{\boldsymbol{0}\}.$$

Conversely, suppose $\boldsymbol{x}^*$ is a arbitrary element of $\mathbb{R}^n_{++} \cup \{\boldsymbol{0}\}$. If $\boldsymbol{x}^* = \boldsymbol{0}$, then obviously:

$$\boldsymbol{x}^* = h(\boldsymbol{p}, 0).$$

On the other hand, if $\boldsymbol{x}^* \in \mathbb{R}^n_{++}$, and we define:

$$p_i^* = a_i/x_i \quad \text{for } i = 1, \ldots, n;$$

then it is easy to show that:

$$\boldsymbol{x}^* = h(\boldsymbol{p}^*, 1).$$

Thus it follows from the arguments of this paragraph that:

$$\mathbb{R}^n_{++} \cup \{\boldsymbol{0}\} \subseteq h(\Omega);$$

and combining this with (18), we then have that:

$$h(\Omega) = \mathbb{R}^n_{++} \cup \{\boldsymbol{0}\}.$$

2. Suppose we change the specification in (3.25) to:

$$a_i \geq 0 \quad \text{for } i = 1, \ldots, n; \ \text{ and } \sum\nolimits_{i=1}^n a_i = 1. \tag{3.19}$$

What is the form of $h(\Omega)$ in this case?

3. Suppose we consider a case in which a consumer has a continuously differentiable and strictly quasi-concave utility function, having the property that:

$$(\forall \boldsymbol{x} \in \mathbb{R}^n_{++}) \colon \nabla u(\boldsymbol{x}) \gg \boldsymbol{0}.$$

Can you then prove that we will have $\mathbb{R}^n_{++} \cup \{\boldsymbol{0}\} \subseteq h(\Omega)$? □

Richter has established the following result.

3.26 Theorem. (Richter [1966]). *Let h be a competitive choice correspondence, suppose that $D(h)$ is a convex set, and that:*

$$\big(\forall(\boldsymbol{p}, w) \in \Omega\big) \colon h(\boldsymbol{p}, w) \textit{ is a closed set.}$$

If h also satisfies the Congruence Axiom, then h is representable.

3.27 Examples/Exercises. Suppose h is a competitive choice function[4] having the property that the proportion of income spent on the ith commodity is equal to some constant $a_i \geq 0$, for $i = 1, \ldots, n$, where:

$$\sum\nolimits_{i=1}^n a_i = 1.$$

Show that h is representable.

[4]Recall the terminology introduced in Section 2. A choice function is a single-valued choice correspondence, but we also think of $h(B)$ as being an element, rather than a subset of B in this case.

3.5 Preferences and Observed Demand Behavior

Let's return to the issue of determining what, exactly, are the implications of preference-maximizing behavor. We'll start by considering the following issue. Suppose we are given a function, $\boldsymbol{h}\colon \Omega \to \mathbb{R}^n_+$. How can we tell if it is consistent with preference maximization; that is, how can we tell whether it *might* be the demand function of a preference-maximizing consumer? Obviously, if it is a demand function, it needs to be positively homogeneous of degree zero in $(\boldsymbol{p}, w)$ and satisfy the condition:

$$(\forall(\boldsymbol{p}, w) \in \Omega)\colon \boldsymbol{p} \cdot \boldsymbol{h}(\boldsymbol{p}, w) \leq w. \tag{3.20}$$

We can take this one step further: consider the following definition.

3.28 Definition. Let X be a non-empty subset of $\mathbb{R}^n$, and let P be a binary relation on X. We shall say that P is **locally non-saturating** iff, given any $\boldsymbol{x} \in X$, and any $\epsilon > 0$, there exists $\boldsymbol{y} \in N(\boldsymbol{x}, \epsilon) \cap X$ such that $\boldsymbol{y}P\boldsymbol{x}$.

It is then easy to prove that if $h\colon \Omega \to X$ is the demand function generated by a locally non-saturating preference relation, P, then h must satisfy the following condition.

3.29 Definition. Let $h\colon \Omega \to \mathbb{R}^n_+$ be a competitive demand function. We shall say that h satisfies the **budget balance condition** iff we have, for all $(\boldsymbol{p}, w) \in \Omega$:

$$(\forall(\boldsymbol{p}, w) \in \Omega)\colon \boldsymbol{p} \cdot \boldsymbol{h}(\boldsymbol{p}, w) = w. \tag{3.21}$$

Thus, to return to our earlier discussion, if there is a $(\boldsymbol{p}, w)$ pair for which:

$$\boldsymbol{p} \cdot \boldsymbol{h}(\boldsymbol{p}, w) < w, \tag{3.22}$$

then this function is not consistent with the maximization of a locally non-saturating preference relation.

So, let's specialize our question a bit, to consider a function, $\boldsymbol{h}\colon \Omega \to \mathbb{R}^n_+$, which is homogeneous of degree zero and satisfies the bundget balance condition. How can we then tell whether or not $\boldsymbol{h}$ is consistent with locally non-saturating-preference-maximizing behavior? To avoid repeating this rather awkward phrase innumerable times in our discussion, let's begin by defining the following.

3.30 Definition. We shall say that a function $\boldsymbol{h}\colon \Omega \to \mathbb{R}^n_+$ is **S-rational** iff it can be rationalized (Definition 3.2) by a locally non-saturating weak order on $\mathbb{R}^n_+$.

Almost 70 years ago, Paul Samuelson provided a partial answer to this question (Samuelson [1938]). To be S-rational, $\boldsymbol{h}$ must satisfy what is now called the Weak Axiom of Revealed Preference (WARP). In order to state this, we begin by defining the relation S on $\mathbb{R}^n_+$ by:

$$\boldsymbol{x}S\boldsymbol{y} \iff \boldsymbol{x} \neq \boldsymbol{y} \text{ and } [(\exists(\boldsymbol{p}, w) \in \Omega)\colon \boldsymbol{x} = \boldsymbol{h}(\boldsymbol{p}, w) \;\&\; \boldsymbol{p} \cdot \boldsymbol{y} \leq w]. \tag{3.23}$$

We can then define the axiom as follows.

3.31 Definition. We say that the function $\boldsymbol{h}\colon \Omega \to \mathbb{R}^n_+$ satisfies the **Weak Axiom of Revealed Preference (WARP)** iff the relation S defined in (3.23) is asymmetric.

Thus, if the function $\boldsymbol{h}$ is S-rational, it must be postively homogeneous of degree zero and satisfy equation 3.21 and WARP. However, this leaves open the question of whether or not these three conditions exhaust the implications of the assumption that $\boldsymbol{h}$ is S-rational. Writing some time after Samuelson, Houthakker [1950] noted, in effect,[5] that an S-rational function must also satisfy what is now known as the Strong Axiom of Revealed Preference. To state this, we begin by defining the relation H as the transitive closure of S; that is, we define H on $\mathbb{R}^n_+$ by:

$$\boldsymbol{x}H\boldsymbol{y} \iff \big[\boldsymbol{x}S\boldsymbol{y} \text{ or } (\exists \boldsymbol{u}_1, \dots, \boldsymbol{u}_s \in \mathbb{R}^n_+)\colon \boldsymbol{x}S\boldsymbol{u}_1S\boldsymbol{u}_2S\dots S\boldsymbol{u}_sS\boldsymbol{y}\big] \tag{3.24}$$

[read: '$\boldsymbol{x}$ is **revealed preferred** to $\boldsymbol{y}$'],

3.32 Definition. We say that the function $\boldsymbol{h}\colon \Omega \to \mathbb{R}^n_+$ satisfies the **Strong Axiom of Revealed Preference (SARP)** iff the relation H defined in (3.24) is asymmetric.

A question which the Houthakker paper left unresolved was whether SARP and WARP were independent conditions. This question was answered in the affirmative by David Gale [1960], who exhibited a function satisfying WARP, but not SARP.[6] Moreover, the question of whether homogeneity, (3.21), and SARP fully exhausted the implications of the assumption that $\boldsymbol{h}$ is S-rational was not definitively answered until the publication of Richter's [1966] paper. The issue here is this: Suppose $\boldsymbol{h}\colon \Omega \to \mathbb{R}^n_+$. Let's agree to call $\boldsymbol{h}$ a **d-function** if it (a) is positively homogeneous of degree zero, (b) satisfies budget balance [equation (3.21)], and (c) satisfies SARP. If we are given a function $\boldsymbol{h}\colon \Omega \to \mathbb{R}^n_+$ which fails any one of conditions (a)–(c), we can be sure that it is *not* S-rational. However, this leaves unanswered the question of whether every d-function is S-rational (that is, whether it *might* be the demand function of a preference-maximizing consumer whose preferences are a locally non-saturating weak order on $\mathbb{R}^n_+$). Richter's article answers this affirmatively and definitively. In order to demonstrate this, we need first to show that, in the present context, SARP and Richter's Congruence Axiom are equivalent. We can do this as follows.

Proof of equivalence, for $\boldsymbol{h}\colon \Omega \to \mathbb{R}^n_+$ satisfying (3.21)

We begin by noting that, under the present conditions, if $\boldsymbol{x}, \boldsymbol{y} \in \mathbb{R}^n_+$ are such that $\boldsymbol{x}W\boldsymbol{y}$, and $\boldsymbol{x} \neq \boldsymbol{y}$, then $\boldsymbol{x}H\boldsymbol{y}$. For, if $\boldsymbol{x}W\boldsymbol{y}$, then there exist $\boldsymbol{u}^1, \dots, \boldsymbol{u}^r \in \mathbb{R}^n_+$ such that, defining $\boldsymbol{u}^0 = \boldsymbol{x}$ and $\boldsymbol{u}^{r+1} = \boldsymbol{y}$, we have:

$$\boldsymbol{u}^iV\boldsymbol{u}^{i+1} \quad \text{for } i = 0, 1, \dots, r. \tag{3.25}$$

[5]Both Samuelson and Houthakker framed their investigations in terms of utility-maximization.

[6]This investigation was extended and expanded by Kihlstrom, Mas-Colell, and Sonnenschein [[1976]. They developed a whole class of functions satisfying WARP, but not SARP; but, more importantly developed necessary, and sufficient conditions for the matrix of substitution terms to be symmetric and negative semi-definite.

Furthermore, if $\boldsymbol{x} \neq \boldsymbol{y}$, then we must have $\boldsymbol{u}^j \neq \boldsymbol{u}^{j+1}$ for at least one $j \in \{0, 1, \ldots, r\}$, and thus:

$$\boldsymbol{u}^j S \boldsymbol{u}^{j+1}.$$

Moreover, for each remaining index, i, for which $\boldsymbol{u}^i = \boldsymbol{u}^{i+1}$, we can eliminate $\boldsymbol{u}^{i+1}$, to obtain a set $\{\boldsymbol{v}^0, \ldots, \boldsymbol{v}^{t+1}\} \subseteq \{\boldsymbol{u}^0, \boldsymbol{u}^1, \ldots, \boldsymbol{u}^{r+1}\}$, with:

$$\boldsymbol{v}^0 = \boldsymbol{x} \ \& \ \boldsymbol{v}^{t+1} = \boldsymbol{y},$$

and satisfying:

$$\boldsymbol{v}^k S \boldsymbol{v}^{k+1} \quad \text{for } k = 0, \ldots, t.$$

Now suppose that $\boldsymbol{h}$ satisfies SARP, and that $(\boldsymbol{p}, w) \in \Omega$ and $\boldsymbol{x}, \boldsymbol{y} \in \mathbb{R}^n_+$ satisfy:

$$\boldsymbol{x} = \boldsymbol{h}(\boldsymbol{p}, w) \ \& \ \boldsymbol{p} \cdot \boldsymbol{y} \leq w.$$

If $\boldsymbol{x} \neq \boldsymbol{y}$, it then follows that $\boldsymbol{x} H \boldsymbol{y}$, and thus by SARP and the argument of the above paragraph we cannot have $\boldsymbol{y} W \boldsymbol{x}$ as well. Therefore, $\boldsymbol{h}$ satisfies the Congruence Axiom.

Conversely, suppose $\boldsymbol{h}$ satisfies the Congruence Axiom, and that $\boldsymbol{x}, \boldsymbol{y} \in \mathbb{R}^n_+$ are such that:

$$\boldsymbol{x} H \boldsymbol{y}.$$

Then we can distinguish two cases.

First, suppose $\boldsymbol{x} S \boldsymbol{y}$. Then there exists $(\boldsymbol{p}, w) \in \Omega$ such that:

$$\boldsymbol{x} = \boldsymbol{h}(\boldsymbol{p}, w) \ \& \ \boldsymbol{p} \cdot \boldsymbol{y} \leq w. \tag{3.26}$$

If we were then also to have $\boldsymbol{y} H \boldsymbol{x}$, we would obviously also have $\boldsymbol{y} W \boldsymbol{x}$, and it would follow from the Congruence Axiom that $\boldsymbol{y} = \boldsymbol{h}(\boldsymbol{p}, w)$, which, since $\boldsymbol{x} S \boldsymbol{y}$ implies $\boldsymbol{x} \neq \boldsymbol{y}$, contradicts (3.26).

Otherwise (that is, if $\neg \boldsymbol{x} S \boldsymbol{y}$), there will exist $\boldsymbol{u}^1, \ldots, \boldsymbol{u}^r \in \mathbb{R}^n_+$ such that:

$$\boldsymbol{x} S \boldsymbol{u}^1, \boldsymbol{u}^1 S \boldsymbol{u}^2, \ldots, \boldsymbol{u}^r S \boldsymbol{y}. \tag{3.27}$$

If we also were to have $\boldsymbol{y} H \boldsymbol{x}$, then there would exist $\boldsymbol{v}^1, \ldots, \boldsymbol{v}^s \in \mathbb{R}^n_+$ such that:[7]

$$\boldsymbol{y} S \boldsymbol{v}^1, \boldsymbol{v}^1 S \boldsymbol{v}^2, \ldots, \boldsymbol{v}^s S \boldsymbol{x}. \tag{3.28}$$

Then, combining (3.27) and (3.28), we see that $\boldsymbol{u}^1 H \boldsymbol{x}$. However, this implies $\boldsymbol{u}^1 W \boldsymbol{x}$, which is impossible; for by the fact that $\boldsymbol{x} S \boldsymbol{u}^1$, there exists $(\boldsymbol{p}, w) \in \Omega$ such that:

$$\boldsymbol{x} = h(\boldsymbol{p}, w) \ \& \ \boldsymbol{p} \cdot \boldsymbol{u}^1 \leq w; \tag{3.29}$$

which, by the Congruence Axiom would imply $\boldsymbol{u}^1 = h(\boldsymbol{p}, w)$, contradicting (3.27). □

Given the equivalence just established, the following is easily established, using Richter's Theorem 3.22. The formal proof will be left as an exercise.

[7]Strictly speaking, we should allow for the case in which $\boldsymbol{y} S \boldsymbol{x}$. However, this leaves the basic argument unaffected.

3.33 Proposition. *Suppose* $h\colon \Omega \to \mathbb{R}^n_+$. *If* $\boldsymbol{h}$ *is S-rational, then it is positively homogeneous of degree zero, and satisfies equation (3.21) (budget balance) and SARP. Conversely, if* $\boldsymbol{h}$ *is positively homogeneous of degree zero, and satisfies equation (3.21) and SARP, then there exists a weak order on* $\mathbb{R}^n_+$, *which is locally non-saturating on:*

$$\boldsymbol{h}(\Omega) = \{\boldsymbol{x} \in \mathbb{R}^n_+ \mid (\exists(\boldsymbol{p}, w) \in \Omega)\colon \boldsymbol{x} = \boldsymbol{h}(\boldsymbol{p}, w)\},$$

and which rationalizes $\boldsymbol{h}$.

The converse statement in the above proposition tells us that the conditions listed exhaust the implications of the assumption that $\boldsymbol{h}$ is S-rational, To put this another way, if we have a function $\boldsymbol{h}\colon \Omega \to \mathbb{R}^n_+$ which is positively homogeneous of degree one and satisfies budget balance and SARP, we can be certain that it *could* be the demand function of a preference-maximizing consumer, whose preference relation is a weak order on $\mathbb{R}^n_+$, and locally non-saturating on the portion of $\mathbb{R}^n_+$ relevant to her or his demand behavior.

However, the proposition just established leaves some issues still unaddressed. First of all, we have been looking at demand *functions;* presuming, in effect, that given the same $(\boldsymbol{p}, w)$ pair in repeated choices, the consumer would always pick the same bundle from the budget set. However, in Chapter 2 we saw that in experimental situations, subjects often varied their choices when given the same budget set in repeated situations. In fact, notice that SARP actually implies that the choice correspondence is single-valued. In the context of consumer demand theory, this means that if, for example a consumer faced the same prices in two successive periods (two successive months in the standard interpretation of the theory), and if the consumer's wealth (money income) were the same in the two preiods, then she/he/it will choose exactly the same commodity bundle in the two periods. It is doubtful whether anyone really believes that this would happen, however. Casual observation suggests that the choice actually made in the two time periods would be influenced by a myriad of factors not taken into account in the standard theory; for example, whether the consumer 'owes' or 'is owed' dinner invitations the season of the year, and so on and so on. We can, of course, treat all such aberrations as 'changes in taste,' but to do so is to imply that the currently received theory of demand has no empirical (predictive) content whatever.

As Richter has pointed out [1966, p. 3a], however, such complications can be allowed for in the following way. Suppose we view the consumer's choice as a two-step process. A 'viable set' of alternatives is chosen from the budget set, and then a final (and unexplained, from the point of view of the standard formal theory) choice is made from this viable set. If the initial choice of a 'viable set' can be regarded as being guided by a weak order, and this viable set is the set of maximal elements of the budget set, then we may still have a theory with empirical content, but one which allows for variations in the final choice. Richter's Congruence Axiom, then provides a complete characterization of the viable set in the context of consumer demand theory. However, there may be a problem with this characterization. Specifically, the difficulty is, that while one can characterize situations in which an investigator may be able to observe values of a consumer's demand function, it is difficult to imagine scenarios in which one can observe the values of the viable sets. In fact, in experimental or statistical studies of actual individual (competitive) choice behavior, all that one can generally hope to observe is a **data set**, $\mathbb{D} = \langle(\boldsymbol{p}^t, w^t), \boldsymbol{x}^t\rangle_{t=1}^T$, where T is a positive integer, and $\boldsymbol{x}^t$ is the bundle chosen at $(\boldsymbol{p}^t, w^t)$, for $t = 1, \ldots, T$. How can we determine whether such a data set is consistent with (one- or two-step) preference maximization?

To see the problem that the difficulty in observing the viable set creates, let's return to Example 3.7; which, as you may recall, exhibited a choice correspondence which could not be rationalized by any *total* binary relation. This time, however, suppose that the individual makes a choice of a viable set (which we will identify with the correspondence h defined in the example initially), and then makes a final choice according to some unknown criterion. We will indicate this final choice by '$d(B)$,' and will suppose that the investigator observes the pairs $\langle B_t, d(B_t)\rangle$, for $t = 1, 2, 3$. Then we may have the situation exhibited in the following example.

3.34 Example. Let $X = \{a, b, c\}, \mathcal{B} = \{B_1, B_2, B_3\}$, and:

$$\begin{array}{lll} B_1 = \{a, c\} & h(B_1) = \{a, c\} & d(B_1) = a, \\ B_2 = \{b, c\} & h(B_2) = \{b, c\} & d(B_2) = b, \\ B_3 = \{a, b, c\} & h(B_3) = \{c\} & d(B_t) = c. \end{array}$$

Given what we are supposing can be observed in this case, we cannot distinguish between the consumer whose choices we have just been describing, and the consumer who maximizes in one step, and whose preference relation is such that a is preferred to b and b is preferred to c. In particular, we could not refute the hypothesis that the consumer's choice of a viable set involves maximization of a weak order. □

This last example shows that problems are created by the fact that we may not observe the entirety of a consumer's viable sets, if our description of the two-step maximization process is a fair description of reality, and probably makes you wonder whether this sort of hypothesis would have any observable implications at all! However. the fact is that in the context of demand theory, the hypothesis does have observable implications, and the Congruence Axiom will help us determine what they are. To see this, suppose that the criterion used to determine the viable set (we will hereafter refer to this as the **first-step criterion** and we will denote the '**viable set correspondence**' by '$\boldsymbol{h}(\cdot)$') is a locally non-saturating weak order. Then when faced with a budget pair $(\boldsymbol{p}, w)$, the consumer's final choice, which we will denote by '$\boldsymbol{d}(\boldsymbol{p}, w)$,' must satisfy:

$$\boldsymbol{p} \cdot \boldsymbol{d}(\boldsymbol{p}, w) = w. \tag{3.30}$$

We will refer to this property of a data set as **budget balance**.

Now suppose that we observe a data set $\mathbb{D} = \langle(\boldsymbol{p}^t, w^t), \boldsymbol{x}^t\rangle_{t=1}^T$, satisfying budget balance, and suppose that for some subset, $\mathbb{D}^* = \langle(\boldsymbol{p}^s, w^s), \boldsymbol{y}^s\rangle_{s=1}^S \subseteq \mathbb{D} = \langle(\boldsymbol{p}^t, w^t), \boldsymbol{x}^t\rangle_{t=1}^T$, we have:

$$\boldsymbol{p}^s \cdot \boldsymbol{y}^{s+1} \leq w^s \quad \text{for } s = 1, \ldots, S-1. \tag{3.31}$$

Then, in terms of the (direct) revealed preference (V) relation corresponding to the viable sets (that is, corresponding to $\boldsymbol{h}$), we have:

$$\boldsymbol{y}^s V \boldsymbol{y}^{s+1} \quad \text{for } s = 1, \ldots, S-1;$$

and therefore:

$$\boldsymbol{y}^1 W \boldsymbol{y}^S. \tag{3.32}$$

Consequently, if $\boldsymbol{y}^1$ is in the budget set $b(\boldsymbol{p}^S, w^S)$, and if we suppose that the first-step criterion is a locally non-saturating weak order, then $\boldsymbol{y}^1$ must be in the viable set, $\boldsymbol{h}(\boldsymbol{p}^S, w^S)$; which implies:

$$\boldsymbol{p}^S \cdot \boldsymbol{y}^1 = w^S.$$

Thus, whether or not $\boldsymbol{y}^1$ is in the final budget set, we must have:

$$\boldsymbol{p}^S \cdot \boldsymbol{y}^1 \geq w^S.$$

This implication is called the **Generalized Axiom of Revealed Preference**. More properly, we state the following definition.

3.35 Definition. We will say the data set, $\mathbb{D} = \langle(\boldsymbol{p}^t, w^t), \boldsymbol{x}^t\rangle_{t=1}^T$ satisfies the **Generalized Axiom of Revealed Preference (GARP)** if, given any subset, $\mathbb{D}^* = \langle(\boldsymbol{p}^s, w^s), \boldsymbol{y}^s\rangle_{s=1}^S \subseteq \mathbb{D} = \langle(\boldsymbol{p}^t, w^t), \boldsymbol{x}^t\rangle_{t=1}^T$ satisfying:

$$\boldsymbol{p}^s \cdot \boldsymbol{y}^{s+1} \leq w^s \quad \text{for } s = 1, \ldots, S-1, \tag{3.33}$$

we have:

$$\boldsymbol{p}^S \cdot \boldsymbol{y}^1 \geq w^S. \tag{3.34}$$

Afriat [1967, 1973] developed the very subtle and insightful theorem which we state as follows.

3.36 Theorem. Afriat *If the data set* $\mathbb{D} = \langle(\boldsymbol{p}^t, w^t), \boldsymbol{x}^t\rangle_{t=1}^T$ *satisfies budget balance and GARP, then there exist real numbers* $u_1, \ldots, u_T$ *and* positive *real numbers* $\lambda_1, \ldots, \lambda_T$ *satisfying:*

$$u_j \leq u_i + \lambda_i(\boldsymbol{p}^i \cdot \boldsymbol{x}^j - w^i) \quad \textit{for } i, j = 1, \ldots, T. \tag{3.35}$$

While we will not provide a proof of Afriat's theorem here, let me recommend that those of you with a particular interest in theory consult the excellent article by Fostel, Scarf, and Todd [2004] in which they provide an elegant, and much shorter and simpler proof than Afriat's original argument.[8]

We can also (and again following Afriat, although not so literally this time) state something which is a sort of converse of the above result. However, we need to begin with some considerations involving the meaning of a preference relation (or utility function) rationalizing demand in the situation under consideration. Since we are supposing that we would not generally observe all of $\boldsymbol{h}(\boldsymbol{p}, w)$, but rather only an element thereof, the following definitions become more important in our current discussion than the definitions of a preference relation (or a utility function) rationalizing $\boldsymbol{h}$.

[8]Such readers should also consult the articles by Diewert [1973] and Varian [1982], who provide alternative arguments and tests for GARP.

3.37 Definition. Let $\mathbb{D} = \langle(\boldsymbol{p}^t, w^t), \boldsymbol{x}^t\rangle_{t=1}^T$ be a data set. We will say that a binary relation, G, on $\mathbb{R}^n_+$ (respectively, a function, $u\colon \mathbb{R}^n_+ \to \mathbb{R}$) **is consistent with** $\mathbb{D}$ iff we have:

$$(\forall \boldsymbol{x} \in \mathbb{R}^n_+)\colon \boldsymbol{p}^t \cdot \boldsymbol{x} \leq w^t \Rightarrow \boldsymbol{x}^t G \boldsymbol{x} \quad \text{for } t = 1, \ldots, T. \tag{3.36}$$

[respectively:

$$(\forall \boldsymbol{x} \in \mathbb{R}^n_+)\colon \boldsymbol{p}^t \cdot \boldsymbol{x} \leq w^t \Rightarrow u(\boldsymbol{x}^t) \geq u(\boldsymbol{x}) \quad \text{for } t = 1, \ldots, T.] \tag{3.37}$$

We can then state a second theorem due to Afriat (albeit our statement is a bit different from Afriat's) as follows.

3.38 Theorem. *Let the data set* $\mathbb{D} = \langle(\boldsymbol{p}^t, w^t), \boldsymbol{x}^t\rangle_{t=1}^T$ *satisfy budget balance, and suppose the real numbers* $u_1, \ldots, u_T$ *and the* positive *real numbers* $\lambda_1, \ldots, \lambda_T$ *satisfy (3.35) of Theorem 3.36. Then the function* $u\colon \mathbb{R}^n_+ \to \mathbb{R}$ *defined by:*

$$u(\boldsymbol{x}) = \min_t [u_t + \lambda_t(\boldsymbol{p}^t \cdot \boldsymbol{x} - w^t)] \tag{3.38}$$

is consistent with $\mathbb{D}$.

Proof. Notice, first of all, that it follows from from budget balance that:

$$u_t + \lambda_t(\boldsymbol{p}^t \cdot \boldsymbol{x}^t - w^t) = u_t;$$

and thus from (3.35) and the defnition of u, we see that:

$$u(\boldsymbol{x}^t) = u_t \quad \text{for } t = 1, \ldots, T.$$

Next we note that if $\boldsymbol{p}^t \cdot \boldsymbol{x} \leq w^t$, then:

$$u(\boldsymbol{x}) \leq u_t + \lambda_t(\boldsymbol{p}^t \cdot \boldsymbol{x} - w^t) \leq u_t = u(\boldsymbol{x}^t). \quad \square$$

Notice that the functions:

$$u_t + \lambda_t(\boldsymbol{p}^t \cdot \boldsymbol{x} - w^t),$$

are continuous, strictly increasing, and concave in $\boldsymbol{x}$. Consequently, the minimum function, u, defined in (3.38) is concave, strictly increasing, and concave as well. Therefore, Theorem 3.36 tells us that GARP and the budget balance condition completely exhaust the observable implications of the assumption that the viable correspondence can be rationalized by a strictly increasing, continuous, and concave utility function.

Interestingly, Matzkin and Richter [1991] have shown that if $\mathbb{D} = \langle(\boldsymbol{p}^t, w^t), \boldsymbol{x}^t\rangle_{t=1}^T$ satisfies budget balance and the *Strong* Axiom of Revealed Preference, then (the consumer's choice correspondence is a function, and) a strengthened version of Afriat's Theorem can be deduced in that all of the inequalities in (3.35) can be taken to be strict, for $\boldsymbol{x}^i \neq \boldsymbol{x}^j$. These inequalities are then used to construct a function, $u(\cdot)$, which is strictly concave, strictly increasing, continuous, and rationalizes the data set.

3.6 The Implications of Asymmetric Orders*

In this section we will take a brief look at some of the implications of the assumption that consumer (strict) preferences can only be assumed to be an asymmetric order. In our treatment here, we will follow Kim and Richter [1986].

3.39 Definitions. If h is a choice correspondence on a budget space $\langle X, \mathcal{B}\rangle$, and $\succ$ is a binary relation on X, then $\succ$ is said to **motivate *h*** iff, for every $B \in \mathcal{B}$:

$$h(B) = \{x \in B \mid (\forall y \in B) \colon y \not\succ x\}. \tag{3.39}$$

Equivalently, we can say that $\succ$ **motivates *h*** iff, for every $B \in \mathcal{B}$:

$$h(B) = \{x \in B \mid (\forall y \in X) \colon y \succ x \Rightarrow y \notin B\}. \tag{3.40}$$

In either case we will say that ***h* is motivated by** $\succ$, and if there exists a binary relation, $\succ$ which motivates h, we will say that ***h* is motivated**. If there exists a binary relation, $\succ$, which motivates h, and which is, respectively: irreflexive, asymmetric, transitive, or asymmetric and transitive, we will say that ***h* is irreflexive-, asymmetric-, transitive-**, or **asymmetric order-motivated**, respectively.

Now, it is easy to show formally that if a choice correspondence, h is motivated by a binary relation, $\succ$, on X, and if we define the binary relation, $\succsim$, on X by:

$$x \succsim y \iff y \not\succ x, \tag{3.41}$$

then h is rationalized by $\succsim$ (as defined in Definition 3.2). Conversely, if h is rationalized by the relation $\succsim$, and we define $\succ$ by:

$$x \succ y \iff \neg[y \succsim x], \tag{3.42}$$

then h is motivated by $\succ$. Thus the proof of the first of the following results is fairly immediate. Similarly, we know that if $\succ$ is asymmetric, and we define $\succsim$ as in (3.41), then $\succsim$ is total and reflexive (Proposition 1.17 of Chapter 1); while if $\succsim$ is total and reflexive, and $\succ$ is defined as in (3.42), then $\succ$ is asymmetric. These considerations provide the basis of the proof of Theorem 3.22.

3.40 Theorem. (Kim and Richter Theorem 3, p. 333) *A choice correspondence, h, is motivated iff h satisfies the V-Axiom. Hence, h is motivated if, and only if, it is rational.*

3.41 Theorem. (Kim and Richter Theorem 5, p. 334) *A choice correspondence, h, is asymmetric-motivated iff it is total-rational, and iff it is total-reflexive-rational.*

3.42 Theorem. (Kim and Richter Theorem 6, p. 334) *Let h be a competitive choice correspondence satisfying the budget balance condition. Then h is asymmetric-motivated if, and only if, h satisfies the V-Axiom.*

Proof. See Kim & Richter [1986, pp. 334–5]. □

3.43 Proposition. *If h satisfies the Congruence Axiom, then h is asymmetric order-motivated.*

Proof. Suppose h satisfies the Congruence Axiom, let W be the revealed preference relation defined by h, and define $\succ$ on X by:

$$x \succ y \iff [xWy \;\&\; \neg yWx]. \tag{3.43}$$

It follows from Theorem 1.13 of Chapter 1 that $\succ$ is an asymmetric order (notice that W is a reflexive and transitive relation), and thus our proof will be complete if we can show that $\succ$ motivates h.

Accordingly, let $B \in \mathcal{B}$, and suppose first that $x \in h(B)$. Then it follows at once from the definition of W that for all $y \in B$, we must have xWy; and from this it is immediate that:

$$(\forall y \in B)\colon y \succ x \Rightarrow y \notin B.$$

Conversely, suppose z is an element of B satisfying:

$$(\forall y \in X)\colon y \succ z \Rightarrow y \notin B.$$

Then in particular, for $x \in h(B)$ [and recall that $h(B)$ must be non-empty, by the definition of a choice correspondence], we must have:

$$x \not\succ z. \tag{3.44}$$

However, since $x \in h(B)$, it follows from the definition of the W relation that we must have xWz. If it were also the case that $\neg zWx$, then it would follow that $x \succ z$; which contradicts (3.44). Thus we must have zWx, and it then follows from the Congruence Axiom that $z \in h(B)$. □

Notice that in the above result we have shown that the satisfaction of the Congruence Axiom is a sufficient condition for the choice function h to be asymmetric order-motivated. Necessary and sufficient conditions for h to be asymmetric order-motivated are apparently not known; however, in the remainder of this section we will investigate some aspects of this question in more detail. We begin with a useful definition which is often used in the revealed preference literature.[9]

3.44 Definition. Let P be an irreflexive binary relation on a non-empty set, X. We shall say that P is **cyclic** iff, for some positive integer, n, there exists points $x_1, x_2, \ldots, x_n \in X$ such that:

$$x_1 P x_2 \;\&\; x_2 P x_3 \;\&\ldots\;\&\; x_{n-1} P x_n,$$

but $x_n P x_1$. If no such cycle exists (that is, if P is *not* cyclic), we shall say that P is **acyclic**.

It is easy to see that if P is acyclic, then it is asymmetric. Conversely, if it is asymmetric and transitive (and thus is an asymmetric order), then it is acyclic (and irreflexive as well). It is, however, easy to construct examples of irreflexive binary relations which are acyclic (and thus are also asymmetric), but which are *not* transitive. For example consider the following.

[9]This definition is also used very frequently in the literature on social choice, as we shall discover in Chapter 14.

3.45 Example. Let X be the three element set, $X = \{a, b, c\}$, and let P be as indicated in the following table.

	a	**b**	**c**
a	...	aPb	...
b	...	...	bPc
c	...	...	....

Then, while this relation is irreflexive and acyclic, it is *not* transitive; transitivity would require that we also have aPc. Notice also that if we were to fill in the lower left cell of the table, specifying that cPa, then the relation would be cyclic. □

We can use the definition of acyclicity to completely characterize choice on finite sets, as follows.

3.46 Proposition. *Suppose $\langle X, \mathcal{B}\rangle$ is a budget space, where X is a finite set, that P is an irreflexive binary relation on X, and define the correspondence, h, on $\mathcal{B}$ by:*

$$h(B) = \{x \in B \mid (\forall y \in X)\colon yPx \Rightarrow y \notin B\}.$$

If P is acyclic, then h is decisive; that is, it is nonempty-valued. Furthermore, if $\mathcal{B}$ includes all subsets of X containing two or more elements, and h is decisive, then P is acyclic.

Proof.

1. Suppose that h is *not* decisive. We wish to prove that it must then be the case that P is cyclic.

Accordingly if h is not decisive, then there exists $B^* \in \mathcal{B}$ such that $h(B^*) = \emptyset$. Since X is finite, we may suppose without loss of generality that $\#B^* = k$, where k is an integer greater than or equal to one. If $k = 1$, that is, if B^* is of the form:

$$B^* = \{x^*\},$$

for some $x^* \in X$, then, since $h(B^*) = \emptyset$, we must have:

$$x^*Px^*;$$

and we see that P is not irreflexive, contrary to our hypothesis. Consequently, we must have $k \geq 2$.

Now, if we choose an arbitrary element of B^* to label 'x_1,' then there must exist an element of B^*, x_2, such that:

$$x_2Px_1.$$

If also x_1Px_2, then we have established that P is cyclic. Otherwise, since $h(B^*) = \emptyset$, there must exist $x_3 \in B^*$, distinct from x_1, such that:

$$x_3Px_2.$$

However, if x_2Px_3, or x_1Px_3, we have established that P is cyclic, and we can stop. Otherwise, since P is irreflexive and, since we have already noted that we must have $x_3 \neq x_1$, it follows that x_1, x_2 and x_3 are distinct elements of B^* satisfying:

$$x_3Px_2 \;\&\; x_2Px_1.$$

Proceeding in this way, suppose we have found m distinct elements, $x_1, \ldots, x_m \in B^*$, where $m \geq 2$, satisfying:

$$x_{j+1} P x_j \quad \text{for } j = 1, \ldots, m-1.$$

Then, since $h(B^*) = \emptyset$, there must exist $x_{m+1} \in B^*$ such that $x_{m+1} P x_m$. However, if $x_{m+1} = x_j$, for some $j \in \{1, \ldots, m\}$, then we have:

$$x_m P x_{m-1} \;\&\; \ldots x_{j+1} P x_j \;\text{ and }\; x_j P x_m,$$

and we have shown that P is cyclic. Consequently, we see that at the mth step, we will either have shown that P is cyclic, or we will obtain an element, $x_{m+1} \in B^*$ such that $x_{m+1}, x_m, \ldots, x_1$ are all distinct elements of B^* and satisfy:

$$x_{j+1} P x_j \quad \text{for } j = 1, \ldots, m. \tag{3.45}$$

However, since there are only k elements in B^*, this process can continue at most until $k = m + 1$. On the other hand, since $h(B^*) = \emptyset$, it then follows that we must have $x_j P x_k$, for some $j \in \{1, \ldots, k-1\}$, and the same basic argument as was presented earlier in this paragraph establishes that P is cyclic.

2. Suppose $\mathcal{B}$ includes all subsets of X containing two or more elements, and suppose P is irreflexive but *not* acyclic; that is, suppose P is irreflexive and cyclic. Then there exists an integer, $n \geq 2$, and elements $x_1, \ldots, x_n \in X$ such that:

$$x_1 P x_2 \;\&\; \ldots \;\&\; x_{n-1} P x_n \;\text{ and }\; x_n P x_1.$$

But then, defining $B^* = \{x_1, \ldots, x_n\}$, we see that for each $j \in \{1, \ldots, n\}$, there exists $k \in \{1, \ldots, n\}$ such that $x_k P x_j$.[10] It follows, therefore, that for each $x \in B^*$, there exists $x' \in B^*$ such that $x' P x$; and thus $h(B^*) = \emptyset$. Since $B^* \in \mathcal{B}$ under the present assumptions, we see that h is not decisive. It therefore follows that if h is decisive (given the extra assumptions on $\mathcal{B}$), then P is acyclic. □

As was noted earlier, full necessary conditions for a choice correspondence to be asymmetric-transitive-motivated are apparently not known.[11] We can, however, make some progress toward the solution of this problem by considering the following examples. In the first of the two, we develop a choice correspondence which can be motivated by a strict preference relation, P, which is acyclic, but such that h does *not* satisfy the congruence axiom;[12] while in the second example, h is asymmetric order-motivated, but nonetheless does not satisfy the congruence axiom.

3.47 Example. Let $X = \{a, b, c\}$, $\mathcal{B} = \{B_1, B_2, B_3, B_4\}$, and:

$$\begin{aligned} B_1 &= \{a, b\} & h(B_1) &= \{a\}, \\ B_2 &= \{a, c\} & h(B_2) &= \{a, c\}, \\ B_3 &= \{b, c\} & h(B_3) &= \{b\}, \\ B_4 &= \{a, b, c\} & h(B_4) &= \{a\}. \end{aligned}$$

[10] If $j \in \{2, \ldots, n\}$, let $k = j - 1$; while if $j = 1$, let $k = n$.

[11] However, see Kim [1987].

[12] It is also true that in this example h is not motivated by the transitive closure of P; which is actually my main reason for presenting it in addition to Example 3.39

Notice that $\mathcal{B}$ includes all subsets of X containing two or more elements. It is easy to prove that the choice correspondence, h is motivated by the following preference, P, wich is obviously irreflexive and acyclic:[13]

	a	**b**	**c**
a	...	aPb	...
b	...	...	bPc
c	...	...	....

In order to show that h does not satisfy the congruence axiom, we begin by noting that the V relation determined by h is as follows (notice that V is total).

	a	**b**	**c**
a	aVa	aVb	aVc
b	...	bVb	bVc
c	cVa	...	cVc.

Thus the W relation is as follows:

	a	**b**	**c**
a	aWa	aWb	aWc
b	bWa	bWb	bWc
c	cWa	cWb	cWc.

But then we have, for example, $b \in B_1$, $a \in h(B_1)$, and bWa, but $b \notin h(B_1)$; which shows that h does not satisfy the congruence axiom. □

3.48 Example. Here we take $X = \{a, b, c, d\}$, $\mathcal{B} = \{B_1, \ldots, B_{11}\}$, and:

$$\begin{aligned}
B_1 &= \{a, b\} & h(B_1) &= \{a\}, \\
B_2 &= \{a, c\} & h(B_2) &= \{a\}, \\
B_3 &= \{a, d\} & h(B_3) &= \{a, d\}, \\
B_4 &= \{b, c\} & h(B_4) &= \{b\}, \\
B_5 &= \{b, d\} & h(B_5) &= \{b, d\}, \\
B_6 &= \{c, d\} & h(B_6) &= \{d\}, \\
B_7 &= \{a, b, c\} & h(B_7) &= \{a\} \\
B_8 &= \{a, b, d\} & h(B_8) &= \{a, d\} \\
B_9 &= \{a, c, d\} & h(B_9) &= \{a, d\} \\
B_{10} &= \{b, c, d\} & h(B_{10}) &= \{b, d\} \\
B_{11} &= \{a, b, c, d\} & h(B_{11}) &= \{a, d\}.
\end{aligned}$$

Notice that, as in the previous example, $\mathcal{B}$ includes all subsets of X which contain at least two elements. Moreover, it is a straightforward exercise to show that the

[13]The transitive closure of P, P^*, is identical to P except that we also have aP^*c. It is easy to see, however, that P^* does not motivate h.

following preference motivates h:

	a	**b**	**c**	**d**
a	...	aPb	aPc	...
b	...	...	bPc	...
c	...	...	...	...
d	...	...	dPc	....

Notice that the relation P defined in the above table is an asymmetric order. It is, however, not negatively transitive; since, for example, we have aPb, but neither aPd nor dPb.[14]

The V relation generated by h is then as follows.

	a	**b**	**c**	**d**
a	aVa	aVb	aVc	aVd
b	...	bVb	bVc	bVd
c	...	...	...	...
d	dVa	dVb	dVc	dVd.

Consequently, the W relation for h is then given by:

	a	**b**	**c**	**d**
a	aWa	aWb	aWc	aWd
b	bWa	bWb	bWc	bWd
c	...	...	...	...
d	dWa	dWb	dWc	dWd.

it is now easy to see that h does not satisfy the congruence axiom; for we have, for example, $b \in B_1$, $a \in h(B_1)$, and bWa, but $b \notin h(B_1)$. □

Exercises.

In each of the following three problems, a choice correspondence is presented. In each case, answer the following questions, and provide a justification for each answer.

Is h (1) rational? (2) total-reflexive-rational? (3) transitive-rational? (4) regular-rational?

1. Let $X = \{a, b, c, d\}, \mathcal{B} = \{B_1, B_2, B_3, B_4\}$, and:

$$B_1 = \{a, b\} \qquad h(B_1) = \{a, b\},$$
$$B_2 = \{b, c\} \qquad h(B_2) = \{b\},$$
$$B_3 = \{a, c\} \qquad h(B_3) = \{c\}.$$
$$B_4 = \{b, d\} \qquad h(B_4) = \{d\}.$$

[14] Kim and Richter [1986] show that if h is asymmetric-negatively transitive-motivated, then it is regular-rational, and thus must satisfy the congruence axiom.

2. Let $X = \{a, b, c, d\}$, and $\mathcal{B}$ and h be defined by:

$$\begin{aligned} B_1 &= \{a, d\} & h(B_1) &= \{a\}, \\ B_2 &= \{b, d\} & h(B_2) &= \{b\}, \\ B_3 &= \{a, c, d\} & h(B_3) &= \{c\}. \\ B_4 &= \{a, b\} & h(B_4) &= \{a\}. \end{aligned}$$

3. Let $X = \{a, b, c, d\}$, and $\mathcal{B}$ and h be defined by:

$$\begin{aligned} B_1 &= \{a, b\} & h(B_1) &= \{a\}, \\ B_2 &= \{b, d\} & h(B_2) &= \{d\}, \\ B_3 &= \{c, d\} & h(B_3) &= \{d\}, \\ B_4 &= \{a, c, d\} & h(B_4) &= \{d\}. \\ B_5 &= \{b, c\} & h(B_5) &= \{b, c\} \\ B_6 &= \{a, c\} & h(B_6) &= \{a, c\}. \end{aligned}$$

4. Can you find the (competitive) choice, or demand function for a consumer having the lexicographic preferences defined in Example 1.28 of Chapter 1? Is the resulting choice function representable?

5. Suppose X is a nonempty set, that P is a binary relation on X, and that $f\colon X \to \mathbb{R}$ is a function satisfying:

$$(\forall x, y \in X)\colon xPy \Rightarrow f(x) > f(y).$$

Is P acyclic? Prove or provide a counterexample. (See also Problem 5, at the end of Chapter 1.)

Chapter 4

Consumer Demand Theory

4.1 Introduction

In this chapter we will add structure to our choice theory model by examining the additional implications which follow from some standard structural/geometric assumptions used in economics. Consequently, much of this chapter will probably be review material. We will suppose throughout the chapter that the ith consumer has a (strict) preference relation, P_i, which is asymmetric and transitive (so that P_i is an asymmetric order); but, as already suggested, we will generally assume that P_i satisfies other assumptions as well. We begin our study by considering the interpretations of the consumers' consumption sets which are used in General Equilibrium Theory.

4.2 The Consumption Set

We will suppose that P_i is defined over the consumer's consumption set, X_i, where $X_i \subseteq \mathbb{R}^n$. In much of this chapter, we will suppose that $X_i \subseteq \mathbb{R}^n_+$; however, we will allow for the more general case when it is convenient, and we will often need to allow for the (mathematically) more general case when we talk about equilibrium in a production economy. We will use the generic notation, '$\boldsymbol{x}_i$,' '$\boldsymbol{x}_i^*$,' etc. to denote the commodity bundle chosen by (and available for consumption by) the ith consumer. Thus we write:

$$\boldsymbol{x}_i = (x_{i1}, \ldots, x_{ij}, \ldots, x_{in}) \in X_i;$$

where 'x_{ij}' denotes the quantity of the jth commodity ($j = 1, \ldots, n$) available to the ith consumer, if $x_{ij} \geq 0$. If $x_{ij} < 0$, then we will take this to mean that the ith consumer is offering to supply the jth commodity in the amount $-x_{ij} = |x_{ij}|$.

There are two basic conventions with respect to the interpretation of the ith consumer's consumption set which are used in general equilibrium theory. The first, which is the one used in the above paragraph, is that the amounts x_{ij} represent the *total* amounts available for consumption, or to be supplied by the ith consumer. The second convention involves the idea of interpreting X_i to be a **trading set** or a **net demand set**. We can relate these two ideas in the following way. Suppose we assume that the consumer's consumption set is necessarily a subset of $\mathbb{R}^n_+$, and let's denote this consumption set by 'C_i;' so that:

$$C_i \subseteq \mathbb{R}^n_+.$$

This convention may necessitate re-defining some of our commodities: for example, if, under the interpretation of the above paragraph, x_{i1} represents the quantity of labor services to be offered by the ith consumer (so that the first commodity is 'labor'), the convention being followed in this second approach is that if $\boldsymbol{c}_i \in C_i$, then c_{i1} will represent the amount of leisure time being enjoyed by the consumer. We then suppose that the consumer has an initial endowment of the n commodities, which we shall denote by $\boldsymbol{r}_i \in \mathbb{R}^n_+$. In particular, '$r_{i1}$' would here denote the total amount of leisure available to the consumer in the time period under consideration, *if no labor services were offered at all.* We would then suppose that any commodity bundle, $\boldsymbol{c}_i$, available to the consumer (that is, any $\boldsymbol{c}_i \in C_i$) would necessarily satisfy the condition:

$$c_{i1} \leq r_{i1};$$

and we would interpret the quantity:

$$\ell_i \stackrel{\text{def}}{=} r_{i1} - c_{i1},$$

to be the quantity of labor services being offered by the consumer, given the total consumption bundle $\boldsymbol{c}_i$.

However, we can conveniently represent the conventions of the above paragraph in a different way, as follows. Let's define the set X_i as:

$$X_i = C_i - \boldsymbol{r}_i.$$

The natural interpretation of X_i is that if $\boldsymbol{x}_i \in X_i$, then the quantity x_{ij} represents the quantity of the jth commodity being demanded from the rest of the economy (if $x_{ij} \geq 0$), or being offered to the rest of the economy (if $x_{ij} < 0$). In particular, recalling our earlier interpretation of the first commodity as representing leisure, notice that if $\boldsymbol{x}_i \in X_i$, then:

$$x_{i1} = -\ell_i.$$

For future reference, notice that with this definition of X_i, it will be the case that X_i will satisfy: for all $\boldsymbol{x}_i \in X_i$:

$$\boldsymbol{x}_i \geq -\boldsymbol{r}_i.$$

That this is so follows from the fact that if $\boldsymbol{x}_i \in X_i$, then the consumer's total consumption (or commodity bundle available for consumption), $\boldsymbol{c}_i$ is given by:

$$\boldsymbol{c}_i = \boldsymbol{x}_i + \boldsymbol{r}_i;$$

and, since $C_i \subseteq \mathbb{R}^n_+$, we necessarily have $\boldsymbol{c}_i \geq \boldsymbol{0}$.

Continuing our discussion of the trading set, notice that if the consumer's preferences can be represented as an asymmetric order, $\succ_i$, on C_i, then we can represent the consumer's preferences on X_i by the relation P_i defined as follows:

$$\boldsymbol{x}_i P_i \boldsymbol{x}'_i \iff (\boldsymbol{x}_i + \boldsymbol{r}_i) \succ_i (\boldsymbol{x}'_i + \boldsymbol{r}_i).$$

It is easy to see that if $\succ_i$ is asymmetric, or transitive, or negatively transitive, then P_i will satisfy exactly the same properties.[1]

Now, having read all of this discussion, you may be inclined to ask this question: "If some elements of X have some negative coordinates, doesn't this mean that we need to use the 'trading set' interepretation of X, and suppose that given a commodity bundle $\boldsymbol{x} \in X$, the consumer's actual commodity bundle available for consumption is given by $\boldsymbol{c} = \boldsymbol{x} + \boldsymbol{r}$, where $\boldsymbol{r}$ is the consumer's initial commodity endowment, and where $\boldsymbol{c} \in \mathbb{R}^n_+$?" Well, the answer to this question is "not necessarily." Suppose we wish to allow for the fact that the consumer may be able to supply two different types of labor, and suppose these two types of labor are commodities one and two (measured in labor hours), that commodity three is, say, 'food,' while for convenience we suppose that there are just these three commodities in the economy. If our consumer needs at least two units of food to survive, and can supply no more than 16 hours of the two types of labor per period then a natural representation of the consumer's consumption set is:

$$X = \{\boldsymbol{x} \in \mathbb{R}^3 \mid 16 + x_1 + x_2 \geq 0, x_j \leq 0 \text{ for } j = 1, 2, \text{ and } x_3 \geq 2\}.$$

The key thing here is that the consumer's choice of leisure (say the quantity $24+x_1+x_2$) does not enable us to determine the quantities of either x_1 of x_2. Consequently, the net trading set representation does not work in this context.

So, the next question is, how do these distinctions affect our analysis. The fact is, that in most of our analysis, we won't need to worry very much about which interpretation of the consumption set should be used. The budget constraint for the consumer will normally be defined by a pair $(\boldsymbol{p}, w)$, where $\boldsymbol{p} \in \mathbb{R}^n_+$ is the vector of prices of the n commodities, and we suppose the consumer's choice is constrainted to be in the set:

$$b(\boldsymbol{p}, w) \stackrel{\text{def}}{=} \{\boldsymbol{x} \in X \mid \boldsymbol{p} \cdot \boldsymbol{x} \leq w\}.$$

Under the 'trading set' interpretation, or under the sort of definition of the consumption set indicated in the preceding paragraph, w is interpreted as 'wealth,' or income from sources other than the supply of labor. On the other hand, in the 'final consumption' interpretation (where we take X_i to be a subset of $\mathbb{R}^n_+$), w will need to include receipts from the 'sale of leisure,' that is, if we return to the case in which we take $X_i \subseteq \mathbb{R}^n_+$, and let the first commodity be the consumer's labor/leisure, with the consumer's total endowment of leisure being given by $r_{i1} > 0$, then the consumer's budget constraint can be expressed as:

$$\boldsymbol{p} \cdot \boldsymbol{x}_i \leq p_1 r_{i1} + w'_i,$$

where now 'w'_i' denotes income from sources other than the sale of labor. Alternatively, in this case we can simply define:

$$w_i = p_1 r_{11} + w'_i;$$

and express the budget constraint exactly as before.

[1] We do, however, need to be careful to note that if $\boldsymbol{r}_i$ should change, then so will P_i; even if $\succ_i$ remains the same!

4.3 Demand Correspondences

Suppose the ith consumer faces the price vector $\boldsymbol{p} = (p_1, \ldots, p_n) \in \mathbb{R}^n_{++}$. Whether one interprets X_i as the consumption set or as a trading set, the consumer's budget set can, in the absence of non-labor income, be represented as the set $\beta_i(\boldsymbol{p})$ defined as:

$$\beta_i(\boldsymbol{p}) = \{\boldsymbol{x}_i \in X_i \mid \boldsymbol{p} \cdot \boldsymbol{x}_i \leq 0\}.$$

Under either the consumption set or the trading set interpretation, however, we will often want to allow for the fact that the consumer may have income or wealth, w_i from other sources; that is, purchasing power which is derived from something (possibly the profits of firms) other than the sale of the consumer's labor services, or initial endowment of commodities. Consequently, we will handle the consumer's budget constraint as follows. We begin by defining the set Ω_i, a subset of R^{n+1}, by:

$$\Omega_i = \{(\boldsymbol{p}, w_i) \in \mathbb{R}^{n+1} \mid \boldsymbol{p} \in \mathbb{R}^n_{++} \ \&\ (\exists \boldsymbol{x}_i \in X_i) \colon \boldsymbol{p} \cdot \boldsymbol{x}_i \leq w_i\}.$$

We then define the consumer's budget set, $b_i(\boldsymbol{p}, w_i)$, for $(\boldsymbol{p}, w_i) \in \Omega_i$ by:

$$b_i(\boldsymbol{p}, w_i) = \{\boldsymbol{x}_i \in X_i \mid \boldsymbol{p} \cdot \boldsymbol{x}_i \leq w_i\}. \tag{4.1}$$

This last equation defines a correspondence, which we define formally in the following.

4.1 Definitions. We define the consumer's **budget correspondence**, $b_i \colon \Omega_i \mapsto X_i$, by equation (4.1), for $(\boldsymbol{p}, w_i) \in \Omega_i$. We then define the consumer's demand correspondence, $\boldsymbol{h}_i$, by:

$$\boldsymbol{h}_i(\boldsymbol{p}, w_i) = \{\boldsymbol{x}_i \in b_i(\boldsymbol{p}, w_i) \mid (\forall \boldsymbol{x}'_i \in X_i) \colon \boldsymbol{x}'_i P_i \boldsymbol{x}_i \Rightarrow \boldsymbol{p} \cdot \boldsymbol{x}'_i > w_i\}, \tag{4.2}$$

for $(\boldsymbol{p}, w_i) \in \Omega_i$. Formally (and sometimes this much formality will be convenient, if not necessary), we shall refer to the correspondence $\boldsymbol{h}_i \colon \Omega_i \mapsto X_i$ defined in (4.2) as the **demand correspondence determined by $\boldsymbol{P}_i$**.[2]

In the remainder of this, and the next four sections, however, we will be dealing with the theory of demand for a single consumer, so that we can drop the subscript 'i' wherever it appears; writing simply '$\boldsymbol{x}$,' '$b(\boldsymbol{p}, w)$,' '$\boldsymbol{h}(\boldsymbol{p}, w)$,' etc.

We begin our investigation of the theory of consumer demand with the most basic consideration of all; namely, under what conditions will the consumer's demand correspondence be well-defined? More precisely, our concern in the remainder of this section is to investigate the conditions under which we will have:

$$\big(\forall (\boldsymbol{p}, w) \in \Omega\big) \colon \boldsymbol{h}(\boldsymbol{p}, w) \neq \emptyset.$$

4.2 Definition. We shall say that a subset, X, of $\mathbb{R}^n$ is **bounded below** iff there exists a point $\boldsymbol{z} \in \mathbb{R}^n$ satisfying:

$$(\forall \boldsymbol{x} \in X) \colon \boldsymbol{x} \geq \boldsymbol{z}. \tag{4.3}$$

[2]It might be objected that the demand correspondence is jointly determined by P_i and X_i as well, but a part of the definition of P_i is a specification of its domain; that is, of X_i.

4.3 Proposition. *Let X be a closed, non-empty subset of $\mathbb{R}^n$, which is bounded below, and let Ω and $b\colon \Omega \mapsto X$ be defined as in Definition 4.1. Then, given any $(\boldsymbol{p}^*, w^*) \in \Omega, b(\boldsymbol{p}^*, w^*)$ is compact and non-empty.*

Proof. It is obvious from the definition of Ω that $b(\boldsymbol{p}^*, w^*)$ is non-empty. To prove that it is also compact, we begin by noting that $b(\boldsymbol{p}^*, w^*)$ is the intersection of the closed half-space:

$$H \stackrel{\text{def}}{=} \{\boldsymbol{x} \in \mathbb{R}^n \mid \boldsymbol{p}^* \cdot \boldsymbol{x} \leq w^*\},$$

with X. Since both of these sets are closed, it follows that $b(\boldsymbol{p}^*, w^*)$ is closed as well.

To prove that $b(\boldsymbol{p}^*, w^*)$ is bounded, we begin by recalling that, since X is bounded below, there exists a point $\boldsymbol{z}$ satisfying (4.3), above. Next define $w^\dagger$ by:

$$\boldsymbol{p}^* \cdot \boldsymbol{z} = w^\dagger, \tag{4.4}$$

and note that it follows from (4.3), the definition of Ω, and the fact that $\boldsymbol{p}^* \geq \boldsymbol{0}$ that:

$$w^\dagger \leq w^*. \tag{4.5}$$

If we now define the vector $\boldsymbol{y} \in \mathbb{R}^n$ by:

$$y_j = z_j + \frac{w^* - w^\dagger}{p_j^*} \quad \text{for } j = 1, \ldots, n, \tag{4.6}$$

it follows from (4.5) that $\boldsymbol{y} \geq \boldsymbol{z}$. We will prove that, defining:

$$Y = \{\boldsymbol{x} \in \mathbb{R}^n \mid \boldsymbol{z} \leq \boldsymbol{x} \leq \boldsymbol{y}\},$$

we must have:

$$b(\boldsymbol{p}^*, w^*) \subseteq Y; \tag{4.7}$$

from which it follows that $b(\boldsymbol{p}^*, w^*)$ is bounded.

To prove (4.7), suppose, by way of obtaining a contradiction, that there exists $\boldsymbol{x}^* \in b(\boldsymbol{p}^*, w^*)$ such that $\boldsymbol{x}^* \notin Y$. Then, in view of (4.3), it must be that $\boldsymbol{y} \not\geq \boldsymbol{x}^*$; so that, for some $j \in \{1, \ldots, n\}$:

$$x_j^* > y_j. \tag{4.8}$$

However, if $\boldsymbol{x}^*$ satisfies (4.8), then we have, making use also of (4.3) and (4.6):

$$\begin{aligned}\boldsymbol{p}^* \cdot \boldsymbol{x}^* = \boldsymbol{p}^* \cdot (\boldsymbol{x}^* - \boldsymbol{z} + \boldsymbol{z}) = \boldsymbol{p}^* \cdot (\boldsymbol{x}^* - \boldsymbol{z}) + \boldsymbol{p}^* \cdot \boldsymbol{z} &\geq p_j^*(x_j^* - z_j) + w^\dagger \\ > p_j^*(y_j - z_j) + w^\dagger = p_j^*(w^* - w^\dagger)/p_j^* &+ w^\dagger = w^*;\end{aligned}$$

that is:

$$\boldsymbol{p}^* \cdot \boldsymbol{x}^* > w^*,$$

contradicting the assumption that $\boldsymbol{x}^* \in b(\boldsymbol{p}^*, w^*)$. Thus we see that (4.7) must hold; and thus that $b(\boldsymbol{p}^*, w^*)$ is bounded. Since we also showed that it was closed, it now follows that $b(\boldsymbol{p}^*, w^*)$ is compact. □

The following is a repetition of Definition 2.14, and is repeated here for the sake of having a convenient reference.

4.4 Definitions. We shall say that an asymmetric binary relation, P, defined on a non-empty subset, X, of $\mathbb{R}^n$, is:

1. **upper semi-continuous** iff, for each $\boldsymbol{x} \in X$, the set $\boldsymbol{x}P$ is open relative to X; that is, for each $\boldsymbol{x}' \in X$ such that $\boldsymbol{x}P\boldsymbol{x}'$, there exists a (Euclidean) neighborhood of $\boldsymbol{x}'$, $N(\boldsymbol{x}')$, such that:

$$\big(\forall \boldsymbol{y} \in N(\boldsymbol{x}') \cap X\big) : \boldsymbol{x}P\boldsymbol{y}.$$

2. **lower semi-continuous** iff, for each $\boldsymbol{x} \in X$, the set $P\boldsymbol{x}$ is open relative to X.

3. **continuous** iff it is both upper and lower semi-continuous.

4. **strongly continuous** iff, for each $\boldsymbol{x}, \boldsymbol{y} \in X$, if $\boldsymbol{x}P\boldsymbol{y}$, then there exist neighborhoods of $\boldsymbol{x}$ and $\boldsymbol{y}$, $N(\boldsymbol{x})$ and $M(\boldsymbol{y})$, respectively such that, for all $\boldsymbol{x}' \in N(\boldsymbol{x}) \cap X$ and all $\boldsymbol{y}' \in M(\boldsymbol{y}) \cap X$, we have $\boldsymbol{x}'P\boldsymbol{y}'$.

The following result is formally proved in the appendix to this chapter. From it we can see that the consumer's demand correspondence is well-defined under very general conditions indeed!

4.5 Theorem. *If X is a non-empty, closed subset of $\mathbb{R}^n$ which is bounded below, and P is an asymmetric ordering on X which is upper semi-continuous, then $\boldsymbol{h}(\cdot)$, the demand correspondence determined by P, satisfies:*

$$\big(\forall (\boldsymbol{p}, w) \in \Omega\big) : \boldsymbol{h}(\boldsymbol{p}, w) \neq \emptyset;$$

that is, for each $(\boldsymbol{p}, w) \in \Omega$, there exists a bundle $\boldsymbol{x} \in b(\boldsymbol{p}, w)$ satisfying:

$$\big(\forall \boldsymbol{x}' \in X\big) : \boldsymbol{x}'P\boldsymbol{x} \Rightarrow \boldsymbol{p} \cdot \boldsymbol{x}' > w.$$

While it has seemed to me to be worthwhile to state and prove (albeit in an appendix) the above result, the fact is that in most of our work with demand correspondences we will be assuming that the consumer's (strict) preference relation satisfies more stringent conditions than are assumed in Theorem 4.5. In fact, more often than not we will be assuming that P is negatively transitive, as well as being asymmetric; in which case, its negation, G, is a weak order. In any case, whether or not P is negatively transitive, the demand correspondence which it determines can equally well be defined by:

$$\boldsymbol{h}(\boldsymbol{p}, w) = \Big\{\boldsymbol{x} \in b(\boldsymbol{p}, w) \mid \big(\forall \boldsymbol{y} \in b(\boldsymbol{p}, w)\big) : \boldsymbol{x}G\boldsymbol{y}\Big\}, \tag{4.9}$$

where G is the negation of P. Since this is the more conventional way of defining demand correspondences in any event, we shall hereafter generally speak of demand correspondences as being determined by a (presumably reflexive) binary relation, G, as per equation (4.9).

4.4 The Budget Balance Condition

A condition which is normally assumed to characterize consumer demand correspondences is budget balance, defined as follows.

4.6 Definition. Let G be a (weak) preference relation, and let $\boldsymbol{h}$ be the demand correspondence determined by G. We shall say that the demand correspondence determined by G, $\boldsymbol{h}(\cdot)$, satisfies the **budget balance condition** (or that **G satisfies the budget balance condition**) if, and only if, for all $(\boldsymbol{p}, w) \in \Omega$, we have:

$$\big(\forall \boldsymbol{x} \in \boldsymbol{h}(\boldsymbol{p}, w)\big) : \boldsymbol{p} \cdot \boldsymbol{x} = w.$$

It is easily seen that the second of the following conditions implies that the consumer's demand correspondence satisfies the budget balance condition. While it is the assumption used most often in economic theory to justify the budget balance condition, we will consider a more general condition shortly.

4.7 Definitions. Let G be a binary relation on $\mathbb{R}^n_+$, and let P be its asymmetric part. We will say that G is:

1. **non-decreasing** iff, given any $\boldsymbol{x}, \boldsymbol{y} \in \mathbb{R}^n_+$:

$$\boldsymbol{x} \geq \boldsymbol{y} \Rightarrow \boldsymbol{x}G\boldsymbol{y},$$

or equivalently, $\boldsymbol{y}P\boldsymbol{x} \Rightarrow \boldsymbol{x} \not\geq \boldsymbol{y}$;

2. **increasing** iff G is non-decreasing and, in addition, satisfies the following condition:

$$(\forall \boldsymbol{x}, \boldsymbol{y} \in \mathbb{R}^n_+) : \boldsymbol{x} \gg \boldsymbol{y} \Rightarrow \boldsymbol{x}P\boldsymbol{y}.$$

3. **strictly increasing** iff, for every $\boldsymbol{x}, \boldsymbol{y} \in \mathbb{R}^n_+$:

$$\boldsymbol{x} > \boldsymbol{y} \Rightarrow \boldsymbol{x}P\boldsymbol{y}.$$

4.8 Definitions. Let X be a non-empty subset of $\mathbb{R}^n$, and let G be a binary relation on X, with P its asymmetric part. We shall say that G is:

1. **non-saturating** iff, given any $\boldsymbol{x} \in X$, there exists $\boldsymbol{y} \in X$ such that $\boldsymbol{y}P\boldsymbol{x}$.
2. **locally non-saturating** iff, given any $\boldsymbol{x} \in X$, and any $\epsilon > 0$, there exists $\boldsymbol{y} \in N(\boldsymbol{x}, \epsilon) \cap X$ such that $\boldsymbol{y}P\boldsymbol{x}$.

Notice that a preference relation which is increasing, as defined in 4.7 above, is locally non-saturating. Moreover, any locally non-saturating binary relation is non-saturating, but the converse is not necessarily true. For instance, Example 1.31.4 of Chapter 1 features a non-saturating binary relation which is not locally non-saturating. Another such example is presented as Example 4.12.1, below.

4.9 Proposition. *Suppose the preference relation, G, is a locally non-saturating weak order, and that $\boldsymbol{x}^* \in \boldsymbol{h}(\boldsymbol{p}^*, w^*)$, for some $(\boldsymbol{p}^*, w^*) \in \Omega$. Then for all $\boldsymbol{x} \in X$:*

$$\boldsymbol{x}G\boldsymbol{x}^* \Rightarrow \boldsymbol{p}^* \cdot \boldsymbol{x} \geq w^*.$$

Proof. Suppose, by way of obtaining a contradiction, that there exists $\bar{\boldsymbol{x}} \in X$ such that:

$$\bar{\boldsymbol{x}}G\boldsymbol{x}^*, \tag{4.10}$$

and:

$$\boldsymbol{p}^* \cdot \bar{\boldsymbol{x}} < w^*.$$

From this last inequality and the continuity of the inner product, there exists $\epsilon > 0$ such that

$$\big(\forall \boldsymbol{x} \in N(\bar{\boldsymbol{x}}, \epsilon)\big) : \boldsymbol{p}^* \cdot \boldsymbol{x} < w^*. \tag{4.11}$$

Now, since G is locally non-saturating, there exists $\boldsymbol{x}' \in N(\bar{\boldsymbol{x}}, \epsilon) \cap X$ such that:

$$\boldsymbol{x}' P \bar{\boldsymbol{x}},$$

where P is the asymmetric part of G; and, making use of (4.10) and the fact that G is a weak order, it follows that $\boldsymbol{x}' P \boldsymbol{x}^*$. However, by (4.11) we also have:

$$\boldsymbol{p}^* \cdot \boldsymbol{x}' < w^*;$$

which contradicts the assumption that $\boldsymbol{x}^* \in \boldsymbol{h}(\boldsymbol{p}^*, w^*)$. □

The next result is easily proved by a modification of the argument just presented. Notice, however, that the assumptions used here are much weaker than those used in 4.9.

4.10 Proposition. *If G is locally non-saturating on X, then G satisfies the budget balance condition.*

It should be apparent that G cannot be locally non-saturating if all commodities are indivisible. However, it is only necessary that one commodity be more or less completely divisible in order that G be locally non-saturating. In our next definition, we will present a condition which will be particularly useful to us, and which implies local non-saturation. In order to present it, however, we need to remind ourselves of a bit of notation. In $\mathbb{R}^n$ we define the n **unit coordinate vectors,** $\boldsymbol{e}_j$ $(j = 1, \ldots, n)$, by:

$$\boldsymbol{e}_j = (\delta_{j1}, \ldots, \delta_{jn}),$$

where δ_{jk} is the Kronecker delta function defined by:

$$\delta_{jk} = \begin{cases} 0 & \text{for } j \neq k, \\ 1 & \text{for } j = k. \end{cases}$$

We can then define the following.

4.11 Definition. Let G be a preference relation on a consumption set, X, with P its asymmetric part. We shall say that the jth commodity is a **numéraire good for G** iff, for all $\boldsymbol{x} \in X$, and all $\theta \in \mathbb{R}_{++}$, we have:

$$\boldsymbol{x} + \theta \boldsymbol{e}_j \in X \text{ and } (\boldsymbol{x} + \theta \boldsymbol{e}_j) P \boldsymbol{x}.$$

We shall say that **G admits a numéraire** iff, for some $j \in \{1, \ldots, n\}$, the jth commodity is a numéraire good for G.

Notice that if G admits a numéraire, then G is locally non-saturating. Notice also that if G is strictly increasing (with $X = \mathbb{R}^n_+$), then all commodities are numéraire goods for G.

4.12 Examples.

1. Let $f\colon \mathbb{R}^n_+ \to \mathbb{R}$ be any non-decreasing function, let δ be a positive constant, and define P on $\mathbb{R}^n_+$ by:

$$\boldsymbol{x}P\boldsymbol{y} \iff f(\boldsymbol{x}) > f(\boldsymbol{y}) + \delta.$$

Here P (or, more correctly, its negation, G) will be non-decreasing, but neither increasing nor locally non-saturating.

2. Let f be defined on $\mathbb{R}^2_+$ by:

$$f(\boldsymbol{x}) = 10x_1 - (x_1)^2 + x_2,$$

and define G on $\mathbb{R}^2_+$ by:

$$\boldsymbol{x}G\boldsymbol{y} \iff f(\boldsymbol{x}) \geq f(\boldsymbol{y}).$$

Here G is locally non-saturating, but is *not* non-decreasing.

3. Let the functions f and g be defined on $\mathbb{R}^2_+$ by:

$$f(\boldsymbol{x}) = x_1 + (x_2)^2 \text{ and } g(\boldsymbol{x}) = (x_1)^2 + x_2,$$

respectively; and define P on $\mathbb{R}^2_+$ by:

$$\boldsymbol{x}P\boldsymbol{y} \iff [f(\boldsymbol{x}) > f(\boldsymbol{y}) \,\&\, g(\boldsymbol{x}) > g(\boldsymbol{y})].$$

In this case, P is a strictly increasing asymmetric order.

4. Define the functions f and g on $\mathbb{R}^2_+$ by:

$$f(\boldsymbol{x}) = 2x_1 - x_2 \text{ and } g(\boldsymbol{x}) = x_2,$$

respectively; and define P on $\mathbb{R}^2_+$ by:

$$\boldsymbol{x}P\boldsymbol{y} \iff [f(\boldsymbol{x}) > f(\boldsymbol{y}) \,\&\, g(\boldsymbol{x}) > g(\boldsymbol{y})].$$

Here P is non-decreasing and locally non-saturating, but not increasing. (To see that P is non-decreasing, notice that if $\boldsymbol{x}P\boldsymbol{y}$, then we must have $g(\boldsymbol{x}) = x_2 > g(\boldsymbol{y}) = y_2$. Thus, obviously, we cannot have $\boldsymbol{y} \geq \boldsymbol{x}$.)

5. Suppose G is representable by the (utility) function, u, defined on $\mathbb{R}^n_+$ by:

$$u(\boldsymbol{x}) = \prod_{j=1}^{n} (x_j + c_j)^{a_j},$$

where:

$$c_j,\ a_j > 0 \text{ for } j = 1, \ldots, n; \text{ and } \sum_{j=1}^{n} a_j = 1.$$

In this case, it is easy to show that G is strictly increasing; probably the easiest way to show this being that the partial derivatives of u are all strictly positive, at any $\boldsymbol{x} \in \mathbb{R}^n_+$. What happens, however, if one of the c_j's is equal to zero? □

4.5 Some Convexity Conditions

In this section we will explore the implications of some convexity conditions which are very often used in general equilibrium theory. A definition from mathematics which will be very useful to us, both in this section and in the remainder of this book is the following.

4.13 Definition. We define the **unit simplex** for $\mathbb{R}^n$, denoted by 'Δ_n,' by:

$$\Delta_n = \left\{\boldsymbol{p} \in \mathbb{R}^n_+ \mid \sum\nolimits_{j=1}^n p_j = 1\right\}.$$

4.14 Definitions. Suppose X is a convex subset of $\mathbb{R}^n$, that G is a weak order on X, and that P is its asymmetric part. We shall say that G is:

1. **weakly convex** iff, for all $\boldsymbol{x} \in X$, $G\boldsymbol{x}$ is a convex set.

2. **convex** iff it is weakly convex, and in addition, for all $\boldsymbol{x}, \boldsymbol{y} \in X$, we have that if $\boldsymbol{x}P\boldsymbol{y}$, then:

$$(\forall \theta \in]0,1[)\colon [\theta\boldsymbol{x} + (1-\theta)\boldsymbol{y}]P\boldsymbol{y}.$$

3. **strictly convex** iff, for all $\boldsymbol{x}, \boldsymbol{y}, \boldsymbol{z} \in X$, we have that if $\boldsymbol{y}G\boldsymbol{x}$, $\boldsymbol{z}G\boldsymbol{x}$, and $\boldsymbol{y} \neq \boldsymbol{z}$, then:

$$(\forall \theta \in]0,1[)\colon [\theta\boldsymbol{y} + (1-\theta)\boldsymbol{z}]P\boldsymbol{x}.$$

Notice that if $\boldsymbol{x}$, $\boldsymbol{y} \in X$, and $\theta \in]0,1[$, then the vector (or 'commodity bundle');

$$\boldsymbol{z} = \theta\boldsymbol{x} + (1-\theta)\boldsymbol{y},$$

can be viewed as a weighted average of the commodity bundles $\boldsymbol{x}$ and $\boldsymbol{y}$. Thus we can see, for example, that weak convexity can be interpreted as stating that if $\boldsymbol{x}$ and $\boldsymbol{y}$ are both considered to be at least as good as some third bundle, $\boldsymbol{z}$, then any weighted average of the two bundles will also be considered to be at least as good as $\boldsymbol{z}$.

4.15 Definitions. Let X be a non-empty and convex subset of $\mathbb{R}^n$, and suppose $f\colon X \to \mathbb{R}$. We shall say that f is:

1. **concave** (respectively, **convex**) iff, for each $\boldsymbol{x}$, $\boldsymbol{y} \in X$, and each $\theta \in]0,1[$, we have:

$$f[\theta\boldsymbol{x} + (1-\theta)\boldsymbol{y}] \geq \theta f(\boldsymbol{x}) + (1-\theta)f(\boldsymbol{y})$$

(respectively, $f[\theta\boldsymbol{x} + (1-\theta)\boldsymbol{y}] \leq \theta f(\boldsymbol{x}) + (1-\theta)f(\boldsymbol{y})$).

2. **strictly concave** (respectively, **strictly convex**) iff, for each $\boldsymbol{x}$, $\boldsymbol{y} \in X$, and each $\theta \in]0,1[$, we have that if $\boldsymbol{x} \neq \boldsymbol{y}$, then:

$$f[\theta\boldsymbol{x} + (1-\theta)\boldsymbol{y}] > \theta f(\boldsymbol{x}) + (1-\theta)f(\boldsymbol{y})$$

(respectively, $f[\theta\boldsymbol{x} + (1-\theta)\boldsymbol{y}] < \theta f(\boldsymbol{x}) + (1-\theta)f(\boldsymbol{y})$).

3. **quasi-concave** (respectively, **quasi-convex**) iff, for each $\boldsymbol{x}$, $\boldsymbol{y} \in X$, and each $\theta \in]0,1[$, we have:

$$f[\theta\boldsymbol{x} + (1-\theta)\boldsymbol{y}] \geq \min\{f(\boldsymbol{x}),\ f(\boldsymbol{y})\}$$

(respectively, $f[\theta\boldsymbol{x} + (1-\theta)\boldsymbol{y}] \leq \max\{f(\boldsymbol{x}),\ f(\boldsymbol{y})\}$).

4. **strictly quasi-concave** (respectively, **strictly quasi-convex**) iff, for each $\boldsymbol{x},\ \boldsymbol{y} \in X$, and each $\theta \in]0,1[$, we have that if $\boldsymbol{x} \neq \boldsymbol{y}$, then:

$$f[\theta \boldsymbol{x} + (1-\theta)\boldsymbol{y}] > \min\{f(\boldsymbol{x}),\ f(\boldsymbol{y})\}$$

$\big($respectively, $f[\theta \boldsymbol{x} + (1-\theta)\boldsymbol{y}] < \max\{f(\boldsymbol{x}),\ f(\boldsymbol{y})\}\big)$.

5. **semi-concave**[3] (respectively, **semi-convex**) iff f is quasi-concave (respectively, quasi-convex) and in addition, for each $\boldsymbol{x},\ \boldsymbol{y} \in X$, and each $\theta \in]0,1[$, we have that if $f(\boldsymbol{x}) > f(\boldsymbol{y})$, then:

$$f[\theta \boldsymbol{x} + (1-\theta)\boldsymbol{y}] > f(\boldsymbol{y})$$

(respectively, $f[\theta \boldsymbol{x} + (1-\theta)\boldsymbol{y}] < f(\boldsymbol{x})$).

Notice that if f is strictly concave, then f is concave and strictly quasi-concave. Similarly, if f is concave, then f is semi-concave; however, it is also true that if f is strictly quasi-concave, then f is semi-concave. It is, of course, obvious that if f is semi-concave, then it is quasi-concave; on the other hand, any strictly increasing transformation of a linear function is semi-concave, but not strictly quasi-concave.

The proof of the following result will be left as an exercise.

4.16 Proposition. *Suppose G is a weak order on a non-empty, convex subset, X, of $\mathbb{R}^n$, and suppose $f: X \to \mathbb{R}$ represents G on X. Then:*
1. *G is weakly convex if, and only if, f is quasi-concave.*
2. *G is strictly convex if, and only if, f is strictly quasi-concave.*
3. *G is convex if, and only if, f is semi-concave.*

4.17 Proposition. *If X is a non-empty convex subset of $\mathbb{R}^n$, and G is a weak order on X which is weakly convex, then for each $(\boldsymbol{p}, w) \in \Omega$, $\boldsymbol{h}(\boldsymbol{p}, w)$ is a convex set.*

Proof. Let $(\boldsymbol{p}^*, w^*) \in \Omega$ be given, let $\boldsymbol{x}$ and $\boldsymbol{x}'$ be elements of $\boldsymbol{h}(\boldsymbol{p}^*, w^*)$, and let $\theta \in [0,1]$ be given. If we then define $\boldsymbol{y} = \theta \boldsymbol{x} + (1-\theta)\boldsymbol{x}'$, we see that:

$$\boldsymbol{p}^* \cdot \boldsymbol{y} = \boldsymbol{p}^* \cdot [\theta \boldsymbol{x} + (1-\theta)\boldsymbol{x}'] = \theta \boldsymbol{p}^* \cdot \boldsymbol{x} + (1-\theta)\boldsymbol{p}^* \cdot \boldsymbol{x}' \leq \theta w^* + (1-\theta) w^* = w^*;$$

where the inequality is by the fact that both $\boldsymbol{x}$ and $\boldsymbol{x}'$ must be in the budget set. Furthermore, by the weak convexity of G and the definition of the consumer's demand correspondence, we see that:

$$\boldsymbol{y} G \boldsymbol{x}. \tag{4.12}$$

Thus, if $\boldsymbol{z} \in b(\boldsymbol{p}^*, w^*)$ it follows from the fact that $\boldsymbol{x} \in \boldsymbol{h}(\boldsymbol{p}^*, w^*)$ that $\boldsymbol{x} G \boldsymbol{z}$. From the transitivity of G and (4.12), it then follows that $\boldsymbol{y} G \boldsymbol{z}$. Therefore, $\boldsymbol{y} \in \boldsymbol{h}(\boldsymbol{p}^*, w^*)$, and it follows that $\boldsymbol{h}(\boldsymbol{p}^*, w^*)$ is a convex set. □

4.18 Proposition. *If X is a non-empty, convex subset of $\mathbb{R}^n$ which is closed and bounded below, and G is a weak order on X which is upper semi-continuous and strictly convex, then for each $(\boldsymbol{p}, w) \in \Omega$, $\boldsymbol{h}(\boldsymbol{p}, w)$ is a singleton; in other words, under these conditions the consumer's demand correspondence is actually a function.*

[3] This is not a standard definition, but it will be useful to us in our remaining work.

Proof. Letting $(\boldsymbol{p}^*, w^*) \in \Omega$ be arbitrary, it follows from Theorem 4.5 of this chapter that $\boldsymbol{h}(\boldsymbol{p}^*, w^*) \neq \emptyset$. Suppose, by way of obtaining a contradiction, that there exist distinct points, $\boldsymbol{x}$ and $\boldsymbol{y}$ which are both elements of $\boldsymbol{h}(\boldsymbol{p}^*, w^*)$. Then $\boldsymbol{y}G\boldsymbol{x}$ and $\boldsymbol{x}G\boldsymbol{x}$, so that by the strict convexity of G we must have:

$$[(1/2)\boldsymbol{x} + (1/2)\boldsymbol{y}]P\boldsymbol{x}. \tag{4.13}$$

However, since both $\boldsymbol{x}$ and $\boldsymbol{y}$ are elements of $b(\boldsymbol{p}^*, w^*)$, it is easy to see that:

$$(1/2)\boldsymbol{x} + (1/2)\boldsymbol{y} \in b(\boldsymbol{p}^*, w^*)$$

as well; and thus (4.13) contradicts the assumption that $\boldsymbol{x} \in \boldsymbol{h}(\boldsymbol{p}^*, w^*)$. □

When $\boldsymbol{h}(\cdot)$ is a function, we will denote the jth coordinate function (the demand function for the jth commodity) by '$h_j(\cdot)$.' Hopefully, you will have no trouble in distinguishing between this and the ith consumer's demand function, which will be denoted by '$\boldsymbol{h}_i(\cdot)$' [the ith consumer's demand function for the jth commodity will be denoted by '$h_{ij}(\cdot)$']. The proof of the following two results will be left as exercises.

4.19 Proposition. *If X is a non-empty, convex subset of $\mathbb{R}^n$, and G is a non-saturating and convex weak order on X, then G is locally non-saturating.*

4.20 Corollary. *If X is a non-empty convex subset of $\mathbb{R}^n$, and G is a non-saturating and convex weak order on X, then the demand correspondence determined by G satisfies the budget balance condition.*

4.6 Wold's Theorem

In this section, we will state and prove the first result to appear in the economics literature which established sufficient conditions for a preference relation to be representable by a real-valued utility function; and which is due to Herman Wold [1943]. It has been generalized since (in particular, by Debreu; see Theorem 1.34); but Wold's original proof is much simpler than the later generalizations, and since we will want to make use of his result in some of our later work, it seems quite appropriate to state and prove his result here.

4.21 Theorem. (Wold [1943]) *Let G be a continuous and increasing weak order on $\mathbb{R}^n_+$. Then there exists a continuous function, $u\colon \mathbb{R}^n_+ \to \mathbb{R}_+$, which represents G on $\mathbb{R}^n_+$.*

Proof. Let $\boldsymbol{x}^* \in \mathbb{R}^n_{++}$ be a (fixed) strictly positive vector in $\mathbb{R}^n_+$, and define:

$$L = \{\boldsymbol{x} \in \mathbb{R}^n_+ \mid (\exists \mu \in \mathbb{R}_+)\colon \boldsymbol{x} = \mu \boldsymbol{x}^*\};$$

in other words, let L be the half-ray determined by $\boldsymbol{x}^*$. We will make use of L to define our utility function in the following way.

Let $\boldsymbol{x} \in \mathbb{R}^n_+$ be arbitrary. Then we note that there exists $\boldsymbol{x}' \in L$ such that:

$$\boldsymbol{x}' \gg \boldsymbol{x},$$

and thus, since G is increasing, $\boldsymbol{x}' P \boldsymbol{x}$. Moreover, we also have $\mathbf{0} \in L$ and $\boldsymbol{x} G \mathbf{0}$. Thus if we define the subsets of $\mathbb{R}_+$, A and B by:

$$A = \{\mu \in \mathbb{R}_+ \mid \boldsymbol{x} G \mu \boldsymbol{x}^*\},$$

and:

$$B = \{\mu \in \mathbb{R}_+ \mid \mu \boldsymbol{x}^* G \boldsymbol{x}\},$$

we see that both A and B are non-empty sets. Obviously the set A is bounded above (by any element of B), and thus A has a least upper bound, call in $u(\boldsymbol{x})$. On the other hand, it is clear that $u(\boldsymbol{x})$ is also the greatest lower bound for the set B. Consequently, since it follows easily from the continuity of G that both sets are closed, we see that:

$$u(\boldsymbol{x}) \in A \cap B;$$

and thus:

$$u(\boldsymbol{x}) \boldsymbol{x}^* I \boldsymbol{x}. \tag{4.14}$$

The argument of the above paragraph establishes the existence of a function $u \colon \mathbb{R}^n_+ \to \mathbb{R}_+$ satisfying (4.14). I will leave as an exercise the task of proving that this function represents G on $\mathbb{R}^n_+$.

To prove that $u(\cdot)$ is continuous, let a be an arbitrary real number. If $a < 0$, then:

$$\{\boldsymbol{x} \in \mathbb{R}^n_+ \mid f(\boldsymbol{x}) \leq a\} = \emptyset \text{ and } \{\boldsymbol{x} \in \mathbb{R}^n_+ \mid f(\boldsymbol{x}) \geq a\} = \mathbb{R}^n_+;$$

both of which are closed sets. On the other hand, if $a \geq 0$, notice that if we define $\boldsymbol{x}_a$ by:

$$\boldsymbol{x}_a = a\boldsymbol{x}^*,$$

we have:

$$u(\boldsymbol{x}_a) = a.$$

Therefore, since $u(\cdot)$ represents G, it follows that:

$$\{\boldsymbol{x} \in \mathbb{R}^n_+ \mid f(\boldsymbol{x}) \leq a\} = \boldsymbol{x}_a G \text{ and } \{\boldsymbol{x} \in \mathbb{R}^n_+ \mid f(\boldsymbol{x}) \geq a\} = G\boldsymbol{x}_a;$$

and since both sets are closed relative to $\mathbb{R}^n_+$ by our continuity assumption, it now follows that $u(\cdot)$ is a continuous function (see Moore [1999], Proposition 3.32, p. 137). □

4.7 Indirect Preferences and Indirect Utility

In this section we will examine some aspects of indirect preferences and indirect utility. We will begin with some very general considerations, and then sharpen our results by considering the implications of some stronger assumptions. Throughout the material to follow, we define the set Z by:

$$Z = \boldsymbol{h}(\Omega) = \big\{\boldsymbol{x} \in X \mid \big(\exists (\boldsymbol{p}, w) \in \Omega\big) \colon \boldsymbol{x} \in \boldsymbol{h}(\boldsymbol{p}, w)\big\},$$

and where we define:

$$\Omega = \{(\boldsymbol{p}, w) \in \mathbb{R}^{n+1} \mid \boldsymbol{p} \in \mathbb{R}^n_{++} \;\&\; (\exists \boldsymbol{x} \in X) \colon \boldsymbol{p} \cdot \boldsymbol{x} \leq w\}; \tag{4.15}$$

We will be denoting the consumer's (weak) preference relation over the consumption set, X, by 'G,' but we will also denote the restriction of G to Z by 'G.' In our initial definition, we suppose the following assumptions hold.

Assumptions I.1. We suppose that the consumer's demand correspondence, $\boldsymbol{h}$, is well-defined on Ω. In other words, we suppose that for each $(\boldsymbol{p}, w) \in \Omega$, there exists $\boldsymbol{x}^* \in b(\boldsymbol{p}, w)$ (where $b\colon \Omega \mapsto X$ is the consumer's budget correspondence), satisfying:

$$(\forall \boldsymbol{x} \in X) \colon \boldsymbol{x} P \boldsymbol{x}^* \Rightarrow \boldsymbol{p} \cdot \boldsymbol{x} > w.$$

We also suppose that the restriction of G to Z is a weak order, with asymmetric and symmetric parts P and I, respectively.

In the context of these assumptions, we then define the following.

4.22 Definition. Given Assumptions I.1, we define the consumer's **indirect preference relation, G^***, on Ω as follows: for $(\boldsymbol{p}, w), (\boldsymbol{p}', w') \in \Omega$:

$$(\boldsymbol{p}, w) G^* (\boldsymbol{p}', w') \iff \big(\exists \boldsymbol{x} \in \boldsymbol{h}(\boldsymbol{p}, w) \;\&\; \boldsymbol{x}' \in \boldsymbol{h}(\boldsymbol{p}', w')\big) \colon \boldsymbol{x} G \boldsymbol{x}'.$$

We then denote the symmetric and asymmetric parts of G^* by 'I^*' and 'P^*,' respectively.

I will leave the proof of the following proposition as an exercise.

4.23 Proposition. *Given Assumptions I.1, the indirect preference relation G^* is a weak order.*

Of course, it follows from the above result and Theorem 1.15 that P^* is negatively transtive (as well as being asymmetric), and that I^* is an equivalence relation.

It can readily be seen that if X is a subset of $\mathbb{R}^n_+$ which contains the origin, then the set Ω defined in (4.15), above, is given by:

$$\Omega = \mathbb{R}^n_{++} \times \mathbb{R}_+.$$

In any case, it will be convenient for us to assume that this condition holds throughout the remainder of this section.

Assumptions I.2. We suppose that $\Omega = \mathbb{R}^n_{++} \times \mathbb{R}_+$ and that h satisfies the **budget balance condition** on Ω; that is, for all $(\boldsymbol{p}, w) \in \Omega$ and all $\boldsymbol{x} \in \boldsymbol{h}(\boldsymbol{p}, w)$, we have $\boldsymbol{p} \cdot \boldsymbol{x} = w$.

4.24 Proposition. *Given Assumptions I.1 and I.2, G^* and P^* satisfy the following conditions (in addition to those set out in Proposition 4.23, above):*

1. given any $(\boldsymbol{p}, w) \in \Omega$, and any $w' \in \mathbb{R}$, we have:

$$w' > w \Rightarrow [(\boldsymbol{p}, w') \in \Omega \;\&\; (\boldsymbol{p}, w') P^* (\boldsymbol{p}, w)].$$

2. given $(\boldsymbol{p}^1, w^1), (\boldsymbol{p}^2, w^2) \in \Omega$ such that $(\boldsymbol{p}^2, w^2) G^* (\boldsymbol{p}^1, w^1)$, any $\theta \in]0, 1[$, and defining $(\boldsymbol{p}^*, w^*)$ by:

$$(\boldsymbol{p}^*, w^*) = \theta(\boldsymbol{p}^1, w^1) + (1 - \theta)(\boldsymbol{p}^2, w^2),$$

we have:

$$(\boldsymbol{p}^2, w^2)G^*(\boldsymbol{p}^*, w^*),$$

and if $(\boldsymbol{p}^2, w^2)P^*(\boldsymbol{p}^1, w^1)$, *then* $(\boldsymbol{p}^2, w^2)P^*(\boldsymbol{p}^*, w^*)$.

Proof. I will leave the proof of part 1 as an exercise. To prove part 2,[4] we begin by showing that:

$$b(\boldsymbol{p}^*, w^*) \subseteq [b(\boldsymbol{p}^1, w^1) \cup b(\boldsymbol{p}^2, w^2)]. \tag{4.16}$$

To prove (4.16), suppose that $\boldsymbol{x} \in X$ is such that:

$$\boldsymbol{x} \notin b(\boldsymbol{p}^1, w^1) \text{ and } \boldsymbol{x} \notin b(\boldsymbol{p}^2, w^2).$$

Then:

$$\boldsymbol{p}^1 \cdot \boldsymbol{x} > w^1 \text{ and } \boldsymbol{p}^2 \cdot \boldsymbol{x} > w^2;$$

and therefore, since $0 < \theta < 1$:

$$\boldsymbol{p}^* \cdot \boldsymbol{x} = \theta \boldsymbol{p}^1 \cdot \boldsymbol{x} + (1-\theta)\boldsymbol{p}^2 \cdot \boldsymbol{x} > \theta w^1 + (1-\theta) w^2 = w^*;$$

and, consequently, $\boldsymbol{x} \notin b(\boldsymbol{p}^*, w^*)$, so we see that (4.16) holds. Having established (4.16), it follows that, since $(\boldsymbol{p}^2, w^2)G^*(\boldsymbol{p}^1, w^1)$, we must have $(\boldsymbol{p}^2, w^2)G^*(\boldsymbol{p}^*, w^*)$.

Now suppose that $(\boldsymbol{p}^2, w^2)P^*(\boldsymbol{p}^1, w^1)$. Then it follows readily from the definition of indirect preferences that there exist $\boldsymbol{x}^t \in h(\boldsymbol{p}^t, w^t)$, for $t = 1, 2$ such that:

$$\boldsymbol{x}^2 P \boldsymbol{x}^1. \tag{4.17}$$

Suppose, then, that $\bar{\boldsymbol{x}} \in X$ is such that $\bar{\boldsymbol{x}} G \boldsymbol{x}^2$. Then by (4.17) and the transitivity of G, $\bar{\boldsymbol{x}} P \boldsymbol{x}^1$; and therefore:

$$\boldsymbol{p}^1 \cdot \bar{\boldsymbol{x}} > w^1. \tag{4.18}$$

On the other hand, if $\boldsymbol{p}^2 \cdot \bar{\boldsymbol{x}} \le w^2$, it follows from the transitivity of G that $\bar{\boldsymbol{x}} \in \boldsymbol{h}(\boldsymbol{p}^2, w^2)$, and thus by budget balance that $\boldsymbol{p}^2 \cdot \bar{\boldsymbol{x}} = w^2$. Thus, in any case we must have:

$$\boldsymbol{p}^2 \cdot \bar{\boldsymbol{x}} \ge w^2;$$

and combining this with (4.18) we see that:

$$\boldsymbol{p}^* \cdot \bar{\boldsymbol{x}} = \theta \boldsymbol{p}^1 \cdot \bar{\boldsymbol{x}} + (1-\theta)\boldsymbol{p}^2 \cdot \bar{\boldsymbol{x}} > \theta w^1 + (1-\theta) w^2 = w^*.$$

It then follows that for all $\boldsymbol{x} \in b(\boldsymbol{p}^*, w^*)$, $\boldsymbol{x}^2 P \boldsymbol{x}$, and therefore:

$$(\boldsymbol{p}^2, w^2)P^*(\boldsymbol{p}^*, w^*). \quad \square$$

An indirect utility function is simply a representation of G^*, as we formally note in the following.

4.25 Definition. We say that a function $V : \Omega \to \mathbb{R}$ is an **indirect utility function representing G^*** (and P^*) iff V satisfies:

$$(\forall (\boldsymbol{p}, w), (\boldsymbol{p}', w') \in \Omega) : V(\boldsymbol{p}, w) \ge V(\boldsymbol{p}', w') \iff (\boldsymbol{p}, w)\, G^*\, (\boldsymbol{p}', w'). \tag{4.19}$$

[4] A careful reading of this part of the proof will reveal that it holds under the weaker assumption that Ω is a convex cone; that is, that Ω is a cone and a convex set as well.

Having now defined what we mean by an indirect utility function, we can obtain the following corollary of 4.24. I will leave the details of the proof as an exercise.

4.26 Corollary. *Given Assumptions I.1 and I.2, if* $V : \Omega \rightarrow \mathbb{R}$ *is an indirect utility function representing* P^*, *then it must be strictly increasing in* w, *for each* $\boldsymbol{p} \in \mathbb{R}^n_{++}$. *Moreover, it must be positively homogeneous of degree zero and semi-convex on* Ω; *that is, it must satisfy the following condition: given any* $(\boldsymbol{p}^t, w^t) \in \Omega$ $(t = 1, 2)$ *such that:*

$$V(\boldsymbol{p}^2, w^2) \geq V(\boldsymbol{p}^1, w^1), \tag{4.20}$$

and any $\theta \in]0, 1[$:

$$V\left[\theta \boldsymbol{p}^1 + (1-\theta)\boldsymbol{p}^2, \theta w^1 + (1-\theta)w^2\right] \leq V(\boldsymbol{p}^2, w^2); \tag{4.21}$$

and if $V(\boldsymbol{p}^2, w^2) > V(\boldsymbol{p}^1, w^1)$, *the strict inequality holds in (4.21).*

It is tempting to replace the second part of the conclusion of the corollary with the statement that $V(\cdot)$ is strictly quasi-convex. However, this is not necessarily the case, as is shown by the first of the following examples.

4.27 Examples.

1. Suppose a consumer's preferences can be represented on $\mathbb{R}^n_+$ by the Leontief utility function:

$$u(\boldsymbol{x}) = \min\left\{\frac{x_j}{a_j}\right\},$$

where $a_j > 0$ for $j = 1, \ldots, n$. Then the consumer's demand function is given by (see Exercise 3, at the end of this chapter):

$$h_j(\boldsymbol{p}, w) = \frac{a_j w}{\sum_{k=1}^n a_k p_k},$$

for $j = 1, \ldots, n$. Consequently, the function:

$$V(\boldsymbol{p}, w) = \frac{w}{\sum_{j=1}^n a_j p_j}, \tag{4.22}$$

is an indirect utility function for the consumer in this case.

Now suppose that $(\boldsymbol{p}^1, w^1)$ and $(\boldsymbol{p}^2, w^2)$ are such that:

$$V(\boldsymbol{p}^1, w^1) = V(\boldsymbol{p}^2, w^2) = \beta > 0, \tag{4.23}$$

and let $\theta \in]0, 1[$. Then, making use of (4.22) and (4.23), we see that:

$$\begin{aligned} V\left[\theta(\boldsymbol{p}^1, w^1) + (1-\theta)(\boldsymbol{p}^2, w^2)\right] &= \frac{\theta w^1 + (1-\theta)w^2}{\sum_{j=1}^n a_j \theta p_j^1 + \sum_{j=1}^n a_j (1-\theta) p_j^2} \\ = \frac{\theta w^1 + (1-\theta)w^2}{\theta \sum_{j=1}^n a_j p_j^1 + (1-\theta)\sum_{j=1}^n a_j p_j^2} &= \frac{\theta w^1 + (1-\theta)w^2}{\theta(1/\beta)w^1 + (1-\theta)(1/\beta)w^2} = \beta; \end{aligned}$$

whether or not $(\boldsymbol{p}^1, w^1) = (\boldsymbol{p}^2, w^2)$. Therefore, we see that $V(\cdot)$ is *not* strictly quasi-convex in this case.

2. Suppose a consumer's preferences can be represented by the Cobb-Douglas utility function:

$$u(\boldsymbol{x}) = \prod_{j=1}^{n} x_j^{a_j},$$

where, as usual, we assume that all of the a_j's are positive and sum to one. Then (see exercise 4, at the end of this chapter), the consumer's indirect preferences can be represented by the function:

$$V(\boldsymbol{p}, w) = \frac{w}{\prod_{j=1}^{n} p_j^{a_j}}.$$

It can be shown that in this case, the indirect utility function is strictly quasi-convex.

Notice that if indirect utility is set equal to some positive value, denoted by v^*, then the associated indifference curve for G^* can be represented by the equation:

$$w = v^* \cdot \prod_{j=1}^{n} p_j^{a_j}; \tag{4.24}$$

and it is of interest to consider the contour curves of this function in the special case in which $n = 2$. Consider first the representation of indirect preferences when we normalize the price of the second commodity; setting $p_2 = 1$. If, for the sake of convenience we then denote the price of the first commodity by p, we see that equation (4.24) reduces to:

$$w = v^* p^{a_1} \stackrel{\text{def}}{=} \phi(p);$$

and, since $0 < a_1 < 1$, $\phi(\cdot)$ is strictly concave. I will leave it to you to consider the shape of the contour curves in this case, as well as what happens when v^* increases. It is also of interest to consider the indifference map for G^* if we fix (or normalize) w, and consider the contour (indifference) curves of G^* in (p_1, p_2)-space. However, I will also leave this as an exercise. □

The following result generalizes propositions established by Antonelli [1886], Allen [1933], and Roy [1942].[5]

4.28 Theorem. *Suppose G satisfies Assumptions I.1 and I.2, that the demand correspondence generated by G is single-valued (and thus is a function), and that $V\colon \Omega \to \mathbb{R}$ is a differentiable indirect utility function representing G^*. Then, if $(\boldsymbol{p}^*, w^*) \in \Omega$ is such that $w^* > 0$, we must have:*

$$\left.\frac{\partial V}{\partial p_k}\right|_{(\boldsymbol{p}^*, w^*)} = -\left(\left.\frac{\partial V}{\partial w}\right|_{(\boldsymbol{p}^*, w^*)}\right) h_k(\boldsymbol{p}^*, w^*) \quad \textit{for } k = 1, \ldots, n. \tag{4.25}$$

Proof. Since V represents G^*, it is clear that, for all $(\boldsymbol{p}, w) \in \Omega$, we must have:

$$\boldsymbol{p} \cdot \boldsymbol{h}(\boldsymbol{p}^*, w^*) \leq w \Rightarrow V(\boldsymbol{p}, w) \geq V(\boldsymbol{p}^*, w^*).$$

[5] In its present form, the result is from Chipman and Moore [1976a, Lemma 3, p. 74]; although the present, simpler, proof is from Chipman and Moore [1990, p. 71]. The assumption that the budget balance condition holds is not needed, as a careful reading of the proof will disclose. Some of the other assumptions can be weakened as well.

Thus we see that $(\boldsymbol{p}^*, w^*)$ *minimizes* $V(\cdot)$ subject to $[(\boldsymbol{p}, w) \in \Omega$ and]:

$$\boldsymbol{p} \cdot \boldsymbol{h}(\boldsymbol{p}^*, w^*) = w.$$

From the classical Lagrangian method, it then follows that there exists a Lagrangian multiplier, $\lambda \in \mathbb{R}$ such that:

$$\left.\frac{\partial V}{\partial p_k}\right|_{(\boldsymbol{p}^*, w^*)} - \lambda h_k(\boldsymbol{p}^*, w^*) = 0 \quad \text{for } k = 1, \ldots, n; \tag{4.26}$$

and:

$$\left.\frac{\partial V}{\partial w}\right|_{(\boldsymbol{p}^*, w^*)} + \lambda = 0. \tag{4.27}$$

Solving for λ in (4.27), and substituting into (4.26), we obtain the desired result. □

Notice that in the result just presented we have not assumed that the indirect utility function was defined as a composite function:

$$V(\boldsymbol{p}, w) = u[\boldsymbol{h}(\boldsymbol{p}, w)],$$

where $u(\cdot)$ is a continuously differentiable direct utility function representing G. There is some question as to whether *every* differentiable indirect utility function representing G^* is obtainable in this way, even when there exists at least one continuously differentiable utility function representing G. More to the point, however, there may exist a continuously differentiable indirect utility function representing G^* even when there exists no utility function (differentiable or not) representing G. Recall, for example, the lexicographic preference relation defined on $\mathbb{R}^2_+$ by:

$$\boldsymbol{x}P\boldsymbol{x}' \iff \begin{cases} x_1 > x_1' & \text{or} \\ x_1 = x_1' \ \& \ x_2 > x_2'. \end{cases}$$

In Chapter 1, we noted that Debreu [1959] has proved that there exists *no* real-valued utility function representing P in this case. However, it is easy to prove that the function $V(\cdot)$ defined by:

$$V(\boldsymbol{p}, w) = w/p_1,$$

is a continuously differentiable indirect utility function representing P^*. In fact, all of the assumptions of the theorem, that is, both of Assumptions I.1 and I.2, are satisfied here.

The discussion of the previous paragraph not withstanding, the most obvious way of defining an indirect utility function is, of course, as a composite function:

$$V(\boldsymbol{p}, w) = u[\boldsymbol{h}(\boldsymbol{p}, w)];$$

at least in the case in which the preference relation, G, is representable by a utility function, and the correspondence h is actually a function. We will, in fact, often make use of this representation in the remainder of this course, but I want to conclude this section by considering a somewhat different way of defining an indirect utility function. The following result is proved in the Appendix.

4.29 Proposition. *Suppose, in addition to Assumptions I.1 and I.2, that the set* $Z = h(\Omega)$ *is a subset of* $\mathbb{R}^n_+$, *and is closed, convex, and contains* $\mathbf{0}$. *Suppose further that the restriction of the consumer's (weak) preference relation to* Z *is a continuous weak order on* Z. *Then, given any* $\bar{\boldsymbol{p}} \in \mathbb{R}^n_{++}$, *and any* $(\mathbf{p}', w') \in \Omega$, *there exists* $\bar{w} \in \mathbb{R}_+$ *such that:*

$$\bar{w} = \min\{w \in \mathbb{R}_+ \mid (\bar{\boldsymbol{p}}, w)G^*(\mathbf{p}', w')\},$$

and we have:

$$(\bar{\boldsymbol{p}}, \bar{w})I^*(\mathbf{p}', w').$$

Because of the above proposition, we see that the following function, which was originally introduced by Hurwicz and Uzawa [1971], is well-defined.

4.30 Definition. Given the assumptions of Proposition 4.29, we define the **income-compensation function**, $\mu\colon \mathbb{R}^n_{++} \times \Omega \to \mathbb{R}_+$ by:

$$\mu(\bar{\boldsymbol{p}}; \boldsymbol{p}, w) = \min\{\overline{w} \in \mathbb{R}_+ \mid (\bar{\boldsymbol{p}}, \overline{w})G^*(\boldsymbol{p}, w)\}.$$

Verbally, the value of $\mu(\bar{\boldsymbol{p}}; \boldsymbol{p}, w)$ is the minimum level of income, or wealth which would leave the consumer exactly as well off, given the price vector $\bar{\boldsymbol{p}}$, as with the price-wealth pair $(\boldsymbol{p}, w)$. The following sets out the basic properties of the income-compensation function.

4.31 Theorem. *The income-compensation function satisfies the following conditions, given the assumptions of Proposition 4.29,:*

1. For all $(\bar{\boldsymbol{p}}; \boldsymbol{p}, w) \in \mathbb{R}^n_{++} \times \Omega$, *and all* $\bar{w} \in \mathbb{R}_+$:

$$(\bar{\boldsymbol{p}}, \bar{w})I^*(\boldsymbol{p}, w) \iff \bar{w} = \mu(\bar{\boldsymbol{p}}; \boldsymbol{p}, w).$$

2. For all $(\boldsymbol{p}, w) \in \Omega$, $\mu(\cdot\,; \boldsymbol{p}, w)$ *is positively homogeneous of degree one in* $\bar{\boldsymbol{p}}$; *that is:*

$$(\forall \lambda \in \mathbb{R}_+)\colon \mu(\lambda\bar{\boldsymbol{p}}; \boldsymbol{p}, w) = \lambda\mu(\bar{\boldsymbol{p}}; \boldsymbol{p}, w).$$

3. For any fixed $\bar{\boldsymbol{p}} \in \mathbb{R}^n_{++}$, $\mu(\bar{\boldsymbol{p}};\ \cdot\ )$ *is an indirect utility function representing* G^*; *that is, for all* $(\boldsymbol{p}, w), (\boldsymbol{p}', w') \in \Omega$:

$$\mu(\bar{\boldsymbol{p}}; \boldsymbol{p}, w) \geq \mu(\bar{\boldsymbol{p}}; \boldsymbol{p}', w') \iff (\boldsymbol{p}, w)G^*(\boldsymbol{p}', w').$$

4. For any fixed $\bar{\boldsymbol{p}} \in \mathbb{R}^n_{++}$, $\mu(\bar{\boldsymbol{p}};\ \cdot\ )$ *is strictly increasing in* w, *and positively homogeneous of degree zero in* $(\boldsymbol{p}, w)$. *Moreover,* $\mu(\bar{\boldsymbol{p}};\ \cdot\ )$ *is semi convex in* $(\boldsymbol{p}, w)$.

Proof.

1. It follows at once from 4.29 that:

$$\big(\bar{\boldsymbol{p}}, \mu(\bar{\boldsymbol{p}}; \boldsymbol{p}, w)\big)I^*(\boldsymbol{p}, w).$$

The converse follows readily from the fact that the budget balance condition implies that G^* must be strictly increasing in w, for fixed $\boldsymbol{p}$.

2. I will leave the proof that $\mu(\cdot; \boldsymbol{p}, w)$ must be positively homogeneous of degree one in $\bar{\boldsymbol{p}}$ as an exercise.

3. Let $\bar{\boldsymbol{p}} \in \mathbb{R}^n_{++}$ be fixed, and suppose $(\boldsymbol{p}, w), (\boldsymbol{p}', w') \in \Omega$ are such that:

$$\mu(\bar{\boldsymbol{p}}; \boldsymbol{p}, w) > \mu(\bar{\boldsymbol{p}}; \boldsymbol{p}', w'). \tag{4.28}$$

Then, since G^* is strictly increasing in w, for fixed $\bar{\boldsymbol{p}} \in \mathbb{R}^n_{++}$, we have:

$$\big(\bar{\boldsymbol{p}}, \mu(\bar{\boldsymbol{p}}; \boldsymbol{p}, w)\big) P^* \big(\bar{\boldsymbol{p}}, \mu(\bar{\boldsymbol{p}}; \boldsymbol{p}', w')\big). \tag{4.29}$$

The fact that $(\boldsymbol{p}, w) P^* (\boldsymbol{p}', w')$ then follows from part 1 and the transitivity of G^*. The proof that (4.29) implies (4.28) can proceed by essentially reversing the above steps.

4. Given Part 3, Part 4 of our conclusion is an immediate consequence of Corollary 4.26. □

4.8 Homothetic Preferences

In this section we will study a special kind of consumer preference relation; the case in which preferences satisfy a condition called 'homotheticity.' Empirical studies have often cast doubt upon the realism of assuming that consumer preferences are homothetic; at least there are reasons to suppose that preferences are not generally homothetic *globally*. On the other hand, almost everyone who has ever written anything for consumers on how to do budget planning has lent support to the belief that there must be some way of aggregating over commmodities which results in a homothetic preference relation; for, as we will see, if a consumer's preferences are homothetic, then the consumer's expenditures on each commodity category (given fixed prices) is a constant percentage of income.

4.32 Definition. Let H be a binary relation on a cone,[6] $X \subseteq \mathbb{R}^n$. We shall say that H is **homothetic** iff for all $\boldsymbol{x}, \boldsymbol{y} \in X$, and every $\theta \in \mathbb{R}_{++}$, we have:

$$\boldsymbol{x} H \boldsymbol{y} \Rightarrow (\theta \boldsymbol{x}) H (\theta \boldsymbol{y}). \tag{4.30}$$

Since we will often be working with strict preference relations for a consumer, we will often be assuming that the consumer's strict preference relation, P, is homothetic. However, if P is homothetic, then its negation is homothetic as well (and conversely), that is, its negation will also satisfy (4.30); as is shown in the following result.

4.33 Proposition. *Suppose H is a binary relation defined on a cone, $X \subseteq \mathbb{R}^n$, and let Q be its negation. Then H is homothetic if, and only if, Q is homothetic as well.*

Proof. Suppose H is homothetic, but, by way of obtaining a contradiction, that there exist $\boldsymbol{x}, \boldsymbol{y} \in X$ and $\theta > 0$ such that:

$$\boldsymbol{x} Q \boldsymbol{y}, \tag{4.31}$$

[6] A set $X \subseteq \mathbb{R}^n$ is said to be a **cone** iff, for each $\boldsymbol{x} \in X$, and every $\theta \in \mathbb{R}_{++}$, $\theta \boldsymbol{x} \in X$.

and yet:

$$\neg[(\theta\boldsymbol{x})Q(\theta\boldsymbol{y})]. \tag{4.32}$$

Then from (4.32) and the fact that Q is the negation of H, it follows that:

$$(\theta\boldsymbol{y})H(\theta\boldsymbol{x}).$$

However, from the homotheticity of H, it then follows that:

$$[(1/\theta)(\theta\boldsymbol{y})]H[(1/\theta)(\theta\boldsymbol{x})];$$

that is, $\boldsymbol{y}H\boldsymbol{x}$, which contradicts (4.31).

By reversing the roles of Q and H in the argument of the above paragraph, it follows that if Q is homothetic, then H is as well. □

4.34 Definition. Suppose $X \subseteq \mathbb{R}^n$ is a cone,[7] and let $f: X \to \mathbb{R}$. We shall say that f is **homothetic** iff there exist functions, $g: X \to \mathbb{R}$ and $F: Y \to \mathbb{R}$, where $g(X) \subseteq Y$, satisfying the following three conditions:

1. g is positively homogeneous of degree one,
2. F is strictly increasing, and:
3. for all $\boldsymbol{x} \in X$, we have: $f(\boldsymbol{x}) = F[g(\boldsymbol{x})]$.

I will leave the proof of the following proposition as an exercise.

4.35 Proposition. *Suppose $X \subseteq \mathbb{R}^n$ is a cone, and that P is an asymmetric and negatively transitive binary relation on X. If there exists a homothetic function, $f: X \to \mathbb{R}$ such that f represents P on X, then P is homothetic.*

As you probably suspect, continuous (and increasing) homothetic preferences can be represented by a utility function which is homogeneous of degree one. What may not be so apparent is that any two such representations of a given preference relation must be scalar multiples of one another. Both facts are established in the following theorem.

4.36 Theorem. *Suppose G is a weak order on $\mathbb{R}^n_+$, and that G is:*

1. *homothetic,*
2. *continuous, and*
3. *increasing.*

Then there exists a function $u: \mathbb{R}^n_+ \to \mathbb{R}_+$ representing G, and such that u is continuous and positively homogeneous of degree one on $\mathbb{R}^n_+$. Moreover, if $\widehat{u}: \mathbb{R}^n_+ \to \mathbb{R}$ is another function representing G which is also positively homogeneous of degree one, then there exists a positive constant $a > 0$ such that:

$$(\forall \boldsymbol{x} \in \mathbb{R}^n_+): \widehat{u}(\boldsymbol{x}) = au(\boldsymbol{x}). \tag{4.33}$$

---

[7]We say that $X \subseteq \mathbb{R}^n$ is a **cone** iff, for each positive real number, $\theta > 0$, and each $\boldsymbol{x} \in X$, we have $\theta\boldsymbol{x} \in X$ as well.

Proof. It follows from the proof of the Wold Representation Theorem (4.21) that if we let $\boldsymbol{x}^* \in \mathbb{R}^n_{++}$ be arbitrary, the function $u \colon \mathbb{R}^n_+ \to \mathbb{R}_+$ defined implicitly by the equation:

$$\boldsymbol{x} I u(\boldsymbol{x})\boldsymbol{x}^*, \tag{4.34}$$

is a continuous utility function representing G.

Now suppose $\boldsymbol{x} \in \mathbb{R}^n_+$ and let $\lambda \in \mathbb{R}_+$. If $\lambda = 0$, then it is clear from the definition of $u(\cdot)$ and (4.34), above, that:

$$u(\lambda\boldsymbol{x}) = u(\mathbf{0}) = 0 = \lambda u(\boldsymbol{x}).$$

On the other hand, if $\lambda > 0$, it follows from (4.34) and the homotheticity of G that:

$$(\lambda\boldsymbol{x})I[\lambda u(\boldsymbol{x})\boldsymbol{x}^*].$$

Thus since $u(\lambda\boldsymbol{x})$ is that unique nonnegative real number such that:

$$(\lambda\boldsymbol{x})I[u(\lambda\boldsymbol{x})\boldsymbol{x}^*],$$

it follows that $u(\lambda\boldsymbol{x}) = \lambda u(\boldsymbol{x})$; and therefore that $u(\cdot)$ is positively homogeneous of degree one.

Finally, suppose $\widehat{u}$ is another function which is both positively homogeneous of degree one and represents G, define the positive real number a by:[8]

$$a = \widehat{u}(\boldsymbol{x}^*),$$

and let $\boldsymbol{x} \in \mathbb{R}^n_+$ be arbitrary. Then, since $\boldsymbol{x}I[u(\boldsymbol{x})\boldsymbol{x}^*]$, and $\widehat{u}$ represents G, we must have:

$$\widehat{u}(\boldsymbol{x}) = \widehat{u}[u(\boldsymbol{x})\boldsymbol{x}^*].$$

However, since $\widehat{u}$ is positively homogeneous of degree one, we also have:

$$\widehat{u}[u(\boldsymbol{x})\boldsymbol{x}^*] = u(\boldsymbol{x})\widehat{u}(\boldsymbol{x}^*) = au(\boldsymbol{x});$$

and equation (4.33) now follows. □

Rather surprisingly we can make use of the theorem just proved to establish sufficient conditions for the existence of a concave utility function representing given preferences.

4.37 Proposition. *If, an addition to the other assumptions of Theorem 4.36, G is weakly convex, then any function which represents G and is positively homogeneous of degree one is also concave.*

Proof. Suppose $f \colon \mathbb{R}^n_+ \to \mathbb{R}$ represents G and is positively homogeneous of degree one. For any $\lambda > 0$, we have:

$$f(\lambda\mathbf{0}) = f(\mathbf{0}) = \lambda f(\mathbf{0});$$

[8] Notice that, since $\widehat{u}$ is positively homogeneous of degree one, we must have $\widehat{u}(\mathbf{0}) = 0$. Consequently, since $\boldsymbol{x}^* P \mathbf{0}$, it follows that $\widehat{u}(\boldsymbol{x}^*) > 0$.

and therefore $f(\mathbf{0}) = 0$. Moreover, since G is increasing and f represents G, it then follows that:

$$(\forall \boldsymbol{x} \in \mathbb{R}^n_{++}) \colon f(\boldsymbol{x}) > 0;$$

while, by the weak convexity of G, we see that f must be quasi-concave (see Proposition 4.15). Consequently, it follows from Corollary 5.101, p. 334, of Moore [1999] that f is concave. □

Homothetic preferences yield demand correspondences having a particularly interesting and tractable form, as is established in the following two results.

4.38 Theorem. *If G is a homothetic preference relation on a cone $X \subseteq \mathbb{R}^n_+$, then the demand correspondence determined by G, $\boldsymbol{h}$, satisfies the following condition: for all $(\boldsymbol{p}, w) \in \Omega$, and all $\lambda \in \mathbb{R}_{++}$:*

$$\boldsymbol{h}(\boldsymbol{p}, \lambda w) = \lambda \boldsymbol{h}(\boldsymbol{p}, w).$$

Proof. Suppose $(\boldsymbol{p}^*, w^*) \in \Omega$, $\lambda > 0$, and that $\boldsymbol{x}^* \in \boldsymbol{h}(\boldsymbol{p}^*, w^*)$. Then:

$$\boldsymbol{p}^* \cdot \boldsymbol{x}^* \leq w^*,$$

and thus:

$$\boldsymbol{p}^* \cdot (\lambda \boldsymbol{x}^*) = \lambda \boldsymbol{p}^* \cdot \boldsymbol{x}^* \leq \lambda w^*.$$

Furthermore, if $\boldsymbol{x} \in X$ is such that $\boldsymbol{x} P (\lambda \boldsymbol{x}^*)$, then, by the homotheticity of P, we have:

$$(1/\lambda)\boldsymbol{x} P \boldsymbol{x}^*.$$

But then, since $\boldsymbol{x}^* \in \boldsymbol{h}(\boldsymbol{p}^*, w^*)$, we have:

$$\boldsymbol{p}^* \cdot (1/\lambda)\boldsymbol{x} > w^*;$$

and therefore:

$$\boldsymbol{p}^* \cdot \boldsymbol{x} > \lambda w^*.$$

It follows that $\lambda \boldsymbol{x}^* \in \boldsymbol{h}(\boldsymbol{p}^*, \lambda w^*)$, and therefore we conclude that:

$$\lambda \boldsymbol{h}(\boldsymbol{p}^*, w^*) \subseteq \boldsymbol{h}(\boldsymbol{p}^*, \lambda w^*).$$

Conversely, suppose $\boldsymbol{x} \in \boldsymbol{h}(\boldsymbol{p}^*, \lambda w^*)$. Then, by reversing the roles of $\lambda \boldsymbol{x}$ and $\boldsymbol{x}$ in the argument of the above paragraph, we can conclude that:

$$(1/\lambda)\boldsymbol{x} \in \boldsymbol{h}(\boldsymbol{p}^*, w^*);$$

and thus that $\boldsymbol{x} \in \lambda \boldsymbol{h}(\boldsymbol{p}^*, w^*)$. Therefore we see that:

$$\boldsymbol{h}(\boldsymbol{p}^*, \lambda w^*) \subseteq \lambda \boldsymbol{h}(\boldsymbol{p}^*, w^*),$$

and our result follows. □

4.39 Theorem. *Suppose G is a homothetic, upper semi-continuous, and strictly convex weak order on a nondegenerate convex cone,[9] $X \subseteq \mathbb{R}^n_+$. Then:*

1. G generates a demand function of the form:

$$\boldsymbol{h}(\boldsymbol{p}, w) = \boldsymbol{g}(\boldsymbol{p})w, \tag{4.35}$$

where $\boldsymbol{g}\colon \mathbb{R}^n_{++} \to X$, and

2. if, in addition, G is increasing, then $\boldsymbol{h}$ satisfies the budget balance condition, and thus:

$$(\forall \boldsymbol{p} \in \mathbb{R}^n_{++})\colon \boldsymbol{p} \cdot \boldsymbol{g}(\boldsymbol{p}) = 1.$$

3. if $\boldsymbol{g}(\cdot)$ is differentiable, then at each $\boldsymbol{p} \in \mathbb{R}^n_{++}$, we have:

$$\frac{\partial g_k}{\partial p_j} = \frac{\partial g_j}{\partial p_k} \quad \text{for } j, k = 1, \ldots, n.$$

Proof. It follows from Proposition 4.18 of this chapter that the demand correspondence generated by G is a function. Furthermore, if we define $g\colon \mathbb{R}^n_{++} \to X$ by:

$$\boldsymbol{g}(\boldsymbol{p}) = \boldsymbol{h}(\boldsymbol{p}, 1) \quad \text{for } \boldsymbol{p} \in \mathbb{R}^n_{++},$$

we have from Theorem 4.38 that, for all $(\boldsymbol{p}, w) \in \Omega$, if $w > 0$, then:

$$\boldsymbol{h}(\boldsymbol{p}, w) = w\boldsymbol{h}(\boldsymbol{p}, 1) = w\boldsymbol{g}(\boldsymbol{p}).$$

If $w = 0$, then the equality in (4.35) obviously holds as well, and thus the first part of our result follows. I will leave the proof of part 2 of the result as an exercise. Part 3 of our result follows readily from the Slutsky symmetry conditions. Details of the argument will be left as an exercise. □

We noted at the beginning of this section that, if a consumer's preference relation is homothetic, then his/her/its expenditure on any given category of commodities was a constant percentage of income (constant, that is, with respect to income). In fact, suppose a consumer's preference relation satisfies the assumptions of Theorem 4.39, and let J be a non-empty subset of $\{1, \ldots, n\}$. Then, according to 4.39, the consumer's expenditures on the commodities corresponding to the set J, as a fraction of income (or wealth) is given by:

$$\left(\frac{1}{w}\right)\left(\sum_{j \in J} p_j x_j\right) = \left(\frac{1}{w}\right)\left(\sum_{j \in J} p_j g_j(\boldsymbol{p}) w\right) = \sum_{j \in J} p_j g_j(\boldsymbol{p});$$

which is clearly independent of w.

4.9 Cost-of-Living Indices

In this section, we shall suppose throughout that $\Omega = \mathbb{R}^n_{++} \times \mathbb{R}_+$, and that the consumer's indirect preference relation, G^*, is a weak order on Ω.

[9]A cone X is non-degenerate if it contains a point $\boldsymbol{x} \neq \boldsymbol{0}$.

4.40 Definition. We shall say that a function $\gamma\colon \mathbb{R}^n_{++} \to \mathbb{R}_{++}$ is a **cost-of-living index for G^*** iff, for all $\boldsymbol{p}, \boldsymbol{p}' \in \mathbb{R}^n_{++}$, and all $w, w' \in \mathbb{R}_{++}$, we have:

$$\frac{w}{\gamma(\boldsymbol{p})} \geq \frac{w'}{\gamma(\boldsymbol{p}')} \iff (\boldsymbol{p}, w)G^*(\boldsymbol{p}', w') \tag{4.36}$$

Notice that $\gamma(\cdot)$ is a cost-of-living index for G^* if, and only if, the function $v\colon \Omega \to \mathbb{R}_+$ defined by:

$$v(\boldsymbol{p}, w) = w/\gamma(\boldsymbol{p}),$$

is an indirect utility function representing G^*. Consequently, it follows that if $\gamma(\cdot)$ is a cost-of-living index for G^*, then it must be positively homogeneous of degree one on $\mathbb{R}^n_{++}$.

Given the assumptions of Theorem 4.36 and 4.39 of this chapter (in particular, that the consumer's preference relation, G is homothetic), there exists a utility function, $u\colon X \to \mathbb{R}_+$ representing G, and such that $u(\cdot)$ is positively homogeneous of degree one on X. By Theorem 4.39, the consumer's demand function, h takes the form:

$$\boldsymbol{h}(\boldsymbol{p}, w) = \boldsymbol{g}(\boldsymbol{p})w;$$

and, as always, the composite function:

$$v(\boldsymbol{p}, w) = u[\boldsymbol{h}(\boldsymbol{p}, w)],$$

is an indirect utility function representing G^* on Ω. However, in this case, we have, for $(\boldsymbol{p}, w) \in \Omega$:

$$v(\boldsymbol{p}, w) = u[\boldsymbol{h}(\boldsymbol{p}, w)] = u[\boldsymbol{g}(\boldsymbol{p})w] = w \cdot u[\boldsymbol{g}(\boldsymbol{p})].$$

Consequently, if we define:

$$\gamma(\boldsymbol{p}) = 1/u[\boldsymbol{g}(\boldsymbol{p})],$$

we see that we can write:

$$v(\boldsymbol{p}, w) = w/\gamma(\boldsymbol{p});$$

and thus $\gamma(\cdot)$ is a cost-of-living index for G^*.

Unfortunately for the generality of the concept of a cost-of-living index, it can be shown that if there exists a cost-of-living index for G^*, then G must be homothetic on $Z = \boldsymbol{h}(\Omega)$; in particular, G^* must be homothetic, where this is defined as follows.

4.41 Definition. We shall say that **an indirect preference relation, G^*, is homothetic** iff, for all $(\boldsymbol{p}', w'), (\boldsymbol{p}, w) \in \Omega$ and all $\lambda > 0$, we have:

$$(\boldsymbol{p}', w')G^*(\boldsymbol{p}, w) \Rightarrow (\boldsymbol{p}', \lambda w')G^*(\boldsymbol{p}, \lambda w).$$

Of course, if G is homothetic, then it follows readily from Theorem 4.38 that G^* is homothetic as well. In principle, however, G^* may be homothetic, as just defined, even though G is not.

There are a few more facts which are of interest regarding cost-of-living indices in the homothetic case, however, and we will investigate some of them in the remainder of this section.

4.42 Proposition. *Suppose G^* is homothetic, and that the income-compensation function for G^* satisfies the condition (see Theorem 4.31): for all $(\bar{\boldsymbol{p}}; \boldsymbol{p}, w) \in \mathbb{R}^n_{++} \times \Omega$, and all $\bar{w} \in \mathbb{R}_+$:*

$$(\bar{\boldsymbol{p}}, \bar{w}) I^*(\boldsymbol{p}, w) \iff \bar{w} = \mu(\bar{\boldsymbol{p}}; \boldsymbol{p}, w). \tag{4.37}$$

Then, given any $(\bar{\boldsymbol{p}}; \boldsymbol{p}, w) \in \mathbb{R}^n_{++} \times \Omega$, and any $\lambda \in \mathbb{R}_{++}$, we have:

$$\mu(\bar{\boldsymbol{p}}; \boldsymbol{p}, \lambda w) = \lambda \cdot \mu(\bar{\boldsymbol{p}}; \boldsymbol{p}, w).$$

Proof. Given $(\bar{\boldsymbol{p}}; \boldsymbol{p}, w) \in \mathbb{R}^n_{++} \times \Omega$, it follows from (4.37) that:

$$(\bar{\boldsymbol{p}}, \mu(\bar{\boldsymbol{p}}; \boldsymbol{p}, w)) I^*(\boldsymbol{p}, w).$$

Thus, it follows from the fact that G^* is homothetic that, for $\lambda \in \mathbb{R}_{++}$:

$$\big(\bar{\boldsymbol{p}}, \lambda\mu(\bar{\boldsymbol{p}}; \boldsymbol{p}, w)\big) I^*(\boldsymbol{p}, \lambda w).$$

Making use of (4.37) once again, it now follows that:

$$\mu(\bar{\boldsymbol{p}}; \boldsymbol{p}, \lambda w) = \lambda\mu(\bar{\boldsymbol{p}}; \boldsymbol{p}, w). \quad \square$$

The above result implies that we can define a cost-of-living index for G^* along the lines of the procedure set out on the previous page, and taking $\mu(\bar{\boldsymbol{p}}; \cdot)$ as the indirect utility function. It turns out, however, that we can turn things around a bit; as is set out in the following proposition.

4.43 Proposition. *If G^* satisfies the hypotheses of Proposition 4.42, then the income-compensation function for G^* satisfies the following condition: for any (fixed) $(\boldsymbol{p}^0, w^0) \in \Omega$, the function $\gamma \colon \mathbb{R}^n_{++} \to \mathbb{R}_{++}$ defined by:*

$$\gamma(\boldsymbol{p}) = \mu(\boldsymbol{p}; \boldsymbol{p}^0, w^0),$$

is a cost-of-living index for G^*.

Proof. Let $(\boldsymbol{p}^0, w^0) \in \Omega$ be fixed, and let $(\boldsymbol{p}, w) \in \Omega$ be arbitrary. Then by (4.37), above, we have:

$$(\boldsymbol{p}^0, w^0) I^*\big(\boldsymbol{p}, \mu(\boldsymbol{p}; \boldsymbol{p}^0, w^0)\big).$$

Therefore, since G^* (and therefore I^*) is homothetic, we have:

$$\left(\boldsymbol{p}^0, \left[\frac{w}{\mu(\boldsymbol{p}; \boldsymbol{p}^0, w^0)}\right] w^0\right) I^* \left(\boldsymbol{p}, \left[\frac{w}{\mu(\boldsymbol{p}; \boldsymbol{p}^0, w^0)}\right] \mu(\boldsymbol{p}; \boldsymbol{p}^0, w^0)\right);$$

or, equivalently:

$$\left(\boldsymbol{p}^0, \frac{w \cdot w^0}{\mu(\boldsymbol{p}; \boldsymbol{p}^0, w^0)}\right) I^*(\boldsymbol{p}, w).$$

Therefore, by (4.37), we have:

$$\mu(\boldsymbol{p}^0; \boldsymbol{p}, w) = w^0 \Big(\frac{w}{\mu(\boldsymbol{p}; \boldsymbol{p}^0, w^0)}\Big).$$

Since $(\boldsymbol{p}, w) \in \Omega$ was arbitrary, and since we know from Theorem 4.31 that $\mu(\boldsymbol{p}^0; \cdot)$ is an indirect utility function representing G^* on Ω, it now follows that the function $v \colon \Omega \to \mathbb{R}_+$ defined by:

$$v(\boldsymbol{p}, w) = \frac{w}{\mu(\boldsymbol{p}; \boldsymbol{p}^0, w^0)},$$

is an indirect utility function representing G^* on Ω, and that:

$$\gamma(\boldsymbol{p}) = \mu(\boldsymbol{p}; \boldsymbol{p}^0, w^0),$$

is a cost-of-living index for G^*. □

4.10 Consumer's Surplus

If I may be allowed the rather egocentric action of quoting an article of which I was a co-author, consumer's surplus has been described as follows [Chipman and Moore (1976a, p.69)]:[10]

> The concept of consumer's surplus is one of the oldest in neoclassical economics, even predating the development of marginal utility theory; and it has proved to be one of the most durable. It has great intuitive appeal to the applied economist, for it promises to provide an objective money measure of a person's satisfaction, in terms of the amount of money he would, as proved by his actions, pay for a thing rather than go without it...

In this section we will begin by following the train of thought set out in the above quotation in that we will seek a measure of the benefit (or loss), B, of a change from one vector of prices $\boldsymbol{p}^1$ to a second vector $\boldsymbol{p}^2$, which is such that, supposing that the change is achievable at a monetary cost C, the change can be regarded as beneficial if, and only if:

$$B \geq C.$$

In terms of indirect preferences, we can express our initial goal as follows: find a benefit function, $\Phi(\boldsymbol{p}^1, \boldsymbol{p}^2; w^1)$ such that:

$$\Phi(\boldsymbol{p}^1, \boldsymbol{p}^2; w^1) \geq w^1 - w^2 \iff (\boldsymbol{p}^2, w^2)G^*(\boldsymbol{p}^1, w^1); \tag{4.38}$$

where, of course, we are interpreting w^2 as $w^1 - C$. In the following example, we will consider the case in which this all works out most nicely.

4.44 Example. In this example, it will be convenient to suppse that there are $n+1$ commodities, and to use the generic notation '$(x_0, \boldsymbol{x})$, $(x_0^*, \boldsymbol{x}^*)$,' etc., to denote commodity bundles. The commodity whose quantity is denoted by 'x_0' we will suppose is a numéraire good, as was defined earlier (Definition 4.11).[11] We will

[10] If I remember correctly, this introduction was actually written by John Chipman, so that I am not being quite as egocentric here as it appears at first glance.

[11] It could also be thought of as 'expenditure on other commodities.' We will return to this interpretation later.

denote price vectors as $(p_0, \boldsymbol{p}) \in \mathbb{R}^{1+n}_{++}$, where p_0 is the price of x_0, and $\boldsymbol{p} \in \mathbb{R}^n_{++}$ is the vector of prices of the remaining n commodities. We will often normalize the price of $\boldsymbol{x}$, however, and define:

$$\boldsymbol{q} = (1/p_0)\boldsymbol{p};$$

referring to q_j as the normalized price of commodity j. Suppose a consumer's preferences can be represented on $\mathbb{R}^{1+n}_{+}$ by the utility function:

$$u(x_0, \boldsymbol{x}) = x_0 + \sum\nolimits_{j=1}^{n} \phi_j(x_j); \tag{4.39}$$

where for each j we have:

$$(\forall x_j \in \mathbb{R}_+)\colon \phi_j'(x_j) > 0 \;\&\; \phi_j''(x_j) \leq 0. \tag{4.40}$$

If you do the mathematics, you can easily verify the fact that the consumer maximizes utility by choosing x_j^* such that:

$$\phi'(x_j^*) = p_j/p_0 \stackrel{\text{def}}{=} q_j \quad \text{for } j = 1, \ldots, n,$$

and:

$$x_0^* = \frac{w - \boldsymbol{p} \cdot \boldsymbol{x}^*}{p_0} = \frac{w}{p_0} - \boldsymbol{q} \cdot \boldsymbol{x}^*.$$

Thus, we can consider the graph of the function $\phi'(\cdot)$ to be the consumer's inverse demand curve for the jth commodity, as a function of the normalized price.

Now suppose that the consumer's price-wealth situation changes from $(p_0^1, \boldsymbol{p}^1, w^1)$ to $(p_0^2, \boldsymbol{p}^2, w^2)$, and define:

$$\boldsymbol{q}^t = (1/p_0^t)\boldsymbol{p}^t \quad \text{for } t = 1, 2.$$

Then, letting:

$$\phi_j'(x_j^t) = q_j^t \quad \text{for } j = 1, \ldots, n; t = 1, 2,$$

the consumer's change in utility is given by:

$$\Delta u = w^2/p_0^2 - \boldsymbol{q}^2 \cdot \boldsymbol{x}^2 + \sum\nolimits_{j=1}^{n} \phi_j(x_j^2) - \left[w^1/p_0^1 - \boldsymbol{q}^1 \cdot \boldsymbol{x}^1 + \sum\nolimits_{j=1}^{n} \phi_j(x_j^1)\right]. \tag{4.41}$$

Suppose we now define:

$$\Delta S_j = (q_j^1 - q_j^2)x_j^1 - q_j^2 \cdot (x_j^2 - x_j^1) + \phi_j(x_j^2) - \phi_j(x_j^1),$$

and:

$$\Delta S = \sum\nolimits_{j=1}^{n} \Delta S_j.$$

Then by (4.41), we have:

$$\Delta u = \frac{w^2}{p_0^2} - \frac{w^1}{p_0^1} + \Delta S.$$

However, we have:

$$\int_{x_j^1}^{x_j^2} \phi_j'(x_j)dx_j = \phi_j(x_j^2) - \phi_j(x_j^1) \quad \text{for } j = 1, \ldots, n; \tag{4.42}$$

and therefore:

$$\begin{aligned}\Delta S &\stackrel{\text{def}}{=} \sum\nolimits_{j=1}^{n} \Delta S_j \stackrel{\text{def}}{=} \sum\nolimits_{j=1}^{n} \Big[(q_j^1 - q_j^2)\cdot x_j^1 - q_j^2 \cdot (x_j^2 - x_j^1) + \int_{x_j^1}^{x_j^2} \phi_j'(x_j)dx_j\Big] \\ &= \boldsymbol{q}^1 \cdot \boldsymbol{x}^1 - \boldsymbol{q}^2 \cdot \boldsymbol{x}^2 + \sum\nolimits_{j=1}^{n} \big[\phi_j(x_j^2) - \phi_j(x_j^1)\big].\end{aligned} \tag{4.43}$$

See Figure 4.1, below, for a graphical depiction of ΔS_j. It may nonetheless not be

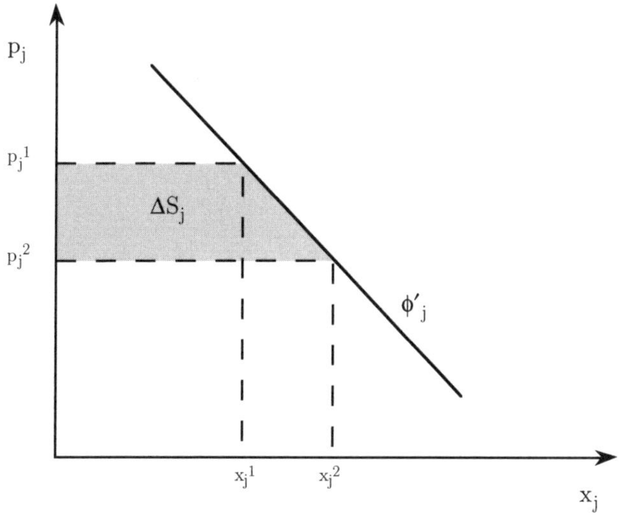

Figure 4.1: Consumer's Surplus in the Simplest Case.

clear why this works out to be a correct measure of the change in the consumer's utility. Perhaps we can clear things up a bit, however, by expressing everything in terms of the demand and indirect utility functions, rather than the inverse demand and direct utility functions. Let's denote the demand functions for the x_j's [which are simply the inverses of the $\phi_j'(\cdot)$'s] by 'δ_j;' so that:

$$\phi_j'[\delta_j(q_j)] = q_j \quad \text{for } j = 1, \ldots, n. \tag{4.44}$$

We will denote the indirect utility function by '$v(p_0, \boldsymbol{p}, w)$;' or, upon normalizing by the price of the numéraire:

$$V(\boldsymbol{q}, w/p_0) \stackrel{\text{def}}{=} v\big[1, (1/p_0)\boldsymbol{p}, w/p_0\big].$$

It is obvious, of course, that what we wish to evaluate in this case is:

$$V(\boldsymbol{q}^2, w^2/p_0^2) - V(\boldsymbol{q}^1, w^1/p_0^1). \tag{4.45}$$

Now, in this case, [using (4.41)] it is easy to see that $V(\boldsymbol{q}, w/p_0)$ takes the form:

$$V(\boldsymbol{q}, w/p_0) = w/p_0 - \boldsymbol{q} \cdot \boldsymbol{\delta}(\boldsymbol{q}) + \sum\nolimits_{j=1}^{n} \psi_j(qp_j) \stackrel{\text{def}}{=} w/p_0 + \sum\nolimits_{j=1}^{n} \varphi_j(q_j), \quad (4.46)$$

where:

$$\psi_j(q_j) = \phi_j[\delta_j(q_j)],$$

and:

$$\varphi(q_j) = \psi_j(q_j) - q_j \cdot \delta_j(q_j) \quad \text{for } j = 1, \ldots, n.$$

Consequently:

$$\frac{\partial V}{\partial q_j} = \varphi_j'(q_j) = \psi'(q_j) - \delta_j(q_j) - q_j \cdot \delta_j'(q_j). \quad (4.47)$$

However, we have:

$$\psi_j'(q_j) = \frac{d}{dq_j}\Big(\phi_j[\delta_j(q_j)]\Big) = \phi_j'[\delta_j(q_j)]\delta_j'(q_j);$$

and from (4.44) and (4.47) it then follows that:

$$\varphi_j'(q_j) = -\delta_j(q_j).$$

Another examination of Figure 4.1 will then make it clear that in the situation under analysis there:

$$\Delta S_j = \int_{q_j^2}^{q_j^1} \delta_j(q_j)dq_j = \int_{q_j^1}^{q_j^2} -\delta_j(q_j)dq_j = \varphi_j(q_j^2) - \varphi_j(q_j^1). \quad (4.48)$$

Notice in particular, that if only the jth (normalized) price is changed (all other prices and income remaining the same); say in the amount Δq_j, then we can unambiguously interpret the benefit (or cost) to the consumer of the price change as the difference:

$$\Delta S_j = \int_{q_j}^{q_j+\Delta q_j} -\delta_j(t)dt = \varphi_j(q_j + \Delta q_j) - \varphi_j(q_j). \quad (4.49)$$

In fact, it is worth noting that in this situation, if only a subset, $J \subseteq \{1, \ldots, n\}$, of prices change, then we will have:

$$\Delta u = \sum_{j \in J}\Big(\int_{q_j}^{q_j+\Delta q_j} -\delta_j(t)dt\Big). \quad \square$$

The situation which we analyzed in the above example is, obviously, very special.[12] If you look back at the analysis we have done to this point, probably the first thing which stands out is that the fact that the demand for the jth commodity was a function of p_j alone meant that we were able to make an unambiguous interpretation of the monetary value of a change in the jth price. Moreover, it allowed us to

[12]However, see also Example 4.47, below.

evaluate an integral of an $(n+1)$-dimensional function as the sum of $n+1$ simple Riemann integrals.

Let's see if we can extend the definitions used in the example to some extent, and seek to find a function, $W(\cdot)$, of $(\boldsymbol{p}^1, w^1)$ and $(\boldsymbol{p}^2, w^2)$ which is such that:

$$W[(\boldsymbol{p}^1, w^1), (\boldsymbol{p}^2, w^2)] \geq 0 \iff (\boldsymbol{p}^2, w^2) G^* (\boldsymbol{p}^1, w^1); \tag{4.50}$$

in which case we will say that $W(\cdot)$ **provides an acceptable indicator of welfare change** for the consumer. We will begin by considering the possibility of obtaining such a function by integrating an observable function of some kind. In this connection, the natural notion of integration to use, since we want to allow for the fact that $\boldsymbol{p}^1$ and $\boldsymbol{p}^2$ may differ in several coordinates, is that of the line integral. Without going into too many technical details,[13] the concept of a line integral works by reducing the integral of a multi-dimensional function to a standard Riemann-Stieljes integral; doing so by making use of the notion of a 'path function;' which, for the situation we are considering here is simply a function $\boldsymbol{\omega}\colon [0,1] \to \Omega$, where we define $\Omega = \mathbb{R}^n_{++} \times \mathbb{R}_+$. We will say that such a path is 'polygonal' if its graph is the union of a finite number of line segments; and that it connects $(\boldsymbol{p}^1, w^1)$ and $(\boldsymbol{p}^2, w^2)$ iff we have:

$$\boldsymbol{\omega}(0) = (\boldsymbol{p}^1, w^1) \text{ and } \boldsymbol{\omega}(1) = (\boldsymbol{p}^2, w^2).$$

(Incidentally, we are going back to our usual supposition that there are n commodities in the remainder of our discussion.)

4.45 Example. Two particularly simple polygonal paths connecting $(\boldsymbol{p}^1, w^1)$ and $(\boldsymbol{p}^2, w^2)$ are given by:

$$\boldsymbol{\omega}(t) = t(\boldsymbol{p}^2, w^2) + (1-t)(\boldsymbol{p}^1, w^1),$$

and, in the case in which $n = 2$:

$$\overline{\boldsymbol{\omega}}(t) = \begin{cases} (\boldsymbol{p}^1, w^1) + 3t[(\boldsymbol{p}^*, w^1) - (\boldsymbol{p}^1, w^1)] & \text{for } 0 \leq t \leq 1/3, \\ (\boldsymbol{p}^*, w^1) + (3t-1)[(\boldsymbol{p}^2, w^1) - (\boldsymbol{p}^*, w^1)] & \text{for } 1/3 < t \leq 2/3, \\ (\boldsymbol{p}^2, w^1) + (3t-2)[(\boldsymbol{p}^2, w^2) - (\boldsymbol{p}^2, w^1)] & \text{for } 2/3 < t \leq 1; \end{cases}$$

where we define $\boldsymbol{p}^*$ by:

$$\boldsymbol{p}^* = (p_1^2, p_2^1). \quad \square$$

To continue, given a function, $\boldsymbol{f}\colon \Omega \to \mathbb{R}^{n+1}$, and a path function, $\boldsymbol{\omega}$, the integral of $\boldsymbol{f}(\cdot)$ from $(\boldsymbol{p}^1, w^1)$ to $(\boldsymbol{p}^2, w^2)$, given a path function $\boldsymbol{\omega}$, satisfying:

$$\boldsymbol{\omega}(0) = (\boldsymbol{p}^1, w^1) \text{ and } \boldsymbol{\omega}(1) = (\boldsymbol{p}^2, w^2),$$

is given by:

$$\int_0^1 \boldsymbol{f}[\boldsymbol{\omega}(t)] \cdot d\boldsymbol{\omega}(t) = \sum\nolimits_{j=1}^{n+1} \int_0^1 f^j[\omega(t)] d\omega_j(t), \tag{4.51}$$

[13] For more detailed analysis, see Chipman and Moore [1976a, 1990].

whenever each of the Riemann-Stieltjes on the right-hand-side of the equation exists.[14] It should also be noted that if $n = 0$; that is, if the integrand function is real-valued, then the line integral reduces to a standard Riemann-Stieljes integral.

Now, the trouble with trying to make use of a line integral to obtain an acceptable indicator of welfare change is that the value of the line integral may be dependent upon the path function chosen; an obviously unacceptable attribute for a measure of consumer welfare. However, if the integrand function, $\boldsymbol{f}(\cdot)$, is continuously differentiable, then given any two polygonal path functions, $\boldsymbol{\omega}$ and $\overline{\boldsymbol{\omega}}$, which connect $(\boldsymbol{p}^1, w^1)$ and $(\boldsymbol{p}^2, w^2)$, we have:

$$\int_0^1 \boldsymbol{f}[\boldsymbol{\omega}(t)] \cdot d\boldsymbol{\omega}(t) = \int_0^1 \boldsymbol{f}[\overline{\boldsymbol{\omega}}(t)] \cdot d\overline{\boldsymbol{\omega}}(t). \tag{4.52}$$

In this case, we will say that $\boldsymbol{f}\colon \Omega \to R^{n+1}$ **provides an acceptable indicator of welfare change on $\boldsymbol{\Omega}$** iff, for all $(\boldsymbol{p}^1, w^1), (\boldsymbol{p}^2, w^2) \in \Omega$ and any polygonal path function connecting the two points, we have:

$$\int_0^1 \boldsymbol{f}[\boldsymbol{\omega}(t)] \cdot d\boldsymbol{\omega}(t) \geq 0 \iff (\boldsymbol{p}^2, w^2) G^*(\boldsymbol{p}^1, w^1); \tag{4.53}$$

where G^* is the consumer's indirect preference relation. The remarkable thing about this is that if the line integral is independent of path on Ω, then there exists a continuously-differentiable function, $V\colon \Omega \to \mathbb{R}$, called a **potential function**, such that, given any polygonal path function, $\boldsymbol{\omega}$ connecting $(\boldsymbol{p}^1, w^1)$ and $(\boldsymbol{p}^2, w^2)$, we have:

$$\int_0^1 \boldsymbol{f}[\boldsymbol{\omega}(t)] \cdot d\boldsymbol{\omega}(t) = V(\boldsymbol{p}^2, w^2) - V(\boldsymbol{p}^1, w^1); \tag{4.54}$$

and for all $(\boldsymbol{p}, w) \in \Omega$:

$$f^j(\boldsymbol{p}, w) = V_j(\boldsymbol{p}, w) \qquad \text{for } j = 1, \ldots, n+1; \tag{4.55}$$

that is, the jth partial derivative of the potential function must equal the jth coordinate function of $\boldsymbol{f}(\cdot)$ at each $(\boldsymbol{p}, w) \in \Omega$, and for each $j = 1, \ldots, n+1$. In fact, the converse is also true: if there exists a continuously differentiable function $V\colon \Omega \to \mathbb{R}$ which satisfies the system (4.55), then (it is a potential function for the integrand $\boldsymbol{f}(\cdot)$ and) equation (4.54) holds for any 'piece-wise smooth' path function connecting $(\boldsymbol{p}^1, w^1)$ and $(\boldsymbol{p}^2, w^2)$. But now if we combine (4.53) and (4.54), we see that if a continuously differentiable function, $\boldsymbol{f}$, provides an acceptable indicator of welfare change on Ω, then the corresponding potential function must be an indirect utility function representing G^*! It then follows from the Antonelli-Allen-Roy theorem (Theorem 4.28) that we must have, for all $(\boldsymbol{p}, w), \in \Omega$:

$$f^j(\boldsymbol{p}, w) = -f^{n+1}(\boldsymbol{p}, w) h^j(\boldsymbol{p}, w), \tag{4.56}$$

where '$h^j(\cdot)$' denotes the demand function for the jth commodity.

[14] If you haven't studied Riemann-Stieltjes integrals, don't worry. We aren't going to do anything technical with such integrals.

Now suppose we look at the functions most often used in this integrand when investigators are doing empirical estimates of consumer's surplus. In such investigations, the integrand most often used is given by:[15]

$$f^j(\boldsymbol{p}, w) = -h^j(\boldsymbol{p}, w) \quad \text{for } j = 1, \ldots, n, \tag{4.57}$$

and:

$$f^{n+1}(\boldsymbol{p}, w) = 1. \tag{4.58}$$

As was pointed out in Chipman and Moore [1976a, p. 79], there exists no preference relation for which this integrand function provides an acceptable indicator of welfare change! Why? because the $(n+1)^{st}$ function must be the marginal utility of income for an indirect preference relation representing the consumer's (indirect) preferences, and as was established more than sixty years ago by Paul Samuelson (Samuelson [1942]), there exists no indirect preference relation yielding a marginal utility of income which is independent of both prices and income.[16]

Incidentally, while we considered an integrand [in equations (4.57) and (4.58)] ostensibly appropriate[17] for analyzing changes in both prices *and* income, the method discussed in the preceding paragraph is no more appropriate if only prices have changed; for, as we have seen, the function corresponding to the jth price *must* take the form of the product of a valid marginal utility of income times the demand function for the jth commodity. However, in Example 4.44 we considered an admittedly very special situation in which something very close to the procedure discussed in the above paragraph worked just fine. There is another case in which a similar integrand defines a valid measure of welfare change; the case in which the consumer's preferences are homothetic. To see this, recall first that if a function $\boldsymbol{f}\colon \Omega \to \mathbb{R}^{n+1}$ is continuously differentiable, then its line integral is independent of path, and there exists a potential function, $V(\cdot)$ satisfying (4.54) and (4.55). Suppose, then, that a consumer's preferences are homothetic, and yield a continuously differentiable demand function. Then, as was noted in Section 4.8, above, we can write this function in the form:

$$\boldsymbol{h}(\boldsymbol{p}, w) = \boldsymbol{g}(\boldsymbol{p})w.$$

Consequently, the integrand function $\boldsymbol{f}\colon \Omega \to \mathbb{R}^{n+1}_+$ defined by:

$$f^j(\boldsymbol{p}, w) = -g^j(\boldsymbol{p}) \quad \text{for } j = 1, \ldots, n, \tag{4.59}$$

and:

$$f^{n+1}(\boldsymbol{p}, w) = 1/w, \tag{4.60}$$

yields line integrals independent of path. Moreover, as noted in the previous section, in the homothetic case, indirect preferences can be represented by a function of the

[15] Typically only a subset of prices, and correspondingly, of demand functions are used in the investigation. This fact does nothing to invalidate the argument presented here, however.

[16] A somewhat simpler argument than Samuelson's is presented in Chipman and Moore [1976, pp. 79–80] who simply note that, since any indirect utility function must be positively homogeneous of degree zero, the marginal utility of income must necessarily be homogeneous of degree minus one.

[17] Since it is $n+1$-dimensional.

form $w/\gamma(\boldsymbol{p})$. Taking the log of this function yields an indirect utility function also representing the consumer's indirect preferences, and is given by:

$$V(\boldsymbol{p}, w) = \log w - \log \gamma(\boldsymbol{p}). \tag{4.61}$$

It is easily shown that this is a potential function for the integrand defined in (4.59) and (4.60); and, consequently, it follows that said integrand function provides an acceptable indicator of welfare change on Ω.

A more detailed discussion of valid possibilities for the Dupuit-Marshall (line integral) type of consumer's surplus can be found in Chipman and Moore [1976, 1990], and we will consider the possibilies for extending the results we've considered here to the multi-consumer case (to obtain consumers' surplus) in Chapter 15. However, before leaving this topic, I should mention the fact that there is a silver lining to these dark clouds. While conventional consumer's surplus analysis is not theoretically correct, this fact is sometimes not as critical as it appears at first glance. The reason is this: if one is estimating the demand function for a consumer (or for a representative consumer), the parameters which must be estimated to define the demand function (more specifically, the vector of demand functions) are often sufficient to define (identify) the corresponding indirect utility function, which can then be used to evaluate the desirability of the change. For example, suppose we have determined that a Cobb-Douglas demand function fits the data well, and are attempting to derive the desirability of a change from $(\boldsymbol{p}^1, w^1)$ to $(\boldsymbol{p}^2, w^2)$. In this situation, the consumer's demand functions are given by:

$$x_j = a_j w/p_j \quad \text{for } j = 1, \ldots, n;$$

and estimating the demand functions amounts to determining the values of the a_j for this consumer. But, the values of the a_j determine an indirect utility function for the consumer; in fact, we know that the function:

$$v(\boldsymbol{p}, w) = \frac{w}{\prod_{j=1}^{n} (p_j/a_j)^{a_j}},$$

is the corresponding indirect utility function. In order to evaluate the desirability of the change, therefore, we need only evaluate:

$$v(\boldsymbol{p}^2, w^2) - v(\boldsymbol{p}^1, w^1).$$

The bad news associated with this point is that it may be necessary to estimate the whole system of demand functions in order to obtain the values of the parameters which determine the indirect utility function. The fact that applied economists often do consumer's surplus analysis for cases where only one or two prices change, and, correspondingly, only one or two demand functions need to be estimated in order to do conventional consumer's surplus estimates, is surely one of the principal reasons that this sort of work appears so often in the literature (and that so many articles have been published analyzing the size of the error involved in using the area under conventional [Marshallian] demand curves as an estimate of 'true' consumer's surplus). However, in some cases the evaluation of changes in indirect utility can be done using only the parameters associated with the demand functions for the commodities whose prices change. This point is pursued further in Example 4.48, below, and in Exercises 8–10 at the end of this chapter.

Let's now turn our attention to two alternative indicators of welfare change which were originally introduced by Hicks [1942]. We will first consider **compensating variation**, which is defined as the amount that would need to be added to a consumer's wealth after a price change in order to make the consumer exactly as well off after the change as before. Having already studied the properties of the income compensation function, it should be apparent that we can define the compensating variation of a proposed change from $(\boldsymbol{p}^1, w^1)$ to $(\boldsymbol{p}^2, w^2)$ as:

$$CV\left[(\boldsymbol{p}^1, w^1), (\boldsymbol{p}^2, w^2)\right] = \mu(\boldsymbol{p}^2; \boldsymbol{p}^1, w^1) - w^2. \tag{4.62}$$

Since this quantity can equivalently be expressed as:

$$\mu(\boldsymbol{p}^2; \boldsymbol{p}^1, w^1) - \mu(\boldsymbol{p}^2; \boldsymbol{p}^2, w^2);$$

it follows from the fact that $\mu(\boldsymbol{p}^2; \cdot)$ is an indirect utility function representing G^* that the change should be undertaken if, and only if:

$$CV\left[(\boldsymbol{p}^1, w^1), (\boldsymbol{p}^2, w^2)\right] \leq 0,$$

although it is probably more natural to turn this around to define the **compensating variation criterion for welfare improvement**, $W^C(\cdot)$, by:

$$W^C\left[(\boldsymbol{p}^1, w^1), (\boldsymbol{p}^2, w^2)\right] = -CV\left[(\boldsymbol{p}^1, w^1), (\boldsymbol{p}^2, w^2)\right] = w^2 - \mu(\boldsymbol{p}^2; \boldsymbol{p}^1, w^1). \tag{4.63}$$

Not only does this provide an acceptable indicator of welfare change, but in the case in which $w^2 = w^1 - C$, with C being the cost of the change, notice that:

$$W^C\left[(\boldsymbol{p}^1, w^1), (\boldsymbol{p}^2, w^2)\right] \geq 0 \iff w^1 - \mu(\boldsymbol{p}^2; \boldsymbol{p}^1, w^1) \geq C;$$

which means that we obtain a valid criterion of the general form of equation (4.38), with $w^1 - \mu(\boldsymbol{p}^2; \boldsymbol{p}^1, w^1)$ as a measure of benfit.

So, we have seen the good news regarding the theory of compensating variation. The question now is, what is the bad news? Well, there are problems with respect to actually estimating $\mu(\boldsymbol{p}^2; \boldsymbol{p}^1, w^1)$ in real-life practical situations; but our concern here will be with the theory, and the theoretical difficulty with this welfare criterion is that it cannot generally be used to rank projects. That is, if two different policies or projects are being considered; with project/policy t yielding the price-wealth combination $(\boldsymbol{p}^t, w^t)$, for $t = 1, 2$, and if the status quo price-wealth combination is $(\boldsymbol{p}^0, w^0)$, it is tempting to say that project/policy 2 is the better one if:

$$W^C\left[(\boldsymbol{p}^0, w^0), (\boldsymbol{p}^2, w^2)\right] > W^C\left[(\boldsymbol{p}^0, w^0), (\boldsymbol{p}^1, w^1)\right]. \tag{4.64}$$

Unfortunately, this inference is not generally valid, as is shown by the following example.

4.46 Example. Let $X = \mathbb{R}^2_+$, and, using the generic notation '(x, y)' to denote points in $\mathbb{R}^2$, suppose a consumer's preferences can be represented by the utility function:

$$u(x, y) = (x + 2) \cdot y.$$

Then (see Exercise 5, at the end of this chapter) the demand functions for x and y are given by:

$$x = \frac{w - 2p_1}{2p_1} \text{ and } y = \frac{w + 2p_1}{2p_2},$$

respectively. Consequently, an indirect utility function for the consumer is given by:

$$v(\boldsymbol{p}, w) = \left(\frac{w - 2p_1}{2p_1}\right)\left(\frac{w + 2p_1}{2p_2}\right) = \frac{w^2 - 4(p_1)^2}{4p_1p_2}.$$

Now suppose the status quo (current) price-wealth situation for the consumer is $(\boldsymbol{p}^0, w^0)$, where:

$$\boldsymbol{p}^0 = (1, 1) \text{ and } w^0 = 2;$$

and suppose project one will result in the price-wealth pair $(\boldsymbol{p}^1, w^1)$, where:

$$\boldsymbol{p}^1 = (2, 1) \text{ and } w^1 = 8.$$

Then, since $v(\boldsymbol{p}^0, w^0) = (4 - 4)/4 = 0$, while:

$$v(\boldsymbol{p}^1, w) = \frac{w^2 - 16}{8} = 0$$

iff $w = 4$, it follows that $\mu(\boldsymbol{p}^1; \boldsymbol{p}^0, w^0) = 4$, and therefore:

$$W^C\left[(\boldsymbol{p}^0, w^0), (\boldsymbol{p}^1, w^1)\right] = 8 - 4 = 4.$$

Now suppose a second project results in the price-wealth pair $(\boldsymbol{p}^2, w^2)$, where:

$$\boldsymbol{p}^2 = (1, 2) \text{ and } w^2 = 7.$$

Then similar considerations to those of the above paragraph establish the fact that $\mu(\boldsymbol{p}^2; \boldsymbol{p}^0, w^0) = 2$, and therefore:

$$W^C(\boldsymbol{p}^0, w^0), (\boldsymbol{p}^2, w^2)] = 7 - 2 = 5 > W^C\left[(\boldsymbol{p}^0, w^0), (\boldsymbol{p}^1, w^1)\right] = 4.$$

However, as you can (and should) readily verify:

$$v(\boldsymbol{p}^1, w^1) = 6 > v(\boldsymbol{p}^2, w^2) = 5\frac{5}{8}.$$

Thus project one should be preferred, despite the contradictory compensating variation comparison. □

Despite the fact that compensating variation comparisons will not generally allow one to rank projects, there are two special cases in which such a ranking is valid. The first case is that in which the consumer's preference relation is homothetic,*if we normalize appropriately.* Suppose project t results in the price-wealth pair $(\boldsymbol{p}^t, w^t)$, for $t = 1, 2$, and that *normalizing by wealth,* we find that:

$$W^C\big[(\boldsymbol{p}^0, w^0), (\boldsymbol{p}^1/w^1, 1)\big] > W^C\big[(\boldsymbol{p}^0, w^0), (\boldsymbol{p}^2/w^2, 1)\big].$$

Then it follows from the definitions that:

$$1 - \mu\big(\boldsymbol{p}^1/w^1; \boldsymbol{p}^0, w^0\big) > 1 - \mu\big(\boldsymbol{p}^2/w^2; \boldsymbol{p}^0, w^0\big);$$

so that $1/\mu\big(\boldsymbol{p}^1/w^1; \boldsymbol{p}^0, w^0\big) > 1/\mu\big(\boldsymbol{p}^2/w^2; \boldsymbol{p}^0, w^0\big)$, and thus, making use of the homogeneity property of the income compensation function:

$$\frac{w^1}{\mu\big(\boldsymbol{p}^1; \boldsymbol{p}^0, w^0\big)} > \frac{w^2}{\mu\big(\boldsymbol{p}^2; \boldsymbol{p}^0, w^0\big)}.$$

But then it follows from Proposition 4.43 that $(\boldsymbol{p}^1, w^1)P^*(\boldsymbol{p}^2, w^2)$. Thus, if indirect preferences are homothetic, the relative values of $W^C\big[(\boldsymbol{p}^0, w^0), (\boldsymbol{p}^t/w^t, 1)\big]$ can be used to rank projects.[18] You can also show that similar considerations allow projects to be ranked by compensating variation comparisons if they result in the same wealth (that is, if $w^1 = w^2$); given that indirect preferences are homothetic. A second situation in which such rankings are valid is set out in the following example.[19]

4.47 Example. Suppose we once again assume that the consumer's consumption set is $\mathbb{R}_+^{n+1}$, and use the generic notation '$(x_0, \boldsymbol{x})$' to denote points (commodity bundles) in $\mathbb{R}^{n+1}$. We then suppose that the consumer's utility function takes the form:

$$u(x_0, \boldsymbol{x}) = x_0 + \varphi(\boldsymbol{x}),$$

where $\varphi\colon \mathbb{R}_+^n \to \mathbb{R}$ is concave, continuous, and positively homogeneous of degree a, where $0 < a < 1$. For future reference, we note that φ can be considered as a utility function on $\mathbb{R}_+^n$, and we will denote the demand function it generates by '$\boldsymbol{h}$,' which can be written in the form:

$$\boldsymbol{h}(\boldsymbol{p}, m) = \boldsymbol{g}(\boldsymbol{p})m.$$

From our work in Section 9 of this chapter, we know that if we define the function ψ by:

$$\psi(\boldsymbol{x}) = \big[\varphi(\boldsymbol{x})\big]^{1/a} \quad \text{for } \boldsymbol{x} \in \mathbb{R}_+^n,$$

then the indirect utility function corresponding to ψ can be written in the form:

$$v^*(\boldsymbol{p}, m) = m/\gamma(\boldsymbol{p}),$$

[18] To the best of my knowledge, this, and the fact that compensating variation cannot generally be used to rank projects, was first pointed out in Chipman and Moore [1980].

[19] I should mention, however, that the next example does not represent the only additional case in which compensating variation can be used to rank projects. If the consumer's utility function is of the form used in Example 4.44, which is not a special case of our next example, the compensating variation criterion can be used to rank projects. See also, Chipman and Moore [1980].

where:

$$\gamma(\boldsymbol{p}) = 1/\psi[g(\boldsymbol{p})].$$

Consequently, it follows that the indirect utility function corresponding to φ can be written in the form:

$$v(\boldsymbol{p}, m) = \big[m/\gamma(\boldsymbol{p})\big]^a. \tag{4.65}$$

Now, given $(p_0^*, \boldsymbol{p}^*, w^*)$, it can be shown that:

$$m^* \stackrel{\text{def}}{=} \min\Big\{w^*, \Big[\frac{ap_0^*}{\gamma(\boldsymbol{p}^*)^a}\Big]^{\frac{1}{1-a}}\Big\},$$

maximizes the function:

$$\frac{w^* - m}{p_o^*} + \Big[\frac{m}{\gamma(\boldsymbol{p}^*)}\Big]^a;$$

and thus we see that, if:

$$w \geq \Big[\frac{ap_0}{\gamma(\boldsymbol{p})^a}\Big]^{\frac{1}{1-a}}, \tag{4.66}$$

then the consumer's demand is given by:

$$x_0 = (w - m)/p_0 \text{ and } \boldsymbol{x} = \boldsymbol{h}(\boldsymbol{p}, m),$$

where:

$$m = \Big[\frac{ap_0}{\gamma(\boldsymbol{p})^a}\Big]^{\frac{1}{1-a}}, \tag{4.67}$$

and $\boldsymbol{h}\colon \Omega \to \mathbb{R}^n_+$ is the demand function for φ. Consequently, we see that the consumer's indirect preferences can be represented by the function V, given by:

$$\begin{aligned} V(p_0, \boldsymbol{p}, w) &= \frac{w - \big[ap_0/\gamma(\boldsymbol{p})^a\big]^{\frac{1}{1-a}}}{p_0} + \frac{\big[ap_0/\gamma(\boldsymbol{p})^a\big]^{\frac{a}{1-a}}}{\gamma(\boldsymbol{p})^a} \\ &= \frac{w}{p_0} + (1-a)\Big(\frac{ap_0}{\gamma(\boldsymbol{p})}\Big)^{\frac{a}{1-a}} \end{aligned}$$

Given that we have found the functional form of the indirect utility function, one can then show by straightforward substitution that:

$$\frac{\mu(p_0, \boldsymbol{p}; \overline{p}_0, \overline{\boldsymbol{p}}, \overline{w})}{p_0} = \frac{\overline{w}}{\overline{p}_0} + (1-a)\bigg(\Big[\frac{a\overline{p}_0}{\gamma(\overline{\boldsymbol{p}})}\Big]^{\frac{a}{1-a}} - \Big[\frac{ap_0}{\gamma(\boldsymbol{p})}\Big]^{\frac{a}{1-a}}\bigg) \tag{4.68}$$

Now suppose the current price-wealth situation for the consumer is $(\overline{p}_0, \overline{\boldsymbol{p}}, \overline{w})$, and that two projects are contemplated which will result in price-wealth situations $(p_0^t, \boldsymbol{p}^t, w^t)$, for $t = 1, 2$. *If we normalize prices and wealth for the new situations*, to obtain:

$$\boldsymbol{q}^t = (1/p_0^t)\boldsymbol{p}^t \text{ and } y^t = w^t/p_0^t \quad \text{for } t = 1, 2;$$

it follows from (4.68) and the homogeneity of the income-compensation function that:

$$\begin{aligned} &W^C\big[(1, \boldsymbol{q}^1, y^1), (\overline{p}_0, \overline{\boldsymbol{p}}, \overline{w})\big] - W^C\big[(1, \boldsymbol{q}^2, y^2), (\overline{p}_0, \overline{\boldsymbol{p}}, \overline{w})\big] \\ &\qquad = y^1 + (1-a)\Big[\frac{a}{\gamma(\boldsymbol{q}^1)}\Big]^{\frac{a}{1-a}} - \bigg(y^2 + (1-a)\Big[\frac{a}{\gamma(\boldsymbol{q}^2)}\Big]^{\frac{a}{1-a}}\bigg). \end{aligned}$$

Thus we see that:

$$W^C\big[(1,\boldsymbol{q}^1,y^1),(\overline{p}_0,\overline{\boldsymbol{p}},\overline{w})\big] \geq W^C\big[(1,\boldsymbol{q}^2,y^2),(\overline{p}_0,\overline{\boldsymbol{p}},\overline{w})\big]$$
$$\iff V(1,\boldsymbol{q}^1,y^1) \geq V(1,\boldsymbol{q}^2,y^2). \quad (4.69)$$

Since:

$$(p_0^t,\boldsymbol{p}^t,w^t)I^*(1,\boldsymbol{q}^t,y^t) \quad \text{for } t=1,2,$$

it follows that project one is preferred if, and only if:

$$W^C\big[(1,\boldsymbol{q}^1,y^1),(\overline{p}_0,\overline{\boldsymbol{p}},\overline{w})\big] > W^C\big[(1,\boldsymbol{q}^2,y^2),(\overline{p}_0,\overline{\boldsymbol{p}},\overline{w})\big] \quad \square$$

In our discussion of Dupuit-Marshall consumer's surplus, we noted that in situations where only a subset of prices change, it may be possible to determine whether indirect utility has increased even if the investigator knows the values of only that subset of parameters corresponding to the commodities whose price has changed. It may also be possible to determine the compensating variation associated with such a change, as the following example demonstrates.

4.48 Example. Suppose a consumer's preferences can be represented by a Cobb-Douglas utility function (with initialy unknown parameter values), and that a proposed policy will change a subset, J, of prices and income/wealth from the status quo, $(\boldsymbol{p}^0,w^0)$ to $(\boldsymbol{p}^1,w^1)$, where:

$$p_j^1 = p_j^0 \;\text{ for all } j \notin J.$$

In order to conduct traditional consumer's surplus analysis in this case, one would need to estimate the Jth demand function:

$$x_j = \frac{a_j w}{p_j} \quad \text{for each } j \in J.$$

Since such estimation involved determining (estimated) values of a_j, for each $j \in J$, for the remainder of our discussion we will assume that these parameter values are known.

To obtain the value of the compensating variation associated with the change, we begin by noting that $\mu(\boldsymbol{p}^1;\boldsymbol{p}^0,w^0)$ is the value of w which solves the equation:

$$\frac{w\prod_{j=1}^n (a_j)^{a_j}}{\prod_{j=1}^n (p_j^1)^{a_j}} = \frac{w^0\prod_{j=1}^n (a_j)^{a_j}}{\prod_{j=1}^n (p_j^0)^{a_j}}.$$

Solving, we find:

$$\mu(\boldsymbol{p}^1;\boldsymbol{p}^0,w^0) = \frac{w^0\prod_{j\in J}(p_j^1)^{a_j}}{\prod_{j\in J}(p_j^0)^{a_j}}.$$

Therefore:

$$W^C\big[(\boldsymbol{p}^0,w^0),(\boldsymbol{p}^1,w^1)\big] = w^0 - \frac{w^0\prod_{j\in J}(p_j^1)^{a_j}}{\prod_{j\in J}(p_j^0)^{a_j}}$$
$$= \frac{w^0}{\prod_{j\in J}(p_j^0)^{a_j}}\Big[\prod\nolimits_{j\in J}(p_j^0)^{a_j} - \prod\nolimits_{j\in J}(p_j^1)^{a_j}\Big].$$

It is also worth noting that if we were allow for a change in all prices (that is, if $J = \{1, \ldots, n\}$), then the above formula becomes:

$$W^C\big[(\boldsymbol{p}^0, w^0), (\boldsymbol{p}^1, w^1)\big] = \frac{w^0}{\prod_{j=1}^n (p_j^0)^{a_j}}\Big[\prod_{j=1}^n (p_j^0)^{a_j} - \prod_{j=1}^n (p_j^1)^{a_j}\Big];$$

that is, the compensating variation criterion is given by the cost-of-living indirect utility function, evaluated at $(\boldsymbol{p}^0, w^0)$, times the change in the cost-of-living index. □

Hicks' second measure of 'consumer surplus,' **equivalent variation**, is defined verbally as the amount which could be added to a consumer's wealth before a price change in order to leave her or him exactly as well off without the change as with it. Thus we can define:

$$EV\big[(\boldsymbol{p}^1, w^1), (\boldsymbol{p}^2, w^2)\big] = \mu(\boldsymbol{p}^1; \boldsymbol{p}^2, w^2) - w^1. \tag{4.70}$$

It is then easy to see that $EV(\cdot)$ is an acceptable indicator of welfare change. Moreover, if we consider the problem of ranking projects/policies as before; with project/policy t yielding the price-wealth combination $(\boldsymbol{p}^t, w^t)$, for $t = 1, 2$, and with the status quo price-wealth combination $(\boldsymbol{p}^0, w^0)$, we see that

$$EV\big[(\boldsymbol{p}^0, w^0), (\boldsymbol{p}^2, w^2)\big] > EV\big[(\boldsymbol{p}^0, w^0), (\boldsymbol{p}^1, w^1)\big], \tag{4.71}$$

if, and only if:

$$\mu(\boldsymbol{p}^0; \boldsymbol{p}^2, w^2) > \mu(\boldsymbol{p}^0; \boldsymbol{p}^1, w^1). \tag{4.72}$$

It then follows at once from the fact that $\mu(\boldsymbol{p}^0; \cdot)$ is a valid indirect utility function (Theorem 4.31) that project two is preferred if inequality (4.71) holds.

So, as we have just seen, there is a real advantage in using equivalent variation, rather than compensating variation as a welfare criterion if we are dealing with a single consumer, or if we are comfortable with the assumption of a 'representative consumer.' However, when dealing with multiple consumers, the advantages are somewhat reversed. For, while neither criterion can generally be used to rank projects which affect multiple consumers, compensating variation provides the better indicator of welfare change in this situation. To see this, suppose a policy is being contemplated which would change the ith consumer's price-wealth situation from $(\boldsymbol{p}^1, w_i^1)$ to $(\boldsymbol{p}^2, w_i^2)$, and suppose:

$$\sum_{i=1}^m W_i^C\big[(\boldsymbol{p}^1, w_i^1), (\boldsymbol{p}^2, w_i^2)\big] \stackrel{\text{def}}{=} \sum_{i=1}^m \big[w_i^2 - \mu_i(\boldsymbol{p}^2; \boldsymbol{p}^1, w_i^1)\big] > 0. \tag{4.73}$$

In this case, it can be shown (and is intuitively apparent) that wealth can be redistributed (and/or the costs of the project can be allocated) in such a way as to make each consumer better off after the change than before. Unfortunately, no such inference follows from the fact that:

$$\sum_{i=1}^m EV_i\big[(\boldsymbol{p}^1, w_i^1), (\boldsymbol{p}^2, w_i^2)\big] = \sum_{i=1}^m \big[\mu_i(\boldsymbol{p}^1; \boldsymbol{p}^2, w_i^2) - w_i^1\big] > 0. \tag{4.74}$$

I will leave this discussion at this point, but we will return to a reconsideration of some of the issues taken up in this last paragraph in a later chapter.

4.11 Appendix

Before presenting a proof of Theorem 4.5, we must first introduce a definition and some facts from Topology.

A.1. Definition. We shall say that a family of sets, $\mathcal{G} = \{G_a \mid a \in A\}$, satisfies the **finite intersection property** iff, for each *finite* subset, $\{G_1, \ldots, G_m\}$, of $\mathcal{G}$, we have:

$$\bigcap\nolimits_{i=1}^{m} G_i \neq \emptyset.$$

A.2. Fact (Theorem) from General Topology. *If F is a compact set, and $\mathcal{G} = \{G_a \mid a \in A\}$ is a family of closed (and non-empty) subsets of F which satisfy the finite intersection property, then:*

$$\bigcap_{a \in A} G_a \neq \emptyset;$$

that is, there must exist at least one element of F which is a member of each G_a.

Proof of Theorem 4.5 Let $(\boldsymbol{p}^*, w^*) \in \Omega$; and, for each $\boldsymbol{x} \in b(\boldsymbol{p}^*, w^*)$, define:

$$G\boldsymbol{x} = \{\boldsymbol{y} \in b(\boldsymbol{p}^*, w^*) \mid \neg \boldsymbol{x} P \boldsymbol{y}\} = [X \setminus \boldsymbol{x}P] \cap b(\boldsymbol{p}^*, w^*).$$

Then we note that, since P is upper semi-continuous, $G\boldsymbol{x}$ is a closed subset of $b(\boldsymbol{p}^*, w^*)$, for each $\boldsymbol{x} \in b(\boldsymbol{p}^*, w^*)$.

Now suppose $\boldsymbol{x}_1, \ldots, \boldsymbol{x}_m$ is a finite subset of $b(\boldsymbol{p}^*, w^*)$, and define:

$$i_1 = 1.$$

Then, since P is asymmetric:

$$\neg \boldsymbol{x}_{i_1} P \boldsymbol{x}_{i_1},$$

and therefore $\boldsymbol{x}_{i_1} \in G\boldsymbol{x}_{i_1}$. Consequently, if:

$$\boldsymbol{x}_{i_1} \notin \bigcap\nolimits_{i=1}^{m} G\boldsymbol{x}_i,$$

then there exists $i_2 \in \{1, \ldots, m\} \setminus \{i_1\}$ such that $\boldsymbol{x}_{i_1} \notin G\boldsymbol{x}_{i_2}$; that is:

$$\boldsymbol{x}_{i_2} P \boldsymbol{x}_{i_1}. \tag{4.75}$$

Therefore, since P is asymmetric and irreflexive:

$$\boldsymbol{x}_{i_2} \in G\boldsymbol{x}_{i_1} \cap G\boldsymbol{x}_{i_2};$$

and thus if:

$$\boldsymbol{x}_{i_2} \notin \bigcap\nolimits_{i=1}^{m} G\boldsymbol{x}_i,$$

then there exists $i_3 \in \{1, \ldots, m\} \setminus \{i_1, i_2\}$ such that:

$$\boldsymbol{x}_{i_3} P \boldsymbol{x}_{i_2}. \tag{4.76}$$

It now follows from (4.75), (4.76), and the transitivity of P that we also have:

$$\boldsymbol{x}_{i_3} P \boldsymbol{x}_{i_1}. \tag{4.77}$$

Now, using (4.76), (4.77), and the asymmetry of P, it now follows that:

$$\boldsymbol{x}_{i_3} \in \bigcap_{j=1}^{3} G\boldsymbol{x}_{i_j};$$

and thus, if

$$\boldsymbol{x}_{i_3} \notin \bigcap_{i=1}^{m} G\boldsymbol{x}_i,$$

there exists $i_4 \in \{1, \ldots, m\} \setminus \{i_1, i_2, i_3\}$ such that:

$$\boldsymbol{x}_{i_4} P \boldsymbol{x}_{i_3}.$$

We can now show, as before, that:

$$\boldsymbol{x}_{i_4} P \boldsymbol{x}_{i_j} \quad \text{for } j = 1, 2, 3;$$

and thus, from the definition of the $G\boldsymbol{x}_j$ sets:

$$\boldsymbol{x}_{i_4} \in \bigcap_{j=1}^{4} G\boldsymbol{x}_{i_j}.$$

Continuing in this fashion, we will, after the kth step ($k \geq 2$), have obtained a finite sequence of distinct points:

$$\{\boldsymbol{x}_{i_1}, \ldots, \boldsymbol{x}_{i_k}\} \subseteq \{\boldsymbol{x}_1, \ldots, \boldsymbol{x}_m\},$$

such that:

$$\boldsymbol{x}_{i_k} P \boldsymbol{x}_{i_j} \quad \text{for } j = 1, \ldots, k-1.$$

Moreover, having obtained $\boldsymbol{x}_{i_k}$, it will either be true that:

$$\boldsymbol{x}_{i_k} \in \bigcap_{i=1}^{m} G\boldsymbol{x}_i,$$

or there exists:

$$\boldsymbol{x}_{i_{k+1}} \notin \{\boldsymbol{x}_{i_1}, \ldots, \boldsymbol{x}_{i_k}\} \tag{4.78}$$

such that:

$$\boldsymbol{x}_{i_{k+1}} P \boldsymbol{x}_{i_k}. \tag{4.79}$$

However, it follows from (4.78), (4.79), and the asymmetry of P that this process must terminate after, at most, m steps. Therefore, there exists $j \in \{1, \ldots, m\}$ such that:

$$\boldsymbol{x}_j \in \bigcap_{i=1}^{m} G\boldsymbol{x}_i.$$

To this point, we see that we have shown that the family of sets:

$$\mathcal{G} = \{G\boldsymbol{x} \mid \boldsymbol{x} \in b(\boldsymbol{p}^*, w^*)\},$$

is a family of closed subsets of $b(\boldsymbol{p}^*, w^*)$ which satisfies the finite intersection property. Since we have from Proposition 4.3 of this chapter that $b(\boldsymbol{p}^*, w^*)$ is compact, it therefore follows from Theorem A.2, above, that there exists $\boldsymbol{x}^* \in b(\boldsymbol{p}^*, w^*)$ such that:

$$\boldsymbol{x}^* \in \bigcap_{\boldsymbol{x} \in b(\boldsymbol{p}^*, w^*)} G\boldsymbol{x};$$

and it is then an immediate consequence of the definition of the $G\boldsymbol{x}$ sets that we must have, for all $\boldsymbol{x} \in b(\boldsymbol{p}^*, w^*)$:

$$\neg \boldsymbol{x} P \boldsymbol{x}^*. \quad \square$$

The following is Proposition 4.29. We repeat its statement here for convenient reference, before providing a proof.

Proposition 4.29. *Suppose, in addition to Assumptions I.1 and I.2, that the set* $Z = h(\Omega)$ *is a subset of* $\mathbb{R}^n_+$, *and is closed, convex, and contains* $\mathbf{0}$. *Suppose further that the restriction of the consumer's (weak) preference relation to* Z *is a continuous weak order on* Z. *Then, given any* $\bar{\boldsymbol{p}} \in \mathbb{R}^n_{++}$, *and any* $(\boldsymbol{p}', w') \in \Omega$, *there exists* $\bar{w} \in \mathbb{R}_+$ *such that:*

$$\bar{w} = \min\{w \in \mathbb{R}_+ \mid (\bar{\boldsymbol{p}}, w) G^* (\boldsymbol{p}', w')\},$$

and we have:

$$(\bar{\boldsymbol{p}}, \bar{w}) I^* (\boldsymbol{p}', w').$$

Proof. Let $\bar{\boldsymbol{p}} \in \mathbb{R}^n_{++}$, let $(\boldsymbol{p}', w') \in \Omega$ be arbitrary, and consider:

$$W = \{w \in \mathbb{R}_+ \mid (\bar{\boldsymbol{p}}, w) G^* (\boldsymbol{p}', w')\}.$$

Clearly, W is a non-empty subset of $\mathbb{R}_+$, so that $\inf W$ exists and is nonnegative. Define:

$$\bar{w} = \inf W = \inf\{w \in \mathbb{R}_+ \mid (\bar{\boldsymbol{p}}, w) G^* (\boldsymbol{p}', w')\}.$$

We wish first to prove that $\bar{w} \in W$.

Accordingly, define the sequence $\langle w_q \rangle$ by:

$$w_q = \bar{w} + 1/q \quad \text{for } q = 1, 2, \ldots,$$

and let $\langle \boldsymbol{x}_q \rangle$ be such that:

$$\boldsymbol{x}_q \in h(\bar{\boldsymbol{p}}, w_q) \quad \text{for } q = 1, 2, \ldots.$$

Since $\langle \boldsymbol{x}_q \rangle$ is contained in $b(\bar{\boldsymbol{p}}, w_1) \cap Z = b(\bar{\boldsymbol{p}}, \bar{w} + 1) \cap Z$, which is a compact set, we may assume, without loss of generality, that there exists $\bar{\boldsymbol{x}} \in Z$ such that:

$$\lim_{q \to \infty} \boldsymbol{x}_q = \bar{\boldsymbol{x}}.$$

Moreover, letting $\boldsymbol{x}' \in h(\boldsymbol{p}', w')$, we see that (since $w_q \in W$ for each q):

$$\boldsymbol{x}_q G \boldsymbol{x}' \quad \text{for } q = 1, 2, \ldots.$$

Therefore, since G is continuous on Z and $\boldsymbol{x}_q \to \bar{\boldsymbol{x}}$, we must have:

$$\bar{\boldsymbol{x}} G \boldsymbol{x}'. \tag{4.80}$$

Now let $\boldsymbol{x} \in b(\bar{\boldsymbol{p}}, \bar{w}) \cap Z$ be arbitrary. Then $\boldsymbol{x} \in b(\bar{\boldsymbol{p}}, w_q)$, for each q, and therefore:

$$\boldsymbol{x}_q G \boldsymbol{x} \quad \text{for } q = 1, 2, \ldots.$$

Once again using the fact that G is continuous on Z, it follows that $\bar{\boldsymbol{x}} G \boldsymbol{x}$; and, since $\bar{\boldsymbol{p}} \cdot \bar{\boldsymbol{x}} = \bar{w}$, we now conclude that $\bar{\boldsymbol{x}} \in h(\bar{\boldsymbol{p}}, \bar{w})$. Making use of (4.80), we can also conclude that:

$$(\bar{\boldsymbol{p}}, \bar{w}) G^* (\boldsymbol{p}', w').$$

Now suppose, by way of obtaining a contradiction, that $\bar{\boldsymbol{x}} P \boldsymbol{x}'$. Then, since h satisfies the budget balance condition, we must have $\bar{\boldsymbol{x}} > \mathbf{0}$. Moreover, since G is lower semi-continuous on Z, there exists a neighborhood, $N(\bar{\boldsymbol{x}})$, such that:

$$\big(\forall \boldsymbol{x} \in N(\bar{\boldsymbol{x}}) \cap Z\big) : \boldsymbol{x} P \boldsymbol{x}'.$$

Furthermore, since $\mathbf{0} \in Z$ and Z is convex, there exists $\theta \in]0, 1[$ such that:

$$\theta \bar{\boldsymbol{x}} \in N(\bar{\boldsymbol{x}}) \cap Z,$$

and thus:

$$\theta \bar{\boldsymbol{x}} P \boldsymbol{x}'.$$

However, this cannot be, for then it follows that:

$$\hat{w} \stackrel{\text{def}}{=} \theta \bar{\boldsymbol{p}} \cdot \bar{\boldsymbol{x}} = \theta \bar{w} < \bar{w},$$

is an element of W; which contradicts the definition of $\bar{w}$. We conclude, therefore, that $\neg \bar{\boldsymbol{x}} P \boldsymbol{x}'$; and thus, since $\bar{\boldsymbol{x}} G \boldsymbol{x}'$, it must be the case that $\boldsymbol{x}' G \bar{\boldsymbol{x}}$ as well. Therefore, $\bar{\boldsymbol{x}} I \boldsymbol{x}'$, and we see that $(\bar{\boldsymbol{p}}, \bar{w}) I^* (\boldsymbol{p}', w')$. □

Exercises.

1. Let $X = \mathbb{R}^n_+$, and let G be the (weak) preference relation representable by:

$$u(\boldsymbol{x}) = \prod_{i=1}^{n} x_i.$$

Is G non-decreasing? increasing? strictly increasing on $\mathbb{R}^n_+$? What is the demand function generated by G? Is G homothetic in this case?

2. Suppose a consumer's (direct) preferences are representable by the function:

$$u(\boldsymbol{x}) = \min_i \left\{ \frac{x_i}{a_i} \right\},$$

where $a_i > 0$ for $i = 1, \ldots, n$. Find the consumer's demand function and a cost-of-living function for the consumer.

3. Follow the instructions for problem 3, except take:

$$u(\boldsymbol{x}) = \prod_{j=1}^{n} x_j^{a_j},$$

with:

$$a_j > 0 \ \text{ for } j = 1, \ldots, n, \qquad \text{and} \qquad \sum_{j=1}^{n} a_j = 1.$$

4. Show that if a function $f\colon \mathbb{R}^n_+ \to \mathbb{R}_+$ is positively homogeneous of degree $\theta > 0$, then it is homothetic.

5. Suppose a consumer's (direct) preferences are representable on $\mathbb{R}^n_+$ by the function:

$$u(\boldsymbol{x}) = \prod_{j=1}^{n} (x_j + c_j)^{a_j},$$

where:

$$a_j > 0 \ \text{ for } j = 1, \ldots, n, \qquad \text{and} \qquad \sum_{j=1}^{n} a_j = 1;$$

and $c_j \geq 0$, for $j = 1, \ldots, n$. Show that, defining:

$$\boldsymbol{c} = (c_1, \ldots, c_n),$$

the consumer's demand functions are given by:

$$h_j(\boldsymbol{p}, w) = \frac{a_j w + \boldsymbol{p} \cdot \boldsymbol{c}}{p_j} - c_j,$$

for $j = 1, \ldots, n$.

6. Show that the function $V(\cdot)$ defined in equation (4.61) is a potential function for the integrand function defined in (4.59) and (4.60).

7. Complete the details of the analysis of Example 4.46.

8. Prove part 1 of Proposition 4.24.

9. Show that if a consumer's preferences can be represented by a Cobb-Douglas utility function, as in Problem 4, above, and, as in Example 4.48, only a subset, J, of prices are affected by a policy change, then the sign of the change in indirect utility can be determined if only the values of a_j, for $j \in J$ (and the original and new values for prices and income) are known. In fact, show that in this situation, the indicated information is sufficient to determine the ratio of $V(\boldsymbol{p}^1, w^1)$ to $V(\boldsymbol{p}^0, w^0)$.

10. Complete the proof of Theorem 4.39

11. Given the conditions of Example 4.48, derive the formula for equivalent variation in this case.

12. Suppose once again that the conditions of Example 4.48 hold, but that in this case the consumer's preferences can be represented by the function:

$$u(\boldsymbol{x}) = \min_j \{x_j / a_j\},$$

where $a_j > 0$ for $j = 1, \ldots, n$. What information is needed to determine equivalent and compensating variation in this case?

Chapter 5

Pure Exchange Economies

5.1 Introduction

In this chapter, we will consider general equilibrium models of 'pure exchange.' Such models were justified in the neo-classical literature by the rationale that what one was doing in such a model was analyzing exchange *after* production had taken place. It may well be that it is more properly portayed as the analysis of the aggregate effects of consumer demand. In any event, this basic model is a fundamental tool in public economics, international trade models and welfare economics. The reason for this is quite simple; a surprising number of fundamental economic principles can be illustrated and analyzed in the context of a pure exchange economy. Moreover, we are able to examine these principles in a context simpler than that of a production economy, and we gain a bonus in that our study of pure exchange economies will make it easier to understand the theoretical analysis of a production economy which we undertake in Chapters 7 and 8.

5.2 The Basic Framework

In dealing with pure exchange economies, we will always suppose, unless otherwise explicitly stated, that X_i, the ith consumer's consumption set, is equal to $\mathbb{R}^n_+$. Thus we can think of a (non-private-ownership) exchange economy with m consumers (agents) as being completely specified by an m-tuple of (strict) preference relations and an aggregate commodity, or resource endowment, $\boldsymbol{r}$. Notationally we will indicate this as follows. When we say that:

$$E = \big(\langle P_i \rangle, \boldsymbol{r}\big),$$

is an **exchange economy**, we will mean that P_i is the ith consumer's (strict) preference relation, for $i = 1, \ldots, m$, and that the total commodity bundle available to the economy, collectively, is given by the **aggregate resource endowment**, $\boldsymbol{r} \in \mathbb{R}^n_+$. In dealing with such an economy, we will always suppose (at least) that each P_i is an irreflexive binary relation on $\mathbb{R}^n_+$. Occasionally (primarily in Chapter 11) we may wish to emphasize the number of consumers in the economy, and we will do so by writing:

$$E = (\langle P_i \rangle_{i=1}^m, \boldsymbol{r}),$$

or:

$$E = (\langle P_i \rangle_{i \in M}, \boldsymbol{r}),$$

where we define $M = \{1, \ldots, m\}$.

5.1 Definition. Let $E = \big(\langle P_i \rangle, \boldsymbol{r}\big)$ be an exchange economy. We shall say that an m-sequence, $\langle \boldsymbol{x}_i \rangle_{i \in M}$ is:[1]

1. an **allocation for E** iff:

$$\boldsymbol{x}_i \in \mathbb{R}_+^n \quad \text{for } i = 1, \ldots, m.$$

2. an **attainable** (or **feasible**) **allocation for E** iff $\langle \boldsymbol{x}_i \rangle_{i \in M}$ is an allocation for E satisfying:

$$\sum\nolimits_{i=1}^m \boldsymbol{x}_i = \boldsymbol{r}.$$

We shall denote the set of all attainable, or feasible allocations for E by '$A(E)$.' As in our definition of an economy, $E = \big(\langle P_i \rangle, \boldsymbol{r}\big)$, however, we will generally not need to exhibit the number of consumers, and thus we will denote allocations simply by '$\langle \boldsymbol{x}_i \rangle$', rather than '$\langle \boldsymbol{x}_i \rangle_{i \in M}$.' Thus we define the **set of attainable allocations for E** by:

$$A(E) = \Big\{ \langle \boldsymbol{x}_i \rangle \in \mathbb{R}_+^{mn} \mid \boldsymbol{x}_i \in \mathbb{R}_+^n, \text{ for } i = 1, \ldots, m, \ \& \ \sum\nolimits_{i=1}^m \boldsymbol{x}_i = \boldsymbol{r} \Big\}.$$

When we are considering competitive equilibria for an exchange economy, however, we will need to specify a distribution of ownership for the aggregate resource endowment, $\boldsymbol{r}$; that is, we will deal with private ownership exchange economies, where individual resource endowments are specified for each of the m consumers. Formally, when we say '$\mathcal{E} = \big(\langle P_i, \boldsymbol{r}_i \rangle\big)$ is a **private ownership exchange economy**,' we shall mean that the associated economy:

$$E = \big(\langle P_i \rangle, \sum\nolimits_{i=1}^m \boldsymbol{r}_i\big),$$

is an exchange economy, and we will let 'r_{ij}' denote the ith consumer's initial endowment of the jth commodity. We will refer to $\boldsymbol{r}_i$ as the ith consumer's **endowment** (or **resource endowment**).

In the remainder of this chapter, we will be concerned much of the time with competitive equilibria for a pure exchange economy. It is assumed that, in a competitive exchange economy, consumers take the vector of prices, $\boldsymbol{p} \in \mathbb{R}_+^n$, as given, and choose the best available commodity bundle, given this price vector and their wealth, which will now be given by:

$$w_i = \boldsymbol{p} \cdot \boldsymbol{r}_i \quad \text{for } i = 1, \ldots, m. \tag{5.1}$$

Thus we will assume that the ith consumer chooses that (or a) bundle, $\boldsymbol{x}_i$ satisfying:

$$\boldsymbol{x}_i \in \mathbb{R}_+^n, \ \boldsymbol{p} \cdot \boldsymbol{x}_i \leq w_i \stackrel{\text{def}}{=} \boldsymbol{p} \cdot \boldsymbol{r}_i, \tag{5.2}$$

[1] The notation '$\langle \boldsymbol{x}_i \rangle$' is, of course, intended to suggest a finite sequence.

and:

$$(\forall \boldsymbol{x}_i' \in \mathbb{R}_+^n)\colon \boldsymbol{x}_i' P_i \boldsymbol{x}_i \Rightarrow \boldsymbol{p} \cdot \boldsymbol{x}_i' > w_i; \tag{5.3}$$

so that $\boldsymbol{x}_i \in \boldsymbol{h}_i(\boldsymbol{p}, w_i)$, where $\boldsymbol{h}_i(\cdot)$ is the ith consumer's demand correspondence.

5.2 Definition. Let $\mathcal{E} = \big(\langle P_i, \boldsymbol{r}_i\rangle\big)$ be a private ownership exchange economy. We shall say that an $(m+1)n$-tuple, $\big(\langle \boldsymbol{x}_i\rangle, \boldsymbol{p}\big)$ is a **competitive** (or **Walrasian**) **equilibrium** iff:

1. $\boldsymbol{p} \in \mathbb{R}_+^n \setminus \{\boldsymbol{0}\}$,
2. $\boldsymbol{x}_i \in \boldsymbol{h}_i(\boldsymbol{p}, w_i)$, where $w_i = \boldsymbol{p} \cdot \boldsymbol{r}_i$ for $i = 1, \ldots, m$,,
3. $\langle \boldsymbol{x}_i \rangle$ is an attainable allocation for $\mathcal{E}$.

Often in the general equilibrium literature, an allocation for a pure exchange economy is said to be 'feasible' if:

$$\sum\nolimits_{i=1}^{m} \boldsymbol{x}_i \leq \boldsymbol{r}.$$

Where this definition of feasibility is used, an additional requirement is added to the definition of a competitive, or Walrasian equilibrium; namely that:

$$\boldsymbol{p} \cdot (\boldsymbol{r} - \boldsymbol{x}) = 0,$$

where:

$$\sum\nolimits_{i=1}^{m} \boldsymbol{x}_i \overset{\text{def}}{=} \boldsymbol{x}.$$

We will discuss this alternative definition in Chapter 7.

5.3 The Edgeworth Box Diagram

Surprisingly enough, a great many of the important results in the theory of pure exchange economies can be illustrated quite handily in the context of a two-consumer, two commodity economy; and, thanks to a very clever invention of the economist F. Y. Edgeworth, we can illustrate much of the analysis diagrammatically. The device in question is the so-called 'Edgeworth Box' diagram, and is developed as follows. In the diagram on the next page, we have supposed that the consumers, Ms. 1 and Mr. 2, have the initial endowments, $\boldsymbol{r}_1$ and $\boldsymbol{r}_2$, respectively, and we have then used the parallelogram law of addition to find the aggregate resource endowment, $\boldsymbol{r}$.

In our diagram we would like to graph the set of all possible allocations of commodity bundles between the two consumers; that is, we would like to graph the set:

$$A(\mathcal{E}) = \{\langle \boldsymbol{x}_i \rangle \in \mathbb{R}_+^4 \mid \boldsymbol{x}_1 + \boldsymbol{x}_2 = \boldsymbol{r}\}.$$

Unfortunately, it is a bit difficult to graph a four-dimensional space, especially on a two-dimensional page. However, Edgeworth developed the basic trick which allows us to construct such a graph; we do this by inserting a second set of coordinate axes in our graph. More specifially, we will indicate the commodity bundle available to Ms. 1 in the usual way; reading the quantitites of the two commodities available to her in the usual way on the axes in Figure 5.1. However we will read the quantities available to Mr. 2 on axes oriented to (that is, with the origin at) the aggregate resource endowment, $\boldsymbol{r}$, and reading from right to left for the quantity of the first commodity

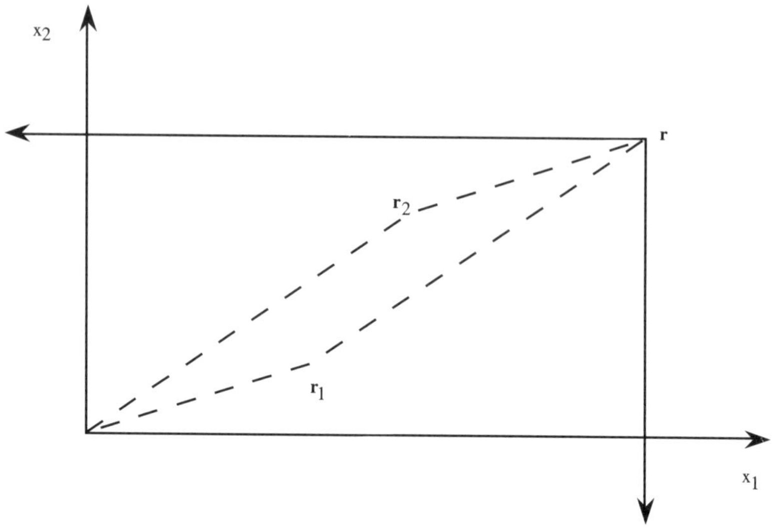

Figure 5.1: The Allocation Space.

(x_{21}) and reading down for the quantity of the second commodity available to Mr. 2 (x_{22}).

Thus, in Figure 5.2, on the next page, the point $\langle \boldsymbol{x}_i^* \rangle$ in the diagram represents an attainable allocation, with quantities as indicated in the diagram; since the quantities of the two commodities going to the two consumers necessarily add up to the totals available in the aggregate resource endowment. Notice also that, using the axes we have constructed for Mr. 2 that his resource endowment will now coincide (reading the quantities along the axes labled 'x_{21}' and 'x_{22}') with $\boldsymbol{r}_1$.

The slightly tricky thing about this sort of diagram is the representation of the consumers' respective indifference maps. Once again there is no particular problem in connection with Ms. 1's indifference map, we can represent it in the usual way; only remembering that Ms. 1's consumption set extends to the 'north' and 'east' of the boundaries of the box. Mr. 2's indifference map may look a bit strange, however, if this is the first time you have enountered an Edgeworth Box diagram, and will, if Mr. 2 has the sort of preference relation favored in textbook diagrams, look something like the curves labled 'I_2,' 'I_2',' and so on, in Figure 5.2. The first thing to keep in mind explaining this graph is that we would expect these indifference curves to be convex to the two axes along which we measure Mr. 2's consumption quantities. Secondly, of course, Mr. 2's prefences will generally increase as we move downward and to the left in our box diagram; and thirdly, we need to keep in mind the fact that Mr. 2's consumption set (and indifference curves) will extend to the south and west beyond the boundaries of the box (as is indicated in Figure 5.2).

Now, "for our next trick," let's see if we can obtain a graphical depiction of a competitive (or Walrasian) equilibrium in such a diagram. Suppose that the price vector $\boldsymbol{p}^*$ prevails in our economy, as indicated in Figure 5.3. Then Ms. 1's budget

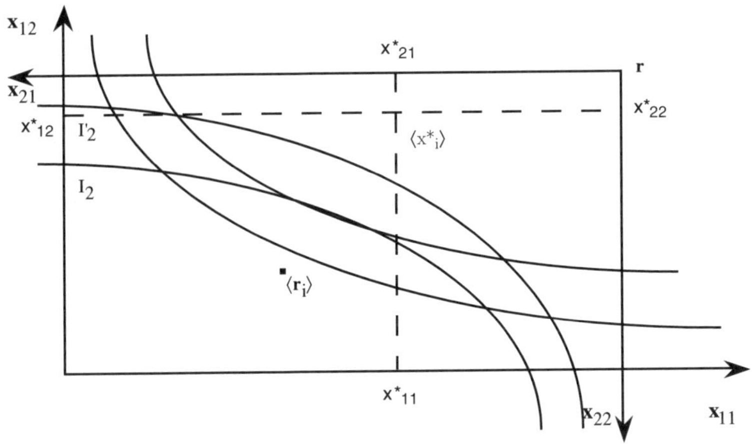

Figure 5.2: The Edgeworth Box.

line will be perpendicular to (the directed line segment) $\boldsymbol{p}^*$ and go through $\boldsymbol{r}_1$.

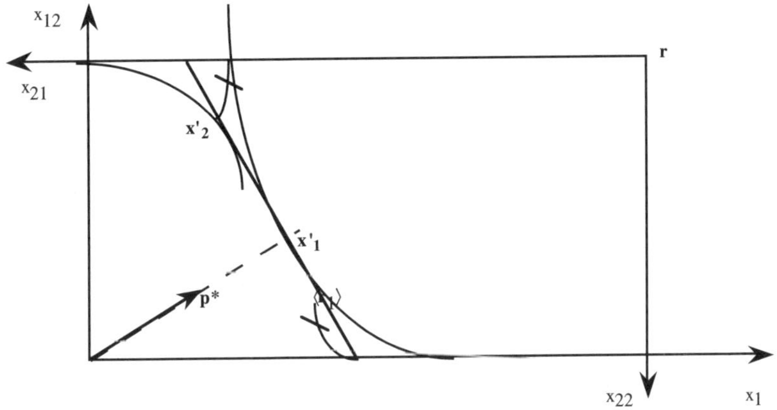

Figure 5.3: A Non-equilibrium Price.

Similarly, 2's budget line will be perpendicular to the price vector $\boldsymbol{p}^*$, and pass through $\boldsymbol{r}_2$. The handy thing about all of this is that if we measure quantities along the new axes we've constructed for Mr. 2, his budget line will coincide with that for Ms. 1. To see this, note first that the line we have constructed for Ms. 1 passes through 2's resource endowment, $\boldsymbol{r}_2$ (when we measure quantities along the new axes). Secondly, recall that the slope of a line is uniquely determined by its angle of incidence with the horizontal axis. Moreover, we know that a transversal between two parallel lines forms equal angles of incidence with the two parallel lines (thus the two angles marked in Figure 5.3 are equal to one another). On the other hand, keep in mind the fact that the consumption bundles chosen by the two consumers need not coincide. Thus, we may have the sort of situation depicted in Figure 5.3, with the bundles demanded by the two consumers denoted by '$\boldsymbol{x}_1'$' and '$\boldsymbol{x}_2'$,' respectively. Notice that we then have excess demand for the first commodity and excess supply of the second.

In order to have a competitive equilibrium, the bundles demanded by the two consumers must then coincide, in order that demands for the two commodities add up to the exact amounts available of the two goods. Thus, with the indifference maps indicated in Figure 5.4, it can easily be seen that both consumers are maximizing preferences, subject to their budget constraint, at $\langle \boldsymbol{x}_i^* \rangle$; which is, therefore, a competitive equilibrium allocation.

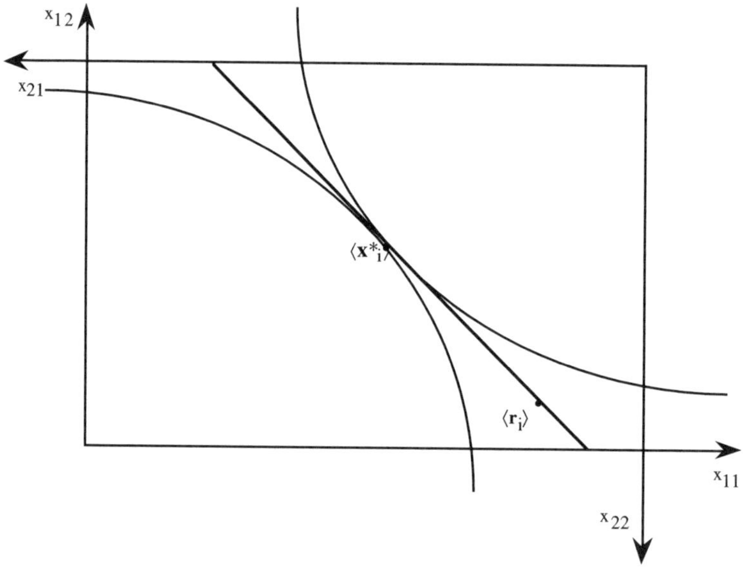

Figure 5.4: A Competitive Equilibrium.

Having examined the graphical depiction of a competitive equilibrium, let's take a look at an algebraic analysis of an example in the two-by-two exchange case. (While we will only go through an analytic solution for this example, you should also try to depict the equilibrium in an Edgeworth Box diagram.)

5.3 Example. Here we will be considering a two-consumer, two-commodity economy, E, in which the consumers' preferences can be represented by the utility functions:

$$u_i(\boldsymbol{x}_1) = x_{11} + x_{12}, \text{ and}$$
$$u_2(\boldsymbol{x}_2) = x_{21}x_{22},$$

respectively. We will consider two situations, and determine whether a competitive equilibrium exists in each case, and if so, what are the competitive allocations and prices. I will list the two specifications, and recommend that you try to determine the answers before reading the analyses which follow.

a. Suppose first that the consumers' initial endowments are given by:

$$\boldsymbol{r}_1 = \boldsymbol{r}_2 = (5, 0).$$

b. Suppose this time that the consumers' initial endowments are given by:

$$\boldsymbol{r}_1 = (16, 4) \text{ and } \boldsymbol{r}_2 = (16, 0),$$

respectively.

Analyses.

a. In this situation, there will be no competitive equilibrium. To see this, notice that if the price of the first good is zero, then consumer one (Ms. 1) will demand an infinite amount of the commodity. On the other hand, if p_1 is positive, then Mr. 2 will have a positive income, and will then demand a positive quantity of good 2, of which there is none available.

b. If there is to be a competitive equilibrium in this situation, then it is clear that the price of the first commodity will have to be positive. Consequently, we can normalize to set $p_1 = 1$. If we then were to have $p_2 < 1$, consumer one would demand more than 4 units of the commodity (consumer one's income/wealth will be equal to $16 + 4p_2$, and given $p_2 < 1 = p_1$, Ms. 1 will demand $16/p_2 + 4$ units of commodity 2).

Now, since comsumer 2 has a Cobb-Douglas utility function, Mr. 2's demand for the second commodity is given by:

$$x_{22} = \frac{\boldsymbol{p} \cdot \boldsymbol{r}_2}{2p_2} = \frac{8}{p_2}.$$

Consequently, if $p_2 = 1$, Mr. 2 will demand 8 units of the second commodity, of which there are only 4 units available. Therefore, we see that if an equilibrium exists, we must have $p_2 > 1$; in which case, Ms. 1 will demand only commodity one, and equilibrium will thus require that:

$$x_{22} = 8/p_2 = 4;$$

that is, $p_2 = 2$.

If we check this out, we see that with $\boldsymbol{p} = (1, 2)$, Mr. 2 demands 8 units of commodity one and 4 units of commodity two. Furthermore, Ms. 1, with linear preferences, will spend all of her income on commodiy one. So, Ms. 1's income is:

$$p_1 \cdot 16 + p_2 \cdot 4 = 16 + 8 = 24,$$

and thus Ms. 1's demand for commodity one is:

$$x_{11} = 24/p_1 = 24.$$

Adding, we then see that demand equals supply for each commodity, and therefore we do have a competitive equilibrium in this case. □

5.4 Demand and Excess Demand Correspondences

In the context of a pure exchange economy, and given a price vector $\boldsymbol{p} \in \mathbb{R}^n_{++}$, a price-taking consumer will choose a commodity bundle, $\boldsymbol{x}_i$, satisfying:

$$\boldsymbol{x}_i \in \boldsymbol{h}_i(\boldsymbol{p}, \boldsymbol{p} \cdot \boldsymbol{r}_i) \stackrel{\text{def}}{=} \boldsymbol{d}_i(\boldsymbol{p}). \tag{5.4}$$

Making use of this, we define the following.

5.4 Definition. Given the ith consumer's demand correspondence, as defined in equation (5.4), we define the ith consumer's **excess demand correspondence**, $\boldsymbol{e}_i \colon \mathbb{R}^n_{++} \mapsto \mathbb{R}^n$, defined by:

$$\boldsymbol{e}_i(\boldsymbol{p}) = \boldsymbol{d}_i(\boldsymbol{p}) - \boldsymbol{r}_i. \tag{5.5}$$

5.5 Definitions. If $\mathcal{E} = \big(\langle P_i, \boldsymbol{r}_i \rangle\big)$ is a private ownership exchange economy, we define the **aggregate demand correspondence**, $\boldsymbol{d}(\cdot)$, for $\mathcal{E}$, by:

$$\boldsymbol{d}(\boldsymbol{p}) = \sum\nolimits_{i=1}^{m} \boldsymbol{d}_i(\boldsymbol{p}), \tag{5.6}$$

and the **aggregate excess demand correspondence** for $\mathcal{E}$, $\boldsymbol{e}(\cdot)$, by:

$$\boldsymbol{e}(\boldsymbol{p}) = \sum\nolimits_{i=1}^{m} [\boldsymbol{d}_i(\boldsymbol{p}) - \boldsymbol{r}_i] = \sum\nolimits_{i=1}^{m} \boldsymbol{d}_i(\boldsymbol{p}) - \sum\nolimits_{i=1}^{m} \boldsymbol{r}_i = \boldsymbol{d}(\boldsymbol{p}) - \boldsymbol{r}. \tag{5.7}$$

The proof of the following facts will be left as an exercise.

Facts regarding the aggregate demand correspondence:

1. The aggregate demand correspondence, $\boldsymbol{d}(\cdot)$, will be positively homogeneous of degree zero in $\boldsymbol{p}$,

2. The aggregate demand correspondence, $\boldsymbol{d}(\cdot)$, will satisfy: for all $\boldsymbol{p} \in \mathbb{R}^n_{++}$:

$$\big(\forall \boldsymbol{x} \in \boldsymbol{d}(\boldsymbol{p})\big) \colon \boldsymbol{p} \cdot \boldsymbol{x} \leq \boldsymbol{p} \cdot \boldsymbol{r},$$

3. The price vector $\boldsymbol{p} \in \mathbb{R}^n_{++}$ defines a competitive equilibrium for $\mathcal{E}$ if, and only if, there exists $\boldsymbol{x} \in \boldsymbol{d}(\boldsymbol{p})$ satisfying:

$$\boldsymbol{x} = \boldsymbol{r}.$$

Making use of the definition of a consumer's excess demand correspondence, you can easily prove the following, very fundamental results.

5.6 Proposition. *Let $\mathcal{E} = \big(\langle P_i, \boldsymbol{r}_i \rangle\big)$ be a private ownership economy. Then for each i, each $\boldsymbol{p} \in \mathbb{R}^n_{++}$, and each $\boldsymbol{z}_i \in \boldsymbol{e}_i(\boldsymbol{p})$:*

$$\boldsymbol{p} \cdot \boldsymbol{z}_i \leq 0. \tag{5.8}$$

Furthermore, if P_i is locally non-saturating, then we will have:

$$\big(\forall \boldsymbol{p} \in \mathbb{R}^n_{++}\big)\big(\forall \boldsymbol{z}_i \in \boldsymbol{e}_i(\boldsymbol{p})\big) \colon \boldsymbol{p} \cdot \boldsymbol{z}_i = 0. \tag{5.9}$$

5.7 Proposition. **[Walras' Law (Weak Form)]** *Let* $\mathcal{E} = (\langle P_i, \boldsymbol{r}_i \rangle)$ *be a private ownership exchange economy, and* $\boldsymbol{e}\colon \mathbb{R}^n_{++} \mapsto \mathbb{R}^n$ *be the aggregate excess demand correspondence for* $\mathcal{E}$. *Then, given any* $\boldsymbol{p} \in \mathbb{R}^n_{++}$, *and any* $\boldsymbol{z} \in \boldsymbol{e}(\boldsymbol{p})$, *we have:*

$$\boldsymbol{p} \cdot \boldsymbol{z} \leq 0. \tag{5.10}$$

5.8 Corollary. **[Walras' Law (Strong Form)]** *If, in addition to the other hypotheses of 5.7 we have:*

$$P_i \text{ is locally nonsaturating, for } i = 1, \ldots, m,$$

then, given any $\boldsymbol{p} \in \mathbb{R}^n_{++}$, *and any* $\boldsymbol{z} \in \boldsymbol{e}(\boldsymbol{p})$:

$$\boldsymbol{p} \cdot \boldsymbol{z} = 0.$$

In our next result, we will say that the $\boldsymbol{j^{th}}$ **market is in equilibrium**, given $\boldsymbol{p} \in \mathbb{R}^n_{++}$ and $\boldsymbol{z} \in \boldsymbol{e}(\boldsymbol{p})$, iff $z_j = 0$.

5.9 Corollary. **[Walras' Law (Original Form)]** *Let* $\mathcal{E} = (\langle P_i, \boldsymbol{r}_i \rangle)$ *be a private ownership exchange economy in which* P_i *is locally non-saturating, for* $i = 1, \ldots, m$; *and suppose that* $\boldsymbol{p}^* \in \mathbb{R}^n_{++}$ *and* $\boldsymbol{z} \in \boldsymbol{e}(\boldsymbol{p}^*)$ *are such that* $n-1$ *of the* n *markets are in equilibrium. Then the* n^{th} *market must be in equilibrium as well.*

Proof. Suppose that for all $j \neq k$, the jth market is in equilibrium. Then by Corollary 5.8, above, we have that:

$$0 = \boldsymbol{p} \cdot \boldsymbol{z} = \sum_{j \neq k} p_j z_j + p_k z_k. \tag{5.11}$$

However, by assumption we have:

$$\sum_{j \neq k} p_j z_j = 0;$$

and thus it follows from (5.11) that $p_k z_k = 0$. Since $p_k > 0$, it now follows that $z_k = 0$. □

Walras' Law is a very useful property of the aggregate excess demand correspondence, and we have shown that it holds under quite general conditions. Moreover, it is also true that this correspondence will be positively homogeneous of degree zero, and because of this, we will often find it convenient to suppose that the domain of the correspondence is Δ_n (as will be done in Theorem 5.9, below). We will later study conditions sufficient to imply that the aggregate excess demand correspondence satisfies sufficiently strong continuity properties as to enable one to prove the existence of a competitive equilibrium. In the meantime, let's consider some additional examples of competitive equilibrium in a pure exchange economy.

5.10 Examples.

1.Once again we consider a 2-person, 2-commodity economy; this time supposing that the ith consumer's preferences can be represented by the utility function:

$$u_i(\boldsymbol{x}_i) = (x_{i1})^{a_i} \cdot (x_{i2})^{1-a_i},$$

with $0 < a_i < 1$, for $i = 1, 2$; and suppose the consumers' initial endowments are given by:

$$\boldsymbol{r}_1 = (r_{11}, 0) \text{ and } \boldsymbol{r}_2 = (0, r_{22}),$$

where:

$$r_{11} > 0 \text{ and } r_{22} > 0.$$

Show that, if we normalize to set $p_2 = 1$, we can find equilibrium p_1 as a function of a_{11}, a_{21}, r_{11} and r_{22}, and that:

$$\frac{\partial p_1}{\partial r_{11}} = -\frac{a_2 r_{22}}{(1 - a_1)(r_{11})^2} \text{ and } \frac{\partial p_1}{\partial r_{22}} = \frac{a_2}{(1 - a_1)(r_{11})}.$$

(And thus $\partial p_1/\partial r_{11} < 0$ and $\partial p_1/\partial r_{22} > 0$.)

Analysis.

Recalling Walras Law (original form), we see that it suffices to find equlibrium in the market for the first commodity. With the price of the second commodity normalized to $p_2 = 1$, and with the given initial endowments, the demands of the two consumers for the first commodity are given by:

$$x_{11} = \frac{a_1 p_1 r_{11}}{p_1} \text{ and } x_{21} = \frac{a_2 r_{22}}{p_1},$$

respectively. Equilibrium in the first market thus requires:

$$\frac{a_1 p_1 r_{11}}{p_1} + \frac{a_2 r_{22}}{p_1} = r_{11}.$$

Solving, we then obtain:

$$p_1 = \frac{a_2 r_{22}}{r_{11}(1 - a_1)};$$

which, when differentiated, yields the indicated values of the partial derivatives.

2. This time we generalize the last example to consider m consumers with initial endowments $\boldsymbol{r}_i$, along with the utility functions used in the last example:

$$u_i(\boldsymbol{x}_i) = (x_{i1})^{a_i} \cdot (x_{i2})^{1-a_i} \quad \text{for } i = 1, \ldots, m.$$

Show that, once again normalizing to set $p_2 = 1$, the equilibrium price for the first commodity is given by:

$$p_1 = \frac{\sum_{i=1}^m a_i r_{i2}}{\sum_{i=1}^m (1 - a_i) r_{i1}}. \quad \square$$

The facts concerning aggregate demand which we noted earlier raise an interesting question, namely: are there further qualitative conditions of aggregate excess demand correspondences which hold under the kinds of assumptions we have been making here and in the previous chapter. Unfortunately, H. Sonnenschein [1973 and 1974] established results which pretty much showed that there are no other qualitative implications for market (aggregate) demand which follow from the standard assumptions about individual preferences. Sonnenschein's original results have been extended and refined in various ways; but our next result is what Shafer and Sonnenschein refer to in their 1982 survey as a 'state of the art' result, and is due to Debreu [1974]. For a proof, consult the original article, or the Shafer and Sonnenschein survey, where a somewhat simpler argument than Debreu's is presented. In the Debreu result, 'Δ_ϵ' will be used to denote that portion of Δ_n satisfying:

$$p_j > \epsilon \quad \text{for } j = 1, \dots, n.$$

5.11 Theorem. (Debreu). *Let $F\colon \Delta_n \to \mathbb{R}^n_+$ be a continuous function satisfying:*

$$(\forall \boldsymbol{p} \in \Delta_n)\colon \boldsymbol{p} \cdot F(\boldsymbol{p}) = 0.$$

Then, for any $\epsilon \in \,]0, 1/n\,[$, there exists an n-consumer exchange economy, $\mathcal{E} = (\langle P_i, \boldsymbol{r}_i \rangle)$, such that each P_i is asymmetric, negatively transitive, continuous, strictly convex, and increasing, and such that F is the aggregate excess demand function for $\mathcal{E}$ on Δ_ϵ.

While the Debreu result stresses the limitations on the qualitative properties of aggregate demand correspondences in general, the structure of a general equilibrium model (in particular, of a pure exchange model) places functional restrictions on incomes. More exactly, in a pure exchange economy, individual wealth is determined solely by prices (given resource endowments), and thus incomes cannot vary independently. Because of this, one can in some cases establish stronger properties for aggregate demand correspondences. Thus, for example, the following result follows from Theorem 4.2 and Example 5.2 of Chipman and Moore [1979].

5.12 Theorem. *Let $\mathcal{E} = (\langle P_i, \boldsymbol{r}_i \rangle)$ be a private ownership exchange economy in which P_i is an asymmetric and negatively transitive binary relation which is:*

1. *continuous,*
2. *non-decreasing, and*
3. *homothetic*

for $i = 1, \dots, m$; and suppose $\boldsymbol{\delta} \in \Delta_m$ and $\boldsymbol{r} \in \mathbb{R}^n_+$ are such that:

$$\boldsymbol{r}_i = \delta_i \boldsymbol{r} \quad \textit{for } i = 1, \dots, m. \tag{5.12}$$

Then the aggregate demand correspondence for the economy is that generated by the utility function, U, given by:

$$U(\boldsymbol{x}) = \max\left\{ \prod_{i=1}^{m} [u_i(\boldsymbol{x}_i)]^{\delta_i} \mid \sum_{i=1}^{m} \boldsymbol{x}_i \leq \boldsymbol{x} \right\}; \tag{5.13}$$

where u_i is any positively homogeneous of degree one utility function representing P_i, for $i = 1, \dots, m$.

5.13 Example. Return to the situation described in Example 5.10.2, except assume now that:

$$\boldsymbol{r}_i = \delta_i \boldsymbol{r} \quad \text{for } i = 1, \dots, m,$$

where:

$$\delta_i \geq 0, \text{ for } i = 1, \dots, m, \ \sum_{i=1}^{m} \delta_i = 1, \tag{5.14}$$

and $\boldsymbol{r} \in \boldsymbol{R}^2_{++}$ is the aggregate resource endowment. Show that in this case, aggregate demand for the first commodity is given by:

$$\sum_{i=1}^m \frac{a_i(\delta_i p_1 r_1 + \delta_i r_2)}{p_1} = \left(\sum_{i=1}^m a_i\delta_i\right)\frac{p_1 r_1 + r_2}{p_1};$$

where we have set $p_2 = 1$. Compare this with the demand function of a single consumer having the resource endowment $\boldsymbol{r}$, and the utility function:

$$u(\boldsymbol{x}) = (x_1)^{\bar{a}} \cdot (x_2)^{1-\bar{a}},$$

where:

$$\bar{a} = \sum_{i=1}^m a_i\delta_i.$$

Notice also that, since $0 < a_i < 1$ for each i, we have:

$$\sum_{i=1}^m a_i\delta_i < \sum_{i=1}^m \delta_i = 1,$$

where the inequality is by (5.14) □

5.5 Pareto Efficiency

In this section, we will be studying some orderings which are intended to be candidates for a 'universally acceptable' criterion for economic improvement for an economy as a whole. Since our concern here is with normative criteria for economic improvement, we will want to abstract from ownership in most of our present considerations, and deal with 'exchange economies' (as opposed to 'private ownership exchange economies').

5.14 Definitions. Let $E = (\langle P_i\rangle, \boldsymbol{r})$ be an exchange economy. We then define:

a. **the unanimity ordering (the strong Pareto ordering)**, $\boldsymbol{Q}$, on $\boldsymbol{X} = \prod_{i=1}^m X_i$ by:

$$\langle \boldsymbol{x}_i\rangle \boldsymbol{Q} \langle \boldsymbol{x}'_i\rangle \iff [\boldsymbol{x}_i P_i \boldsymbol{x}'_i \quad \text{for } i = 1, \ldots, m]. \tag{5.15}$$

b. **the Pareto (at-least-as-good-as) ordering**, $\boldsymbol{R}$, on $\boldsymbol{X}$ by:

$$\langle \boldsymbol{x}_i\rangle \boldsymbol{R} \langle \boldsymbol{x}'_i\rangle \iff [\boldsymbol{x}_i G_i \boldsymbol{x}'_i \quad \text{for } i = 1, \ldots, m]. \tag{5.16}$$

c. **the strict Pareto ordering**, $\boldsymbol{P}$, on $\boldsymbol{X}$ by:

$$\langle \boldsymbol{x}_i\rangle \boldsymbol{P} \langle \boldsymbol{x}'_i\rangle \iff [(\langle \boldsymbol{x}_i\rangle \boldsymbol{R} \langle \boldsymbol{x}'_i\rangle \text{ and } \neg\langle \boldsymbol{x}'_i\rangle \boldsymbol{R} \langle \boldsymbol{x}_i\rangle]. \tag{5.17}$$

In dealing with these three orderings, we will use the following terminology.

If $\langle \boldsymbol{x}_i\rangle \boldsymbol{Q} \langle \boldsymbol{x}'_i\rangle$, we will say that $(\boldsymbol{x}_i)$ is **unanimously preferred to** $(\boldsymbol{x}'_i)$.

If $\langle \boldsymbol{x}_i\rangle \boldsymbol{R} \langle \boldsymbol{x}'_i\rangle$, we will say that $(\boldsymbol{x}_i)$ is **Pareto non-inferior to** $\langle \boldsymbol{x}'_i\rangle$ (or that $\langle \boldsymbol{x}'_i\rangle$ is **no better than** $\langle \boldsymbol{x}_i\rangle$ **in the Pareto sense**).

If $\langle \boldsymbol{x}_i\rangle \boldsymbol{P} \langle \boldsymbol{x}'_i\rangle$, we will say that $\langle \boldsymbol{x}_i\rangle$ is **Pareto superior to** $\langle \boldsymbol{x}'_i\rangle$ (or that $\langle \boldsymbol{x}_i\rangle$ **Pareto dominates** $\langle \boldsymbol{x}'_i\rangle$).

5.15 Proposition. *If each P_i is an asymmetric order, then $\boldsymbol{Q}$, as defined in 5.14.a, above, is also an asymmetric order; and $\boldsymbol{R}$, as defined in 5.14.b, is reflexive. If each P_i is negatively transitive, then $\boldsymbol{R}$ is also transitive; and its asymmetric part, $\boldsymbol{P}$, is transitive as well (in addition to being asymmetric). [Neither $\boldsymbol{Q}$ nor $\boldsymbol{P}$ will generally be negatively transitive, however.]*

Proof. I will prove that if each P_i is negatively transitive, then P is transitive (in fact, I will prove a slightly stronger statement, as you will see). I will leave the remainder of the proof as an exercise (see also Exercise 7); although we will look at an example to show that neither $\boldsymbol{Q}$ nor $\boldsymbol{P}$ is necessarily negatively transitive following this proof.

Suppose $\langle \boldsymbol{x}_i \rangle$, $\langle \boldsymbol{x}_i^* \rangle$, and $\langle \boldsymbol{x}_i' \rangle$ are such that:

$$\langle \boldsymbol{x}_i \rangle \boldsymbol{R} \langle \boldsymbol{x}_i^* \rangle \text{ and } \langle \boldsymbol{x}_i^* \rangle \boldsymbol{P} \langle \boldsymbol{x}_i' \rangle. \tag{5.18}$$

Then from the first relation in (5.18), we have:

$$\boldsymbol{x}_i G_i \boldsymbol{x}_i^* \quad \text{for } i = 1, \ldots, m, \tag{5.19}$$

where G_i is the negation of P_i; while from the second relationship:

$$\boldsymbol{x}_i^* G_i \boldsymbol{x}_i' \quad \text{for } i = 1, \ldots, m, \tag{5.20}$$

and, for some $h \in \{1, \ldots, m\}$, we have:

$$\boldsymbol{x}_h^* P_h \boldsymbol{x}_h'. \tag{5.21}$$

But then from (5.19), (5.20), and the fact that each P_i is negatively transitive:

$$\boldsymbol{x}_i G_i \boldsymbol{x}_i' \quad \text{for } i =, \ldots, m;$$

while from (5.19) and (5.21):

$$\boldsymbol{x}_h P_h \boldsymbol{x}_i'.$$

Therefore $\langle \boldsymbol{x}_i \rangle \boldsymbol{P} \langle \boldsymbol{x}_i' \rangle$. □

Generally economists have proceeded as if the following is a universally acceptable value judgment

A.1. If $\langle \boldsymbol{x}_i \rangle \boldsymbol{Q} \langle \boldsymbol{x}_i' \rangle$, for two feasible allocations, $\langle \boldsymbol{x}_i \rangle$ and $\langle \boldsymbol{x}_i' \rangle$, then society should choose $\langle \boldsymbol{x}_i \rangle$ over $\langle \boldsymbol{x}_i' \rangle$.

We will refer to the acceptance of A.1, above, as a criterion for economic improvement as the '**unanimity principle**;' and to the acceptance of the corresponding statement when the strict Pareto criterion, $\boldsymbol{P}$, is substituted for $\boldsymbol{Q}$ in A.1 as the '**Pareto principle**.'

Earlier I emphasized the fact that in dealing with welfare economics, we would make every effort to develop as much of the material as possible assuming only that consumers' strict preference relations were asymmetric orders (and not necessarily negatively transitive). It is probably already apparent that under these assumptions the strict Pareto ordering defined in 5.14.c will be of somewhat dubious value. The following example illustrates the problem.

5.16 Example. Consider the exchange economy with 3 consumers ($m = 3$), 2 commodities ($n = 2$), and individualistic preferences, and the allocations:

$$\begin{aligned}\langle \boldsymbol{x}_i^1 \rangle &= \big((4,1),(5,1),(6,1)\big),\\ \langle \boldsymbol{x}_i^2 \rangle &= \big((1/10,50),(3,2),(2,2)\big),\\ \langle \boldsymbol{x}_i^3 \rangle &= \big((120,1/20),(40,1/10),(100,1/20)\big).\end{aligned} \tag{5.22}$$

Clearly, it may be difficult for the consumers to make comparisons between these, quite widely dispersed, commodity bundles. Suppose, in fact that the consumers' preferences are defined by:

$$\boldsymbol{x}_i P_i \boldsymbol{x}_i' \iff u(\boldsymbol{x}_i) > u(\boldsymbol{x}_i') + 1 \quad \text{for } i = 1,2,3; \tag{5.23}$$

where:

$$u(\boldsymbol{x}_i) = x_{i1} \cdot x_{i2}. \tag{5.24}$$

If we then denote the values of this 'utility function' at the three allocations by:

$$\boldsymbol{u}(\langle \boldsymbol{x}_i^t \rangle) \stackrel{\text{def}}{=} \big(u(\boldsymbol{x}_1^t), u(\boldsymbol{x}_2^t), u(\boldsymbol{x}_3^t)\big) \quad \text{for } t = 1,2,3, \tag{5.25}$$

we have:

$$\begin{aligned}\boldsymbol{u}(\langle \boldsymbol{x}_i^1 \rangle) &= (4,5,6),\\ \boldsymbol{u}(\langle \boldsymbol{x}_i^2 \rangle) &= (5,6,4),\\ \boldsymbol{u}(\langle \boldsymbol{x}_i^3 \rangle) &= (6,4,5).\end{aligned} \tag{5.26}$$

Consequently, according to the strict Pareto ordering defined in Definition 5.14, we have:

$$\langle \boldsymbol{x}_i^1 \rangle P \langle \boldsymbol{x}_i^2 \rangle \ \& \ \langle \boldsymbol{x}_i^2 \rangle P \langle \boldsymbol{x}_i^3 \rangle \ \& \ \langle \boldsymbol{x}_i^3 \rangle P \langle \boldsymbol{x}_i^1 \rangle; \tag{5.27}$$

in other words, the strict Pareto relation is cyclic (Definition 3.44) in this case. □

5.17 Definition. Let $E = \big(\langle P_i \rangle, \boldsymbol{r}\big)$ be an exchange economy. We shall say that a feasible allocation for E, $\langle x_i^* \rangle$, is **Pareto efficient for E** (respectively, **strongly Pareto efficient for E**) iff there exists no alternative feasible allocation for E, $\langle \boldsymbol{x}_i \rangle$, satisfying:

$$\langle \boldsymbol{x}_i \rangle \boldsymbol{Q} \langle \boldsymbol{x}_i^* \rangle \text{ [respectively, } \langle \boldsymbol{x}_i \rangle \boldsymbol{P} \langle \boldsymbol{x}_i^* \rangle],$$

or alternatively:

$$\langle \boldsymbol{x}_i \rangle Q \langle \boldsymbol{x}_i^* \rangle \Rightarrow \langle \boldsymbol{x}_i \rangle \notin \mathcal{X}^*(E);$$

where the orderings $\boldsymbol{Q}$ and $\boldsymbol{P}$ are the unanimity and strict Pareto orderings, respectively, and '$\mathcal{X}^*(E)$' denotes the set of feasible consumption allocations for E.

The favorite textbook picture of Pareto efficiency in an Edgeworth Box diagram tends to look like Figure 5.5, on the next page. In the diagram, the heavy curve running from the southwest corner to the northeast corner of the box is (usually) called the contract curve,[2] and is the locus of the Pareto efficient allocations for the

[2]We will use a somewhat more restrictive definition of the contract curve in Chapter 11.

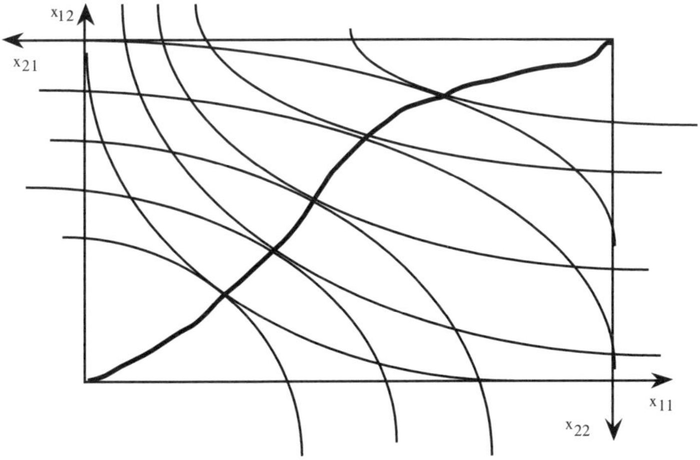

Figure 5.5: The Contract Curve.

economy. Notice that it is the locus of the tangency points of the indifference curves of the two consumers. We can show that this is necessarily the case in two ways.

First we'll consider a mathematical development. We can characterize a Pareto efficient allocation (for the two-consumer, two-commodity exchange case) as the solution of the problem:

$$\max_{w.r.t.x_{11},x_{12}} u_1(x_{11}, x_{12}), \tag{5.28}$$

subject to:

$$u_2(x_{21}, x_{22}) \geq u^*, \tag{5.29}$$

and:

$$\boldsymbol{x}_1 + \boldsymbol{x}_2 \leq \boldsymbol{r}, \tag{5.30}$$

for some feasible value of u^*.[3] If we assume that these utility functions are increasing, as well as being differentiable, then we can simplify things a bit because we can then replace the inequalities in the two constraints [equations (5.29) and (5.30)] by equalities. This in turn will enable us to use the classical Lagrangian multiplier method to derive the necessary conditions for an interior solution (that is, a solution in which both consumers receive positive quantities of both commodities). I will then leave it to you to verify the fact that the necessary conditions imply that at an interior Pareto efficient allocation, we must have:

$$\frac{\partial u_1}{\partial x_{11}} \Big/ \frac{\partial u_1}{\partial x_{12}} = \frac{\partial u_2}{\partial x_{21}} \Big/ \frac{\partial u_2}{\partial x_{22}}. \tag{5.31}$$

Which verifies the fact that the slopes of the two individuals' indifference curves have to be equal at a Pareto (actually strongly Pareto) efficient allocation in this case.

[3]Insofar as the derivation of necessary conditions are concerned, the numerical value of u^* is not important. Consequently, for this demonstration, we need not worry about which values of u^* are 'feasible.'

Moreover, if u_1 and u_2 are both quasi-concave, then any (interior) allocation at which (5.31) is satisfied is strongly Pareto efficient, given the assumptions we have been making here.

Our second development is geometric, and in a sense is much more general than the mathematical development which we have just presented; although we will be considering the solution to the same problem as before, that is, the problem set out in equations (5.28)–(5.30), above. Moreover, once again we will assume that the consumers' utility functions are increasing, so that we can replace the inequalities in the two constraints by equalities. Now, the question is, how do we handle these constraints in our geometric development?

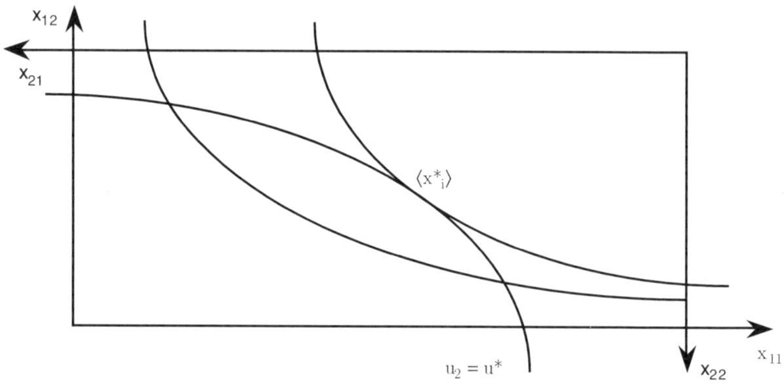

Figure 5.6: An Interior Pareto Efficient Allocation.

Considering the constraints in reverse order, we can see that the second constraint is automatically satisfied within the confines of the Edgeworth Box, for every point in the box satisfies this constraint. Therefore we can concentrate our attention upon the problem of maximizing u_1 on the set of points satisfying the first constraint; maximizing 1's utility on 2's indifference curve for the value $u_2 = u^*$. Thus, it is easy to see that the allocation $\langle \boldsymbol{x}_i^* \rangle$ in Figure 5.6 is Pareto efficient.

In Figure 5.6 we have shown for a second time that an interior Pareto efficient allocation occurs at a point of tangency of the two individuals' indifference curves. What happens, however, if the utility functions are not differentiable? It is in its ability to handle this contingency that the geometric method of analysis is more general (in a sense) than the mathematical development which we went through earlier. For example, in Figure 5.7 we have illustrated a case in which both individuals have Leontief-type utility functions, which are, of course, not differentiable. However, by concentrating our attention on the problem of maximizing u_1 subject to being on 2's indifference curve for $u_2 = u^*$, it is easy to see that any allocation on the heavily-shaded line segment between $\langle \boldsymbol{x}_i \rangle$ and $\langle \boldsymbol{x}_i^* \rangle$ is Pareto efficient.

In the following examples, we present two illustrative cases in which there are allocations which are Pareto efficient for E, but which are not strongly Pareto efficient.

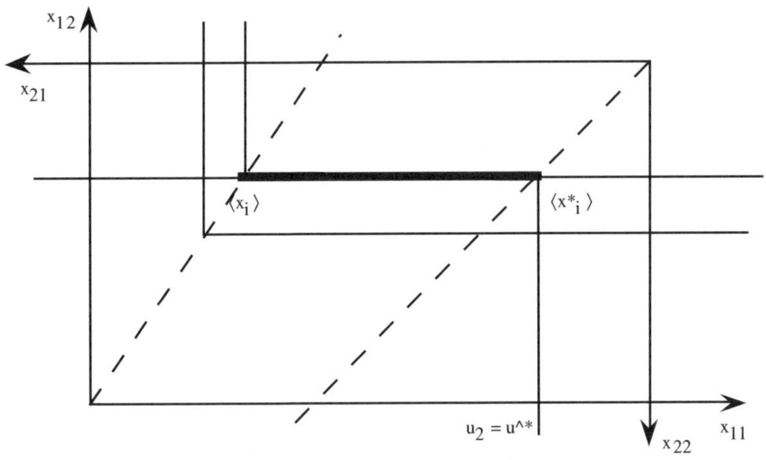

Figure 5.7: Pareto Efficiency in the Leontief Case.

5.18 Examples.

1. Define:

$$Z = \{\boldsymbol{x} \in \mathbb{R}^2 \mid x_j \text{ is an integer, for } j = 1, 2\};$$

and let:

$$X_i = \mathbb{R}^2_+ \cap Z \quad \text{for } i = 1, 2;$$

that is, X_i is the set of all vectors in $\mathbb{R}^2$ having each coordinate a nonnegative integer, for $i = 1, 2$. We also suppose that the initial endowments of the two consumers are given by:

$$\boldsymbol{r}_1 = (1, 0) \text{ and } \boldsymbol{r}_2 = (1, 1);$$

and that the two consumers have preference relations which can be represented on:

$$X_i^* \stackrel{\text{def}}{=} \{\boldsymbol{x}_i \in X_i \mid \boldsymbol{0} \leq \boldsymbol{x}_i \leq \boldsymbol{r}\}$$

by the functions f_1 and f_2 defined in the table below.

$\boldsymbol{x}_i$	f_1	f_2
$(0, 0)$	1	1
$(1, 0)$	3	3
$(0, 1)$	3	2
$(1, 1)$	5	4
$(2, 0)$	5	5
$(2, 1)$	6	6

We can show in this case that the two allocations $\big((1, 0), (1, 1)\big)$ and $\big((2, 0), (0, 1)\big)$ are both Pareto efficient, but are *not* strongly Pareto efficient.

2. Consider the same example as before, except that the preference relations P_i can be represented by the functions f_1 and f_2 set out in the following.

$\boldsymbol{x}_i$	f_1	f_2
$(0,0)$	1	1
$(1,0)$	2	3
$(0,1)$	4	2
$(1,1)$	5	4
$(2,0)$	3	5
$(2,1)$	6	6

In this example, all of the Pareto efficient allocations will be strongly Pareto efficient.

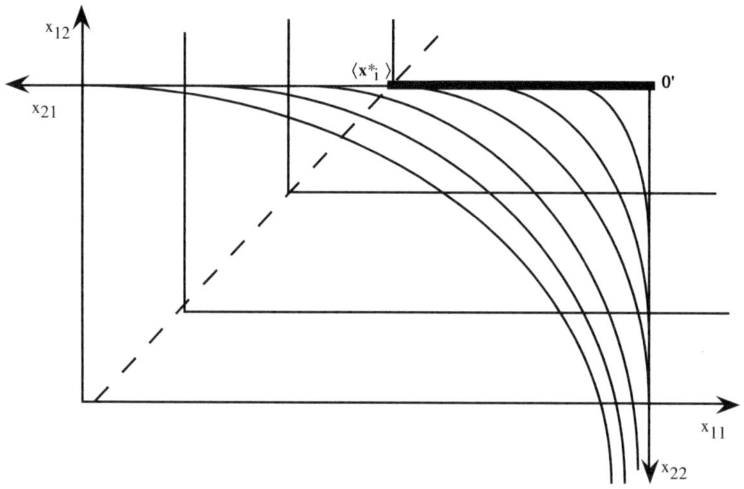

Figure 5.8: Pareto Efficiency without Strong Pareto Efficiency.

3. Lest you get the idea that the divergence between Pareto efficiency and strong Pareto efficiency can only occur when consumers have discrete consumption sets (commodities only available in discrete quantities), consider the example presented in Figure 5.8, above. In this example, we have a classic Edgeworth Box case with both consumers having increasing utility functions. However, you should have no difficulty in verifying the fact that all of the allocations lying on the open line segment connecting $\langle \boldsymbol{x}_i^* \rangle$ and $0'$ are Pareto efficient, but not strongly Pareto efficient. □

The significance of our next result resides in the fact that it provides suffient conditions for Pareto efficiency to imply strong Pareto efficiency.

5.19 Proposition. *Suppose $\mathcal{E} = (\langle P_i, \boldsymbol{r}_i \rangle)$ is such that P_i is lower semi-continuous, strictly increasing, asymmetric, and negatively transitive, for $i = 1, \ldots, m$. Then for all $\langle \boldsymbol{x}_i \rangle, \langle \boldsymbol{x}_i^* \rangle \in \mathbb{R}_+^{mn}$ such that $\langle \boldsymbol{x}_i \rangle \boldsymbol{P} \langle \boldsymbol{x}_i^* \rangle$, there exists $\langle \boldsymbol{x}_i' \rangle \in \mathbb{R}_+^{mn}$ such that:*

$$\sum_{i=1}^{m} \boldsymbol{x}_i' = \sum_{i=1}^{m} \boldsymbol{x}_i \text{ and } \langle \boldsymbol{x}_i' \rangle \boldsymbol{Q} \langle \boldsymbol{x}_i^* \rangle.$$

Proof. Suppose that $\langle \boldsymbol{x}_i \rangle \boldsymbol{P} \langle \boldsymbol{x}_i^* \rangle$. Then by definition of the strict Pareto ordering, we have:

$$\boldsymbol{x}_i G_i \boldsymbol{x}_i^* \quad \text{for } i = 1, \ldots, m; \tag{5.32}$$

and, for some $k \in \{1, \ldots, m\}$:

$$\boldsymbol{x}_k P_k \boldsymbol{x}_k^*. \tag{5.33}$$

Now, since P_k is strictly increasing, we have:

$$(\forall \boldsymbol{x}'_k \in \mathbb{R}^n_+) \colon \boldsymbol{x}'_k G_k \mathbf{0},$$

where we have denoted the origin in $\mathbb{R}^n$ by '$\mathbf{0}$;' and thus from (5.33) we see that:

$$\boldsymbol{x}_k > \mathbf{0}. \tag{5.34}$$

Furthermore, since P_k is lower semi-continuous, it also follows that there exists $\theta \in]0, 1[$ satisfying:

$$\theta \boldsymbol{x}_k P_k \boldsymbol{x}_k^*. \tag{5.35}$$

But now consider the allocation $(\boldsymbol{x}'_i) \in \mathbb{R}^{mn}_+$ defined by:

$$\boldsymbol{x}'_k = \theta \boldsymbol{x}'_k, \tag{5.36}$$

and:

$$\boldsymbol{x}'_i = \boldsymbol{x}_i + [(1 - \theta)/(m - 1)]\boldsymbol{x}_k \quad \text{for all } i \neq k.$$

From (5.34) we have, for each $i \neq k$, $\boldsymbol{x}'_i > \boldsymbol{x}_i$; and thus, since each P_i is strictly increasing:

$$\boldsymbol{x}'_i P_i \boldsymbol{x}_i \quad \text{for all } i \neq k. \tag{5.37}$$

Therefore, using (5.35) and (5.37), we see that:

$$\langle \boldsymbol{x}'_i \rangle \boldsymbol{Q} \langle \boldsymbol{x}_i^* \rangle.$$

Furthermore, from the definition of $\langle \boldsymbol{x}'_i \rangle$, we have:

$$\sum_{i=1}^{m} \boldsymbol{x}'_i = \theta \boldsymbol{x}_k + \sum_{i \neq k} \Big(\boldsymbol{x}_i + [(1 - \theta)/(m - 1)]\boldsymbol{x}_k \Big)$$
$$= \theta \boldsymbol{x}_k + \sum_{i \neq k} \boldsymbol{x}_i + (m - 1)[(1 - \theta)/(m - 1)]\boldsymbol{x}_k = \sum_{i=1}^{m} \boldsymbol{x}_i. \quad \square$$

Our next proposition is now an easy consequence of the result just proved. Details will be left as an exercise.

5.20 Proposition. *If $\mathcal{E} = \big(\langle P_i, \boldsymbol{r}_i \rangle\big)$ is such that P_i is lower semi-continuous, strictly increasing, asymmetric, and negatively transitive, for $i = 1, \ldots, m$, then an allocation $\langle \boldsymbol{x}_i^* \rangle$ is Pareto efficient for E if, and only if, it is strongly Pareto efficient for E.*

5.6 Pareto Efficiency and 'Non-Wastefulness'

In the terminology introduced by Hurwicz [1960], we will demonstrate that, loosely speaking:

1. the competitive mechanism is **non-wasteful**, in the sense that any competitive equilibrium is Pareto efficient.

2. the competitive mechanism is **unbiased**, in the sense that (given some additional assumptions) any Pareto efficient allocation can be made a competitive equilibrium.

The first of the above two results is often called the 'First Fundamental Theorem of Welfare Economics,' and seems to have been originally established by Enrico Barone [1908]. The second result is the 'Second Fundamental Theorem of Welfare Economics,' and was originally formulated and proved by Kenneth Arrow [1951a]. We will conclude this chapter with two versions of the 'First Fundamental Theorem,' but we will postpone our study of the 'Second Fundamental Theorem' until Chapter 7.

5.21 Theorem. *If $\big(\langle \boldsymbol{x}_i^* \rangle, \boldsymbol{p}^*\big)$ is a competitive equilibrium for a private ownership economy, $\mathcal{E}$, then $\langle \boldsymbol{x}_i^* \rangle$ is Pareto efficient for $\mathcal{E}$.*

Proof. Suppose $\langle \boldsymbol{x}_i \rangle \in \mathbb{R}_+^{mn}$ is such that $\langle \boldsymbol{x}_i \rangle \boldsymbol{Q} \langle \boldsymbol{x}_i^* \rangle$, so that:

$$\boldsymbol{x}_i P_i \boldsymbol{x}_i^* \quad \text{for } i = 1, \ldots, m.$$

Then, since $\boldsymbol{x}_i^* \in \boldsymbol{h}_i(\boldsymbol{p}^*, \boldsymbol{p}^* \cdot \boldsymbol{r}_i)$, for each i, we must have:

$$\boldsymbol{p}^* \cdot \boldsymbol{x}_i > \boldsymbol{p}^* \cdot \boldsymbol{r}_i \quad \text{for } i = 1, \ldots, m.$$

But then, by summing over i, we see that we must have:

$$0 < \sum_{i=1}^m (\boldsymbol{p}^* \cdot \boldsymbol{x}_i - \boldsymbol{p}^* \cdot \boldsymbol{r}_i) = \boldsymbol{p}^* \cdot \Big(\sum_{i=1}^m \boldsymbol{x}_i - \boldsymbol{r}\Big).$$

However, it then follows that we cannot have:

$$\sum_{i=1}^m \boldsymbol{x}_i = \boldsymbol{r};$$

and thus $\langle \boldsymbol{x}_i \rangle$ is not feasible for $\mathcal{E}$. Therefore, $\langle \boldsymbol{x}_i^* \rangle$ is Pareto efficient for $\mathcal{E}$. □

Our alternative version of the 'First Fundamental Theorem' is as follows.

5.22 Theorem. *Suppose $\big(\langle \boldsymbol{x}_i^* \rangle, \boldsymbol{p}^*\big)$ is a competitive equilibrium for a private ownership economy, $\mathcal{E}$, and that each P_i is locally non-saturating, asymmetric, and negatively transitive. Then $\langle \boldsymbol{x}_i^* \rangle$ is strongly Pareto efficient for $\mathcal{E}$*

Proof. Suppose $\langle \boldsymbol{x}_i \rangle \in \mathbb{R}_+^{mn}$ is such that $\langle \boldsymbol{x}_i \rangle \boldsymbol{P} \langle \boldsymbol{x}_i^* \rangle$. Then:

$$\boldsymbol{x}_i G_i \boldsymbol{x}_i^* \quad \text{for } i = 1, \ldots, m;$$

and, for some $k \in \{1, \ldots, m\}$:

$$\boldsymbol{x}_k P_k \boldsymbol{x}_k^*.$$

It then follows from Proposition 4.9 and the definition of $\boldsymbol{h}_i(\cdot)$, respectively, that:

$$\boldsymbol{p}^* \cdot \boldsymbol{x}_i \geq \boldsymbol{p}^* \cdot \boldsymbol{r}_i \quad \text{for } i = 1, \ldots, m, \tag{5.38}$$

and:

$$\boldsymbol{p}^* \cdot \boldsymbol{x}_k > \boldsymbol{p}^* \cdot \boldsymbol{r}_k. \tag{5.39}$$

Adding (5.38) and (5.39) over all i, we then see that:

$$0 < \sum_{i=1}^{m} (\boldsymbol{p}^* \cdot \boldsymbol{x}_i - \boldsymbol{p}^* \cdot \boldsymbol{r}_i) = \boldsymbol{p}^* \cdot \left(\sum_{i=1}^{m} \boldsymbol{x}_i - \boldsymbol{r}\right);$$

from which we see that we cannot have:

$$\sum_{i=1}^{m} \boldsymbol{x}_i = \boldsymbol{r}.$$

Therefore, $(\boldsymbol{x}_i) \notin A(\mathcal{E})$, and we conclude that $\langle \boldsymbol{x}_i^* \rangle$ is strongly Pareto efficient for $\mathcal{E}$. □

Exercises

Each of the following four problems deals with a two-person, two-commodity exchange economy, in which we suppose the consumers' preferences can be represented by the utility functions given in the problem.

1. Suppose the consumers' utility functions are given by:

$$u_1(\boldsymbol{x}_1) = x_{11}^{3/4} x_{12}^{1/4},$$

and:

$$u_2(\boldsymbol{x}_2) = x_{21}^{1/4} x_{22}^{3/4},$$

respectively; and suppose the income distribution in the economy is given by:

$$w_i = (1/2)W \quad \text{for } i = 1, 2,$$

where $W = W(\boldsymbol{p}) = \boldsymbol{p} \cdot \boldsymbol{r}$ for $\boldsymbol{p} \in \mathbb{R}^2_{++}$.

a. Find the aggregate demand function in this case, if one exists.

b. Does the market demand behave as if there is a single utility-maximizing individual in the economy in this case? Explain.

2. Suppose the consumers' utility functions are given by:

$$u_i(\boldsymbol{x}_i) = \min\left\{\frac{x_{i1}}{a_{i1}}, \frac{x_{i2}}{a_{i2}}\right\} \quad \text{for } i = 1, 2,$$

where $a_{ij} > 0 \quad \text{for } i, j = 1, 2.$

a. Supposing that the ith consumer's initial resource endowment is given by $\boldsymbol{r} = (r_{i1}, r_{i2})$, find the ith consumer's demand function, $d_i(\boldsymbol{p})$.

b. If we now suppose that:

$$a_{11} = a_{12} = a_{21} = 1, a_{22} = 4, r_{12} = r_{21} = 0, r_{11} = 5, \text{ and } r_{22} = 10,$$

find the (or a) competitive equilibrium for the economy, or show that no competitive equilibrium exists in this case.

c. Now suppose that all the data assumed in part b still holds, except that we now have $a_{22} = 2$, and find the (or a) competitive equilibrium for the economy, or show that no competitive equilibrium exists.

3. Suppose the consumers' utility functions are given by:

$$u_i(\boldsymbol{x}_i) = x_{i1}^{a_i} \cdot x_{i2}^{1-a_i}, \quad \text{for } i = 1, 2,$$

where:

$$0 < a_i < 1, \quad \text{for } i = 1, 2.$$

a. Supposing that the ith consumer's initial resource endowment is given by $\boldsymbol{r} = (r_{i1}, r_{i2})$, find the ith consumer's demand function for the first commodity, $d_{i1}(\boldsymbol{p})$.

b. If we now suppose that:

$$r_{12} = r_{21} = 0, \quad \text{while } r_{11} > 0 \;\&\; r_{22} > 0,$$

find the (or a) competitive equilibrium for the economy.

c. With the values for $\boldsymbol{r}_i$ as specified in part b, can you find $\partial p_1 / \partial r_{22}$? Does its value make sense to you? Explain.

4. Suppose the consumers' utility functions are given by:

$$u_1(\boldsymbol{x}_1) = \min\left\{\frac{x_{11}}{2}, x_{12}\right\} \text{ and } u_2(\boldsymbol{x}_2) = \min\left\{x_{21}, \frac{x_{22}}{2}\right\},$$

while the initial resource endowments are given by:

$$\boldsymbol{r}_1 = (3, 0) \text{ and } \boldsymbol{r}_2 = (0, 3).$$

On the basis of this information, answer the following questions.

a. Is the first consumer's preference relation homothetic?

b. Is the first consumer's preference relation weakly convex? convex? strictly convex?

c. Find the ith consumer's demand function for the first commodity, $d_{i1}(\boldsymbol{p})$; normalizing the price of the second commodity to equal one, that is, setting $p_2 = 1$.

d. Find the (or a) competitive equilibrium for the economy, or show that no competitive equilibrium exists in this case.

5. Consider a pure exchange economy in which $X_i = \mathbb{R}^2_+ \cap Z$, for $i = 1, 2$; where:

$$Z = \{\boldsymbol{x} \in \mathbb{R}^2 \mid x_j \text{ is an integer, for } j = 1, 2\},$$

that is, X_i is the set of all vectors in $\mathbb{R}^2$ having each coordinate a nonnegative integer. Suppose that the initial endowments are given by $\boldsymbol{r}_1 = (1, 0)$ and $\boldsymbol{r}_2 = (1, 1)$; and that the two consumers have preference relations which can be represented on X_i^* by the functions f_1 and f_2, respectively, given by Table 1, below.

$\boldsymbol{x}_i$	f_1	f_2
$(0,0)$	1	1
$(1,0)$	3	3
$(0,1)$	3	2
$(0,2)$	4	3
$(1,1)$	5	4
$(2,0)$	5	5
$(2,1)$	6	6

Table 1.

Show that $\big((\boldsymbol{x}_i^*), \boldsymbol{p}^*\big)$ is a competitive equilibrium for this economy, where:

$$\boldsymbol{x}_1^* = (1,0),\ \boldsymbol{x}_2^* = (1,1),\ \text{and}\ \boldsymbol{p}^* = (3/5, 2/5).$$

Is this allocation Pareto efficient? Is it strongly Pareto efficient? Explain your answers briefly.

6. Complete the proof of Proposition 5.14.

7. Show that neither the unanimity relationship, nor the strict Pareto ordering is necessarily negatively transitive, even if individual preferences are negatively transitive. (Note: it suffices to produce an example for each [or possibly one dual-purpose example] in which negative transitivity fails. This can be done with simple Edgeworth Box diagrams.)

8. Suppose P_i is asymmetric and negatively transitive for each i, and let $\boldsymbol{R}$ and $\boldsymbol{P}$ be the Pareto at-least-as-good-as, and strict Pareto dominance relation. Show that if $\langle \boldsymbol{x}_i \rangle$, $\langle \boldsymbol{x}_i^* \rangle$, $\langle \boldsymbol{x}_i' \rangle$. and $\langle \boldsymbol{x}_i'' \rangle$ are such that:

$$\langle \boldsymbol{x}_i \rangle \boldsymbol{R} \langle \boldsymbol{x}_i^* \rangle, \langle \boldsymbol{x}_i^* \rangle \boldsymbol{P} \langle \boldsymbol{x}_i' \rangle,\ \text{and}\ \langle \boldsymbol{x}_i' \rangle \boldsymbol{R} \langle \boldsymbol{x}_i'' \rangle, \tag{5.40}$$

then $\langle \boldsymbol{x}_i \rangle \boldsymbol{P} \langle \boldsymbol{x}_i'' \rangle$.

Chapter 6

Production Theory

6.1 Introduction

In this chapter we will develop the 'bare bones' of production theory as it is utilized in the remaining chapters of this book. The next section covers the most basic topics, while in section 3 we consider the special case of linear production sets. Linear production sets play a major role in many portions of applied general equilibrium analysis; particularly in the area of public economics, where the assumption of linear production sets greatly simplifies the analysis in the literature on 'optimal commodity taxation,' for example. Input-Output analysis, which has played a key role in development and planning models, also utilizes the assumption of an aggregate linear production set; and we will undertake a *very* brief study of one version of this model in Section 4.

In Section 5 we will examine the issue of profit maximization for a competitive firm, as well as some of the properties of the firm's profit function in this case. We then move on in Section 6 to a consideration of the specifically general equilibrium development of production theory; in particular, the relationship of the aggregate production set and the aggregate profit function to the individual production sets, and individual profit functions, respectively. Finally, in Section 6 we present a brief development of the theory of 'activity analysis.' This is, in effect, a special case of a linear production model, and is a topic which can easily be skipped on one's first passage through this text; although some of the mathematical results in this section will be utilized in our study of general equilibrium under uncertainty.

6.2 Basic Concepts of Production Theory

In our development of the theory of the firm in a general equilibrium context, we suppose that the set of technologically feasible production vectors for the firm is a subset, Y, of $\mathbb{R}^n$, which we shall call a **production set**. The producer chooses a production plan (or **production vector**, or '**netput**' vector):

$$\boldsymbol{y} = (y_1, \ldots, y_n),$$

from Y, with the interpretation: if y_j is:

positive, then the vector $\boldsymbol{y}$ specifies a net production of the jth commodity in the amount $y_j > 0$;

negative, then the jth commodity is being used as an input in the amount:

$$-y_j = |y_j| > 0;$$

and, of course, if $y_j = 0$, then there is neither net production nor a net useage of commodity j as an input.

In this section, we will examine some of the implications of several assumptions which are commonly-used in connection with production in general equilibrium models, as well as some definitions we will use a great deal in our discussion of production. In intermediate theory courses one of the most important and significant assumptions which one examines is the assumption of decreasing returns to scale; or, in the short-run, diminishing returns. In general equilibrium developments, the corresponding condition is that the production set is convex. However, this is getting a bit ahead of our story; let's begin by considering the following conditions.

6.1 Definitions. A production set, $Y \subseteq \mathbb{R}^n$ is said to satisfy (or to exhibit):

a. **non-increasing returns to scale** iff, for any $\boldsymbol{y} \in Y$ and any $\theta \in]0, 1]$, we have $\theta\boldsymbol{y} \in Y$ as well.

b. **non-decreasing returns to scale** iff, for any $\boldsymbol{y} \in Y$ and any $\theta \geq 1$, we have $\theta\boldsymbol{y} \in Y$ as well.

c. **constant returns to scale** iff Y is a cone; that is, for all $\boldsymbol{y} \in Y$ and all $\theta > 0$, we have $\theta\boldsymbol{y} \in Y$.

d. **increasing returns to scale** iff Y satisfies non-decreasing returns to scale, and does *not* satisfy non-increasing returns to scale.

e. **decreasing returns to scale** iff Y does *not* satisfy non-decreasing returns to scale, and does satisfy non-increasing returns to scale.

In words, a production set satisfies non-increasing returns if whenever we decrease all input and output quantities of a feasible production vector in the same proportion, we arrive at another feasible production vector. I will leave it to you to develop analagous verbal statements for the other properties set out in the above definition. In the figures on the next page, we present examples satisfying non-increasing and non-decreasing returns to scale, respectively, for production sets in $\mathbb{R}^2$. In fact, in Figure 6.1.a, Y satisfies decreasing returns; while in Figure 6.1.b, Y exhibits increasing returns.

While the definitions presented in 6.1 are frequently seen in the literature, definitions (d) and (e) (increasing and decreasing returns to scale, respectively) are not very satisfactory. For example the production set in Figure 6.2.a satisfies the definition of increasing returns, while that in 6.2.b satisfies decreasing returns. The essential relationship between non-increasing returns and convexity is set out in the following proposition.

6.2 Proposition. *If $Y \subseteq \mathbb{R}^n$ is convex, and contains the origin (that is, $\mathbf{0} \in Y$), then Y satisfies non-increasing returns to scale.*

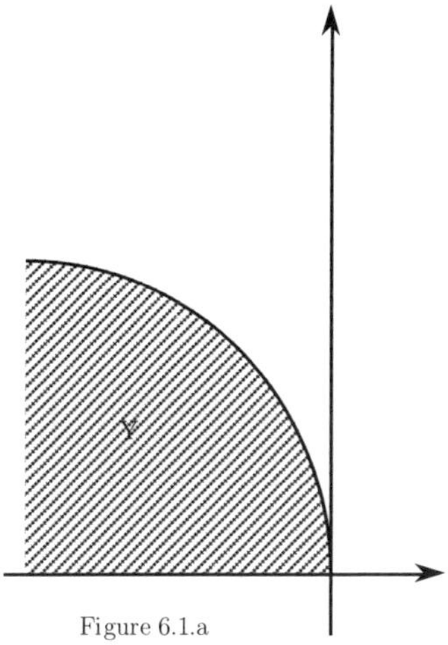

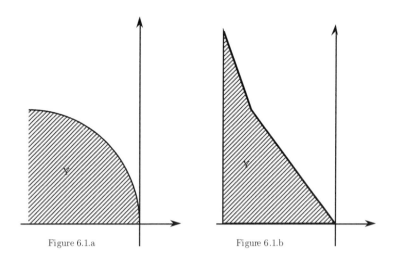

Figure 6.1: Decreasing and Increasing Returns.

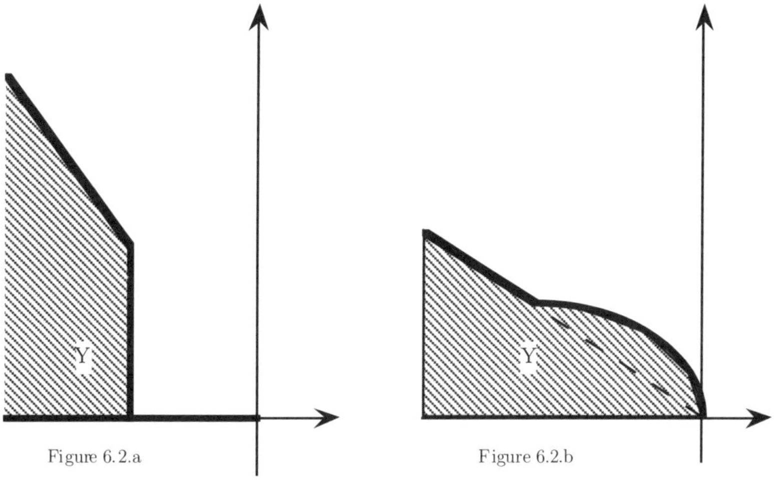

Figure 6.2: Increasing and Decreasing Returns: Another Example.

Proof. If $\boldsymbol{y} \in Y$ and $\theta \in]0, 1]$, then making use of the fact that $\mathbf{0} \in Y$, and the convexity of Y, we have:

$$\theta \boldsymbol{y} + (1 - \theta)\mathbf{0} = \theta \boldsymbol{y} \in Y. \quad \square$$

While the above result shows that convexity, together with the assumption that $\mathbf{0} \in Y$ is sufficient to imply non-increasing returns to scale, these conditions are not necessary for same. In fact, in Figure 6.3, Y contains the origin and satisfies decreasing returns to scale even though it is not convex. On the other hand, convexity of the production set can also be viewed as the counterpart of the assumption that the short-run production function satisfies diminishing returns; as is illustrated in the examples which follow.

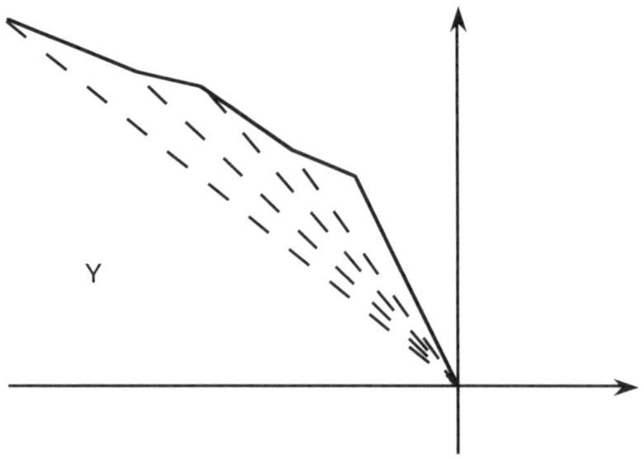

Figure 6.3: Decreasing Returns without Convexity.

6.3 Examples.
1. Let $\varphi\colon \mathbb{R}^{n-1}_+ \to \mathbb{R}_+$ be a production function, and define $Y \subseteq \mathbb{R}^n$ by:

$$Y = \{\boldsymbol{y} = (\boldsymbol{v}, x) \in \mathbb{R}^{(n-1)+1} \mid \boldsymbol{v} \in -\mathbb{R}^{n-1}_+ \;\&\; 0 \le x \le \varphi(-\boldsymbol{v})\}.$$

If φ is concave, then Y is a convex set. Moreover, if $\varphi(\mathbf{0}) = 0$, then (as established by Proposition 6.2, above) Y satisfies non-increasing returns to scale.

2. As above, let $\varphi\colon \mathbb{R}^{n-1}_+ \to \mathbb{R}_+$ be a production function, but this time suppose that φ is positively homogeneous of degree one. If we then define $Y \subseteq \mathbb{R}^n$ by:

$$Y = \{\boldsymbol{y} = (\boldsymbol{v}, x) \in \mathbb{R}^{(n-1)+1} \mid \boldsymbol{v} \in -\mathbb{R}^{n-1}_+ \;\&\; 0 \le x \le \varphi(-\boldsymbol{v})\}, \tag{6.1}$$

it is easy to show that Y satisfies constant returns to scale.

3. Suppose once again that φ is concave, but that, say v_1 is fixed at the level $v_1 = v_1^*$ in the short run. In this case the set Y defined in (1), above, will be convex; however, the short-run production set will be given by:

$$\widehat{Y} = \{\boldsymbol{y} = (\boldsymbol{v}, x) \in \mathbb{R}^{(n-1)+1} \mid \boldsymbol{v} \in -\mathbb{R}^{n-1}_+ \;\&\; 0 \le x \le \varphi(-\boldsymbol{v}) \;\&\; v_1 = v_1^*\}.$$

On the other hand, we can define the set $\widehat{Y}$ by:

$$\widehat{Y} = Y \cap \{\boldsymbol{y} \in \mathbb{R}^n \mid y_1 = v_1^*\};$$

so that we see that $\widehat{Y}$ is the intersection of two convex sets, and is, therefore, also convex. $\square$

The following definition sets out some further conditions which are often used in the general equilibrium theory of production.

6.4 Definitions.

1. **Impossibility of Free Production:** $Y \cap \mathbb{R}^n_+ \subseteq \{\mathbf{0}\}$.
2. **Irreversibility:** $Y \cap (-Y) \subseteq \{\mathbf{0}\}$; that is, if $\boldsymbol{y} \in Y$, and $-\boldsymbol{y} \in Y$ as well, then $\boldsymbol{y} = \mathbf{0}$.
3. **Additivity:** if $\boldsymbol{y} \in Y$ and $\boldsymbol{y}^* \in Y$, then $\boldsymbol{y} + \boldsymbol{y}^* \in Y$ as well.
4. **Possibility of inaction:** $\mathbf{0} \in Y$.
5. **Disposability:**
   a. **Limited:** If $\boldsymbol{y} \in Y, \boldsymbol{y}^* \in \mathbb{R}^n$, and for all j we have:

$$0 \leq y_j^* \leq y_j \text{ or } y_j^* \leq y_j < 0,$$

then $\boldsymbol{y}^* \in Y$.
   b. **Semi-Free:** $Y - \mathbb{R}^n_+ \subseteq Y$; that is, if $\boldsymbol{y} \in Y$ and $\boldsymbol{y}' \in \mathbb{R}^n$ are such that $\boldsymbol{y}' \leq \boldsymbol{y}$, then $\boldsymbol{y}' \in Y$.
   c. **Free:** $-\mathbb{R}^n_+ \subseteq Y$.[1]

Let's begin our consideration of these definitions by noting some relationships among the disposability conditions. The proofs of the first and third of the following facts we will leave as exercises. Fact 2 is demonstrated by Figures 6.4.a and 6.4.b. (In Figure 6.4.a, Y is the set $Y = \{\boldsymbol{y} \in \mathbb{R}^2 \mid \boldsymbol{y} \leq \boldsymbol{y}^*\}$; while in Figure 6.4.b, Y is the union of the third quadrant and the ray indicated.)

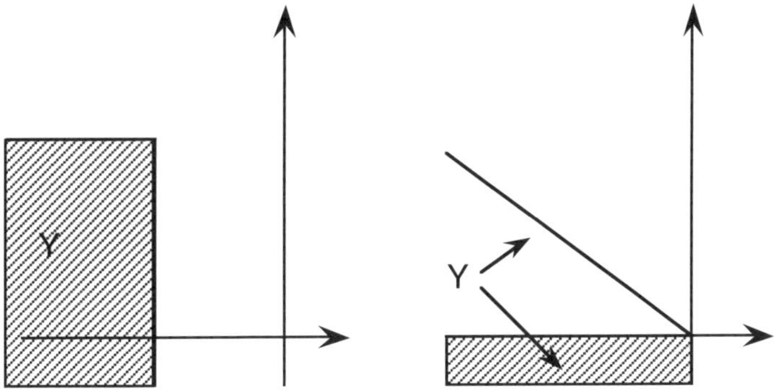

Figure 6.4.a Figure 6.4.b

Figure 6.4: Semi-Free and Free Disposability.

Facts Regarding Disposability.

1. If Y satisfies 'semi-free disposability,' then it satisfies limited disposability. However, the converse is not true.

[1] This is an assumption which is usually applied only to the aggregate production set.

2. A production set may satisfy semi-free disposability, but not satisfy free disposability. Furthermore, a production set may satisfy free disposability but not limited disposability (and hence not semi-free disposability).

3. If Y satisfies semi-free disposability and $\mathbf{0} \in Y$, then Y satisfies free disposability.

4. If Y satisfies free disposability and is closed and convex, then Y also satisfies semi-free disposability.

Proof of Fact 4. Let $\boldsymbol{y}^* \in Y$ and $\boldsymbol{y}' \in \mathbb{R}^n$ be such that $\boldsymbol{y}' \leq \boldsymbol{y}^*$. Then:

$$\bar{\boldsymbol{y}} \stackrel{\text{def}}{=} \boldsymbol{y}' - \boldsymbol{y}^* \leq \mathbf{0};$$

and thus $\bar{\boldsymbol{y}} \in Y$, by free disposability. But then, since $-\mathbb{R}^n_+ \subseteq Y$ and $-\mathbb{R}^n_+$ is a cone, we see that for all $\mu > 0$ (in particular, for $\mu > 1$), we must have $\mu\bar{\boldsymbol{y}} \in Y$. Therefore, given an arbitrary $\mu > 1$, if we let $\theta = 1/\mu$, it follows from the convexity of Y that:

$$\theta(\mu\bar{\boldsymbol{y}}) + (1-\theta)\boldsymbol{y}^* = \bar{\boldsymbol{y}} + \boldsymbol{y}^* - \Big(\frac{1}{\mu}\Big)\boldsymbol{y}^* = \boldsymbol{y}' - \boldsymbol{y}^* + \boldsymbol{y}^* - \Big(\frac{1}{\mu}\Big)\boldsymbol{y}^* = \boldsymbol{y}' - \Big(\frac{1}{\mu}\Big)\boldsymbol{y}^* \in Y.$$

But:

$$\boldsymbol{y}' - \Big(\frac{1}{\mu}\Big)\boldsymbol{y}^* \to \boldsymbol{y}' \quad \text{as} \quad \mu \to +\infty;$$

and, since Y is closed, it follows that $\boldsymbol{y}' \in Y$. □

6.5 Definition. We shall say that a set $Y \subseteq \mathbb{R}^n$ is a **convex cone** iff Y is a cone which is also a convex set.

Within the context of production theory, an interesting property of convex cones is set out in the following.

6.6 Proposition. *If a set $Y \subseteq \mathbb{R}^n$ is a convex cone, then Y is additive.*

Proof. Suppose $\boldsymbol{y}$ and $\boldsymbol{y}'$ are elements of Y. Then, since Y is convex:

$$(1/2)\boldsymbol{y} + (1/2)\boldsymbol{y}' \in Y.$$

But then, since Y is also a cone:

$$2\big[(1/2)\boldsymbol{y} + (1/2)\boldsymbol{y}'\big] = \boldsymbol{y} + \boldsymbol{y}' \in Y. \quad □$$

In the next section, we will investigate the properties of a particular kind of production set in detail. In the meantime, we close this section by presenting some alternative useful methods for characterizing a production set.

6.7 Example. Let $P\colon \mathbb{R}^q_+ \mapsto \mathbb{R}^m_+$ be a correspondence; where, for $\boldsymbol{v} \in \mathbb{R}^q_+$ we interpret $P(\boldsymbol{v})$ as the set of all output vectors, $\boldsymbol{x} \in \mathbb{R}^m_+$ such that $\boldsymbol{v}$ can produce $\boldsymbol{x}$. In fact, in this case, for $\boldsymbol{v} \in \mathbb{R}^q_+$, we call $P(\boldsymbol{v})$ the **production possibility set for** $\boldsymbol{v}$ (and we will refer to such a correspondence as a **production possibility correspondence**). In this case, if $n = q + m$, then one approach to the task of defining Y from P is to more or less define Y as the graph of P; more exactly, we can define:

$$Y = \{(\boldsymbol{v}, \boldsymbol{x}) \in \mathbb{R}^{q+m} \mid \boldsymbol{v} \in -\mathbb{R}^q_+ \ \&\ \boldsymbol{x} \in P(-\boldsymbol{v})\}.$$

A better approach, however, from the standpoint of general equilibrium theory is to do the following. We can always assume, without loss of generality, that $q = m = n$. With this convention, we can then define Y by:

$$Y = \{\boldsymbol{y} \in \mathbb{R}^n \mid (\exists \boldsymbol{v}, \boldsymbol{x} \in \mathbb{R}^n_+) \colon \boldsymbol{x} \in P(\boldsymbol{v}) \ \&\ \boldsymbol{y} = \boldsymbol{x} - \boldsymbol{v}\}. \quad \square$$

6.8 Example. Let $V \colon \mathbb{R}^m_+ \mapsto \mathbb{R}^q_+$ be a correspondence, where for $\boldsymbol{x} \in \mathbb{R}^m_+$ we interpret the set $V(\boldsymbol{x})$ as the set of input vectors, $\boldsymbol{v} \in \mathbb{R}^q_+$, such that $\boldsymbol{v}$ can produce $\boldsymbol{x}$, and we call the set $V(\boldsymbol{x})$ the **input-requirement set for** $\boldsymbol{x}$. In this case the correspondence, V, is called the **input-requirement correspondence**, and the corresponding production set, Y, is often defined by:

$$Y = \{(\boldsymbol{v}, \boldsymbol{x}) \in \mathbb{R}^{q+m} \mid \boldsymbol{x} \in \mathbb{R}^m_+ \ \&\ -\boldsymbol{v} \in V(\boldsymbol{x})\}.$$

However, once again a better approach is to suppose, without loss of generality, that $q = m = n$, and to define Y by:

$$Y = \{\boldsymbol{y} \in \mathbb{R}^n \mid (\exists \boldsymbol{v}, \boldsymbol{x} \in \mathbb{R}^n_+) \colon \boldsymbol{v} \in V(\boldsymbol{x}) \ \&\ \boldsymbol{y} = \boldsymbol{x} - \boldsymbol{v}\}. \quad \square$$

6.3 Linear Production Sets

In order to get a better feel for the meaning of the conditions set out in Definitions 6.4, we will examine them within the context of a particular type of production set; one which we will often be considering in the material to follow.

6.9 Definition. We shall say that a production set, $Y \subseteq \mathbb{R}^n$ is **linear** iff there exists a non-zero $m \times n$ matrix, $\boldsymbol{A}$ such that:

$$Y = \{\boldsymbol{y} \in \mathbb{R}^n \mid \boldsymbol{A}\boldsymbol{y} \leq \boldsymbol{0}\}. \tag{6.2}$$

Before considering the general properties of linear production sets, let's take a look at what is apparently a quite different sort of linear production relationship.

6.10 Example. Let $\boldsymbol{B}$ be a semi-positive $q \times r$ matrix. Define a pair $(\boldsymbol{v}, \boldsymbol{x}) \in \mathbb{R}^{q+r}_+$ to be technologically feasible iff:

$$\boldsymbol{v} \geq \boldsymbol{B}\boldsymbol{x}.$$

The situation here is that we suppose that $\boldsymbol{B}$ is an **input-requirement matrix**, with the interpretation that the matrix-vector product $\boldsymbol{B}\boldsymbol{x}$ gives the minimal amounts of the q inputs which are needed to produce $\boldsymbol{x}$. Notice that with this specification, the corresponding input-requirement correspondence (Example 6.8 of the previous section) is given by:

$$V(\boldsymbol{x}) = \{\boldsymbol{v} \in \mathbb{R}^q_+ \mid \boldsymbol{v} \geq \boldsymbol{B}\boldsymbol{x}\} \quad \text{for } \boldsymbol{x} \in \mathbb{R}^r_+.$$

If we now suppose that $n = q + r$, we can define a production set for this case by (assuming that the first q commodities are the inputs used in the production process, and that the last $n - q = r$ commodities are the outputs):

$$Y = \{(\boldsymbol{v}, \boldsymbol{x}) \in \mathbb{R}^n \mid -\boldsymbol{v} \geq \boldsymbol{B}\boldsymbol{x} \ \&\ \boldsymbol{x} \in \mathbb{R}^r_+\} = \{(\boldsymbol{v}, \boldsymbol{x}) \in \mathbb{R}^n \mid \boldsymbol{v} + \boldsymbol{B}\boldsymbol{x} \leq \boldsymbol{0} \ \&\ \boldsymbol{x} \in \mathbb{R}^r_+\}.$$

Alternatively, we can express this production set in the general form of equation (6.2) above, by defining the $n \times n$ matrix $\boldsymbol{A}$ as:

$$\boldsymbol{A} = \begin{pmatrix} \boldsymbol{I}_q & \boldsymbol{B} \\ \boldsymbol{O} & -\boldsymbol{I}_r \end{pmatrix},$$

where I_q and I_r are the $q \times q$ and $r \times r$ identity matrices, respectively; and then defining Y by:

$$Y = \left\{ \boldsymbol{y} = \begin{pmatrix} \boldsymbol{v} \\ \boldsymbol{x} \end{pmatrix} \in \mathbb{R}^n \mid \boldsymbol{A} \begin{pmatrix} \boldsymbol{v} \\ \boldsymbol{x} \end{pmatrix} \leq \boldsymbol{0} \right\}. \quad \square$$

Returning to the general definition of a linear production set, we can prove the following.

6.11 Proposition. *If Y is a linear production set, then $\boldsymbol{0} \in Y$, and Y is a closed convex cone. Thus, in particular, Y is additive.*

Proof. Suppose Y is linear; so that there exists an $m \times n$ non-zero matrix, $\boldsymbol{A}$, such that:

$$Y = \{\boldsymbol{y} \in \mathbb{R}^n \mid \boldsymbol{A}\boldsymbol{y} \leq \boldsymbol{0}\}.$$

Obviously we then have $\boldsymbol{0} \in Y$. To prove that Y is convex, let $\boldsymbol{y}, \boldsymbol{y}' \in Y$ and $\theta \in [0, 1]$. We then have:

$$\boldsymbol{A}\boldsymbol{y} \leq \boldsymbol{0} \ \&\ \boldsymbol{A}\boldsymbol{y}' \leq \boldsymbol{0},$$

and thus:

$$\boldsymbol{A}[\theta \boldsymbol{y} + (1 - \theta)\boldsymbol{y}'] = \theta \boldsymbol{A}\boldsymbol{y} + (1 - \theta)\boldsymbol{A}\boldsymbol{y}' \leq \boldsymbol{0},$$

and it follows that:

$$\theta \boldsymbol{y} + (1 - \theta)\boldsymbol{y}' \in Y.$$

Similarly, if $\boldsymbol{y} \in Y$ and $\lambda > 0$, we have $\boldsymbol{A}\boldsymbol{y} \leq \boldsymbol{0}$, and therefore:

$$\boldsymbol{A}\lambda \boldsymbol{y} = \lambda \boldsymbol{A}\boldsymbol{y} \leq \boldsymbol{0}.$$

To prove that Y is closed, denote the ith row of $\boldsymbol{A}$ by '$\boldsymbol{a}_{i\cdot}$,' for $i = 1, \ldots, m$, and define $H_i \subseteq \mathbb{R}^n$ by:

$$H_i = \{\boldsymbol{y} \in \mathbb{R}^n \mid \boldsymbol{a}_{i\cdot} \cdot \boldsymbol{y} \leq 0\} \quad \text{for } i = 1, \ldots, m.$$

Then we note that:

$$Y = \bigcap_{i=1}^{m} H_i;$$

and, since each H_i is a closed lower half-space, it follows that Y is closed as well.

Finally, the fact that Y is additive follows from Proposition 6.6, now that we have shown Y to be a convex cone. $\square$

The generic example presented in Example 6.10, above, is obviously very special in at least three senses (in addition to the linearity of the technology). First, it takes the first q commodities as inputs and produces the last $n-q$ commodities as outputs. Secondly, every commodity is either an input or an output of the production process. Third, each production process uses the inputs in fixed proportions; there is no substitution possible between inputs. In principle, however, all of these deficiencies are readily correctible while retaining the assumption of linearity, as is shown in the following examples.

6.12 Examples.

1. Suppose $n = 6$, and that a firm operates two production processes. Process 1, we will suppose, produces commodity two using the first and third commodities as inputs, with input-output combinations feasible if, and only if, they satisfy the production constraint:

$$y_2 \leq -a_{11}y_1 - a_{13}y_3,$$

where $a_{11} > 0$ and $a_{13} > 0$. Process 2 produces the fourth commodity using the sixth commodity as an input; with production constraint:

$$y_4 \leq -a_{26}y_6,$$

where $a_{26} > 0$. Commodity five, we will suppose, does not enter into this firm's production processes at all. Notice that in this case the first production process allows possible substitution between the first and the third commodities as inputs.

We can characterize this firm's technology in the form of equation (6.2) by defining the 9×6 matrix $\boldsymbol{A}$ by:

$$\boldsymbol{A} = \begin{pmatrix} a_{11} & 1 & a_{13} & 0 & 0 & 0 \\ 0 & 0 & 0 & 1 & 0 & a_{26} \\ 1 & 0 & 0 & 0 & 0 & 0 \\ 0 & -1 & 0 & 0 & 0 & 0 \\ 0 & 0 & 1 & 0 & 0 & 0 \\ 0 & 0 & 0 & -1 & 0 & 0 \\ 0 & 0 & 0 & 0 & 1 & 0 \\ 0 & 0 & 0 & 0 & -1 & 0 \\ 0 & 0 & 0 & 0 & 0 & 1 \end{pmatrix}. \tag{6.3}$$

I will leave it to you to verify that the firm's production set, Y, can be represented as:

$$Y = \{\boldsymbol{y} \in \mathbb{R}^6 \mid \boldsymbol{Ay} \leq \boldsymbol{0}\},$$

and that Y satisfies the properties set out at the beginning of this paragraph.

In this example, we have only one commodity produced by each of the production processes, but this is not at all necessary for the type of representation being considered here; we can allow for joint production of two or more commodities in a single production process in much the same way that we allow for more than one input.

2. In connection with the preceding example, notice that if we define the matrix $\boldsymbol{B}$ by:

$$\boldsymbol{B} = \begin{pmatrix} a_{11} & 1 & a_{13} & 0 & 0 & 0 \\ 0 & 0 & 0 & 1 & 0 & a_{26} \end{pmatrix},$$

then we can equally well define the production set by:

$$Y = \{\boldsymbol{y} \in \mathbb{R}^6 \mid \boldsymbol{B}\boldsymbol{y} \leq \boldsymbol{0},\ y_1 \leq 0,\ y_2 \geq 0,\ y_3 \leq 0,\ y_4 \geq 0,\ y_5 = 0 \ \&\ y_6 \leq 0\}.$$

In fact, if one is actually to construct a detailed or numerical example, it is usually more convenient to represent the production set by defining a $q \times n$ semi-positive matrix, $\boldsymbol{B}$ (where q is the number of production processes), and nonempty sets I and J, where, defining $N = \{1, \ldots, n\}$, we have:

$$I \cup J = N, \tag{6.4}$$

in such a way that Y can be defined as:

$$Y = \{\boldsymbol{y} \in \mathbb{R}^n \mid \boldsymbol{B}\boldsymbol{y} \leq \boldsymbol{0} \ \&\ (\forall i \in I)\colon y_i \leq 0 \ \&\ (\forall j \in J)\colon y_j \geq 0\}. \tag{6.5}$$

Letting $K = I \cap J$, we then see that these sets have the following interpretations:

$I \setminus J = \{i \in \{1, \ldots, n\}\}$ such that commodity i is used as an input by the technology,

$J \setminus I = \{j \in \{1, \ldots, n\}\}$ such that commodity j is produced by the technology, and

$K = \{k \in \{1, \ldots, n\}\}$ such that commodity k does not enter the technology.

In the present example:

$$I = \{1, 3, 5, 6\} \ \&\ J = \{2, 4, 5\}.$$

While the representation in (6.5) is often more convenient than the representation in (6.2), and is probably more intuitive as well, it is important to notice that if a production set Y can be defined as in (6.5), then one can define an $m \times n$ matrix, $\boldsymbol{A}$, using the example developed in (6.3) as a model, to equivalently define Y in the form of equation (6.2). Thus, in particular, if a production set can be defined as in equation (6.5), then it is linear, as we have defined the term.

3. Suppose this time that $n = 5$, and consider a somewhat more complicated example, as follows. Once again we suppose that the firm operates two production processes. Process one produces the second commodity, using the first and third commodities as inputs, with production constraint:

$$y_2 \leq -b_{11}y_1 - b_{13}z_1, \tag{6.6}$$

where b_{11} and b_{13} are both positive, and we are denoting the quantity of the third commodity used in the first production process by 'z_1.' We will suppose that the second process produces the fifth commmodity, using the third commodity as input, with the production constraint given by:

$$y_5 \leq -b_{23}z_2, \tag{6.7}$$

where $b_{23} > 0$, and $z_2 \leq 0$ is the quantity of the third commodity used as an input in the second process. Commodity four, we then suppose, does not enter into this firm's technology at all. On the other hand, the total amount of commodity three used as an input by this firm must satisfy the constraint:

$$y_3 = z_1 + z_2, \tag{6.8}$$

if the firm is paying a positive price for commodity three, and if the firm is to maximize profits.

Here, in order to express this production set in the form of equation (6.5), we can proceed as follows. Define $b_{15} = (b_{13}/b_{23})$, the matrix $\boldsymbol{B}$ by:

$$\boldsymbol{B} = \begin{pmatrix} b_{11} & 1 & b_{13} & 0 & b_{15} \\ 0 & 0 & b_{23} & 0 & 1 \end{pmatrix}, \tag{6.9}$$

and the sets I and J by:

$$I = \{1,3,4\} \ \& \ J = \{2,4,5\}. \tag{6.10}$$

(See Exercise 3, at the end of this chapter.)

4. An interesting oddity of the definitions being used here stems from the fact that, given a set Y of the form indicated in (6.5), we can define an *almost* equivalent production set as follows. Define the set J^* as the set of indices of commodities which are produced by the technology; that is, in the notation of the previous two examples, let:

$$J^* = J \setminus I,$$

and define $I^* = N \setminus J^*$. Now consider the set Y^* defined by:

$$Y^* = \{\boldsymbol{y} \in \mathbb{R}^n \mid \boldsymbol{By} \leq \boldsymbol{0} \ \& \ (\forall i \in I^*)\colon y_i \leq 0\}.$$

Is the production set Y^* the same as the set Y defined in (6.5)? It is probably pretty obvious that in general, these two sets will not be the same. However, what happens if, say, $n = 2$, $I = I^* = \{1\}$, and $J = J^* = \{2\}$?

Alternatively, consider the set $Y^\dagger$ defined by:

$$Y^\dagger = \{\boldsymbol{y} \in \mathbb{R}^n \mid \boldsymbol{By} \leq \boldsymbol{0} \ \& \ (\forall i \in J^*)\colon y_i \geq 0\}.$$

Does the set $Y^\dagger = Y$? Does $Y^\dagger = Y^*$? □

Turning now to the issue of which of the remaining conditions set out in Definition 6.4 are satisfied by linear technologies, I will leave it as an easy exercise to show that if Y is specified as in equations (6.4) and (6.5), above, then Y will satisfy Irreversibility:

$$Y \cap (-Y) \subseteq \{\boldsymbol{0}\}.$$

The question of whether Y will satisfy 6.4.2, Impossibility of Free Production, in this case is a little more complicated, however, as is shown in the following example.

6.13 Example. In the notation of Example 6.12.2, let $n = 4$, $I = \{1, 2\}$, $J = \{3, 4\}$, and suppose the matrix $\boldsymbol{B}$ is given by:

$$\boldsymbol{B} = \begin{pmatrix} 1 & 0 & 1 & 0 \\ 0 & 1 & 3 & 0 \end{pmatrix}$$

In this case, will Y satisfy Impossibility of Free Production? What if $\boldsymbol{B}$ is given by the following matrix?

$$\boldsymbol{B} = \begin{pmatrix} 1 & 0 & 1 & 2 \\ 0 & 0 & 0 & 0 \end{pmatrix} \quad \square$$

At the risk of sounding rather too English, let's coin the following definition.

6.14 Definition. We shall say that Y is a **proper linear technology** iff there exist a semi-positive $m \times n$ matrix, $\boldsymbol{B}$, and sets $I,\ J \subseteq N \equiv \{1, \dots, n\}$ such that:

1. $I \cup J = N$, $I \setminus J \neq \emptyset$, $J \setminus I \neq \emptyset$,
2. defining $K = I \cap J$, we have:

$$(\forall j \in N \setminus K)(\exists i \in \{1, \dots, m\})\colon b_{ij} > 0,$$

and

3. $Y = \{\boldsymbol{y} \in \mathbb{R}^n \mid \boldsymbol{By} \leq \boldsymbol{0} \;\&\; (\forall i \in I)\colon y_i \leq 0 \;\&\; (\forall j \in J)\colon y_j \geq 0\}$.

I will leave the proof of the following result as an exercise.

6.15 Proposition. *If $Y \subseteq \mathbb{R}^n$ is a proper linear technology, then Y satisfies Impossibility of Free Production, Irreversibility, and Limited Disposability, in addition to the properties set out in Proposition 6.11.*

Suppose a production set Y takes the form:

$$Y = \{\boldsymbol{y} \in \mathbb{R}^n \mid (\exists \boldsymbol{z} \in \mathbb{R}^k_+)\colon \boldsymbol{y} = \boldsymbol{Bz}\},$$

where k is a positive integer, and $\boldsymbol{B}$ is an $n \times k$ matrix. Obviously it would make sense to call Y a linear production set in this case, but this seems to be a very different specification of technology than that which we have termed 'linear' in this section. One can show, however, that this new specification of the production set is linear, as we have defined the term in this section. While the proof of this is deferred until Section 8 of this chapter (a starred section), we will devote the next section to a discussion of a very important example of this second formulation of a linear technology.

6.4 Input-Output Analysis

Input-Output analysis has been a primary tool of applied economics and planning models for the past 60 years. It was developed by W. Leontief during the 1930's, and is still a basic component of many computational general equilibrium models today. In applications it may be used to model the aggregate production possibilities of a nation or a region, or even a large firm; however, hereafter in this discussion we will refer to the entity being analyzed as an 'economy,' and proceed as if we are attempting to characterize the aggregate production set of the economy.

The basic assumption needed is that the products produced in the economy can be classified into n non-overlapping and exhaustive categories. In some applications, this categorization might be very broad: for example Agricultural Products, Manufactured Goods, and Services; while in other applications the categorization may be much finer. In any case, given a categorization of products, the production sector of the economy can be divided into n corresponding production sectors, with each producing exactly one (aggregated) commodity. It is then postulated that the technology of the jth production sector can be characterized by the (Leontief) production function:

$$y_j = \min\left\{\frac{y_{1j}}{z_{1j}}, \dots \frac{y_{nj}}{a_{nj}}, \frac{z_{1j}}{b_{1j}}, \dots, \frac{z_{mj}}{b_{mj}}\right\}; \tag{6.11}$$

where:

y_{ij} is the amount of the ith sector's output used as an input in the jth sector ($i = 1, \dots, n$),

z_{kj} is the amount of the kth primary (non-produced) input used in the jth sector ($k = 1, \dots, m$),

and:[2]

$$a_{ij} \geq 0 \quad \text{for } i = 1, \dots, n, \text{ and}$$
$$b_{kj} \geq 0 \quad \text{for } k = 1, \dots, m.$$

If no goods are free, then efficient production in the jth sector will require that:

$$y_{ij} = a_{ij}x_j \text{ for } i = 1, \dots, n, \text{ and } z_{kj} = b_{kj}x_j \text{ for } k = 1, \dots, m. \tag{6.12}$$

On the other side of the coin, the ith sector's output may be used as in input in the production of any of the the other $n-1$ sectors' outputs, or it may be delivered to the consumer sector, or to government, or exported as a final good. In our discussion we will lump these three sectors together as (exogenous) 'final demand,' and we will denote the quantity of this final demand by 'c_j.'

Now, if we ignore primary factor inputs for the moment, we can picture the structure of the aggregate production technology as in the following table.

Delivering Sector	**Receiving Sector**				
1 (y_1)	y_{11}	y_{12}	...	y_{1n}	c_1
2 (y_2)	y_{21}	y_{22}	...	y_{2n}	c_2
...			...		
n (y_n)	y_{n1}	y_{n2}	...	y_{nn}	c_n

[2]If, for some i, j, we have $a_{ij} = 0$, we define $x_y/a_{ij} = +\infty$, and similarly if, for some k, j, $b_{kj} = 0$. Notice that in this eventuality, for example, if $a_{hj} = 0$, then $\min\{x_j/a_{ij}\}$ is never equal to x_j/a_{hj}.

Consequently, we see that the net output, that is, the quantity of the ith good available for final consumption, is given by:

$$y_i - \sum_{j=1}^{n} a_{ij} y_j;$$

and if the vector, $\boldsymbol{c}$, of final demands is feasible, it must be that:

$$c_i = y_i - \sum_{j=1}^{n} a_{ij} y_j \quad \text{for } i = 1, \ldots, n. \tag{6.13}$$

Alternatively, if we define the $n \times n$ matrix $\boldsymbol{A} = [a_{ij}]$ (the **technology matrix**), we can express this feasibility requirement as:

$$\boldsymbol{y} - \boldsymbol{A}\boldsymbol{y} = \boldsymbol{c};$$

or, denoting the $n \times n$ identity matrix by '$\boldsymbol{I}$:'

$$(\boldsymbol{I} - \boldsymbol{A})\boldsymbol{y} = \boldsymbol{c}.$$

(The matrix $\boldsymbol{I} - \boldsymbol{A}$ is called the **Leontief matrix**.) Since we are treating final demand as exogenous, we can express the basic problem with which input-output analysis deals as:

Problem A. Given $\boldsymbol{c} \in \mathbb{R}^n_+$, does there exist $\boldsymbol{y} \in \mathbb{R}^n_+$ such that:

$$(\boldsymbol{I} - \boldsymbol{A})\boldsymbol{y} = \boldsymbol{c}, \tag{6.14}$$

and:

$$\boldsymbol{B}\boldsymbol{y} \leq \overline{\boldsymbol{z}}, \tag{6.15}$$

where '$\overline{\boldsymbol{z}}$' denotes the vector of available primary input quantities, and $\boldsymbol{B}$ is the $m \times n$ matrix, $\boldsymbol{B} = [b_{kj}]$

In the remainder of our discussion, however, we will concentrate our attention upon the equality (6.14), and ignore inequality (6.15), for the following reason. Suppose that for some $\boldsymbol{c}^*$ and $\boldsymbol{y}^*$, equation (6.14) is satisfied;, but that for some subset, K, of the primary resources, we have:

$$(\forall k \in K)\colon \boldsymbol{b}_{k\cdot} \cdot \boldsymbol{y}^* > \overline{z}_k;$$

where '$b_{k\cdot}$' denotes the kth row of the matrix $\boldsymbol{B}$. If we then define the numbers μ_k by:

$$\mu_k = \frac{\overline{z}_k}{b_{k\cdot} \cdot \boldsymbol{y}^*} \quad \text{for } k \in K,$$

let:

$$\mu = \min_{k \in K} \mu_k,$$

and define:

$$\boldsymbol{y}' = \mu \boldsymbol{y}^* \text{ and } \boldsymbol{c}' = \mu \boldsymbol{x}^*,$$

we will have:

$$(\boldsymbol{I} - \boldsymbol{A})\boldsymbol{y}' = \mu(\boldsymbol{I} - \boldsymbol{A})\boldsymbol{y}^* = \mu \boldsymbol{c}^* = \boldsymbol{c}',$$

while:[3]

$$By' \leq \overline{z}. \tag{6.16}$$

Because of this, the key problem is to find a solution to (6.15). If this can be done, it may then be necessary to scale back demand in order that the resource constaints are satisfied, but the primary problem is to determine whether the proportions involved in a vector of final demands is or is not feasible. Consequently, instead of Problem A, we will be concentrating our attention upon the following.

Problem B. Given $\boldsymbol{c} \in \mathbb{R}^n_+$, does there exist $\boldsymbol{y} \in \mathbb{R}^n_+$ such that:

$$(\boldsymbol{I} - \boldsymbol{A})\boldsymbol{y} = \boldsymbol{c}?$$

For anyone who has taken a course in linear algebra, Problem B may appear rather trivial, at least at first glance. One's inclination is probably to simply remark that, in order to solve Problem B, we can simply let:

$$\boldsymbol{y} = (\boldsymbol{I} - \boldsymbol{A})^{-1}\boldsymbol{c},$$

and move on to other things. However, there are two major difficulties here. First the Leontief matrix, $(\boldsymbol{I} - \boldsymbol{A})$ may be singular. Secondly, even if the Leontief matrix is non-singular, it may be that the vector:

$$(\boldsymbol{I} - \boldsymbol{A})\boldsymbol{c},$$

is not nonnegative; and thus the mathematical solution found may not be economically meaningful. In connection with these two points, consider the following conditions regarding equation (6.15) and the Leontief matrix.

Condition I. There exists $\boldsymbol{y} \in \mathbb{R}^n_+$ such that:

$$(\boldsymbol{I} - \boldsymbol{A})\boldsymbol{y} > \boldsymbol{0}.$$

Condition II. For every $\boldsymbol{c} \in \mathbb{R}^n_+$, there exists $\boldsymbol{y} \in \mathbb{R}^n_+$ such that:

$$(\boldsymbol{I} - \boldsymbol{A}) = \boldsymbol{c}.$$

Condition III. The n upper left-hand corner principal minors of the matrix $(\boldsymbol{I}-\boldsymbol{A})$ are all positive; that is:

$$\begin{vmatrix} 1 - a_{11} & \dots & -a_{1k} \\ & \dots & \\ -a_{k1} & \dots & 1 - a_{kk} \end{vmatrix} > 0 \quad \text{for } k = 1, \dots, n.$$

[3]The student should verify this inequality: see Excercise 7, at the end of this chapter.

We will say that the technology matrix is **productive** if Condition (I) holds.

Now consider what is, in effect, the dual of the problem we have been examining. Suppose we consider the jth production sector as a potentially profit-maximizing entity. If a vector of prices for the n outputs, $\boldsymbol{p} \in \mathbb{R}^n_+$, is given, and the jth sector's output is y_j, its input cost for intermediate goods inputs (the other y_i's) is given by:

$$C_j(y_j) = \sum\nolimits_{i=1}^{n} p_i a_{ij} y_j.$$

Consequently, value added in the jth sector, given the price vector $\boldsymbol{p}$ is given by:

$$\left(p_j - \sum\nolimits_{i=1}^{n} p_i a_{ij}\right) y_j = \boldsymbol{p}^\top (\boldsymbol{I} - \boldsymbol{A})_j y_j,$$

where I am denoting the jth column of the Leontief matrix by '$(\boldsymbol{I} - \boldsymbol{A})_j$; and '$\boldsymbol{p}^\top$' denotes the transpose of the vector $\boldsymbol{p}$; that is, the row vector:

$$\boldsymbol{p}^\top = (p_1, \ldots, p_n).$$

In relation to this issue, consider the following conditions.

Condition I'. There exists $\boldsymbol{p} \in \mathbb{R}^n_+$ such that:

$$\boldsymbol{p}^\top (\boldsymbol{I} - \boldsymbol{A}) > \boldsymbol{0}. \tag{6.17}$$

Condition II'. Given any $\boldsymbol{v} \in \mathbb{R}^n_+$, there exists $\boldsymbol{p} \in \mathbb{R}^n_+$ such that:

$$\boldsymbol{p}^\top (\boldsymbol{I} - \boldsymbol{A}) = \boldsymbol{v}^\top.$$

We will say that the jth production sector is **viable**, given the price vector $\boldsymbol{p} \in \mathbb{R}^n_+$ iff:

$$\boldsymbol{p}^\top (\boldsymbol{I} - \boldsymbol{A})_j \geq 0;$$

and that the production sector is **sustainable**, iff there exists a price vector $\boldsymbol{p} \in \mathbb{R}^n_+$ such that:

$$\boldsymbol{p}^\top (\boldsymbol{I} - \boldsymbol{A}) > \boldsymbol{0}.$$

In other words, if the production sector is sustainable then there exists a price vector, $\boldsymbol{p} \in \mathbb{R}^n_+$ such that the jth production sector is viable, for each $j = 1, \ldots, n$; more simply, the production sector is sustainable iff Condition (I') holds.

One can prove the following, quite remarkable theorem; which establishes the aptness of the terminology we have introduced in this section. (For a proof, see Nikaido [1968, pp. 90–4].)

6.16 Theorem. *Given that the technology matrix, $\boldsymbol{A}$, is nonnegative, Conditions (I)–(III) and (I') and (II') are mutually equivalent.*

Thus if the technology matrix is productive, it is also sustainable; in fact, it will satisfy Condition (II'). Conversely, if the technology is sustainable, then it will satisfy Condition (II) as well. Because of this, it turns out that there is a very simple pair of conditions, either of which is sufficient to insure that all of Conditions (I)–(III) and (I') and (II') hold. Define:

$$r_i = \sum\nolimits_{j=1}^{n} a_{ij} \quad \text{for } i = 1, \ldots, n,$$

and:

$$s_j = \sum\nolimits_{i=1}^{n} a_{ij} \quad \text{for } j = 1, \ldots, n.$$

6.17 Theorem. (Brauer-Solow Conditions) Either of the following implies Conditions (I)–(III), as well as Conditions (I') and (II')

$$r_i < 1, \text{ for } i = 1, \ldots, n, \text{or}: \tag{6.18}$$

$$s_j < 1, \text{for } j = 1, \ldots, n. \tag{6.19}$$

Proof. Let $\boldsymbol{y}^* = (1, 1, \ldots, 1)^\top$. Then if (6.18) holds;

$$\left[(\boldsymbol{I} - \boldsymbol{A})\boldsymbol{y}^*\right]_i = 1 - \sum\nolimits_{j=1}^{n} a_{ij} = 1 - r_i > 0 \quad \text{for } i = 1, \ldots, n.$$

Thus Condition (I) holds, and it follows at once from Theorems 6.16 that Conditions (II), (III), (I') and (II') hold as well. A similar argument proves that (6.19) implies that Condition (I') holds. □

In the proof of the above result we have shown that the technology matrix is productive if (6.18) holds. A similar argument establishes the fact that the technology matrix is sustainable if (6.19) holds.

Before concluding this section, let's return to the formulation of 'Problem A,' which was set out earlier in this section. In particular, recall (6.14) and (6.15) in the specification of the problem. Let's change notation a bit here to require that the matrices $\boldsymbol{I}$ and $\boldsymbol{A}$ in (6.14) are both $r \times r$, where r is a positive integer; while the matrix $\boldsymbol{B}$ in (6.15) which defines the requirement of primary inputs is $s \times r$, where s is a positive integer, and where $n = r + s$. Finally, we note that the aggregate production set assumed to hold in input-output analysis is given by:

$$Y = \left\{ \boldsymbol{y} \in \mathbb{R}^n \mid (\exists \boldsymbol{z} \in \mathbb{R}^r_+) \colon \boldsymbol{y} = \begin{bmatrix} \boldsymbol{I} - \boldsymbol{A} \\ -\boldsymbol{B} \end{bmatrix} \boldsymbol{z} \right\}.$$

6.5 Profit Maximization

In this section, we will consider some basic results concerning profit maximization, and the relationship between profit maximization and efficiency. I will leave the proof of the result following the definition as an exercise.

6.18 Definition. Given a production set, $Y \subseteq \mathbb{R}^n$, we shall say that $\boldsymbol{y}^* \in Y$ is **efficient** (in Y) iff there exists no $\boldsymbol{y} \in Y$ satisfying $\boldsymbol{y} > \boldsymbol{y}^*$.

6.19 Proposition. *If $\boldsymbol{y}^* \in Y$ and $\boldsymbol{p}^* \in \mathbb{R}^n_{++}$ are such that:*

$$(\forall \boldsymbol{y} \in Y) \colon \boldsymbol{p}^* \cdot \boldsymbol{y}^* \geq \boldsymbol{p}^* \cdot \boldsymbol{y},$$

then $\boldsymbol{y}^*$ is efficient in Y.

6.20 Example. Suppose $f\colon -\mathbb{R}_+ \to \mathbb{R}_+$, and define $Y \subseteq \mathbb{R}^2$ by:

$$Y = \{\boldsymbol{y} \in \mathbb{R}^2 \mid y_1 \leq 0 \;\&\; 0 \leq y_2 \leq f(y_1)\}.$$

While the definition just presented yields a quite conventional production set, it follows from Proposition 6.19 that if we are analyzing the behavior of a profit-maximizing firm, then we can confine our attention to *efficient* points in the production set. Obviously, a production vector $\boldsymbol{y} \in Y$ is efficient if, and only if $y_2 = f(y_1)$. Consequently, we can analyze this sort of situation with the simplest sort of production function; one showing a single output as a function of a simple input. The only unfamiliar aspect of this example is that in order to insure that Y is convex, it is usual to assume that for all $y_1 \in -\mathbb{R}_+$:

$$f'(y_1) \leq 0 \;\;\&\;\; f''(y_1) \leq 0. \quad \square$$

The following two mathematical results will be used several times in the remainder of this book, but will be stated here without proof. For those interested, proofs are given in, for example, Moore [1999, pp. 297–300].

6.21 Theorem. *If A and B are disjoint and non-empty convex subsets of $\mathbb{R}^n$, then there exists a hyperplane separating A and B; that is, there exists $\mathbf{p}^* \in \mathbb{R}^n \setminus \{\mathbf{0}\}$ such that:*

$$\sup_{\boldsymbol{a}\in A} \boldsymbol{p}^* \cdot \boldsymbol{a} \leq \inf_{\boldsymbol{b}\in B} \boldsymbol{p}^* \cdot \boldsymbol{b}.$$

In other words, there exists a non-zero $\mathbf{p}^* \in \mathbb{R}^n$ and $\alpha \in \mathbb{R}$ such that, for all $\mathbf{a} \in A$ and all $\mathbf{b} \in B$:

$$\mathbf{p}^* \cdot \boldsymbol{a} \leq \alpha \leq \boldsymbol{p}^* \cdot \boldsymbol{b}.$$

6.22 Theorem. *If A and B are nonempty, disjoint, closed convex sets, at least one of which is bounded, then there exists a hyperplane strongly separating A and B; that is, there exists a nonzero $\mathbf{p} \in \mathbb{R}^n$ and $\beta \in \mathbb{R}$ such that:*

$$\sup_{\boldsymbol{a}\in A} \boldsymbol{p} \cdot \boldsymbol{a} < \beta < \inf_{\boldsymbol{b}\in B} \boldsymbol{p} \cdot \boldsymbol{b}.$$

6.23 Theorem. *If Y is a convex (production) subset of $\mathbb{R}^n$, and $\boldsymbol{y}^* \in Y$ is efficient, then there exists $\mathbf{p}^* \in \mathbb{R}^n_+ \setminus \{\mathbf{0}\}$ such that:*

$$(\forall \boldsymbol{y} \in Y)\colon \boldsymbol{p}^* \cdot \boldsymbol{y} \leq \boldsymbol{p}^* \cdot \boldsymbol{y}^*.$$

Proof. Define the set B by:

$$B = \{\boldsymbol{y} \in \mathbb{R}^n \mid \boldsymbol{y} > \boldsymbol{y}^*\}.$$

Then B is a convex set, and, since $\boldsymbol{y}^*$ is efficient in Y, $Y \cap B = \emptyset$. It then follows from Theorem 6.21 that there exists a non-zero (price) vector in $\mathbb{R}^n$ and a real number α satisfying:

$$(\forall \boldsymbol{y} \in Y)(\forall \boldsymbol{z} \in B)\colon \boldsymbol{p}^* \cdot \boldsymbol{y} \leq \alpha \leq \boldsymbol{p}^* \cdot \boldsymbol{z}. \tag{6.20}$$

Our proof will therefore be complete if we can show that:

$$\boldsymbol{p}^* > \mathbf{0}, \tag{6.21}$$

and:

$$\boldsymbol{p}^* \cdot \boldsymbol{y}^* = \alpha. \tag{6.22}$$

To prove (6.21), suppose, by way of obtaining a contradiction, that for some $h \in \{1, \ldots, n\}$, we have $p_h^* < 0$; and define $\boldsymbol{z}^* \in \mathbb{R}^n$ by:

$$\boldsymbol{z}^* = \boldsymbol{y}^* - p_h^* \boldsymbol{e}_h,$$

where $\boldsymbol{e}_h$ is the hth unit coordinate vector. Then, since $p_h^* < 0$, it follows that $\boldsymbol{z}^* > \boldsymbol{y}^*$, and thus $\boldsymbol{z}^* \in B$. However,

$$\alpha - \boldsymbol{p}^* \cdot \boldsymbol{z}^* \geq \boldsymbol{p}^* \cdot \boldsymbol{y}^* - \boldsymbol{p}^* \cdot \boldsymbol{z}^* = \boldsymbol{p}^* \cdot (\boldsymbol{y}^* - \boldsymbol{z}^*) = \boldsymbol{p}^* \cdot (p_h^* \boldsymbol{e}_h) = (p_h^*)^2 > 0.$$

Therefore, $\alpha - \boldsymbol{p}^* \cdot \boldsymbol{z}^* > 0$, that is:

$$\alpha > \boldsymbol{p}^* \cdot \boldsymbol{z}^*;$$

which, since $\boldsymbol{z}^* \in B$, contradicts (6.20). Consequently, we see that (6.21) must hold.

To prove (6.22), let $\epsilon > 0$ be given, and let i be such that $p_i^* > 0$. Defining:

$$\boldsymbol{z} = \boldsymbol{y}^* + (\epsilon / p_i^*) \boldsymbol{e}_i,$$

we have $\boldsymbol{z} > \boldsymbol{y}^*$, so that $\boldsymbol{z} \in B$, and thus:

$$\boldsymbol{p}^* \cdot \boldsymbol{z}^* \geq \alpha.$$

However, we then have:

$$\boldsymbol{p}^* \cdot \boldsymbol{y}^* \leq \alpha \leq \boldsymbol{p}^* \cdot \boldsymbol{z} = \boldsymbol{p}^* \cdot [\boldsymbol{y}^* + (\epsilon / p_i^*) \boldsymbol{e}_i] = \boldsymbol{p}^* \cdot \boldsymbol{y}^* + (\epsilon / p_i^*) p_i^* = \boldsymbol{p}^* \cdot \boldsymbol{y}^* + \epsilon.$$

Thus we see that, for all $\epsilon > 0$, we have:

$$\boldsymbol{p}^* \cdot \boldsymbol{y}^* \leq \alpha \leq \boldsymbol{p}^* \cdot \boldsymbol{y}^* + \epsilon,$$

and it follows that $\boldsymbol{p}^* \cdot \boldsymbol{y}^* = \alpha$. □

Given an arbitrary price vector, $\boldsymbol{p}$, there may or may not exist a production vector, $\boldsymbol{y}^* \in Y$ which maximizes profits on Y. Obviously, however, we have a particular interest in those price vectors for which such a profit-maximizing output vector exists.

6.24 Definitions. For a production set, Y, we define:

1. $\Pi = \Pi(Y) = \{\boldsymbol{p} \in \mathbb{R}^n \mid (\exists \boldsymbol{y}^* \in Y)(\forall \boldsymbol{y} \in Y) \colon \boldsymbol{p} \cdot \boldsymbol{y}^* \geq \boldsymbol{p} \cdot \boldsymbol{y}\}$,
2. and, for $\boldsymbol{p} \in \Pi$, we then define:

$$\pi(\boldsymbol{p}) = \max_{\boldsymbol{y} \in Y} \boldsymbol{p} \cdot \boldsymbol{y},$$

and:

$$\sigma(\boldsymbol{p}) = \{\boldsymbol{y} \in Y \mid \boldsymbol{p} \cdot \boldsymbol{y} = \pi(\boldsymbol{p})\}.$$

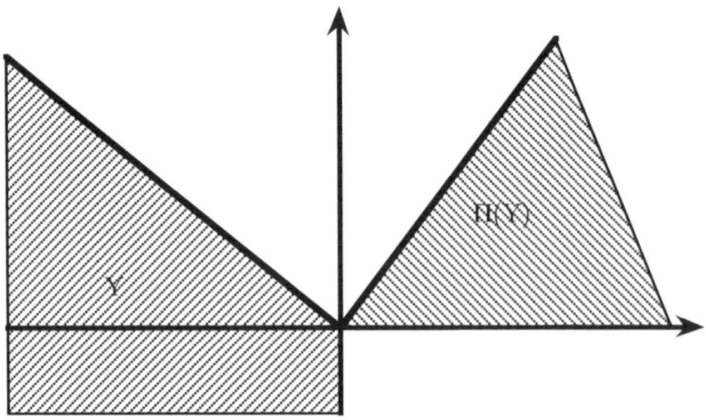

Figure 6.5: $\Pi(Y)$**, for linear** Y**.**

Figure 6.5, above, illustrates the relationship between Y and $\Pi(Y)$ for a linear production set.

I will leave the proof of the first of the following two results as an exercise.

6.25 Proposition. *If Y satisfies free or semi-free disposability, and $\boldsymbol{p}^* \in \Pi(Y)$, then $\boldsymbol{p}^* \geq \boldsymbol{0}$.*

6.26 Proposition. *Whatever the form of Y, $\Pi = \Pi(Y)$ will be a non-empty cone; and $\pi(\cdot)$ and $\sigma(\cdot)$ will be positively homogeneous of degrees one and zero, respectively. Furthermore, if Π is convex, then the profit function, $\pi(\cdot)$ is convex on Π.*

Proof. I will only prove the last part of this result here; leaving the remainder of the proof as an exercise.

Let $\boldsymbol{p}', \boldsymbol{p}^* \in \Pi(Y)$, and let $\theta \in [0,1]$. Then, given an arbitrary $\boldsymbol{y} \in Y$, we have, since both θ and $1-\theta$ are nonnegative:

$$\big[\theta\boldsymbol{p}' + (1-\theta)\boldsymbol{p}^*\big]\cdot\boldsymbol{y} = \theta\boldsymbol{p}'\cdot\boldsymbol{y} + (1-\theta)\boldsymbol{p}^*\cdot\boldsymbol{y} \leq \theta\pi(\boldsymbol{p}') + (1-\theta)\pi(\boldsymbol{p}^*).$$

Since $\Pi(Y)$ is convex:

$$\theta\boldsymbol{p}' + (1-\theta)\boldsymbol{p}^* \in \Pi(Y),$$

and since $\boldsymbol{y}$ was an arbitrary element of Y, it follows that:

$$\pi\big[\theta\boldsymbol{p}' + (1-\theta)\boldsymbol{p}^*\big] \leq \theta\pi(\boldsymbol{p}') + (1-\theta)\pi(\boldsymbol{p}^*). \quad \square$$

The following rather remarkable result first appeared in Debreu [1959, p. 47]. Interestingly enough, however, Debreu credits the result to Samuelson [1947, Chapter 4]. (Samuelson's Chapter 4 is concerned with the theory of revealed preference for consumer demand.)

6.27 Theorem. *If $\boldsymbol{p}', \boldsymbol{p}'' \in \Pi(Y)$, $\boldsymbol{y}' \in \sigma(\boldsymbol{p}')$, and $\boldsymbol{y}'' \in \sigma(\boldsymbol{p}'')$, then:*

$$\Delta\boldsymbol{p}\cdot\Delta\boldsymbol{y} \geq 0;$$

where we define:

$$\Delta \boldsymbol{p} = \boldsymbol{p}'' - \boldsymbol{p}' \ \text{ and } \ \Delta \boldsymbol{y} = \boldsymbol{y}'' - \boldsymbol{y}'.$$

Furthermore, if $\mathbf{p}''$ *takes the form:*

$$\boldsymbol{p}'' = \boldsymbol{p}' + \Delta p_j \boldsymbol{e}_j,$$

where Δp_j *is a non-zero real number, and* $\mathbf{e}_j$ *is the* j^{th} *unit coordinate vector, then:*

$$\Delta y_j / \Delta p_j \geq 0. \tag{6.23}$$

Proof. By profit maximization, we have:

$$\boldsymbol{p}'' \cdot \boldsymbol{y}'' \geq \boldsymbol{p}'' \cdot \boldsymbol{y}',$$

and thus:

$$\boldsymbol{p}'' \cdot (\boldsymbol{y}'' - \boldsymbol{y}') = \boldsymbol{p}'' \cdot \Delta \boldsymbol{y} \geq 0. \tag{6.24}$$

Similarly, $\boldsymbol{p}' \cdot \boldsymbol{y}' \geq \boldsymbol{p}' \cdot \boldsymbol{y}''$, and thus:

$$0 \geq \boldsymbol{p}' \cdot \boldsymbol{y}'' - \boldsymbol{p}' \cdot \boldsymbol{y}' = \boldsymbol{p}' \cdot \Delta \boldsymbol{y}. \tag{6.25}$$

Combining (6.24) and (6.25), we have:

$$\boldsymbol{p}'' \cdot \Delta \boldsymbol{y} \geq 0 \geq \boldsymbol{p}' \cdot \Delta \boldsymbol{y};$$

and thus:

$$0 \leq \boldsymbol{p}'' \cdot \Delta \boldsymbol{y} - \boldsymbol{p}' \cdot \Delta \boldsymbol{y} = \Delta \boldsymbol{p} \cdot \Delta \boldsymbol{y}. \tag{6.26}$$

Now suppose that $\boldsymbol{p}''$ takes the form:

$$\boldsymbol{p}'' = \boldsymbol{p}' + \Delta p_j \boldsymbol{e}_j,$$

where Δp_j is non-zero. Then in this case, $\Delta \boldsymbol{p}$ takes the form:

$$\Delta \boldsymbol{p} = \Delta p_j \boldsymbol{e}_j = (0, \ldots, 0, \Delta p_j, 0, \ldots, 0);$$

and thus, using (6.26):

$$\Delta \boldsymbol{p} \cdot \Delta \boldsymbol{y} = \Delta p_j \Delta y_j \geq 0 \tag{6.27}$$

(where $\Delta y_j \stackrel{\text{def}}{=} y_j'' - y_j'$). Dividing both sides of (6.27) by $(\Delta p_j)^2$ yields (6.23). □

The following simple little result is useful surprisingly often in examples and applications of general equilibrium theory. I will leave the proof as an exercise.

6.28 Proposition. *If* $\mathbf{0} \in Y$, *and* $\boldsymbol{p} \in \mathbb{R}^n$ *satisfies:*

$$(\forall \boldsymbol{y} \in Y) \colon \boldsymbol{p} \cdot \boldsymbol{y} \leq 0,$$

then $\boldsymbol{p} \in \Pi(Y)$, *and* $\pi(\boldsymbol{p}) = 0$.

By making use of the preceding result, we can give a complete characterization of $\Pi(Y)$ and $\pi(\cdot)$ for production sets containing the origin and satisfying non-decreasing returns to scale; as follows.

6.29 Proposition. *If $Y \subseteq \mathbb{R}^n$ satisfies non-decreasing returns to scale and $\mathbf{0} \in Y$, then:*

$$\Pi(Y) = \{\boldsymbol{p} \in \mathbb{R}^n \mid (\forall \boldsymbol{y} \in Y)\colon \boldsymbol{p} \cdot \boldsymbol{y} \leq 0\}.$$

Furthermore, if $\boldsymbol{p} \in \Pi(Y)$ and $\boldsymbol{y} \in \sigma(\boldsymbol{p})$, then $\boldsymbol{p} \cdot \boldsymbol{y} = 0$. Thus we have:

$$\big(\forall \boldsymbol{p} \in \Pi(Y)\big)\colon \pi(\boldsymbol{p}) = 0.$$

Proof. Suppose $\boldsymbol{p} \in \Pi(Y)$ and $\boldsymbol{y}^* \in \sigma(\boldsymbol{p})$. Then, since $\mathbf{0} \in Y$, we must have $\boldsymbol{p} \cdot \boldsymbol{y}^* \geq 0$ Suppose, by way of obtaining a contradiction, that $\boldsymbol{p} \cdot \boldsymbol{y}^* > 0$. Since Y satisfies non-decreasing returns to scale, $2\boldsymbol{y}^* \in Y$. However, since $\boldsymbol{p} \cdot \boldsymbol{y}^* > 0$:

$$\boldsymbol{p} \cdot (2\boldsymbol{y}^*) = 2\boldsymbol{p} \cdot \boldsymbol{y}^* > \boldsymbol{p} \cdot \boldsymbol{y}^*;$$

contradicting the assumption that $\boldsymbol{y}^* \in \sigma(\boldsymbol{p})$. Thus we see that we must have $\boldsymbol{p} \cdot \boldsymbol{y}^* = \pi(\boldsymbol{p}) \leq 0$; and thus:

$$(\forall \boldsymbol{y} \in Y)\colon \boldsymbol{p} \cdot \boldsymbol{y} \leq 0.$$

Combining this with Proposition 6.28 completes our proof. □

6.6 Profit Maximizing with Constant Returns to Scale*

Recall that, in terms of the definitions in this book, a production set satisfies constant returns to scale if, and only if, it is a cone; that is, if and only if for all $\boldsymbol{y} \in Y$, and all $\theta > 0$, we have $\theta \boldsymbol{y} \in Y$. The case in which Y is linear is a special case of this, and our first result of this section deals with a special sort of cone.

6.30 Proposition. *If $Y \subseteq \mathbb{R}^n$ is a proper linear production set; so that there exists a semi-positive $m \times n$ matrix, $\boldsymbol{B}$, and sets I and J satisfying the conditions of Definition 6.14, then defining $K = I \cap J$, we have the following. Given any $\boldsymbol{a} \in \Delta_m \cap \mathbb{R}^n_{++}$ and any $\boldsymbol{p} \in \mathbb{R}^n$ satisfying:*

$$(\forall i \in N \setminus K)\colon p_i = 0 \;\&\; (\forall k \in K)\colon p_k > 0,$$

the price vector $\boldsymbol{p}^* \in \mathbb{R}^n$ defined by:[4]

$$\boldsymbol{p}^* = \boldsymbol{B}^\top \boldsymbol{a} + \boldsymbol{p},$$

is a strictly positive element of Π.

Proof. Suppose $\boldsymbol{p}^*$ has the indicated form, and let $\boldsymbol{y} \in Y$ be arbitrary. Then:

$$\boldsymbol{p}^* \cdot \boldsymbol{y} = (\boldsymbol{p}^*)^\top \boldsymbol{y} = (\boldsymbol{a}^\top \boldsymbol{B} + \boldsymbol{p}^\top)\boldsymbol{y} = \boldsymbol{a}^\top \boldsymbol{B}\boldsymbol{y} + \boldsymbol{p}^\top \boldsymbol{y} = \boldsymbol{a}^\top \boldsymbol{B}\boldsymbol{y};$$

where the last equality is by the definition of K and the fact that $p_j = 0$ for all $j \notin K$. However, since $\boldsymbol{a} \gg \mathbf{0}$ and $\boldsymbol{B}\boldsymbol{y} \leq \mathbf{0}$ by the fact that $\boldsymbol{y} \in Y$, we then see that $\boldsymbol{p} \cdot \boldsymbol{y} \leq 0$. Thus it follows from Proposition 6.28 that $\boldsymbol{p}^* \in \Pi(Y)$. The fact

[4] We denote the transpose of a matrix, $\boldsymbol{B}$, by '$\boldsymbol{B}^\top$.'

that $\boldsymbol{p}^* \gg \mathbf{0}$ follows from the definition of a proper linear technology, the fact that $\boldsymbol{a} \gg \mathbf{0}$ and the specification of $\boldsymbol{p}$. Details will be left as an exercise. □

A type of cone which is of particular interest for us is the cone dual to another cone, where we define this as follows.[5]

6.31 Definition. If $K \subseteq \mathbb{R}^n$ is a cone, we define the **dual cone for $\boldsymbol{K}$, $\boldsymbol{K^*}$**, by:

$$K^* = \{\boldsymbol{y} \in K \mid (\forall \boldsymbol{x} \in K) \colon \boldsymbol{y} \cdot \boldsymbol{x} \leq 0\}.$$

A reason for our particular interest in the dual cone is that if Y is a cone which contains the origin in $\mathbb{R}^n$, then $\Pi(Y) = Y^*$, a fact whose proof I will leave as an exercise. Having established a reason for an interest in the dual cone, let's take a look at some of the properties of same. The proof of the first of the following two results will be left as an exercise. A closed cone, incidentally, is simply a cone which is also a closed set. Notice that, while a cone does not necessarily contain the origin, a closed cone necessarily does.

6.32 Proposition. *If $Y \subseteq \mathbb{R}^n$ is a cone, its dual, Y^*, is a closed, convex cone.*[6]

If K^* is the dual cone to a cone, K, K^* itself has a dual cone, which we denote by 'K^{**};' that is:

$$K^{**} \stackrel{\text{def}}{=} (K^*)^*.$$

The following result sets out the basic properties of this second dual.

6.33 Proposition. *If $K \subseteq \mathbb{R}^n$ is a cone, then:*
i. we have $K \subseteq K^{**}$, and:
ii. $K = K^{**}$ if, and only if, K is closed and convex.

Proof. I will leave the proof of part (i) of this result and the of the fact that $K^{**} = K$ *only if* K is a closed convex coneas an exercise. To prove that $K^{**} \subseteq K$ if K is a closed convex cone, suppose K is a closed convex cone, and that $\boldsymbol{a} \in \mathbb{R}^n$ is a point which is *not* an element of K. Then by Theorem 6.22 of the previous section, there exists a nonzero $\boldsymbol{p} \in \mathbb{R}^n$, and $\beta \in \mathbb{R}$ such that:

$$\boldsymbol{p} \cdot \boldsymbol{a} > \beta = \sup_{\boldsymbol{x} \in K} \boldsymbol{p} \cdot \boldsymbol{x}. \tag{6.28}$$

Now, we must then have:

$$(\forall \boldsymbol{x} \in K) \colon \boldsymbol{p} \cdot \boldsymbol{x} \leq 0;$$

for, suppose that for some $\boldsymbol{x}^* \in K$ we have $\boldsymbol{p} \cdot \boldsymbol{x}^* > 0$. Then for all $\theta \in \mathbb{R}_{++}$, $\theta \boldsymbol{x}^* \in K$; and:

$$\lim_{\theta \to +\infty} \boldsymbol{p} \cdot \boldsymbol{\theta} \boldsymbol{x}^* = \Big[\lim_{\theta \to +\infty} \theta\Big] \boldsymbol{p} \cdot \boldsymbol{x}^* = +\infty,$$

[5] In the mathematical literature, it is quite common to see the dual cone defined as the set of points having a nonnegative inner product with each element of Y, rather than, as here, those having a nonpositive inner product with each element. The mathematical properties of the two types of dual are, however, precisely the same, and for us the more directly useful definition is the one given here.

[6] Remember that a convex cone is simply a cone which is also a convex set. The result stated here does not, as you may have noticed, really need the assumption that Y is a cone.

which contradicts (6.28). Since $\mathbf{0} \in K$ by virtue of the fact that K is closed, it now follows that $\beta = 0$; and, therefore, that $\boldsymbol{p} \in K^*$. However, since $\boldsymbol{p} \cdot \boldsymbol{a} > 0$, we can now see that $\boldsymbol{a} \notin K^{**}$. Therefore, if $\boldsymbol{y} \in K^{**}$, we must have $\boldsymbol{y} \in K$; that is, $K^{**} \subseteq K$ Since, by part (i) we always have $K \subseteq K^{**}$, it now follows that the two sets are equal in this case. □

This last result is the basis of one type of duality result in production theory; namely, if Y is a closed convex cone, then $Y = [\Pi(Y)]^*$. In other words, if we specify, or have empirical data providing a reason to believe that a production set Y is a closed convex cone, then Y is completely determined by $\Pi(Y)$, the set of prices for which a maximum profit output exists. As one simple application: in such a situation, if we assume or specify that a maximum profit level exists for each $\boldsymbol{p} \in \mathbb{R}^n_+$, then Y can only be $-\mathbb{R}^n_+$; in other words, no positive production can take place.

Another useful mathematical result concerning cones and their duals is the following theorem; a proof of which can be found in Nikaido [1968, pp. 35–6].

6.34 Theorem. *Let $K \subseteq \mathbb{R}^n$ be a closed convex cone. If K contains no semi-positive element, then K^* contains a positive element, and vice-versa.*

Thus, if Y is a closed convex production set satisfying constant returns to scale, and also satisfies 6.4.1 (Impossibility of Free Production), then $\Pi(Y)$ contains a strictly postive price, $\boldsymbol{p} \gg \mathbf{0}$.

6.7 Production in General Equilibrium Theory

In our basic general equlibrium model of the next chapter, we suppose that the number of producers is a given positive integer, ℓ; and producers are indexed by k ($k = 1, \ldots, \ell$). The kth producer chooses a production plan (or **production vector**, or **netput vector**), $\boldsymbol{y}_k$, from some non-empty subset of $\mathbb{R}^n, Y_k$. We refer to the set Y_k as the kth producer's **production set**.

6.35 Definitions. If $\boldsymbol{y}_k \in Y_k$ for $k = 1, \ldots, \ell$, then the vector:

$$\boldsymbol{y} = \sum\nolimits_{k=1}^{\ell} \boldsymbol{y}_k,$$

is called the **aggregate** (or **total**) **production vector**; and:

$$Y \stackrel{\text{def}}{=} \sum\nolimits_{k=1}^{\ell} Y_k,$$

is called the **aggregate** (or **total**) **production set**.

The notation of the previous sections is, in this context, extended as follows.

Notation/Definitions. For $k = 1, \ldots, \ell$, we define:
1. $\Pi_k = \{\boldsymbol{p} \in \mathbb{R}^n \setminus \{\mathbf{0}\} \mid (\exists \boldsymbol{y}^* \in Y_k)(\forall \boldsymbol{y} \in Y_k) \colon \boldsymbol{p} \cdot \boldsymbol{y}^* \geq \boldsymbol{p} \cdot \boldsymbol{y}\}$,
2. and for $\boldsymbol{p} \in \Pi_k$, we then define:

$$\pi_k(\boldsymbol{p}) = \max_{\boldsymbol{y} \in Y_k} \boldsymbol{p} \cdot \boldsymbol{y} \quad \text{and} \quad \sigma_k(\boldsymbol{p}) = \{\boldsymbol{y} \in Y_k \mid \boldsymbol{p} \cdot \boldsymbol{y} = \pi_k(\boldsymbol{p})\}.$$

3. We use a similar notation, simply dropping the subscript 'k,' to denote the corresponding concepts for the aggregate production set, Y.

If one is given ℓ production sets, Y_k, it may be very difficult to characterize the aggregate production set, Y. Fortunately, we will generally not need to do so; it is usually sufficient to use the formal definition of the summation set. There are some cases, however, in which it is easy and may be useful to characterize the set Y. One interesting example, which corresponds to a model used quite frequently in the theoretical public economics literature, is presented in the following.

6.36 Example. Suppose there are $n-1$ firms (that is, $\ell = n-1$); with the kth firm's production set given by:

$$Y_k = \{\boldsymbol{y}_k \in \mathbb{R}^n \mid y_{kk} \geq 0, y_{kn} \leq 0, c_k y_{kk} + y_{kn} \leq 0 \ \& \ y_{kj} = 0 \text{ for } j \notin \{k, n\}\}; \quad (6.29)$$

where we suppose:

$$c_k > 0 \text{ for } k = 1, \dots, n-1.$$

In other words, we suppose the kth firm produces only the kth commodity, and uses only the nth commodity (which we generally suppose to be labor) as an input. In this case, the aggregate production set is given by:

$$Y = \Big\{\boldsymbol{y} \in \mathbb{R}^n \mid \sum_{k=1}^{\ell-1} c_k y_k + y_n \leq 0, y_k \geq 0, \text{ for } k = 1, \dots, n-1, \& \ y_n \leq 0\Big\}. \quad (6.30)$$

In other words, in the framework of the terminology introduced in Definition 6.14, the aggregate production set is a proper linear technology, with $I = \{n\}$ and $J = \{1, \dots, n-1\}$.

We can prove this for the special case in which $\ell = 2$ as follows (the basic argument for the case in which ℓ is an arbitrary positive integer is basically the same; the notation is just messier).

Suppose first that $\boldsymbol{y}_k \in Y_k$, for $k = 1, 2$. Then $\boldsymbol{y}_1$ and $\boldsymbol{y}_2$ are of the form:

$$\boldsymbol{y}_1 = (y_{11}, 0, y_{13}) \text{ and } \boldsymbol{y}_2 = (0, y_{22}, y_{23}),$$

with:

$$y_{11} \geq 0, y_{13} \leq 0, c_1 y_{11} + y_{13} \leq 0, \quad (6.31)$$

and:

$$y_{22} \geq 0, y_{23} \leq 0, c_2 y_{22} + y_{23} \leq 0. \quad (6.32)$$

It is then easy to check (I will leave it to you to verify the details) that:

$$\boldsymbol{y}_1 + \boldsymbol{y}_2 = (y_{11}, y_{22}, y_{13} + y_{23}),$$

is an element of the set Y defined in (6.30).

Conversely, suppose $\boldsymbol{y} \in Y$. In this case, if we define $\boldsymbol{y}_1$ and $\boldsymbol{y}_2$ by:

$$y_{11} = y_1, y_{12} = 0, y_{13} = -c_1 y_1,$$

and:

$$y_{21} = 0, y_{22} = y_2, y_{23} = y_3 + c_1 y_1,$$

it is easy to show that $\boldsymbol{y}_k \in Y_k$, for $k = 1, 2$, and that:

$$\boldsymbol{y}_1 + \boldsymbol{y}_2 = \boldsymbol{y}. \quad \square$$

I will leave the proof of the following mathematical result as an exercise. It will often be useful in our work in general equilibrium theory.

6.37 Proposition. *If A_j is a convex subset of $\mathbb{R}^n$, for $j = 1, \ldots, m$, then the set A defined by:*

$$A = \sum\nolimits_{j=1}^{m} A_j,$$

is a convex set.

One implication of this last result is this: if each production set, Y_k, is convex, then so is the aggregate production set. This implication is particularly interesting in view of our next result.

Suppose we actually knew the exact form of the aggregate production set in an economy, and also knew the sum of the individual production vectors chosen by the ℓ firms in the economy, given a price vector, $\boldsymbol{p}^*$. Could we then tell if the individual firms were maximizing profits, given $\boldsymbol{p}^*$, even given that we did not know the production vectors chosen by the individual firms? The answer is, yes, we could; as is established in the following theorem.

6.38 Theorem. *Suppose $\boldsymbol{p}^*$ is a non-zero price vector, let Y_k be a production set (a non-empty subset of $\mathbb{R}^n$), for $k = 1, \ldots, \ell$, and let:*

$$Y \stackrel{def}{=} \sum\nolimits_{k=1}^{\ell} Y_k,$$

be the corresponding aggregate production set. Then:

1. if $\boldsymbol{y}^* \in Y$ is such that:

$$(\forall \boldsymbol{y} \in Y)\colon \boldsymbol{p}^* \cdot \boldsymbol{y} \leq \boldsymbol{p}^* \cdot \boldsymbol{y}^*, \tag{6.33}$$

and $\boldsymbol{y}'_k \in Y_k$ $(k = 1, \ldots, \ell)$ are such that:

$$\boldsymbol{y}^* = \sum\nolimits_{k=1}^{\ell} \boldsymbol{y}'_k,$$

then for each k, we have:

$$(\forall \boldsymbol{y}_k \in Y_k)\colon \boldsymbol{p}^* \cdot \boldsymbol{y}_k \leq \boldsymbol{p}^* \cdot \boldsymbol{y}'_k. \tag{6.34}$$

2. Conversely, if (6.34) holds, for $\boldsymbol{y}'_k \in Y_k$ $(k = 1, \ldots, \ell)$, and we define:

$$\boldsymbol{y}^* = \sum\nolimits_{k=1}^{\ell} \boldsymbol{y}'_k,$$

then $\boldsymbol{y}^*$ will satisfy (6.33).

Proof. Suppose $\boldsymbol{y}^* \in Y$ is such that:

$$(\forall \boldsymbol{y} \in Y)\colon \boldsymbol{p}^* \cdot \boldsymbol{y} \leq \boldsymbol{p}^* \cdot \boldsymbol{y}^*, \tag{6.35}$$

and that:

$$\boldsymbol{y}^* = \sum_{k=1}^{\ell} \boldsymbol{y}_k',$$

with $\boldsymbol{y}_k' \in Y_k$, for $k = 1, \dots, \ell$. We will prove that $\boldsymbol{y}_1'$ maximizes profits over Y_1. A similar argument then establishes the general case.

Accordingly, let $\boldsymbol{y}_1 \in Y_1$ be arbitrary. Then if we define $\boldsymbol{y}'' \in Y$ by:

$$\boldsymbol{y}'' = \boldsymbol{y}_1 + \sum_{k=2}^{\ell} \boldsymbol{y}_k',$$

we see from (6.34) that we must have:

$$\boldsymbol{p}^* \cdot \boldsymbol{y}^* \geq \boldsymbol{p}^* \cdot \boldsymbol{y}''. \tag{6.36}$$

However:

$$\boldsymbol{p}^* \cdot \boldsymbol{y}^* = \boldsymbol{p}^* \cdot \left(\sum_{k=1}^{\ell} \boldsymbol{y}_k\right) = \boldsymbol{p}^* \cdot \boldsymbol{y}_1' + \sum_{k=2}^{\ell} \boldsymbol{p}^* \cdot \boldsymbol{y}_k', \tag{6.37}$$

while:

$$\boldsymbol{p}^* \cdot \boldsymbol{y}'' = \boldsymbol{p}^* \cdot \left(\boldsymbol{y}_1 + \sum_{k=2}^{\ell} \boldsymbol{y}_k\right) = \boldsymbol{p}^* \cdot \boldsymbol{y}_1 + \sum_{k=2}^{\ell} \boldsymbol{p}^* \cdot \boldsymbol{y}_k', \tag{6.38}$$

and thus, from (6.36)–(6.38), we see that;

$$\boldsymbol{p}^* \cdot \boldsymbol{y}_1' \geq \boldsymbol{p}^* \cdot \boldsymbol{y}_1.$$

Since $\boldsymbol{y}_1$ was an arbitrary element of Y_1, it now follows that $\boldsymbol{y}_1'$ maximizes profits on Y_1.

Now suppose that:

$$\boldsymbol{y}^* = \sum_{k=1}^{\ell} \boldsymbol{y}_k', \tag{6.39}$$

where $\boldsymbol{y}_k'$ maximizes profits on Y_k, for $k = 1, \dots, \ell$, and let $\boldsymbol{y} \in Y$ be arbitrary. Then there exist $\boldsymbol{y}_k \in Y_k$, for $k = 1, \dots, \ell$, such that:

$$\boldsymbol{y} = \sum_{k=1}^{\ell} \boldsymbol{y}_k.$$

Since $\boldsymbol{y}_k'$ maximizes profits over Y_k, for each k, it then follows easily from (6.39) that:

$$\boldsymbol{p}^* \cdot \boldsymbol{y}^* \geq \boldsymbol{p}^* \cdot \boldsymbol{y}. \quad \square$$

The following is a more or less immediate implication of Theorem 6.38.

6.39 Corollary. *Given the ℓ production sets, $Y_1, \dots, Y_\ell$ and corresponding aggregate production set, Y, we have:*

$$\Pi = \bigcap_{k=1}^{\ell} \Pi_k;$$

and, for all $\boldsymbol{p} \in \Pi$:

$$\pi(\boldsymbol{p}) = \sum_{k=1}^{\ell} \pi_k(\boldsymbol{p}) \ \ \textit{and} \ \ \sigma(\boldsymbol{p}) = \sum_{k=1}^{\ell} \sigma_k(\boldsymbol{p}).$$

6.8 Activity Analysis*

In this section, we will discuss another method of specifying a linear production set; namely Activity Analysis. Activity Analysis no longer commands the attention of economists to the extent which it did thirty-five years or so ago, but it still can be quite useful in applied work, as well as involving some interesting theoretical problems. We will, however, consider only a very abbreviated presentation of this theory here.

Consider a production process, which we will refer to as a **productive activity**, and suppose that if the activity is operated at the level $z \in \mathbb{R}_+$, then the input requirements are given by:

$$\boldsymbol{u} = \boldsymbol{a}z,$$

while output is given by:

$$\boldsymbol{x} = \boldsymbol{b}z,$$

where $\boldsymbol{a}$ and $\boldsymbol{b}$ are semi-positive q and m-vectors, respectively. We can express the corresponding production set, Y, as:

$$Y = \{(\boldsymbol{v}, \boldsymbol{x}) \in \mathbb{R}^{q+m} \mid (\exists z \in \mathbb{R}_+)\colon (\boldsymbol{v}, \boldsymbol{x}) = (-\boldsymbol{a}, \boldsymbol{b})z\}. \tag{6.40}$$

For example, suppose that $q = m = 1, \boldsymbol{a} = 2$, and $\boldsymbol{b} = 1$. Then:

$$\begin{aligned} Y &= \{(v, x) \in \mathbb{R}^2 \mid (\exists z \in \mathbb{R}_+)\colon (v, x) = (-2, 1)z\} \\ &= \{(v, x) \in \mathbb{R}^2 \mid v \leq 0 \;\&\; x = -(1/2)v\}. \end{aligned}$$

We can usefully modify and generalize this example in two directions, as follows. Suppose first that there are ℓ such productive activities, and denote the input-requirement and output vectors for the kth such activity by '$\boldsymbol{a}_k$' and '$\boldsymbol{b}_k$,' respectively; with the level at which the kth activity is being operated being denoted by 'z_k,' for $k = 1, \ldots, \ell$. If we assume that all activities can be operated simultaneously with no loss of productive efficiency (that is, if there are no externalities in production), then with the activity levels given by the vector $\boldsymbol{z} = (z_1, \ldots, z_\ell)$, output, $\boldsymbol{x}$, is given by:

$$\boldsymbol{x} = \sum_{k=1}^{\ell} \boldsymbol{b}_k z_k,$$

while the required input vector is given by:

$$\boldsymbol{u} = \sum_{k=1}^{\ell} \boldsymbol{a}_k z_k.$$

More compactly, if we define the $q \times \ell$ matrix $\boldsymbol{A}$, and the $m \times \ell$ matrix $\boldsymbol{B}$ by:

$$\boldsymbol{A} = \begin{bmatrix} -\boldsymbol{a}_1 & -\boldsymbol{a}_2 & \ldots & -\boldsymbol{a}_\ell \end{bmatrix} \text{ and } \boldsymbol{B} = \begin{bmatrix} \boldsymbol{b}_1 & \boldsymbol{b}_2 & \ldots & \boldsymbol{b}_\ell \end{bmatrix},$$

respectively; then we see that the input-output vector $(\boldsymbol{v}, \boldsymbol{x})$ is feasible (where we are now returning to the general equilibrium convention of denoting input quantities by non-positive numbers) if, and only if, there exists $\boldsymbol{z} \in \mathbb{R}^\ell_+$ such that:

$$Y = \{(\boldsymbol{v}, \boldsymbol{x}) \in \mathbb{R}^{q+m} \mid (\exists \boldsymbol{z} \in \mathbb{R}^\ell_+)\colon \boldsymbol{v} = \boldsymbol{A}\boldsymbol{z} \;\&\; \boldsymbol{x} = \boldsymbol{B}\boldsymbol{z}\}. \tag{6.41}$$

The formulation of the above paragraph is very convenient and useful in a great many contexts. In the standard general equilibrium model which we are in the process of considering, there is a better way of proceeding, however. We begin by noting that we can assume here, without loss of generality, that $q = m = n$, where n is the total number of commodities available. Thus, suppose that a productive activity uses two inputs to produce one output; with $\boldsymbol{a}$ and $\boldsymbol{b}$ given by:

$$\boldsymbol{a} = (2, 3) \text{ and } \boldsymbol{b} = 1,$$

respectively. Suppose further that there are only four commodities available in a given economy, and that the two inputs being used here are the first and the third commodities, while the commodity being produced is the second commodity. Let's then define the new input-requirement and output vectors, $\boldsymbol{a}^*$ and $\boldsymbol{b}^*$ by:

$$\boldsymbol{a}^* = (2, 0, 3, 0) \text{ and } \boldsymbol{b}^* = (0, 1, 0, 0),$$

respectively. Next, define the vector $\boldsymbol{c}$ as:

$$\boldsymbol{c} = \boldsymbol{b}^* - \boldsymbol{a}^*,$$

and notice that the feasible production set can now be defined as:

$$Y = \{\boldsymbol{y} \in \mathbb{R}^4 \mid (\exists z \in \mathbb{R}_+) \colon \boldsymbol{y} = \boldsymbol{c}z\}.$$

We can complete this line of generalization by allowing for several productive activities once again, leading to a production set of the form:

$$Y = \{\boldsymbol{y} \in \mathbb{R}^n \mid (\exists \boldsymbol{z} \in \mathbb{R}^\ell_+) \colon \boldsymbol{y} = \boldsymbol{C}\boldsymbol{z}\}, \tag{6.42}$$

where $\boldsymbol{C}$ is taken to be a non-zero $n \times \ell$ matrix. By allowing for disposability, we are led to the two further generic examples:

$$Y = \{\boldsymbol{y} \in \mathbb{R}^n \mid (\exists z \in \mathbb{R}_+) \colon \boldsymbol{y} \leq \boldsymbol{c}z\}, \tag{6.43}$$

or, for the ℓ activity case, we can consider:

$$Y = \{\boldsymbol{y} \in \mathbb{R}^n \mid (\exists \boldsymbol{z} \in \mathbb{R}^\ell_+) \colon \boldsymbol{y} \leq \boldsymbol{C}\boldsymbol{z}\}. \tag{6.44}$$

Denoting the ith row of the matrix $\boldsymbol{C}$ by '$\boldsymbol{c}_{i\cdot}$,' for $i = 1, \ldots, n$; we generally assume in this context that there exists $h \in \{1, \ldots, n\}$ such that:

$$\boldsymbol{c}_{h\cdot} < \boldsymbol{0}. \tag{6.45}$$

Notice that in the case of a single activity, and where $\boldsymbol{c}$ is given by:

$$\boldsymbol{c} = \boldsymbol{b} - \boldsymbol{a},$$

condition (6.45) will be guaranteed by the requirement:

$$\boldsymbol{a} \cdot \boldsymbol{b} = 0 \tag{6.46}$$

(remember that we are assuming that both $\boldsymbol{a}$ and $\boldsymbol{b}$ are semi-positive vectors). Verbally interpret this condition. If (6.45) [or the special case in (6.46)] holds, which of the conditions of Definition 6.4 will Y satisfy?

Clearly the sort of production set being considered here, in particular, as specified by (6.42), above, is in some sense linear. On the other hand, this seems to be quite a different sort of set from our definition of a linear production set, as presented in Definition 6.9. Remarkably enough, however, the two specifications are, in a formal mathematical sense, equivalent. In order to demonstrate this, we will, in the remainder of this section, consider some results from linear algebra which are of interest in their own right. The first three of these results are known collectively, as 'theorems of the alternative,' and our proofs of them are adapted from Nikaido [1968, pp. 36–9].

6.40 Theorem. (Stiemke, 1915). If $\boldsymbol{A}$ is an $m \times n$ matrix, then *exactly one* of the following holds. The equation:

$$\boldsymbol{A}\boldsymbol{x} = \boldsymbol{0}, \tag{6.47}$$

has a (strictly) positive solution, or the inequality:

$$\boldsymbol{p}^\top \boldsymbol{A} > \boldsymbol{0}, \tag{6.48}$$

has a solution.

Proof. Suppose, by way of obtaining a contradiction, that there exist $\bar{\boldsymbol{x}} \in \mathbb{R}^n_{++}$ and $\bar{\boldsymbol{p}} \in \mathbb{R}^m$ satisfying (6.47) and (6.48), respectively. Then we have:

$$\bar{\boldsymbol{p}}^\top \boldsymbol{A}\bar{\boldsymbol{x}} = \bar{\boldsymbol{p}}^\top \big(\boldsymbol{A}\bar{\boldsymbol{x}}\big) = \bar{\boldsymbol{p}}^\top \cdot \boldsymbol{0} = 0.$$

On the other hand, by (6.48) and the fact that $\bar{\boldsymbol{x}} \gg \boldsymbol{0}$:

$$\bar{\boldsymbol{p}}^\top \boldsymbol{A}\bar{\boldsymbol{x}} = \big(\bar{\boldsymbol{p}}^\top \boldsymbol{A}\big)\bar{\boldsymbol{x}} > 0;$$

giving a contradiction.

Now suppose there exists no $\boldsymbol{p} \in \mathbb{R}^m$ which satisfies (6.48), and define:

$$L = \{\boldsymbol{y} \in \mathbb{R}^n \mid (\exists \boldsymbol{p} \in \mathbb{R}^m)\colon \boldsymbol{A}^\top \boldsymbol{p} = \boldsymbol{y}\}.$$

Then, by hypothesis, L does *not* contain a semi-positive point. But then, by Theorem 6.34, L^* contains a strictly positive element, $\boldsymbol{x}^*$. Moreover, since L is a linear subspace, $L^* = L^\perp$; and, given our definition of L, it is clear that:

$$L^\perp = \{\boldsymbol{x} \in \mathbb{R}^n \mid \boldsymbol{A}\boldsymbol{x} = \boldsymbol{0}\};$$

which establishes the desired result. □

The nifty thing about Stiemke's theorem is that it often provides a much simpler way of establishing whether or not a system of linear equations has a positive solution than trying to solve it directly. Our next result is probably even more interesting, from a theoretical point of view.

6.41 Theorem. (Tucker, 1956) If $\boldsymbol{A}$ is an $m \times n$ matrix, then the system of linear inequalities:

$$\boldsymbol{A}^\top \boldsymbol{p} \geq \boldsymbol{0},$$

and the homogeneous linear equation:

$$\boldsymbol{A}\boldsymbol{x} = \boldsymbol{0},$$

always have a pair of solutions, $\boldsymbol{p}^* \in \mathbb{R}^m$ and $\boldsymbol{x}^* \in \mathbb{R}^n$ such that:

$$\boldsymbol{x}^* \geq \boldsymbol{0} \;\&\; \boldsymbol{A}^\top \boldsymbol{p}^* + \boldsymbol{x}^* \gg \boldsymbol{0}.$$

Proof. For any $\boldsymbol{p} \in \mathbb{R}^m$, let '$\left[\boldsymbol{A}^\top \boldsymbol{p}\right]_j$' denote the jth coordinate of $\boldsymbol{A}^\top \boldsymbol{p}$, and define:

$$N(\boldsymbol{p}) = \left\{j \in \{1, \ldots, n\} \mid \left[\boldsymbol{A}^\top \boldsymbol{p}\right]_j > 0\right\}.$$

Next, define the set P by:

$$P = \{\boldsymbol{p} \in \mathbb{R}^m \mid \boldsymbol{A}^\top \boldsymbol{p} \geq \boldsymbol{0}\};$$

and note that, since for all $\boldsymbol{p} \in P$, we have $N(\boldsymbol{p}) \subseteq \{1, \ldots, n\}$, the number of elements in $N(\boldsymbol{p})$, $\#N(\boldsymbol{p})$, is maximized at some point $\widehat{\boldsymbol{p}} \in P$. We distinguish several cases, depending upon the value of $\#N(\widehat{\boldsymbol{p}})$.

1. $\#N(\widehat{\boldsymbol{p}}) = 0$. Here it follows from Stiemke's Theorem that there exists $\boldsymbol{x}^* \in \mathbb{R}^n_{++}$ such that $\boldsymbol{x}^* \gg \boldsymbol{0}$, and $\boldsymbol{A}\boldsymbol{x}^* = \boldsymbol{0}$. Consequently, if we define $\boldsymbol{p}^* = \boldsymbol{0} \in \mathbb{R}^m$, we have:

$$\boldsymbol{A}^\top \boldsymbol{p}^* \geq \boldsymbol{0}, \boldsymbol{A}\boldsymbol{x}^* = \boldsymbol{0}, \boldsymbol{x}^* \geq \boldsymbol{0}, \text{and } \boldsymbol{A}^\top \boldsymbol{p}^* + \boldsymbol{x}^* \gg \boldsymbol{0},$$

as desired.

2. $\#N(\widehat{\boldsymbol{p}}) = n$. Here we must have $\boldsymbol{A}^\top \widehat{\boldsymbol{p}} \gg \boldsymbol{0}$, and setting $\widehat{\boldsymbol{x}} = \boldsymbol{0} \in \mathbb{R}^n$, we have:

$$\boldsymbol{A}\widehat{\boldsymbol{x}} = \boldsymbol{0} \text{ and } \boldsymbol{A}^\top \widehat{\boldsymbol{p}} + \widehat{\boldsymbol{x}} \gg \boldsymbol{0},$$

once again.

3. $1 \leq \#N(\widehat{\boldsymbol{p}}) \stackrel{\text{def}}{=} k < n$. Here we can assume, without loss of generality, that:

$$[\boldsymbol{A}^\top \widehat{\boldsymbol{p}}]_j > 0 \text{ for } j = 1, \ldots, k, \quad \text{and } [\boldsymbol{A}^\top \widehat{\boldsymbol{p}}]_j = 0 \text{ for } j = k+1, \ldots, n.$$

Define $\boldsymbol{B}$ as the submatrix consisting of the first k columns of $\boldsymbol{A}$, and $\boldsymbol{C}$ as the submatrix consisting of columns $k+1, \ldots, n$. Thus we can write:

$$\boldsymbol{A} = [\boldsymbol{B}\ \boldsymbol{C}],$$

and we have:

$$\boldsymbol{B}^\top \widehat{\boldsymbol{p}} \gg \boldsymbol{0} \text{ and } \boldsymbol{C}^\top \widehat{\boldsymbol{p}} = \boldsymbol{0}.$$

Now suppose, by way of obtaining a contradiction, that there exists $\boldsymbol{y} \in \mathbb{R}^m$ such that:

$$\boldsymbol{C}^\top \boldsymbol{y} > \boldsymbol{0},$$

and define $\boldsymbol{p} \in \mathbb{R}^m$ by:

$$\boldsymbol{p} = \theta \widehat{\boldsymbol{p}} + \boldsymbol{y},$$

where:

$$\theta = \max_{1 \le j \le k} \left(\frac{-[\boldsymbol{B}^\top \boldsymbol{y}]_j}{[\boldsymbol{B}^\top \widehat{\boldsymbol{p}}]_j} \right) + 1.$$

Then, for each $j \in \{1, \ldots, n\}$, we have:

$$[\boldsymbol{B}^\top (\theta \widehat{\boldsymbol{p}})]_j = \theta [\boldsymbol{B}^\top \widehat{\boldsymbol{p}}]_j \ge -[\boldsymbol{B}^\top \boldsymbol{y}]_j + [\boldsymbol{B}^\top \widehat{\boldsymbol{p}}]_j > -[\boldsymbol{B}^\top \boldsymbol{y}]_j;$$

so that:

$$\boldsymbol{B}^\top (\theta \widehat{\boldsymbol{p}} + \boldsymbol{y}) = \boldsymbol{B}^\top (\theta \widehat{\boldsymbol{p}}) + \boldsymbol{B}^\top \boldsymbol{y} > \boldsymbol{0}.$$

Moreover:

$$\boldsymbol{C}^\top (\theta \widehat{\boldsymbol{p}} + \boldsymbol{y}) = \theta \boldsymbol{C}^\top \widehat{\boldsymbol{p}} + \boldsymbol{C}^\top \boldsymbol{y} = \boldsymbol{C}^\top \boldsymbol{y} \gg \boldsymbol{0},$$

so that $\boldsymbol{p} \in P$. But this is impossible, since $\#N(\boldsymbol{p}) > \#N(\widehat{\boldsymbol{p}})$.

From the argument of the preceding paragraph we conclude that there exists no $\boldsymbol{y} \in \mathbb{R}^m$ satisfying:

$$\boldsymbol{C}^\top \boldsymbol{y} > \boldsymbol{0}.$$

Consequently, it follows from Stiemke's Theorem that there exists $\boldsymbol{z} \in R^{n-k}_{++}$ such that:

$$\boldsymbol{C}\boldsymbol{z} = \boldsymbol{0}.$$

If we now define $\boldsymbol{x}^* \in \mathbb{R}^n_+$ by:

$$\boldsymbol{x}^* = \begin{pmatrix} \boldsymbol{0} \\ \boldsymbol{z} \end{pmatrix},$$

we see that:

$$\boldsymbol{A}\boldsymbol{x}^* = [\boldsymbol{B}\ \boldsymbol{C}] = [\boldsymbol{B}\ \boldsymbol{C}] \begin{pmatrix} \boldsymbol{0} \\ \boldsymbol{z} \end{pmatrix} = \boldsymbol{B}\boldsymbol{0} + \boldsymbol{C}\boldsymbol{z} = \boldsymbol{0},$$

and:

$$\boldsymbol{A}^\top \widehat{\boldsymbol{p}} + \boldsymbol{x}^* = \begin{pmatrix} \boldsymbol{B}^\top \\ \boldsymbol{C}^\top \end{pmatrix} \widehat{\boldsymbol{p}} + \begin{pmatrix} \boldsymbol{0} \\ \boldsymbol{z} \end{pmatrix} = \begin{pmatrix} \boldsymbol{B}^\top \widehat{\boldsymbol{p}} \\ \boldsymbol{C}^\top \widehat{\boldsymbol{p}} \end{pmatrix} + \begin{pmatrix} \boldsymbol{0} \\ \boldsymbol{z} \end{pmatrix} = \begin{pmatrix} \boldsymbol{B}^\top \widehat{\boldsymbol{p}} \\ \boldsymbol{z} \end{pmatrix} \gg \boldsymbol{0},$$

as desired. □

By making use of Tucker's Theorem, we can prove yet another sort of 'theorem of the alternative,' which in this case is known as the 'Minkowski-Farkas Lemma.'[7]

6.42 Theorem. (Farkas, 1902; Minkowski, 1910) If $\boldsymbol{A}$ is an $m \times n$ matrix and $\boldsymbol{b} \in \mathbb{R}^m$, then exactly one of the following holds. Either the equation:

$$\boldsymbol{A}\boldsymbol{x} = \boldsymbol{b}, \tag{6.49}$$

has a nonnegative solution, or the inequalities:

$$\boldsymbol{p}^\top \boldsymbol{A} \ge \boldsymbol{0} \ \ \& \ \ \boldsymbol{p}^\top \boldsymbol{b} = \boldsymbol{p} \cdot \boldsymbol{b} < 0, \tag{6.50}$$

have a solution.

[7]For yet additional results of this sort, see Nikaido [1968, pp. 38–9], or Mangasarian [1969, Chapter 2].

Proof. This time I will leave as an exercise the proof that there cannot exist both $\boldsymbol{x}^* \in \mathbb{R}^n_+$ satisfying (6.49) and $\boldsymbol{p}^* \in \mathbb{R}^m$ satisfying (6.50).

To show that one of the two must hold, we note that by Tucker's Theorem, there exist $\boldsymbol{p}^* \in \mathbb{R}^m$ and $\boldsymbol{z}^* \in \mathbb{R}^{n+1}_{++}$ such that:

$$\begin{bmatrix} \boldsymbol{A}^\top \\ -\boldsymbol{b}^\top \end{bmatrix} \boldsymbol{p}^* = \begin{bmatrix} \boldsymbol{A}^\top \boldsymbol{p}^* \\ -\boldsymbol{b}^\top \boldsymbol{p}^* \end{bmatrix} \geq \boldsymbol{0}, \tag{6.51}$$

$$\begin{bmatrix} \boldsymbol{A} - \boldsymbol{b} \end{bmatrix} \boldsymbol{z}^* = \boldsymbol{0}, \tag{6.52}$$

and:

$$\begin{bmatrix} \boldsymbol{A}^\top \\ -\boldsymbol{b}^\top \end{bmatrix} \boldsymbol{p}^* + \boldsymbol{z}^* = \begin{bmatrix} \boldsymbol{A}^\top \boldsymbol{p}^* \\ -\boldsymbol{b}^\top \boldsymbol{p}^* \end{bmatrix} + \boldsymbol{z}^* \gg \boldsymbol{0}. \tag{6.53}$$

Now, if in fact there exists no $\boldsymbol{p} \in \mathbb{R}^m$ such that:

$$\boldsymbol{A}^\top \boldsymbol{p} \geq \boldsymbol{0} \text{ and } \boldsymbol{p}^\top \boldsymbol{b} < 0,$$

then it follows from (6.51) that:

$$\boldsymbol{p}^* \cdot \boldsymbol{b} = 0;$$

in which case we have from (6.53) that $z_{n+1} > 0$; so that, defining $\boldsymbol{x}^* \in \mathbb{R}^n_+$ by:

$$x_j^* = \frac{z_j}{z_{n+1}} \quad \text{for } j = 1, \ldots, n,$$

it follows from (6.52) that:

$$\boldsymbol{A}\boldsymbol{x}^* = \boldsymbol{b}. \quad \square$$

6.43 Definition. If a vector $\boldsymbol{a}$ is an element of $\mathbb{R}^n$, we define the **ray generated by $\boldsymbol{a}$**, denoted by '$(\boldsymbol{a})$,' by:

$$(\boldsymbol{a}) = \{\boldsymbol{x} \in \mathbb{R}^n \mid (\exists \theta \in \mathbb{R}_+) \colon \boldsymbol{x} = \theta \boldsymbol{a}\}.$$

Given $\boldsymbol{a}_j \in \mathbb{R}^m$, for $j = 1, \ldots, n$, we follow the usual set summation rule in defining $(\boldsymbol{a}_1) + (\boldsymbol{a}_2) + \ldots (\boldsymbol{a}_n)$ as the set of all $\boldsymbol{y} \in \mathbb{R}^m$ such that there exist $\boldsymbol{y}_j \in (\boldsymbol{a}_j)$, for $j = 1, \ldots, n$, such that:

$$\boldsymbol{y} = \sum\nolimits_{j=1}^n \boldsymbol{y}_j.$$

The definition toward which I have been aiming can now be set forth as follows.

6.44 Definition. A set $K \subseteq \mathbb{R}^m$ is said to be a **polyhedral cone** iff there exist vectors $\boldsymbol{a}_1, \ldots, \boldsymbol{a}_n \in \mathbb{R}^m$ such that:

$$K = (\boldsymbol{a}_1) + \cdots + (\boldsymbol{a}_n).$$

I will leave it to you to prove that a polyhedral cone is a convex cone. I will also leave the proof of the following proposition as an exercise.

6.45 Proposition. *A set $K \subseteq \mathbb{R}^m$ is a polyhedral cone if, and only if there exists an $m \times n$ matrix, $\boldsymbol{A}$ such that:*

$$K = \big\{\boldsymbol{y} \in \mathbb{R}^m \mid (\exists \boldsymbol{x} \in \mathbb{R}^n_+) \colon \boldsymbol{y} = \boldsymbol{A}\boldsymbol{x}\big\}.$$

Now, having established Proposition 6.45, let's return to a consideration of the general activity analysis model developed at the beginning of this section. Recall that, given the assumptions of that model, the production set took the form:

$$Y = \{\boldsymbol{y} \in \mathbb{R}^n \mid (\exists \boldsymbol{x} \in \mathbb{R}^n_+) \colon \boldsymbol{y} = \boldsymbol{C}\boldsymbol{x}\};$$

which we now see is a polyhedral cone. In fact, every one of the 'theorem of the alternative' type results which we have just established has an economically meaningful application/interpretation in terms of the activity analysis model of production; although I will leave as an exercise the task of verifying this statement. I will close this section with some results which effectively define the relationship between the activity analysis model and the linear production set definitions set forth earlier in this chapter.

6.46 Theorem. *If $K \subseteq \mathbb{R}^m$ is a polyhedral cone, then $K^{**} = K$.*

Proof. We have already noted the fact that, for any cone, K, we must necessarily have:

$$K \subseteq K^{**}.$$

To complete our proof, let the $m \times n$ matrix $\boldsymbol{A}$ be such that:

$$K = \{\boldsymbol{y} \in \mathbb{R}^m \mid (\exists \boldsymbol{x} \in \mathbb{R}^n_+) \colon \boldsymbol{y} = \boldsymbol{A}\boldsymbol{x}\}. \tag{6.54}$$

Then it is easy to prove that:

$$K^* = \{\boldsymbol{z} \in \mathbb{R}^m \mid \boldsymbol{z}^\top \boldsymbol{A} \leq \boldsymbol{0}\} = \{\boldsymbol{z} \in \mathbb{R}^m \mid \boldsymbol{z}^\top(-\boldsymbol{A}) \geq \boldsymbol{0}\}.$$

Now, suppose $\boldsymbol{b} \notin K$. Then it follows from (6.54) that there exists no nonnegative $\boldsymbol{x}$ such that:

$$\boldsymbol{A}\boldsymbol{x} = \boldsymbol{b}, \text{ or } (-\boldsymbol{A})\boldsymbol{x} = -\boldsymbol{b};$$

and thus by the Minkowski-Farkas Lemma, there exists $\bar{\boldsymbol{z}} \in \mathbb{R}^m$ such that:

$$\bar{\boldsymbol{z}}^\top(-\boldsymbol{A}) \geq \boldsymbol{0} \text{ and } \bar{\boldsymbol{z}} \cdot (-\boldsymbol{b}) < 0.$$

By the first of these two inequalities, we see that $\bar{\boldsymbol{z}} \in K^*$, while by the second, it then follows that $\boldsymbol{b} \notin K^{**}$. Consequently, if $\boldsymbol{b} \in K^{**}$, then $\boldsymbol{b} \in K$. □

The following is now an immediate implication of Proposition 6.33 and Theorem 6.46.

6.47 Corollary. *A polyhedral cone is a closed set.*

For a proof of our next, and final result of this section, see, for example, Nikaido [1968, p. 42].

6.48 Theorem. *If $\boldsymbol{A}$ is an $m \times n$ matrix, the set $Y \subseteq \mathbb{R}^n$ defined by:*

$$Y = \{\boldsymbol{y} \in \mathbb{R}^n \mid \boldsymbol{A}\boldsymbol{y} \leq \boldsymbol{0}\},$$

is a polyhedral cone.

As you have probably already noticed, it follows immediately from Theorem 6.45 that any linear production set can be generated by an activity analysis model. The converse is also true. In this case, the proof of the statement is a little more tricky, but we can proceed as follows. Let $Y \subseteq \mathbb{R}^n$ be generated by an activity analysis model, so that there exists an $n \times m$ matrix, $\boldsymbol{A}$, such that:

$$Y = \{\boldsymbol{y} \in \mathbb{R}^n \mid (\exists \boldsymbol{x} \in \mathbb{R}^m_+) \colon \boldsymbol{y} = \boldsymbol{A}\boldsymbol{x}\}.$$

Then it is easy to prove (although I will leave this as an exercise) that:

$$Y^* = \Pi(Y) = \{\boldsymbol{p} \in \mathbb{R}^n \mid \boldsymbol{A}^\top \boldsymbol{p} \leq \boldsymbol{0}\}. \tag{6.55}$$

But then it follows from Theorem 6.48 that Y^* is a polyhedral cone; so that there exists a positive integer, q, and an $n \times q$ matrix, $\boldsymbol{B}$, such that:

$$Y^* = \{\boldsymbol{p} \in \mathbb{R}^n \mid (\exists \boldsymbol{z} \in \mathbb{R}^q_+) \colon \boldsymbol{p} = \boldsymbol{B}\boldsymbol{z}\}.$$

However, by the same reasoning as yielded (6.55), we can see that:

$$Y^{**} = (Y^*)^* = \{\boldsymbol{y} \in \mathbb{R}^n \mid \boldsymbol{B}^\top \boldsymbol{y} \leq \boldsymbol{0}\}.$$

Since Theorem 6.46 tells us that $Y^{**} = Y$, we can now see that Y is a linear production set, as per Definition 6.9.

Exercises. 1. Let $\boldsymbol{A}$ be a nonnegative $m \times q$ matrix; and define the production set, T, by:

$$T = \{(\boldsymbol{v}, \boldsymbol{x}) \in \mathbb{R}^{m+q} \mid \boldsymbol{v} + \boldsymbol{A}\boldsymbol{x} \leq \boldsymbol{0} \; \& \; \boldsymbol{x} \in \mathbb{R}^q_+\}$$

On the basis of this information, answer the following five questions. In each case, you should try to prove the property directly, and *not* by appealing to results established in this chapter.

a. Does this production process satisfy non-decreasing returns to scale? Explain briefly.

b. Is this production set convex? Is it additive? Explain your answer.

c. Suppose now that the matrix, $\boldsymbol{A}$, satisfies the following condition: for each $j \in \{1, \ldots, q\}$, there exists $i \in \{1, \ldots, m\}$ such that $a_{ij} > 0$. Can you characterize the efficient pairs, $(\boldsymbol{v}, \boldsymbol{x})$ for this production set?

d. Suppose now that $\boldsymbol{w} \in \mathbb{R}^m_{++}$ is the vector of prices for the inputs, $\boldsymbol{v}$, and that $\boldsymbol{p} \in \mathbb{R}^q_{++}$ is the vector of prices of the outputs, $\boldsymbol{x}$. What must be the relationship between $\boldsymbol{w}$ and $\boldsymbol{p}$ if a price-taking (and profit-maximizing) firm operating this technology is to produce a non-zero (but finite) output?

e. Given that $\boldsymbol{A}$ satisfies the condition defined in part (c), above, does T satisfy Impossibility of Free Production? Does it satisfy Irreversibility? Explain your answers.

2. Prove the first and third of the 'Facts Regarding Disposability' set out in Section 2.

3. Show that production constraints (6.6)–(6.8) are equivalent to (6.9)–(6.10) in the example in Section 3.

4. Prove Proposition 6.15.

5. Complete the proof outlined in Example 6.36.

6. Prove Proposition 6.37

7. Verify inequality (6.16) in Section 4.

Chapter 7

Fundamental Welfare Theorems

7.1 Introduction

In this chapter, we will be extending the development of the 'Fundamental Theorems of Welfare Economics' from the pure exchange economy case which we discussed in Chapter 5 to a production economy. Much of the analysis will be almost unchanged from the corresponding material set out in Chapter 5. The main difference is that we will present here a detailed proof and discussion of the 'Second Fundamental Theorem of Welfare Economics' (the unbaisedness result), neither of which was included in Chapter 5.

7.2 Competitive Equilibrium with Production

We will suppose in our discussion here that there are given finite (integer) numbers of commodities, consumers, and firms; and we will denote these quantities by 'n, m,' and 'ℓ,' respectively. The **commodity space** then becomes $\mathbb{R}^n$, and we will employ the following system of notation. First, we will denote the set of consumers by 'M,' and will use 'K' to denote the set of producers; in other words:

$$M = \{1, \ldots, m\} \text{ and } K = \{1, \ldots, \ell\}.$$

The remaining basic notation is as follows:

$X_i \subseteq \mathbb{R}^n$ denotes the ith consumer's consumption set, and we will assume throughout that $X_i \neq \emptyset$, for $i = 1, \ldots, m$ (that is, $X_i \neq \emptyset$, for all $i \in M$);

'P_i' denotes the ith consumer's (strict) preference relation on X_i, and *we will assume throughout that P_i is irreflexive, for $i = 1, \ldots, m$;*

$Y_k \subseteq \mathbb{R}^n$ denotes the kth firm's production set, and we will asssume that $Y_k \neq \emptyset$, for $k = 1, \ldots, \ell$;

$\boldsymbol{r} \in \mathbb{R}^n$ will denote the aggregate resource endowment of the economy.

7.1 Definition. When we write 'E is an economy,' we will mean that E is a tuple of the form:

$$E = (\langle X_i, P_i \rangle, \langle Y_k \rangle, \boldsymbol{r}),$$

where $\langle X_i, P_i\rangle$ $(i = 1, \ldots, m)$, $\langle Y_k\rangle$ $(k = 1, \ldots, \ell)$, and $\boldsymbol{r} \in \mathbb{R}^n$ satisfy the above conditions.

In dealing with allocations for an economy, we will denote the coordinates of the ith consumer's commodity bundle, $\boldsymbol{x}_i$, by 'x_{ij}', $(j = 1, \ldots, n)$; that is, we write:

$$\boldsymbol{x}_i = (x_{i1}, \ldots, x_{in}),$$

where 'x_{ij}' denotes the quantity of the jth commodity available to (or being made available by) the ith consumer. We then follow the convention that if:

a. $x_{ij} \geq 0$, then the jth commodity is available for i's consumption in the amount x_{ij}, while if:

b. $x_{ij} < 0$, then the consumer is offering to supply the jth commodity (or service), in the amount $|x_{ij}| = -x_{ij}$.

With this convention, notice that if commodity prices are given by the vector $\boldsymbol{p} \in \mathbb{R}^n_+$, then the net expenditure necessary for the consumer to obtain the bundle $\boldsymbol{x}_i \in X_i$ is given by the inner product of $\boldsymbol{p}$ and $\boldsymbol{x}_i$, that is, $\boldsymbol{p} \cdot \boldsymbol{x}_i$.

A similar convention will be followed with respect to production vectors $\boldsymbol{y}_k \in Y_k$; which, where necessary for clarity, we will write out as:

$$\boldsymbol{y}_k = (y_{k1}, \ldots, y_{kn}).$$

In this case, if:

a. $y_{kj} \geq 0$, then the kth producer is producing (or planning to produce) the jth commodity in the net amount y_{kj}; while if

b. $y_{kj} < 0$, then the producer is using the jth commodity as an input[1] in the amount $|y_{kj}| = -y_{kj}$.

Consequently, given prices $\boldsymbol{p} \in \mathbb{R}^n_+$, the profit to the kth producer yielded by the choice of $\boldsymbol{y}_k \in Y_k$ is given by $\boldsymbol{p} \cdot \boldsymbol{y}_k$. We will, in fact, define the profit function π_k on $\mathbb{R}^n_+$ by:

$$\pi_k(\boldsymbol{p}) = \max\{\boldsymbol{p} \cdot \boldsymbol{y}_k \mid \boldsymbol{y}_k \in Y_k\},$$

and assume that the producer attempts to maximize profits, taking the price vector as given.[2]

In dealing with an economy, $E = (\langle X_i, P_i\rangle, \langle Y_k\rangle, \boldsymbol{r})$, we denote the cartesian product of the X_i's, the **consumption allocation space**, by '$\mathcal{X}$', or by '$\mathcal{X}(E)$,' if it appears that a reminder might be needed as to the association of the set with the economy; that is,

$$\mathcal{X} \equiv \mathcal{X}(E) = \prod_{i=1}^{m} X_i = \prod_{i \in M} X_i.$$

Similarly, we denote the product of the Y_k's, the **production allocation space** by $\mathcal{Y}$, or $\mathcal{Y}(E)$:

$$\mathcal{Y} \equiv \mathcal{Y}(E) = \prod_{k=1}^{\ell} Y_k = \prod_{k \in K} Y_k.$$

[1]For this particular production plan; other technologically feasible production plans for the producer may have $y_{kj} = 0$, or, indeed, have $y_{kj} > 0$.

[2]So that the producers are taken to be pure competitors, that is 'price-takers.'

We use the generic notation, '$\langle \boldsymbol{x}_i \rangle, \langle \boldsymbol{x}'_i \rangle, \langle \boldsymbol{x}^*_i \rangle,$[3]' and so on to denote elements of $\mathcal{X}$; and, similarly, '$\langle \boldsymbol{y}_k \rangle, \langle \boldsymbol{y}'_k \rangle, \langle \boldsymbol{y}^*_k \rangle$,' will be used to denote elements of $\mathcal{Y}$. Combining these, we will use '$(\langle \boldsymbol{x}_i \rangle, \langle \boldsymbol{y}_k \rangle), (\langle \boldsymbol{x}'_i \rangle, \langle \boldsymbol{y}'_k \rangle), (\langle \boldsymbol{x}^*_i \rangle, \langle \boldsymbol{y}^*_k \rangle)$,' and so on, to denote elements of the **allocation space**:

$$\mathcal{X} \times \mathcal{Y} \equiv \left(\prod_{i=1}^{m} X_i \right) \times \left(\prod_{k=1}^{\ell} Y_k \right).$$

7.2 Definitions. Let E be an economy. An $(m+\ell)\cdot n$-tuple, $\big(\langle \boldsymbol{x}_i \rangle, \langle \boldsymbol{y}_k \rangle\big) \in \mathbb{R}^{(m+\ell)n}$ will be said to be a **feasible** (or **attainable**) **allocation for *E*** iff:

1. $\boldsymbol{x}_i \in X_i \quad$ for $i = 1, \ldots, m$,
2. $\boldsymbol{y}_k \in Y_k \quad$ for $k = 1, \ldots, \ell$, and:
3. $\sum_{i=1}^{m} \boldsymbol{x}_i = \boldsymbol{r} + \sum_{k=1}^{\ell} \boldsymbol{y}_k$

In other words, $\big(\langle \boldsymbol{x}_i \rangle, \langle \boldsymbol{y}_k \rangle\big)$is feasible for E iff $\big(\langle \boldsymbol{x}_i \rangle, \langle \boldsymbol{y}_k \rangle\big) \in \mathcal{X} \times \mathcal{Y}$, and:

$$\sum_{i \in M} \boldsymbol{x}_i = \boldsymbol{r} + \sum_{k \in K} \boldsymbol{y}_k.$$

We will denote the set of all feasible or attainable allocations for E by '$A(E)$'.

We are going to want to discuss competitive, or Walrasian equilibria[4] for an economy, E; but of course we cannot define such an equilibrium without specifying what the consumers have to spend in such an economy. Our next definition will provide us with a great deal of flexibility in this respect.

7.3 Definition. Given an economy, $E = (\langle X_i, P_i \rangle, \langle Y_k \rangle, \boldsymbol{r})$, we shall say that a vector $\boldsymbol{w} = (w_1, \ldots, w_m)$ is an **assignment of wealth** (or a **wealth-assignment vector**) **for *E*, given the price vector $\boldsymbol{p}^* \in \mathbb{R}^n$**, iff for each i we have:

$$(\exists \bar{\boldsymbol{x}}_i \in X_i) \colon \boldsymbol{p}^* \cdot \bar{\boldsymbol{x}}_i \leq w_i,$$

and $\boldsymbol{w}$ satisfies:

$$\sum_{i=1}^{m} w_i = \boldsymbol{p}^* \cdot \boldsymbol{r} + \sum_{k=1}^{\ell} \pi_k(\boldsymbol{p}^*). \tag{7.1}$$

A wealth assignment vector, given a vector of prices, $\boldsymbol{p}^*$, must provide each consumer with enough wealth to purchase something in its consumption set, and must also exhaust the sum of the value of resources plus aggregate profits, given $\boldsymbol{p}^*$. We then use the concept of a wealth assignment vector to define a competitive (or Walrasian) equilibrium for an economy, E, as follows.

7.4 Definitions. An $(m + \ell + 1) \cdot n$-tuple, $(\langle \boldsymbol{x}^*_i \rangle, \langle \boldsymbol{y}^*_k \rangle, \boldsymbol{p}^*)$, is a **competitive** (or **Walrasian**) **equilibrium for the economy** $E = (\langle X_i, P_i \rangle, \langle Y_k \rangle, \boldsymbol{r})$, iff there is an assignment of wealth for E, given $\boldsymbol{p}^*$, $\boldsymbol{w} = (w_1, \ldots, w_m)$, such that:

1. $\boldsymbol{p}^* \neq \boldsymbol{0}$,
2. $\big(\langle \boldsymbol{x}^*_i \rangle, \langle \boldsymbol{y}^*_k \rangle\big) \in A(E)$,
3. for each $k \in K$, we have: $\boldsymbol{p}^* \cdot \boldsymbol{y}^*_k = \pi_k(\boldsymbol{p}^*)$, and

[3] Or sometimes, for example, $\langle \boldsymbol{x}_i \rangle_{i=1}^{m}$, or $\langle \boldsymbol{x}_i \rangle_{i \in M}$, as was indicated in Chapter 5.

[4] We will use these two terms, competitive equilibrium and Walrasian equilibrium, as synonyms throughout the remainder of this book; except in Chapter 8, as will be explained there.

4. for each $i \in M$, we have:
   a. $\boldsymbol{p}^* \cdot \boldsymbol{x}_i^* \leq w_i$, and:
   b. $(\forall \boldsymbol{x}_i \in X_i)\colon \boldsymbol{x}_i P_i \boldsymbol{x}_i^* \Rightarrow \boldsymbol{p}^* \cdot \boldsymbol{x}_i > w_i$.

In this case we shall also say that $(\langle \boldsymbol{x}_i^* \rangle, \langle \boldsymbol{y}_k^* \rangle, \boldsymbol{p}^*)$ is a **competitive** (or **Walrasian**) **equilibrium for the economy E, given the wealth assignment**, $\boldsymbol{w} = (w_1, \ldots, w_m)$.

I should at this point take a moment to explain the inclusion of the condition $\boldsymbol{p}^* \neq \boldsymbol{0}$ in the above definition. It should be clear that if at each attainable consumption allocation, $\langle \boldsymbol{x}_i \rangle \in \mathcal{X}(E)$, at least one consumer is not satiated, then the satisfaction of condition 4 of Definition 7.4 implies $\boldsymbol{p}^* \neq \boldsymbol{0}$. Consequently, the requirement that $\boldsymbol{p}^* \neq \boldsymbol{0}$ is almost redundant in our definition of a competitive equilibrium. However, we will need this last condition as a part of our definition of a 'quasi-competitive equilibrium,' a concept we will be defining later in this chapter; and in order to ensure that a competitive equilibrium is always a special case of a quasi-competitive equilibrium, I have included condition 1 in our definition.

The most usual way of specifying a wealth assignment for an economy is based upon the following definition.

7.5 Definitions. Let $E = (\langle X_i, P_i \rangle, \langle Y_k \rangle, \boldsymbol{r})$ be an economy. We shall say that $(\langle \boldsymbol{r}_i \rangle, [s_{ik}])$ is a **distribution of ownership for E** iff:

1. $\boldsymbol{r}_i \in \mathbb{R}^n$, for $i = 1, \ldots, m$, and $\sum_{i=1}^m \boldsymbol{r}_i = \boldsymbol{r}$,
2. $s_{ik} \geq 0$, for $i = 1, \ldots, m, k = 1, \ldots, \ell$, and
3. $\sum_{i=1}^m s_{ik} = 1$, for $k = 1, \ldots, \ell$.

We shall then say that $\mathcal{E} = (\langle X_i, P_i \rangle, \langle Y_k \rangle, \langle \boldsymbol{r}_i \rangle, [s_{ik}])$ is a **private ownership economy** iff $E = (\langle X_i, P_i \rangle, \langle Y_k \rangle, \boldsymbol{r})$ is an economy, and $(\langle \boldsymbol{r}_i \rangle, [s_{ik}])$ is a distribution of ownership for E.

Notice that the above definition allows for the possibility that the kth firm is a sole proprietorship (in which case, there exists some i such that $s_{ik} = 1$), an equal-shares partnership (in which case there would exist some h, i such that $s_{hk} = s_{ik} = 1/2$), or a publicly-traded corporation (in which we would have $s_{ik} > 0$ for many of the consumers).

If $\mathcal{E} = (\langle X_i, P_i \rangle, \langle Y_k \rangle, \langle \boldsymbol{r}_i \rangle, [s_{ik}])$ is a private ownership economy, and a price vector, $\boldsymbol{p} \in \mathbb{R}^n$ is given, the usual definition of the **i^{th} consumer's income** (or **wealth**) is given by:

$$w_i(\boldsymbol{p}) = \boldsymbol{p} \cdot \boldsymbol{r}_i + \sum_{k=1}^{\ell} s_{ik} \pi_k(\boldsymbol{p}). \tag{7.2}$$

Notice that if $\boldsymbol{0} \in Y_k$ (although in this chapter we will not usually be requiring this condition), then for any price vector, $\boldsymbol{p}$, we will have $w_i(\boldsymbol{p}) \geq \boldsymbol{p} \cdot \boldsymbol{r}_i$.

The definition of a private ownership economy is due to Debreu, and private ownership economies, as thus defined, are the principal subjects of investigation in Debreu's *Theory of Value* (Debreu [1959]). As we have just indicated, Debreu treated individual resource endowments as something distinct from the production sets in the economy. From a formal mathematical point of view, however, we could easily eliminate the explicit inclusion of individual resource endowments by defining m new production sets:

$$Y_k = \{\boldsymbol{r}_{k-\ell}\} \quad \text{for } k = \ell + 1, \ldots, \ell + m;$$

and correspondingly adding m shares of ownership, with:

$$s_{ik} = \begin{cases} 1 & \text{for } i = k - \ell, \\ 0 & \text{otherwise,} \end{cases}$$

for $i = 1, \ldots, m; k = \ell + 1, \ldots \ell + m$. (Or alternatively, by making use of the consumption sets X_i' defined as $X_i' = X_i - \boldsymbol{r}_i$.) Why then, you may well ask, do we complicate our notation by explicitly including the individual resource endowments? Well, the discussion in the above paragraph indicates one reason; if we add to the assumption that $\mathbf{0} \in Y_k$, for each K, the condition $\boldsymbol{r}_i \in X_i$, for each i, then given any price vector, $\boldsymbol{p}$, each consumer will have a non-empty budget set. A second reason is that the inclusion of individual resource endowments allows us to consider private ownership pure exchange economies as the special case of a private ownership economy in which $\ell = 1$, and $Y = \{\mathbf{0}\}$. On the other hand we obviously gain no generality in our model by including individual resource endowments in the specification of an economy, and in later chapters we will often simplify our notation by not explicitly taking such endowments into account.

As we have seen, the definition of a private ownership economy provides a natural definition of a wealth assignment for the economy, given a price vector, $\boldsymbol{p}$. However, the more abstract notion of a wealth-assignment for E provides us with much more flexibility in our analysis, and allows us to deal with many different situations with more or less the same arguments. Some of those different situations are set out in the following examples.

7.6 Examples.

1. Let $E = (\langle X_i, P_i \rangle, \langle Y_k \rangle, \boldsymbol{r})$ be an economy, let $(\langle \boldsymbol{r}_i \rangle, [s_{ik}])$ be a distribution of ownership for E, and denote the resultant private ownership economy by '$\mathcal{E}$.' We will say that $(\langle \boldsymbol{x}_i^* \rangle, \langle \boldsymbol{y}_k^* \rangle, \boldsymbol{p}^*)$ is a competitive equilibrium for $\mathcal{E}$ if it satisfies Definition 7.4 with the wealth assignment:

$$w_i = w_i(\boldsymbol{p}^*) = \boldsymbol{p}^* \cdot \boldsymbol{r}_i + \sum\nolimits_{k=1}^{\ell} s_{ik} \pi_k(\boldsymbol{p}^*), \text{ for } i = 1, \ldots, m.$$

However, we will sometimes wish to modify this arrangement, as follows. Define a vector $\boldsymbol{t} = (t_1, \ldots, t_m)$ to be a **system of lump-sum transfers for $\mathcal{E}$** iff:

$$\sum\nolimits_{i=1}^{m} t_i = 0.$$

We then say that $(\langle \boldsymbol{x}_i^* \rangle, \langle \boldsymbol{y}_k^* \rangle, \boldsymbol{p}^*)$ is a **competitive equilibrium for $\mathcal{E}$, given the system of lump-sum transfers, t**, iff, defining the assignment of wealth $\boldsymbol{w} = (w_1, \ldots, w_m)$ by:

$$w_i = \boldsymbol{p}^* \cdot \boldsymbol{r}_i + \sum\nolimits_{k=1}^{\ell} s_{ik} \pi_k(\boldsymbol{p}^*) + t_i, \quad \text{for } i = 1, \ldots, m,$$

it is true that $(\langle \boldsymbol{x}_i^* \rangle, \langle \boldsymbol{y}_k^* \rangle, \boldsymbol{p}^*)$ is a competitive equilibrium for E, given the assignment of wealth $\boldsymbol{w}$,

2. Let $E = (\langle X_i, P_i \rangle, \langle Y_k \rangle, \boldsymbol{r})$ be an economy, and suppose $(\langle \boldsymbol{x}_i^* \rangle, \langle \boldsymbol{y}_k^* \rangle, \boldsymbol{p}^*)$ satisfies conditions 1–3 of Definition 7.4, above; and, in addition, that for each i we have:

$$(\forall \boldsymbol{x}_i \in X_i) \colon \boldsymbol{x}_i P_i \boldsymbol{x}_i^* \Rightarrow \boldsymbol{p}^* \cdot \boldsymbol{x}_i > \boldsymbol{p}^* \cdot \boldsymbol{x}_i^*.$$

Then if we assign the wealth levels:

$$w_i^* = \boldsymbol{p}^* \cdot \boldsymbol{x}_i^* \quad \text{for } i = 1, \ldots, m;$$

it follows that $(\langle \boldsymbol{x}_i^* \rangle, \langle \boldsymbol{y}_k^* \rangle, \boldsymbol{p}^*)$ is a competitive equilibrium for E with the wealth assignment vector $\boldsymbol{w}^* = (w_1^*, \ldots, w_m^*)$.

3. **Exercise.** Show that in the above examples, the vectors $\boldsymbol{w}$ and $\boldsymbol{w}^*$ are wealth assignments for E, respectively, given the price vector $\boldsymbol{p}^*$; that is, that they satisfy Definition 7.3 □

7.3 Some Diagrammatic Techniques

In this and the next chapter, we will be looking at a great many simple examples of competitive or Walrasian equilibria. Our examples will deal primarily with two special cases. One is that in which $m = n = 2$, $\ell = 1$, and $Y = \{\mathbf{0}\}$ or $Y = -\mathbb{R}_+^2$. Either of these cases can be interpreted as the classical two-person, two-commodity pure exchange model, and the primary diagrammatic technique which we will use in the analysis of such examples will be the Edgeworth Box diagram with which you are already familiar. The other special case with which we shall be dealing is that in which:

$$m = \ell = 1, \; n = 2,$$

and in which there is non-zero production. In this case, it will be convenient to use some diagrammatic techniques with which you may not be familiar.[5]

Consider the economy, E, in which $m = \ell = 1$, $n = 2$, and:

$$X = \{\boldsymbol{x} \in \mathbb{R}^2 \mid (-2, 1) \leq \boldsymbol{x} \;\&\; x_1 \leq 0\}, Y = \{\boldsymbol{y} \in \mathbb{R}^2 \mid y_1 + y_2 \leq 0 \;\&\; y_1 \leq 0\},$$

and $\boldsymbol{r} = (0, 3/2)$. We can graph this production set and consumption set as in Figure 7.1.a, on the next page.

In the right-hand diagram (Figure 7.1.b) on the next page, we have indicated the **attainable consumption set**, which we will denote by 'X^*,' and which is simply the set of all consumption bundles which correspond to an attainable allocation for the economy. In other words, it is the set of all consumption bundles which can be attained with the resources and production technology available in the economy. Formally, we define X^* as follows. s

Suppose a pair $(\boldsymbol{x}, \boldsymbol{y})$ is in $A(E)$ for this economy, so that $(\boldsymbol{x}, \boldsymbol{y}) \in X \times Y$, and satisfies:

$$\boldsymbol{x} = \boldsymbol{r} + \boldsymbol{y}.$$

Then $\boldsymbol{x}$ must be a member of the set:

$$\boldsymbol{r} + Y \stackrel{\text{def}}{=} \{\boldsymbol{z} \in \mathbb{R}^2 \mid (\exists \boldsymbol{y} \in Y) \colon \boldsymbol{z} = \boldsymbol{r} + \boldsymbol{y}\}.$$

Consequently, defining X^*, the attainable consumption set, by:

$$X^* = \{\boldsymbol{x} \in X \mid (\exists \boldsymbol{y} \in Y) \colon (\boldsymbol{x}, \boldsymbol{y}) \in A(E)\},$$

[5] This type of diagram was introduced, and used extensively in Koopmans [1957].

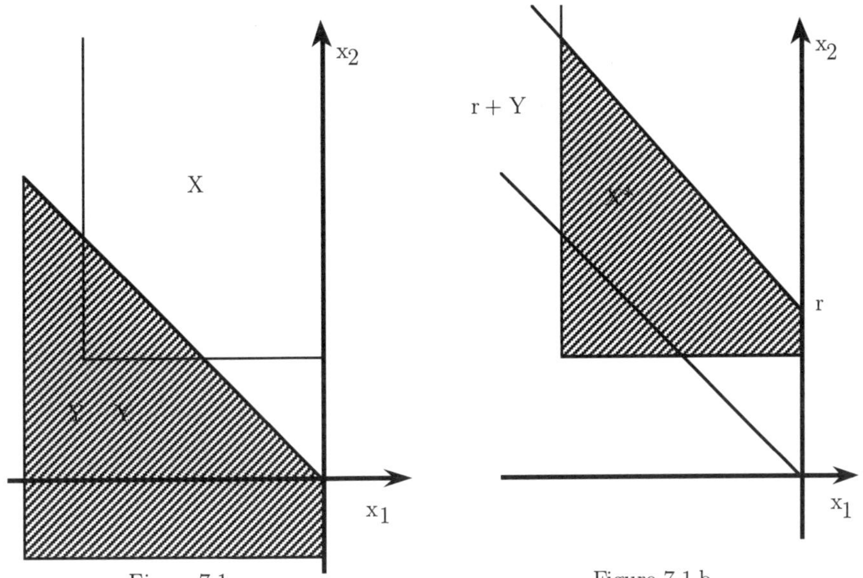

Figure 7.1.a

Figure 7.1.b

Figure 7.1: A Production Economy.

we see that we must have:

$$X^* \subseteq X \cap [\boldsymbol{r} + Y]. \tag{7.3}$$

Conversely, if:

$$\boldsymbol{x} \in X \cap [\boldsymbol{r} + Y], \tag{7.4}$$

then $\boldsymbol{x} \in X$, and, by definition of $\boldsymbol{r} + Y$, there exists $\boldsymbol{y} \in Y$ such that $\boldsymbol{x} = \boldsymbol{r} + \boldsymbol{y}$. We see, therefore, that if (7.4) holds, then $(\boldsymbol{x}, \boldsymbol{y}) \in A(E)$, and it follows that:

$$X \cap [\boldsymbol{r} + Y] \subseteq X^*. \tag{7.5}$$

Combining (7.3) and (7.5), we see, therefore, that:

$$X^* = X \cap [\boldsymbol{r} + Y]. \tag{7.6}$$

It is also easy to see that in our example, the set $\boldsymbol{r} + Y$ will be given by:

$$\boldsymbol{r} + Y = \{\boldsymbol{z} \in \mathbb{R}^2 \mid z_1 \leq 0 \;\&\; z_1 + z_2 \leq 3/2\},$$

as is indicated in Figure 7.1.b; and thus that the attainable consumption set will be as shown in that diagram.

Since the attainable consumption set is given by (7.6), it is clear that the set $\boldsymbol{r} + Y$ will be quite important in the analysis of our examples. In the examples we will present, the set $\boldsymbol{r} + Y$ will also be quite easy to represent graphically. For example, it is easy to see that if the commodity space is $\mathbb{R}^2$, $\boldsymbol{0} \in Y$, and $\boldsymbol{r}$ is of the form $\boldsymbol{r} = (0, r_2)$, where $r_2 > 0$, then $\boldsymbol{r} + Y$ will simply be the translate of Y

obtained by sliding the production set upward along the vertical axis until its vertex coincides with $\boldsymbol{r}$, rather than the origin. In fact, it is easy to prove the following (and it would be a good exercise to do so).

Suppose Y is a subset of $\mathbb{R}^2$ of the form:

$$Y = \{\boldsymbol{y} \in \mathbb{R}^2 \mid y_1 \leq 0 \;\&\; a_1 y_1 + a_2 y_2 \leq 0\},$$

where a_1 and a_2 are positive constants. Then $\boldsymbol{r}+Y$ is the set whose upper boundary is given by:

$$y_2 = (-a_1/a_2)y_1, \tag{7.7}$$

whose vertical intercept is at the origin; and if $\boldsymbol{r}$ is of the form:

$$\boldsymbol{r} = (0, r_2),$$

then:

$$\boldsymbol{r} + Y = \{\boldsymbol{z} \in \mathbb{R}^2 \mid z_1 \leq 0 \;\&\; a_1 z_1 + a_2 z_2 \leq a_2 r_2\}.$$

The upper boundary of $\boldsymbol{r} + Y$ is therefore given by the line:

$$z_2 = (-a_1/a_2)z_1 + r_2, \tag{7.8}$$

which is parallel to the line defined in (7.7), and which has the vertical intercept r_2.

The set $\boldsymbol{r} + Y$ can, for much of our analysis of this type of example, actually be thought of as a kind of production set. In fact, if we are trying to determine the level of supply, given a price vector $\boldsymbol{p}$, we can actually effectively ignore the set Y, and search for a 'profit-maximizing' vector, $\boldsymbol{z}$, in the set $Z = \boldsymbol{r} + Y$. That this is so stems from the following fact, which is actually a special case of Theorem 6.38, from Chapter 6; however, it might nonetheless be a good exercise to prove this fact directly.

7.7 Proposition. *If $\boldsymbol{z}^* = \boldsymbol{r} + \boldsymbol{y}^*$ maximizes 'profits' on Z, given the price vector $\boldsymbol{p}^*$, then $\boldsymbol{y}^*$ maximizes $\boldsymbol{p}^* \cdot \boldsymbol{y}$ on Y. Conversely, if $\boldsymbol{y}'$ maximizes $\boldsymbol{p}^* \cdot \boldsymbol{y}$ on Y, then $\boldsymbol{z}' = \boldsymbol{r} + \boldsymbol{y}'$ maximizes 'profits,' $\boldsymbol{p}^* \cdot \boldsymbol{z}$, on Z.*

Now consider a private ownership economy, $\mathcal{E}$, where $m = \ell = 1$. Since there is only one consumer and one firm, we have:

$$s \equiv s_{11} = \text{the share of the first consumer in the profits of the first firm} = 1,$$

and:

$$\boldsymbol{r}_1 \equiv \text{the first consumer's resource endowment} = \boldsymbol{r},$$

the aggregate resource endowment. Therefore, if $\boldsymbol{z}^* \in Z$ maximizes 'profits' on Z, given the price vector $\boldsymbol{p}^*$, the consumer's budget line (or hyperplane), given $\boldsymbol{p}^*$, is given by:

$$b(\boldsymbol{p}^*) = \{\boldsymbol{x} \in X \mid \boldsymbol{p}^* \cdot \boldsymbol{x} = \boldsymbol{p}^* \cdot \boldsymbol{z}^*\}.$$

Thus, in the case where $n = 2$, we have the sort of situation illustrated in Figure 7.2, below, which depicts a Walrasian equilibrium at $(\boldsymbol{x}^*, \boldsymbol{y}^*, \boldsymbol{p}^*)$.

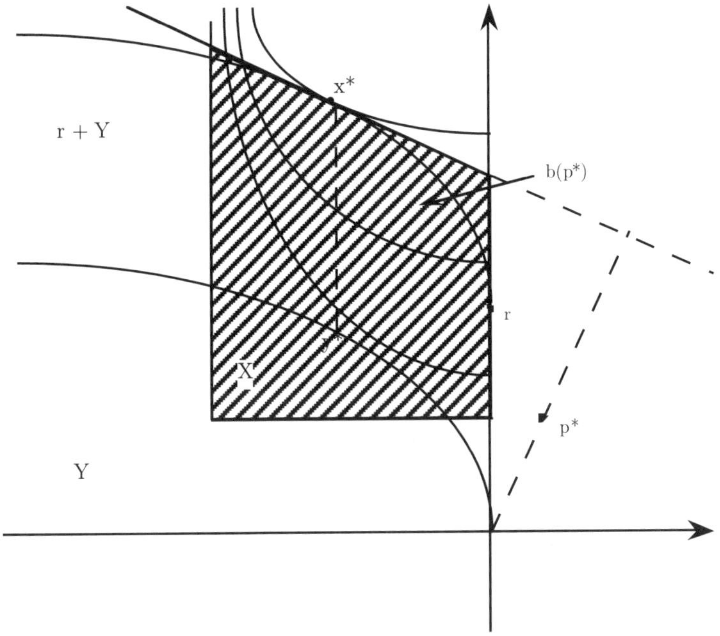

Figure 7.2: A Competitive Equilibrium.

We can simplify our discussion of examples, and gain some further understanding of a Walrasian equilibrium, by consideration of the following material. Earlier we defined the sets Π_k and Π by:

$$\Pi_k = \{\boldsymbol{p} \in \mathbb{R}^n \mid (\exists \boldsymbol{y}_k^* \in Y_k)(\forall \boldsymbol{y}_k \in Y_k) \colon \boldsymbol{p} \cdot \boldsymbol{y}_k^* \geq \boldsymbol{p} \cdot \boldsymbol{y}_k\} \quad \text{for } k = 1, \ldots, \ell;$$

and:

$$\Pi = \{\boldsymbol{p} \in \mathbb{R}^n \mid (\exists \boldsymbol{y}^* \in Y)(\forall \boldsymbol{y} \in Y) \colon \boldsymbol{p} \cdot \boldsymbol{y}^* \geq \boldsymbol{p} \cdot \boldsymbol{y}\}, \tag{7.9}$$

where 'Y' denotes the aggregate production set. We then defined the aggregate profit function, $\pi(\cdot)$, and the aggregate supply correspondence, $\sigma(\cdot)$:

$$\pi(\boldsymbol{p}) = \max_{\boldsymbol{y} \in Y} \boldsymbol{p} \cdot \boldsymbol{y} \quad \text{for } \boldsymbol{p} \in \Pi, \tag{7.10}$$

and:

$$\sigma(\boldsymbol{p}) = \{\boldsymbol{y} \in Y \mid \boldsymbol{p} \cdot \boldsymbol{y} = \pi(\boldsymbol{p})\} \quad \text{for } \boldsymbol{p} \in \Pi. \tag{7.11}$$

Recall also that it follows from Theorem 6.38 that:

$$\mathbf{0} \in \Pi \ \& \ \Pi = \bigcap\nolimits_{k=1}^{\ell} \Pi_k; \tag{7.12}$$

and, for $\boldsymbol{p} \in \Pi$:

$$\pi(\boldsymbol{p}) = \sum\nolimits_{k=1}^{\ell} \pi_k(\boldsymbol{p}) \ \& \ \sigma(\boldsymbol{p}) = \sum\nolimits_{k=1}^{\ell} \sigma_k(\boldsymbol{p}). \tag{7.13}$$

Furthermore, it follows easily from the same theorem that if $(\langle \boldsymbol{x}_i^* \rangle, \langle \boldsymbol{y}_k^* \rangle, \boldsymbol{p}^*)$ is a competitive equilibrium for an economy, E, then we must have $\boldsymbol{p}^* \in \Pi$, and, defining:

$$\boldsymbol{y}^* = \sum_{k=1}^{\ell} \boldsymbol{y}_k^*,$$

we must have:

$$\boldsymbol{p}^* \cdot \boldsymbol{y}^* = \pi(\boldsymbol{p}^*) \;\;\&\;\; \boldsymbol{y}^* \in \sigma(\boldsymbol{p}^*).$$

Now define the set Π^* by:

$$\Pi^* = \Pi \cap \Delta_n,$$

where 'Δ_n' denotes the unit simplex in $\mathbb{R}^n$:

$$\Delta_n = \Big\{\boldsymbol{p} \in \mathbb{R}_+^n \mid \sum_{j=1}^{n} p_j = 1\Big\}.$$

The homogeneity of the consumers' demand correspondences and the producers' supply correspondences, together with Proposition 6.25 and Theorem 6.38 of Chapter 6, imply the following; the proof of which will be left as an exercise.

7.8 Proposition. *Suppose the economy E satisfies either:*

$$Y - \mathbb{R}_+^n \subseteq Y, \tag{7.14}$$

or:

$$-\mathbb{R}_+^n \subseteq Y, \tag{7.15}$$

and that $(\langle \boldsymbol{x}_i^* \rangle, \langle \boldsymbol{y}_k^* \rangle, \boldsymbol{p}^*)$ is a competitive equilibrium for E. If we define:

$$\bar{\boldsymbol{p}} = \left(\frac{1}{\sum_{j=1}^{n} p_j^*}\right) \boldsymbol{p}^*,$$

then $\bar{\boldsymbol{p}} \in \Pi^*$ and $\big(\langle \boldsymbol{x}_i^* \rangle, \langle \boldsymbol{y}_k^* \rangle, \bar{\boldsymbol{p}}\big)$ is also a competitive equilibrium for E.

We will often employ a variant of the model we have been discussing here, one which also deals with the case in which there is one producer, one consumer, and two commodities as before, but makes use of the assumption that $X = \mathbb{R}_+^2$. Here we will also generally suppose that $\boldsymbol{r}$ takes the form:

$$\boldsymbol{r} = (r, 0),$$

and we will use the generic notation:

$$\boldsymbol{p} = (w, p),$$

to denote price vectors; although we will frequently normalize the price of the first commodity; that is, set $w = 1$. In this case we interpret 'x_1' as the quantity of leisure demanded by the consumer; while:

$$\ell = x_1 - r,$$

is the quantity of labor offered. A typical competitive equilibrium in this case might look something like the following, in diagrammatic presentation.

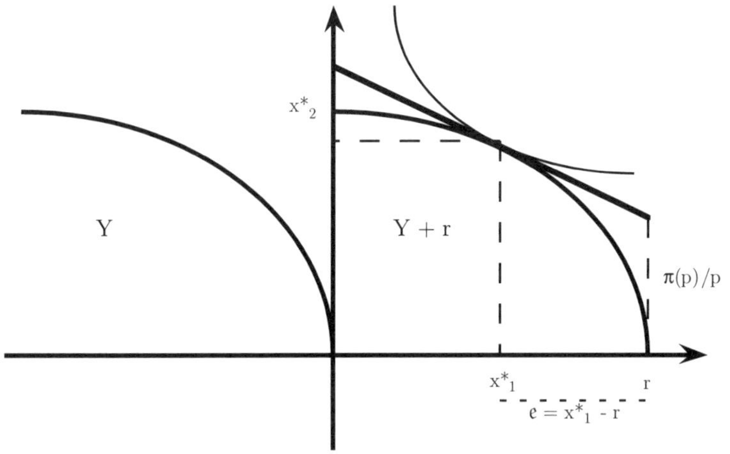

Figure 7.3: Competitive Equilibrium for the Labor/Leisure Model.

7.4 Walras' Law with Production

We begin this section by extending our definition of a wealth-assignment vector to a wealth-assignment function. In the definition, recall that we have defined Π for an economy as the collection of all price vectors for which a maximum profit exists on *each* production set, and, for each $i \in M$:

$$\Omega_i = \{(\boldsymbol{p}, w) \in \mathbb{R}^{n+1} \mid (\exists \boldsymbol{x} \in X_i) \colon \boldsymbol{p} \cdot \boldsymbol{x} \le w\}.$$

7.9 Definition. Let $E = (\langle X_i, P_i \rangle, \langle Y_k \rangle, \boldsymbol{r})$ be an economy. We shall say that a function, $\boldsymbol{w} \colon \Pi \to \mathbb{R}^m$ is a **feasible wealth-assignment function for $\boldsymbol{E}$** iff, for each $\boldsymbol{p} \in \Pi$, we have:

$$\big(\boldsymbol{p}, w_i(\boldsymbol{p})\big) \in \Omega_i \quad \text{for } i = 1, \ldots, m,$$

and:

$$\sum_{i=1}^{m} w_i(\boldsymbol{p}) = \boldsymbol{p} \cdot \boldsymbol{r} + \sum_{k=1}^{\ell} \pi_k(\boldsymbol{p}).$$

We have already considered competitive equilibria with an arbitrary assignment of wealth. The notion of a feasible wealth-assignment function simply extends this basic idea by allowing for a wealth assignment which is a function of prices.[6]

7.10 Examples.

1. Let $\mathcal{E} = (\langle X_i, P_i \rangle, \langle Y_k \rangle, \langle \boldsymbol{r}_i \rangle, [s_{ik}])$ be a private ownership economy satisfying:

$$\boldsymbol{r}_i \in X_i \quad \text{for } i = 1, \ldots, m,$$

and:

$$\boldsymbol{0} \in Y_k \quad \text{for } k = 1, \ldots, \ell;$$

[6] The very useful notion of a wealth-assignment function was introduced in Gale and Mas-Colell [1975].

and, as we have done previously, define $\boldsymbol{w}\colon \Pi \to \mathbb{R}^m$ by:

$$w_i(\boldsymbol{p}) = \boldsymbol{p}\cdot\boldsymbol{r}_i + \sum_{k=1}^{\ell} s_{ik}\pi_k(\boldsymbol{p}) \quad \text{for } i = 1,\ldots,m. \tag{7.16}$$

Then, since $\mathbf{0} \in Y_k$ for each k, we see that for any $\boldsymbol{p} \in \Pi$;

$$\pi_k(\boldsymbol{p}) \geq 0 \quad \text{for } k = 1,\ldots,\ell.$$

Therefore, for each $\boldsymbol{p} \in \Pi$ and each i:

$$w_i(\boldsymbol{p}) = \boldsymbol{p}\cdot\boldsymbol{r}_i + \sum_{k=1}^{\ell} s_{ik}\pi_k(\boldsymbol{p}) \geq \boldsymbol{p}\cdot\boldsymbol{r}_i.$$

Since $\boldsymbol{r}_i \in X_i$, and (as I will leave you to verify), for each $\boldsymbol{p} \in \Pi$:

$$\sum_{i=1}^{m} w_i(\boldsymbol{p}) = \boldsymbol{p}\cdot\boldsymbol{r} + \sum_{k=1}^{\ell} \pi_k(\boldsymbol{p}),$$

it follows that $\boldsymbol{w}(\cdot)$ is a feasible wealth-assignment function for $\mathcal{E}$.

2. It will sometimes be useful to make use of the idea of a wealth-assignment function to formally relate results for pure exchange economies to production economies with linear technologies, as follows. Let $E = (\langle X_i, P_i\rangle, \langle Y_k\rangle, \boldsymbol{r})$ be an economy in which Y_k is linear, for $k = 1,\ldots,\ell$. We can then define the function $\boldsymbol{w}\colon \Pi \to \mathbb{R}^m$ by:

$$w_i(\boldsymbol{p}) = \boldsymbol{p}\cdot\boldsymbol{r}_i \quad \text{for } i = 1,\ldots,m. \tag{7.17}$$

If $(\boldsymbol{r}_i)$ satisfies the condition:

$$\boldsymbol{r}_i \in X_i \quad \text{for } i = 1,\ldots,m;$$

then, since we showed in Chapter 6 that we must have $\pi_k(\boldsymbol{p}) = 0$ for each $\boldsymbol{p} \in \Pi$ and each k, it follows that $\boldsymbol{w}(\cdot)$ is a feasible wealth-assignment function for E.

3. Suppose once again that $\mathcal{E} = (\langle X_i, P_i\rangle, \langle Y_k\rangle, \langle \boldsymbol{r}_i\rangle, [s_{ik}])$ satisfies:

$$\boldsymbol{r}_i \in X_i \quad \text{for } i = 1,\ldots,m;$$

and that:

$$\mathbf{0} \in Y_k \quad \text{for } k = 1,\ldots,\ell;$$

and let's introduce government as an $(m+1)^{st}$ consumer; using the subscript '0' to denote government's income and consumption. We will then suppose that $X_0 = \mathbb{R}^n_+$, so that:

$$\Omega_0 = \{(\boldsymbol{p}, w) \in \mathbb{R}^{n+1} \mid \boldsymbol{p} \in \mathbb{R}^n_{++} \;\&\; w \geq 0\}.$$

We will suppose that the ith consumer pays the tax:

$$t_i = \tau_i \cdot \Big(\sum_{k=1}^{\ell} s_{ik}\pi_k(\boldsymbol{p})\Big),$$

where:[7]

$$0 \leq \tau_i < 1 \quad \text{for } i = 1,\ldots,m.$$

[7] In practice, there would typically be only three or four different tax rates; with the tax rate paid by the ith consumer depending upon her or his income. However, the formulation here incorporates this situation as a special case, and is simpler to deal with in this example.

We then define $\langle s_{0k} \rangle$ by:

$$s_{0k} = \sum_{i=1}^{m} \tau_i \cdot s_{ik} \quad \text{for } k = 1, \ldots, \ell;$$

and $\boldsymbol{w} \colon \Pi \to \mathbb{R}^{1+m}$ by:

$$w_0(\boldsymbol{p}) = \sum_{k=1}^{\ell} s_{0k}\pi_k(\boldsymbol{p}), \text{ and:}$$
$$w_i(\boldsymbol{p}) = \boldsymbol{p} \cdot \boldsymbol{r}_i + \sum_{k=1}^{\ell} s_{ik}\pi_k(\boldsymbol{p}) - t_i \quad \text{for } i = 1, \ldots, m.$$

I will leave it to you to verify that this is a feasible wealth assignment function for $\mathcal{E}$. □

If a feasible wealth-assignment function is defined for an economy, individual and aggregate demand and excess demand become functions of prices alone; as is set out formally in the following definition.

7.11 Definition. Let $E = (\langle X_i, P_i \rangle, \langle Y_k \rangle, \boldsymbol{r})$, and let $\boldsymbol{w} \colon \Pi \to \mathbb{R}^m$ be a feasible wealth-assignment function for E. We define the **excess demand correspondence for $\boldsymbol{E}$, given $\boldsymbol{w}, \boldsymbol{\eta} \colon \Pi \mapsto \mathbb{R}^n$**, by:

$$\boldsymbol{\eta}(\boldsymbol{p}) = \sum_{i=1}^{m} \delta_i(\boldsymbol{p}) - \boldsymbol{r} - \sum_{k=1}^{\ell} \sigma_k(\boldsymbol{p}), \tag{7.18}$$

where we define $\boldsymbol{\delta}_i(\cdot)$ by:

$$\boldsymbol{\delta}_i(\boldsymbol{p}) = \boldsymbol{h}_i[\boldsymbol{p}, w_i(\boldsymbol{p})] \quad \text{for } i = 1, \ldots, m;$$

where $\boldsymbol{h}_i(\cdot)$ is the ith consumer's demand correspondence.

7.12 Proposition. [**Walras' Law (Weak Form)**] *Let $E = (\langle X_i, P_i \rangle, \langle Y_k \rangle, \boldsymbol{r})$ be an economy, let $\boldsymbol{w} \colon \Pi \to \mathbb{R}^m$ be a feasible wealth-assignment function for E, and let $\boldsymbol{\eta} \colon \Pi \mapsto \mathbb{R}^n$ be the aggregate excess demand correspondence for E, given $\boldsymbol{w}$. Then for any $\boldsymbol{p} \in \Pi$ and any $\boldsymbol{z} \in \boldsymbol{\eta}(\boldsymbol{p})$ we have $\boldsymbol{p} \cdot \boldsymbol{z} \leq 0$.*

Proof. Under the stated conditions, there exist $\big(\langle \boldsymbol{x}_i \rangle, \langle \boldsymbol{y}_k \rangle\big)$ such that:

$$\boldsymbol{x}_i \in \boldsymbol{h}_i[\boldsymbol{p}, w_i(\boldsymbol{p})] \quad \text{for } i = 1, \ldots, m, \tag{7.19}$$
$$\boldsymbol{y}_k \in \boldsymbol{\sigma}_k(\boldsymbol{p}) \quad \text{for } k = 1, \ldots, \ell, \tag{7.20}$$

and:

$$\boldsymbol{z} = \sum_{i=1}^{m} \boldsymbol{x}_i - \boldsymbol{r} - \sum_{k=1}^{\ell} \boldsymbol{y}_k. \tag{7.21}$$

Since $\boldsymbol{y}_k \in \boldsymbol{\sigma}_k(\boldsymbol{p})$, for each k, we also have:

$$\boldsymbol{p} \cdot \boldsymbol{y}_k = \pi_k(\boldsymbol{p}) \quad \text{for } k = 1, \ldots, \ell; \tag{7.22}$$

while from (7.30), we have:

$$\boldsymbol{p} \cdot \boldsymbol{x}_i - w_i(\boldsymbol{p}) \leq 0 \quad \text{for } i = 1, \ldots, m. \tag{7.23}$$

From (7.23) we have, upon adding over i, and making use of (7.31), (7.21), and the definition of a feasible wealth-assignment function:

$$0 \geq \sum_{i=1}^{m} \boldsymbol{p} \cdot \boldsymbol{x}_i - \sum_{i=1}^{m} w_i(\boldsymbol{p}) = \sum_{i=1}^{m} \boldsymbol{p} \cdot \boldsymbol{x}_i - \Big(\boldsymbol{p} \cdot \boldsymbol{r} + \sum_{k=1}^{\ell} \pi_k(\boldsymbol{p})\Big)$$
$$= \boldsymbol{p} \cdot \Big(\sum_{i=1}^{m} \boldsymbol{x}_i - \boldsymbol{r} - \sum_{k=1}^{\ell} \boldsymbol{y}_k\Big) = \boldsymbol{p} \cdot \boldsymbol{z}. \quad \square$$

The proof of the next two results will be left as exercises.

7.13 Proposition. [**Walras' Law (Strong Form)**] *Let $E = (\langle X_i, P_i\rangle, \langle Y_k\rangle, \boldsymbol{r})$ be an economy, let $\boldsymbol{w}\colon \Pi \to \mathbb{R}^m$ be a feasible wealth-assignment function for E, let $\boldsymbol{\eta}\colon \Pi \mapsto \mathbb{R}^n$ be the aggregate excess demand correspondence for E, given $\boldsymbol{w}$, and suppose that:*

$$P_i \text{ is locally non-saturating, for } i = 1, \ldots, m.$$

Then for any $\boldsymbol{p} \in \Pi$ and any $\boldsymbol{z} \in \boldsymbol{\eta}(\boldsymbol{p})$ we have $\boldsymbol{p} \cdot \boldsymbol{z} = 0$.

7.14 Corollary. [**Walras' Law (Original Form)**] *Let $E = (\langle X_i, P_i\rangle, \langle Y_k\rangle, \boldsymbol{r})$ be an economy, let $\boldsymbol{w}\colon \Pi \to \mathbb{R}^m$ be a feasible wealth-assignment function for E, let $\boldsymbol{\eta}\colon \Pi \mapsto \mathbb{R}^n$ be the aggregate excess demand correspondence for E, given $\boldsymbol{w}$, and suppose that P_i is locally non-saturating, for $i = 1, \ldots, m$. Then if $\boldsymbol{p}^* \in \Pi \cap \mathbb{R}^n_{++}$ and $\boldsymbol{z}^* \in \boldsymbol{\eta}(\boldsymbol{p}^*)$ are such that for some $k \in \{1, \ldots, n\}$, we have*

$$(\forall j \in \{1, \ldots, n\} \setminus \{k\})\colon z_j^* = 0,$$

we must have $z_k^* = 0$ as well. In other words if $n-1$ of the markets are in equilibrium, then the remaining market must be in equilibrium as well.

7.15 Example. Suppose that in the economy, $\mathcal{E} = (\langle X_i, P_i\rangle, \langle Y_k\rangle, \langle \boldsymbol{r}_i\rangle, [s_{ik}])$, the ℓth firm represents government production. We are not excluding the possibility that we may have $Y_\ell \subseteq -\mathbb{R}^n_+$; that is, that government is simply a consumer, as was the case in Example 7.10.3; however, we can equally well suppose that some elements $\boldsymbol{y}_\ell \in Y_\ell$ have some positive coordinates. In any case, an allocation, $\big(\langle \boldsymbol{x}_i\rangle, \langle \boldsymbol{y}_k\rangle\big)$ will be feasible iff:

$$\sum_{i=1}^{m} \boldsymbol{x}_i = \sum_{i=1}^{m} \boldsymbol{r}_i + \sum_{k=1}^{\ell} \boldsymbol{y}_k. \tag{7.24}$$

We define:

$$\Pi = \bigcap_{k=1}^{\ell-1} \Pi_k;$$

and, defining the function $\boldsymbol{w}\colon \Pi \to \mathbb{R}^m$ by:

$$w_i(\boldsymbol{p}) = \boldsymbol{p} \cdot \boldsymbol{r}_i + \sum_{k=1}^{\ell-1} s_{ik}\pi_k(\boldsymbol{p}) - t_i(\boldsymbol{p}) \quad \text{for } i = 1, \ldots, m, \tag{7.25}$$

we suppose that the tax/transfer function $\boldsymbol{t}\colon \Pi \to \mathbb{R}^m$ has been defined in such a way that, for each $\boldsymbol{p} \in \Pi$:

$$\big(\boldsymbol{p}, w_i(\boldsymbol{p})\big) \in \Omega_i \quad \text{for } i = 1, \ldots, m. \tag{7.26}$$

We will suppose that the government's choice of 'production' can be characterized by a function of price, $\boldsymbol{\sigma}_\ell\colon \Pi \to Y_\ell$. We will then say that $(\langle \boldsymbol{x}_i^*\rangle, \langle \boldsymbol{y}_k^*\rangle, \boldsymbol{p}^*)$ is a **competitive equilibrium, given the governmental policy $(\boldsymbol{\sigma}_\ell, \boldsymbol{t})$** iff:

1. $\boldsymbol{p}^* \neq \mathbf{0}$,
2. $(\langle \boldsymbol{x}_i^* \rangle, \langle \boldsymbol{y}_k^* \rangle)$ is feasible,
3. $\boldsymbol{y}_k^* \in \boldsymbol{\sigma}_k(\boldsymbol{p}^*)$, for $k = 1, \ldots, \ell$, and:
4. $\boldsymbol{x}_i^* \in \boldsymbol{h}_i[\boldsymbol{p}^*, w_i(\boldsymbol{p}^*)]$, for $i = 1, \ldots, m$.

Now suppose that each consumer's demand correspondence satisfies the budget balance condition (Definition 4.6). We can then derive a rather interesting conclusion, as follows. By the budget balance condition, we have:

$$\boldsymbol{p}^* \cdot \boldsymbol{x}_i^* = w_i(\boldsymbol{p}^*) = \boldsymbol{p}^* \cdot \boldsymbol{r}_i + \sum\nolimits_{k=1}^{\ell-1} s_{ik}\pi_k(\boldsymbol{p}^*) - t_i(\boldsymbol{p}^*), \text{ for } i = 1, \ldots, m.$$

Thus, adding over i, and defining $\boldsymbol{x}^* = \sum_{k=1}^{m} \boldsymbol{x}_i^*$ and $\tau(\boldsymbol{p}) = \sum_{i=1}^{m} t_i(\boldsymbol{p})$, we obtain:

$$\boldsymbol{p}^* \cdot \boldsymbol{x}^* = \boldsymbol{p}^* \cdot \boldsymbol{r} + \sum\nolimits_{k=1}^{\ell-1} \boldsymbol{p}^* \cdot \boldsymbol{y}_k^* - \tau(\boldsymbol{p}^*). \tag{7.27}$$

On the other hand, since $(\langle \boldsymbol{x}_i^* \rangle, \langle \boldsymbol{y}_k^* \rangle)$ is feasible, we obtain from (7.24) that:

$$\boldsymbol{p}^* \cdot \boldsymbol{x}^* = \boldsymbol{p}^* \cdot \boldsymbol{r} + \sum\nolimits_{k=1}^{\ell-1} \boldsymbol{p}^* \cdot \boldsymbol{y}_k^* + \boldsymbol{p}^* \cdot \boldsymbol{y}_\ell^* \tag{7.28}$$

Combining (7.27) and (7.28), we see that:

$$\boldsymbol{p}^* \cdot \boldsymbol{y}_\ell^* = -\tau(\boldsymbol{p}^*);$$

that is, the government's budget is necessarily balanced!

How is it that this conclusion, which is so very unlike our recent experience in the U. S., can be reached? Well, it is rather a variant of Walras' Law; and comes about essentially because the value of demand must be equal to the value of supply. To put this another way, we cannot have an unbalanced governmental budget in this sort of model unless we introduce a financial sector into the model.

Notice finally that it follows from (7.27) and (7.28) that the function defined in (7.24) is a feasible wealth-assignment function for $\mathcal{E}$. □

7.5 The 'First Fundamental Theorem'

The following definitions are essentially unchanged from those presented in Chapter 5. I repeat them here largely for the sake of providing a convenient reference.

7.16 Definitions. Let E be an economy. We then define:

1. the **unanimity ordering** (or the **strong Pareto ordering**), $\boldsymbol{Q}$, on $\mathcal{X}$ by:

$$\langle \boldsymbol{x}_i \rangle \boldsymbol{Q} \langle \boldsymbol{x}'_i \rangle \iff [\boldsymbol{x}_i P_i \boldsymbol{x}'_i \quad \text{for } i = 1, \ldots, m]. \tag{7.29}$$

2. the **Pareto (at-least-as-good-as) ordering**, $\boldsymbol{R}$, on $\mathcal{X}$, by:

$$\langle \boldsymbol{x}_i \rangle \boldsymbol{R} \langle \boldsymbol{x}'_i \rangle \iff [\neg \boldsymbol{x}'_i P_i \boldsymbol{x}_i, \text{ for } i = 1, \ldots, m]. \tag{7.30}$$

3. the **strict Pareto ordering**, $\boldsymbol{P}$, on $\mathcal{X}$, by:

$$\langle \boldsymbol{x}_i \rangle \boldsymbol{P} \langle \boldsymbol{x}'_i \rangle \iff [\langle \boldsymbol{x}_i \rangle \boldsymbol{R} \langle \boldsymbol{x}'_i \rangle \;\&\; \neg \langle \boldsymbol{x}'_i \rangle \boldsymbol{R} \langle \boldsymbol{x}_i \rangle]. \tag{7.31}$$

We will use the following terminology in dealing with these three orderings. If:
$\langle \boldsymbol{x}_i^* \rangle \boldsymbol{Q} \langle \boldsymbol{x}_i \rangle$, we shall say that $\langle \boldsymbol{x}_i^* \rangle$ is **unanimously preferred to** $\langle \boldsymbol{x}_i \rangle$,
$\langle \boldsymbol{x}_i^* \rangle \boldsymbol{R} \langle \boldsymbol{x}_i \rangle$, we shall say that $\langle \boldsymbol{x}_i^* \rangle$ **(weakly) Pareto dominates** $\langle \boldsymbol{x}_i \rangle$,
$\langle \boldsymbol{x}_i^* \rangle \boldsymbol{P} \langle \boldsymbol{x}_i \rangle$, we shall say that $\langle \boldsymbol{x}_i^* \rangle$ **strictly Pareto dominates** $\langle \boldsymbol{x}_i \rangle$.

7.17 Definitions. Let $E = (\langle X_i, P_i \rangle, \langle Y_k \rangle, \boldsymbol{r})$ be an economy. We shall say that a feasible allocation for E, $\big(\langle \boldsymbol{x}_i^* \rangle, \langle \boldsymbol{y}_k^* \rangle\big)$ is **Pareto efficient for $\boldsymbol{E}$** [respectively, **strongly Pareto efficient for $\boldsymbol{E}$**] iff there exists no alternative feasible allocation for E, $\big(\langle \boldsymbol{x}_i \rangle, \langle \boldsymbol{y}_k \rangle\big)$, satisfying:

$$\langle \boldsymbol{x}_i \rangle \boldsymbol{Q} \langle \boldsymbol{x}_i^* \rangle [\text{respectively, } \langle \boldsymbol{x}_i \rangle \boldsymbol{P} \langle \boldsymbol{x}_i^* \rangle];$$

where the orderings $\boldsymbol{Q}$ and $\boldsymbol{P}$ are defined in equations (7.29) and (7.31), above.

While the above definitions are stated for an economy, the corresponding definitions for a private ownership economy, $\mathcal{E} = (\langle X_i, P_i \rangle, \langle Y_k \rangle, \langle \boldsymbol{r}_i \rangle, [s_{ik}])$, are obvious, and will be used where needed without further comment.

In the terminology just introduced, a feasible allocation, $\big(\langle \boldsymbol{x}_i^* \rangle, \langle \boldsymbol{y}_k^* \rangle\big)$, will be said to be Pareto efficient for E iff there exists no alternative *feasible* allocation which all consumers prefer to $\big(\langle \boldsymbol{x}_i^* \rangle, \langle \boldsymbol{y}_k^* \rangle\big)$. The feasible allocation $\big(\langle \boldsymbol{x}_i^* \rangle, \langle \boldsymbol{y}_k^* \rangle\big)$ is strongly Pareto efficient for E iff there exists no alternative *feasible* allocation which strictly Pareto dominates $\big(\langle \boldsymbol{x}_i^* \rangle, \langle \boldsymbol{y}_k^* \rangle\big)$. Since $\big(\langle \boldsymbol{x}_i^* \rangle, \langle \boldsymbol{y}_k^* \rangle\big)$ may be such that, while no feasible alternative allocation is unanimously preferred, there nonetheless is another feasible allocation, $\big(\langle \boldsymbol{x}_i \rangle, \langle \boldsymbol{y}_k \rangle\big)$, where no consumer is worse off, and at least one consumer is better off than at $\big(\langle \boldsymbol{x}_i^* \rangle, \langle \boldsymbol{y}_k^* \rangle\big)$, there are in principle more Pareto efficient allocations than there are strongly Pareto efficient allocations for a given economy, E. This is the reason for the terminology used here.

Making use of the terminology introduced by Hurwicz [1960], we will demonstrate that, loosely speaking:

1. the competitive mechanism is **non-wasteful**, in the sense that any competitive

equilibrium is Pareto efficient, and

2. the competitive mechanism is **unbiased**, in the sense that (given some additional

assumptions) any Pareto efficient allocation can be made a competitive equilibrium.

Roughly speaking, these two results respectively constitute what are known as the 'First' and 'Second Fundamental Theorems of Welfare Economics.'

In the material to be presented here, we will concentrate on Pareto efficient allocations, as opposed to strongly Pareto efficient allocations; for the reasons set out in Chapter 5. However, the following result provides sufficient conditions for Pareto efficient allocations to be strongly Pareto efficient. I will leave the proof as an exercise, since it can be done in essentially the same way as we proved Proposition 5.19.

7.18 Proposition. *If $E = (\langle X_i, P_i \rangle, \langle Y_k \rangle, \boldsymbol{r})$ is an economy in which $X_i = \mathbb{R}^n_+$ and P_i is asymmetric, negatively transitive, lower semi-continuous, and strictly increasing, for $i = 1, \ldots, m$; then an allocation $\big(\langle \boldsymbol{x}_i^* \rangle, \langle \boldsymbol{y}_k^* \rangle\big)$ is Pareto efficient for E if, and only if, it is strongly Pareto efficient for E.*

Turning now to the 'non-wastefulness' property of the competitive mechanism, we begin by defining a slight generalization of the idea of a competitive, or Walrasian equilibrium for an economy. The definition presented here will be critical to our initial development of the 'unbiasedness' result; in fact, the usual statement of the 'Second Fundamental Theorem' is something like: 'given the appropriate convexity conditions it is the case that, given any Pareto efficient allocation $\big(\langle \boldsymbol{x}_i^* \rangle, \langle \boldsymbol{y}_k^* \rangle\big)$, there exists a price vector, $\boldsymbol{p}^*$, and a wealth assignment, $\boldsymbol{w}^*$, such that $(\langle \boldsymbol{x}_i^* \rangle, \langle \boldsymbol{y}_k^* \rangle, \boldsymbol{p}^*)$ is a quasi-competitive equilibrium for E, given $\boldsymbol{w}^*$. In the definition we make use of a bit of notation which we will continue to use throughout the remainder of this book; given a vector $\boldsymbol{p}^* \in \mathbb{R}^n$ and a set $Z \subseteq \mathbb{R}^n$, we use the expression '$\min \boldsymbol{p}^* \cdot Z$' as shorthand for:

$$\min\{\boldsymbol{p}^* \cdot \boldsymbol{z} \mid \boldsymbol{z} \in Z\}.$$

7.19 Definition. If E is an economy, we shall say that $(\langle \boldsymbol{x}_i^* \rangle, \langle \boldsymbol{y}_k^* \rangle, \boldsymbol{p}^*)$ is a **quasi-competitive equilibrium for $\boldsymbol{E}$**, iff there exists a wealth-assignment for E, given $\boldsymbol{p}^*$, $\boldsymbol{w} = (w_1, \ldots, w_m)$, such that:
1. $\boldsymbol{p}^* \neq \boldsymbol{0}$,
2. $\big(\langle \boldsymbol{x}_i^* \rangle, \langle \boldsymbol{y}_k^* \rangle\big) \in A(E)$,
3. $\boldsymbol{p}^* \cdot \boldsymbol{y}_k^* = \pi_k(\boldsymbol{p}^*)$, for $k = 1, \ldots, \ell$,
4. for each i ($i = 1, \ldots, m$), we have $\boldsymbol{p}^* \cdot \boldsymbol{x}_i^* \leq w_i$, and either:

$$w_i == \min \boldsymbol{p}^* \cdot X_i, \tag{7.32}$$

or:

$$(\forall \boldsymbol{x}_i \in X_i) \colon \boldsymbol{x}_i P_i \boldsymbol{x}_i^* \Rightarrow \boldsymbol{p}^* \cdot \boldsymbol{x}_i > w_i \tag{7.33}$$

(or both). In this case, we shall also say that $(\langle \boldsymbol{x}_i^* \rangle, \langle \boldsymbol{y}_k^* \rangle, \boldsymbol{p}^*)$ is a quasi-competitive equilibrium for E, given the wealth-assignment, $\boldsymbol{w}$.

If you compare Definition 7.19 with the definition of a competitive equilibrium for E (Definition 7.4), you will see that the two definitions differ only in condition 4: condition 4 of Definition 7.4 says that any commodity bundle $\boldsymbol{x}_i$ which is preferred to $\boldsymbol{x}_i^*$ must cost more than w_i (given the price vector, $\boldsymbol{p}^*$) while condition 4 of 7.19 says that the former condition can fail only if every commodity bundle in X_i costs at least as much, given the price vector $\boldsymbol{p}^*$, as does $\boldsymbol{x}_i^*$.

The following presents the properties upon which our proof of the 'First Fundamental Theorem' is based.

7.20 Proposition. *If $(\langle \boldsymbol{x}_i^* \rangle, \langle \boldsymbol{y}_k^* \rangle, \boldsymbol{p}^*)$ is a quasi-competitive equilibrium for an economy, E, given the wealth-assignment, $\boldsymbol{w}^* = (w_1^*, \ldots, w_m^*)$, then:*

$$\boldsymbol{p}^* \cdot \boldsymbol{x}_i^* = w_i^* \quad \textit{for } i = 1, \ldots, m; \tag{7.34}$$

and, for any feasible allocation, $\big(\langle \boldsymbol{x}_i \rangle, \langle \boldsymbol{y}_k \rangle\big)$, we have:

$$\sum_{i=1}^m w_i^* = \boldsymbol{p}^* \cdot \Big(\sum_{i=1}^m \boldsymbol{x}_i^*\Big) = \sum_{i=1}^m \boldsymbol{p}^* \cdot \boldsymbol{x}_i^* \geq \boldsymbol{p}^* \cdot \Big(\sum_{i=1}^m \boldsymbol{x}_i\Big) = \sum_{i=1}^m \boldsymbol{p}^* \cdot \boldsymbol{x}_i. \tag{7.35}$$

Proof. Since $(\langle \boldsymbol{x}_i^* \rangle, \langle \boldsymbol{y}_k^* \rangle, \boldsymbol{p}^*)$ is a quasi-competitive equilibrium for E, $\big(\langle \boldsymbol{x}_i^* \rangle, \langle \boldsymbol{y}_k^* \rangle\big)$ is feasible for E, and thus:

$$0 = \boldsymbol{p}^* \cdot \Big(\sum_{i=1}^m \boldsymbol{r}_i + \sum_{k=1}^\ell \boldsymbol{y}_k^* - \sum_{i=1}^m \boldsymbol{x}_i^*\Big) = \sum_{i=1}^m \boldsymbol{p}^* \cdot \boldsymbol{r}_i + \sum_{k=1}^\ell \boldsymbol{p}^* \cdot \boldsymbol{y}_k^* - \sum_{i=1}^m \boldsymbol{p}^* \cdot \boldsymbol{x}_i^*. \quad (7.36)$$

However, by the definition of a quasi-competitive equilibrium, we have that:

$$w_i^* - \boldsymbol{p}^* \cdot \boldsymbol{x}_i^* \geq 0 \quad \text{for } i = 1, \dots, m, \quad (7.37)$$

$$\sum_{i=1}^m w_i^* = \boldsymbol{p}^* \cdot \boldsymbol{r} + \sum_{k=1}^\ell \pi_k(\boldsymbol{p}^*), \quad (7.38)$$

and:

$$\boldsymbol{p}^* \cdot \boldsymbol{y}_k^* = \pi_k(\boldsymbol{p}^*) \quad \text{for } k = 1, \dots, \ell. \quad (7.39)$$

Adding the terms in (7.37) and making use of (7.38) and (7.39), we have:

$$\begin{aligned}\sum_{i=1}^m (w_i^* - \boldsymbol{p}^* \cdot \boldsymbol{x}_i^*) &= \sum_{i=1}^m w_i^* - \boldsymbol{p}^* \cdot \Big(\sum_{i=1}^m \boldsymbol{x}_i^*\Big) \\ &= \boldsymbol{p}^* \cdot \boldsymbol{r} + \sum_{k=1}^\ell \pi_k(\boldsymbol{p}^*) - \sum_{i=1}^m \boldsymbol{p}^* \cdot \boldsymbol{x}_i^* \\ &= \boldsymbol{p}^* \cdot \boldsymbol{r} + \sum_{k=1}^\ell \boldsymbol{p}^* \cdot \boldsymbol{y}_k^* - \sum_{i=1}^m \boldsymbol{p}^* \cdot \boldsymbol{x}_i^* \\ &= \boldsymbol{p}^* \cdot \Big(\boldsymbol{r} + \sum_{k=1}^\ell \boldsymbol{y}_k^* - \sum_{i=1}^m \boldsymbol{x}_i^*\Big) = 0.\end{aligned}$$

Since the sum of nonnegative terms can only be zero if all of these terms are zero, we can then conclude that $\boldsymbol{p}^* \cdot \boldsymbol{x}_i^* = w_i^*$, for $i = 1, \dots, m$; and furthermore:

$$\sum_{i=1}^m \boldsymbol{p}^* \cdot \boldsymbol{x}_i^* = \boldsymbol{p}^* \cdot \Big(\sum_{i=1}^m \boldsymbol{x}_i^*\Big) = \boldsymbol{p}^* \cdot \boldsymbol{r} + \sum_{k=1}^\ell \boldsymbol{p}^* \boldsymbol{y}_k^*. \quad (7.40)$$

Now suppose that $\big(\langle \boldsymbol{x}_i \rangle, \langle \boldsymbol{y}_k \rangle\big)$ is feasible for E. Then we have:

$$\boldsymbol{p}^* \cdot \sum_{i=1}^m \boldsymbol{x}_i = \boldsymbol{p}^* \cdot \sum_{i=1}^m \boldsymbol{r}_i + \boldsymbol{p}^* \cdot \sum_{k=1}^\ell \boldsymbol{y}_k = \boldsymbol{p}^* \cdot \boldsymbol{r} + \sum_{k=1}^\ell \boldsymbol{p}^* \cdot \boldsymbol{y}_k. \quad (7.41)$$

However, since $(\langle \boldsymbol{x}_i^* \rangle, \langle \boldsymbol{y}_k^* \rangle, \boldsymbol{p}^*)$ is a quasi-competitive equilibrium for E, we have:

$$\sum_{k=1}^\ell \boldsymbol{p}^* \cdot \boldsymbol{y}_k \leq \sum_{k=1}^\ell \boldsymbol{p}^* \cdot \boldsymbol{y}_k^*; \quad (7.42)$$

and combining (7.40)–(7.42), we then obtain:

$$\boldsymbol{p}^* \cdot \sum_{i=1}^m \boldsymbol{x}_i \leq \boldsymbol{p}^* \cdot \sum_{i=1}^m \boldsymbol{x}_i^*. \quad \square$$

The following is our first version of the 'First Fundamental Theorem.'

7.21 Theorem. *If $(\langle \boldsymbol{x}_i^* \rangle, \langle \boldsymbol{y}_k^* \rangle, \boldsymbol{p}^*)$ is a quasi-competitive equilibrium for an economy, E, and:*

$$\boldsymbol{p}^* \cdot \Big(\sum_{i=1}^m \boldsymbol{x}_i^*\Big) > \min \boldsymbol{p}^* \cdot X, \quad (7.43)$$

then $\big(\langle \boldsymbol{x}_i^* \rangle, \langle \boldsymbol{y}_k^* \rangle\big)$ is Pareto efficient for E.

Proof. Suppose $(\langle \boldsymbol{x}_i^* \rangle, \langle \boldsymbol{y}_k^* \rangle, \boldsymbol{p}^*)$ is a quasi-competitive equilibrium for E, given the wealth-assignment $\boldsymbol{w}^*$, and that $(\langle \boldsymbol{x}_i \rangle, \langle \boldsymbol{y}_k \rangle)$ is such that $\langle \boldsymbol{x}_i \rangle \boldsymbol{Q} \langle \boldsymbol{x}_i^* \rangle$. By (7.43), there exists $h \in \{1, \ldots, m\}$ such that:

$$\boldsymbol{p}^* \cdot \boldsymbol{x}_h^* > \min \boldsymbol{p}^* \cdot X_h; \tag{7.44}$$

and, for any $h \in \{1, \ldots, m\}$ satisfying (7.44), we must have, by the definition of a quasi-competitive equilibrium and Proposition 7.20:

$$\boldsymbol{p}^* \cdot x_h > w_h^* = \boldsymbol{p}^* \cdot \boldsymbol{x}_h^*. \tag{7.45}$$

On the other hand, for $i \in \{1, \ldots, m\}$ not satisfying (7.44), we obviously have:

$$\boldsymbol{p}^* \cdot \boldsymbol{x}_i \geq \min \boldsymbol{p}^* \cdot X_i = \boldsymbol{p}^* \cdot \boldsymbol{x}_i^*. \tag{7.46}$$

Combining (7.45) and (7.46), we see that:

$$\boldsymbol{p}^* \cdot \Big(\sum\nolimits_{i=1}^m \boldsymbol{x}_i\Big) > \boldsymbol{p}^* \cdot \Big(\sum\nolimits_{i=1}^m \boldsymbol{x}_i^*\Big);$$

and it follows from Proposition 7.20 that $(\langle \boldsymbol{x}_i \rangle, \langle \boldsymbol{y}_k \rangle)$ is not an attainable allocation for E □

In Chapter 5, we showed that there are undesirable properties of the strict Pareto ordering, $\boldsymbol{P}$ in the case where each P_i is an asymmetric order, but not necessarily negatively transitive (Example 5.16). Correspondingly, it seems that there is a case to be made for concentrating on Pareto efficient allocations, as opposed to strongly Pareto efficient allocations for an economy. It is only fair, therefore, that we present Theorem 7.21 as our first version of the 'First Fundamental Theorem;' for clearly a quasi-competitive equilibrium can involve a very undesirable allocation, in that any consumer minimizing expenditure over X_i at $\boldsymbol{x}_i^*$ can be very badly off indeed! Of course, from our discussion in Chapter 5 we already knew that some Pareto efficient allocations may be quite undesirable.

Since a competitive equilibrium for E is necessarily also a quasi-competitive equilibrium for E, it is immediately apparent that Theorem 7.21 remains correct if we substitute 'competitive equilibrium for E' for 'quasi-competitive equilibrium for E' in its statement. However, we can prove a somewhat stronger result for competitive equilibria, as follows. The proof will be left as an (easy) exercise, since it is an almost immediate corollary of the *proof* of Theorem 7.21.

7.22 Theorem. *If $(\langle \boldsymbol{x}_i^* \rangle, \langle \boldsymbol{y}_k^* \rangle, \boldsymbol{p}^*)$ is a competitive equilibrium for an economy, E, then $(\langle \boldsymbol{x}_i^* \rangle, \langle \boldsymbol{y}_k^* \rangle)$ is Pareto efficient for E.*

Our third (and basic alternative) version of the 'First Fundamental Theorem' is a little more complicated, but still fairly simple to state and prove.

7.23 Theorem. *If $(\langle \boldsymbol{x}_i^* \rangle, \langle \boldsymbol{y}_k^* \rangle, \boldsymbol{p}^*)$ is a competitive equilibrium for an economy, E, and each P_i is asymmetric, negatively transitive, and locally non-saturating, then $(\langle \boldsymbol{x}_i^* \rangle, \langle \boldsymbol{y}_k^* \rangle)$ is strongly Pareto efficient for E.*

Proof. Suppose $(\langle \boldsymbol{x}_i^* \rangle, \langle \boldsymbol{y}_k^* \rangle, \boldsymbol{p}^*)$ is a competitive equilibrium for E, given the assignment of wealth levels $\boldsymbol{w} = (w_1, \ldots, w_m)$, and that $\langle \boldsymbol{x}_i \rangle$ is a consumption allocation such that $\langle \boldsymbol{x}_i \rangle \boldsymbol{P}xalas$. Then, by definition of the strict Pareto dominance relation:

$$\boldsymbol{x}_i G_i \boldsymbol{x}_i^* \quad \text{for } i = 1, \ldots, m, \tag{7.47}$$

and, for some $h \in \{1, \ldots, m\}$:

$$\boldsymbol{x}_h P_h \boldsymbol{x}_h^*. \tag{7.48}$$

However, by (7.47) and Proposition 4.9, we have:

$$\boldsymbol{p}^* \cdot \boldsymbol{x}_i \geq w_i \quad \text{for } i = 1, \ldots, m; \tag{7.49}$$

while by (7.48) and the definition of a competitive equilibrium:

$$\boldsymbol{p}^* \cdot \boldsymbol{x}_h > w_h. \tag{7.50}$$

Adding over equations (7.49) and (7.50), we have:

$$\sum_{i=1}^m \boldsymbol{p}^* \cdot \boldsymbol{x}_i > \sum_{i=1}^m w_i;$$

and it then follows from Proposition 7.20 that $\langle \boldsymbol{x}_i \rangle$ cannot be feasible for E. Therefore, $\big(\langle \boldsymbol{x}_i^* \rangle, \langle \boldsymbol{y}_k^* \rangle\big)$ is strongly Pareto efficient for E. □

As we showed earlier, a quasi-competitive equilibrium is Pareto efficient if the aggregate non-minimum expenditure condition [equation (7.43)] is satisfied. However, a quasi-competitive equilibrium is not necessarily strongly Pareto efficient, even if (7.43) is satisfied and, in addition, the individual preference relations satisfy the assumptions of the above result. This is shown by the following example.

7.24 Example. Consider the two-person, two-commodity exchange economy in which the two consumers' preferences can be represented by the utility functions:

$$u_1(\boldsymbol{x}_1) = x_{11} + x_{12},$$

and:

$$u_2(\boldsymbol{x}_2) = \min\{x_{21}, x_{22}\},$$

respectively; and let:

$$\boldsymbol{r}_1 = (1, 0) \quad \text{and} \quad \boldsymbol{r}_2 = (2, 1).$$

Then, as you can easily show, if we define:

$$\boldsymbol{x}_i^* = \boldsymbol{r}_i \ \text{ for } i = 1, 2, \ \text{ and } \boldsymbol{p}^* = (0, 1),$$

then $\big(\langle \boldsymbol{x}_i^* \rangle, \boldsymbol{p}^*\big)$ is a quasi-competitive equilibrium for E. Moreover, the two consumers' preferences satisfy the assumptions of 7.23, while:

$$\boldsymbol{p}^* \cdot \boldsymbol{x}_1^* + \boldsymbol{p}^* \cdot \boldsymbol{x}_2^* = 1 > \min \boldsymbol{p}^* \cdot X = 0.$$

Nonetheless, $\langle \boldsymbol{x}_i^* \rangle$ is not strongly Pareto efficient for E. □

This last example also illustrates a problem involved in using the word 'equilibrium' as a part of the phrase 'quasi-competitive equilibrium.' The 'equilibrium' defined in the example is not a situation from which there is 'no net tendency to change;' in fact, consumer one will demand an indefinitely large quantity of the first commodity, given its zero price, despite having zero income. Nonetheless, the concept of a quasi-competitive equilibrium will provide a useful 'stepping stone' in developing the results of the next section.

The three versions of the 'First Fundamental Theorem' which were presented here, in particular, the last two, state conditions under which the competitive mechanism is 'non-wasteful.' While the assumptions used in the two results are quite general (particularly in the case of Theorem 7.22), it should be noted that two assumptions which were used implicitly, but not stated as explicit hypotheses are:

1. P_i is individualistic, for $i = 1, \ldots, m$, and:
2. there are no external effects in production, in the sense that if $\boldsymbol{y}_k \in Y_k$, for $k = 1, \ldots, \ell$, then the aggregate production level:

$$\boldsymbol{y} = \sum_{k=1}^{\ell} \boldsymbol{y}_k,$$

can be achieved.

These two assumptions were not stated explicitly for the simple reason that we have been implicitly maintaining them throughout our discussion of competitive equilibrium; and in fact, we would have to re-define what we mean by such an equilibrium if either of these conditions fails. It is nonetheless important to keep in mind that these two assumptions are implicitly used in the result.

It is also worth noting that we can construct an example, based on Example 5.18 of Chapter 5, of a competitive equilibrium which is *not* strongly Pareto efficient, as follows. Let:

$$\boldsymbol{r}_1 = (1, 0) \text{ and } \boldsymbol{r}_2 = (1, 1),$$

and let P_1 and P_2 be as set out in the example, with $f_2(0, 2) = 3$ (that is, we define 2's utility at the bundle $(0, 2)$ to be 3). If we let:

$$\boldsymbol{p}^* = (3/2, 1), \ \boldsymbol{x}_1^* = \boldsymbol{r}_1, \text{ and } \boldsymbol{x}_2^* = \boldsymbol{r}_2,$$

then $(\langle \boldsymbol{x}_i^* \rangle, \boldsymbol{p}^*)$ is a competitive equilibrium for $\mathcal{E}$, but $\langle \boldsymbol{x}_i^* \rangle$ is not strongly Pareto efficient for $\mathcal{E}$.

7.6 'Unbiasedness' of the Competitive Mechanism

While our proof of the 'First Fundamental Theorem' was straightforward indeed, establishing the 'Second Fundamental Theorem' is a bit more complicated. A first difficulty is that a reallocation of initial endowments and shares of ownership in firms may be necessary in order to make a competitive equilibrium of a Pareto efficient allocation. One way of dealing with this difficulty, which is the approach we will follow here, is to seek a wealth-assignment vector for E, which enables equilibrium to be achieved. A second difficulty arises in that standard assumptions do not imply that an arbitrary Pareto efficient allocation can actually be made a competitive equilibrium. The results that people loosely interpret as establishing this implication usually actually establish the existence of a weakened form of competitive equilibrium. We will follow this pattern initially; making use of the concept of a quasi-competitive equilibrium, as defined in the previous section.

In doing our first version of the 'Second Fundamental Theorem,' we will need a supporting result and one further definition, as follows.

7.25 Proposition. *If P_i is a lower semi-continuous binary relation on a convex set, X_i, and $\boldsymbol{x}_i^* \in X_i$ and $\boldsymbol{p}^* \in \mathbb{R}^n$ satisfy:*

$$(\forall \boldsymbol{x}_i \in X_i) \colon \boldsymbol{x}_i P_i \boldsymbol{x}_i^* \Rightarrow \boldsymbol{p}^* \cdot \boldsymbol{x}_i \geq \boldsymbol{p}^* \cdot \boldsymbol{x}_i^*, \tag{7.51}$$

and

$$\boldsymbol{p}^* \cdot \boldsymbol{x}_i^* > \min \boldsymbol{p}^* \cdot X_i, \tag{7.52}$$

then:

$$(\forall \boldsymbol{x}_i \in X_i) \colon \boldsymbol{x}_i P_i \boldsymbol{x}_i^* \Rightarrow \boldsymbol{p}^* \cdot \boldsymbol{x}_i > \boldsymbol{p}^* \cdot \boldsymbol{x}_i^*.$$

Proof. Suppose, by way of obtaining a contradiction, that there exists $\boldsymbol{x}_i' \in X_i$ such that $\boldsymbol{x}_i' P_i \boldsymbol{x}_i^*$, but:

$$\boldsymbol{p}^* \cdot \boldsymbol{x}_i' \leq \boldsymbol{p}^* \cdot \boldsymbol{x}_i^*.$$

Since P_i is lower semi-continuous, there exists a neighborhood, $N(\boldsymbol{x}_i')$ such that:

$$(\forall \boldsymbol{x}_i \in N(\boldsymbol{x}_i') \cap X_i) \colon \boldsymbol{x}_i P_i \boldsymbol{x}_i^*. \tag{7.53}$$

Now, by (7.52) there exists $\overline{\boldsymbol{x}}_i \in X_i$ such that $\boldsymbol{p}^* \cdot \overline{\boldsymbol{x}}_i < \boldsymbol{p}^* \cdot \boldsymbol{x}_i^*$, and we then have:

$$(\forall \theta \in]0,1]) \colon \boldsymbol{p}^* \cdot [\theta \overline{\boldsymbol{x}}_i + (1-\theta)\boldsymbol{x}_i'] < \boldsymbol{p}^* \cdot \boldsymbol{x}_i^*. \tag{7.54}$$

However, it is clear that, since X_i is convex, there exists a value of $\theta > 0$ and small enough so that:

$$\theta \overline{\boldsymbol{x}}_i + (1-\theta)\boldsymbol{x}_i' \in N(\boldsymbol{x}_i') \cap X_i;$$

which, given (7.53) and (7.54), contradicts (7.51). □

7.26 Definition. Let X_i be a convex subset of $\mathbb{R}^n$, and let P_i be an irreflexive binary relation on X_i. We shall say that ***P_i* is weakly convex** iff, for each $\boldsymbol{x}_i^* \in X_i$, the set $P_i \boldsymbol{x}_i^*$ defined by:

$$P_i \boldsymbol{x}_i^* = \{\boldsymbol{x}_i \in X_i \mid \boldsymbol{x}_i P_i \boldsymbol{x}_i^*\},$$

is convex.

The initial version of the 'Second Fundamental Theorem' which we will consider is a generalization of the theorem originally developed by Arrow [1951]. Early variations of Arrow's result were published by Debreu [1954] and Koopmans [1957].

7.27 Theorem. *Let $E = (\langle X_i, P_i\rangle, \langle Y_k\rangle, \boldsymbol{r})$ be an economy such that:*

a. X_i is convex,

b. P_i is weakly convex, locally non-saturating, and lower semi-continuous, for each $i = 1, \ldots, m$; and suppose that:

c. $Y \stackrel{\text{def}}{=} \sum_{k=1}^{\ell} Y_k$ is a convex set.

Then if $(\langle \boldsymbol{x}_i^*\rangle, \langle \boldsymbol{y}_k^*\rangle)$ is Pareto efficient for E, there exists a price vector, $\boldsymbol{p}^* \in \mathbb{R}^n$ such that $(\langle \boldsymbol{x}_i^*\rangle, \langle \boldsymbol{y}_k^*\rangle, \boldsymbol{p}^*)$ is a quasi competitive equilibrium for E given the assignment of wealth $\boldsymbol{w}^*$ defined by:

$$w_i^* = \boldsymbol{p}^* \cdot \boldsymbol{x}_i^* \quad \text{for } i = 1, \ldots, m. \tag{7.55}$$

Proof. We note first that, since $(\langle \boldsymbol{x}_i^*\rangle, \langle \boldsymbol{y}_k^*\rangle)$ is Pareto efficient for E, we must have:

$$\sum_{i=1}^{m} \boldsymbol{x}_i^* = \sum_{i=1}^{m} \boldsymbol{r}_i + \sum_{k=1}^{\ell} \boldsymbol{y}_k^*. \tag{7.56}$$

Define:

$$\boldsymbol{x}^* = \sum_{i=1}^{m} \boldsymbol{x}_i^*, \;\; \boldsymbol{r} = \sum_{i=1}^{m} \boldsymbol{r}_i;$$

and the subset, $\mathbb{P}$, of $\mathbb{R}^n$, by:

$$\mathbb{P} = \sum_{i=1}^{m} P_i \boldsymbol{x}_i^*;$$

where we recall the notation:

$$P_i \boldsymbol{x}_i^* = \{\boldsymbol{x}_i \in X_i \mid \boldsymbol{x}_i P_i \boldsymbol{x}_i^*\} \quad \text{for } i = 1, \ldots, m.$$

By way of completing our preliminaries, we note also that it follows immediately from the assumption that each P_i is weakly convex and the fact that the sum of convex sets is convex, that $\mathbb{P}$ is a convex set.

Next, we note that, by assumption (c) and using Proposition 6.37 once again, the set $\boldsymbol{r} + Y$ is convex; and it is easy to see that, since $(\langle \boldsymbol{x}_i^*\rangle, \langle \boldsymbol{y}_k^*\rangle)$ is Pareto efficient for E, we must have:

$$\mathbb{P} \cap [\boldsymbol{r} + Y] = \emptyset.$$

Thus, by Theorem 6.21 (the Separating Hyperplane Theorem), there exists a non-zero $\boldsymbol{p}^* \in \mathbb{R}^n$ such that:

$$\alpha \stackrel{\text{def}}{=} \sup\{\boldsymbol{p}^* \cdot \boldsymbol{z} \mid \boldsymbol{z} \in \boldsymbol{r} + Y\} \leq \beta \stackrel{\text{def}}{=} \inf\{\boldsymbol{p}^* \cdot \boldsymbol{x} \mid \boldsymbol{x} \in \mathbb{P}\}. \tag{7.57}$$

Now, it follows at once from (7.56), (7.57), and our definition of $\boldsymbol{x}^*$, that:

$$\boldsymbol{p}^* \cdot \boldsymbol{x}^* \leq \alpha. \tag{7.58}$$

We are going to prove that we also must have $\boldsymbol{p}^* \cdot \boldsymbol{x}^* \geq \beta$. To do this, let $\epsilon > 0$ be given. Then, using the continuity of the inner product function and the fact that each P_i is locally non-saturating, we see that, for each i, there exists $\boldsymbol{x}_i^\dagger$ satisfying:

$$\boldsymbol{x}_i^\dagger P_i \boldsymbol{x}_i^* \;\;\&\;\; \boldsymbol{p}^* \cdot \boldsymbol{x}_i^\dagger < \boldsymbol{p}^* \cdot \boldsymbol{x}_i^* + \epsilon/m \quad \text{for } i = 1, \ldots, m. \tag{7.59}$$

Adding the inequalities on the right in (7.59), we then obtain:

$$\beta \leq \boldsymbol{p}^* \cdot \left(\sum_{i=1}^{m} \boldsymbol{x}_i^\dagger\right) = \sum_{i=1}^{m} \boldsymbol{p}^* \cdot \boldsymbol{x}_i^\dagger < \sum_{i=1}^{m} (\boldsymbol{p}^* \cdot \boldsymbol{x}_i^* + \epsilon/m) = \boldsymbol{p}^* \cdot \boldsymbol{x}^* + \epsilon, \tag{7.60}$$

where the first inequality in (7.60) is by the definitions of β and $\mathbb{P}$ and the left-hand part of (7.59). However, since (7.60) has been shown to hold for any positive real number, ϵ, it follows that $\beta \leq \boldsymbol{p}^* \cdot \boldsymbol{x}^*$; and, combining this with (7.57) and (7.58), we see that:

$$\alpha = \beta = \boldsymbol{p}^* \cdot \boldsymbol{x}^*. \tag{7.61}$$

Now let $j \in K$, let $\boldsymbol{y}_j$ be an arbitrary element of Y_j, and consider the production allocation $\langle \boldsymbol{y}_k^\dagger \rangle \in \mathcal{Y}$ defined by:

$$\langle \boldsymbol{y}_k^\dagger \rangle = \begin{cases} \boldsymbol{y}_j & \text{for } k = j, \\ \boldsymbol{y}_k^* & \text{for } k \neq j. \end{cases}$$

From (7.56), (7.57), and (7.61), we have:

$$\boldsymbol{p}^* \cdot \boldsymbol{r} + \sum_{k \neq j} \boldsymbol{p}^* \cdot \boldsymbol{y}_k^* + \boldsymbol{p}^* \cdot \boldsymbol{y}_j \leq \boldsymbol{p}^* \cdot \boldsymbol{r} + \sum\nolimits_{k=1}^{\ell} \boldsymbol{p}^* \cdot \boldsymbol{y}_k^*;$$

from which we obtain:

$$\boldsymbol{p}^* \cdot \boldsymbol{y}_j \leq \boldsymbol{p}^* \cdot \boldsymbol{y}_j^*;$$

and we conclude that:

$$\boldsymbol{p}^* \cdot \boldsymbol{y}_k^* = \pi_k(\boldsymbol{p}^*) \quad \text{for } k = 1, \ldots, \ell. \tag{7.62}$$

Next, defining $\boldsymbol{w}^* = (w_1^*, \ldots, w_m^*)$ by:

$$w_i^* = \boldsymbol{p}^* \cdot \boldsymbol{x}_i^* \quad \text{for } i = 1, \ldots, m,$$

we note that it follows from (7.55) and (7.56) that $\boldsymbol{w}^*$ is a wealth-assignment for E, given $\boldsymbol{p}^*$; and thus we have shown that $(\langle \boldsymbol{x}_i^* \rangle, \langle \boldsymbol{y}_k^* \rangle, \boldsymbol{p}^*)$ satisfies the first three of the conditions defining a quasi-competitive equilibrium for E, given the wealth assignment $\boldsymbol{w}^*$; and the first part of Condition 4 as well. Therefore, to complete our proof, we need only establish that, for each i, either (7.32) or (7.33) of Definition 7.19 must hold.

Accordingly, let $i \in M$ be arbitrary, suppose $\boldsymbol{x}_i^\dagger \in X_i$ is such that $\boldsymbol{x}_i^\dagger P_i \boldsymbol{x}_i^*$, and let $\epsilon > 0$ be given. Since each P_h is locally non-saturating, and making use of the continuity of the inner product, we see that for each $h \neq i$, there exists $\bar{\boldsymbol{x}}_h \in X_h$ such that:

$$\bar{\boldsymbol{x}}_h P_h \boldsymbol{x}_h^* \text{ and } \boldsymbol{p}^* \cdot \bar{\boldsymbol{x}}_h < \boldsymbol{p}^* \cdot \boldsymbol{x}_h^* + \epsilon/(m-1). \tag{7.63}$$

If we then define $(\widehat{\boldsymbol{x}}_h)$ by:

$$\widehat{\boldsymbol{x}}_h = \begin{cases} \bar{\boldsymbol{x}}_h & \text{for } h \neq i, \\ \boldsymbol{x}^\dagger & \text{for } h = i, \end{cases}$$

we see that $\sum_{h=1}^m \widehat{\boldsymbol{x}}_h \in \boldsymbol{P}$, and thus by (7.57) and (7.61):

$$\boldsymbol{p}^* \cdot \Big(\sum\nolimits_{h=1}^m \widehat{\boldsymbol{x}}_h \Big) = \sum_{h \neq i} \boldsymbol{p}^* \cdot \bar{\boldsymbol{x}}_h + \boldsymbol{p}^* \cdot \boldsymbol{x}_i^\dagger \geq \beta = \sum\nolimits_{h=1}^m \boldsymbol{p}^* \cdot \boldsymbol{x}_h^*$$

However, by (7.63) we have:

$$\sum_{h \neq i} \boldsymbol{p}^* \cdot \bar{\boldsymbol{x}}_h < \sum_{h \neq i} \left[\boldsymbol{p}^* \cdot \boldsymbol{x}_h^* + \epsilon/(m-1)\right] = \sum_{h \neq i} \boldsymbol{p}^* \cdot \boldsymbol{x}_h^* + \epsilon; \tag{7.64}$$

and from (7.63) and (7.64), we then obtain:

$$\sum_{h \neq i} \boldsymbol{p}^* \cdot \boldsymbol{x}_h^* + \epsilon + \boldsymbol{p}^* \cdot \boldsymbol{x}_i^{\dagger} \geq \sum\nolimits_{h=1}^{m} \boldsymbol{p}^* \cdot \boldsymbol{x}_h^*;$$

so that:

$$\boldsymbol{p}^* \cdot \boldsymbol{x}_i^{\dagger} \geq \boldsymbol{p}^* \cdot \boldsymbol{x}_i^* - \epsilon.$$

Since $\epsilon > 0$ was arbitrary, we can now conclude that:

$$\boldsymbol{p}^* \cdot \boldsymbol{x}_i^{\dagger} \geq \boldsymbol{p}^* \cdot \boldsymbol{x}_i^*.$$

From the argument of the above paragraph, we conclude that for all $\boldsymbol{x}_i \in X_i$:

$$\boldsymbol{x}_i P_i \boldsymbol{x}_i^* \Rightarrow \boldsymbol{p}^* \cdot \boldsymbol{x}_i \geq \boldsymbol{p}^* \cdot \boldsymbol{x}_i^* \quad \text{for } i = 1, \ldots, m. \tag{7.65}$$

Condition 4 of Definition 7.19 now follows from Proposition 7.25. □

This last result can be generalized to the extent of allowing for the commodity space to be infinite-dimensional.[8] It is also possible that it could be generalized within the context of $\mathbb{R}^n$. However, the conclusion of the result does not hold if any one of the conditions of Theorem 7.27 is simply dropped. In fact, a series of examples presented in the next chapter show that none of the assumptions of Theorem 7.27 can be dispensed with:

1. Example 8.3 shows that we cannot drop the assumption that each of the consumption sets, X_i, is convex.
2. Example 8.7 shows that we cannot drop the assumption that each P_i is locally non-saturating.
3. Example 8.8 shows that we cannot drop the assumption that each P_i is weakly convex.
4. Example 8.9 shows that we cannot drop the assumption that each P_i is lower semi-continuous.
5. Example 8.14 shows that we cannot drop the assumption that Y is convex.

Under the hypotheses of Theorem 7.27, we have shown that given any Pareto efficient allocation, $\big(\langle \boldsymbol{x}_i^* \rangle, \langle \boldsymbol{y}_k^* \rangle\big)$, there exists a price vector, $\boldsymbol{p}^*$, such that $(\langle \boldsymbol{x}_i^* \rangle, \langle \boldsymbol{y}_k^* \rangle, \boldsymbol{p}^*)$ is a quasi-competitive equilibrium. Unfortunately, the hypotheses of that result are not sufficiently strong to imply that we can obtain a competitive equilibrium. In fact, under the hypotheses of Theorem 7.27 an allocation may be strongly Pareto efficient, yet there may nonetheless be no price vector, $\boldsymbol{p}^*$, such that the allocation becomes a competitive equilibrium with this price vector. That this is so is demonstrated by the following, which is based upon a famous example by Arrow [1951].

[8] Although some of the other assumptions then need to be strengthened.

7.28 Example. Let $\mathcal{E}$ be a private ownership economy in which $m = n = 2$, $\ell = 1$, and $Y = -\mathbb{R}^2_+$. If the preference relations P_i are such that either P_1 or P_2 is strictly increasing, then it is clear that any competitive equilibrium, $(\langle \boldsymbol{x}_i^* \rangle, \boldsymbol{y}^*, \boldsymbol{p}^*)$, if one exists, must satisfy $\boldsymbol{p}^* \gg \boldsymbol{0}$. However, this in turn implies that in any such competitive equilibrium, we must have $\boldsymbol{y}^* = \boldsymbol{0}$. Consequently, in this case we can analyze the possible existence of a competitive equilibrium within the context of a traditional Edgeworth Box diagram. Suppose, then, that P_1, P_2, $\boldsymbol{r}_1$, and $\boldsymbol{r}_2$ are as in Figure 7.4, on the next page, and that $X_1 = X_2 = \mathbb{R}^2_+$, and consider the price vector $\boldsymbol{p}^* = (0, 1)$. It is easy to see that $(\langle \boldsymbol{x}_i^* \rangle, \boldsymbol{y}^*, \boldsymbol{p}^*)$ is a quasi-competitive equilibrium. However, consumer 1's demand for the first commodity is unbounded, given a zero price for that commodity. Therefore, $(\langle \boldsymbol{x}_i^* \rangle, \boldsymbol{y}^*, \boldsymbol{p}^*)$ is not a competitive equilibrium. Furthermore, it is easily seen in this case that if a nonnegative price vector defines a line separating $P_1\boldsymbol{x}_1^*$ and $P_2\boldsymbol{x}_2^*$, it must be a scalar multiple of $\boldsymbol{p}^* = (0, 1)$; and thus it follows that no price vector, $\boldsymbol{p}^*$, exists which is such that $(\langle \boldsymbol{x}_i^* \rangle, \boldsymbol{y}^*, \boldsymbol{p}^*)$ is a competitive equilibrium. □

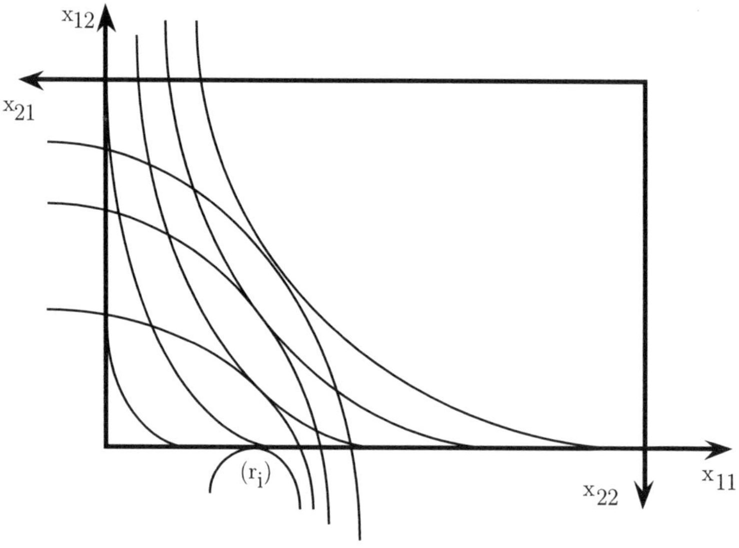

Figure 7.4: Arrow's 'Exceptional Case.'

Now let's turn our attention to finding conditions ensuring that $(\langle \boldsymbol{x}_i^* \rangle, \langle \boldsymbol{y}_k^* \rangle, \boldsymbol{p}^*)$, as obtained in the conclusion of Theorem 7.27, is a competitive equilibrium with the given wealth assignment; as opposed to the weaker conclusion established in the theorem. Actually, one can easily obtain the stronger conclusion by making some rather stringent additional assumptions; the simplest of which makes use of the following mathematical condition.

7.29 Definition. The **interior of a set**, $A \subseteq \mathbb{R}^n$, denoted $int(A)$ is defined as the set of all $\boldsymbol{x} \in A$ such that there exists $\epsilon > 0$ such that $N(\boldsymbol{x}, \epsilon) \subseteq A$.

While the following corollary is very easy to prove, it is not a very satisfactory result, for reasons we will discuss shortly.

7.30 Corollary. *Suppose, in addition to the assumptions and conditions of Theorem 7.27, that the allocation* $(\langle \boldsymbol{x}_i^* \rangle, \langle \boldsymbol{y}_k^* \rangle)$ *satisfies:*

$$\boldsymbol{x}_i^* \in int X_i \quad \text{for } i = 1, \ldots, m. \tag{7.66}$$

Then there exists $\boldsymbol{p}^* \in \mathbb{R}^n$ *such that* $(\langle \boldsymbol{x}_i^* \rangle, \langle \boldsymbol{y}_k^* \rangle, \boldsymbol{p}^*)$ *is a competitive equilibrium for* E *given the assignment of wealth* $\boldsymbol{w}^*$ *defined by:*

$$w_i^* = \boldsymbol{p}^* \cdot \boldsymbol{x}_i^* \quad \text{for } i = 1, \ldots, m.$$

In order to prove the corollary, we need only note that a linear function, in this case $f_i(\boldsymbol{x}_i) \equiv \boldsymbol{p}^* \cdot \boldsymbol{x}_i$, can only be minimized at an interior point of a set if it is identically zero on the space. However, since we know that $\boldsymbol{p}^* \neq \boldsymbol{0}$, we see that this cannot be the case; that is, we must have:

$$\boldsymbol{p}^* \cdot \boldsymbol{x}_i^* > \min \boldsymbol{p}^* \cdot X_i \quad \text{for } i = 1, \ldots, m.$$

It then follows from the definitions that the quasi-competitive equilibrium obtained in the theorem must, with this additional assumption, actually be a competitive equilibrium (with the assignment of wealth, $\boldsymbol{w}^*$).

The problem with this result is, of course, that the sort of Pareto efficient allocation which would satisfy this assumption is strange and rare indeed! Notice, in particular, equation (7.66) implies that, for each consumer, i, there exists a point, $\boldsymbol{x}_i' \in X_i$ satisfying:

$$\boldsymbol{x}_i' \ll \boldsymbol{x}_i^*.$$

Consequently, it follows that at the allocation $(\boldsymbol{x}_i^*)$, and given any commodity, j, each consumer must either possess a strictly positive quantity of the jth commodity, or be supplying less of it than he or she is capable of supplying. It is not at all clear that any such point could ever be a Pareto efficient allocation;[9] and in any case, it is clear that most of the efficient allocations of interest will not satisfy this property.

In fact, to carry this argument one step further, remember that we would like to be able to claim that the model allows for a finite number of time periods; with commodities being differentiated by time of availability as well as physical characteristics. In this context, it is worth noting that in most of our work in this and the next chapter the assumptions of the model incorporate as a special case the situation in which there are T time periods, and G physically distinguishable commodites, so that $n = G \times T$; and where, for each i there exist positive integers t_i and t_i' such that:

$$1 \leq t_i < t_i' \leq T$$

and such that X_i takes the form:

$$X_i = \boldsymbol{0}_{n_i} \times C_i \times \boldsymbol{0}_{n_i'},$$

[9]Unless consumer preferences are identical, the satisfaction of equation (7.66) will almost certainly indicate that Pareto-improving trades among consumers are possible; and thus that $((\boldsymbol{x}_i^*), (\boldsymbol{y}_k^*))$ cannot be Pareto efficient.

where:

$$n_i = t_i \cdot G,\ n'_i = (T - t'_i) \cdot G,$$

and, defining $n''_i = t'_i - t_i$:

$$C_i \subseteq \mathbb{R}^{n''_i \cdot G}.$$

The idea here is that, for a consumer who is born in the t_i^{th} period and dies in period t'_i, only its consumption in periods $t_i + 1, \ldots, t'_i$ affects its survival or preferences. Under these conditions, the sets X_i do not even possess interiors!

In the next section we will consider a different sort of strengthening of Theorem 7.27, but before ending this discussion it may be of some interest to show how easily we can now obtain a version of the 'Second Fundamental Theorem' for private ownership economies. In doing this we will define a quasi-competitive equilibrium for $\mathcal{E}$, given a system of lump-sum transfers; where we shall say that $\boldsymbol{t} \in \mathbb{R}^m$ is a **system of lump-sum transfers** for $\mathcal{E}$ iff:

$$\sum_{i=1}^{m} t_i = 0. \tag{7.67}$$

7.31 Definition. Let $\mathcal{E} = (\langle X_i, P_i \rangle, \langle Y_k \rangle, \langle \boldsymbol{r}_i \rangle, [s_{ik}])$ be a private ownership economy. We shall say that $(\langle \boldsymbol{x}_i^* \rangle, \langle \boldsymbol{y}_k^* \rangle, \boldsymbol{p}^*)$ is a **quasi-competitive equilibrium for $\mathcal{E}$ with** lump-sum transfers $\boldsymbol{t}$ iff:

1. $\boldsymbol{p}^* \neq \boldsymbol{0}$,
2. $((\boldsymbol{x}_i^*), (\boldsymbol{y}_k^*)) \in A(\mathcal{E})$,
3. $\boldsymbol{p}^* \cdot \boldsymbol{y}_k^* = \pi_k(\boldsymbol{p}^*)$, for $k = 1, \ldots, \ell$,
4. for each i ($i = 1, \ldots, m$), we have $\boldsymbol{p}^* \cdot \boldsymbol{x}_i^* \leq w_i(\boldsymbol{p}^*)$, and either:

$$w_i(\boldsymbol{p}^*) = \min\{\boldsymbol{p}^* \cdot \boldsymbol{x}_i \mid \boldsymbol{x}_i \in X_i\} \stackrel{\text{def}}{=} \min \boldsymbol{p}^* \cdot X_i, \tag{7.68}$$

or:

$$(\forall \boldsymbol{x}_i \in X_i)\colon \boldsymbol{x}_i P_i \boldsymbol{x}_i^* \Rightarrow \boldsymbol{p}^* \cdot \boldsymbol{x}_i > w_i(\boldsymbol{p}^*) \tag{7.69}$$

(or both), where:

$$w_i(\boldsymbol{p}^*) = \boldsymbol{p}^* \cdot \boldsymbol{r}_i + \sum_{k=1}^{\ell} s_{ik} \pi_k(\boldsymbol{p}^*) + t_i \quad \text{for } i = 1, \ldots, m;$$

I will leave as an exercise the following corollary of Theorem 7.27.

7.32 Corollary. *Let $\mathcal{E} = (\langle X_i, P_i \rangle, \langle Y_k \rangle, \langle \boldsymbol{r}_i \rangle, [s_{ik}])$ be a private ownership economy such that:*

a. X_i is convex,

b. P_i is weakly convex, locally non-saturating, and lower semi-continuous, for each $i = 1, \ldots, m$; and suppose that:

c. $Y \stackrel{def}{=} \sum_{k=1}^{\ell} Y_k$ is a convex set.

Then if $\big(\langle \boldsymbol{x}_i^* \rangle, \langle \boldsymbol{y}_k^* \rangle\big)$ is Pareto efficient for $\mathcal{E}$, there exists a price vector, $\boldsymbol{p}^* \in \mathbb{R}^n$ and a vector $\boldsymbol{t}^* \in \mathbb{R}^m$ such that $(\langle \boldsymbol{x}_i^* \rangle, \langle \boldsymbol{y}_k^* \rangle, \boldsymbol{p}^*)$ is a quasi-competitive equilibrium for $\mathcal{E}$ with the lump-sum transfers $\boldsymbol{t}^*$.

7.7 A Stronger Version of 'The Second Theorem'

In order to develop a better version of the 'Second Fundamental Theorem,' we will make use of two principal conditions, the first of which is defined as follows.

7.33 Definition. We shall say that the economy, $E = (\langle X_i, P_i \rangle, \langle Y_k \rangle, \boldsymbol{r})$ is **irreducible at the allocation $(\langle \boldsymbol{x}_i^* \rangle, \langle \boldsymbol{y}_k^* \rangle)$** iff, given any partition of the consumers, $\{I_1, I_2\}$,[10] there exists $\langle \boldsymbol{x}_i \rangle \in \mathfrak{X}$ and $\boldsymbol{z} \in \mathbb{R}^n$ such that:

$$\boldsymbol{z} \in \boldsymbol{r} + Y, \tag{7.70}$$

$$\sum_{i=1}^{m} \boldsymbol{x}_i = \boldsymbol{z}, \tag{7.71}$$

and:

$$(\forall i \in I_1) \colon \boldsymbol{x}_i P_i \boldsymbol{x}_i^*. \tag{7.72}$$

The condition just defined is developed from the 'irreducibility condition' introduced by L. McKenzie [1959, 1961], and was generalized somewhat in Moore [1970, 1975]. Effectively, it implies that at the allocation $((\boldsymbol{x}_i^*), (\boldsymbol{y}_k^*))$, there is no definable subgroup of consumers, I_1, who could not make themselves better off, collectively, if they were simply allowed to exploit the remaining consumers, and, possibly, to re-organize the means of production. In the simplest situation, a two-person, two-commodity exchange economy, E is irreducible at an allocation in the Edgeworth box if, and only if, each consumer prefers some other attainable allocation (not, of course, necessarily the same one).[11] A similar interpretation applies in the case of a pure exchange economy with an arbitrary finite number of consumers; and thus, in this case the satisfaction of the irreducibility condition at $\langle \boldsymbol{x}_i^* \rangle$ implies that there is no subgroup of consumers (I_2) so poor that the remaining consumers could not (if allowed to do so) exploit them in such a way as to make themselves better off. In other words, and somewhat loosely interpreting, the condition guarantees that no subgroup of consumers is so poor as to have nothing which is valued by the remaining consumers. We will consider the meaning of the condition in more detail shortly, but first let me introduce the second of the key conditions mentioned earlier.

7.34 Definition. Let $E = (\langle X_i, P_i \rangle, \langle Y_k \rangle, \boldsymbol{r})$ and $\mathbb{E} = (\langle X_i', P_i' \rangle, T)$ be economies. We will say that $\mathbb{E}$ **is aggregatively similar to $\boldsymbol{E}$ at an allocation** $(\langle \boldsymbol{x}_i \rangle, \langle \boldsymbol{y}_k \rangle) \in A(E)$ iff:

$$\langle X_i, P_i \rangle = \langle X_i', P_i' \rangle \text{ for } i = 1, \ldots, m, \tag{7.73}$$

$$\boldsymbol{r} + \sum_{k=1}^{\ell} Y_k \subseteq T, \tag{7.74}$$

and the allocation $(\langle \boldsymbol{x}_i^* \rangle, \boldsymbol{z}^*)$ is Pareto efficient for $\mathbb{E}$.

Clearly if $\mathbb{E} = (\langle X_i', P_i' \rangle, T)$ is aggregatively similar to $\boldsymbol{E}$ at $(\langle \boldsymbol{x}_i^* \rangle, \langle \boldsymbol{y}_k^* \rangle) \in A(E)$, then $(\langle \boldsymbol{x}_i^* \rangle, \langle \boldsymbol{y}_k^* \rangle)$ is Pareto efficient for E. The following presents what are probably the simplest examples of this aggregatively similar relationship.

[10] By a partition of the consumers, $\{I_1, I_2\}$, we mean $I_j \subseteq I$ & $I_j \neq \emptyset$, for $i = 1, 2$, $I_1 \cap I_2 = \emptyset$, and $I_1 \cup I_2 = I$.

[11] Notice that this condition fails at the allocation $\langle \boldsymbol{x}_i^* \rangle$ in Example 7.28.

7.35 Examples.

1. Let $E = (\langle X_i, P_i\rangle, \langle Y_k\rangle, \boldsymbol{r})$ be an economy, $\big(\langle \boldsymbol{x}_i^*\rangle, \langle \boldsymbol{y}_k^*\rangle\big)$ be a Pareto efficient allocation for E, and define the set T by:

$$T = \boldsymbol{r} + \sum\nolimits_{k=1}^{\ell} Y_k.$$

Then $\mathbb{E} = \big(\langle X_i, P_i\rangle, T\big)$ is aggregatively similar to E at $\big(\langle \boldsymbol{x}_i^*\rangle, \langle \boldsymbol{y}_k^*\rangle\big)$.

2. Let $E = (\langle X_i, P_i\rangle, \langle Y_k\rangle, \boldsymbol{r})$ be an economy, suppose $(\langle \boldsymbol{x}_i^*\rangle, \langle \boldsymbol{y}_k^*\rangle, \boldsymbol{p}^*)$ is a competitive equilibrium for E, and define:

$$T = \big\{\boldsymbol{z} \in \mathbb{R}^n \mid \boldsymbol{p}^* \cdot \boldsymbol{z} \leq \boldsymbol{p}^* \cdot \boldsymbol{r} + \sum_{k=1}^{\ell} \pi_k(\boldsymbol{p}^*)\big\}.$$

Then, as you can easily prove, $\mathbb{E} = \big(\langle X_i, P_i\rangle, T\big)$ is aggregatively similar to E at $\big(\langle \boldsymbol{x}_i^*\rangle, \langle \boldsymbol{y}_k^*\rangle\big)$. □

7.36 Theorem. *Let $E = (\langle X_i, P_i\rangle, \langle Y_k\rangle, \boldsymbol{r})$ be an economy such that:*

a. X_i is convex,

b. P_i is weakly convex, locally non-saturating, and lower semi-continuous, for each $i = 1, \ldots, m$; suppose $\big(\langle \boldsymbol{x}_i^*\rangle, \langle \boldsymbol{y}_k^*\rangle\big)$ is Pareto efficient for E, and suppose there exists a convex set $T \subseteq \mathbb{R}^n$ such that $\mathbb{E} = \big(\langle X_i, P_i\rangle, T\big)$ is aggregatively similar to E at $\big(\langle \boldsymbol{x}_i^*\rangle, \langle \boldsymbol{y}_k^*\rangle\big)$,

c. $int(X) \cap T \neq \emptyset$, and

d. $\mathbb{E}$ is irreducible at $(\langle \boldsymbol{x}_i^*\rangle, \boldsymbol{z}^*)$, where:

$$\boldsymbol{z}^* = \boldsymbol{r} + \sum\nolimits_{k=1}^{\ell} \boldsymbol{y}_k^*.$$

Then there exists a price vector, $\boldsymbol{p}^* \in \mathbb{R}^n$ such that $(\langle \boldsymbol{x}_i^*\rangle, \langle \boldsymbol{y}_k^*\rangle, \boldsymbol{p}^*)$ is a competitive equilibrium for E given the assignment of wealth $\boldsymbol{w}^*$, where:

$$w_i^* = \boldsymbol{p}^* \cdot \boldsymbol{x}_i^* \quad \mathit{for}\, i = 1, \ldots, m; \tag{7.75}$$

and we have:

$$w_i^* > \min \boldsymbol{p}^* \cdot X_i \quad \mathit{for}\, i = 1, \ldots, m. \tag{7.76}$$

Proof. It follows from Theorem 7.27 that there exists a price vector, $\boldsymbol{p}^* \neq \boldsymbol{0}$ such that $(\langle \boldsymbol{x}_i^*\rangle, \boldsymbol{z}^*, \boldsymbol{p}^*)$ is a quasi-competitive equilibrium for $\mathbb{E}$, given the wealth assignment:

$$w_i^* = \boldsymbol{p}^* \cdot \boldsymbol{x}_i^* \quad \text{for } i = 1, \ldots, m. \tag{7.77}$$

Moreover, from Assumption c, we see that there exists $\widehat{\boldsymbol{x}} \in X \cap T$ and $\theta \in \mathbb{R}_{++}$ such that:

$$\boldsymbol{x}^\dagger \stackrel{\text{def}}{=} \widehat{\boldsymbol{x}} - \theta \boldsymbol{p}^* \in X, \tag{7.78}$$

and thus:

$$\boldsymbol{p}^* \cdot \boldsymbol{x}^\dagger = \boldsymbol{p}^* \cdot [\widehat{\boldsymbol{x}} - \theta \boldsymbol{p}^*] = \boldsymbol{p}^* \cdot \widehat{\boldsymbol{x}} - \theta \boldsymbol{p}^* \cdot \boldsymbol{p}^* < \boldsymbol{p}^* \cdot \widehat{\boldsymbol{x}} \leq \boldsymbol{p}^* \cdot \boldsymbol{x}^*,$$

where the last inequality is from Proposition 7.20. Therefore since (again by 7.20) at a quasi-competitive equilibrium, each consumer's consumption expenditure must be equal to wealth, it must be the case that for some $i \in M$:

$$w_i^* = \boldsymbol{p}^* \cdot \boldsymbol{x}_i^* > \min \boldsymbol{p}^* \cdot X_i. \tag{7.79}$$

Now defining the sets of consumers $I_h \subseteq M$ $(h = 1, 2)$ by:

$$I_1 = \{i \in I \mid w_i^* > \min \boldsymbol{p}^* \cdot X_i\},$$

and:

$$I_2 = \{i \in I \mid w_i^* = \min \boldsymbol{p}^* \cdot X_i\},$$

respectively, it follows from (7.79) that $I_1 \neq \emptyset$. Suppose by way of obtaining a contradiction, that $I_2 \neq \emptyset$ as well. Then, since $\mathbb{E}$ is irreducible at $(\langle \boldsymbol{x}_i^* \rangle, \boldsymbol{z}^*)$, there exists $(\langle \boldsymbol{x}_i \rangle, \boldsymbol{z}) \in A(\mathbb{E})$ satisfying:

$$\boldsymbol{z} \in T, \tag{7.80}$$

$$\sum_{i \in I_2} x_i = \boldsymbol{z} - \sum_{i \in I_1} \boldsymbol{x}_i, \tag{7.81}$$

and:

$$(\forall i \in I_1) \colon \boldsymbol{x}_i P_i \boldsymbol{x}_i^*. \tag{7.82}$$

For future reference, we note that it follows from the definition of I_1 that for each $i \in I_1$:

$$\boldsymbol{p}^* \cdot \boldsymbol{x}_i > w_i^* = \boldsymbol{p}^* \cdot \boldsymbol{x}_i^*. \tag{7.83}$$

Now, from (7.81) and (7.83), we have

$$\sum_{i \in I_2} \boldsymbol{p}^* \cdot \boldsymbol{x}_i = \boldsymbol{p}^* \cdot \boldsymbol{z} - \sum_{i \in I_1} \boldsymbol{p}^* \cdot \boldsymbol{x}_i < \boldsymbol{p}^* \cdot \boldsymbol{z} - \sum_{i \in I_1} w_i^*. \tag{7.84}$$

Moreover, since $\boldsymbol{z} \in T$, we see that we must have $\boldsymbol{p}^* \cdot \boldsymbol{z} \leq \boldsymbol{p}^* \cdot \boldsymbol{z}^*$; and, since $\boldsymbol{w}^*$ is a feasible wealth assignment for $\mathbb{E}$, we also have:

$$\sum_{i \in M} w_i^* = \boldsymbol{p}^* \cdot \boldsymbol{z}^*. \tag{7.85}$$

Therefore, it now follows from (7.84) that:

$$\sum_{i \in I_2} \boldsymbol{p}^* \cdot \boldsymbol{x}_i < \sum_{i \in M} w_i^* - \sum_{i \in I_1} w_i^* = \sum_{i \in I_2} w_i^*.$$

But this contradicts the definition of I_2. It follows, therefore, that I_2 is empty, and, consequently, that (7.76) holds and that $(\langle \boldsymbol{x}_i^* \rangle, \boldsymbol{z}^*, \boldsymbol{p}^*)$ is a competitive equilibrium for $\mathbb{E}$. Since:

$$\boldsymbol{r} + \sum\nolimits_{k=1}^{\ell} Y_k \subseteq T,$$

you can now (making use of Theorem 6.38) easily prove that $(\langle \boldsymbol{x}_i^* \rangle, \langle \boldsymbol{y}_k^* \rangle, \boldsymbol{p}^*)$ is a competitive equilibrium for E as well, given the wealth distribution, $\boldsymbol{w}^*$. □

One might question whether Theorem 7.36 is of much significance, even from a purely theoretical point of view. After all, under the assumptions of Theorem 7.27, we have shown that if $\langle \boldsymbol{x}_i^* \rangle$ is Pareto efficient, then there exists $\boldsymbol{p}^*$ such that $(\langle \boldsymbol{x}_i^* \rangle, \langle \boldsymbol{y}_k^* \rangle, \boldsymbol{p}^*)$ is a quasi-competitive equilibrium. Moreover, $(\langle \boldsymbol{x}_i^* \rangle, \langle \boldsymbol{y}_k^* \rangle, \boldsymbol{p}^*)$ will be a competitive equilibrium unless for one or more consumeres, we have $\boldsymbol{p}^* \cdot \boldsymbol{x}_i^* = \min \boldsymbol{p}^* \cdot X_i$; and, since $\boldsymbol{p}^* \neq \boldsymbol{0}$, the only way this can happen is if $\boldsymbol{x}_i^*$ is on the boundary of X_i. However, as I suggested in the discussion at the end of the previous section, if we define commodities finely (as we usually specify that we are when dealing with competitive behavior), all reasonable allocations would result in each consumer's commodity bundle being on the boundary of its consumption set. After all, does anyone consume a positive quantity of each commodity available in the U. S. each month? I think not! Consequently, even though we tend to think of boundary values as being a very special case, they are the norm in reality. Perhaps it is because we use Edgeworth Box diagrams so frequently in our analysis that we think of consumption values on the boundary of X_i as being an 'exceptional case,' and if it were reasonable to assume that in reality there are only two or three commodities available in an economy, this attitude would probably be correct. However, one needs to allow for a large number of commodities in order to justify our competitive assumptions; and, correspondingly, we need to treat boundary values as the norm.

The assumption that there exists a *convex* set, T, such that $\mathbb{E} = (\langle X_i, P_i \rangle, T)$ is aggregatively similar to $E = (\langle X_i, P_i \rangle, \langle Y_k \rangle, \boldsymbol{r})$ at $\big(\langle \boldsymbol{x}_i^* \rangle, \langle \boldsymbol{y}_k^* \rangle\big)$ is obviously critical in the proof of Theorem 7.36. Figure 7.5, on the next page, in which the set $\boldsymbol{P}$ is intended to represent the set:

$$\boldsymbol{P} = \sum\nolimits_{i=1}^{m} P_i \boldsymbol{x}_i^*,$$

and Y represents the aggregate production set, conveys some idea of the generality of the assumption (in the diagram, we are supposing that $\boldsymbol{r} = \boldsymbol{0}$). We can obtain another strengthened version of the Second Fundamental Theorem by making use of the following definition (which is partially repeated from Chapter 4).

7.37 Definitions. We will say that the jth commodity is a **numéraire good for $\boldsymbol{P}_i$** iff for all $\boldsymbol{x} \in X_i$ and all $\theta \in \mathbb{R}_{++}$, we have:

$$\boldsymbol{x} + \theta \boldsymbol{e}_j \in X_i \ \text{ and } \ (\boldsymbol{x} + \theta \boldsymbol{e}_j) P_i \boldsymbol{x},$$

where $\boldsymbol{e}_j$ is the jth unit coordinate vector. We shall say that the jth commodity is a **numéraire good for the economy**, E, at an allocation $\big(\langle \boldsymbol{x}_i^* \rangle, \langle \boldsymbol{y}_k^* \rangle\big) \in A(E)$ iff it is a numéraire good for each $i \in M$, and for each $i \in M$ there exists $\theta_i > 0$ such that:

$$\boldsymbol{x}_i^* - \theta_i \boldsymbol{e}_j \in X_i.$$

Since it is easily seen that the numéraire good assumption in the following implies that E is irreducible at $\big(\langle \boldsymbol{x}_i^* \rangle, \langle \boldsymbol{y}_k^* \rangle\big)$, and that each preference relation is locally non-saturating, this next result is a corollary of 7.36. Details of the proof will be left as an exercise.

7.38 Theorem. *Let $E = (\langle X_i, P_i \rangle, \langle Y_k \rangle, \boldsymbol{r})$ be an economy such that:*

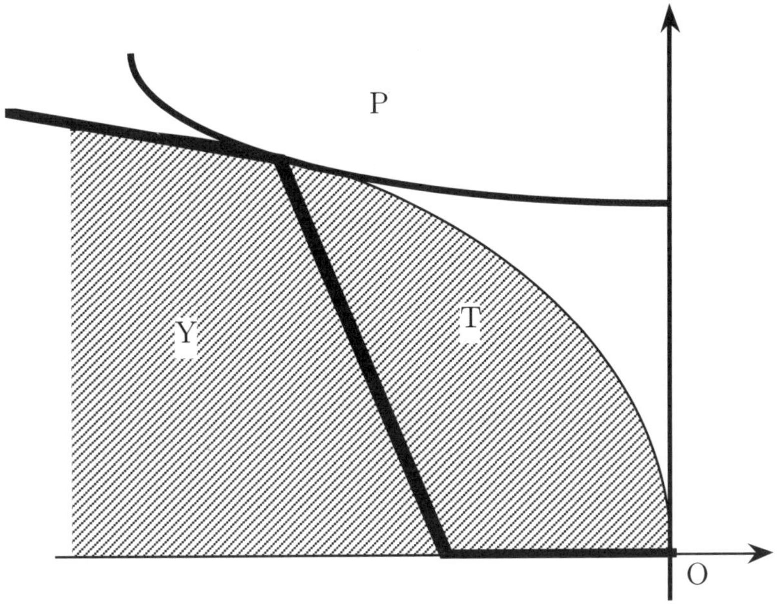

Figure 7.5: Aggregatively Similar Economies.

a. X_i *is convex,*

b. P_i *is weakly convex and lower semi-continuous,*

for each $i = 1, \ldots, m$*;*

c. $Y \stackrel{def}{=} \sum_{k=1}^{\ell} Y_k$ *is a convex set, and suppose that:*

d. $int(X) \cap (\boldsymbol{r} + Y) \neq \emptyset$.

Then we have the following.

If $(\langle \boldsymbol{x}_i^* \rangle, \langle \boldsymbol{y}_k^* \rangle)$ *is Pareto efficient for* E *and for some* $\bar{\jmath} \in \{1, \ldots, n\}$*, the* $\bar{\jmath}^{th}$ *commodity is a numéraire good for the economy at* $(\langle \boldsymbol{x}_i^* \rangle, \langle \boldsymbol{y}_k^* \rangle)$*, then there exists a price vector,* $\boldsymbol{p}^* \in \mathbb{R}^n$ *such that* $(\langle \boldsymbol{x}_i^* \rangle, \langle \boldsymbol{y}_k^* \rangle, \boldsymbol{p}^*)$ *is a competitive equilibrium for* E *given the assignment of wealth* $\boldsymbol{w}^*$*, where:*

$$w_i^* = \boldsymbol{p}^* \cdot \boldsymbol{x}_i^* \quad for\, i = 1, \ldots, m; \tag{7.86}$$

and we have:

$$w_i^* > \min \boldsymbol{p}^* \cdot X_i \quad for\, i = 1, \ldots, m. \tag{7.87}$$

Notes on the Literature.

The notion of Pareto dominance was apparently formally introduced into economics in Pareto [1894]; although it seems to have been Pareto's friend and colleague, Enrico Barone who first stated and proved a version of the 'First Fundamental Theorem' (Barone [1908]). Kenneth Arrow first stated and proved a version of the 'Second Fundamental Theorem' (Arrow [1951a]): although Gerard Debreu independently published a closely-related result

(Debreu [1951]). Early refinements/generalizations of the 'Second Fundamental Theorem' were done by Debreu [1954], Koopmans [1957], and Koopmans and Bausch [1959]. Hurwicz [1960] was the first to examine the issue of non-wastefulness and unbiasedness of abstract resource allocation mechanisms generally.

As mentioned in the text, the irreducibility condition used in Theorem 7.36, while adapted from an earlier condition introduced by Lionel McKenzie. It was generalized in various ways in Moore[1970], and further generalized and refined in Moore [1973, 1975]. However, the irreducibility condition introduced here incorporates and generalizes all of these conditions.

Exercises.

1. Consider the economy, E, in which we have one consumer, one producer, and two commodities, and where X, $\mathbf{r}$ and Y are given by:

$$\begin{aligned} X =& \{\mathbf{x} \in \mathbb{R}^2 \mid -4 \leq x_1 \leq 0 \; \& \; x_2 \geq 3\}, \\ \mathbf{r} =& (0,2), \text{ and} \\ Y =& \{\mathbf{y} \in \mathbb{R}^2 \mid y_1 \leq 0 \leq y_2 \; \& \; y_1 + y_2 \leq 0\}, \end{aligned}$$

respectively; and suppose the consumer's preferences can be represented by the utility function:

$$u(\mathbf{x}) = \min\{8 + 2x_1, x_2\}.$$

(a) Show that the allocation $(\mathbf{x}^*, \mathbf{y}^*)$ is Pareto efficient for E, where:

$$\mathbf{x}^* = (-2,4) \text{ and } \mathbf{y}^* = (-2,2).$$

[Note: There are various hard ways to verify this answer, as well as an easy way. Try to find the easy way, but before concluding that you have found it, answer part (b).]

(b) Is the allocation $(\mathbf{x}', \mathbf{y}')$, where:

$$\mathbf{x}' = (-4,6) \text{ and } \mathbf{y}' = (-4,4),$$

Pareto efficient for E?

2. Consider the pure exchange economy, E, in which $m = n = 2$, and in which the ith consumer's preference relation, P_i, is representable by the utility function:

$$u_i(\mathbf{x}_i) = x_{i1} \cdot x_{i2} \quad \text{for } i = 1,2;$$

and where $\mathbf{r} = (1,1)$. Show that an allocation, $(\mathbf{x}_i)$, is Pareto efficient for E if, and only if, there exists some $\theta \in [0,1]$ such that:

$$\mathbf{x}_1 = (\theta, \theta) \text{ and } \mathbf{x}_2 = (1-\theta, 1-\theta).$$

Once again there is an easy way to do this.

3. Consider the two-consumer, two-commodity exchange economy in which the consumer's preferences can be represented by the utility functions:

$$u_1(\mathbf{x}_1) = x_{11} + x_{12},$$

and:

$$u_2(\boldsymbol{x}_2) = \min\{x_{21}, x_{22}\},$$

respectively; and with $\boldsymbol{r}$, the aggregate resource endowment, given by $\boldsymbol{r} = (10, 10)$.

a. Find, either graphically or algebraically, all Pareto efficient allocations for this economy.

b. Let $\theta \in]0,1[$, and consider the function $\boldsymbol{w}\colon \mathbb{R}^2_{++} \to \mathbb{R}^2_+$ defined by:

$$w_1(\boldsymbol{p}) = 10\theta(p_1 + p_2) \text{ and } w_2(\boldsymbol{p}) = 10(1-\theta)(p_1 + p_2),$$

respectively. Is $\boldsymbol{w}(\cdot)$ a feasible wealth-assignment funtion for E?

c. Given the wealth-assignment function defined in part b, above, can you find a Walrasian equilibrium for E, given $\boldsymbol{w}(\cdot)$ and an arbitrary $\theta \in]0,1[$?

4. Prove Proposition 7.13.

5. Prove Corollary 7.14.

6. Prove Corollary 7.32

7. In the literature on general competitive equilibrium, it is quite common to find the condition:

$$\sum\nolimits_{i=1}^{m} \boldsymbol{x}_i \leq \boldsymbol{r} + \sum\nolimits_{k=1}^{\ell} \boldsymbol{y}_k\,, \tag{7.88}$$

used in place of Condition 3 in the definition of an attainable allocation for an economy, E. The tuple $(\langle \boldsymbol{x}_i^* \rangle, \langle \boldsymbol{y}_k^* \rangle, \boldsymbol{p}^*)$ is then said to be a competitive equilibrium for E if it satisfies this modified attainability condition; and, in addition to the other conditions of Definition 7.4, it satisfies:

$$\boldsymbol{p}^* \cdot \left(\boldsymbol{r} + \sum\nolimits_{l=1}^{\ell} y_k^* - \sum\nolimits_{i=1}^{m} \boldsymbol{x}_i^*\right) = 0.$$

Show that this approach is equivalent to using the definitions in the present chapter, while maintaining the assumption that Y satisfies 'semi-free disposability' (Definition 6.4.5.b)

Chapter 8

The Existence of Competitive Equilibrium

8.1 Introduction

In this chapter, our main concern is to analyze the following theorem concerning the existence of Walrasian equilibrium for a private ownership economy. It is based upon (and is a special case of) Gale and Mas-Colell [1975], and is a generalization of Theorem 5.7.1, pp. 83–4 of Debreu [1959].

8.1 Theorem. *The private ownership economy,* $\mathcal{E} = (\langle X_i, P_i \rangle, \langle Y_k \rangle, \langle \boldsymbol{r}_i \rangle, [s_{ik}])$, *has a Walrasian equilibrium if:*

for each i $(i = 1, \ldots, m)$*:*

a. X_i *is closed, convex, and bounded below,*

b. P_i *is (irreflexive and):*

1. non-saturating,

2. weakly convex, and:

3. strongly continuous;

c. $(\exists \bar{\boldsymbol{x}}_i \in X_i) \colon \bar{\boldsymbol{x}}_i \ll \boldsymbol{r}_i$.

d.1. $\mathbf{0} \in Y_k$, *for* $k = 1, \ldots, \ell$,

d.2. $Y \equiv \sum_{k=1}^{\ell} Y_k$ *is closed and convex,*

d.3. $Y \cap (-Y) \subseteq \{\mathbf{0}\}$, *and:*

d.4. $-\mathbb{R}^n_+ \subseteq Y$.

In the Sections 2–4 of this chapter, we will go through the assumptions of this theorem one by one; showing in each case that the assumption cannot simply be dispensed with. In Section 4 we will present at statement and brief discussion of the original Gale and Mas-Colell theorem; and in Section 5 we will prove an especially simple version of an existence theorem.

Returning to our discussion of Theorem 8.1, let me explain that by the statement, "an assumption cannot be dispensed with," in a given theorem, I mean the following. Suppose we have a theorem of the form:

$$A_1 \ \& \ A_2 \ \& \ A_3 \Rightarrow C,$$

where 'A_i' denotes the statement of an assumption, for $i = 1, 2, 3$; and 'C' denotes the statement of the conclusion. We will say that, for example, A_1 cannot be dispensed with in this result if we can find an example satisfying A_2 and A_3, but where the conclusion, C, does not hold. Notice that the existence of such an example does not preclude the possibility of their being an assumption A_1^* such that:

$$A_1^* \,\&\, A_2 \,\&\, A_3 \Rightarrow C,$$

and where A_1^* generalizes A_1 (that is, $A_1 \Rightarrow A_1^*$, but not conversely). Thus, to show, for example, that assumption (b.1) cannot be dispensed with in Theorem 8.1, we need to find an example of a private ownership economy satisfing all of the remaining assumptions of the theorem, but for which no Walrasian equilibrium exists.

Our examples will deal primarily with two special cases introduced in the previous chapter: the classical two-person, two-commodity pure exchange model; and the one-person, one producer, two commodity model.

Recall the notation used in the previous chapter: we define Π by:

$$\Pi = \{\boldsymbol{p} \in \mathbb{R}^n \mid (\exists \bar{\boldsymbol{y}} \in Y)(\forall \boldsymbol{y} \in Y)\colon \boldsymbol{p} \cdot \bar{\boldsymbol{y}} \geq \boldsymbol{p} \cdot \boldsymbol{y}\}, \tag{8.1}$$

where 'Y' denotes the aggregate production set. Furthermore, just as we defined the profit functions, $\pi_k(\cdot)$ and the supply correspondences, $\sigma_k(\cdot)$ on Π_k, we defined an aggregate profit function, $\pi(\cdot)$, and an aggregate supply correspondence, $\sigma(\cdot)$, on Π by:

$$\pi(\boldsymbol{p}) = \max_{\boldsymbol{y} \in Y} \boldsymbol{p} \cdot \boldsymbol{y} \text{ and } \sigma(\boldsymbol{p}) = \{\boldsymbol{y} \in Y \mid \boldsymbol{p} \cdot \boldsymbol{y} = \pi(\boldsymbol{p})\} \qquad \text{for } \boldsymbol{p} \in \Pi, \tag{8.2}$$

respectively. Recall also that it follows immediately from Proposition 7.8, that, under the assumptions of the present Theorem 8.1:

$$\Pi \subseteq \mathbb{R}^n_+, \tag{8.3}$$

so that, if we define the set Π^* by:

$$\Pi^* = \Pi \cap \Delta_n, \tag{8.4}$$

where 'Δ_n' denotes the unit simplex in $\mathbb{R}^n$:

$$\Delta_n = \left\{\boldsymbol{p} \in \mathbb{R}^n_+ \mid \sum\nolimits_{j=1}^n p_j = 1\right\}, \tag{8.5}$$

we can make use of the homogeneity of the producers' supply- and consumers' demand correspondences to confine our search for equilibrium prices to the set Π^*.

In this chapter, although nowhere else in this book, we will distinguish between competitive and Walrasian equilibria. A Walrasian equilibrium will be one which satisfies Definition 7.4; while a competitive equilibrium will be defined as is set out in Exercise 7, at the end of Chapter 7. In the examples to follow, it will be shown that no Walrasian equilibrium exists; however, it can also be shown that no competitive equilibrium (as just defined) exists either. We will discuss this issue further in the Appendix to this chapter.

8.2 Examples, Part 1

In this section we will develop a number of examples in which we drop only one of the assumptions of Theorem 8.1, and then show that a Walrasian equilibrium does not exist. One particular thing to keep in mind as you study the examples to follow is this: we know from our work in the previous chapter that if a Walrasian equilibrium exists, then the allocation involved must be Pareto efficient. Consequently, if there is only one consumer involved in the example, then the only commodity bundle which could be involved in a Walrasian equilibrium is the one which maximizes the consumer's preferences over the attainable consumption set (or a member of the maximal set, if the preference-maximizing bundle is not unique). Correspondingly, if there is no bundle in the attainable, or feasible consumption set at which the consumer's preferences are maximized, then there will be no Walrasian equilbrium in the example. This fact is the basis of our first example, which shows that the closure of each X_i is an assumption which cannot be dispensed with in Theorem 8.1).

8.2 Example. Let $m = \ell = 1$, $n = 2$, and suppose X, Y and $\boldsymbol{r}$ are as indicated in Figure 8.1, below. While the consumer can almost achieve the level of satisfaction (or utility) corresponding to the indifference curve I_1, we are supposing that the left-most boundary of the consumption set (the dashed vertical line) is not contained in the consumption set. Thus, in particular, the point at which I_1 and the upper boundary of $\boldsymbol{r} + Y$ appear to intersect is not an element of the consumptions set, and thus no maximal point exists within the attainable consumption set. □

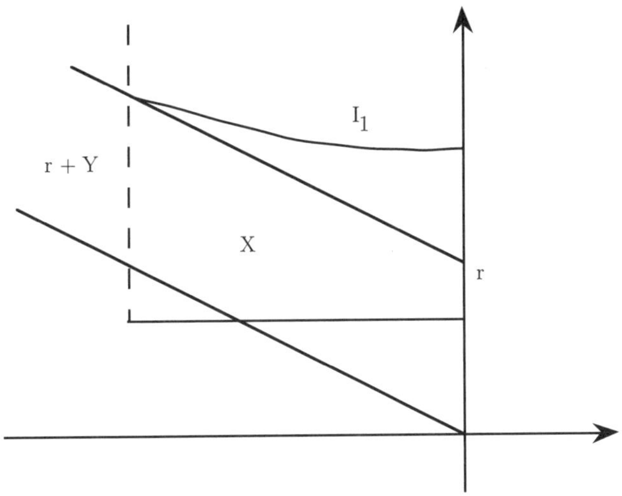

Figure 8.1: Consumption Set not Closed.

The example just presented is, admittedly, highly artificial in appearance. This is an inevitable consequence of the fact that continuity and set closure are not generally assumptions which have any empirical content; that is, in general closure and continuity cannot be directly refuted by empirical observation. For example, in $\mathbb{R}^n$ a set is closed if, given any convergent sequence of points from the set, the limit of the sequence is also an element of the set. Since we cannot observe all the terms in an infinite sequence, we cannot determine whether a finite sequence of empirical observations is or is not drawn from a convergent infinite sequence. Consequently, we cannot generally refute the assumption that individual consumption sets are closed.

The reason which I have slightly qualified my statements in the above paragraph is that there may be real choice situations in which one can see that the choice set (or the attainable consumption set in our general equilibrium examples) is not closed. One such example, which is, I believe, due to Marcel K. Richter, runs as follows. Suppose I am given a gold bar, and told that I need to cut it into two parts, giving one part to you, and keeping the other for myself—with no stipulations about relative size, and no choice of part by yourself. In this case, I, being the greedy person that I am, will try to cut off and give you as thin a slice of the bar as possible in order that I maximize the amount of gold which I get to keep for myself. Well, as you can see, there is no maximal point in this problem; no matter how thin the slice I give you, I might have been able to cut off a still thinner slice. Formally, this is an example in which the attainable consumption set is not closed, and, correspondingly, in which no preference-maximizing choice exists for me.

8.3 Example. (Showing that the convexity of each X_i is an assumption which cannot be dispensed with in 8.1.) Let $m = \ell = 1$, $n = 2$, and suppose X, Y and $\boldsymbol{r}$ are as indicated in Figure 8.2, below.

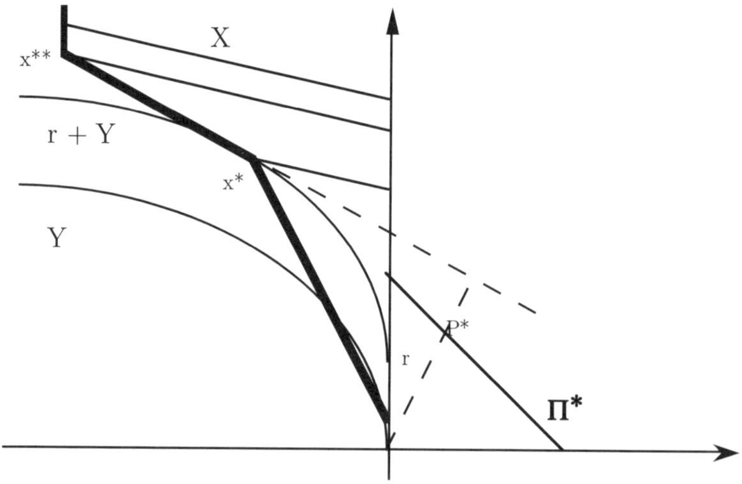

Figure 8.2: Non-convexity of the Consumption Set.

In Figure 8.2, we can see that if $\boldsymbol{p} \in \Pi^*$ is such that $p_1 \geq p_1^*$, then the consumer's preference-maximizing commodity bundle will be on the left-hand boundary of the consumption set (at some $\boldsymbol{x}$ for which $x_1 = x_1^{**}$, which is outside the attainable consumption set, X^*). On the other hand, if $\boldsymbol{p} \in \Pi^*$ is such that $p_1 < p_1^*$, then the producer will maximize profits at some point $\boldsymbol{y} \in Y$ where $\boldsymbol{r} + \boldsymbol{y} \notin X$ (in particular, for any such $\boldsymbol{p}$, we will have $y_1 < x_1^*$, for the profit-maximizing value of $\boldsymbol{y}$). Therefore, no Walrasian equilibrium exists for $\mathcal{E}$ in this case. □

8.4 Example. (Showing that the assumption that X_i is bounded below for each i cannot be dispensed with in Theorem 8.1) Let $m = \ell = 1$, $n = 2$, and suppose X, Y and $\boldsymbol{r}$ are as indicated in Figure 8.3, below; where X is the set of all points in the second quadrant lying above the heavy horizontal line, and the consumer's indifference curves are the family of parallel lines drawn with slope flatter than the upper boundary of Y (which is indicated by the heavy upward-sloping line in the figure).

We can see from the figure that if $\boldsymbol{p} \in \Pi^*$ is such that $p_1 > p_1^*$, then the producer will maximize profits at the origin (that is, at $\boldsymbol{y} = \boldsymbol{0}$); while if $\boldsymbol{p} = \boldsymbol{p}^*$, then the producer's profits are maximized at all points $\boldsymbol{y}$ on the upper boundary of the production set. However, in either of these cases, the consumer's demand is unbounded. On the other hand, if $\boldsymbol{p} \in \Delta_n$ is such that $p_1 < p_1^*$, then the producer's profits are unbounded (and no profit-maximizing production vector exists). Therefore, no Walrasian equilibrium exists in this case. □

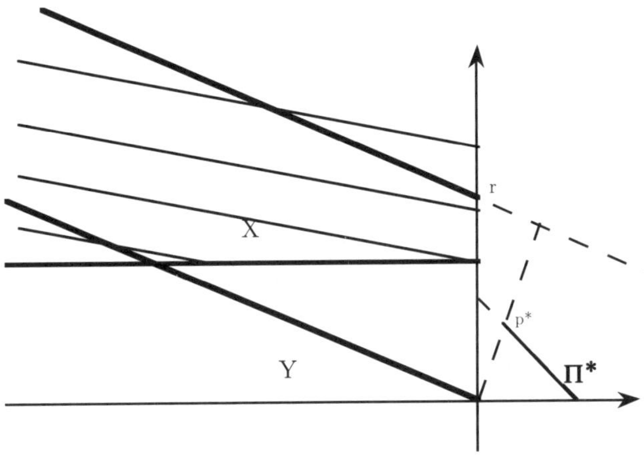

Figure 8.3: An Unbounded Consumption Set.

Notice that the attainable consumption set, X^*, is unbounded in Example 8.4. Intuitively, we would probably suspect that the attainable consumption set is going to have to be bounded if there is to exist a Walrasian equilibrium. As a matter of fact, however, it is possible for an equilibrium to exist even if X^* is unbounded. On the other hand, it is very difficult to specify non-trivial, yet meaningful conditions which are sufficient to guarantee that an equilibrium exists in such a case; in fact, nearly all of the existence proofs with which I am acquainted make use of assumptions which insure that X^* is bounded. Our next example shows an even more insurmountably difficult situation which can arise when X^* is unbounded, and notice that this time X is bounded below.

8.5 Example. (Showing that (d.3), the aggregate irreversibility of production assumption, cannot be dispensed with in 8.1) Let $m = \ell = 1$, $n = 2$, and suppose X, Y and $\boldsymbol{r}$ are as indicated in Figure 8.4, below; where Y is the half-space consisting of all points on or to the left of the vertical axis. Notice that, since Y is a convex cone in this case, and $\boldsymbol{r} \in Y$, we will have $\boldsymbol{r} + Y = Y$. You should have no difficulty in establishing the fact that no Walrasian equilibrium exists in this case. □

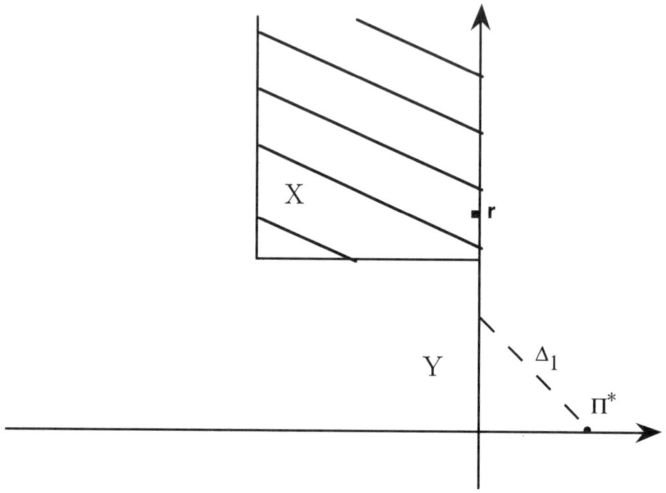

Figure 8.4: Reversible Production.

Debreu has proved the following result, which I will simply state without proof. It is proved on p. 77 of Debreu [1959], for those of you who might be interested.

8.6 Proposition. (Debreu) Let E be an economy n which X is bounded below, Y is closed and convex, and $Y \cap \mathbb{R}^n_+ = \{\mathbf{0}\}$. If, in addition, $\ell = 1$, and/or $Y \cap (-Y) = \{\mathbf{0}\}$, then $A(E)$ is bounded; as then are X^* and Y^* as well.

This result, together with Examples 4 and 5, show the role played by the assumptions:

$$X_i \text{ is bounded below, for } i = 1, \ldots, m,$$

and:

$$Y \cap (-Y) \subseteq \{\mathbf{0}\}.$$

In Theorem 8.1; as well as indicating one of the roles played by each of the assumptions (d.1), (d.2), and (d.4) in the result. Notice, incidentally, that if Y satisfies:

$$-\mathbb{R}^n_+ \subseteq Y \text{ and } Y \cap (-Y) \subseteq \{\mathbf{0}\},$$

then we will also have:

$$\mathbb{R}^n_+ \cap Y = \{\mathbf{0}\}.$$

8.7 Example. (Showing that the assumption that each P_i is non-saturating cannot be dispensed with in Theorem 8.1) Once again we consider a case in which there is one consumer, two commodities, and one producer; this time with the consumption and production sets as indicated in Figure 8.5, below. The concentric ovals represent the consumer's indifference map in this case; with the consumer essentially ordering bundles in terms of their distance from his or her 'bliss point,' which is $\boldsymbol{x}^*$. If there were a Walrasian equilibrium in this situation, we would need to be able to find a non-null price vector, $\boldsymbol{p}$, such that the producer maximizes profits at $\boldsymbol{y}^* \stackrel{\text{def}}{=} \boldsymbol{x}^* - \boldsymbol{r}$. Obviously, however, there is no such price vector.

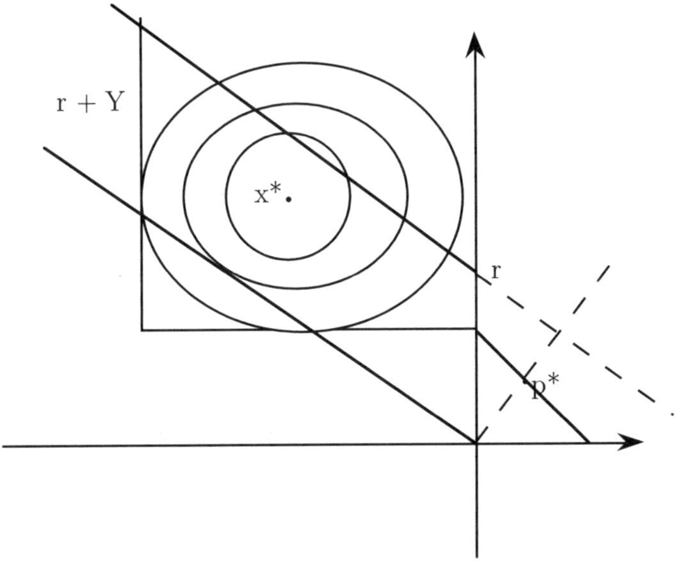

Figure 8.5: Saturating Preferences.

8.8 Example. (Showing that the assumption that each P_i is weakly convex cannot be dispensed with in 8.1). Let $m = \ell = 1$, $n = 2$, and suppose X, Y and $\boldsymbol{r}$ are as indicated in Figure 8.6, on the next page. The indifference curves for P, the consumer's preference relation, are the kinked lines with vertices on the upward-sloping dashed line emanating from the lower left corner of X. Notice that for $\boldsymbol{p} \in \Pi^*$ such that $p_1 > p_1^*$, the consumer's optimal commodity bundle will have $x_1 = x_1^*$ and $x_2 \geq x_2^*$. On the other hand, for $\boldsymbol{p} \in \Pi^*$ satisfying $0 < p_1 < p_1^*$, the consumer's optimal commodity bundle, $\boldsymbol{x}$, will satisfy $x_1 = 0$. Finally, for $\boldsymbol{p} = \boldsymbol{p}^*$, $h(\boldsymbol{p}^*, w) = \{\boldsymbol{x}^*, \boldsymbol{x}^\dagger\}$. Thus we can see that no Walrasian Equilibrium exists in this case. □

8.9 Example. (Showing that the assumption that each P_i is strongly continuous cannot be dispensed with in Theorem 8.1) In Figure 8.7, on the next page, the line with the arrows represents the consumer's 'behavior line' through $\boldsymbol{x}^*$; that is, everything above the line is preferred to anything on or below the line, while anything

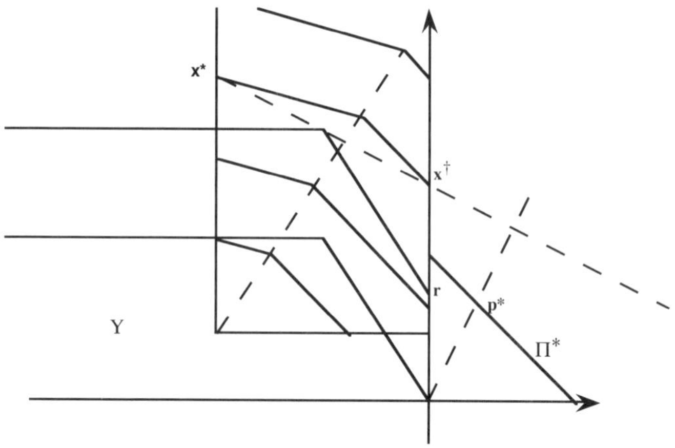

Figure 8.6: Non-convex Upper ontour Sets.

on the line is preferred to anything below the line. On the line itself, however, there is an ordering, as indicated by the direction of the arrows.

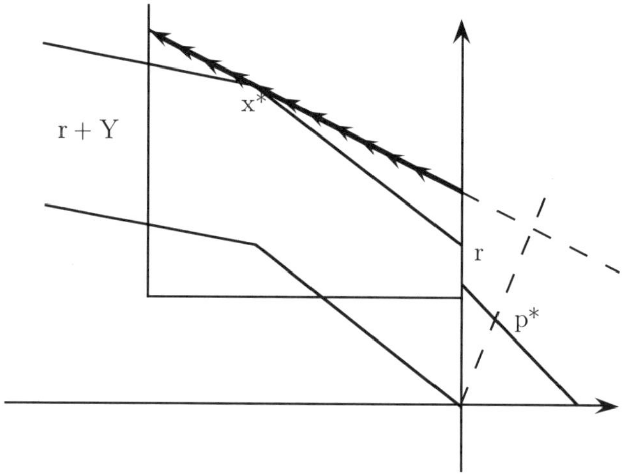

Figure 8.7: Non-continuous Preferences.

In this case, there is a maximal consumption bundle (Pareto efficient consumption allocation), namely $\boldsymbol{x}^*$. Moreover, there is a wide range of price vectors, among them $\boldsymbol{p}^*$, such that the producer will maximize profits at $\boldsymbol{x}^* - \boldsymbol{r}$. However, if the price vector is anything (in Δ_2) other than $\boldsymbol{p}^*$, the consumer will maximize satisfaction at a point on one of the boundaries of X. On the other hand, with $\boldsymbol{p} = \boldsymbol{p}^*$, the consumer will maximize satisfaction at the point where the 'behavior line' intersects the left-hand boundary of X. Therefore, no Walrasian equilibrium exists in this case. □

8.3 Assumption (c) and the Attainable Set

A condition which we have not as yet considered, but which is obviously a necessary condition for the existence of a Walrasian equilibrium for an economy, E, is that:

$$A(E) \neq \emptyset.$$

Our next result shows how this property is guaranteed under the hypotheses of Theorem 8.1.

8.10 Proposition. *If E satisfies:*
c′. $(\exists \bar{\boldsymbol{x}} \in X)\colon \bar{\boldsymbol{x}} \leq \boldsymbol{r}$,
and (d.4):

$$-\mathbb{R}^n_+ \subseteq Y,$$

then $A(E) \neq \emptyset$.

Proof. Let $\bar{\boldsymbol{x}}$ satisfy (c'). Then there exist $\bar{\boldsymbol{x}}_i \in X_i$, for $i = 1, \dots, m$, satisfying:

$$\bar{\boldsymbol{x}} = \sum_{i=1}^{m} \bar{\boldsymbol{x}}_i \leq \boldsymbol{r}. \tag{8.6}$$

Defining:

$$\bar{\boldsymbol{y}} = \bar{\boldsymbol{x}} - \boldsymbol{r},$$

we see by (8.6) that $\bar{\boldsymbol{y}} \in -\mathbb{R}^n_+$. Therefore, by (d.4) there exists $\bar{\boldsymbol{y}}_k \in Y_k$, for $k = 1, \dots, \ell$, satisfying:

$$\bar{\boldsymbol{x}} - \boldsymbol{r} = \bar{\boldsymbol{y}} = \sum_{k=1}^{\ell} \bar{\boldsymbol{y}}_k. \tag{8.7}$$

It follows at once from (8.7) that $\big((\bar{\boldsymbol{x}}_i), (\bar{\boldsymbol{y}}_k)\big) \in A(E)$. □

The above proposition indicates another role which assumption (d.4) plays in Theorem 8.1, as well as one role played by assumption (c).[1] The role played by assumption (c) in Theorem 8.1 is much more complicated and more subtle than this, however. Consider, for instance, the following example.

8.11 Example. Let $\mathcal{E}$ be the private ownership economy in which $\ell = 1,\ m = n = 2$, and:

$$X_i = \{\boldsymbol{x}_i \in \mathbb{R}^2 \mid -1 \leq x_{i1} \leq 0\ \&\ x_{i2} \geq 2\} \quad \text{for } i = 1, 2;$$
$$Y = \{\boldsymbol{y} \in \mathbb{R}^2 \mid y_1 + y_2 \leq 0\ \&\ y_1 \leq 0\},$$

let:

$$\boldsymbol{r}_1 = (0, 5), \qquad s_1 = 1,$$
$$\boldsymbol{r}_2 = (0, 0), \qquad s_2 = 0;$$

and consider the commodity bundles:

$$\boldsymbol{x}_1^* = \boldsymbol{x}_2^* = (-1, 2) \in X_i \quad \text{for } i = 1, 2.$$

[1] Obviously (c) implies (c').

We have:

$$\boldsymbol{x}^* \stackrel{\text{def}}{=} \boldsymbol{x}_1^* + \boldsymbol{x}_2^* = (-2, 4) \ll \boldsymbol{r} \stackrel{\text{def}}{=} \boldsymbol{r}_1 + \boldsymbol{r}_2 = (0, 5),$$

and thus we see that $\mathcal{E}$ satisfies the condition:

$$c.'' \qquad (\exists \bar{\boldsymbol{x}} \in X) \colon \bar{\boldsymbol{x}} \ll \boldsymbol{r}.$$

Moreover, it is easy to show that $\mathcal{E}$ satisfies (a), (d.1), (d.2), (d.3), and (d.4) of Theorem 8.1. We will demonstrate, however, that, if P_1 is a locally non-saturating binary relation, then whatever the form of P_2, no Walrasian equilibrium exists in this case.

We begin by noting that in this case the set Π^* will be given by:

$$\Pi^* = \{\boldsymbol{p} \in \mathbb{R}^2_+ \mid 1/2 \leq p_1 \leq 1 \;\&\; p_2 = 1 - p_1\}.$$

Now, suppose $\boldsymbol{p} \in \Pi^*$ is such that $1/2 < p_1 \leq 1$. Then the producer will maximize profits at $\boldsymbol{y} = \boldsymbol{0}$, and the first consumer's budget constraint will be of the form:

$$p_1 x_{11} + p_2 x_{12} \leq w_1(\boldsymbol{p}) = \boldsymbol{p} \cdot \boldsymbol{r}_1 + s_1 \pi(\boldsymbol{p}) = 5p_2. \tag{8.8}$$

However, if P_1 is any locally non-saturating preference relation, then any $\boldsymbol{x}_1^* \in X_1$ which maximizes P_1, given $\boldsymbol{p}$ and $w_1(\boldsymbol{p})$ will satisfy (8.8) with an equality:

$$p_1 x_{11}^* = (5 - x_{12}^*) p_2. \tag{8.9}$$

Since $\boldsymbol{x}^* \in X_1$ implies $x_{11}^* \leq 0$, we see that (8.9) implies that $x_{12}^* \geq 5$. However, in order that $\boldsymbol{x}_2$ be an element of X_2, we must have $x_{22} \geq 2$, and thus it is clear that there exists no $\boldsymbol{x}_2 \in X_2$ satisfying:

$$\boldsymbol{x}_1^* + \boldsymbol{x}_2 = \boldsymbol{r} + \boldsymbol{y} = (0, 5) + \boldsymbol{0} = (0, 5);$$

and, consequently, that no Walrasian equilibrium exists in which

$$1/2 < p_1 \leq 1. \tag{8.10}$$

Now, if $\boldsymbol{p} \in \Pi^*$, the only alternative to its satisfying (8.10) is that $\boldsymbol{p} = (1/2, 1/2)$. However, for this value of $\boldsymbol{p}$, the second consumer's budget constraint is give by:

$$\boldsymbol{p} \cdot \boldsymbol{x}_2 = (1/2)(x_{21} + x_{22}) \leq w_2(\boldsymbol{p}) = 0. \tag{8.11}$$

On the other hand, if $\boldsymbol{x}_2 \in X_2$, we have $x_{21} \geq -1$ and $x_{22} \geq 2$, so that:

$$(\forall \boldsymbol{x}_2 \in X_2) \colon \boldsymbol{p} \cdot \boldsymbol{x}_2 = (1/2)(x_{21} + x_{22}) \geq (1/2)(-1 + 2) = 1/2. \tag{8.12}$$

Upon comparing (8.11) and (8.12), we see that the second consumer's budget set is empty if $\boldsymbol{p} = (1/2, 1/2)$; and since we have now considered all values of $\boldsymbol{p}$ which are consistent with profit-maximization for the producer, it follows that no Walrasian equilibrium exists for this economy. □

As is made clear by the last example, one of the functions of assumption (c) in Theorem 8.1 is that it [together with (d.1)] guarantees that each consumer will have sufficient wealth to participate in a market economy. However, consider the following result.

8.12 Proposition. *Suppose $\mathcal{E}$ satisfies:*
c.‴ $(\exists \bar{\boldsymbol{x}}_i \in X_i)\colon \bar{\boldsymbol{x}}_i \leq \boldsymbol{r}_i$ *for* $i = 1, \ldots, m$;
and (d.1):

$$\mathbf{0} \in Y_k \quad \textit{for } i = 1, \ldots, \ell.$$

Then we have:

$$(\forall \boldsymbol{p} \in \Pi^*)\colon B_i(\boldsymbol{p}) \stackrel{def}{=} \{\boldsymbol{x} \in X_i \mid \boldsymbol{p} \cdot \boldsymbol{x} \leq w_i(\boldsymbol{p})\} \neq \emptyset \quad \textit{for } i = 1, \ldots, m.$$

Proof. Let $\boldsymbol{p}^*$ be an arbitrary element of Π^*. Then, by (d.1) and the definition of Π^*, it follows that for each k:

$$\pi_k(\boldsymbol{p}^*) \geq 0. \tag{8.13}$$

Thus, if $\bar{\boldsymbol{x}}_i \in X_i$ satisfies (c'''), it follows from (8.13) and the fact that $\boldsymbol{p}^* \geq \mathbf{0}$ that:

$$\boldsymbol{p} \cdot \bar{\boldsymbol{x}}_i \leq \boldsymbol{p}^* \cdot \boldsymbol{r}_i \leq \boldsymbol{p}^* \cdot \boldsymbol{r}_i + \sum_{k=1}^{\ell} s_{ik}\pi_k(\boldsymbol{p}^*) = w_i(\boldsymbol{p}^*).$$

Since $\boldsymbol{p}^*$ was an arbitrary element of Π^*, and i was arbitrary, our result follows. □

Propositions 8.10 and 8.12 together suggest that we might be able to generalize Theorem 8.1 by replacing hypothesis (c) with (c'''). However, this in not the case; in fact consider the somewhat stronger assumption:

c^*. $\quad (\exists \bar{\boldsymbol{x}}_i \in X_i)\colon \bar{\boldsymbol{x}}_i < \boldsymbol{r}_i$, for $i = 1, \ldots, m$; and $\sum_{i=1}^m \bar{\boldsymbol{x}}_i \ll \boldsymbol{r} \stackrel{\text{def}}{=} \sum_{i=1}^m \boldsymbol{r}_i$.

That Theorem 8.1 does not remain correct if (c^*) is substituted for (c) is demonstrated by the following.[2]

8.13 Example. Consider the economy, $\mathcal{E}$, in which $m = n = 2$, $X_1 = X_2 = \mathbb{R}^2_+$, and the two preferences can be represented by the utility functions:

$$u_1(\boldsymbol{x}_1) = \min\{x_{11}/2, x_{12}\},$$

and:

$$u_2(\boldsymbol{x}_2) = x_{22},$$

respectively. Suppose further that:

$$\boldsymbol{r}_1 = (4, 4), \boldsymbol{r}_2 = (0, 4), s_1 = s_2 = 1/2, \ \& \ Y = -\mathbb{R}^2_+.$$

You can easily confirm the fact that $\mathcal{E}$ satisfies all of the assumptions of Theorem 8.1, *except* assumption (c), and that it does satisfy assumption (c*).[3] We will show that no Walrasian equilibrium exists for $\mathcal{E}$ in this case.

Accordingly, suppose, by way of obtaining a contradiction, that $\big((\boldsymbol{x}_i^*), \boldsymbol{y}^*, \boldsymbol{p}^*\big)$ is a Walrasian equilibrium for $\mathcal{E}$. Then, since Y satisfies (d.4) and both consumers' preferences are increasing, it must be the case that $\boldsymbol{p}^* > \mathbf{0}$; and we may therefore

[2]See also Example 7.28, which provides a bit different insight into the role played by assumption (c) in the theorem.

[3]And yes, it is essentially a pure exchange economy.

suppose that $\boldsymbol{p}^* \in \Delta_2$. If we suppose that $\boldsymbol{p}^* \gg \mathbf{0}$, then the first consumer's demand for the first commodity is given by (verify this):

$$x_{11}^* = \frac{8}{2p_1^* + p_2^*}.$$

Now, given the form of Y, it then follows that we must have:

$$\frac{8}{2p_1^* + p_2^*} \leq 4;$$

and, since $p_2^* = 1 - p_1^*$, this requires $p_1^* = 1$, and, correspondingly, $p_2^* = 0$. However, with $p_2^* = 0$, the second consumer has unbounded demand for the second commodity! Therefore, no Walrasian equilibrium exists in this case.

What the simple analytics of this example brings out in sharp relief is this: given any finite price ratio of p_1/p_2, the first consumer has strictly positive excess demand for the first commodity; which means that this consumer's demand for the first commodity exceeds the total quantity available. If, on the other hand, p_1 rises to $p_1 = 1$ (thereby making $p_1/p_2 = +\infty$), consumer one no longer has positive excess demand for the first commodity, but the second consumer now has unbounded demand for the second commodity. On the other hand, if we change the second consumer's endowment to set $r_{21} > 0$,[4] then, however small the quantity of the first commodity we add to his endowment, consumer one's excess demand for the first commodity can be accommodated with a finite ratio of p_1/p_2, and a Walrasian equilibrium exists for $\mathcal{E}$ (as I will leave you to verify). □

Insofar as the remaining hypotheses of 8.1 are concerned, our final example of this section shows that the convexity of Y cannot be dispensed with. We will not include an example showing that the closure of Y cannot be dispensed with, but such examples are easy to construct. Bergstrom [1976] has shown that the free disposal assumption can actually be dispensed with. (See also Shafer [1976].)

8.14 Example. Let $m = \ell = 1$, $n = 2$, and suppose X, Y $\boldsymbol{r}$, and P are as indicated in Figure 8.8, on the next page. Notice that if $\boldsymbol{p} \in \Pi^*$ satisfies:

$$p_1^* < p_1 \leq 1, \tag{8.14}$$

then the producer will maximize profits at $\boldsymbol{y} = \mathbf{0}$; whereas the consumer's preference-maximizing commmodity bundle will not equal $\boldsymbol{r}$. Hence, no Walrasian equilibrium can exist for any $\boldsymbol{p} \in \Pi^*$ satisfying (8.14). On the other hand, for $\boldsymbol{p} \in \Pi^*$ satisfying:

$$0 \leq p_1 \leq p_1^*,$$

the producer will maximize profits at some $\boldsymbol{y} \in Y$ such that $\boldsymbol{r} + \boldsymbol{y} \notin X$. Therefore, no Walrasian equilibrium exists in this case.

This example shows the way in which the competitive pricing system can break down in the presence of non-convexity. Notice that, from the consumer's point of view, a best consumption vector exists in X^*; namely at $\boldsymbol{x} = \boldsymbol{x}^*$. However, no price vector exists such that the producer will maximize his profits at $\boldsymbol{y} = \boldsymbol{x}^* - \boldsymbol{r}$. □

---

[4]Thereby satisfying assumption (c).

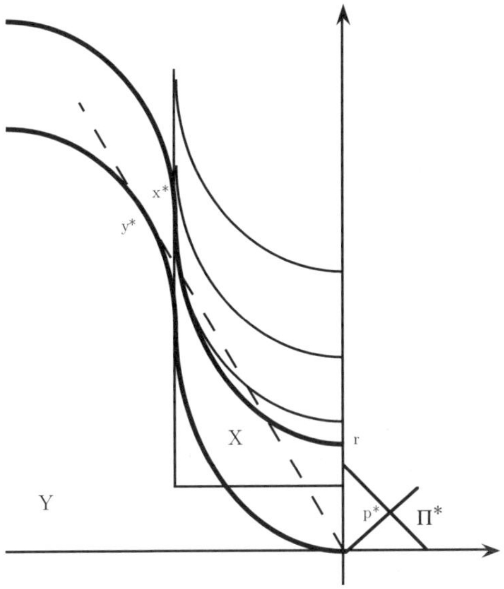

Figure 8.8: Non-convexity of the Production Set.

8.4 The Gale and Mas-Colell Theorem

As indicated in the introduction, Theorem 8.1 is a special case of the main theorem in Gale and Mas-Colell [1975]. Their theorem makes use of a wealth-assignment function, as per Definition 7.9, and is stated as follows.

8.15 Theorem. *The economy E has a Walrasian equilibrium if:*
for each i ($i = 1, \ldots, m$):
a. X_i is closed, convex, and bounded below,
b. P_i is (irreflexive and):
1. non-saturating,
2. weakly convex, and:
3. strongly continuous;
c. the feasible wealth assignment function, $\boldsymbol{w}\colon \Pi \cap \Delta_n \to \mathbb{R}^m$ is continuous [that is each $w_i(\cdot)$ is a continuous real-valued function], and satisfies:

$$(\forall \boldsymbol{p} \in \Pi^*)\colon w_i(\boldsymbol{p}) > \min \boldsymbol{p} \cdot \boldsymbol{X}, \tag{8.15}$$

where $\Pi^* = \Pi \cap \Delta_n$,
and the aggregate production set, $Y \equiv \sum_{k=1}^{\ell} Y_k$:
d.1. is closed and convex,
d.2. has a bounded intersection with $\mathbb{R}^n_+$, and:
d.3. contains $-\mathbb{R}^n_+$.

While I don't propose to go through the assumptions of this result, showing each assumption cannnot be dispensed with, in the way that we did in connection with Theorem 8.1, a few comments may be in order.

First of all, if you compare the statements of this and the earlier result, you will see that the differences occur in assumption (c) and (d) in the respective statements. Assumptions (d.1) and (d.3) of the present theorem are used in the earlier result as well. Moreover, it is obvious that the assumption $Y \cap (-Y) \subseteq \mathbf{0}$, together with assumption (d.3) of the present result, implies that Y has a bounded intersection with $\mathbb{R}^n_+$ (in fact, that $Y \cap \mathbb{R}^n_+ \subseteq \mathbf{0}$) Consequently, the present assumptions regarding the production sector generalize those in the earlier result. On the other hand, in Example 8.5 we have already shown that the present assumption (d.2) cannot be dispensed with; indeed that it cannot be weakened to:

$$Y \cap \mathbb{R}^n_{++} = \emptyset.$$

The relationship between the two different assumptions (c) is more intriguing, however. We saw in Section 8.3 that one of the roles played by Assumption (c) in Theorem 8.1 is to guarantee that the set of attainable allocations is non-empty. One can be forgiven a little head-scratching over the puzzle of what it is that Gale and Mas-Colell have assumed which guarantees this same condition. However, it is their assumption (c) which does the trick here as well, for notice that if $X \cap Y = \emptyset$,[5] then it follows from the 'separating hyperplane theorem' (Theorem 6.21) that there exists $\boldsymbol{p}^\dagger \in \Delta_n$ such that:

$$\sup_{\boldsymbol{y} \in Y} \boldsymbol{p}^\dagger \cdot \boldsymbol{y} \leq \inf_{\boldsymbol{x} \in X} \boldsymbol{p}^\dagger \cdot \boldsymbol{x}. \tag{8.16}$$

However, since $\boldsymbol{w}(\cdot)$ is required to be a feasible wealth assignment function, we must have:

$$\sum_{i=1}^{m} w_i(\boldsymbol{p}^\dagger) = \pi(\boldsymbol{p}^\dagger) = \sup_{\boldsymbol{y} \in Y} \boldsymbol{p}^\dagger \cdot \boldsymbol{y}.$$

Combining this with (8.16), we see that we must have:

$$\sum_{i=1}^{m} w_i(\boldsymbol{p}^\dagger) \leq \inf_{\boldsymbol{x} \in X} \boldsymbol{p} \cdot \boldsymbol{x};$$

which clearly contradicts equation (8.15) in the statement of the Gale and Mas-Colell theorem. Consequently, we see that if E satisfies the present assumption (c), then we must have $X \cap Y \neq \emptyset$.

We saw in Example 8.13 that we cannot weaken the present assumption (c) by replacing equation (8.15) with:

$$(\forall \boldsymbol{p} \in \Pi^*) \colon w_i(\boldsymbol{p}) \geq \min \boldsymbol{p} \cdot \boldsymbol{X}_i, \tag{8.17}$$

Our next example, with which we will close this section, shows that we cannot dispense with the assumption that $\boldsymbol{w}(\cdot)$ is continuous.

8.16 Example. Let E be an economy in which $m = n = 2$, $\ell = 1$, $X_i = \mathbb{R}^2_+$, for $i =, 2$, and:

$$Y = \{\boldsymbol{y} \in \mathbb{R}^2 \mid \boldsymbol{y} \leq (6, 6)\}.$$

[5] We will follow Gale and Mas-Colell in dispensing with aggregate resource endowment, $\boldsymbol{r}$, in this discussion. One can, of course, allow for such an endowment by defining an 'extra' production set, $Y_0 = \{\boldsymbol{r}\}$.

We suppose also that the consumers' preferences can be represented by the utility functions:

$$u_1(\boldsymbol{x}_1) = (x_{11})^{1/3} \cdot (x_{12})^{2/3}, \text{ and } u_2(\boldsymbol{x}_2) = (x_{21})^{2/3} \cdot (x_{22})^{1/3},$$

respectively. Since we know that we can, without loss of generality, confine our attention to price vectors $\boldsymbol{p} \in \Delta_2$, we can define the wealth function to be used as a function of p_1 alone (implicitly assuming that $p_2 = 1 - p_1$). We then we suppose that the wealth-assignment function, $\boldsymbol{w}\colon [0,1] \to \mathbb{R}^2_+$, is given by:

$$w_1(p_1) = \begin{cases} 4 & \text{for } 1 \geq p_1 \geq 1/2, \\ 2 & \text{for } 1/2 > p_1 \geq 0; \end{cases}$$

and:

$$w_2(p_1) = \begin{cases} 2 & \text{for } 1 \geq p_1 \geq 1/2, \\ 4 & \text{for } 1/2 > p_1 \geq 0. \end{cases}$$

It is easy to show that $\boldsymbol{w}(\cdot)$ is a feasible wealth-assignment function for E, and it satisfies:

$$(\forall p_1 \in [0,1])\colon w_i(p_1) > \min \boldsymbol{p} \cdot \boldsymbol{X}_i = 0,$$

for $i = 1, 2$. However, for this wealth-assignment function, aggregate demand for the first commodity, given $p_1 \geq 1/2$, satisfies:

$$\delta(p_1) \leq \delta(p_1) = \frac{(4/3 + (2/3) \cdot 2)}{1/2} = 16/3 = 5\,\frac{1}{3};$$

while for $0 < p_1 < 1/2$:

$$\delta(p_1) > \lim_{p_1 \nearrow 1/2} \delta(p_1) = 6\,\frac{2}{3}.$$

Consequently, no Walrasian equilibrium exists in this case. □

8.5 An (Especially) Simple Existence Theorem

In this section we will study a very simple theorem establishing the existence of a Walrasian equilibrium. While the result assumes a very special case of a private ownership economy, it incorporates one which is often used in the Public Economics literature (although often only implicitly); and, I believe that working through the proof of the result which we are going to study may help you to attain some valuable insights into the meaning and nature of a general competitive equilibrium. The basic model which we are going to be studying here builds upon the model presented at the end of Section 7.3, in that we take the consumers' consumption sets to be a subset of the nonnegative orthant of the commodity space, and suppose that only the initial endowments of leisure are positive.

We will deal with a private ownership economy in which there are m consumers and $n+1$ commodities, with the 0th commodity being leisure/labor. The production sector will be characterized by an aggegate Leontief technology, as described in Section 6.4. Thus, the aggregate production set can be expressed as:

$$Y = \left\{ \boldsymbol{y} \in \mathbb{R}^{n+1} \middle| (\exists \boldsymbol{z} \in \mathbb{R}^n_+) \colon \boldsymbol{y} = \begin{pmatrix} -\boldsymbol{c} \\ \boldsymbol{I} - \boldsymbol{A} \end{pmatrix} \boldsymbol{z} \right\}, \tag{8.18}$$

where $\boldsymbol{A}$ is an $n \times n$ semi-positive matrix, and $\boldsymbol{c}$ is a strictly positive n-vector. We suppose that each consumer's initial endowment takes the form:

$$\boldsymbol{r}_i = (r_{i0}, \boldsymbol{0}),$$

where $r_{i0} > 0$ and '$\boldsymbol{0}$' denotes the origin in $\mathbb{R}^n$, and we will use the generic notation '$(x_{i0}, \boldsymbol{x}_i)$' to denote the ith consumer's commodity bundle. Using this notation, we suppose that the ith consumer's consumption set is given by:

$$X_i = \{(x_{i0}, \boldsymbol{x}_i) \in \mathbb{R}^{n+1}_+ \mid 0 \leq x_{i0} \leq r_{i0}\}; \tag{8.19}$$

and, given $(x_{i0}, \boldsymbol{x}_i) \in X_i$, the ith consumer's labor offer is:

$$\ell_i = x_{i0} - r_{i0}.$$

We will suppose that **leisure is a numéraire for $\mathcal{E}$**, a condition we define as follows; for each $i \in M$, each $(x_{i0}, \boldsymbol{x}_i) \in X_i$ and each $\Delta x_0 \in \mathbb{R}_{++}$, we have:

$$0 \leq x_{i0} + \Delta x_0 < r_{i0} \Rightarrow (x_{i0} + \Delta x_0, \boldsymbol{x}_i) P_i (x_{i0}, \boldsymbol{x}_i). \tag{8.20}$$

The definitions of feasible allocations and competitive equilibria adapt easily to this context, although it is easier and more natural to identify the production level with the n-vector $\boldsymbol{z}$ than with the $(n+1)$-vector $\boldsymbol{y}$. Accordingly, we will say that an allocation, $\big(\langle (x_{i0}, \boldsymbol{x}_i) \rangle, \boldsymbol{z}\big)$ **is feasible for $\mathcal{E}$** iff:

$$(x_{i0}, \boldsymbol{x}_i) \in X_i, \text{for } i = 1, \ldots, m; \ \boldsymbol{z} \in \mathbb{R}^n_+,$$

and:

$$\sum_{i=1}^m \begin{pmatrix} \ell_i \\ \boldsymbol{x}_i \end{pmatrix} = \begin{pmatrix} -\boldsymbol{c} \\ \boldsymbol{I} - \boldsymbol{A} \end{pmatrix} \boldsymbol{z};$$

that is:

$$\sum_{i=1}^m (x_{i0} - r_{i0}) \equiv \sum_{i=1}^m \ell_i = -\boldsymbol{c} \cdot \boldsymbol{z} \text{ and } \sum_{i=1}^m \boldsymbol{x}_i = (\boldsymbol{I} - \boldsymbol{A})\boldsymbol{z}.$$

Turning now to the issue of defining a Walrasian equilibrium for this economy, let me begin by noting that we will always normalize to set the price of leisure (the wage rate) equal to one, so that prices are completely determined by the n-vector, $\boldsymbol{p} \in \mathbb{R}^n_+$, consisting of the prices of the n produced goods. Thus, we will say that a tuple $\big(\langle (x^*_{i0}, \boldsymbol{x}^*_i) \rangle, \boldsymbol{z}^*, \boldsymbol{p}^*\big)$ is a **Walrasian equilibrium for $\mathcal{E}$** iff (a) $\big(\langle (x^*_{i0}, \boldsymbol{x}^*_i) \rangle, \boldsymbol{z}^*\big)$ is feasible for $\mathcal{E}$, (b) z^*_j maximizes profits in the jth sector, given $\boldsymbol{p}^*$, and (c) for each $i \in M$, $(x^*_{i0}, \boldsymbol{x}^*_i)$ maximizes P_i, given:

$$x^*_{i0} + \boldsymbol{p}^* \cdot \boldsymbol{x}^*_i \leq r_{i0} \tag{8.21}$$

Since the production technology is linear, there will be a profit-maximizing output in sector j only if there is zero profit in producing commodity j; and in order that a non-zero net output of commodity j be produced, it is necessary that its price be equal to the unit cost of production; that is, $\boldsymbol{p}^*$ must satisfy $p_j^* z_j - C(z_j) = 0$, so that:

$$p_j^* z_j - \sum_{k=1}^n p_j^* a_{kj} z_j - c_j z_j = \Big(p_j^* - \sum_{k=1}^n p_j^* a_{kj} - c_j\Big) z_j = 0 \tag{8.22}$$

With these considerations in mind, we can turn to our existence theorem.

8.17 Theorem. *Suppose the private ownership economy,* $\mathcal{E} = \big(\langle X_i, P_i\rangle, \langle \boldsymbol{r}_i\rangle, Y\big)$, *satisfies the following conditions:*

for each i $(i = 1, \ldots, m)$;

a. $X_i = \{(x_{i0}, \boldsymbol{x}_i) \in \mathbb{R}_+^{n+1} \mid 0 \leq x_{i0} \leq r_{i0}\}$,

b. P_i *is:*

1. asymmetric,

2. transitive,

3. locally non-saturating, and:

4. upper semi-continuous;

c. $\boldsymbol{r}_i$ *is of the form;*

$$\boldsymbol{r}_i = (r_{i0}, \mathbf{0}),$$

where $r_{i0} > 0$,

d. the aggregate production set takes the form set out in (8.18), where $\mathbf{c} \gg \mathbf{0}$, *and the matrix* $\boldsymbol{A}$ *is semipositive, and satisfies:*

$$\sum_{k=1}^n a_{kj} < 1 \quad \textit{for } j = 1, \ldots, n, \tag{8.23}$$

and:

e. leisure is a numéraire for $\mathcal{E}$.

Then $\mathcal{E}$ *has a Walrasian equilibrium..*

Proof. Since $\boldsymbol{A}$ satisfies (8.23), it follows from Theorem 6.17 that there exists a unique vector $\boldsymbol{p}^* \in \mathbb{R}_+^n$ satisfying:

$$(\boldsymbol{p}^*)^\top(\boldsymbol{I} - \boldsymbol{A}) = \mathbf{c}^\top. \tag{8.24}$$

Moreover, since $\mathbf{c} \gg \mathbf{0}$, we must have $\boldsymbol{p}^* \gg \mathbf{0}$ as well, for we can write the jth equation in (8.24 as:

$$p_j^* - \sum_{k=1}^n p^* a_{kj} = c_j;$$

and, since $a_{kj} \geq 0$ for all k, j, it follows that we must have $p_j^* > 0$. Since $\boldsymbol{p}^* \gg \mathbf{0}$, it now follows from Theorem 4.5 that, for each i, there exists $(x_{i0}^*, \boldsymbol{x}_i^*) \in X_i$ satisfying:

$$x_{i0}^* + \boldsymbol{p}^* \cdot \boldsymbol{x}_i^* = r_{i0}, \tag{8.25}$$

and, for all $(x_{i0}, \boldsymbol{x}_i) \in X_i$:

$$(x_{i0}, \boldsymbol{x}_i) P_i (x_{i0}^*, \boldsymbol{x}_i^*) \Rightarrow x_{i0} + \boldsymbol{p}^* \cdot \boldsymbol{x}_i > r_{i0}.$$

Notice also that it follows from (8.22) and Theorem 6.17 that, defining $\boldsymbol{x}^* \in \mathbb{R}^n_+$ by:

$$\boldsymbol{x}^* = \sum_{i=1}^m x_i^*,$$

there exists $\boldsymbol{z}^* \in \mathbb{R}^n_+$ satisfying:

$$\boldsymbol{x}^* = (\boldsymbol{I} - \boldsymbol{A})\boldsymbol{z}^*. \tag{8.26}$$

Now, it follows from (8.25) that:

$$\sum_{i=1}^m \boldsymbol{p}^* \cdot \boldsymbol{x}_i^* = \boldsymbol{p}^* \cdot \boldsymbol{x}^* = \sum_{i=1}^m (r_{i0} - x_{i0}^*) = -\sum_{i=1}^m \ell_i^*. \tag{8.27}$$

On the other hand, from the definition of $\boldsymbol{z}^*$ and (8.24), we have:

$$\boldsymbol{p}^* \cdot \boldsymbol{x}^* = (\boldsymbol{p}^*)^\top (\boldsymbol{I} - \boldsymbol{A})\boldsymbol{z}^* = \boldsymbol{c} \cdot \boldsymbol{z}^*. \tag{8.28}$$

Combining (8.26)–(8.28) with the fact that, for each $i \in M$, $(x_{i0}^*, \boldsymbol{x}_i^*) \in X_i$, we see that $\big(\langle (x_{i0}^*, \boldsymbol{x}_i^*)\rangle, \boldsymbol{z}^*\big)$ is feasible for $\mathcal{E}$. Finally, we note that for each j:

$$p_j^* - \sum_{k=1}^n p_k^* a_{kj} - c_j = 0,$$

so that profits in the jth sector are zero (and thus maximized) when $z_j = z_j^*$. Therefore $\big(\langle (x_{i0}^*, \boldsymbol{x}_i^*)\rangle, \boldsymbol{z}^*, \boldsymbol{p}^*\big)$ is a Walrasian equilibrium for $\mathcal{E}$. □

It is worth noting that if we strengthen the hypotheses of this existence theorem by requiring that, in addition to the hypotheses of Theorem 8.17, each P_i is negatively transitive and strictly convex, then the Walrasian equilibrium established in our proof is unique. In the special case of the model used here which is often used in public economics literature, the aggregate production set takes the form:

$$Y = \{\boldsymbol{y} \in \mathbb{R}^n \mid \boldsymbol{c} \cdot \boldsymbol{y} \leq 0 \ \&\ y_j \geq 0, \ \text{for } j = 1, \ldots, n-1\}.$$

See Exercise 7, at the end of this chapter.

8.6 Appendix

Making use of the distinction introduced in Section 1, we have been showing that no Walrasian equilibrium exists for the economies in the examples presented in this chapter. This raises the question of whether we might have been able to find a competitive (free disposal) equilibrium in some cases. However, recall that in most of our examples, Y satisfied the semi-free disposability condition:

$$\text{if } \boldsymbol{y}, \boldsymbol{y}' \in \mathbb{R}^n \text{ are such that } \boldsymbol{y} \in Y \text{ and } \boldsymbol{y}' \leq \boldsymbol{y}, \text{ then } \boldsymbol{y}' \in Y;$$

and, in addition, in nearly all of our examples, we assumed $\boldsymbol{0} \in Y$, so that:

$$-\mathbb{R}^n_+ \subseteq Y. \tag{8.29}$$

Under these conditions, suppose $(\langle \boldsymbol{x}_i^* \rangle, \langle \boldsymbol{y}_k^* \rangle, \boldsymbol{p}^*)$ is a competitive equilibrium for the economy, but that:

$$\sum_{i=1}^{m} \boldsymbol{x}_i^* < \boldsymbol{r} + \sum_{k=1}^{\ell} \boldsymbol{y}_k^*. \tag{8.30}$$

From (8.29) and Proposition 7.8 of Chapter 7, it follows that we must have $\boldsymbol{p}^* > \boldsymbol{0}$. However, if in fact, $\boldsymbol{p}^* \gg \boldsymbol{0}$, then from (8.30) we have:

$$\boldsymbol{p}^* \cdot \left[\boldsymbol{r} + \sum_{k=1}^{\ell} \boldsymbol{y}_k^* - \sum_{i=1}^{m} \boldsymbol{x}_i^*\right] > 0;$$

contradicting the definition of a competitive equilibrium. If, on the other hand, $\boldsymbol{p}^* > \boldsymbol{0}$, but one or more $p_i^* = 0$, then it will typically be the case that the consumer(s) could not be maximizing preferences at $\boldsymbol{x}_i^*$.

The example in which it may seem most likely that a competitive (free disposal) equilibrium, as opposed to a Walrasian equilibrium, may exist is Example 8.7. If you return to Figure 8.5, you can easily verify the fact that we can find a price vector such that the producer maximizes profits at a point $\boldsymbol{y}$ satisfying:

$$\boldsymbol{r} + \boldsymbol{y} \geq \boldsymbol{x}^*. \tag{8.31}$$

However, notice that the only price vector yielding this relationship (that is, the only one in Δ_2) is $\boldsymbol{p}^*$; and, since $\boldsymbol{p}^* \gg \boldsymbol{0}$, any $\boldsymbol{y} \in Y$ and satisfying equation (8.31) is such that:

$$\boldsymbol{p}^* \cdot (\boldsymbol{r} + \boldsymbol{y} - \boldsymbol{x}^*) > 0.$$

Therefore, no competitive equilibrium exists in this case either. □

Exercises.

1. Suppose there are two commodities, and a consumer has the consumption set:

$$X_i = \{\boldsymbol{x}_i \in \mathbb{R}^2 \mid -2 \leq x_{i1} \;\&\; x_{i2} \geq 2\}.$$

Answer the following questions.

a. Show that the consumer's demand correspondence can be defined only for pairs $(\boldsymbol{p}, w) \in \mathbb{R}^3_+$ satisfying $w \geq \mu(\boldsymbol{p})$, where:

$$\mu(\boldsymbol{p}) = 2(p_2 - p_1),$$

for $\boldsymbol{p} \in \mathbb{R}^2_{++}$.

b. Given that the consumer's preferences can be represented by the utility function $u_i(\boldsymbol{x}_i) = x_{i1}$, find the consumer's demand correspondence.

c. Given that the consumer's preferences can be represented by the utility function $u_i(\boldsymbol{x}_i) = x_{i2}$, find the consumer's demand corresponcence.

d. Consider the private ownership economy, $\mathcal{E}$, in which we have one producer, two consumers and two commodities, and where X_i, $\boldsymbol{r}_i$, s_i, and Y are given by:

$$\begin{aligned} X_i &= \{\boldsymbol{x}_i \in \mathbb{R}^2 \mid -2 \leq x_{i1} \;\&\; x_{i2} \geq 2\} \text{ and } \boldsymbol{r}_i = (0,1) \quad \text{for } i = 1,2; \\ s_1 &= s_2 = 1/2, \text{ and} \\ Y &= \{\boldsymbol{y} \in \mathbb{R}^2 \mid y_1 \leq 0 \leq y_2 \;\&\; y_1 + y_2 \leq 0\}, \end{aligned}$$

respectively; and suppose that the consumers' preferences can be represented by the utility functions:

$$u_1(\boldsymbol{x}_1) = x_{11} \text{ and } u_2(\boldsymbol{x}_2) = x_{22},$$

respectively.

Find the (a) competitive equilibrium for $\mathcal{E}$, if one exists, or show that no competitive equilibrium exists for $\mathcal{E}$ in this case; and (b) answer the following question: Are all of the assumptions of Theorem 8.1 satisfied by the economy presented in this problem? If not, which assumptions are violated?

2. Follow the same sequence of questions as in problem 1 for the private ownership economy, $\mathcal{E}$ having one consumer, one producer, two commodities, and where $X, \boldsymbol{r}$ and Y are given by:

$$\begin{aligned} X &= \{\boldsymbol{x} \in \mathbb{R}^2 \mid -2 \le x_1 \;\&\; 2 \le x_2\}, \\ \boldsymbol{r} &= (0, 1), \text{ and} \\ Y &= \{\boldsymbol{y} \in \mathbb{R}^2 \mid y_1 \le 0 \le y_2 \;\&\; y_1 + 2y_2 \le 0\}, \end{aligned}$$

and where the consumer's preferences can be represented by the utility function:

$$u(\boldsymbol{x}) = \min\{4 + x_1, x_2\}.$$

3. Consider the private ownership economy, $\mathcal{E}$, in which we have one consumer, one producer, two commodities, and in which X, $\boldsymbol{r}$, and Y are given by:

$$\begin{aligned} X &= \{\boldsymbol{x} \in \mathbb{R}^2 \mid -4 \le x_1 \;\&\; x_2 \ge 4\}, \\ \boldsymbol{r} &= (0, 2), \text{ and} \\ Y &= \{\boldsymbol{y} \in \mathbb{R}^2 \mid y_1 \le 0 \;\&\; 3y_1 + y_2 \le 0\}, \end{aligned}$$

respectively; and suppose the consumer's preferences can be represented by the utility function:

$$u(\boldsymbol{x}) = \min\{2x_1 + 12, x_2\}.$$

On the basis of this information,

a. Find the consumer's demand function (correspondence) as a function of $\boldsymbol{p} \in \mathbb{R}^2_{++}$ and $w \ge 0$.

b. Find the (or a) Walrasian equilibrium for this economy, or show that no Walrasian equilibrium exists in this case.

4. Consider the private ownership economy, $\mathcal{E}$, in which we have one consumer, one producer, two commodities, and where X, $\boldsymbol{r}$, and Y are given by:

$$\begin{aligned} X &= \mathbb{R}^2_+, \\ \boldsymbol{r} &= (24, 0), \text{ and} \\ Y &= \{\boldsymbol{y} \in \mathbb{R}^2 \mid y_1 \le 0 \;\&\; y_1 + y_2 \le 0\}, \end{aligned}$$

respectively; and suppose the consumer's preferences can be represented by the utility function:

$$u(\boldsymbol{x}) = (x_1)^2 \cdot (x_2).$$

On the basis of this information, answer the following questions.

a. Find the consumer's demand function (correspondence) as a function of $\boldsymbol{p} \in \mathbb{R}^2_{++}$ and $w \geq 0$.

b. Find the (or a) Walrasian equilibrium for this economy.

c. If we interpret the first commodity as the consumer's leisure, how much labor is being offered in the Walrasian equilibrium?

5. Here we will consider a case in which we have two consumers, two commodities, and one producer. We will depart from our usual notation to denote quantities of the first commodity by 'x' (interpreted as 'leisure'), and the second (produced) commodity by 'y.' We will suppose the two consumers' preferences can be represented by the utility functions:

$$u_1(x_1, y_1) = (x_1)^{1/4} \cdot (y_1)^{3/4} \text{ and } u_2(x_2, y_2) = (x_2)^{3/4} \cdot (y_2)^{1/4},$$

respectively; and have the initial endowments:

$$\boldsymbol{r}_i = (24, 0),$$

for $i = 1, 2$. Finally, we suppose that the producer's production function is given by:

$$y = 2\sqrt{-z},$$

where 'z' denotes the aggregate labor supplied by the consumers, and we suppose that the consumers' shares of ownership in the firm are given by:

$$s_i = 1/2,$$

for $i = 1, 2$. Given this information, find the competitive equilbrium for this economy, or show that none exists.

6. In this question, we will be considering a two-consumer, two-commodity economy, E, in which the consumers' preferences can be represented by the utility functions:

$$u_i(\boldsymbol{x}_i) = x_{i1} + x_{i2} \quad \text{for } i = 1, 2,$$

with the initial endowments:

$$\boldsymbol{r}_1 = (16, 4) \text{ and } \boldsymbol{r}_2 = (16, 0),$$

respectively. On the basis of this information, answer the following two questions.

a. Suppose $Y = \{\mathbf{0}\}$; that is, that this is a pure exchange economy. Find the (or the set of) Walrasian equilibrium (or equilibria) in this case, or show that no such equilibrium exists.

b. Now suppose there is one producer, whose production set is given by:

$$Y = \{\boldsymbol{y} \in \mathbb{R}^2 \mid y_1 \leq 0 \;\&\; y_1 + 2y_2 \leq 0\},$$

and that the shares of ownership in the firm are given by:

$$s_1 = s_2 = 1/2.$$

Find the (or the set of) Walrasian equilibrium (or equilibria) in this case, or show that no such equilibrium exists.

7. Show that the production set Y defined by:

$$Y = \{\boldsymbol{y} \in \mathbb{R}^n \mid \boldsymbol{c} \cdot \boldsymbol{y} \leq 0 \;\&\; y_j \geq 0, \text{ for } j = 1, \ldots, n-1\},$$

where $\boldsymbol{c} \in \mathbb{R}^n_{++}$, is a special case of the aggregate production set specified in Theorem 8.17.

Chapter 9

Examples of General Equilibrium Analyses

9.1 Introduction

In this chapter, we are going to consider some applications of general equilibrium theory to policy analysis. While our treatment here will stop far short of the current frontiers of the related policy analyses, I hope that it will illustrate some of the flavor of such analysis, and the usefulness of general equilibrium models therein. We will begin by presenting the elements of the basic theory of 'optimal taxation.' We will first take up the theory of 'optimal commodity taxation,' and we will then take an even briefer look at the theory of 'optimal income taxation.' While neither discussion will take us very far toward the current frontier of research in the respective fields, the analysis to follow should provide a bit of insight into the role of general equilibrium theory in current policy analysis. Moreover, in the discussion of optimal income taxation, we will for the first time in this book encounter the problem of 'incentive compatibility,' an issue which will play a key role in much of our work in Chapters 16–18.

After our consideration of optimal taxation, we will examine some extensive examples incorporating monopoly, and then money in a general equilibrium model. We then conclude the chapter with an example incorporating indivisible commodities into a general equilibrium model.

9.2 Optimal Commodity Taxation: Initial Formulation

The problem which we will be examining initially is the 'efficiency aspect' of optimal commodity taxation. For this, it is customary to suppose that there is only one consumer, or that the consumption sector of the economy as a whole behaves as if it were a single consumer. In principle, this enables us to separate the efficiency aspect from equity considerations, which occur because there is more than one consumer in the economy. We will first set out the standard 'text book model' used in this literature, we will then analyze the workings of this model in the simplest case possible, and finally, we will set out a few of the conclusions reached in this literature.

The standard 'text book model' assumes one consumer, $n+1$ commodities (n produced commodities plus labor), and constant returns to scale in production, with each commodity produced with the use of labor alone.

In each industry, it is assumed that the input-requirement function is given by:

$$\ell_j = -c_j y_j, \tag{9.1}$$

where $c_j > 0$, for each j. Consequently, given the price p_j for the jth commodity, and a wage w for labor, profit-maximization at non-zero production will require that:

$$p_j y_j + w\ell_j = p_j y_j - wc_j y_j = (p_j - wc_j)y_j \equiv 0; \tag{9.2}$$

and thus, normalizing to set $w = 1$, we must have:

$$p_j = c_j \quad \text{for } j = 1, \ldots, n \tag{9.3}$$

Moreover, for future reference, notice that the aggregate production set is here given by:

$$Y = \{(y_0, \boldsymbol{y}) \in \mathbb{R}^{1+n} \mid \boldsymbol{y} \in \mathbb{R}^n_+ \ \&\ y_0 + \boldsymbol{c} \cdot \boldsymbol{y} \leq 0\} \tag{9.4}$$

(compare Example 6.27 of Chapter 6); and notice that, if $(y_0, \boldsymbol{y}) \in Y$, and $(w, \boldsymbol{p})$ satisfies (9.3) (and with $w = 1$), we have:

$$\pi(y_0, \boldsymbol{y}) = (w, \boldsymbol{p}) \cdot (y_0, \boldsymbol{y}) = wy_0 + \boldsymbol{p} \cdot \boldsymbol{y} = y_0 + \boldsymbol{c} \cdot \boldsymbol{y} \leq 0.$$

Now, the consumer can (and will) pay a positive tax on the jth commodity, and we will denote the price paid by the consumer by 'q_j,' where:

$$q_j = p_j + t_j \quad \text{for } j = 1, \ldots, n,$$

and where t_j is the tax levied on the jth commodity, for each j. Denoting the consumer's consumption bundle, generically, by '$(x_0, \boldsymbol{x})$,' where $0 \leq x_0 \leq r$ and $\boldsymbol{x} \in \mathbb{R}^n_+$, with $r > 0$ representing the consumer's endowment of leisure, and:

$$\ell = x_0 - r, \tag{9.5}$$

denoting the consumer's offer of labor; the tax revenue raised by the government, given the tax vector $\boldsymbol{t} \in \mathbb{R}^n_+$, is given by:

$$T = \boldsymbol{t} \cdot \boldsymbol{x} = \sum\nolimits_{j=1}^{n} t_j x_j. \tag{9.6}$$

Notice that our formulation allows some coordinates of $\boldsymbol{t}$ to be negative; on the other hand, the problem becomes analytically somewhat more tractable in some ways if we require that:

$$t_j \geq 0 \quad \text{for } j = 1, \ldots, n.$$

However, we will ignore this complication for the moment; coming back to it at the end of the next section.

Initially, we will suppose that the government needs to raise a given amount, R, of funds; and thus we require that:

$$T = \boldsymbol{t} \cdot \boldsymbol{x} \geq R. \tag{9.7}$$

However, we will also sometimes specialize this requirement to assume that the government uses the tax revenue to purchase a commodity bundle, $\boldsymbol{x}^g \in \mathbb{R}^n_+$ (which may be used as an input in the production of 'governmental services'); in which case the government's budget constraint becomes:

$$T = \boldsymbol{t} \cdot \boldsymbol{x} \geq \boldsymbol{p} \cdot \boldsymbol{x}^g. \tag{9.8}$$

We suppose the (representative) consumer maximizes a utility function, $u(x_0, \boldsymbol{x})$ subject to the budget constraint:

$$\boldsymbol{q} \cdot \boldsymbol{x} + wx_0 = wr;$$

or, equivalently:

$$\boldsymbol{q} \cdot \boldsymbol{x} = w(r - x_0) = -w\ell;$$

which, since we are normalizing throughout with $w = 1$, becomes:

$$\boldsymbol{q} \cdot \boldsymbol{x} = (r - x_0) = -\ell. \tag{9.9}$$

In our treatment, we will suppose that for each vector of commodity prices, $\boldsymbol{q} \in \mathbb{R}^n_{++}$,[1] there exists a unique utility-maximizing bundle, $\big(h_0(\boldsymbol{q}), \boldsymbol{h}(\boldsymbol{q})\big)$.[2] Finally, we denote the consumer's indirect utility function by '$v(\boldsymbol{q})$;' and we note that in this case, we can take $v(\cdot)$ to be given by:

$$v(\boldsymbol{q}) = u\big[\big(h_0(\boldsymbol{q}), \boldsymbol{h}(\boldsymbol{q})\big)\big].$$

Notice that there is no income term in our expression for the consumer's demand function and indirect utility function. This is because in our analysis of optimal commodity taxation, we must necessarily take the consumer's non-labor income to be zero; profits from the production sector are necessarily zero because of our linearity assumption regarding production, and we are assuming a closed general equilibrium system. Given our assumptions about the consumer, however, we can certainly define a demand function which allows income to vary, and an indirect utility function which treats income as an independent variable as well. We will denote these functions by '$\big(H_0(\boldsymbol{q}, I), \boldsymbol{H}(\boldsymbol{q}, I)\big)$' and '$V(\boldsymbol{q}, I)$,' respectively (where '$I$' denotes non-labor income). We then have the following relationships: for any $\boldsymbol{q} \in \mathbb{R}^n_{++}$:

$$\big(H_0(\boldsymbol{q}, 0), \boldsymbol{H}(\boldsymbol{q}, 0)\big) = \big(h_0(\boldsymbol{q}), \boldsymbol{h}(\boldsymbol{q})\big) \text{ and } V(\boldsymbol{q}, 0) = v(\boldsymbol{q}). \tag{9.10}$$

The usual formulation of the optimal commodity taxation problem is then to maximize the consumer's utility (with respect to $\boldsymbol{t}$), given R. Thus we can formulate the problem as:

$$\max_{w.r.t.\ \boldsymbol{t}} u\big[\big(h_0(\boldsymbol{p} + \boldsymbol{t}), \boldsymbol{h}(\boldsymbol{p} + \boldsymbol{t})\big)\big] \text{ subject to } \boldsymbol{t} \cdot \boldsymbol{h}(\boldsymbol{p} + \boldsymbol{t}) \geq R;$$

or, equivalently:

$$\max_{w.r.t.\ \boldsymbol{t}} v(\boldsymbol{p} + \boldsymbol{t}) \text{ subject to } \boldsymbol{t} \cdot \boldsymbol{h}(\boldsymbol{p} + \boldsymbol{t}) \geq R \tag{9.11}$$

While this is the normal statement of the problem which is used in the literature, the first question which I want to investigate here is 'what happened to the production constraint?' We will pursue an answer to this question in the next section.

---

[1]Because of our normalization, we are suppressing the variable w throughout our treatment.

[2]In other words, we suppose that the consumer's demand correspondence is a function.

9.3 A Reconsideration of the Problem

In the previous sections of this chapter we have been considering a one-consumer, one-producer economy with $n+1$ commodities; in this section we will generalize this model to the extent of allowing an arbitrary finite number, m, of consumers. Thus we consider a private ownership economy:

$$\mathcal{E} = \big(\langle \mathbb{R}^{1+n}_{+}, G_i\rangle, \langle \boldsymbol{r}_i\rangle, Y\big),$$

where G_i is a continuous weak order on $\mathbb{R}^{n+1}_{+}$, and $\boldsymbol{r}_i$ is of the form:

$$\boldsymbol{r}_i = r_{i0}, \boldsymbol{0}),$$

where $r_{i0} > 0$. We shall also suppose that each G_i is strictly convex; so that, for each $(w, \boldsymbol{q}) \in \mathbb{R}^{1+n}_{++}$, there exists a unique vector $(x^*_{i0}, \boldsymbol{x}^*_i) \in \mathbb{R}^{1+n}_{+}$ satisfying $wx^*_{i0} + \boldsymbol{q}\cdot\boldsymbol{x}^*_i \leq wr_{i0}$ and:

$$\big(\forall (x_0, \boldsymbol{x}) \in \mathbb{R}^{1+n}_{+}\big) \colon (x_0, \boldsymbol{x}) P_i (x^*_{i0}, \boldsymbol{x}^*_i) \Rightarrow wx_0 + \boldsymbol{q}\cdot\boldsymbol{x} > w^* r_{i0}.$$

Continuing as per the discussion in the previous section, we will always normalize to set $w = 1$; and, given $\boldsymbol{q} \in \mathbb{R}^n_{++}$, we denote the values of x^*_{i0} and $\boldsymbol{x}^*_i$ which satisfy the above conditions (with $w = 1$) by '$h_{i0}(\boldsymbol{q})$' and '$\boldsymbol{h}_i(\boldsymbol{q})$,' respectively.

9.1 Definition. We shall say that $\big(\langle (x^*_{i0}, \boldsymbol{x}^*_i)\rangle, (y^*_0, \boldsymbol{y}^*)\big)$ is **feasible for $\mathcal{E}$, given the governmental demand $\boldsymbol{x}^g \in \mathbb{R}^n_{+}$**, iff:

1. $(x^*_{i0}, \boldsymbol{x}^*_i) \in \mathbb{R}^{1+n}_{+}$, for $i = 1, \ldots, n$, $(y^*_0, \boldsymbol{y}^*) \in Y$,
2. $\sum_{i=1}^m (x^*_{i0} - r_{i0}) = y^*_0$ and $\sum_{i=1}^m \boldsymbol{x}^*_i + \boldsymbol{x}^g = \boldsymbol{y}^*$.

We then make use of this definition of feasibility to introduce the following equilibrium concept; where, as in the previous section, we normalize to set wages (the price of the 0th commodity) equal to one; and where we say that a pair $(\boldsymbol{t}, \boldsymbol{x}^g) \in \mathbb{R}^{2n}$ is a **level of governmental activity in $\mathcal{E}$** iff $\boldsymbol{x}^g \in \mathbb{R}^n_{+}$ and $\boldsymbol{t} \in \mathbb{R}^n$.

9.2 Definition. If $\boldsymbol{x}^g \in \mathbb{R}^n_{+}$ and $\boldsymbol{t} \in \mathbb{R}^n$, we shall say that $\big(\langle (x^*_{i0}, \boldsymbol{x}^*_i)\rangle, (y^*_0, \boldsymbol{y}^*), \boldsymbol{p}^*\big)$ is a **competitive equilibrium for $\mathcal{E}$, given the governmental activity $(\boldsymbol{t}, \boldsymbol{x}^g)$**, iff:

1. $\boldsymbol{p}^* \in \mathbb{R}^n_{+} \setminus \{\boldsymbol{0}\}$ and $\boldsymbol{q}^* \stackrel{\text{def}}{=} \boldsymbol{p}^* + \boldsymbol{t} \in \mathbb{R}^n_{++}$,
2. $\big(\langle (x^*_{i0}, \boldsymbol{x}^*_i)\rangle, (y^*_0, \boldsymbol{y}^*)\big)$ is feasible for $\mathcal{E}$, given $\boldsymbol{x}^g$,
3. $(y^*_0, \boldsymbol{y}^*)$ maximizes profits on Y, given $(1, \boldsymbol{p}^*)$,
4. $(x^*_{i0}, \boldsymbol{x}^*_i) = \big(h_{i0}(\boldsymbol{q}^*), \boldsymbol{h}_i(\boldsymbol{q}^*)\big)$, for $i = 1, \ldots, m$, and
5. $\boldsymbol{t}\cdot\boldsymbol{x}^* \geq \boldsymbol{p}^*\cdot\boldsymbol{x}^g$, where $\boldsymbol{x}^* = \sum_{i=1}^m \boldsymbol{x}^*_i$.

9.3 Proposition. *Suppose that Y is linear; that is, that there exists $\boldsymbol{c} \in \mathbb{R}^n_{+} \setminus \{\boldsymbol{0}\}$ such that:*

$$Y = \{(y_0, \boldsymbol{y}) \in \mathbb{R}^{1+n} \mid \boldsymbol{y} \in \mathbb{R}^n_{+} \ \&\ y_0 + \boldsymbol{c}\cdot\boldsymbol{y} \leq 0\}, \tag{9.12}$$

and that $\langle (x^*_{i0}, \boldsymbol{x}^*_i)\rangle$ and the level of governmental activity $(\boldsymbol{t}, \boldsymbol{x}^g) \in \mathbb{R}^{2n}$ satisfy:

$$\boldsymbol{c} + \boldsymbol{t} \in \mathbb{R}^n_{++}, \tag{9.13}$$

$$(x^*_{i0}, \boldsymbol{x}^*_i) = \big(h_{i0}(\boldsymbol{c}+\boldsymbol{t}), \boldsymbol{h}_i(\boldsymbol{c}+\boldsymbol{t})\big) \text{ for each } i \text{ and } \boldsymbol{t}\cdot\boldsymbol{x}^* = \boldsymbol{c}\cdot\boldsymbol{x}^g. \tag{9.14}$$

Then, defining:

$$\boldsymbol{p}^* = \boldsymbol{c}, \tag{9.15}$$

$$\boldsymbol{y}^* = \boldsymbol{x}^* + \boldsymbol{x}^g, \tag{9.16}$$

$$\ell_j = -c_j y_j^* \quad \text{for } j = 1, \ldots, n, \text{ and} \tag{9.17}$$

$$y_0^* = \sum_{j=1}^{n} \ell_j; \tag{9.18}$$

$\big(\langle(x_{i0}^*, \boldsymbol{x}_i^*)\rangle, (y_0^*, \boldsymbol{y}^*), \boldsymbol{p}^*\big)$ *is a competitive equilibrium for* $\mathcal{E}$, *given the level of governmental activity* $(\boldsymbol{t}, \boldsymbol{x}^g)$.

Proof. If $\langle(x_{i0}^*, \boldsymbol{x}_i^*)\rangle$ and the level of governmental activity $(\boldsymbol{t}, \boldsymbol{x}^g) \in \mathbb{R}^{2n}$ satisfy (9.13) and (9.14), and we define $\boldsymbol{p}^* = \boldsymbol{c}$ and $\boldsymbol{q}^* = \boldsymbol{p}^* + \boldsymbol{t}$, it follows immediately that $\big((x_0^*, \boldsymbol{x}^*), (y_0^*, \boldsymbol{y}^*), \boldsymbol{p}^*\big)$ and $(\boldsymbol{t}, \boldsymbol{x}^g)$ satisfy conditions 1, 4, and 5 of Definition 9.2. Furthermore, defining $(y_0^*, \boldsymbol{y}^*)$ as in (9.16) and (9.18), we have:

$$y_0^* + \boldsymbol{c} \cdot \boldsymbol{y}^* = \sum_{j=1}^{n} \ell_j + \boldsymbol{c} \cdot \boldsymbol{y}^* = -\boldsymbol{c} \cdot \boldsymbol{y}^* + \boldsymbol{c} \cdot \boldsymbol{y}^* = 0;$$

so that $(y_0, \boldsymbol{y}) \in Y$. Moreover, since $w = 1$ and $\boldsymbol{p}^* = \boldsymbol{c}$, we have:

$$w y_0^* + \boldsymbol{p}^* \cdot \boldsymbol{y}^* = y_0^* + \boldsymbol{c} \cdot \boldsymbol{y}^* = 0;$$

and it also follows that $(y_0^*, \boldsymbol{y}^*)$ maximizes profits over Y, given $(w, \boldsymbol{p}^*) = (1, \boldsymbol{c})$.

Now, since, for each i, $(x_{i0}^*, \boldsymbol{x}_i^*) = \big(h_{i0}(\boldsymbol{q}^*), \boldsymbol{h}_i(\boldsymbol{q}^*)\big)$, we have:

$$x_{i0}^* = r_{i0} - \boldsymbol{q}^* \cdot \boldsymbol{x}_i^* \quad \text{for } i = 1, \ldots, m.$$

Therefore:

$$\begin{aligned}\sum_{i=1}^{m} (x_{i0}^* - r_{i0}) = -\sum_{i=1} \boldsymbol{q}^* \cdot \boldsymbol{x}_i^* &= -\boldsymbol{q}^* \cdot \boldsymbol{x}^* = -\boldsymbol{c} \cdot \boldsymbol{x}^* - \boldsymbol{t} \cdot \boldsymbol{x}^* \\ &= -\boldsymbol{c} \cdot \boldsymbol{x}^* - \boldsymbol{c} \cdot \boldsymbol{x}^g = -\boldsymbol{c} \cdot \boldsymbol{y}^* = y_0^*; \quad \square\end{aligned}$$

and it now follows that $\big(\langle(x_{i0}^*, \boldsymbol{x}_i^*)\rangle, (y_0^*, \boldsymbol{y}^*), \boldsymbol{p}^*\big)$ satisfies condition 2 of Definition 9.2. □

Our next result is stated in a bit more generality than is needed; that is, it is somewhat more general than the context in which we have been working. In it, we will drop our assumption that Y is linear; which means in turn that we will have to allow for the possibility that the firm may make a profit. As usual, we will denote the maximum profit achievable in Y, given a price vector, $\boldsymbol{p} \in \mathbb{R}_+^n$ by '$\pi(\boldsymbol{p})$,' and denote the ith consumer's share of these profits by 's_i.' In our model, this profit becomes the non-labor income that we considered at the end of the previous section; and, in terms of the notation introduced there, the ith consumer's preference-maximizing commodity bundle at consumer prices $\boldsymbol{q} \in \mathbb{R}_{++}^n$ and producer prices $\boldsymbol{p} \in \mathbb{R}_+^n$ will be given by:

$$(x_{i0}, \boldsymbol{x}_i) = \big(H_{i0}[\boldsymbol{q}, s_i\pi(\boldsymbol{p})], \boldsymbol{H}_i[\boldsymbol{q}, s_i\pi(\boldsymbol{p})]\big).$$

This is the notation utilized in our next result.

9.4 Proposition. *Suppose* $(\langle\langle(x_{i0}^*, \boldsymbol{x}_i^*)\rangle\rangle, (y_0^*, \boldsymbol{y}^*), \boldsymbol{p}^*)$ *and the level of government activity* $(\boldsymbol{t}, \boldsymbol{x}^g)$ *are such that:*

1. $\boldsymbol{p}^* \in \mathbb{R}_+^n \setminus \{\boldsymbol{0}\}$ *and* $\boldsymbol{q}^* \stackrel{def}{=} \boldsymbol{p}^* + \boldsymbol{t} \in \mathbb{R}_{++}^n$,
2. $(\langle\langle(x_{i0}^*, \boldsymbol{x}_i^*)\rangle\rangle, (y_0^*, \boldsymbol{y}^*))$ *is feasible for* $\mathcal{E}$*, given* $\boldsymbol{x}^g$,
3. $(y_0^*, \boldsymbol{y}^*)$ *maximizes profits on* Y*, given* $\boldsymbol{p}^*$ *(and* $w = 1$*),*
4. $(x_{i0}^*, \boldsymbol{x}_i^*) = (H_{i0}[\boldsymbol{q}^*, s_i\pi(\boldsymbol{p}^*)], \boldsymbol{H}_i[\boldsymbol{q}^*, s_i\pi(\boldsymbol{p}^*)])$*, for* $i = 1, \ldots, m$.

Then $\boldsymbol{t} \cdot \boldsymbol{x}^* = \boldsymbol{p}^* \cdot \boldsymbol{x}^g$*, and* $(\langle\langle(x_{i0}^*, \boldsymbol{x}_i^*)\rangle\rangle, (y_0^*, \boldsymbol{y}^*), \boldsymbol{p}^*)$ *is a competitive equilibrium for* $\mathcal{E}$*, given the level of governmental activity* $(\boldsymbol{t}, \boldsymbol{x}^g)$.

Proof. Upon re-checking Definition 9.2, we see that it suffices to prove that $\boldsymbol{t} \cdot \boldsymbol{x}^* = \boldsymbol{p}^* \cdot \boldsymbol{x}^g$. To establish this equality, we make use of the definitions and conditions 1–4 to obtain:

$$\begin{aligned}\boldsymbol{t} \cdot \boldsymbol{x}^* &= (\boldsymbol{q}^* - \boldsymbol{p}^*) \cdot \boldsymbol{x}^* = \boldsymbol{q}^* \cdot \boldsymbol{x}^* - \boldsymbol{p}^* \cdot (\boldsymbol{y}^* - \boldsymbol{x}^g) \\ &= \boldsymbol{q}^* \cdot \boldsymbol{x}^* - \boldsymbol{p}^* \cdot \boldsymbol{y}^* + \boldsymbol{p}^* \cdot \boldsymbol{x}^g = \sum_{i=1}^m (r_{i0} - x_{i0}^* + s_i\pi(\boldsymbol{p}^*)) - \boldsymbol{p}^* \cdot \boldsymbol{y}^* + \boldsymbol{p}^* \cdot \boldsymbol{x}^g \\ &= -y_0^* + [y_0^* + \boldsymbol{p}^* \cdot \boldsymbol{y}^*] - \boldsymbol{p}^* \cdot \boldsymbol{y}^* + \boldsymbol{p}^* \cdot \boldsymbol{x}^g = \boldsymbol{p}^* \cdot \boldsymbol{x}^g. \quad \square\end{aligned}$$

Returning to the one consumer case, the thrust of the above two results is that we can equivalently formulate our optimal commodity taxation problem either as (going back to the assumption that Y is linear [that is, satisfies (9.12)], and setting $\boldsymbol{p} = \boldsymbol{c}$):

$$\max_{w.r.t.\ \boldsymbol{t}} v(\boldsymbol{p} + \boldsymbol{t}) \text{ subject to } \boldsymbol{t} \cdot \boldsymbol{h}(\boldsymbol{p} + \boldsymbol{t}) = \boldsymbol{c} \cdot \boldsymbol{x}^g \text{ (and } \boldsymbol{p} + \boldsymbol{t} \in \mathbb{R}_{++}^n\text{);} \tag{9.19}$$

or as:

$$\max_{w.r.t.\ \boldsymbol{t}} v(\boldsymbol{c} + \boldsymbol{t}) \text{ subject to:}$$
$$(h_0(\boldsymbol{q}) - r, \boldsymbol{h}(\boldsymbol{q}) + \boldsymbol{x}^g) \in \sigma(\boldsymbol{c}) \text{ (and } \boldsymbol{q} \equiv \boldsymbol{c} + \boldsymbol{t} \in \mathbb{R}_{++}^n\text{);} \tag{9.20}$$

which, as is easily verified, given the form of Y, is equivalent to:

$$\max_{w.r.t.\ \boldsymbol{t}} v(\boldsymbol{c}+\boldsymbol{t}) \text{ subject to: } h_0(\boldsymbol{q}) - r + \boldsymbol{c} \cdot [\boldsymbol{h}(\boldsymbol{q}) + \boldsymbol{x}^g] = 0 \text{ (\& } \boldsymbol{q} \equiv \boldsymbol{c}+\boldsymbol{t} \in \mathbb{R}_{++}^n\text{).} \tag{9.21}$$

Formally, it will sometimes be more convenient to state the problem as follows:

$$\max_{w.r.t.\ \boldsymbol{t}} v(\boldsymbol{c} + \boldsymbol{t}) \tag{9.22}$$

subject to:

$$x_0 - r + \boldsymbol{c} \cdot [\boldsymbol{x} + \boldsymbol{x}^g] = 0, \tag{9.23}$$
$$(x_0, \boldsymbol{x}) = (h_0(\boldsymbol{c} + \boldsymbol{t}), \boldsymbol{h}(\boldsymbol{c} + \boldsymbol{t})), \text{ and:} \tag{9.24}$$
$$\boldsymbol{q} \equiv \boldsymbol{c} + \boldsymbol{t} \in \mathbb{R}_{++}^n. \tag{9.25}$$

It should be apparent that the problem stated in equations (9.22)–(9.25) is equivalent to both (9.20) and (9.21). On the other hand, the longer formulation perhaps makes some aspects of the problem stand out in sharper relief. In particular, consider the last constraint, (9.25). Since we are assuming that $\boldsymbol{c} \gg \mathbf{0}$, it will automatically be satisfied if we require that:

$$\boldsymbol{t} \geq \mathbf{0}. \tag{9.26}$$

Moreover, if we substitute (9.26) for (9.25), the constraint set becomes closed (in fact, it is also bounded, and thus is compact); and therefore a solution will necessarily exist. On the other hand, using (9.25), rather than (9.26) allows us to characterize the solution via calculus, which is done in most of the literature on optimal commodity taxation. We will find the long statement of the problem useful in our discussion of the next section as well.

9.4 The Simplest Model of Optimal Commodity Taxation

In this section, we will examine the simplest special case of the optimal commodity taxation problem formulated in the two previous sections; one in which we have one consumer, one produced commodity and labor, and where the commodity is produced under conditions of constant returns to scale. The consumer will choose a bundle, $(x_0, x) \in \mathbb{R}^2_+$, where:

'x_0' denotes the quantity of leisure, and

'x' denotes the quantity of the produced good,

chosen by the consumer. We then denote the initial endowment of leisure by 'r;' and the quantity of labor supplied by the consumer is given by:

$$\ell = x_0 - r$$

(thus the quantity of labor supplied is given by a negative number).

The production set will be given by:

$$Y = \{\boldsymbol{y} \in \mathbb{R}^2 \mid 0 \leq y_1 \leq 0 \;\&\; y_0 + cy_1 \leq 0\},$$

where $c > 0$ is a constant. Thus the 'production function' is given by:

$$y_1 = -y_0/c,$$

for $y_0 \in \mathbb{R}_-$.

Now, in fact, we will always be concerned with allocations in which:

$$y_0 = \ell = x_0 - r;$$

and hereafter we will write production vectors as a pair (ℓ, y), where $\ell \in \mathbb{R}_-$, and $y \in \mathbb{R}_+$ denotes the quantity of the good being produced. Correspondingly, we can write the production set as:

$$Y = \{(\ell, y) \in \mathbb{R}^2 \mid 0 \leq y \leq -\ell/c\} = \{(\ell, y) \in \mathbb{R}^2 \mid y \geq 0 \;\&\; \ell + cy \leq 0\}.$$

We will denote the vector of prices faced by the producer by '(w, p),' where $(w, p) \in \mathbb{R}^2_+$; and we will normalize to set $w = 1$. The price vector faced by the consumer will

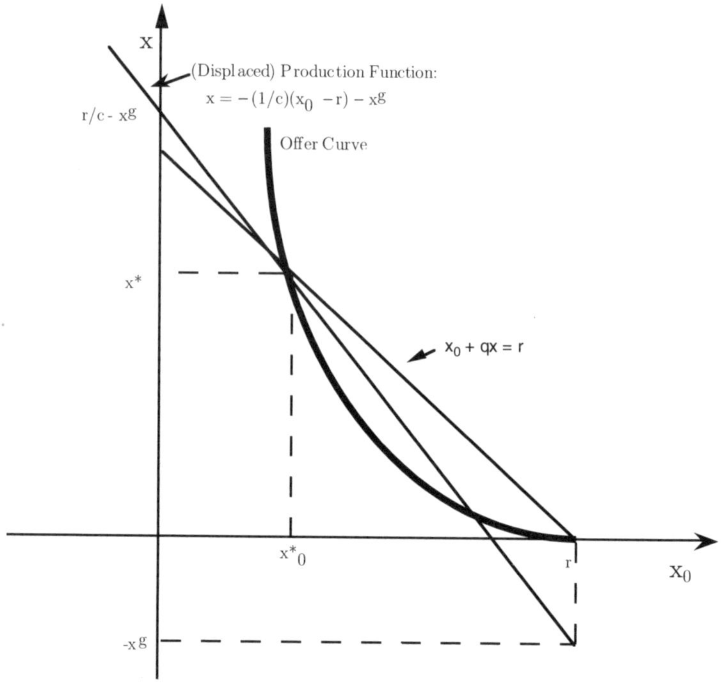

Figure 9.1: Basic Solution.

be denoted by $\boldsymbol{q} = (q_1, q_2)$; although in the present analysis we will be supposing that $q_1 = w = 1$, so that we can (and will) denote the vector of prices faced by the consumer by '$(1, q)$.' As before, we suppose that the government is to purchase an amount x^g of the produced good, and will be levying a tax, $t \in \mathbb{R}_+$ in such a way that:

$$px^g = tx = (q - p)x.$$

The basic diagrammatic solution to the problem:

$$\max_{w.r.t.\ t} v(c + t), \tag{9.27}$$

subject to:

$$x_0 - r + c \cdot (x + x^g) = 0, \tag{9.28}$$

$$(x_0, x) = \big(h_0(c + t), h(c + t)\big), \text{ and:} \tag{9.29}$$

$$c + t > 0, \tag{9.30}$$

is indicated in the diagram above.

9.5 Some Results

In this section we will derive two of the more famous results concerning optimal commodity taxation in the one-consumer (efficiency) case (for a development of [equity] results for the multiple-consumer case, see Myles [1995, pp. 108–14[). We will suppose throughout that the consumer's indirect utility function and demand functions are all continuously differentiable. We will say that $\boldsymbol{t}^* \in \mathbb{R}^n$ is an **optimal tax vector** if $\boldsymbol{t}^*$ solves the problem stated in (9.22)–(9.25), above.

9.5 Proposition. (Diamond-Mirlees [1971]) *If $\boldsymbol{t}^*$ is an optimal tax vector, then there exists $\beta \in \mathbb{R}$ such that:*

$$\sum_{j=1}^{n} \frac{\partial h_j}{\partial q_i} t_j^* = -\beta h_i(\boldsymbol{p} + \boldsymbol{t}^*) \quad \textit{for } i = 1, \ldots, n; \tag{9.31}$$

where the partial derivatives are evaluated at $\boldsymbol{p} + \boldsymbol{t}^*$.

Proof. Since $\boldsymbol{t}^*$ is optimal, it solves the problem stated as equation (9.19) in the previous section; or, equivalently, recalling the discussion in Section 1 of this chapter:

$$\max_{w.r.t.\ \boldsymbol{t}} V(\boldsymbol{p} + \boldsymbol{t}, 0) \text{ subject to } \boldsymbol{t} \cdot \boldsymbol{h}(\boldsymbol{p} + \boldsymbol{t}) = \boldsymbol{p} \cdot \boldsymbol{x}^g \text{ (and } \boldsymbol{p} + \boldsymbol{t} \in \mathbb{R}^n_{++}).$$

Forming the appropriate Lagrangian:

$$\varphi(\boldsymbol{t}, \lambda) = V(\boldsymbol{p} + \boldsymbol{t}, 0) + \lambda\big(\boldsymbol{p} \cdot \boldsymbol{x}^g - \boldsymbol{t} \cdot \boldsymbol{h}(\boldsymbol{p} + \boldsymbol{t})\big),$$

and setting the partial derivatives equal to zero, we have:

$$\frac{\partial V}{\partial q_i} - \lambda\Big(h_i(\boldsymbol{p} + \boldsymbol{t}^*) + \sum_{j=1}^{n} t_j \frac{\partial h_j}{\partial q_i}\Big) = 0 \quad \text{for } i = 1, \ldots, n,$$

with all partial derivatives evaluated at $(\boldsymbol{p} + \boldsymbol{t}^*, 0)$; from which we obtain:

$$\lambda \cdot \sum_{j=1}^{n} t_j \frac{\partial h_j}{\partial q_i} = \lambda h_i(\boldsymbol{p} + \boldsymbol{t}^*) - \frac{\partial V}{\partial q_i} \quad \text{for } i = 1, \ldots, n. \tag{9.32}$$

However, by the Antonelli-Allen-Roy conditions (see Theorem 4.28 of Chapter 4) we have, evaluating all partial derivatives at $(\boldsymbol{p} + \boldsymbol{t}^*, 0)$:

$$\frac{\partial V}{\partial q_i} = -\frac{\partial V}{\partial I} h_i(\boldsymbol{q}) \quad \text{for } i = 1, \ldots, n. \tag{9.33}$$

Defining:

$$\alpha = \frac{\partial V}{\partial I},$$

and substituting into (9.32), we obtain:

$$\lambda \cdot \sum_{j=1}^{n} t_j \frac{\partial h_j}{\partial q_i} = \lambda h_i(\boldsymbol{p} + \boldsymbol{t}^*) + \alpha \cdot h_i(\boldsymbol{p} + \boldsymbol{t}^*);$$

or, defining:

$$\beta = (\lambda + \alpha)/\lambda,$$

we obtain:

$$\sum_{j=1}^{n} t_j \frac{\partial h_j}{\partial q_i} = \beta \cdot h_i(\boldsymbol{p} + \boldsymbol{t}^*) \quad \text{for } i = 1, \ldots, n. \quad \square \tag{9.34}$$

In our next result, we make use of the compensated demand functions:

$$g_j(\boldsymbol{q}, u) \quad \text{for } j = 1, \dots, n,$$

and we recall that:

$$S_{ij}(\boldsymbol{q}, u) \stackrel{\text{def}}{=} \frac{\partial h_i}{\partial q_j} + h_j(\boldsymbol{q})\frac{\partial H_i}{\partial I} = \frac{\partial h_j}{\partial q_i} + h_i(\boldsymbol{q})\frac{\partial H_j}{\partial I} \stackrel{\text{def}}{=} S_{ji}(\boldsymbol{q}, u) \quad \text{for } i, j = 1, \dots, n; \tag{9.35}$$

and that, if we define $u^* = V(\boldsymbol{c} + \boldsymbol{t}^*, 0)$, we must have:

$$g_j(\boldsymbol{c} + \boldsymbol{t}^*, u^*) = h_j(\boldsymbol{c} + \boldsymbol{t}^*) \quad \text{for } j = 1, \dots, n. \tag{9.36}$$

9.6 Proposition. (Ramsey [1927]) *If $\boldsymbol{t}^*$ is an optimal tax vector, then there exists a real number, θ, such that:*

$$\sum_{j=1}^n t_j^* \frac{\partial g_j}{\partial q_i} = -\theta \cdot g_i(\boldsymbol{c} + \boldsymbol{t}^*, u^*) \quad \textit{for } i = 1, \dots, n; \tag{9.37}$$

where all partial derivatives are evaluated at $\boldsymbol{c} + \boldsymbol{t}^*$. Moreover, we have $\theta T^* \geq 0$; where $T^* = \boldsymbol{t}^* \cdot h(\boldsymbol{c} + \boldsymbol{t}^*) = \boldsymbol{t}^* \cdot g(\boldsymbol{c} + \boldsymbol{t}^*, u^*)$.

Proof. If we substitute from (9.35) and (9.36) into (9.34), we obtain:

$$\sum_{j=1}^n t_j^* \Big(S_{ji} - g_i(\boldsymbol{c} + \boldsymbol{t}^*, u^*)\frac{\partial H_j}{\partial I}\Big) = -(1 + \alpha/\lambda) g_i(\boldsymbol{c} + \boldsymbol{t}^*, u^*) \quad \text{for} i = 1, \dots, n. \tag{9.38}$$

Rearranging, we have:

$$\sum_{j=1}^n t_j^* S_{ji} = -\Big(1 + \alpha/\lambda + \sum_{j=1}^n t_j^* \frac{\partial H_j}{\partial I}\Big) g_i(\boldsymbol{c} + \boldsymbol{t}^*, u^*) \quad \text{for } i = 1, \dots, n; \tag{9.39}$$

which, defining:

$$\theta = 1 + \alpha/\lambda + \sum_{j=1}^n t_j^* \frac{\partial H_j}{\partial I};$$

establishes the first part of our result.

To prove the 'moreover' portion of our result, we begin by defining $t_0^* = 0$, to obtain from (9.39), making use of the symmetry of the Slutsky matrix, $[S_{ij}]$:

$$\sum_{j=0}^n S_{ij} t_j^* = -\theta \cdot g_i(\boldsymbol{c} + \boldsymbol{t}^*, u^*) \quad \text{for } i = 1, \dots, n; \tag{9.40}$$

But then, multiplying both sides of (9.40) by t_i^*, and adding over i:

$$\sum_{i=1}^n \sum_{j=0}^n S_{ij} t_i^* t_j^* = -\theta \sum_{i=1}^n g_i(\boldsymbol{c} + \boldsymbol{t}^*, u^*) t_i^*. \tag{9.41}$$

However, with $t_0^* \equiv 0$, we have:

$$\sum_{i=1}^n \sum_{j=0}^n S_{ij} t_i^* t_j^* = \sum_{i=0}^n \sum_{j=0}^n S_{ij} t_i^* t_j^*;$$

so that:

$$\sum_{i=0}^n \sum_{j=0}^n S_{ij} t_i^* t_j^* = -\theta \sum_{i=1}^n g_i(\boldsymbol{c} + \boldsymbol{t}^*, u^*) t_i^* = -\theta \cdot T^*. \tag{9.42}$$

Since $[S_{ij}]$ is negative semi-definite, our result follows. □

Going back to (9.39) of the above proof, and making use of the symmetry of $[S_{ij}]$, we obtain:

$$\frac{\sum_{j=1}^{n} S_{ij}t_j^*}{g_i(\boldsymbol{c}+\boldsymbol{t}^*, u^*)} \equiv \frac{\sum_{j=1}^{n} S_{ij}t_j^*}{x_i^*} = -\theta \quad \text{for } i = 1, \ldots, n. \tag{9.43}$$

Roughly speaking:

$$\sum\nolimits_{j=1}^{n} S_{ij}t_j^* = \sum\nolimits_{j=1}^{n} \frac{\partial g_i}{\partial q_j} t_j^* = \sum\nolimits_{j=1}^{n} \frac{\partial g_i}{\partial q_j} \Delta q_j,$$

is the total differential of the compensated demand function for the ith commodity; so that the basic interpretation of (9.43) is that the proportionate reduction in compensated demand which results from the imposition of the commodity tax scheme $\boldsymbol{t}^*$ should be the same for all commodities.

9.6 Optimal Income Taxation

The literature on this topic begins with the seminal work of Mirrlees [1971], and the model to be presented here is an adaptation of the one originally developed by him. It is typical of the framework used throughout much of the recent discussion of this topic.[3]

The fundamental assumption is that there are a finite number of consumer types in the economy; all of whom have the same (continuously differentiable) utility function, but who have differing labor productivities. We also follow the vast majority of recent articles by supposing that there are only two goods in the economy, a consumption good and labor/leisure. Thus consumer i has the utility function:

$$U_i = u(\boldsymbol{x}_i),$$

where $u\colon \mathbb{R}^2_+ \to \mathbb{R}$; and we suppose each consumer has the same initial endowment:

$$\boldsymbol{r} = (r, 0),$$

where $r > 0$. As already mentioned, however, the consumers differ in their labor productivity, with this productivity being indexed by s_i, where we assume, without loss of generality, that:

$$1 = s_1 \leq s_2 \leq \cdots \leq s_m.$$

The (aggregate) production set is assumed to be linear:

$$Y = \{\boldsymbol{y} \in \mathbb{R}^2 \mid y_1 + y_2 \leq 0 \;\&\; y_1 \leq 0\},$$

where in equilibrium (that is, with labor used in production equal to the labor offered by consumers):

$$y_1 = \sum\nolimits_{i=1}^{m} s_i(x_{i1} - r).$$

[3]For surveys of the literature on this topic, see Stiglitz [1987], Mirrlees [1986] and Auerbach and Hines [2002].

Thus efficient (equilibrium) production requires:

$$y_2 = \sum_{i=1}^{m} s_i(r - x_{i1}).$$

If we normalize to set the price of the produced good equal to one, then it follows from our previous work with linear production sets that profit maximization at positive output can only occur if w_i, the wage rate paid the ith consumer, is equal to s_i.

We suppose that government plans to purchase x_g units of the second good, to be paid for with taxes:

$$t = \sum_{i=1}^{m} t_i,$$

with t_i the tax to be levied on the ith consumer. A balanced budget for the government then requires that:

$$x_g = \sum_{i=1}^{m} t_i.$$

The utility maximization problem for the ith consumer thus becomes:

$$\max u(x_{i1}, x_{i2}) \quad \text{subject to} : x_{i2} = s_i(r - x_{i1}) - t_i.$$

The ith consumer's income (before taxes) is given by:

$$z_i = w_i(r - x_{i1}) = s_i(r - x_{i1}).$$

The standard assumption in this literature is that government can observe z_i, for each i, but cannot observe either x_{i1} or s_i. We will not attempt a full-scale analysis of this problem here, but will be content to consider the case in which there are only two consumer types, with m_h consumers of each type, where

$$m_1 + m_2 = m, \text{ and } m_h \geq 1, \text{ for } h = 1, 2.$$

Many, if not most of the key issues raised in the recent literature arise even in this simple context.

In fact, we will begin by considering a particular example which illustrates many of the basic issues.[4] In our example, we will suppose that the consumers' utility function is additively separable; taking the form:

$$u(\boldsymbol{x}_i) = \phi(x_{i1}) + \psi(x_{i2}), \tag{9.44}$$

where, for all $\boldsymbol{x}_i \in \mathbb{R}^2_+$:

$$\phi'(x_{i1}) > 0, \psi'(x_{i2}) > 0, \phi''(x_{i1}) < 0 \text{ and } \psi''(x_{i2}) < 0. \tag{9.45}$$

Now, if all consumers of a given type face the same price for x_2, and are paid the same wage, they will each make the same consumption choice (and labor offer) as every other consumer of the same type. Moreover, as per the general assumptions

[4]This example is an adaptation of one preseented in Stiglitz [1987].

stated earlier, we will assume that the marginal products of labor are constant, and equal to 1 and $s > 1$, for the two types. Thus the production function is given by:[5]

$$y = m_1(r - x_{11}) + m_2 s(r - x_{21}) \stackrel{\text{def}}{=} m_1\ell_1 + m_2 s\ell_2, \tag{9.46}$$

where x_{i1} denotes the consumption of leisure by type i; with equality of supply and demand requiring that:

$$y = m_1 x_{12} + m_2 x_{22} + x_g. \tag{9.47}$$

However, if the government's budget balance condition:

$$m_1 t_1 + m_2 t_2 = x_g, \tag{9.48}$$

is satisfied, and if each consumer is paid the value of her/his marginal product, we will have:

$$\begin{aligned} m_1 x_{12} + m_2 x_{22} &= m_1\big[(r - x_{11}) - t_1\big] + m_2\big[s(r - x_{21}) - t_2\big] \\ &= m_1\ell_1 + m_2 s\ell_2 - m_1 t_1 - m_2 t_2 = m_1\ell_1 + m_2 s\ell_2 - x_g; \end{aligned} \tag{9.49}$$

and thus:

$$m_1 x_{12} + m_2 x_{22} + x_g = m_1\ell_1 + m_2 s\ell_2.$$

We will suppose that the government (IRS, or whatever; hereafter we will refer to this entity as 'the policy-maker') wishes to maximize the sum of utilities (which, as we will see in Chapter 15, amounts to maximizing a utilitarian social welfare function). Consequently, given price-taking behavior by the consumers, the policy-maker's optimization problem reduces to the following:

$$\max_{w.r.t.\ x_{11}, x_{21}, t_1, t_2} m_1\big[\phi(x_{11}) + \psi(r - x_{11} - t_1)\big] + m_2\big[\phi(x_{21}) + \psi[s(r - x_{21}) - t_2]\big], \tag{9.50}$$

subject to:

$$m_1 t_1 + m_2 t_2 - x_g = 0. \tag{9.51}$$

If we write out the standard Lagrangian function, and take first-order conditions, we see that the optimal values must satisfy:

$$m_1\big[\phi'(x_{11}^*) - \psi'(r - x_{11}^* - t_1^*)\big] = 0 \tag{9.52}$$

$$m_2\big[\phi'(x_{21}^*) - s\psi'[(s(r - x_{21}^*) - t_2^*]\big] = 0 \tag{9.53}$$

$$-m_1\psi'(r - x_{11}^* - t_1^*) + m_1\lambda = 0, \tag{9.54}$$

and:

$$-m_2\psi'[s(r - x_{21}^*) - t_2^*] + m_2\lambda = 0 \tag{9.55}$$

From (9.54) and (9.55), we have:

$$\psi'[s(r - x_{21}^*) - t_2^*] = \psi'(r - x_{11}^* - t_1^*);$$

[5] In this discussion, we will simplify our notation by denoting the production sector's output of the produced good by 'y,' rather than 'y_2.'

which, since $\psi'' < 0$ everywhere, implies:

$$x_2^* \stackrel{\text{def}}{=} s(r - x_{21}^*) - t_2^* = r - x_{11}^* - t_1^*; \tag{9.56}$$

and thus all consumers have the same consumption of good 2 at the optimum. We then have from (9.52) and (9.53) that:

$$\phi'(x_{11}^*) = \psi'(x_2^*), \tag{9.57}$$

and:

$$\phi'(x_{21}^*) = s\psi'(x_2^*), \tag{9.58}$$

respectively. Consequently, it follows that:

$$\phi'(x_{21}^*) = s\phi'(x_{11}^*) > \phi'(x_{11}^*).$$

But, since $\phi'' < 0$, this means that:

$$x_{21}^* < x_{11}^*;$$

which, in turn implies that:

$$u_2(\boldsymbol{x}_2^*) = \phi(x_{21}^*) + \psi(x_2^*) < \phi(x_{11}^*) + \psi(x_2^*) = u(\boldsymbol{x}_1^*);$$

and thus that consumers of type 2 are worse off at the optimum than are consumers of type 1.

Now, suppose for the moment that the policy-maker is able to observe a consumer's type. In this event, the policy-maker could simply levy a tax of t_i^* on each consumer of type i ($i = 1, 2$). Then, for example, consumers of type 2 would seek to:

$$\max_{w.r.t.\ x_{21}} \phi(x_{21}) + \psi[s(r - x_{21}) - t_2^*];$$

which we see, upon taking derivatives, would result in a choice of x_{21}, call it '$\widehat{x}_{21}$,' satisfying:

$$\phi'(\widehat{x}_{21}) = s\psi'[s(r - \widehat{x}_{21}) - t_2^*]. \tag{9.59}$$

Diagrammatically, the consumer's solution would look like the situation illustrated in Figure 9.2, on the next page. However, given our assumptions, there is a unique point of intersection of the curves $\phi'(x_{21})$ and $s\psi'[s(r - x_{21}) - t_2^*]$. Consequently, if we return to (9.58), we see that $\widehat{x}_{21} = x_{21}^*$; that is, the consumers' choices, once their tax was announced, would coincide with the optimal quantities.

Unfortunately for the policy-maker, however, in our scenario we are supposing that, while the policy-maker can observe consumers' incomes, it cannot observe either consumers' types or their choice of leisure ($\widehat{x}_{i1}$). Consequently, if, for example, the policy-maker were to announce the tax rule:

$$t = \begin{cases} t_1 & \text{if you are type 1} \\ t_2 & \text{if you are type 2,} \end{cases} \tag{9.60}$$

consumers of type 2 would have every incentive to lie, and claim to be of type 1. Notice that this is true even if a large penalty is assessed for lying, assuming that

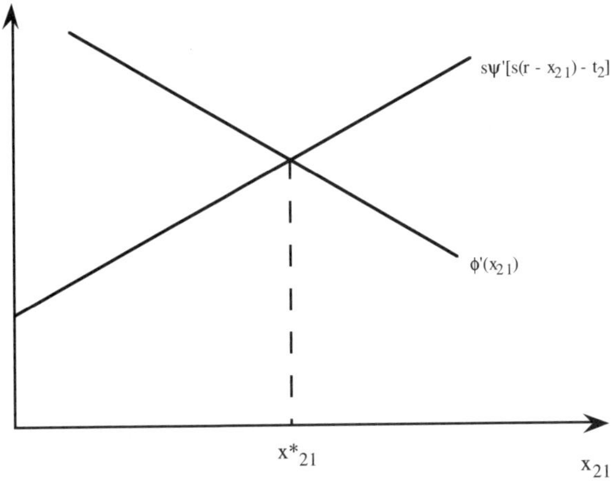

Figure 9.2: Type 2 Consumers' Optimum.

the policy-maker can only observe income and tax paid; for, as you can easily prove, consumers of type 2 are better off setting:

$$x_{21} = \Big(\frac{s-1}{s}\Big)r + \frac{x^*_{11}}{s},$$

thereby accepting the income and consumption level of good 2 enjoyed by consumers of type 1, rather than admitting to be of type 2, and paying the higher tax, t^*_2. Thus, as we say, the mechanism being employed here is not *incentive-compatible*; consumers of type 2 will find it preferable to deny being of type 2. This is, of course, disastrous for the policy-maker's plans, for the situation which will result will not yield a balanced budget, and will not maximize the sum of utilities.

Now let's generalize this example a bit. We will retain the assumption that there are only two types of consumers; however, we will drop the assumption that their common utility function is additively separable. We will suppose instead that $u(\cdot)$ is continuously differentiable, satisfies, for all $\boldsymbol{x} \in \mathbb{R}^2_+$:

$$u_1(\boldsymbol{x}) \stackrel{\text{def}}{=} \frac{\partial u}{\partial x_1}\Big|_{\boldsymbol{x}} > 0, u_2(\boldsymbol{x}) \stackrel{\text{def}}{=} \frac{\partial u}{\partial x_2}\Big|_{\boldsymbol{x}} > 0,$$

and that $u(\cdot)$ is strictly quasi-concave. We will also suppose that $u(\cdot)$ satisfies a fourth assumption, but this last condition requires a little explanation.

First of all, in our remaining discussion, it will be convenient to concentrate our attention upon the income-x_2-space. However, in dealing with this space, let's modify our previous notation slightly to use the generic notation '(z, x)' to denote points in this space, where 'x' denotes the quantity of the produced good. In this space the utility-maximization problem faced by a consumer having an index of labor productivity s, and facing a tax of t can be expressed as:

$$\max U(z, x) \stackrel{\text{def}}{=} u(r - z/s, x), \tag{9.61}$$

subject to:

$$x = z - t. \tag{9.62}$$

If we form the appropriate Lagrangian expression and take first-order conditions, we find that the optimizing values, (z^*, x^*) must satisfy:

$$\frac{u_1(r - z^*/s, x^*)}{su_2(r - z^*/s, x^*)} = 1 \tag{9.63}$$

Since it is easily shown that the slope of the indifference curve through an arbitrary point (z, x) is given by:

$$-\frac{U_1(z,x)}{U_2(z,x)} = \frac{u_1(r - z/s, x)}{su_2(r - z/s, x)}, \tag{9.64}$$

it follows that the slope of the indfference curve must equal one at the optimal point, (z^*, x^*). This brings us to our fourth assumption, a condition which is standard in the optimal income taxation literature, and which is called 'agent monotonicity.'

9.7 Definition. The utility function $u(\cdot)$ satisfies **agent monotonicity** iff the marginal rate of substitution;

$$-\frac{U_1(z,x)}{U_2(z,x)} = \frac{u_1(r - z/s, x)}{su_2(r - z/s, x)},$$

is a decreasing function of s.

In other words, the consumer's indifference curves in (z, x)-space will be flatter, the higher is the agent's productivity index. Suppose, for example, that:

$$u(x_0, x) = (x_0)^a (x)^{1-a},$$

for some real number, a, satisfying:

$$0 < a < 1,$$

and where we are using 'x_0' and 'x' to denote the quantities of leisure and the produced good, respectively. Then, as you can easily verify:

$$-\frac{U_1(z,x)}{U_2(z,x)} = \frac{u_1(r - z/s, x)}{su_2(r - z/s, x)} = \frac{ax}{(1-a)(sr - z)};$$

which is obviously decreasing in s. Therefore, agent monotonicity is satisfied in the Cobb-Douglas case.

Suppose, that the policy-maker has calculated optimal values, $x_{10}^*, x_{20}^*, x_1^*, x_2^*, t_1^*$, and t_2^*. It can then calculate the optimal (pre-tax) income values:

$$z_1^* = r - x_{10}^* \text{ and } z_2^* = s(r - x_{20}^*),$$

and we will suppose that:[6]

$$t_1^* < t_2^* \;\&\; z_1^* < z_2^*. \tag{9.65}$$

[6] We will return to a discussion of this assumption later.

Suppose our policy-maker now chooses any number (income), $z^\dagger$, satisfying:

$$z_1^* < z^\dagger < z_2^*, \tag{9.66}$$

and then attempts to implement the following tax schedule:

$$t = \begin{cases} t_1^* & \text{for } 0 \leq z \leq z^\dagger \\ t_2^* & \text{for } z^\dagger < z. \end{cases} \tag{9.67}$$

In order to analyze the consumers' choices here, let's begin by considering the optimization problem for a consumer of type 1, and supposing for the moment, that the tax schedule is simply $t = t_1$, for all z. In this case consumers of type one face the optimization problem:

$$\max_{w.r.t.\ z_1,\ x_1} u(r - z_1, x_1) \text{ subject to: } x_1 = z_1 - t_1.$$

This yields the first-order necessary conditions:

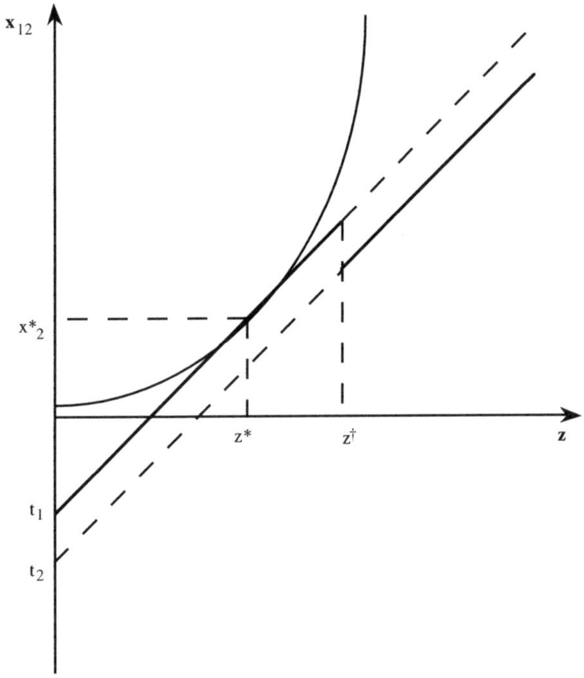

Figure 9.3: Type 1 Consumers' Optimum.

$$\frac{u_1(r - z_1^*, x_1^*)}{u_2(r - z_1^*, x_1^*)} = 1, \tag{9.68}$$

where 'z_1^*' and 'x_1^*' denote the optimal quantities of income and consumption of the produced good, respectively. Recalling that the left-hand fraction is the slope of the indifference curve through (z_1^*, x_1^*), we see that the utility-maximizing solution looks like that shown in Figure 9.3, on the previous page.

As the diagram makes clear, this is the optimizing value for consumers of type one; that is, we do not have to consider the portion of the consumption schedule corresponding to $z > z^\dagger$ in determining the consumption-income choice for consumers of type one. However, what about consumers of type 2? By the assumption of agent monotonicity, consumers of type 2 will have an indifference curve through (z_1^*, x_1^*) which looks something like that shown in Figure 9.4. Consequently, the difference between the tax levied on the productive group (t_2) and that levied on the lower productivity group (t_1) can be no greater than that indicated in Figure 9.4, since a higher value for $t_2 - t_1$ (which would move the t_2-schedule downward parallel to itself) would result in the consumers of type 2 achieving a higher utility by mimicking the behavior of type 1 consumers, than they would by achieving the higher income associated with t_2. In fact, of course, with the tax differential and schedules shown in Figure 9.4, consumers of type 2 would maximize utility by setting $z_2 = z^\dagger$. The policy-maker can correct for this by charging a tax of t_2 for all persons with an income greater than z_1^*; however, this amounts to choosing the tax schedule:

$$t = \begin{cases} t_1 & \text{for } 0 \le z \le z_1^*, \\ t_2 & \text{for } z_1^* < z; \end{cases} \tag{9.69}$$

which would leave consumers of type 2 indifferent between z_1^* and z_2^*. Moreover, the policy maker can set this schedule only if she knows the exact value of z_1^*.

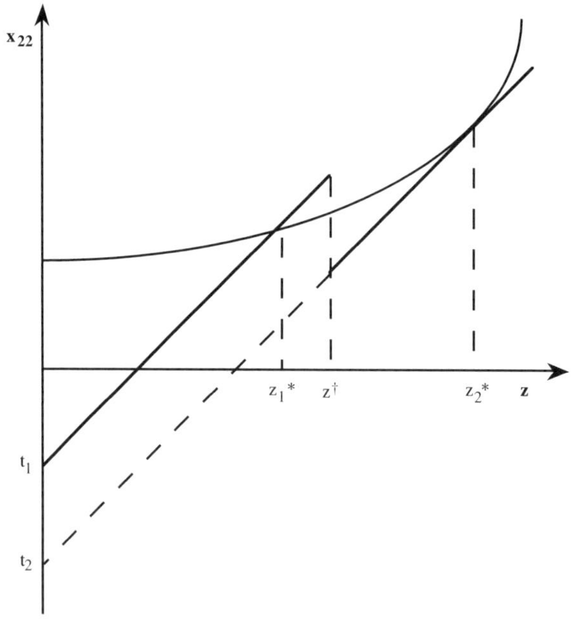

Figure 9.4: Limiting Tax Value for Consumer 2.

Let's see if we can extend this analysis a bit. In our example, we supposed that the policy-maker wished to maximize the sum of the consumers' utilities. As one would expect, however, maximizing according to a different objective function will yield different optimal consumption and tax rates for the two consumer types. Let's take a look at what sorts of solutions will be self-enforcing, in the sense that, if the policy-maker determines an optimal tax and income, (t_i^*, z_i^*), for consumers of type i, an announcement of an appropriate tax schedule will lead to consumers maximizing utility at the policy-maker's optimal values. If we go back to take another look at Figure 9.3, it becomes apparent that the policy-maker's tax choices must satisfy $t_2^* \geq t_1^*$; since in the opposite case, consumers of type 1 would maximize utility at some point along the type 2 schedule. Thus, whatever the objective function which the policy-maker is attempting to maximize, the tax for the type 2 consumers needs to be at least as high as for type one consumers if the solution is to be consistent with utility maximization; that is, if it is to be 'incentive compatible.'

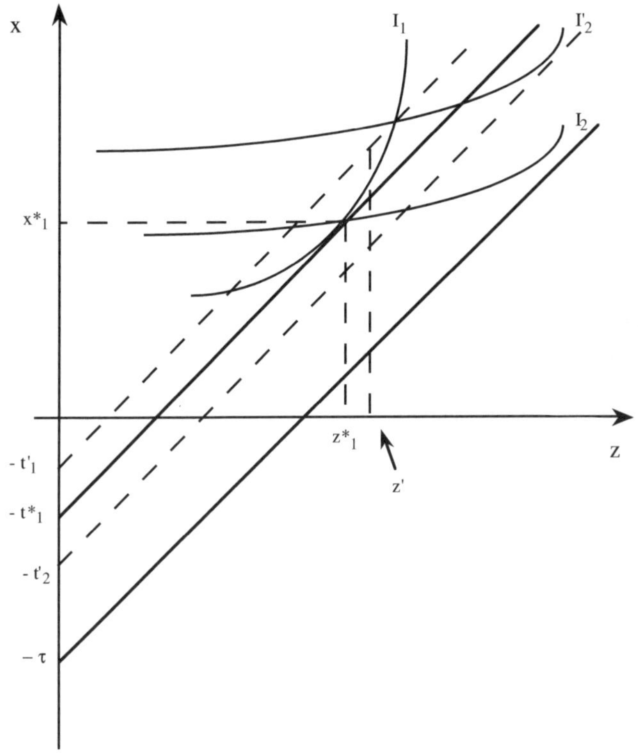

Figure 9.5: A Pareto Inefficient Solution.

The next question is, what must be true of z_2^* *vis-a-vis* z_1^*? A moment's study of Figure 9.3 should suffice to convince you that, given that the utility function satisfies agent monotonicity (and since we must have $t_2^* \geq t_1^*$), the policy-maker's choice of an optimal income for type 2 consumers must be at least as high as for type one consumers. In fact, consider the limiting case in which the policy-maker's choice of (t_1^*, z_1^*) and (t_2^*, z_2^*) satisfy:

$$t_1^* = t_2^* \text{ and } z_1^* = z_2^*.$$

Then a tax schedule of the form:

$$T = \begin{cases} t_1^* & \text{for } 0 \leq z \leq z_1^*, \text{ and} \\ \tau & \text{for } z_1^* < z, \end{cases} \tag{9.70}$$

with τ as indicated in Figure 9.5 will induce the consumers to choose the income-consumption pairs which the policy-maker views to be optimal. *However, the resulting situation cannot be optimal for any objective function which is an increasing function of consumers' utilities*! To see this, notice that the tax schedule indicated will result in that utility for type 1 consumers corresponding to the the indifference curve I_1, while type 2 consumers will achieve the utility associated with indifference curve I_2. However, if the tax schedule:

$$T = \begin{cases} t_1' & \text{for } 0 \leq z \leq z', \text{ and} \\ t_2' & \text{for } z' < z, \end{cases} \tag{9.71}$$

is instituted instead, the same tax revenue will be raised, and both consumers will be better off!

The perceptive reader may have noticed from the outset that the type of taxation we have been analyzing in this section is not a conventional sort of income tax at all, but is effectively a lump-sum tax. That is, in our discussion we have, for all practical purposes, been assuming that the policy-maker wishes to assess a tax of t_i on consumers of type i. What happens if we instead consider a tax schedule more typical of that used in actual practice? In particular, suppose we consider the simplest sort of income tax schedule; one defined by the function:

$$T = tz,$$

where $0 < t < 1$.[7]

In this case, the consumers' consumption schedule (for the produced good) takes the form:

$$x = z - T = (1 - t)z.$$

Consequently, as you can easily demonstrate, consumers of the two types will choose incomes z_1^* and z_2^* satisfying:

$$-\frac{U_1}{U_2} = \frac{u_1[r - z_1^*, (1 - t)z_1^*]}{u_2[r - z_1^*, (1 - t)z_1^*]} = 1 - t,$$

and:

$$-\frac{U_1}{U_2} = \frac{u_1[r - z_2^*/x, (1 - t)z_2^*]}{su_2[r - z_2^*/s, (1 - t)z_1^* s]} = 1 - t,$$

[7]For discussion of the 'optimal' tax rates for schedules of this type, see Hellwig[1986] and Strawczynski [1998].

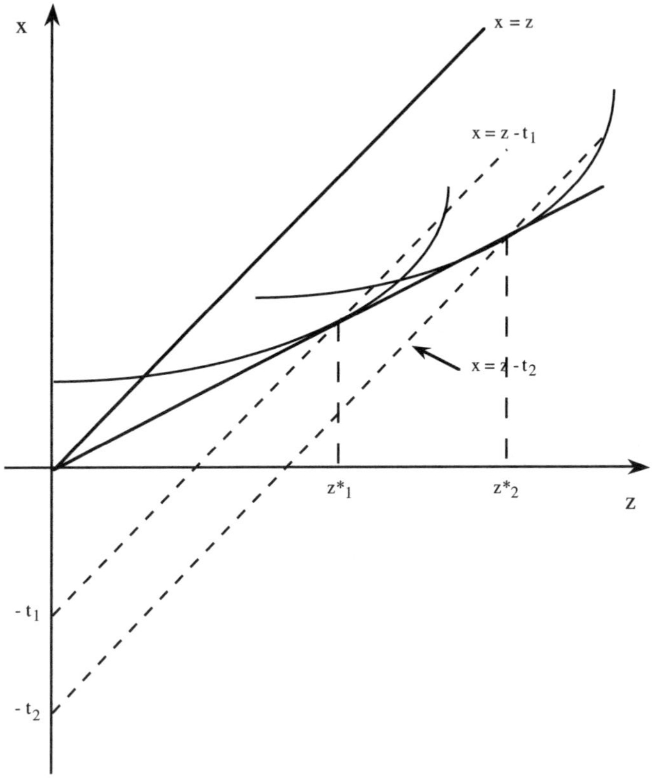

Figure 9.6: Inefficiency of the Conventional Tax Schedule.

respectively. Thus, utility-maximization by the two consumers will result in the sort of situation depicted in Figure 9.6, above. However, the solution, while simple and incentive-compatible, is not efficient. As indicated in the diagram, the same tax revenue would be raised if consumers of type 1 were to pay a tax of t_1, while consumers of type 2 pay a tax of t_2 (I will leave the details of the reasoning to you). Unfortunately, of course, this second solution may not be implementable, and certainly will not be practicable!

9.7 Monopoly in a General Equilibrium Model

In this section we'll look at a very simple general equilibrium model with three commodities (two produced goods and leisure), two consumers, and two firms. My goal here is to develop a bit different perspective as regards the 'First Fundamental Theorem,' and the second-best optimality of the optimal commodity tax solution. Consumer (agent) one ('the worker') will be assumed to have preferences representable by the utility function:

$$u_1(\boldsymbol{x}_1) = (x_{11}) \cdot (x_{12} + 1)(x_{13}),$$

while consumer (agent) two ('the capitalist') has preferences representable by:

$$u_2(\boldsymbol{x}_2) = x_{21} \cdot x_{23}.$$

Thus, consumer two cares nothing for the second commodity. We will suppose that consumer one faces the usual budget constraint:

$$p_1 x_{11} + p_2 x_{12} + w x_{13} \leq w r_1, \text{ and } 0 \leq x_{13} \leq r_1,$$

where we have denoted the price of the third commodity by 'w,' to suggest 'wages,' and 'r_1' denotes the consumer's initial endowment of leisure. Consumer 2 ('the capitalist'), on the other hand, has the budget constraint:

$$p_1 x_{21} \leq \pi_2, \text{ and } 0 \leq x_{23} \leq r_2,$$

where 'π_2' denotes the profits of the second firm, and r_2 is the consumer's endowment of leisure.

I will leave to you the task of showing that, in the absence of government intervention, with leisure chosen as the numéraire ($w = 1$), and setting:

$$r_1 = r_2 = 32,$$

the consumers' demand functions are given by (see Exercise 5, at the end of Chapter 4):

$$x_{11} = \frac{32 + p_2}{3p_1}, x_{12} = \frac{32 + p_2}{3p_2} - 1, x_{13} = \frac{32 + p_2}{3}, \tag{9.72}$$

and:

$$x_{21} = \frac{\pi_2}{p_1}. \tag{9.73}$$

Finally, we will suppose that both production technologies are linear, with labor requirement functions given by:

$$\ell_j = x_j \quad \text{for } j = 1, 2. \tag{9.74}$$

We will first consider the case in which Firm 2 behaves as a monopolist; while Firm 1 behaves as a price-taker (setting price equal to marginal cost).

Firm 2's profit function is given by:

$$\pi_2 = \frac{32 + p_2}{3} - p_2 - \frac{32 + p_2}{3p_2} + 1,$$

and we see that profits (and consumer 2's utility) are maximized when $p_2 = 4$, resulting in $\pi_2 = 6$. Since Firm 2 sets price equal to marginal cost, we must have (given that $w = 1$):

$$p_1 = 1;$$

and thus:

$$x_1 = x_{11} + x_{21} = \frac{36}{3} + 6 = 18,$$
$$x_{12} = \frac{36}{12} - 1 = 2,$$
$$x_{13} = \frac{36}{3} = 12$$
$$\ell_1 + \ell_2 = x_1 + x_2 = 20,$$

so that:

$$r_1 - x_{13} = 32 - 12 = 20 = \ell_1 + \ell_2.$$

Thus, we have found an equilibrium for the economy, and in this equilibrium the consumers' utilities are given by:

$$u_1 = 12 \times 3 \times 12 = 432, \tag{9.75}$$

and:

$$u_2 = 32 \times 6 = 192. \tag{9.76}$$

Now suppose government regulates the monopolist; requiring that p_2 = marginal cost = 1, while compensating consumer 2, and paying for this compensation by taxing consumer one's consumption of the first commodity with a tax of t per unit. Then consumer one pays a price of $1+t$ per unit of good one, while consumer 2 pays a price of 1 per unit. In this case, the demand for the first commodity becomes:

$$x_1 = x_{11} + x_{21} = \frac{32 + p_2}{3(p_1 + t)} + \frac{t \cdot x_{11}}{p_1} = \frac{11}{1+t}(1+t) = 11. \tag{9.77}$$

In order that agent two achieve her/his previous utility level, we must have $x_{21} = 6$, or:

$$t \cdot x_{11}/p_1 = t\left(\frac{11}{1+t}\right) = 6;$$

so that $t = 6/5$. We then have:

$$x_{11} = 5, \; x_{12} = 10, \text{ and } x_{13} = 11.$$

I will leave it to you to verify that this is indeed an equilibrium, and that the first consumer's utility is $u_1 = 550$; which is considerably higher than the utility of 432 which the first consumer achieved in the monopoly situation. Thus the second-best situation with a commodity tax and subsidy strictly Pareto dominates the original unregulated monopoly equilibrium. As a matter of fact, it can easily be shown (although I will leave it as an exercise) that if government were to set $t = 7/4$, then the consumers' utilities at the new equilibrium are:

$$u_1 = 440 \text{ and } u_2 = 224;$$

so that both consumers are better off than in the monopoly situation!

9.8 Money in a General Equilibrium Model

It is often postulated that individual's have a 'transactions demand' for money balances. Typically this would mean that, for each i, there exists $\alpha_i \in]0, 1[$ such that $i's$ demand for money balances is related to commodity demands by the equation

$$m_i = \alpha_i \boldsymbol{p} \cdot \boldsymbol{x}_i \quad \text{for } i = 1, \ldots, I; \tag{9.78}$$

where $\boldsymbol{x}_i$ is the vector of consumer i's commodity demands. In this section we will consider two questions. First, is (9.78) consistent with utility maximization? If so, what sort of utility function yields the relation set out in (9.78)? Secondly, what are some of the implications of (9.78) for the economy as a whole?

Turning our attention to the first question, suppose the ith consumer's utility function takes the form:

$$u_i(\boldsymbol{x}_i, m_i) = \phi^i(\boldsymbol{x}_i) \cdot (m_i)^{\alpha_i}, \tag{9.79}$$

where $\alpha_i \in]0, 1[$, and $\phi^i \colon \mathbb{R}^n_+ \to \mathbb{R}_+$ is an increasing function which is continuously differentiable and positively homogeneous of degree one, and 'm_i'denotes agent i's desired money balance, or the quantity of gold to be held, if you prefer. We suppose that the ith consumer maximizes utility subject to:

$$\sum_{j=1}^{n} p_j x_{ij} + m_i = \boldsymbol{p} \cdot \boldsymbol{x}_i + m_i = w_i(\boldsymbol{p}) + m_{i0}, \tag{9.80}$$

where, as usual 'w_i' demotes the ith consumer's wealth, and 'm_{i0}' denotes its initial money (or gold) balance. [The assumption that ϕ^i is increasing allows us to use an equality sign, rather than inequality in (9.80).] We will also suppose that $\boldsymbol{w} \colon \mathbb{R}^n_+ \to \mathbb{R}^I_+$ is a feasible wealth-assignment function.

Now, suppose that $(\boldsymbol{x}_i^*, m_i^*)$ is the $(n+1)$-tuple which maximizes i's utility, given $(\boldsymbol{p}^*, m_{i0}^*)$. If we form the relevant Lagrangian for i's utility maximization problem, we see that there must exist a scalar, μ such that at $(\boldsymbol{x}_i^*, m_i^*)$ we will have:

$$\phi^i_j(\boldsymbol{x}_i^*) \cdot (m_i^*)^{\alpha_i} = \mu p_j^* \quad \text{for } j = 1, \ldots, n, \tag{9.81}$$

and:

$$\alpha_i \phi^i(\boldsymbol{x}_i^*)(m_i^*)^{\alpha_i - 1} = \mu. \tag{9.82}$$

Multiplying each of the equations in (9.81) by the corresponding x_{ij}^* and adding, we have:

$$(m_i^*)^{\alpha_i} \sum_{j=1}^{n} x_{ij}^* \phi^i_j(\boldsymbol{x}_i^*) = \mu \boldsymbol{p}^* \cdot \boldsymbol{x}_i^*. \tag{9.83}$$

Substituting (9.82) into (9.83), we then obtain:

$$(m_i^*)^{\alpha_i} \sum_{j=1}^{n} x_{ij}^* \phi^i_j(\boldsymbol{x}_i^*) = \alpha_i \phi^i(\boldsymbol{x}_i^*)(m_i^*)^{\alpha_i - 1} \boldsymbol{p}^* \cdot \boldsymbol{x}_i^*,$$

or:

$$m_i^* \Big[\sum_{j=1}^{n} x_{ij}^* \phi^i_j(\boldsymbol{x}_i^*) \Big] = \alpha_i \phi^i(\boldsymbol{x}_i^*) \boldsymbol{p}^* \cdot \boldsymbol{x}_i^*. \tag{9.84}$$

However, by Euler's theorem, we have:

$$\sum_{j=1}^{n} x_{ij}^* \phi_j^i(\boldsymbol{x}_i^*) = \phi^i(\boldsymbol{x}_i^*);$$

and thus we obtain from (9.84):

$$m_i^* = \alpha_i \boldsymbol{p}^* \cdot \boldsymbol{x}_i^*. \tag{9.85}$$

Thus, transactions demand for money of the form (9.78) is consistent with maximization of a utility function of the form (9.79). However, in this case, we can express the transactions demand for money balances in a somewhat more useful way as follows. From the budget constraint and (9.85), we have:

$$m_i = \alpha_i[w_i(\boldsymbol{p}) + m_{i0} - m_i];$$

so that:

$$m_i = \Big(\frac{\alpha_i}{1+\alpha_i}\Big)\big[w_i(\boldsymbol{p}) + m_{i0}\big],$$

or:

$$m_i = \beta_i\big[w_i(\boldsymbol{p}) + m_{i0}\big], \tag{9.86}$$

where we have defined β_i as:

$$\beta_i = \frac{\alpha_i}{1+\alpha_i}. \tag{9.87}$$

Notice that we necessarily have $0 < \beta_i < 1$.

Now suppose we have an equilibrium at $\big(\langle(\boldsymbol{x}_i^*, m_i^*)\rangle, \boldsymbol{p}^*, \boldsymbol{m}_0\big)$. Then we have:

$$\begin{aligned} \sum_{i=1}^{I} \boldsymbol{x}_i^* &= \sum_{k=1}^{K} \sigma_k(\boldsymbol{p}^*), \\ \sum_{i=1}^{I} m_i^* &= M \\ \sum_{i=1}^{I} w_i(\boldsymbol{p}^*) &= \sum_{k=1}^{K} \boldsymbol{p}^* \cdot \sigma_k(\boldsymbol{p}^*); \end{aligned} \tag{9.88}$$

where σ_k is the kth firm's supply correspondence, 'M' denotes the aggregate money supply, and:

$$\boldsymbol{m}_0 = (m_{10}, \ldots, m_{i0}, \ldots, m_{I0}).$$

Notice that it follows from the last relation in (9.88) that $w_i(\cdot)$ is positively homogeneous of degree one, for each i. For future reference, we also note that, adding the budget constraints over i:

$$\sum_{i=1}^{I} \boldsymbol{p}^* \cdot \boldsymbol{x}_i^* + \sum_{i=1}^{I} m_i^* = \sum_{i=1}^{I} w_i(\boldsymbol{p}^*) + \sum_{i=1}^{I} m_{i0}, \tag{9.89}$$

and using the last equation in (9.88), we have:

$$\sum_{i=1}^{I} w_i(\boldsymbol{p}^*) + \sum_{i=1}^{I} m_{i0} = \sum_{k=1}^{K} \boldsymbol{p}^* \cdot \boldsymbol{\sigma}_k(\boldsymbol{p}^*) + \sum_{i=1}^{I} m_{i0} \tag{9.90}$$

Furthermore, from the first and second equations in (9.88), we have:

$$\sum_{i=1}^{I} \boldsymbol{p}^* \cdot \boldsymbol{x}_i^* + \sum_{i=1}^{I} m_i^* = \sum_{k=1}^{K} \boldsymbol{p}^* \cdot \boldsymbol{\sigma}_k(\boldsymbol{p}^*) + M. \tag{9.91}$$

Combining (9.89)–(9.91), we have, therefore:

$$\sum\nolimits_{i=1}^{I} m_{i0} = M. \tag{9.92}$$

Now suppose the money supply is multiplied by $\lambda > 0$, and that this is done by multiplying each consumer's initial money balance by λ. Then from (9.86), we have:

$$m_i^\dagger = \beta_i\big[w_i(\boldsymbol{p}^\dagger) + \lambda m_{i0}\big],$$

where $\boldsymbol{p}^\dagger$ is the new price vector, and $m_i^\dagger$ is consumer i's new demand for money balances. However, if we set $\boldsymbol{p}^\dagger = \lambda\boldsymbol{p}^*$, then:

$$m_i = \beta_i[w_i(\lambda\boldsymbol{p}^*) + \lambda m_{i0}] = \lambda[w_i(\boldsymbol{p}^*) + m_{i0}];$$

and thus:

$$\sum\nolimits_{i=1}^{I} m_i = \sum\nolimits_{i=1}^{I} \lambda m_i^* = \lambda M,$$

where the last inequality frollows from (9.88). Furthermore, since consumer commodity demand and producer's supply functions are positively homogeneous of degree zero in $(\boldsymbol{p}, w_i)$ and $\boldsymbol{p}$, respectively, it now follows from (9.88) that the commodity market is also in equilibrium with $\boldsymbol{p}^\dagger = \lambda\boldsymbol{p}^*$. Thus, in this new equilibrium, all commodity prices are multiplied by λ.

Now let's see if we can generalize this a bit. Suppose now that the ith consumer's utility function can be written in the form (9.79), as before, but let's now drop the assumption that ϕ^i is positively homogeneous and continuously differentable; supposing only that it is increasing. Using the same notation as before, suppose we have an equilibrium at $\big(\langle(\boldsymbol{x}_i^*, m_i^*)\rangle, \boldsymbol{p}^*, \boldsymbol{m}_0\big)$. Then (9.88) and (9.89) are satisfied; and, for each i, $(\boldsymbol{x}_i^*, m_i^*)$ maximizes u_i, given $(\boldsymbol{p}^*, w_i(\boldsymbol{p}^*), m_{i0}^*)$. Now let $\lambda > 0$. Then we note that:

$$(\lambda\boldsymbol{p}^*)\cdot\boldsymbol{x}_i^* + \lambda m_i^* = \lambda(\boldsymbol{p}^*\cdot\boldsymbol{x}_i + m_i^*) = \lambda\cdot[w_i(\boldsymbol{p}^*) + m_{i0}^*] = w_i(\lambda\boldsymbol{p}^*) + \lambda m_{i0}^*,$$

where the last inequality follows from the fact that a feasible wealth-assignment function must be positively homogeneous of degree one in $\boldsymbol{p}$. Thus we see that $(\boldsymbol{x}_i^*, \lambda m_i^*)$ is in the budget set defined by $(\lambda\boldsymbol{p}^*, \lambda m_{i0}^*)$:

$$B(\lambda\boldsymbol{p}^*, \lambda m_{i0}^*) = \big\{(\boldsymbol{x}_i, m_i) \in R_+^{n+1} \mid \boldsymbol{p}^*\cdot\boldsymbol{x}_i + m_i \le w_i(\lambda\,\boldsymbol{p}^*) + \lambda m_{i0}^*\big\}.$$

Now suppose $(\boldsymbol{x}_i, m_{i0})$ is such that:

$$(\lambda\boldsymbol{p}^*)\cdot\boldsymbol{x}_i + m_i \le w_i(\boldsymbol{p}^*) + \lambda m_{i0}^*.$$

Then we have:

$$\boldsymbol{p}^*\cdot\boldsymbol{x}_i + m_i/\lambda \le w_i(\boldsymbol{p}^*) + m_{i0}^*.$$

Therefore, since $(\boldsymbol{x}_i^*, m_i^*)$ maximizes utility subject to:

$$\boldsymbol{p}^*\cdot\boldsymbol{x}_i + m_i \le w_i(\boldsymbol{p}^*) + m_{i0}^*,$$

it follows that:

$$u_i(\boldsymbol{x}_i, m_i/\lambda) = \phi^i(\boldsymbol{x}_i) \cdot \big(m_i/\lambda\big)^{\alpha_i} \le \phi^i(\boldsymbol{x}_i^*) \cdot (m_i^*)^{\alpha_i};$$

or:

$$\phi^i(\boldsymbol{x}_i) \cdot (m_i^*)^{\alpha_i} \le \phi^i(\boldsymbol{x}_i^*) \cdot (\lambda m_i^*)^{\alpha_i}.$$

Therefore, denoting i's demand function for money balances by '$h_{i,n+1}(\boldsymbol{p}, m_{io})$,' we have:

$$h_{i,n+1}(\lambda \boldsymbol{p}, \lambda m_{io}) = \lambda h_{i,n+1}(\boldsymbol{p}, m_{io}). \tag{9.93}$$

I will leave it as an exercise to show that if $\big(\langle(\boldsymbol{x}_i^*, m_i^*)\rangle, \boldsymbol{p}^*, \boldsymbol{m}_0\big)$ is an equilibrium for the economy, and the money supply is multiplied by a positive number, λ, by means of multiplying each consumer's initial money balance by λ, then $\big(\langle(\boldsymbol{x}_i^*, \lambda m_i^*)\rangle, \lambda\boldsymbol{p}^*, \lambda\boldsymbol{m}_0\big)$ will be a new competitive equilibrium.

One thing that may be slightly troubling about the above analysis is the assumption that each individual's initial money balance is increased (or decreased) in exactly the same proportion. This is consistent with government's changing the official exchange rate, or the amount of gold backing, for example; but is not consistent with the way that the money supply is adjusted by the Federal Reserve in the U. S., for example. So, it would seem to be of interest to investigate the question of what happens to the price level if the aggregate money supply is changed, but individual initial holdings are not. (This may be a situation in which it is more realistic to suppose that we are dealing with the total supply of and demand for gold balances, rather than money balances *per se*.)

Suppose, then, that the aggregate money (or gold) supply, M, is multiplied by a positive number λ, which for definiteness we will suppose is greater than one. In order to simplify our analysis we will also suppose that all the β_i's are equal:

$$\beta_i = \beta \quad \text{for } i = 1, \dots, I,$$

where $0 < \beta < 1$. If, in fact, the vector of commodity prices, $\boldsymbol{p}$ is changed in the proportion μ, then equilibrium requires that:

$$\lambda M = \sum_{i=1}^{I} m_i = \sum_{i=1}^{I} \beta\big[w_i(\mu\boldsymbol{p}) + m_{i0}\big] = \beta\mu\Big(\sum_{i=1}^{I} w_i(\boldsymbol{p})\Big) + \beta\Big(\sum_{i=1}^{I} m_{i0}\Big). \tag{9.94}$$

However, by (9.92) we have:

$$\lambda M = \sum\nolimits_{i=1}^{I} m_{i0};$$

and if the economy was initially in equilibrium, then:

$$M = \beta \sum\nolimits_{i=1}^{I} \Big(w_i(\boldsymbol{p}) + m_{i0}\Big).$$

Substituting these last two equations into (9.94), we obtain:

$$\beta\lambda \sum\nolimits_{i=1}^{I} \Big(w_i(\boldsymbol{p}) + m_{i0}\Big) = \beta\mu\Big(\sum_{i=1}^{I} w_i(\boldsymbol{p})\Big) + \beta\Big(\sum_{i=1}^{I} m_{i0}\Big),$$

or:

$$\mu = \lambda + (\lambda - 1)\left(\frac{M}{\sum_{i=1}^{I} w_i(\boldsymbol{p})}\right);$$

and thus we see that prices in the new equilibrium have gone up more than proportionately to the increase in the money supply.

9.9 Indivisible Commodities

In this section I will simply present an example of general equilibrium analysis with indivisible commodities. The example itself was inspired by Ellickson [1993, pp. 111–14].

We consider an economy in which we have four commodities and one hundred consumers. The fourth commodity will be assumed to be perfectly divisible (perhaps a 'composite commodity'); while the first three commodities are types of rental properties (hereafter called "apartments") of low (L), medium (M), and high (H) quality, respectively. We will suppose that each of the consumers has preferences representable by the utility function:

$$u(\boldsymbol{x}_i) = (1 + x_{i1} + 2x_{i2} + 5x_{i3})x_{i4}.$$

While this utility function is quite conventional, we are going to suppose that consumers rent (consume) at most one apartment. We do this by defining their (common) consumption set in the following way. We first define the set C by:

$$C = \{\boldsymbol{x} \in \mathbb{R}^4_+ \mid x_j \in \{0,1\} \text{ for } j = 1,2,3\},$$

and then define X, the consumption set, by:

$$X = \{\boldsymbol{x} \in C \mid x_1 + x_2 + x_3 \leq 1\}.$$

We will label the individuals from lowest to highest wealth; supposing the individual commodity endowments, $\boldsymbol{\omega}_i$, are given by:

$$\boldsymbol{\omega}_i = \begin{cases} (0,0,0,199+5i) & \text{for } i = 1,\ldots,80, \\ (2,0,0,5i) & \text{for } i = 81,\ldots,90, \\ (2,4,2,2i) & \text{for } i = 91,\ldots,100. \end{cases}$$

Obviously the fourth commodity is a numéraire for the economy in this case, and consequently, in considering competitive allocations we can normalize to set $p_4 = 1$. Thus consumer i's wealth, w_i will be given by:

$$w_i = p_1\omega_{i1} + p_2\omega_{i2} + p_3\omega_{i3} + \omega_{i4};$$

so that, for example consumer one's wealth is, w_1, is equal to 204 (units of commodity 4).

To analyze the workings of this type of example, notice that, for example, a consumer will prefer to be homeless, as opposed to renting a low-quality apartment if, and only if the utility obtained with no housing exceeds that obtained with one unit of good 1:

$$w_i > 2(w_i - p_1);$$

that is:

$$2p_1 > w_i. \tag{9.95}$$

Similarly, consumer i will prefer to rent low-quality housing rather than medium iff:

$$2(w_i - p_1) > 3(w_i - p_2),$$

or:

$$3p_2 - 2p_1 > w_i. \tag{9.96}$$

Finally, consumer i will prefer to rent a medium, as opposed to a high quality apartment, if and only if:

$$3(w_i - p_2) > 6(w_i - p_3),$$

or:

$$2p_3 - p_2 > w_i. \tag{9.97}$$

We will first consider a competitive equilibrium, and then examine a disequilibrium situation.

Consider the price vector $\boldsymbol{p}^* = (100, 200, 400, 1)$. Since our poorest consumer has a wealth equal to 204, it follows from (9.95) that consumer 1 will prefer to rent a low-quality apartment to the alternative of being homeless. On the other hand:

$$w_{40} = 199 + 5 \cdot 40 = 399 < 3p_2 - 2p_1 = 400 < w_{41} = 404;$$

so we see that the first forty consumers will choose to rent low-quality housing, while the forty-first consumer will prefer to rent a medium-quality apartment. Since each consumer with a label larger than 41 will have a higher wealth, it follows that exactly 40 consumers will demand a low quality apartment, and this is exactly equal to the supply of same. Similarly, we find that:

$$w_{80} = 599 < 2p_3 - p_2 = 600 < w_{81} = 2 \cdot 100 + 5 \cdot 81 = 605;$$

so that only consumers 41–80 will rent medium quality apartments, and since there are 40 such apartments available, we once again have demand equal to supply. Finally, it is easy to show that consumers 81–100 will demand high-quality apartments, and since there are 20 such available, we have found a competitive equilibrium.

However, suppose a (bad) mistake is made in pricing the high-quality apartments, so that p_3 is set equal to 500, instead of 400. Then it is easy to see that, with p_1 and p_2 set at the levels just considered, we will have:

$$w_{90} = 650 < 2p_3 - p_2 = 800;$$

so that consumers 81–90 will now opt for medium-, rather than high-quality apartments. However, since there are only 40 such units available, the price of medium-quality housing will have to rise by enough to induce consumers 41–50 to choose low-quality housing. But this in turn means that the price of low-quality housing will have to rise by enough to induce consumers 1–10 to prefer homelessness to renting a low-quality apartment. Are there prices which will accomplish all of this?

In order to answer this question, let's begin by considering the price of low-quality housing. In order to induce 10 consumers to opt for homelessness, we need to have p_1 satisfy:

$$2p_1 > w_{10} = 249.$$

Consequently, the desired result will be achieved with $p_1 = 125$. In order to induce consumers 41–50 to choose low-quality housing, we see from (9.96) that we need p_2 to satisfy:

$$3p_2 - 2p_1 = 3p_2 - 250 > w_{50} = 449,$$

or:

$$p_2 > 233.$$

However, we also need to have the fifty-first consumer choose medium-quality housing, so that we need:

$$w_{51} = 454 > 3p_2 - 2p_1 = 3p_2 - 250,$$

or:

$$3p_2 < 704.$$

Thus, supply for medium-quality housing will be equal to demand if $p_2 = 234$. I will leave it to you to show that consumers 81–90 will continue to opt for medium-quality housing, despite their increased income.

Are there then market forces which will tend to move prices back toward the equilibrium levels which we found earlier? Before considering this question as such, let's take a look at the utility levels of these 100 consumers when prices are at the (non-equilibrium) level $\boldsymbol{p}' = (125, 234, 500, 1)$. This means that, for example, consumers 1-10 in the new situation attain a utility of:

$$u(\boldsymbol{x}_i) = w_i = 199 + 5i;$$

whereas with $\boldsymbol{p}^*$, the corresponding utility values were:

$$u(\boldsymbol{x}_i) = 2(w_i - 100) = 2(99 + 5i) = 198 + 5i;$$

so that each of these consumers was strictly better off in the equilibrium situation, as compared to that with $\boldsymbol{p} = \boldsymbol{p}'$. In fact, although I will leave you to show this (Exercise 3), *every consumer has a higher utility at the equilibrium prices* than in the disequilibrium situation!

Returning to the issue of whether there are forces tending to push this market toward equilibrium, as you have probably already noticed, with $\boldsymbol{p} = \boldsymbol{p}'$, there are 10 high-quality apartments standing empty. I will leave it to you to trace out the forces which will then move this market toward equilibrium.

Exercises.

1. Let E be an economy with one consumer, whose preferences can be represented by the utility function:

$$u(x_0, x_1) = \sqrt{x_0 \cdot x_1},$$

and whose initial endowment is given by $\boldsymbol{r} = (20, 0)$; and suppose there is one producer whose production set is given by:

$$Y = \{\boldsymbol{y} \in \mathbb{R}^2 \mid y_0 \leq 0 \;\&\; y_0 + y_1 \leq 0\}.$$

Find the competitive equilibrium for E, given the level of governmental activity $(\boldsymbol{t}, \boldsymbol{x}^g)$, where:

$$\boldsymbol{t} = (0, 1) \text{ and } \boldsymbol{x}^g = (0, 5);$$

or show that no such equilibrium exists in this case.

2. Let E be an economy with one consumer, whose preferences can be represented by the utility function:

$$u(x_0, x_1, x_2) = x_0 \cdot x_1 \cdot x_2,$$

and whose initial endowment is given by $\boldsymbol{r} = (24, 0, 0)$; and suppose there is one producer whose production set is given by:

$$Y = \{\boldsymbol{y} \in \mathbb{R}^3 \mid y_1, y_2 \geq 0 \;\&\; y_0 + 2y_1 + y_2 \leq 0\}.$$

Find the optimal commodity tax vector, $\boldsymbol{t} = (t_1, t_2)$; supposing that government demand, $\boldsymbol{x}^g$, is given by:

$$\boldsymbol{x}^g = (0, 3, 2).$$

3. Verify the claims made in the Example of Section 9.7 regarding the situation in which government sets a tax of $t = 7/4$.

4. Show that, in the Example of Section 9, every consumer has a higher utility at the equilibrium prices than in the disequilibrium situation.

Chapter 10

Comparative Statics and Stability

10.1 Introduction

We are accustomed to saying that if a commodity is a normal good (positive income effect), and given a normal supply curve (increasing with respect to price), an increase in the demand for the good will result in an increase in both the price of the good and the quantity of the good traded. However, this is a partial equilibrium analysis. In a general equilibrium context, the increased demand for good i must have repercussions for, or come about because of, changes in excess demand for one or more other goods. Moreover, as the price of the ith good increases, there will be changes in the quantity demanded of other commodities, which in turn will have feedback effects on the market for the ith good. Consequently, can we still be sure the the new equilibrium price for the ith good will be higher? You're probably thinking that it surely will be higher; it is, after all, only 'common sense.' Unfortunately, the theoretical conditions under which one can verify this simple analysis in the context of a general equilibrium model are rather more restrictive than one would like, although I am speaking here of known sufficient conditions for the analysis to hold; the known necessary conditions are less discouraging.

In this chapter we will investigate a portion of what is known of comparative statics in the context of general equilibrium, as well as giving brief consideration to the issues of the uniqueness of equilibrium and the stability of general competitive equilibrium. It is particularly reasonable to combine these topics; for, first of all, even in the context of partial equilibrium analysis, comparative statics analysis is likely to become meaningless unless the equilibria in a market (both before and after a demand change, say) are unique. In turn, comparative statics analysis will also break down if the equilibria in a market are not stable; after all, comparative statics analysis proceeds by comparing equilibria before and after a change has taken place. There is little point in such a comparison unless the market price and quantities approach the new equilibrium. Moreover, there is another aspect of this relationship; the sufficient conditions for stability of general competitive equilbria may enable us to deduce the sign of price changes following a change in underlying demand or supply conditions. This is the 'correspondence principle,' first stated and discussed by Samuelson [1947].

Comparative statics analysis in general equilibrium proceeds along essentially two different lines; global analysis, and local analysis. We will begin by considering some global analysis; which as we will develop it here, is based upon a variation of the weak axiom of revealed preference for aggregate excess demand. This approach has admitted weaknesses, and we will consider some of these as well. However, the analysis is, I believe, interesting, intuitive, and the conclusions seem to be in accord with the stylized facts of real economies. In any event, we will present the basic approach and the fundamental analysis based upon this approach in the next section. We will then look at two conditions which imply this form of the weak axiom; the 'law of demand,' and gross substitutability. We will discuss the 'law of demand' in Section 3, where we will show that it is implied by homothetic preferences. We will also show that it is unfortunately difficult to justify the assumption that aggregate *excess* demand functions satisfy the law of demand, even if preferences are homothetic. On the other hand, we will show that this law will hold under plausible empirical conditions, and as such it has strong and interesting implications. In Section 4 we will discuss the assumption of 'gross substitutability,' and we will show that it implies the weak axiom condition of Section 2. In Section 5, we will look at an alternative approach to comparative statics analysis; the local (differential) approach constituting a portion of what is called 'qualitative economics.' Finally, in the last two sections of this chapter, we provide a very brief introduction to the literature on the stability of general competitive equilibrium.

10.2 Aggregate Excess Demand

We will assume throughout this and the next two sections that individual preferences are continuous, strictly convex, and that the nth commodity is a numéraire good for the economy (and thus consumers will have demand *functions* satisfying the budget balance condition); in fact, we will suppose that the aggregate excess demand correspondence for the economy is single-valued (that is, is a function).

More formally, we suppose that the producers in the economy are price-takers, and that the aggregate supply correspondence is single-valued, and thus is a function, $\boldsymbol{s}(\boldsymbol{p})$. We denote the portion of the domain of this function which lies within $\mathbb{R}^n_{++}$ by Π, we let $\pi\colon \Pi \to \mathbb{R}$ be the aggregate profit function, and we will *assume throughout this, and the next two sections of this chapter, that* Π *is a convex cone.*[1] We further suppose that consumer wealth is defined by a feasible wealth-assignment function, $\boldsymbol{w}\colon \Pi \to \mathbb{R}^m$ (see Definition 7.9). Thus, for $\boldsymbol{p} \in \Pi$, we define the **aggregate excess demand function** for the economy, $\boldsymbol{z}\colon \Pi \to \mathbb{R}^n$. by:

$$\boldsymbol{z}(\boldsymbol{p}) = \sum_{i \in M} h_i[\boldsymbol{p}, w_i(\boldsymbol{p})] - \boldsymbol{r} - \boldsymbol{s}(\boldsymbol{p}), \tag{10.1}$$

where $\boldsymbol{r} \in \mathbb{R}^n$ is the aggregate resource endowment for the economy.

[1]The fact that Π is a cone follows from the definitions without any special assumptions. The issue of whether or not Π is also convex is discussed in Chapter 6.

A particularly important special case of this occurs when E is a private ownership, pure exchange economy. In this situation, both the supply and profit functions are null,

$$\boldsymbol{r} = \sum\nolimits_{i \in M} \boldsymbol{r}_i,$$

where $\boldsymbol{r}_i$ is the ith consumer's initial resource endowment, and the feasible wealth-assignment function is given by:

$$w_i(\boldsymbol{p}) = \boldsymbol{p} \cdot \boldsymbol{r}_i \quad \text{for } \boldsymbol{p} \in \Pi, \text{ and } i = 1, \ldots, m.$$

In this case, it is also useful to define the **individual excess demand functions**, $\boldsymbol{z}_i \colon \Pi \to \mathbb{R}^n$ by:

$$\boldsymbol{z}_i(\boldsymbol{p}) = \boldsymbol{h}_i[\boldsymbol{p}, \boldsymbol{p} \cdot \boldsymbol{r}_i] - \boldsymbol{r}_i \quad \text{for } i = 1, \ldots, m.^2$$

Returning to the general case, notice that under the assumptions being utilized here, $\boldsymbol{p}^* \in \Pi$ is an equilibrium price vector if, and only if:

$$\boldsymbol{z}(\boldsymbol{p}^*) = \boldsymbol{0}.$$

It is also important to notice that, given our assumptions, the excess demand function will be positively homogeneous of degree zero in $\boldsymbol{p}$ (recall that Π will be a cone), and will satisfy the strong form of Walras Law, which can be expressed as the condition:

$$(\forall \boldsymbol{p} \in \Pi) \colon \boldsymbol{p} \cdot \boldsymbol{z}(\boldsymbol{p}) = 0.$$

A key consideration in many investigations of conditions under which an economy will have a unique competitive equilibrium revolves around the question of whether or not the aggregate excess demand function satisfies the same properties as are satisfied by individual excess demand functions; in particular, the following condition.

10.1 Definition. The aggregate excess demand function, $\boldsymbol{z}(\cdot)$, satisfies the **Weak Axiom of Revealed Preference** (abbreviated **WA**) iff, for any pair of price vectors, $\boldsymbol{p}$ and $\boldsymbol{p}'$, we have:

$$\left[\boldsymbol{z}(\boldsymbol{p}) \neq \boldsymbol{z}(\boldsymbol{p}') \;\&\; \boldsymbol{p} \cdot \boldsymbol{z}(\boldsymbol{p}') \leq 0\right] \Rightarrow \boldsymbol{p}' \cdot \boldsymbol{z}(\boldsymbol{p}) > 0.$$

In fact, Mas-Colell, Whinston, and Green prove that the satisfaction of WA by the aggregate excess demand function is, in a sense, both necessary and sufficient for the uniqueness of competitive equilibrium in the situation in which the aggregate production set is a convex cone (see Proposition 17.F.2, p. 609, of Mas-Colell, Whinston, and Green [1995]). However, in our analysis we will make use of the slightly weaker conditon defined as follows.

10.2 Definition. The aggregate excess demand function, $\boldsymbol{z}(\cdot)$, satisfies the **Weak* Axiom of Revealed Preference** (abbreviated **WA***) iff, given any equilibrium price vector, $\boldsymbol{p}^*$, and any second price vector, $\boldsymbol{p} \in \Pi$ such that $\boldsymbol{z}(\boldsymbol{p}) \neq \boldsymbol{0}$, we have:

$$\boldsymbol{p}^* \cdot \boldsymbol{z}(\boldsymbol{p}) > 0.$$

[2]Notice also that in this case we can take Π to be equal to $\mathbb{R}^n_{++}$.

It is an easy exercise to show that if the aggregate demand function satisfies the Weak Axiom (WA), then it satisfies WA* (see Exercise 1, at the end of this chapter). We can then prove the following variant of the sufficiency portion of the MWG result.

10.3 Proposition. *Suppose the aggregate excess demand function, $\boldsymbol{z}(\cdot)$, satisfies WA*. Then the set of equilibrium price vectors is convex.*

Proof. Suppose $\boldsymbol{p}$ and $\boldsymbol{p}'$ are both equilibrium price vectors for the economy E, let $\theta \in [0,1]$, and define:

$$\boldsymbol{p}^* = \theta \boldsymbol{p} + (1-\theta)\boldsymbol{p}'.$$

Suppose, by way of obtaining a contradiction, that $\boldsymbol{p}^*$ is not a competitive equilibrium price. Then $\boldsymbol{z}(\boldsymbol{p}^*) \neq \boldsymbol{0}$, and it follows from WA* that:

$$\boldsymbol{p} \cdot \boldsymbol{z}(\boldsymbol{p}^*) > 0 \text{ and } \boldsymbol{p}' \cdot \boldsymbol{z}(\boldsymbol{p}^*) > 0. \tag{10.2}$$

But this is impossible, for by Walras' Law:

$$0 = \boldsymbol{p}^* \cdot \boldsymbol{z}(\boldsymbol{p}^*) = [\theta \boldsymbol{p} + (1-\theta)\boldsymbol{p}'] \cdot \boldsymbol{z}(\boldsymbol{p}^*) = \theta \boldsymbol{p} \cdot \boldsymbol{z}(\boldsymbol{p}^*) + (1-\theta)\boldsymbol{p}' \cdot \boldsymbol{z}(\boldsymbol{p}^*);$$

which contradicts (10.2) □

As noted in Mas-Colell, Whinston, and Green [1995], if the set of normalized equilibria is finite, then the above result implies that, given WA*, the set of normalized equilibrium prices is a singleton (that is, there is a unique normalized equilibrium price). When the aggregate excess demand satisfies WA*, increases in the demand for a given commodity will result in an increase in the price of that commodity. In demonstrating this, we will more or less follow McKenzie [2002, pp. 143–4]. We begin with the following definition.

10.4 Definition. Let $\boldsymbol{p}^*$ be an equilibrium price vector, given the aggregate excess demand function $\boldsymbol{z}(\cdot)$, let $\widehat{\boldsymbol{z}}(\cdot)$ be a second excess demand function, and let $j \in \{1, \ldots, n-1\}$ be arbitrary. We will say that $\widehat{\boldsymbol{z}}$ exhibits **increased excess demand for the jth commodity** (alone), as compared with $\boldsymbol{z}(\cdot)$ iff we have:

$$\widehat{z}_j(\boldsymbol{p}^*) > 0,$$

and, for all $k \in \{1, \ldots, n-1\} \setminus \{j\}$:

$$\widehat{z}_k(\boldsymbol{p}^*) = 0.$$

The idea of the above definition corresponds to the situation described in the introduction to this chapter. We begin with a situation in which the economy is in equilibrium, given the excess demand function $\boldsymbol{z}$, and then we suppose that one or more consumers' tastes change in favor of the jth commodity (relative to the numéraire), or perhaps there has been a wealth transfer from a given consumer to a second consumer who values the jth commodity (relative to the numéraire) more than did the first.[3] Given such a change, we will have a new excess demand function for the economy which exhibits increased excess demand for the jth commodity. For instance, consider the following examples.

[3]Given the other conditions of the definition, we will necessarily have $\widehat{z}_n(\boldsymbol{p}^*) < 0$. Why?

10.5 Examples. We consider a pure exchange economy with $m = 2$, $n = 3$, and suppose initially that both consumers have the utility functions:

$$u_i(\boldsymbol{x}_i) = \left[x_{i1} \cdot x_{i2} \cdot x_{i3}\right]^{1/3}. \tag{10.3}$$

Suppose also that initial endowments are given by:

$$\boldsymbol{r}_1 = (4, 0, 2) \text{ and } \boldsymbol{r}_2 = (0, 4, 2). \tag{10.4}$$

In this case, as I will leave it to you to demonstrate, if we set $\boldsymbol{p} = \boldsymbol{p}^* = (1, 1, 1)$, then $\boldsymbol{z}(\boldsymbol{p}^*) = \boldsymbol{0}$; so that $\boldsymbol{p}^*$ is an equilibrium price, given $\boldsymbol{z}$. Now suppose that the first consumer's utility changes to:

$$u_1(\boldsymbol{x}_1) = x_{11}^{1/3} \cdot x_{12}^{1/2} \cdot x_{13}^{1/6}, \tag{10.5}$$

while consumer 2's utility function remains as in (10.3). Then, as I will leave for you to demonstrate, the aggregate excess demand function for the economy changes to (normalizing to set $p_3 = 1$):

$$\widehat{\boldsymbol{z}}(\boldsymbol{p}) = \Big(\frac{4p_1 + 4p_2 + 4}{3p_1} - 4, \frac{6p_1 + 4p_2 + 5}{3p_2} - 4, \frac{2p_1 + 4p_2 + 3}{3} - 4\Big);$$

so that:

$$\widehat{\boldsymbol{z}}(\boldsymbol{p}^*) = (0, 1, -1);$$

and we see that $\widehat{\boldsymbol{z}}$ exhibits increased demand for the second commodity. I will leave it to you to construct a similar example in which a transfer of a quantity of the numéraire good (the third commodity in this case) from one consumer to the other results in an increase in demand for the second commodity (see Exercise 4, at the end of this chapter). □

Making use of WA* and our (McKenzie's) definition of an increase in excess demand for the jth commodity, we can establish the following.[4]

10.6 Proposition. *Suppose* $\mathbf{p}^*$ *is an equilibrium price for an economy, given the excess demand function* $\boldsymbol{z}(\cdot)$, *and that excess demand changes to* $\widehat{\boldsymbol{z}}(\cdot)$, *which exhibits increased demand for commodity* j, *and satisfies WA*. Then if* $\widehat{\mathbf{p}}$ *is the equilibrium price for* E, *given* $\widehat{\boldsymbol{z}}$, *we will have* $\widehat{p}_j > p_j^*$.

Proof. Since $\widehat{\boldsymbol{z}}$ exhibits increased excess demand for commodity j, and satisfies WA*, we have:

$$0 < \widehat{\boldsymbol{p}} \cdot \widehat{\boldsymbol{z}}(\boldsymbol{p}^*) = \widehat{p}_j \widehat{z}_j(\boldsymbol{p}^*) + \widehat{z}_n(\boldsymbol{p}^*).$$

However, we also have, by (the strong form of) Walras' Law:

$$0 = \boldsymbol{p}^* \cdot \widehat{\boldsymbol{z}}(\boldsymbol{p}^*) = p_j^* \widehat{z}_j(\boldsymbol{p}^*) + \widehat{z}_n(\boldsymbol{p}^*);$$

so that:

$$p_j^* \widehat{z}_j(\boldsymbol{p}^*) < \widehat{p}_j \widehat{z}_j(\boldsymbol{p}^*),$$

and thus:

$$(\widehat{p}_j - p_j^*) \cdot \widehat{z}_j(\boldsymbol{p}^*) > 0.$$

Since $\widehat{z}_j(\boldsymbol{p}^*) > 0$, our result follows. □

[4] The statement and proof of which are lifted almost verbatim from McKenzie [2002, Theorem 10, p. 144].

Thus, the satisfaction of WA* by the aggregate excess demand function has quite strong and useful implications. Unfortunately, even in the case of a pure exchange economy, and with homothetic preferences, the aggregate excess demand function does not necessarily satisfy WA*, as is demonstrated by the following example.

10.7 Example. Suppose $m = n = 2$, that $X_1 = X_2 = \mathbb{R}^2_+$, and that the consumers' preferences are represented by the utility functions:

$$\begin{aligned} u_1(\boldsymbol{x}_1) &= \min\{x_{11}, x_{12}\}, \text{and} \\ u_2(\boldsymbol{x}_2 &= \min\{4x_{21}, x_{22}\} \end{aligned} \tag{10.6}$$

respectively; and suppose:

$$\boldsymbol{r}_1 = (2, 0) \text{ and } \boldsymbol{r}_2 = (0, 5).$$

If $\boldsymbol{p} = (1, 1)$, then:

$$x_{11} + x_{21} = 1 + 1 = r_{11},$$

and thus (by Walras' Law) $\boldsymbol{p}$ is a competitive equilibrium price.

Now consider the price vector $\boldsymbol{p}' = (1, 1/4)$. We have:

$$\boldsymbol{z}(\boldsymbol{p}') = \left(\frac{8}{5} + \frac{5}{8} - 2, \frac{8}{5} + \frac{5}{2} - 5\right) = \left(\frac{9}{40}, \frac{-9}{10}\right).$$

Therefore:

$$\boldsymbol{p} \cdot \boldsymbol{z}(\boldsymbol{p}') = 9/40 - 9/10 = -27/40 < 0;$$

which violates WA*. □

It is quite reasonable at this point for you to be wondering just why it is that we are bothering with the condition WA* if there is such a simple example, with preferences so well-behaved, in which the condition is violated. Well, there are two cases of some interest in which WA* is satisfied.[5] The first of these two cases is that in which aggregate excess demand satisfies the 'Law of Demand.' We will study this case in the next section, and while we will find that this condition is not satisfied in general, we can specify intuitive empirical conditions in which it is satisfied. The second case in which WA* is satisfied is that in which the commodities are all gross substitutes. We will consider this condition in section 4 of this chapter.

10.3 The 'Law of Demand'

One often sees the phrase 'the law of demand' used in the economics literature to mean that the aggregate demand function for a commodity is downward-sloping. The following condition reduces to this in the situation in which only one price has changed.[6] In the second part of the following definition, we will use the notation $H(\cdot)$ to denote the aggregate demand function defined by:

$$\boldsymbol{H}(\boldsymbol{p}, \boldsymbol{w}) = \sum_{i=1}^{m} h_i(\boldsymbol{p}, w_i).$$

[5]It should also be noted that, insofar as I am aware, no one has proved that these are the *only* two conditions implying WA*.

[6]I believe that this label was first attached to this condition by J. R. Hicks [1956]. The treatment here owes much to Section 4.C of Mas-Colell, Whinston, and Green [1995], however.

10.8 Definition. We will say that the ith consumer's demand function satisfies the **law of demand** iff, given any $\boldsymbol{p}, \boldsymbol{p}' \in \mathbb{R}^n_{++}$, and any $w_i \in \mathbb{R}_+$ such that $\boldsymbol{h}_i(\boldsymbol{p}, w_i) \neq \boldsymbol{h}_i(\boldsymbol{p}', w_i)$, we have:

$$[\boldsymbol{p} - \boldsymbol{p}'] \cdot [\boldsymbol{h}_i(\boldsymbol{p}, w_i) - \boldsymbol{h}_i(\boldsymbol{p}', w_i)] < 0. \tag{10.7}$$

Similarly, we will say that **aggregate demand satisfies the law of demand** iff given any $\boldsymbol{p}, \boldsymbol{p}' \in \mathbb{R}^n_{++}$, and any $\boldsymbol{w} \in \mathbb{R}^m_+$ such that $\boldsymbol{H}(\boldsymbol{p}, \boldsymbol{w}) \neq \boldsymbol{H}(\boldsymbol{p}', \boldsymbol{w})$, we have:

$$[\boldsymbol{p} - \boldsymbol{p}'] \cdot [\boldsymbol{H}(\boldsymbol{p}, \boldsymbol{w}) - \boldsymbol{H}(\boldsymbol{p}', \boldsymbol{w})] < 0. \tag{10.8}$$

As already suggested, the above condition implies that the demand function for each commodity is downward-sloping. As you know, an individual demand function does not necessarily satisfy the law of demand; there is always the possibility that the infamous Giffen good case may arise. However, if an individual's preferences are homothetic, then the consumer's demand function will satisfy the law of demand, and thus the demand function for each commodity is necessarily downward-sloping in this case.

10.9 Theorem. *Suppose G is homothetic, continuous, strictly convex, and locally non-saturating on $\mathbb{R}^n_+$. Then the demand function determined by G satisfies the law of demand on Ω.*

Proof. Recall from Theorem 4.39 that, given the present assumptions, the demand function takes the form:

$$\boldsymbol{h}(\boldsymbol{p}, w) = \boldsymbol{g}(\boldsymbol{p})w.$$

Now, let $\boldsymbol{p}, \boldsymbol{p}' \in \mathbb{R}^n_{++}$, and define:

$$\mu = \boldsymbol{p}' \cdot \boldsymbol{g}(\boldsymbol{p}).$$

Then we note that it follows from the fact that $\boldsymbol{p}' \cdot \boldsymbol{g}(\boldsymbol{p}) \leq \mu$, that:

$$\boldsymbol{p} \cdot \boldsymbol{g}(\boldsymbol{p}')\mu \geq 1$$

[remember that $\boldsymbol{h}(\boldsymbol{p}', \mu) = \boldsymbol{g}(\boldsymbol{p}')\mu$]. Therefore, we see that:

$$\boldsymbol{p}' \cdot \boldsymbol{g}(\boldsymbol{p}) + \boldsymbol{p} \cdot \boldsymbol{g}(\boldsymbol{p}') \geq \mu + 1/\mu. \tag{10.9}$$

Now consider the function $f\colon \mathbb{R}_{++} \to \mathbb{R}_{++}$ defined by:

$$f(\mu) = \mu + 1/\mu. \tag{10.10}$$

If you check first- and second-order conditions for an extremum, it is easy to show that f has a unique minimum at $\mu = 1$; and that f is strictly convex, so that for all $\mu \in \mathbb{R}_{++}$:

$$\mu \neq 1 \Rightarrow f(\mu) > f(1) = 2. \tag{10.11}$$

Thus it follows from (10.9)–(10.11) that if $\boldsymbol{p}' \cdot \boldsymbol{g}(\boldsymbol{p}) \neq 1$, then:[7]

$$\boldsymbol{p}' \cdot \boldsymbol{g}(\boldsymbol{p}) + \boldsymbol{p} \cdot \boldsymbol{g}(\boldsymbol{p}') > 2. \tag{10.12}$$

On the other hand, if $\mu = \boldsymbol{p}' \cdot \boldsymbol{g}(\boldsymbol{p}) = 1$, and $\boldsymbol{g}(\boldsymbol{p}) \neq \boldsymbol{g}(\boldsymbol{p}')$, then it follows from WA that $\boldsymbol{p} \cdot \boldsymbol{g}(\boldsymbol{p}') > 1$, so that (10.12) holds in this case as well.

Now let w be arbitrary, and let $\boldsymbol{p}, \boldsymbol{p}' \in \mathbb{R}^n_{++}$ be such that:

$$\boldsymbol{h}(\boldsymbol{p}, w) = \boldsymbol{g}(\boldsymbol{p})w \neq \boldsymbol{h}(\boldsymbol{p}', w) = \boldsymbol{g}(\boldsymbol{p}')w.$$

Then obviously, $\boldsymbol{g}(\boldsymbol{p}) \neq \boldsymbol{g}(\boldsymbol{p}')$, so it follows from equation (10.12) that:

$$\boldsymbol{p}' \cdot \boldsymbol{g}(\boldsymbol{p})w + \boldsymbol{p} \cdot \boldsymbol{g}(\boldsymbol{p}')w > 2w,$$

so that:

$$w - \boldsymbol{p}' \cdot \boldsymbol{g}(\boldsymbol{p})w + w - \boldsymbol{p} \cdot \boldsymbol{g}(\boldsymbol{p}')w < 0.$$

But then we see that:

$$\begin{aligned} 0 > \boldsymbol{p}' \cdot \boldsymbol{g}(\boldsymbol{p}')w - \boldsymbol{p}' \cdot \boldsymbol{g}(\boldsymbol{p})w + \boldsymbol{p} \cdot \boldsymbol{g}(\boldsymbol{p})w - \boldsymbol{p} \cdot \boldsymbol{g}(\boldsymbol{p}')w \\ = \boldsymbol{p}' \cdot \big[\boldsymbol{h}(\boldsymbol{p}', w) - \boldsymbol{h}(\boldsymbol{p}, w)\big] - \boldsymbol{p} \cdot \big[\boldsymbol{h}(\boldsymbol{p}', w) - \boldsymbol{h}(\boldsymbol{p}, w)\big] \\ = (\boldsymbol{p}' - \boldsymbol{p}) \cdot \big[\boldsymbol{h}(\boldsymbol{p}', w) - \boldsymbol{h}(\boldsymbol{p}, w)\big]. \quad \square \end{aligned}$$

The good news about the Law of Demand is that if it holds for each individual consumer's demand function, then it holds for the aggregate demand function as well. Thus we obtain the following as an easy corollary of Theorem 10.9. I will leave the details of the proof as an exercise.

10.10 Corollary. *If each individual demand function, $h_i(\cdot)$ saisfies the Law of Demand, then the aggregate demand function $\boldsymbol{H} \colon \mathbb{R}^n_{++} \times \mathbb{R}^m_+ \to \mathbb{R}^n_+$ also satisfies the Law of Demand.*

While the aggregate demand function, $\boldsymbol{H}$, is well-defined whenever all individual demand functions are well-defined, economists often (especially in applied work) assume that aggregate demand can be expressed as a function of price and aggregate income. The following condition provides one method for justifying such an assumption.

10.11 Definition. We will say that a function $\boldsymbol{\omega} \colon \mathbb{R}_+ \to \mathbb{R}^m_+$ is an **income distribution function** iff it is positively homogeneous of degree one, and satisfies:

$$(\forall w \in \mathbb{R}_+) \colon \sum\nolimits_{i=1}^m \omega_i(w) = w.$$

Given such an income distribution function, we can define an aggregate demand function, $\boldsymbol{h}$, on $\mathbb{R}^n_{++} \times \mathbb{R}_+$ by:

$$\boldsymbol{h}(\boldsymbol{p}, w) = \sum\nolimits_{i=1}^m \boldsymbol{h}_i[\boldsymbol{p}, \omega_i(w)].$$

[7]Notice that if $\boldsymbol{g}(\boldsymbol{p}) = \boldsymbol{g}(\boldsymbol{p}')$, then $\boldsymbol{p}' \cdot \boldsymbol{g}(\boldsymbol{p}) = 1$.

It is easy to show that, given any such $\boldsymbol{\omega}$, the function $\boldsymbol{h}(\boldsymbol{p}, w)$ will be positively homogeneous of degree zero in $(\boldsymbol{p}, w)$; and if each individual demand function satisfies the budget balance condition (as we are assuming is the case throughout this section), the aggregate demand function will satisfy this condition as well, that is:

$$(\forall(\boldsymbol{p}, w) \in \mathbb{R}^n_{++} \times \mathbb{R}_+)\colon \boldsymbol{p} \cdot \boldsymbol{h}(\boldsymbol{p}, w) = w.$$

It is also easy to show that if each individual demand function, $\boldsymbol{h}_i$ satisfies the Law of Demand, then so will the aggregate demand function, $\boldsymbol{h}$. Of course, the most interesting special case of this is where we have the following

10.12 Definition. Let $E = (\langle X_i, P_i\rangle, \langle Y_k\rangle, \boldsymbol{r})$ be an economy, suppose aggregate supply is well-defined[8] on the set $\Pi \subseteq \mathbb{R}^n_{++}$, and let $\boldsymbol{w}\colon \Pi \to \mathbb{R}^m_+$ be a feasible wealth function for E. We will say that E and $\boldsymbol{w}$ satisfy the **income distribution condition** iff, there exists an income distribution function, $\boldsymbol{\omega}\colon \mathbb{R}_+ \to \mathbb{R}^m_+$ such that, for all $\boldsymbol{p} \in \Pi$ we have:

$$w_i(\boldsymbol{p}) = \omega_i[w(\boldsymbol{p})] \quad \text{for } i = 1, \ldots, n,$$

where $w\colon \Pi \to \mathbb{R}_+$ is defined by:

$$w(\boldsymbol{p}) = \boldsymbol{p} \cdot \boldsymbol{r} + \boldsymbol{p} \cdot \sigma(\boldsymbol{p}).$$

10.13 Example. Suppose $\mathcal{E} = (\langle X_i, P_i\rangle, \langle Y_k\rangle, \langle \boldsymbol{r}_i\rangle, [s_{ik}])$ is a private ownership economy, and let $\boldsymbol{a} \in \Delta_m$ be such that for each $i \in \{1, \ldots, m\}$:

$$\boldsymbol{r}_i = a_i\Big(\sum\nolimits_{j=1}^m \boldsymbol{r}_j\Big) \equiv a_i\boldsymbol{r} \text{ and } s_{ik} = a_i, \text{ for } k = 1, \ldots, \ell.$$

If we now define $\boldsymbol{w}(\cdot)$ in the usual way:

$$w_i(\boldsymbol{p}) = \boldsymbol{p} \cdot \boldsymbol{r}_i + \sum\nolimits_{k=1}^{\ell} s_{ik}\pi_k(\boldsymbol{p}) \quad \text{for } i = 1, \ldots, m,$$

it is easy to show that $\boldsymbol{w}$ and $\mathcal{E}$ satisfy the incoe distribution condition. □

The following result sets forth an interesting property guaranteed by the Law of Demand.

10.14 Proposition. *Let $\boldsymbol{\omega}\colon \mathbb{R}_+ \to \mathbb{R}^m_+$ be an income distribution function, let $\boldsymbol{h}$ be the corresponding aggregate demand function, and suppose $\boldsymbol{h}$ satisfies the Law of Demand.*[9] *Then $\boldsymbol{h}$ satisfies the weak axiom (WA) on $\Omega \stackrel{def}{=} \mathbb{R}^n_{++} \times \mathbb{R}_+$.*

Proof. Suppose $(\boldsymbol{p}, w)$, and $(\boldsymbol{p}', w')$ are elements of Ω such that:

$$\boldsymbol{h}(\boldsymbol{p}, w) \neq \boldsymbol{h}(\boldsymbol{p}', w') \text{ and } \boldsymbol{p} \cdot \boldsymbol{h}(\boldsymbol{p}', w') \leq w.$$

[8]That is, for each $\boldsymbol{p} \in \Pi$, there exists a unique $\boldsymbol{y} \in Y$ which maximizes profits on Y, given $\boldsymbol{p}$.

[9]As will be the case, remember, if each $\boldsymbol{h}_i$ satisfies the Law of Demand.

Then, defining $\boldsymbol{p}'' = (w/w')\boldsymbol{p}'$, we have, by the homogeneity of degree zero of the aggregate demand function, that $\boldsymbol{h}(\boldsymbol{p}'', w) = \boldsymbol{h}(\boldsymbol{p}', w')$. It then follows from the fact that aggregate demand satisfies the law of demand on Ω, that:

$$\begin{aligned} 0 > [\boldsymbol{p}'' - \boldsymbol{p}] \cdot [\boldsymbol{h}(\boldsymbol{p}'', w) - \boldsymbol{h}(\boldsymbol{p}, w)] &= \boldsymbol{p}'' \cdot \boldsymbol{h}(\boldsymbol{p}'', w) - \boldsymbol{p}'' \cdot \boldsymbol{h}(\boldsymbol{p}, w) \\ - \boldsymbol{p} \cdot \boldsymbol{h}(\boldsymbol{p}'', w) + \boldsymbol{p} \cdot \boldsymbol{h}(\boldsymbol{p}, w) &= w - \boldsymbol{p}'' \cdot \boldsymbol{h}(\boldsymbol{p}, w) + w - \boldsymbol{p} \cdot \boldsymbol{h}(\boldsymbol{p}'', w), \end{aligned} \tag{10.13}$$

where the last equality is by the fact that each $\boldsymbol{h}_i$ satisfies the budget balance condition. Moreover, by the fact that $\boldsymbol{h}(\boldsymbol{p}'', w) = \boldsymbol{h}(\boldsymbol{p}', w')$, and our hypothesis, we have that:

$$w - \boldsymbol{p} \cdot \boldsymbol{h}(\boldsymbol{p}'', w) \geq 0.$$

Thus it follows from (10.13) that

$$\boldsymbol{p}'' \cdot \boldsymbol{h}(\boldsymbol{p}, w) > w;$$

and using the definition of $\boldsymbol{p}''$, it now follows that:

$$\boldsymbol{p}' \cdot \boldsymbol{h}(\boldsymbol{p}, w) > w'. \quad \square$$

Returning to the not-so-good news about the Law of Demand, notice that, under the assumptions which we're employing, we can write aggregate excess demand as:

$$\boldsymbol{z}(\boldsymbol{p}) = \sum\nolimits_{i=1}^{m} \boldsymbol{z}_i(\boldsymbol{p}) - \boldsymbol{s}(\boldsymbol{p}),$$

where $\boldsymbol{z}_i(\cdot)$ is the ith consumer's excess demand function, for $i = 1, \ldots, m$.[10] Thus, given $\boldsymbol{p}, \boldsymbol{p}' \in \Pi$, we will have:

$$(\boldsymbol{p}'-\boldsymbol{p}) \cdot \big[\boldsymbol{z}(\boldsymbol{p}')-\boldsymbol{z}(\boldsymbol{p})\big] = \sum_{i=1}^{m} (\boldsymbol{p}'-\boldsymbol{p}) \cdot \big[\boldsymbol{z}_i(\boldsymbol{p}')-\boldsymbol{z}_i(\boldsymbol{p})\big] - (\boldsymbol{p}'-\boldsymbol{p}) \cdot \big[\boldsymbol{s}(\boldsymbol{p}')-\boldsymbol{s}(\boldsymbol{p})\big] \tag{10.14}$$

Now, it follows from Theorem 6.27 that we necessarily have:

$$-(\boldsymbol{p}' - \boldsymbol{p}) \cdot \big[\boldsymbol{s}(\boldsymbol{p}') - \boldsymbol{s}(\boldsymbol{p})\big] \leq 0;$$

with strict inequality if $\boldsymbol{s}(\boldsymbol{p}) \neq \boldsymbol{s}(\boldsymbol{p}')$. (See Exercise 6, at the end of this chapter.) Consequently, if each individual excess demand function satisfies the Law of Demand, then the aggregate excess demand function will satisfy the Law as well. Unfortunately, individual excess demand functions do not necessarily satisfy the Law of Demand even if the consumer's preferences are homothetic, as is shown by the following example.

10.15 Example. Let $n = 2, X_i = \mathbb{R}^2_+$, and $\boldsymbol{r}_i = (1, 0)$; while the consumer's utility function is given by:

$$u_i(\boldsymbol{x}_i) = x_{i1}^{\alpha} \cdot x_{i2}^{1-\alpha},$$

[10]If we are allowing for individually owned resource endowments, $\boldsymbol{r}_i \in \mathbb{R}^n$, then $\boldsymbol{z}_i(\boldsymbol{p}) = h[\boldsymbol{p}, w_i(\boldsymbol{p})] - \boldsymbol{r}_i$. If we are not allowing for such ownership, then we can add $\boldsymbol{r}$ to $\boldsymbol{s}(\boldsymbol{p})$.

where $0 < \alpha < 1$, and let:

$$\boldsymbol{p}' = (1,1) \text{ and } \boldsymbol{p}'' = \Big(\frac{4}{(1-\alpha)}, 2\Big)$$

Then we have:

$$\boldsymbol{z}_i(\boldsymbol{p}') = (\alpha - 1, 1-\alpha),$$

while:

$$\boldsymbol{z}_i(\boldsymbol{p}'') = \Big(\frac{\alpha \times \frac{4}{(1-\alpha)}}{\frac{4}{1-\alpha}} - 1, \frac{(1-\alpha)\frac{4}{(1-\alpha)}}{2}\Big) = (\alpha - 1, 2).$$

Therefore:

$$(\boldsymbol{p}'' - \boldsymbol{p}') \cdot \big[\boldsymbol{z}_i(\boldsymbol{p}'') - \boldsymbol{z}_i(\boldsymbol{p}')\big] = \Big(\frac{3+\alpha}{1-\alpha}, 1\Big) \cdot (0, 1+\alpha) = 1 + \alpha > 0.$$

Thus we see that $\boldsymbol{z}_i(\cdot)$ does *not* satisfy the Law of Demand. □

As the above example demonstrates, the reason that the ith consumer's excess demand function may not satisfy the Law of Demand is that, in general, when $\boldsymbol{p}$ changes, $w_i(\boldsymbol{p})$ changes as well. In fact, while the above example is discouraging to be sure, notice that to achieve the violation we changed the consumer's income from 1 to $4/(1-\alpha)$. Since $0 < \alpha < 1$, this is an extremely large change, in percentage terms.

In order to consider this issue further, suppose $\boldsymbol{p}$ changes from $\boldsymbol{p}'$ to $\boldsymbol{p}''$, and consider the ith consumer's demand and excess demand functions. We have:

$$\begin{aligned}
(\boldsymbol{p}'' - \boldsymbol{p}') \cdot \big[\boldsymbol{z}_i(\boldsymbol{p}'') - \boldsymbol{z}_i(\boldsymbol{p}')\big] &= (\boldsymbol{p}'' - \boldsymbol{p}') \cdot \big(\boldsymbol{h}_i\big[\boldsymbol{p}'', w_i(\boldsymbol{p}'')\big] - \boldsymbol{h}_i\big[\boldsymbol{p}', w_i(\boldsymbol{p}')\big]\big) \\
= (\boldsymbol{p}'' - \boldsymbol{p}') \cdot \Big(\boldsymbol{h}_i\big[\boldsymbol{p}'', w_i(\boldsymbol{p}'')\big] &- \boldsymbol{h}_i\big[\boldsymbol{p}', w_i(\boldsymbol{p}'')\big] + \boldsymbol{h}_i\big[\boldsymbol{p}', w_i(\boldsymbol{p}'')\big] - \boldsymbol{h}_i\big[\boldsymbol{p}', w_i(\boldsymbol{p}')\big]\Big) \\
&= (\boldsymbol{p}'' - \boldsymbol{p}') \cdot \Big(\boldsymbol{h}_i\big[\boldsymbol{p}'', w_i(\boldsymbol{p}'')\big] - \boldsymbol{h}_i\big[\boldsymbol{p}', w_i(\boldsymbol{p}'')\big]\Big) \\
&\quad + (\boldsymbol{p}'' - \boldsymbol{p}') \cdot \Big(\boldsymbol{h}_i\big[\boldsymbol{p}', w_i(\boldsymbol{p}'')\big] - \boldsymbol{h}_i\big[\boldsymbol{p}', w_i(\boldsymbol{p}')\big]\Big) \qquad (10.15)
\end{aligned}$$

If the consumer's demand function satisfies the Law of Demand, and if:

$$\boldsymbol{h}_i\big[\boldsymbol{p}'', w_i(\boldsymbol{p}'')\big] \neq \boldsymbol{h}_i\big[\boldsymbol{p}', w_i(\boldsymbol{p}'')\big],$$

then the first inner product on the right-hand-side of (10.15) is negative. Consequently, if the income change, $w_i(\boldsymbol{p}'') - w_i(\boldsymbol{p}')$, is sufficiently small, the inner product on the left-hand-side of (10.15) will be negative as well. Thus, to oversimplify things just a bit, we can say that if individual consumers' demand functions satisfy the Law of Demand, then the aggregate excess demand will also satisfy the Law of Demand for price changes which do not induce 'large' income changes.

10.4 Gross Substitutes

The idea behind the formal definition of gross substitutes is exactly the intuitive idea of substitute commodities; we say that commodities i and j are gross substitutes if, whenever the price of i goes up, other things being equal, the demand for j goes up as well. This is as opposed to the more sophisticated (Hicksian) idea of substitutes, which says that commmodities i and j are substitutes iff:

$$S_{ij} \equiv \frac{\partial h_i}{\partial p_j} + h_j \frac{\partial h_i}{\partial w} > 0.$$

The formal definition which we will use is as follows.

10.16 Definition. Given the excess demand function, $\boldsymbol{z} \colon \Pi \to \mathbb{R}^n$, we shall say that the **commodities i and j are gross substitutes (finite increment form)**[11] iff, for any $\boldsymbol{p}^* \in \Pi$ and any *positive* real numbers, Δp_i and Δp_j we have:

$$z_i(\boldsymbol{p}^* + \Delta p_j \boldsymbol{e}_j) > z_i(\boldsymbol{p}^*) \text{ and } z_j(\boldsymbol{p}^* + \Delta p_i \boldsymbol{e}_i) > z_j(\boldsymbol{p}^*), \tag{10.16}$$

where '$\boldsymbol{e}_i$' and '$\boldsymbol{e}_j$' denote the ith and j^{th} unit coordinate vectors, respectively. We will say that **$\boldsymbol{z}(\cdot)$ satisfies (S)** iff commodities j and k are gross substitutes, for each j, k such that $k \neq j$.

We will also be interested in a differential version of Definition 10.16, stated as follows.

10.17 Definition. Given the differentiable excess demand function, $\boldsymbol{z} \colon \Pi \to \mathbb{R}^n$, we shall say that the **commodities i and j are gross substitutes** iff, for any $\boldsymbol{p}^* \in \Pi$, we have:

$$\frac{\partial}{\partial p_j}[z_i(\boldsymbol{p}^*)] > 0 \text{ and } \frac{\partial}{\partial p_i}[z_j(\boldsymbol{p}^*)] > 0. \tag{10.17}$$

Clearly, if $\boldsymbol{z}(\cdot)$ is differentiable and satisfies Definition 10.17, it also satisfies 10.16; although the converse is not quite true, even for differentiable excess demand functions.[12] The following example shows that our definitions are not vacuous.

10.18 Example. Suppose $\mathcal{E}$ is a pure exchange economy in which the ith consumer has the Cobb-Douglas utility function:

$$u_i(\boldsymbol{x}_i) = \prod_{j=1}^n x_{ij}^{a_{ij}}, \text{ where } a_{ij} > 0 \text{ for all } i, j, \text{ and } \sum_{j=1}^n a_{ij} = 1, \text{ for } i = 1, \ldots, m.$$

Then the ith consumer's excess demand function for the jth commodity is given by:

$$z_{ij}(\boldsymbol{p}) = \frac{a_{ij} \boldsymbol{p} \cdot \boldsymbol{r}_i}{p_j} - r_{ij}.$$

Thus we see that if $k \neq j$, then:

$$\frac{\partial}{\partial p_k}[z_{ij}(\boldsymbol{p})] = \frac{a_{ij} r_{ij}}{p_j} \geq 0.$$

[11] While we will not be dealing with the alternative defintion here, gross substitutes are sometimes defined by substituting weak inequalities for the strict inequalities we have used in (10.16). Usually, however, i and j are then said to be 'weak gross substitutes.'

[12] If $\boldsymbol{z}(\cdot)$ is differentiable and satisfies Definition 10.17, it may nonetheless be true that there exist price vectors at which the partial derivatives appearing in 10.17 are zero. There cannot, however, be any neighborhoods in which the partials are zero throughout the neighborhood.

Therefore, if:

$$\boldsymbol{r} \stackrel{\text{def}}{=} \sum_{i=1}^{m} \boldsymbol{r}_i \gg \mathbf{0},$$

it follows that, for $k \neq j$, we will have:

$$\frac{\partial}{\partial p_k}[z_j(\boldsymbol{p})] = \frac{\partial}{\partial p_k}\Big[\sum_{i=1}^{m} z_{ij}(\boldsymbol{p})\Big] = \Big(\frac{1}{p_j}\Big)\sum_{i=1}^{m} a_{ij} r_{ij} > 0, \text{ for } j = 1, \ldots, n. \quad \square$$

It is easily shown that if each consumer's excess demand function satisfies condition (S), then the aggregate excess demand function for the economy will satisfy condition (S) as well. Unfortunately, in an economy with production each individual consumer's *demand* function may satisfy condition (S), while the aggregate *excess* demand function fails to satisfy the condition.[13]

It can be shown that, under the assumptions which we have been employing, if $\boldsymbol{z}(\cdot)$ satisfies (S), and $\boldsymbol{p}^* \in \Pi$ is an equilibrium for $\boldsymbol{z}(\cdot)$ [so that $\boldsymbol{z}(\boldsymbol{p}^*) = \mathbf{0}$], then $\boldsymbol{p}^* \gg \mathbf{0}$ (for a proof, see Arrow, Block, and Hurwicz [1959]). This fact is employed in the next result.

10.19 Proposition. *If $\boldsymbol{z}(\cdot)$ is a continuous excess demand function satisfying (S), and if $\mathbf{p}^*$ and $\mathbf{p}'$ are equilibria for $\boldsymbol{z}(\cdot)$, then there exists $\theta \in \mathbb{R}_{++}$ such that $\mathbf{p}' = \theta\mathbf{p}^*$.*

Proof. Let:

$$\mu = \min\{p_1'/p_1^*, \ldots, p_n'/p_n^*\}.$$

By the homogeneity of $\boldsymbol{z}$, we have:

$$\boldsymbol{z}(\mu\boldsymbol{p}^*) = \boldsymbol{z}(\boldsymbol{p}^*) = \mathbf{0}. \tag{10.18}$$

If we suppose, by way of obtaining a contradiction, that $\boldsymbol{p}' \neq \mu\boldsymbol{p}^*$, then we must have, for some $i, k \in \{1, \ldots, n\}$:

$$p_i' = \mu p_i^*,\ p_k' > \mu p_k^*, \text{ and } p_j' \geq \mu p_j^* \text{ for all } j \neq i.$$

But then it follows from (S) that:

$$z_i(\boldsymbol{p}') > z_i(\mu\boldsymbol{p}^*);$$

which, together with (10.18), contradicts the assumption that $\boldsymbol{p}'$ is an equilibrium for $\boldsymbol{z}(\cdot)$. $\square$

The following result is of particular interest in connection with stability analysis. (See particularly, Proposition 10.24 of Section 7.) While the conclusion of the following result holds in an economy with any finite number of commodities, we will confine our argument to the case in which $n = 2$. (For a proof for the case of an arbitrary finite number of economies, see Arrow, Block, and Hurwicz [1959].)

10.20 Proposition. *If $\boldsymbol{z}(\cdot)$ is a continuous excess demand function satisfying (S) and the strong form of Walras' Law, and if $\mathbf{p}^*$ is an equilibrium for $\boldsymbol{z}(\cdot)$, then for any $\boldsymbol{p} \in \mathbb{R}_{++}$ which is not a scalar multiple of $\mathbf{p}^*$, we must have $\mathbf{p}^* \cdot \boldsymbol{z}(\boldsymbol{p}) > 0$.*

[13] For a more detailed discussion of this difficulty, see Mas-Colell, Whinston, and Green [1995], pp. 612–14.

Proof. (For the case in which $n = 2$). Suppose:

$$p_2^*/p_1^* > p_2/p_1,$$

and define $\boldsymbol{p}^\dagger$ and $\boldsymbol{p}'$ by:

$$\boldsymbol{p}^\dagger = (1, p_2^*/p_1^*) \text{ and } \boldsymbol{p}' = (1, p_2/p_1),$$

respectively. Then by homogeneity and (S), we have:

$$0 = z_1(\boldsymbol{p}^*) = z_1(\boldsymbol{p}^\dagger) > z_1(\boldsymbol{p}') = z_1(\boldsymbol{p}).$$

Since $z_1(\boldsymbol{p}) < 0$, it follows readily from (the strong form of) Walras' Law that we also have $z_2(\boldsymbol{p}) > 0$. Therefore:

$$\begin{aligned}(1/p_1^*)\boldsymbol{p}^* \cdot \boldsymbol{z}(\boldsymbol{p}) = \boldsymbol{p}^\dagger \cdot \boldsymbol{z}(\boldsymbol{p}') &= z_1(\boldsymbol{p}') + (p_2^*/p_1^*)z_2(\boldsymbol{p}') \\ &> z_1(\boldsymbol{p}') + (p_2/p_1)z_2(\boldsymbol{p}') = \boldsymbol{p}' \cdot \boldsymbol{z}(\boldsymbol{p}') = 0;\end{aligned}$$

where the first equality is by the homogeneity of $\boldsymbol{z}$, and the last is by Walras' Law. Consequently:

$$(1/p_1^*)\boldsymbol{p}^* \cdot \boldsymbol{z}(\boldsymbol{p}) > 0,$$

and thus $\boldsymbol{p}^* \cdot \boldsymbol{z}(\boldsymbol{p}) > 0$.

On the other hand, if:

$$p_2^*/p_1^* < p_2/p_1,$$

then:

$$p_1^*/p_2^* > p_1/p_2;$$

so that, if we define:

$$\boldsymbol{p}^\dagger = (p_1^*/p_2^*, 1) \text{ and } \boldsymbol{p}' = (p_1/p_2, 1),$$

it follows from the homogeneity of $\boldsymbol{z}$ and (S) that:

$$0 = z_2(\boldsymbol{p}^\dagger) > z_2(\boldsymbol{p}') = z_2(\boldsymbol{p}).$$

Proceeding as in the argument of the previous paragraph, we can then show that $z_1(\boldsymbol{p}') > 0$, and thus that:

$$\boldsymbol{p}^* \cdot \boldsymbol{z}(\boldsymbol{p}) > 0. \quad \square$$

10.5 Qualitative Economics

We will confine our investigation of the 'local' approach to the study of comparative statics in a general equilibrium context to a discussion of what is known as 'qualitative economics.' This is an area of investigation which was originally proposed and given its initial development by Samuelson [1947, Chapters 2 and 3], and further developed in its early stages by Gorman [1964], Lancaster [1962], and James Quirk and various collaborators (for example, Bassett, Maybee, and Quirk [1968]).

In our present discussion, we will borrow heavily from Lang, Moore, and Whinston [1995].[14]

The basic issue with which qualitative economics is concerned is the development of comparative statics results in an essentialy general equilibrium context, while making use of only purely qualitative information. We can illustrate the basic idea which motivated this work with a very simple illustration. Suppose in a given market, we can write the demand and supply functions as:

$$\delta(p, I) \text{ and } \sigma(p),$$

respectively, where 'p' and 'I' denote the price of the product and consumer income, respectively; and suppose further that for all relevant values of the variables, we have:

$$\frac{\partial \delta}{\partial p} < 0, \ \frac{\partial \delta}{\partial I} > 0, \ \text{ and } \sigma'(p) > 0.$$

As usual, we suppose the market is in equilibrium at (p^*, I^*) if:

$$\delta(p^*, I^*) - \sigma(p^*) = 0.$$

What I want to do now is to see whether we can deduce the effect on p of an increase in I, making use only of the information given above. Of course, I know that you probably already know perfectly well how to do this, but bear with me; I want to do this in a way which fairly naturally extends to a system of equations (and a general equilbrium system).

Formally, we know from the Implicit Function Theorem that if:

$$\frac{\partial}{\partial p}\big[\delta(p, I) - \sigma(p)\big]\Big|_{(p^*, I^*)} \neq 0, \tag{10.19}$$

then we can solve for p as a function of I in a neighborhood of (p^*, I^*). Moreover, if we denote this function by '$\rho(I)$,' then the functional solution obtained will satisfy the identity:

$$\delta\big[\rho(I), I\big] - \sigma\big[\rho(I)\big] \equiv 0. \tag{10.20}$$

I will leave it to you to verify that (10.19) is necessarily satisfied at an equilibrium. Differentiating the identity in equation (10.20), we have:

$$\frac{d}{dI}\Big(\delta[\rho(I), I] - \sigma[\rho(I)]\Big) = \frac{\partial \delta}{\partial p} \cdot \rho'(I) + \frac{\partial \delta}{\partial I} - \sigma' \cdot \rho'(I) = 0; \tag{10.21}$$

from which we obtain:

$$\rho'(I) = \frac{-\partial\delta/\partial I}{\partial\delta/\partial p - \sigma'}. \tag{10.22}$$

From the assumptions we have made regarding the signs of the relevant derivatives, it now follows that $\rho'(I) > 0$; that is, that the effect of an increase in I will be to increase the equilibrium price.

[14]For a more complete review of most aspects of this research, see Quirk and Saposnik [1968, Chapter 6], and for an alternative recent development of this type of material, see Fontaine, Garbely, and Gilli [1991], as well as McKenzie [2002, Chapter 2].2

Now, the intriguing thing about the derivation which we have just gone through is that we deduced the direction of change which will take place in equilibrium price solely on the basis of qualitative information and/or assumptions about the demand and supply functions. Given the difficulty in obtaining reliable numerical estimates in economics, this is obviously very important. So now the issue is, how well can we do in extending this mode of analysis to a general equilibrium context.

In general equilibrium analysis, we often deal with equilibrium conditions of the form:

$$f^i(x_1, \ldots, x_m; z_1, \ldots, z_n) = 0 \quad \text{for } i = 1, \ldots, m; \tag{10.23}$$

or, more compactly:

$$\boldsymbol{f}(\boldsymbol{x}; \boldsymbol{z}) = \boldsymbol{0}, \tag{10.24}$$

where:

$$\boldsymbol{f}\colon X \times Z \to \mathbb{R}^m, X \subseteq \mathbb{R}^m, \text{ and } Z \subseteq \mathbb{R}^n.$$

In this context, the vector $\boldsymbol{x}$ would generally represent *endogenous* variables, while the vector $\boldsymbol{z}$ would consist of n *exogenous* (possibly governmental *policy variables*); and we will say that $(\boldsymbol{x}^*, \boldsymbol{z}^*) \in X \times Z$ is an **equilibrium of the system** iff $\boldsymbol{f}(\boldsymbol{x}^*; \boldsymbol{z}^*) = \boldsymbol{0}$. Writing:

$$f^i_j(\boldsymbol{x}^*; \boldsymbol{z}^*) = \left.\frac{\partial f^i}{\partial x_j}\right|_{(\boldsymbol{x}^*, \boldsymbol{z}^*)} \quad \text{for } i, j = 1, \ldots, m,$$

and, similarly:

$$f^i_k(\boldsymbol{x}^*; \boldsymbol{z}^*) = \left.\frac{\partial f^i}{\partial z_k}\right|_{(\boldsymbol{x}^*; \boldsymbol{z}^*)} \quad \text{for } i = 1, \ldots, m, \text{ and } k = m+1, \ldots, m+n,$$

(for the sake of convenience, we will label the coordinates of vectors $\boldsymbol{z}$ by $m + 1, \ldots, m+n$)[15] it is common in applied general equilibrium analysis to specify the sign of each of these derivatives, or to specify that one or more of these derivatives is identically zero.

Now, if at an equilibrium $(\boldsymbol{x}^*; \boldsymbol{z}^*) \in X \times Z$, we have:

$$|\boldsymbol{J}| \neq 0,$$

where:

$$\boldsymbol{J} = \begin{bmatrix} f^1_1(\boldsymbol{x}^*; \boldsymbol{z}^*) & f^1_2(\boldsymbol{x}^*; \boldsymbol{z}^*) & \ldots & f^1_m(\boldsymbol{x}^*; \boldsymbol{z}^*) \\ f^2_1(\boldsymbol{x}^*; \boldsymbol{z}^*) & f^2_2(\boldsymbol{x}^*; \boldsymbol{z}^*) & \ldots & f^2_m(\boldsymbol{x}^*; \boldsymbol{z}^*) \\ \ldots & \ldots & \ldots & \ldots \\ f^m_1(\boldsymbol{x}^*; \boldsymbol{z}^*) & f^m_2(\boldsymbol{x}^*; \boldsymbol{z}^*) & \ldots & f^m_m(\boldsymbol{x}^*; \boldsymbol{z}^*) \end{bmatrix},$$

and '$|\boldsymbol{J}|$' denotes the determinant of $\boldsymbol{J}$, it follows from the Implicit Function Theorem that there exists a neighborhood, N of $\boldsymbol{z}^*$ (contained in Z), and a function $\boldsymbol{g}\colon N \to X$, such that $\boldsymbol{g}$ has continuous first partials, and satisfies:

$$(\forall \boldsymbol{z} \in N)\colon \boldsymbol{f}\big[\boldsymbol{g}(\boldsymbol{z}); \boldsymbol{z}\big] = \boldsymbol{0}. \tag{10.25}$$

[15] Thus, in particular, we write:

$$\boldsymbol{z}^* = (z^*_{m+1}, \ldots, z^*_{m+n}).$$

From (10.25) and the differentiability of $\boldsymbol{f}$ and $\boldsymbol{g}$, we then have, for each $\boldsymbol{z} \in N$:

$$\sum\nolimits_{j=1}^{m} f_j^i[\boldsymbol{g}(\boldsymbol{z}); \boldsymbol{z}] g_k^j(\boldsymbol{z}) + f_k^i[\boldsymbol{g}(\boldsymbol{z}), \boldsymbol{z}] = 0 \quad \text{for } i = 1, \ldots, m; k = m+1, \ldots, m+n.$$

Thus, we have, in particular:

$$\sum\nolimits_{j=1}^{m} f_j^i(\boldsymbol{x}^*; \boldsymbol{z}^*) g_k^j(\boldsymbol{z}^*) = -f_k^i(\boldsymbol{x}^*; \boldsymbol{z}^*) \quad \text{for } i = 1, \ldots, m; k = m+1; \ldots, m+n;$$

so that, defining:

$$\boldsymbol{g}_k(\boldsymbol{z}^*) = \big(g_k^1(\boldsymbol{z}^*), \ldots, g_k^n(\boldsymbol{z}^*)\big) \text{ and } \boldsymbol{f}_k(\boldsymbol{x}^*; \boldsymbol{z}^*) = \big(f_k^1(\boldsymbol{x}^*; \boldsymbol{z}^*), \ldots, f_k^n(\boldsymbol{x}^*; \boldsymbol{z}^*)\big)$$

we obtain the system:

$$\boldsymbol{J}\boldsymbol{g}_k(\boldsymbol{z}^*) = -\boldsymbol{f}_k(\boldsymbol{z}^*) \quad \text{for } k = m+1, \ldots, m+n. \tag{10.26}$$

From (10.26), we then obtain:

$$\boldsymbol{g}_k(\boldsymbol{z}^*) = -\boldsymbol{J}^{-1}\boldsymbol{f}_k(\boldsymbol{z}^*) \quad \text{for } k = m+1, \ldots, m+n. \tag{10.27}$$

Now, it is often possible to determine whether $|\boldsymbol{J}| \neq 0$ (and thus whether $\boldsymbol{J}^{-1}$ exists); in fact, to determine the sign of $|\boldsymbol{J}|$, solely on the basis of qualitative assumptions regarding $\boldsymbol{f}$; that is, solely on the basis of a specification of the signs of the partial derivatives, $f_j^i = \partial f^i / \partial x_j$ and $f_k^i = \partial f^i / \partial z_k$. Sometimes, though less often, it is actually possible to determine the signs of $g_k^i = \partial g^i / \partial z_k$ solely on the basis of this sort of qualitative information. The idea is this: let us specify only that each f_ℓ^i (globally) takes on one of the three values:

$$f_\ell^i = \begin{cases} + \\ 0 \\ - \end{cases} ; \tag{10.28}$$

and we can agree to use the following arithmetic for these symbols:

$$\begin{gathered}
(+) \cdot (+) = (-) \cdot (-) = +, \\
(+) \cdot (-) = (-) \cdot (+) = -, \\
(+) \cdot (0) = (-) \cdot (0) = (0) \cdot (+) = (0) \cdot (-) = 0, \\
(+) + (+) = (+) - (-) = +, \\
(-) + (-) = (-) - (+) = - \\
(+) - (+) = (+) + (-) = (-) + (+) = (-) - (-) = ?, \\
(+) + (0) = (0) + (+) = +, \\
(0) + (0) = (0) - (0) = 0, \\
(0) - (+) = (-) + (0) = -
\end{gathered} \tag{10.29}$$

We can illustrate the principles involved here with an example.

10.21 Example. Consider the system:

$$f^i(x_1, x_2; z_1, z_2) = 0 \quad \text{for } i = 1, 2; \tag{10.30}$$

and suppose:

$$\begin{array}{llll} f_1^1 = + & f_2^1 = - & f_3^1 = + & f_4^1 = -, \\ f_1^2 = + & f_2^2 = + & f_3^2 = - & f_4^2 = 0. \end{array} \tag{10.31}$$

Then:

$$|\boldsymbol{J}| = \begin{vmatrix} f_1^1 & f_2^1 \\ f_1^2 & f_2^2 \end{vmatrix} = \begin{vmatrix} + & - \\ + & + \end{vmatrix} = (+)\cdot(+) - (+)\cdot(-) = +. \tag{10.32}$$

Consequently we know that the function $\boldsymbol{g}\colon N \to X$ (giving $\boldsymbol{x}$ as a function of $\boldsymbol{z}$) will exist, for any system of functions satisfying the qualitative specification in (10.31).

For this example, and for $k = 1$ (or $k = m + 1 = 2 + 1 = 3$) equation (10.27) becomes:

$$\begin{aligned} \begin{pmatrix} g_1^1 \\ g_1^2 \end{pmatrix} &= \boldsymbol{J}^{-1} \cdot \begin{pmatrix} -f_3^1 \\ -f_3^2 \end{pmatrix} = \frac{-1}{|\boldsymbol{J}|} \begin{bmatrix} f_2^2 & -f_2^1 \\ -f_1^2 & f_1^1 \end{bmatrix} \begin{pmatrix} f_3^1 \\ f_3^2 \end{pmatrix} = \frac{-1}{|\boldsymbol{J}|} \begin{pmatrix} f_2^2 f_3^1 - f_2^1 f_3^2 \\ -f_1^2 f_3^1 + f_1^1 f_3^2 \end{pmatrix} \\ &= (-)\cdot \begin{pmatrix} (+)\cdot(+) - (-)\cdot(-) \\ (-)\cdot(+)\cdot(+) + (+)\cdot(-) \end{pmatrix} = \begin{pmatrix} (-)\cdot(?) \\ (-)\cdot(-) \end{pmatrix} = \begin{pmatrix} ? \\ + \end{pmatrix}. \end{aligned} \tag{10.33}$$

Similarly:

$$\begin{aligned} \begin{pmatrix} g_2^1 \\ g_2^2 \end{pmatrix} &= \boldsymbol{J}^{-1} \cdot \begin{pmatrix} -f_4^1 \\ -f_4^2 \end{pmatrix} = \frac{-1}{|\boldsymbol{J}|} \begin{bmatrix} f_2^2 & -f_2^1 \\ -f_1^2 & f_1^1 \end{bmatrix} \begin{pmatrix} f_4^1 \\ f_4^2 \end{pmatrix} = \frac{-1}{|\boldsymbol{J}|} \begin{pmatrix} f_2^2 f_4^1 - f_2^1 f_4^2 \\ -f_1^2 f_4^1 + f_1^1 f_4^2 \end{pmatrix} \\ &= (-)\cdot \begin{pmatrix} (+)\cdot(-) - (-)\cdot(0) \\ (-)\cdot(+)\cdot(-) + (+)\cdot(0) \end{pmatrix} = \begin{pmatrix} (-)\cdot(-) \\ (-)\cdot(+) \end{pmatrix} = \begin{pmatrix} + \\ - \end{pmatrix}. \end{aligned} \tag{10.34}$$

Thus, in this case we see that only $\partial x_2/\partial z_1 = g_1^2$ has an indeterminate sign. □

I will not pursue this topic further here. For those interested, it is probably still true that the best survey of this material, and introduction to the correspondence principle, is contained in Quirk and Saposnik [1968, Chapter 6]; and for a different, and quite promising approach, see Milgrom and Roberts [1994].

10.6 Stability in a Single Market

Turning our attention to the stability of competitive equilibrium, in this section we will introduce our topic by considering a markeet for a single commodity; which we suppose is characterized by demand and supply curves $D(p)$ and $S(p)$, respectively, where here 'p' denotes the (scalar) price of the good in question. We then denote the excess demand function by '$E(p)$;' that is:

$$E(p) \stackrel{\text{def}}{=} D(p) - S(p).$$

Stability (of competitive equilibrium) analysis is concerned with two related issues: (1) how does the market (or markets) behave out of equilibrium? and (2) does the market (or markets) approach equilibrium over time? The two classic (continuous time) adjustment mechanisms developed for the analysis of stability in a single market are due to Walras and Marshall.

The Walrasian adjustment mechanism, in its simplest formulation, postulates:

$$\frac{dp}{dt} \equiv \dot{p} = k \cdot [E(p)], \tag{10.35}$$

where k is a positive constant. The Marshallian adjustment mechanism is a bit more complicated to formulate, however we proceed as follows. First, we will suppose that both $D(\cdot)$ and $S(\cdot)$ are invertible; which, of course, they will be if D is everywhere downward-sloping, and S is everywhere upward-sloping. We can then interpret the values of $D^{-1}(x)$, for a given quantity of the commodity, x, as being the **demand price of the quantity x**; that is, the (maximum) price at which the quantity x can be sold. Similarly, for a given quantity, x, $S^{-1}(x)$, the **supply price of x** is the (minimum) price sufficient to bring the quantity x onto the market. The Marshallian adjustment mechanism is then, in its simplest form, given by:

$$\frac{dx}{dt} = \mu \cdot [D^{-1}(x) - S^{-1}(x)], \tag{10.36}$$

where μ is a positive constant.

10.22 Definitions. Suppose $p^* \in \mathbb{R}_{++}$ is such that $D(p^*) = S(p^*)$ and let $x^* = D(p^*) = S(p^*)$. We shall say that the equilibrium (x^*, p^*) is:

1. **Walrasian Stable** iff, for each $p > p^*$, $S(p) > D(p)$, and for each $p < p^*$, $S(p) < D(p)$.
2. **Marshallian stable** iff, for each $x > x^*, S^{-1}(x) > D^{-1}(x)$, and for each $x < x^*, S^{-1}(x) < D^{-1}(x)$.

If we compare these definitions with the adjustment mechanisms defined in (10.35) and (10.36), respectively, the logic of the definitions should be clear enough: the equilibrium is said to be Walrasian stable iff whenever $p > p^*$, $dp/dt < 0$, and whenever $p < p^*$, $dp/dt > 0$. Marshallian stability has an analogous interpretation in terms of changes in the quantity of the commodity on the market.

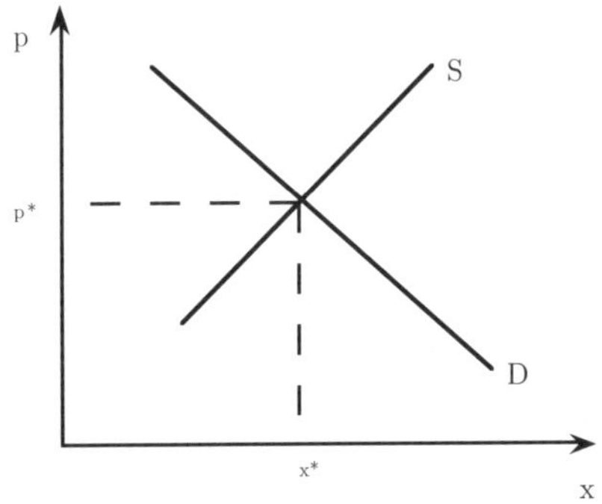

Figure 10.1: Demand and Supply: The Textbook Case.

In Figure 1, on the previous page, the market is both Marshallian stable and Walrasian stable. In Figure 2a, below, we have Marshallian, but not Walrasian stability; while in Figure 2b, we have Walrasian, but not Marshallian stability.

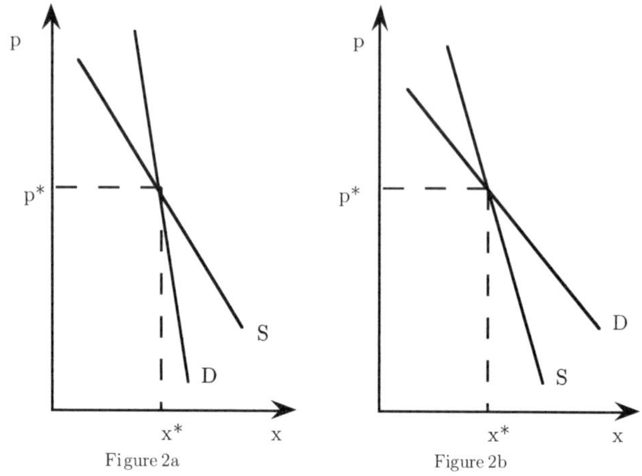

Figure 10.2: Stability and Instability.

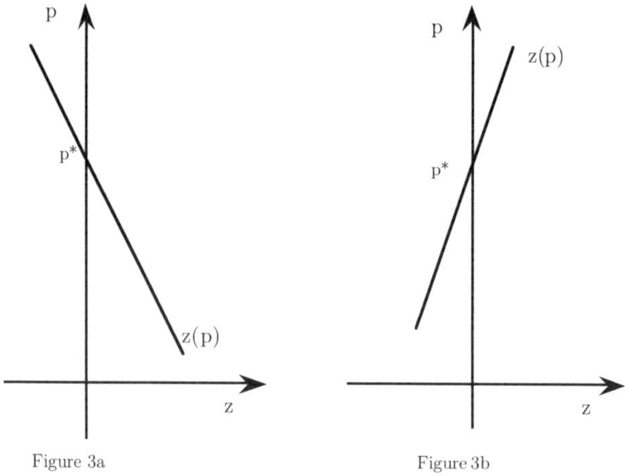

Figure 10.3: Excess Demand Functions.

Notice, however, that both Figures 1 and 2b correspond to an excess demand configuration like that depicted in Figure 3a; while Figure 2b corresponds to the sort of excess demand function shown in Figure 3b. (**Note:** in these two diagrams, we have used '$z(\cdot)$' to denote the excess demand function for this market; that is, for each p, we define $z(p) = D(p) - S(p)$.]

In order to study this question of stability, or lack thereof, in greater generality, we will need to set out our relationships more formally.

We will begin our analysis with a brief consideration of a more formal development of stability analysis and the Walrasian adjustment mechanism along the lines just discussed, but for an economy with only two goods. In such an economy it suffices to look at stability in one of the two markets, because, by Walras' law, if one market is in equilibrium, the other must be also. In fact, it can be shown that (in an exchange economy) if we are moving toward equilibrium in one market, we have to be moving toward equilibrium in the other as well.

In dealing with stability of competitive equilibrium, it is very convenient to normalize prices; in fact, many of our definitions will become simpler, and it is much easier to actually work through the mathematics in this case. Moreover, since excess demand functions are necessarily positively homogeneous of degree zero, it suffices to study normalized prices; which is what we will do in the remainder of our present discussion. We consider an economy, $\mathcal{E}$, in which $n = 2$, and normalize prices by setting $p_2 = 1$, and define:

$$p = p_1/p_2 = p_1, \tag{10.37}$$

where the second equality arises from our nornalization. Suppose further that the rules of change in market one are given by:

$$\dot{p} = dp/dt = f[\zeta(p)]; \tag{10.38}$$

where $f\colon \mathbb{R} \to \mathbb{R}$, and where '$\zeta(p)$' denotes excess demand in market one as a function of p; that is, we define:

$$\zeta(p) = z_1(p, 1). \tag{10.39}$$

Question for Discussion. If we define *equilibrium* in a market as a situation from which there is *no net tendency to change*, and if the rules of change in market one are given by (10.38), then consider the folllowing question (due to L. Hurwicz): "Under what conditions is it true that the market is in equilibrium if, and only if, supply equals demand?"

10.23 Definition. Let $f\colon \mathbb{R} \to \mathbb{R}$. We will say that $\boldsymbol{f}$ **is sign-preserving** iff, for all $z \in \mathbb{R}$:

1. $f(z) = 0$ iff $z = 0$, and:
2. $zf(z) > 0$ for $z \neq 0$.

A particular example of a sign-preserving function is the identity function $\boldsymbol{i}(\cdot)$, defined on $\mathbb{R}$ by:

$$\boldsymbol{i}(z) = z \quad \text{for } z \in \mathbb{R}. \tag{10.40}$$

Other examples are:

$$f(z) = \arctan z,$$

and

$$f(z) = \exp(z) - 1.$$

Of course, if we take any sign-preserving function, and multiply it by a positive constant, we get another sign-preserving function. Thus, if $\alpha > 0$ is a positive constant, the function:

$$f(z) = \alpha[\exp(z) - 1],$$

is also a sign-preserving function.

Now, if the rules of change in market one are given by (10.38), where f is a sign-preserving function, then:

$$\dot{p}\zeta(p) > 0, \tag{10.41}$$

for all $p \in \mathbb{R}_+$ such that $\zeta(p) \neq 0$ Thus, if $p^* \in \mathbb{R}_{++}$ is such that $\zeta(p^*) = 0$, and if $\zeta(\cdot)$ is downward-sloping, as in Figure 10.3.a, above (and thus the market is Walrasian Stable), then we will have:

$$\begin{cases} \dot{p} > 0 \iff p < p^*, \\ \dot{p} = 0 \iff p = p^*, \quad \text{and} \\ \dot{p} < 0 \iff p > p^*; \end{cases} \tag{10.42}$$

that is, price will increase over time if, and only if, price is initially below the equilibrium level, and so on.

Thus, in the case of just two markets, there isn't much to (*tâtonnement*) stability analysis. Essentiallly, if price in market one changes according to a rule of the form of (10.38), above, where f is a sign-preserving function, and if the excess demand function is downward-sloping, then an equilibrium of the economy must be (at least locally) stable. Complications arise, however, in connection with the parenthetic insertions in the statement just made, and in connection with intermarket reactions when there are more than two commodities, as we will see in the next section.[16]

10.7 Multi-Market Stability

The material in this section is only a sampling of results conerning multimarket stability, and is largely drawn from the work of Arrow and Hurwicz [1958] and Arrow, Block, and Hurwicz [1959]. More complete surveys are provided by Negishi [1962], Quirk and Saposnik [1968], Hahn [1982], Takayama [1985], and McKenzie [2002,Chapter 2].

In our discussion, we will use the same normalization that was introduced in the beginning of this section, except that we will suppose for notational convenience that there are $n+1$ commodities. In fact, we will normalize to consider price vectors of the form:

$$\boldsymbol{P} = (\boldsymbol{p}, 1),$$

where $\boldsymbol{p} \in \mathbb{R}^n_{++}$. We will assume throughout the remainder of this section that the excess demand function, $\boldsymbol{z}(\cdot)$ satisfies the properties which were introduced in Section 2; so that it satisfies the strong form of Walras' Law (W), homogeneity of degree zero (H), and continuity (C). It follows from Walras' Law (W) that we

[16] Complications also arise in connection with the treatment of time, continuous or discrete, but we will consider only continuous time models here.

need only consider equilibrium and price adjustments for the first n commoditities. Consequently, in considering stability issues we will consider a price adjustment mechanism of the form:

$$\frac{dp_j}{dt} = f_j[\zeta_j(\boldsymbol{p})] \quad \text{for } j = 1, \ldots, n; \tag{10.43}$$

where we will take $f_j(\cdot)$ to be a continous and sign-preserving function, and where we define $\boldsymbol{\zeta}\colon \mathbb{R}^n_{++} \to \mathbb{R}^n_+$ by:

$$\boldsymbol{\zeta}(\boldsymbol{p}) = \big(\zeta_i(\boldsymbol{p}), \ldots, \zeta_n(\boldsymbol{p})\big) = \big(z_i(\boldsymbol{p}, 1), \ldots, z_n(\boldsymbol{p}, 1)\big) = \boldsymbol{z}(\boldsymbol{p}, 1). \tag{10.44}$$

The system of equations (10.43) is a system of differential equations;[17] a solution of which is a function $\boldsymbol{\rho}\colon \mathbb{R}_+ \to \mathbb{R}^n_{++}$, satisfying:

$$\frac{d\rho_j}{dt} = f_j\big(z_j[\boldsymbol{\rho}(t)]\big) \quad \text{for } j = 1, \ldots, n. \tag{10.45}$$

Under the conditions we will be assuming to hold here, the system (10.43) will always possess a solution. However, the solution will not generally be unique unless we specify a starting value for $\boldsymbol{p}$; that is, a value $\boldsymbol{p}^0 \in \mathbb{R}^n_{++}$, which we take to be $\boldsymbol{\rho}(0)$. Thus a function $\boldsymbol{\rho}\colon \mathbb{R}_+ \to \mathbb{R}^n_{++}$ is said to be a **solution of (10.43), given the initial value $\boldsymbol{p}^0$**, iff:

$$\frac{d\rho_j}{dt} = f_j\big(z_j[\boldsymbol{\rho}(t)]\big) \quad \text{for } j = 1, \ldots, n, \text{ and all } t \in \mathbb{R}_+, \tag{10.46}$$

and:

$$\boldsymbol{\rho}(0) = \boldsymbol{p}^0. \tag{10.47}$$

Under the conditions which we are assuming to hold here, and subject to some mild technical qualifications (which we will ignore in this discussion), the system (10.43) will have a unique solution, for each initial value $\boldsymbol{p}^0 \in \Pi$.

For purposes of the present discussion, let us agree to call a pair $\langle \boldsymbol{\zeta}, \boldsymbol{f} \rangle$ a **price adjustment mechanism** iff $\boldsymbol{\zeta}(\cdot)$ is a continuous function, and $\boldsymbol{f}\colon \mathbb{R}^n \to \mathbb{R}^n$ is continuous and sign-preserving (that is, each f_j is a sign-preserving function). Given a pair, $\langle \boldsymbol{\zeta}, \boldsymbol{f} \rangle$, we shall say that a function $\boldsymbol{\rho}$ is a **solution for the mechanism, given the initial value $\boldsymbol{p}^0 \in \mathbb{R}^n_{++}$** iff ρ satisfies (10.46) and (10.47). We shall say that $\boldsymbol{p}^* \in \mathbb{R}^n_{++}$ is an **equilibrim for $\langle \boldsymbol{\zeta}, \boldsymbol{f} \rangle$** iff $\boldsymbol{\zeta}(\boldsymbol{p}^*) = \boldsymbol{0}$. Notice that, since $\boldsymbol{f}$ is sign-preserving, this last condition is equivalent to requiring that:

$$f_j[\zeta_j(\boldsymbol{p}^*)] = 0 \quad \text{for } j = 1, \ldots, n. \tag{10.48}$$

10.24 Definitions. If $\langle \boldsymbol{\zeta}, \boldsymbol{f} \rangle$ is a price adjustment mechanism, we shall say that:

1. an equilibrium for $\langle \boldsymbol{\zeta}, \boldsymbol{f} \rangle$, $\boldsymbol{p}^*$, is **globally stable for $\langle \boldsymbol{\zeta}, \boldsymbol{f} \rangle$** iff we have:

$$\lim_{t \to \infty} \boldsymbol{\rho}(t) = \boldsymbol{p}^*, \tag{10.49}$$

[17] The system corresponds to the Walrasian idea of *tâtonnement*. We will briefly consider the idea of a *non-tâtonnement* process in Section 8, below.

given any $\boldsymbol{p}^0 \in \mathbb{R}^n_{++}$, and any $\boldsymbol{\rho}(\cdot)$ which is a solution for $\langle \boldsymbol{\zeta}, \boldsymbol{f} \rangle$, given the initial value $\boldsymbol{p}^0$.

2. an equilibrium for $\langle \boldsymbol{\zeta}, \boldsymbol{f} \rangle$, $\boldsymbol{p}^*$, is **locally stable for** $\langle \boldsymbol{\zeta}, \boldsymbol{f} \rangle$ iff there exists a neighborhood of $\boldsymbol{p}^*$, $N(\boldsymbol{p}^*)$, such that for all $\boldsymbol{p}^0 \in N(\boldsymbol{p}^*)$, and any $\boldsymbol{\rho}(\cdot)$ which is a solution for $\langle \boldsymbol{\zeta}, \boldsymbol{f} \rangle$, given the initial value $\boldsymbol{p}^0$, we have:

$$\lim_{t \to \infty} \boldsymbol{\rho}(t) = \boldsymbol{p}^*.$$

3. **the mechanism** $\langle \boldsymbol{\zeta}, \boldsymbol{f} \rangle$ **is stable**, or that $\langle \boldsymbol{\zeta}, \boldsymbol{f} \rangle$ **posesses system stability**, iff for each $\boldsymbol{p}^0 \in \mathbb{R}^n_{++}$, and any $\boldsymbol{\rho}(\cdot)$ which is a solution for $\langle \boldsymbol{\zeta}, \boldsymbol{f} \rangle$, given the initial value $\boldsymbol{p}^0$, there exists an equilibrium for $\langle \boldsymbol{\zeta}, \boldsymbol{f} \rangle$, $\boldsymbol{p}^*$,, such that:

$$\lim_{t \to \infty} \boldsymbol{\rho}(t) = \boldsymbol{p}^*.$$

In connection with Definitions 10.21.1 and 10.21.2, it is worth noting that if $\boldsymbol{P}^*$ is an equilibrium for $\boldsymbol{z}(\cdot)$, then it follows from (H) that, for all $\lambda \in \mathbb{R}_{++}$, $\lambda \boldsymbol{P}^*$ is also an equilibrium for $\boldsymbol{z}(\cdot)$. Consequently, if we were to define concepts here for non-normalized [and $(n+1)$-dimensional] prices, we cannot have a unique equilibrium price vector; the most we can have is uniqueness up to a scalar multiple. Correspondingly, if we were to define concepts here for non-normalized price vectors, we could never have global stability in the sense of Definition 10.21.1.

In the analysis to follow, we will work with a special case of a price adjustment mechanism: namely those $\langle \boldsymbol{\zeta}, \boldsymbol{f} \rangle$ for which $\boldsymbol{f}$ is the identity function:

$$\boldsymbol{i}(\boldsymbol{\zeta}) = \boldsymbol{\zeta} \quad \text{for all } \boldsymbol{\zeta} \in \mathbb{R}^n. \tag{10.50}$$

Before proceeding further, let's take a moment to consider the form which Walras' Law, in the strong version, will take here. If we denote the excess demand function for the $(n+1)^{st}$ commodity by '$\zeta_{n+1}(\boldsymbol{p})$;' then, remembering that we are taking p_{n+1} to be identically equal to one, we will have, for all $\boldsymbol{p} \in \Pi$:

$$0 = (\boldsymbol{p}, 1) \cdot \big(\boldsymbol{\zeta}(\boldsymbol{p}), \zeta_{n+1}(\boldsymbol{p})\big) = \boldsymbol{p} \cdot \boldsymbol{\zeta}(\boldsymbol{p}) + \zeta_{n+1}(\boldsymbol{p});$$

or:

$$\boldsymbol{p} \cdot \boldsymbol{\zeta}(\boldsymbol{p}) = -\zeta_{n+1}(\boldsymbol{p}). \tag{10.51}$$

As a special, but particularly useful case of this, notice that if $\boldsymbol{\rho}(\cdot)$ is a solution for a mechanism, $\langle \boldsymbol{\zeta}, \boldsymbol{f} \rangle$, given $\boldsymbol{p}^0 \in \Pi$, then we will have, for all $t \in \mathbb{R}_+$:

$$\boldsymbol{\rho}(t) \cdot \boldsymbol{\zeta}[\boldsymbol{\rho}(t)] = -\zeta_{n+1}[\boldsymbol{\rho}(t)]. \tag{10.52}$$

Now, if $\boldsymbol{p}^*$ is an equilibrium for $\langle \boldsymbol{z}_i \rangle_{i \in M}$, and $\boldsymbol{\rho}(\cdot)$ is a solution for $\langle \boldsymbol{z}_i \rangle_{i \in M}$, given $\boldsymbol{p}^0 \in \Pi$, the distance of $\boldsymbol{\rho}(t)$ from $\boldsymbol{p}^*$ will be given by:

$$\|\boldsymbol{\rho}(t) - \boldsymbol{p}^*\| = \Big(\sum\nolimits_{j=1}^n [\rho_j(t) - p_j^*]^2\Big)^{1/2}. \tag{10.53}$$

It can be shown that:

$$\lim_{t \to \infty} \rho(t) = \boldsymbol{p}^*,$$

if, for all $t, t' \in \mathbb{R}_+$ such that $0 < t < t'$, we have:

$$\|\boldsymbol{\rho}(t') - \mathbf{p}^*\| < \|\boldsymbol{\rho}(t) - \mathbf{p}^*\|. \tag{10.54}$$

Inequality (10.54) will hold in turn if the function $V(\cdot)$ defined:

$$V(t) = (1/2)\|\boldsymbol{\rho}(t) - \mathbf{p}^*\|^2 = (1/2)\sum_{j=1}^{n}[\rho_j(t) - p_j^*]^2, \tag{10.55}$$

is strictly decreasing. Because of this relationship, the basic tool which we will use in our analysis is the following result.

10.25 Proposition. *Suppose $\boldsymbol{\zeta}(\cdot)$ satisfies (W) and (C), and that $\mathbf{p}^*$ is an equilibrium for $\langle \boldsymbol{z}_i \rangle_{i\in M}$ which satisfies the following condition (WA*): for all $\mathbf{p} \in \Pi$ such that $\mathbf{p} \neq \mathbf{p}^*$:*

$$\mathbf{p}^* \cdot \boldsymbol{\zeta}(\mathbf{p}) + \zeta_{n+1}(\mathbf{p}) > 0. \tag{10.56}$$

Then, given any $\mathbf{p}^0 \in \Pi$, if $\boldsymbol{\rho}(\cdot)$ is a solution for $\langle \boldsymbol{z}_i \rangle_{i\in M}$, given the initial value $\mathbf{p}^0$:

$$\frac{dV}{dt} = -\big(\mathbf{p}^* \cdot \boldsymbol{\zeta}[\boldsymbol{\rho}(t)] + \zeta_{n+1}[\boldsymbol{\rho}(t)]\big) < 0. \tag{10.57}$$

Proof. Since $\boldsymbol{\rho}(\cdot)$ satisfies (10.46) and (10.47) for the mechanism $\langle \boldsymbol{z}_i \rangle_{i\in M}$, the function $V(\cdot)$ is differentiable in t, and we have:

$$\begin{aligned}\frac{dV}{dt} &= (1/2)\Big(2\sum_{j=1}^{n}[\rho_j(t) - p_j^*]\cdot[d\rho_j/dt]\Big) = \sum_{j=1}^{n}[\rho_j(t) - p_j^*]\cdot\zeta_j[\boldsymbol{\rho}(t)] \\ &= \sum_{j=1}^{n}\rho_j(t)\cdot\zeta_j[\boldsymbol{\rho}(t)] - \sum_{j=1}^{n}p_j^*\cdot\zeta_j[\boldsymbol{\rho}(t)] \\ &= -\zeta_{n+1}[\boldsymbol{\rho}(t)] - \sum_{j=1}^{n}p_j^*\cdot\zeta_j[\boldsymbol{\rho}(t)]\end{aligned} \tag{10.58}$$

or:

$$\frac{dV}{dt} = -\big(\mathbf{p}^* \cdot \boldsymbol{\zeta}[\boldsymbol{\rho}(t)] + \zeta_{n+1}[\boldsymbol{\rho}(t)]\big); \tag{10.59}$$

where the second and last equalities in (10.58) are by (10.46) and (10.52), respectively. Our conclusion is then an immediate consequence of (10.59), given (10.56). □

The last part of the following result is a more or less immediate consequence of Proposition 10.22, above. We will simply accept the first part of the result without a formal proof; hopefully, however, our earlier discussion at least renders it intuitively plausible.

10.26 Theorem. (Arrow and Hurwicz [1958]) *If $\boldsymbol{z}(\cdot)$ satisfies (W) and (C), then the mechanism $\langle \boldsymbol{z}_i \rangle_{i\in M}$ posesses a unique solution, $\boldsymbol{\rho}(\cdot)$, for each initial value, $\mathbf{p}^0 \in \Pi$. Furthermore, if there exists an equilibrium for $\langle \boldsymbol{z}_i \rangle_{i\in M}$, $\mathbf{p}^*$, which satisfies (10.56) of Proposition 10.22, then $\mathbf{p}^*$ is globally stable for $\langle \boldsymbol{z}_i \rangle_{i\in M}$.*

The following result is now a more or less immediate consequence of the above Theorem and Proposition 10.20.

10.27 Theorem. (Arrow, Block, and Hurwicz [1959]) *If $\boldsymbol{z}(\cdot)$ satisfies (H), (W), (C), and the differential version of gross substitutability, as defined in Section 4, then the mechanism $\langle \boldsymbol{z}_i \rangle_{i\in M}$ posesses a unique solution, $\rho(\cdot)$, for each initial value, $\mathbf{p}^0 \in \Pi$. Furthermore, if $\mathbf{p}^*$ is an equilibrium for $\boldsymbol{\zeta}(\cdot)$, then given any $\mathbf{p}^0 \in \Pi$, the solution $\rho(t;\boldsymbol{p}^*)$ for $\langle \boldsymbol{z}_i \rangle_{i\in M}$, given $\mathbf{p}^0$, satisfies:*

$$\lim_{t\to+\infty} \rho(t;\boldsymbol{p}^0) = \boldsymbol{p}^*.$$

Unfortunately, not all aggregate excess demand functions satisfy condition (S), nor do they necessarily posess an equilibrium price satisfying (10.56). As a matter of fact, not all mechanisms are stable, as the following example demonstrates.

10.28 Example. (Scarf [1960]) Consider the pure exchange economy in which $m = 3, n = 3$, and the consumers have the respective utility functions:

$$\begin{aligned} u_1(\boldsymbol{x}_1) &= \min\{x_{11}, x_{12}\}, \\ u_2(\boldsymbol{x}_2) &= \min\{x_{22}, x_{23}\}, \text{ and} \\ u_3(\boldsymbol{x}_3) &= \min\{x_{31}, x_{33}\}; \end{aligned} \tag{10.60}$$

and that:

$$\boldsymbol{r}_1 = (1,0,0), \boldsymbol{r}_2 = (0,1,0), \text{ and } \boldsymbol{r}_3 = (0,0,1).$$

In this example, we will let the vector $\mathbf{p}$ *be three-dimensional.* Thus, for example, the first consumer's excess demand function will be given by:

$$\boldsymbol{z}_1(\boldsymbol{p}) = \Big(\frac{p_1}{p_1+p_2}, \frac{-p_2}{p_1+p_2}, 0\Big);$$

while the aggregate excess demand equations will be given by:

$$\begin{aligned} z_1(\boldsymbol{p}) &= \frac{p_3}{p_1+p_3} - \frac{p_2}{p_1+p_2} = \frac{p_1(p_3-p_2)}{(p_1+p_2)(p_1+p_3)}, \\ z_2(\boldsymbol{p}) &= \frac{p_1}{p_1+p_2} - \frac{p_3}{p_2+p_3} = \frac{p_2(p_1-p_3)}{(p_1+p_2)(p_2+p_3)}, \\ z_3(\boldsymbol{p}) &= \frac{p_2}{p_2+p_3} - \frac{p_1}{p_1+p_3} = \frac{p_3(p_2-p_1)}{(p_1+p_3)(p_2+p_3)}, \end{aligned} \tag{10.61}$$

respectively; and it is then easy to show that if $\boldsymbol{p}^*$ is an equilibrium for $\boldsymbol{z}(\cdot)$, we must have:

$$p_1^* = p_2^* = p_3^*. \tag{10.62}$$

Now suppose we have an initial value, $\boldsymbol{p}^0 \in \Pi$ satisfying:

$$\|\boldsymbol{p}^0\|^2 = 3 \text{ and } p_1^0 \cdot p_2^0 \cdot p_3^0 \neq 1 \tag{10.63}$$

[for example, $\boldsymbol{p}^0 = (1/\sqrt{3}, \sqrt{5/3}, 1)$]. Then we note first that if a solution, $\boldsymbol{\rho}(\cdot)$, exists for $\langle \boldsymbol{z}_i \rangle_{i\in M}$, then we must have:

$$\frac{d}{dt}\|\boldsymbol{\rho}(t)\|^2 = \frac{d}{dt}\Big[\sum_{j=1}^{3} \rho_j(t)^2\Big] = 2\Big(\rho_1(t)z_1[\boldsymbol{\rho}(t)] + \rho_2(t)z_2[\boldsymbol{\rho}(t)] + \rho_3(t)z_3[\boldsymbol{\rho}(t)]\Big) = 0, \tag{10.64}$$

for all t; where the last equality is by Walras' Law. Thus we see that the norm of $\boldsymbol{\rho}$ must remain constant, and it then follows from (10.62) and (10.63) that if $\boldsymbol{\rho}(\cdot)$ converges to an equilibrium, it must converge to the vector:

$$\boldsymbol{p}^* \stackrel{\text{def}}{=} (1,1,1).$$

However, again supposing that $\boldsymbol{\rho}(\cdot)$ is a solution for $\langle \boldsymbol{z}_i \rangle_{i \in M}$, given $\boldsymbol{p}^0$, consider the expression $\prod_{j=1}^3 \rho_j(t;\boldsymbol{p}^0)$. We have:

$$\begin{aligned}\frac{d}{dt}\Big[\prod\nolimits_{j=1}^3 \rho_j(t;\boldsymbol{p}^0)\Big] &= \frac{d}{dt}\Big[\prod\nolimits_{j=1}^3 p_j\Big] \\ &= \frac{dp_1}{dt}\cdot(p_2p_3) + \frac{dp_2}{dt}\cdot(p_1p_3) + \frac{dp_3}{dt}\cdot(p_1p_2) \\ &= z_1(\boldsymbol{p})\cdot(p_2p_3) + z_2(\boldsymbol{p})\cdot(p_1p_3) + z_3(\boldsymbol{p})\cdot(p_1p_2);\end{aligned} \tag{10.65}$$

where the last equation in (10.65) is by the assumption that $\boldsymbol{\rho}(\cdot)$ is a solution for $\langle \boldsymbol{z}_i \rangle_{i \in M}$, given $\boldsymbol{p}^0$. Substituting from (10.61), we then obtain:

$$\begin{aligned}&\frac{d}{dt}\Big[\prod\nolimits_{j=1}^3 \rho_j(t;\boldsymbol{p}^0)\Big] \\ &\quad= \frac{p_1p_2p_3\big[(p_3-p_2)(p_2+p_3)+(p_1-p_3)(p_1+p_3)+(p_2-p_1)(p_1+p_2)\big]}{(p_1+p_2)(p_1+p_3)(p_2+p_3)} \\ &= \frac{p_1p_2p_3}{(p_1+p_2)(p_1+p_3)(p_2+p_3)}\big[(p_3)^2-(p_2)^2+(p_1)^2-(p_3)^2+(p_2)^2-(p_1)^2\big] = 0.\end{aligned}$$

Thus we see that if $\boldsymbol{\rho}(\cdot)$ is a solution for $\langle \boldsymbol{z}_i \rangle_{i \in M}$, given $\boldsymbol{p}^0$, we must have:

$$\prod\nolimits_{j=1}^3 \rho_j(t;\boldsymbol{p}^0) = \prod\nolimits_{j=1}^3 p_j^0 \neq 1;$$

where the last equality is by our choice of $\boldsymbol{p}^0$. We see, therefore, that no such solution can converge to $\boldsymbol{p}^*$. □

10.8 A Note on Non-Tâtonnement Processes

The idea of tâtonnement, as originally set forth by Walras, is this. Imagine that all individuals interested in trading in a given commodity gather in a room at an appointed time. An official ('auctioneer' or 'referee') then announces a price for the commodity, and each individual writes on a card the amount he or she would buy (a positive number) or sell (a negative number) at that price. These cards are then passed in to the auctioneer, who adds up the totals. If the excess demand is positive, then the auctioneer announces a new price higher than the original, and if excess demand is negative, price is lowered. This process continues until excess demand is zero at some price; and then, and only then, does *any* trading actually take place.

In our study of stability in the preceding sections, we can be said to have studied whether this tâtonnement process will ever come to a halt. Of course, no real market functions exactly like the process just described; and, fortunately, our analysis did not need explicit assumptions about the institutional structure of the markets' functionings. However, we did assume throughout our analysis that the excess demand functions remained fixed for the duration of the adjustment process; an assumption which is difficult to justify unless we assume that no trading actually takes place until an equilibrium is achieved.

To see the significance of this last point, consider a pure exchange economy in which the *ith* individual's excess demand function is given by:

$$\boldsymbol{z}_i(\boldsymbol{p}) = \boldsymbol{d}_i(\boldsymbol{p}) - \boldsymbol{r}_i. \tag{10.66}$$

Agent i's initial resource endowment obviously affects this excess demand function: in fact, it does so in two ways: (a) directly, as it enters equation (10.66), and (b) indirectly, since in a pure exchange economy, $w_i(\boldsymbol{p}) = \boldsymbol{p} \cdot \boldsymbol{r}_i$, and:

$$\boldsymbol{d}_i(\boldsymbol{p}) = \boldsymbol{h}_i[\boldsymbol{p}, w_i(\boldsymbol{p})]. \tag{10.67}$$

We can formally indicate this dependence by writing agent i's excess demand function as:

$$\boldsymbol{z}_i = \boldsymbol{z}_i(\boldsymbol{p}; \boldsymbol{r}_i) \quad \text{for } i = 1, \ldots, m. \tag{10.68}$$

The basic idea of a non-tâtonnement process is simply that trade is allowed in the process of achieving equilibrium. Thus, our tâtonnement-type adjustment equation [(10.43) of Section 7):

$$\frac{dp_j}{dt} = f_j[z_j(\boldsymbol{p})] \quad \text{for } j = 1, \ldots, n; \tag{10.69}$$

is no longer valid, and must b e modified to reflect trades. Accordingly, we might replace (10.69) by:

$$\begin{aligned} \frac{dp_j}{dt} &= f_j\big(z_j[\boldsymbol{p}, (\boldsymbol{r}_i)]\big) && \text{for } j = 1, \ldots, n, \\ \frac{dr_{ij}}{dt} &= g_{ij}[\boldsymbol{p}, (\boldsymbol{r}_i)] && \text{for } i = 1, \ldots, m; j = 1, \ldots, n; \end{aligned} \tag{10.70}$$

where once again each f_j is taken to be a sign-preserving function, while $g_{ij}(\cdot)$ reflects the transaction rules governing trade out of equilibrium, and thus must satisfy:

$$\sum\nolimits_{i=1}^{m} g_{ij}[\boldsymbol{p}, (\boldsymbol{r}_i)] = 0 \quad \text{for } j = 1, \ldots, n; \text{ and all } (\boldsymbol{p}, (\boldsymbol{r}_i)). \tag{10.71}$$

A solution, given the initial values $\boldsymbol{p}^0$ and $(\boldsymbol{r}_i)$, is a function $(\boldsymbol{\rho}, \boldsymbol{\gamma})$ mapping $\mathbb{R}_+$ into $\mathbb{R}^{n(1+m)}$ satisfying:

$$\begin{aligned} \frac{d\rho_j}{dt} &= f_j\big(z_j[\boldsymbol{\rho}(t), \boldsymbol{\gamma}(t)]\big) && \text{for } j = 1, \ldots, n, \\ \frac{d\gamma_{ij}}{dt} &= g_{ij}[\boldsymbol{\rho}(t), \boldsymbol{\gamma}(t)] && \text{for } i = 1, \ldots, m; j = 1, \ldots, n; \end{aligned} \tag{10.72}$$

and:

$$\boldsymbol{\rho}(0) = \boldsymbol{p}^0 \quad \text{and} \quad \boldsymbol{\gamma}(0) = (\boldsymbol{r}_i). \tag{10.73}$$

While we will not pursue this topic further here, good introductions to non-tâtonnement process are provided in Arrow and Hahn [1971, Chapter 13], Negishi [1962, Sections 8 - 10], and Quirk and Saposnik [1968, pp. 191–3].

Exercises.

1. Show that if the aggregate demand function satisfies WA, then it satisfies WA*.

2. Verify the details of Example 10.5.

3. Suppose a consumer's preferences can be represented by the utility function:

$$u(\boldsymbol{x}) = \min\left\{\frac{x_1}{a_1}, \frac{x_2}{a_2}\right\}.$$

where $a_1, a_2 > 0$.

a. Find the consumer's demand functions for the two commodities.

b. Are the commodities gross substitutes in this case?

4. Construct an example with two consumers and three commodities in which a transfer of a unit of the numéraire (the third commodity) from the first consumer to the second results in an increase in excess demand for the second commodity.

5. Prove Corollary 10.10.

6. Show that if the aggregate supply correspondence is single-valued, so that we can consider it to be a function, $\boldsymbol{s}\colon \Pi \to \mathbb{R}^n$, then for all $\boldsymbol{p}, \boldsymbol{p}' \in \Pi$, we have:

$$-(\boldsymbol{p}' - \boldsymbol{p}) \cdot \left[\boldsymbol{s}(\boldsymbol{p}') - \boldsymbol{s}(\boldsymbol{p})\right] \leq 0.$$

Furthermore, if $\boldsymbol{s}(\boldsymbol{p}) \neq \boldsymbol{s}(\boldsymbol{p}')$, then the above inequality is strict.

7. Verify the details of Example 10.22

Chapter 11

The Core of an Economy

11.1 Introduction

In this chapter, we will be concerned with the core of a production economy. Some of the results and concepts become somewhat more difficult to deal with in this context than would be the case if we confined our attention to a pure exchange economy, but the generality we will gain, and the additional insights obtained in this context more than compensate for the slight added difficulty.

We will begin our discussion by considering a private ownership economy. In our discussion here, however, we will suppose that $\mathcal{E}$ takes the form:

$$\mathcal{E} = \big(\langle X_i, P_i\rangle, \langle Y_k\rangle, [s_{ik}]\big).$$

that is, we will dispense with the explicit display of the consumers' resourse endowments.[1] We will be assuming thoughout that, for each $i \in M$: (Please note that I am changing the notation slightly here; 'I,' rather than 'M' was used to denote this set in earlier chapters.)

$$P_i \text{ is irreflexive,}$$

where:

$$M \stackrel{\text{def}}{=} \{1, \ldots, m\}.$$

We will often be concerned with the allocation of consumption bundles to the consumers; denoting such an allocation by, for example, '$\langle \boldsymbol{x}_i\rangle_{i\in M}$,' which we can think of as vectors in $\mathbb{R}^{mn}$, or as a sequence of m vectors from $\mathbb{R}^n$. Formally, we define:

11.1 Definitions. If $\mathcal{E} = \big(\langle X_i, P_i\rangle, \langle Y_k\rangle, [s_{ik}]\big)$ is a private ownership economy, we will say that $\langle \boldsymbol{x}_i\rangle_{i\in M} \in \mathbb{R}^{mn}$ is a **consumption allocation for $\mathcal{E}$** iff:

$$\boldsymbol{x}_i \in X_i \quad \text{for } i = 1, \ldots, m;$$

and is an **attainable consumption allocation for $\mathcal{E}$** iff there exist $\boldsymbol{y}_k \in Y_k$ $(k = 1, \ldots, \ell)$ such that:

$$\sum\nolimits_{i=1}^{m} \boldsymbol{x}_i = \sum\nolimits_{k=1}^{\ell} \boldsymbol{y}_k. \tag{11.1}$$

[1] We can account for such endowments either by supposing that the first m production sets are of the form $Y_i = \{\boldsymbol{r}_i\}$, with $s_{ih} = 1$ for $i = h$ and 0 for $i \neq h$; or by interpreting the X_i as 'trading sets,' à la Chapter 4.

In other words, $\langle \boldsymbol{x}_i \rangle_{i\in M}$ is an attainable consumption allocation for $\mathcal{E}$ iff there exists $\langle \boldsymbol{y}_k \rangle_{k\in L}$ such that $\big(\langle \boldsymbol{x}_i \rangle_{i\in M}, \langle \boldsymbol{y}_k \rangle_{k\in L}\big) \in A(\mathcal{E})$. We will denote the set of all attainable consumption allocations for $\mathcal{E}$ by '$\boldsymbol{X}^*(\mathcal{E})$,' or simply by '$\boldsymbol{X}^*$' if the type of economy is understood.

In this chapter, we will generally express equation (11.1) as:

$$\sum_{i\in M} \boldsymbol{x}_i = \sum_{k\in L} \boldsymbol{y}_k, \tag{11.2}$$

where we define:

$$L = \{1, \ldots, \ell\}.$$

We will be considering possible actions of coalitions of consumers, where a coalition of consumers can be identified with a subset, S, of M; the idea here, of course, being that the coalition $S \subseteq M$ consists of those agents (consumers), i, such that $i \in S$. We will denote the collection of all such coalitions, that is, the collection of all non-empty subsets of M, by '$\mathbf{S}$.'

When dealing with coalitions, we will need to concern ourselves with the issue of what they could accomplish as a group if they operated as a separate sub-economy, independently of the other consumers. In doing this, we will take a bit different approach to the idea of ownership of firms than has been our custom. Up to this point, when we have considered a private ownership economy, $\mathcal{E} = (\langle X_i, P_i \rangle, \langle Y_k \rangle, \langle \boldsymbol{r}_i \rangle, [s_{ik}])$, we have supposed that s_{ik} $(i = 1, \ldots, m; k = 1, \ldots, \ell)$ represented the ith consumer's (proportionate) share of the profits of the kth firm. In this chapter, however, we suppose instead that the ith consumer controls the production set Z_{ik}, which, in the case of a private ownership economy would generally be defined as:

$$Z_{ik} = s_{ik} Y_k \stackrel{\text{def}}{=} \{\boldsymbol{z} \in \mathbb{R}^n \mid (\exists \boldsymbol{y}_k \in Y_k) \colon \boldsymbol{z} = s_{ik} \boldsymbol{y}_k\}. \tag{11.3}$$

With this definition, it is easy to prove the following (see the exercises at the end of this chapter).

11.2 Proposition. *Let* $\mathcal{E} = \big(\langle X_i, P_i \rangle, \langle Y_k \rangle, [s_{ik}]\big)$ *be an economy, and* $\mathbf{p}^* \in \mathbb{R}^n \setminus \{\mathbf{0}\}$. *Then:*

1. *if* $\boldsymbol{y}_k^*$ *maximizes* $\mathbf{p}^* \cdot \boldsymbol{y}$ *on* Y_k, *then:*

$$\boldsymbol{z}_{ik}^* \stackrel{def}{=} s_{ik} \boldsymbol{y}_k^*,$$

maximizes $\mathbf{p}^* \cdot \boldsymbol{z}$ *on* Z_{ik}*; and:*

2. *if* $s_{ik} > 0$, *and* $\boldsymbol{z}_{ik} \in Z_{ik}$ *maximizes* $\mathbf{p}^* \cdot \boldsymbol{z}$ *on* Z_{ik}, *then:*

$$\boldsymbol{y}_k \stackrel{def}{=} (1/s_{ik}) \boldsymbol{z}_{ik},$$

maximizes profits on Y_k. *Moreover,*

3. *if we define, for* $\mathbf{p} \in \Pi(Y_k) \equiv \Pi_k$*:*

$$\widehat{\pi}_{ik}(\boldsymbol{p}) = \max_{\boldsymbol{z}\in Z_{ik}} \boldsymbol{p} \cdot \boldsymbol{z}, \tag{11.4}$$

then (a) $\Pi_k \subseteq \Pi(Z_{ik})$, *(b) if* $s_{ik} > 0$, *then* $\Pi(Z_{ik}) = \Pi_k$, *and (c) for any* $\boldsymbol{p} \in \Pi_k$*:*

$$\widehat{\pi}_{ik}(\boldsymbol{p}) = s_{ik}\pi_k(\boldsymbol{p}) \equiv s_{ik} \max_{\boldsymbol{y} \in Y_k} \boldsymbol{p} \cdot \boldsymbol{y}.$$

With the ideas of the above paragraph and proposition in mind, we can define the ith consumer's production set, Z_i, for a private ownership economy, $\mathcal{E}$ as:

$$Z_i = \sum\nolimits_{k=1}^{\ell} Z_{ik} = \sum\nolimits_{k=1}^{\ell} s_{ik} Y_k. \tag{11.5}$$

It must be confessed at the outset that the ideas just presented are only completely consistent with our definition of the attainable set for the economy, $A(\mathcal{E})$, if each Y_k is convex and contains $\mathbf{0}$. We will come back to this point shortly; in the meantime, the following result notes how neatly things do work out if each Y_k is convex.

11.3 Proposition. *Suppose* $\mathcal{E} = \big(\langle X_i, P_i\rangle, \langle Y_k\rangle, [s_{ik}]\big)$ *is such that* Y_k *is convex, for* $k = 1, \ldots, \ell$, *and that* Z_i, $(i = 1, \ldots, m)$ *is defined as in equation (12.5), above. Then the following holds: If* $\boldsymbol{z}_i \in Z_i$, *for* $i = 1, \ldots, m$, *then:*

$$\sum_{i \in M} \boldsymbol{z}_i \in Y \equiv \sum_{k \in L} Y_k;$$

and conversely, if $\boldsymbol{y}^* \in Y$, *then there exist* $\boldsymbol{z}_i^* \in Z_i$ *for* $i = 1, \ldots, m$, *such that:*

$$\boldsymbol{y}^* = \sum_{i \in M} \boldsymbol{z}_i^*.$$

Proof. Suppose first that $\boldsymbol{z}_i \in Z_i$, for $i = 1, \ldots, m$. Then, by the definitions of Z_i and Z_{ik}, for each i there exist $\boldsymbol{y}_k^i$, for $k = 1, \ldots, \ell$, such that:

$$\boldsymbol{z}_i = \sum_{k \in L} s_{ik} \boldsymbol{y}_k^i. \tag{11.6}$$

However, since each Y_k is convex, and $\sum_{i \in M} s_{ik} = 1$:

$$\boldsymbol{y}_k \stackrel{\text{def}}{=} \sum_{i \in M} s_{ik} \boldsymbol{y}_k^i,$$

is an element of Y_k, for each $k \in L$. Moreover:

$$\sum_{i \in M} \boldsymbol{z}_i = \sum_{i \in M} \sum_{k \in L} s_{ik} \boldsymbol{y}_k^i = \sum_{k \in L} \sum_{i \in M} s_{ik} \boldsymbol{y}_k^i = \sum_{k \in L} \boldsymbol{y}_k \Big(\sum_{i \in M} s_{ik} \Big) = \sum_{k \in L} \boldsymbol{y}_k.$$

Conversely, suppose $\boldsymbol{y}^* \in Y$. Then there exist $\boldsymbol{y}_k^* \in Y_k$, for $k = 1, \ldots, \ell$, such that:

$$\boldsymbol{y}^* = \sum_{k \in L} \boldsymbol{y}_k^*.$$

But then, if for each i, we define $\boldsymbol{z}_{ik}^*$ by:

$$\boldsymbol{z}_{ik}^* = s_{ik} \boldsymbol{y}_k^*,$$

it follows from the definition of Z_{ik} that $\boldsymbol{z}_{ik}^* \in Z_{ik}$, for $k = 1, \ldots, \ell$. Furthermore, we have:

$$\sum_{i \in M} \sum_{k \in L} \boldsymbol{z}_{ik}^* = \sum_{k \in L} \sum_{i \in M} \boldsymbol{z}_{ik}^* = \sum_{k \in L} \sum_{i \in M} s_{ik} \boldsymbol{y}_k^* = \sum_{k \in L} \boldsymbol{y}_k^* \sum_{i \in M} s_{ik} = \sum_{k \in L} \boldsymbol{y}_k^* \equiv \boldsymbol{y}^*. \quad \square$$

We will explore these relationships in more depth in the next section.

11.2 Convexity and the Attainable Consumption Set

If you go back over the proof of Proposition 11.3, you can readily verify the fact that the convexity of the Y_k sets was not used in the second part of the proof. Consequently, the following corollary follows easily for those who are comfortable with the idea of set summation.

11.4 Corollary. *If $\mathcal{E} = \big(\langle X_i, P_i\rangle, \langle Y_k\rangle, [s_{ik}]\big)$ is a private ownership economy, then, given the definitions of the previous section:*

$$Y \equiv \sum_{k\in L} Y_k \subseteq \sum_{i\in M} Z_i. \tag{11.7}$$

Moreover, if Y_k is convex, for each $k \in L$, then we also have:

$$\sum_{i\in M} Z_i \subseteq Y. \tag{11.8}$$

Now let's return to the issue of why it is that our definitions are a bit inconsistent unless the Y_k are both convex and contain $\mathbf{0}$. First of all, without convexity of the individual production sets, Y_k, the inclusion in equation (12.8) will not necessarily hold; which means that without convexity, we cannot suppose consumer i can produce whatever net output vector $\boldsymbol{z}_i \in Z_i$ is desired, for $i = 1, \ldots, m$. That is, the combined results of such production will not necessarily be feasible, in the aggregate. There is a further difficulty, however. In our treatment of production, we have supposed that Y_k contains all of the production vectors which can be produced by the kth firm, and only those production vectors. In other words, Y_k contains all feasible production vectors, given the technology available to the kth firm, and given any fixed factors embodied in the firm's production facilities. Suppose now that there exists a production vector $\bar{\boldsymbol{y}}_k \in Y_k$ such that for all $\theta \in]0,1[$, we have:

$$\theta\bar{\boldsymbol{y}}_k \notin Y_k;$$

in other words, suppose Y_k does *not* satisfy non-increasing returns to scale. Then, while $s_{ik}\bar{\boldsymbol{y}}_k \in Z_{ik}$, for each i, $s_{ik}\bar{\boldsymbol{y}}_k \notin Y_k$. Thus in this case there will be elements of Z_{ik} which cannot actually be produced; for, given the definition of the sets Z_{ik} which was presented in the previous section, it would seem to be logical to define:

$$Z_{ik}^* = Z_{ik} \cap Y_k; \tag{11.9}$$

as the kth production set actually feasible for i, and the ith consumer's production set as:

$$Z_i^* = \sum\nolimits_{k=1}^{\ell} Z_{ik}^*. \tag{11.10}$$

If this definition is reasonable, then it presents some interesting insights into the potential gains from cooperation by the consumers. Consider, for instance, the following example.

11.5 Example. Let $\mathcal{E}$ be the private ownership economy in which $m = n = 2$, and in which there is one firm whose production set is given by:

$$Y = \{\boldsymbol{y} \in \mathbb{R}^2 \mid 0 \leq y_2 \leq -y_1 \text{ \& } y_1 \leq -4\}.$$

Suppose further that:

$$s_1 = s_2 = 1/2,$$

and that:

$$X_1 = X_2 = \{\boldsymbol{x}_i \in \mathbb{R}^2 \mid -3 \leq x_{i1} \leq 0 \text{ \& } x_{i2} \geq 1\}.$$

Then we have:

$$(-2, 2) \in X_i \cap Z_i,$$

if we simply define $Z_i = s_i Y$. However, if Z_i^* is defined as in equation (11.10), then it is easy to see that:

$$Z_i^* = Y \quad \text{for } i = 1, 2;$$

which means that, for each i:

$$X_i \cap Z_i^* = \emptyset.$$

In other words, neither consumer can survive utilizing only her or his own resources! On the other hand, suppose the consumers combine forces to form a firm with production set Y as originally given, so that:

$$Y = Z_1 + Z_2 = (1/2)Y + (1/2)Y_1 = Y.$$

Then, for example the consumption bundles $\boldsymbol{x}_i$ given by:

$$\boldsymbol{x}_1 = \boldsymbol{x}_2 = (-2, 2),$$

are such that $\boldsymbol{x}_i \in X_i$ for each i, and:

$$\boldsymbol{x}_1 + \boldsymbol{x}_2 = (-4, 4) \in Y;$$

so that $\langle(\boldsymbol{x}_i, \boldsymbol{z}_i)\rangle_{i \in M}$ is attainable. □

If, however, $\mathbf{0} \in Y_k$ and Y_k is convex (in other words, if Y_k satisfies non-increasing returns, see Proposition 6.2), then the distinction between Z_{ik} and Z_{ik}^* disappears; in fact, we have the following, the proof of which I will leave as an exercise.

11.6 Proposition. *If Y_k is convex and contains the origin, and Z_{ik} and Z_{ik}^* are defined as in (11.3) and (11.9), respectively, then $Z_{ik} = Z_{ik}^*$.*

In the remainder of this chapter, we consider economies $\mathbb{E} = \big\langle(X_i, P_i, Z_i)\big\rangle$, where X_i and Z_i are nonempty subsets of $\mathbb{R}^n$, and P_i is an irreflexive binary relation on X_i, for $i = 1, \ldots, m$. We will also assume that, for each i:

$$X_i \cap Z_i \neq \emptyset.$$

We will not be assuming that the Z_i sets are necessarily of the form:

$$Z_i = \sum_{k \in L} Z_{ik} = \sum_{k \in L} s_{ik} Y_k, \tag{11.11}$$

and thus that the economy $\mathbb{E}$ is derived from a private ownership economy, $\mathcal{E}$; however, we will not be ruling this possibility out either. Consequently, it will be worth our while to consider some further aspects of the relationship between $\mathbb{E}$ and $\mathcal{E}$ if the former *is* derived from the latter.

We can begin by introducing the definition we will use for attainable allocations and competitive equilibria in economies $\mathbb{E} = \langle (X_i, P_i, Z_i) \rangle$.

11.7 Definitions. We will say that $\langle (\boldsymbol{x}_i, \boldsymbol{z}_i) \rangle_{i \in M} \in \mathbb{R}^{2mn}$ is a **feasible allocation** for $\mathbb{E} = \langle (X_i, P_i, Z_i) \rangle$ iff:

$$(\boldsymbol{x}_i, \boldsymbol{z}_i) \in X_i \times Z_i \quad \text{for } i = 1, \ldots, m,$$

and:

$$\sum_{i \in M} (\boldsymbol{x}_i - \boldsymbol{z}_i) = \mathbf{0}.$$

As we did in the case of private ownership economies, we will denote the set of all attainable consumption allocations for $\mathbb{E}$ by '$\boldsymbol{X}^*(\mathbb{E})$.'

11.8 Definition. We will say that a tuple, $(\langle \boldsymbol{x}_i^*, \boldsymbol{z}_i^* \rangle, \boldsymbol{p}^*)$ is a **competitive** (or **Walrasian**) **equilibrium** for $\mathbb{E} = \langle (X_i, P_i, Z_i) \rangle$ iff

1. $\boldsymbol{p}^* \neq \mathbf{0}$,
2. $\langle (\boldsymbol{x}_i^*, \boldsymbol{z}_i^*) \rangle_{i \in M}$ is an attainable allocation for $\mathbb{E}$,
3. for each i, we have:

$$(\forall \boldsymbol{z}_i \in Z_i) \colon \boldsymbol{p}^* \cdot \boldsymbol{z}_i \leq \boldsymbol{p}^* \cdot \boldsymbol{z}_i^*,$$
$$\boldsymbol{p}^* \cdot \boldsymbol{x}_i^* \leq \boldsymbol{p}^* \cdot \boldsymbol{z}_i^*$$

and:

$$(\forall \boldsymbol{x}_i \in X_i) \colon \boldsymbol{x} P_i \boldsymbol{x}_i^* \Rightarrow \boldsymbol{p}^* \cdot \boldsymbol{x}_i > \boldsymbol{p}^* \cdot \boldsymbol{z}_i^*.$$

The following proposition, the proof of which I will leave as an exercise, sets forth the relationship between competitive equilibria for a private ownership economy, $\mathcal{E}$, and an economy $\mathbb{E} = \langle (X_i, P_i, Z_i) \rangle$, assuming that the latter is derived from $\mathcal{E}$ as per equation (11.11).

11.9 Proposition. *Suppose $\mathcal{E} = (\langle X_i, P_i \rangle, \langle Y_k \rangle, [s_{ik}])$ is a private ownership economy, and that $\mathbb{E} = \langle (X_i, P_i, Z_i) \rangle$ is derived from $\mathcal{E}$ as per equation (11.5). Then we have the following.*

1. If $(\langle \boldsymbol{x}_i^* \rangle, \langle \boldsymbol{y}_k^* \rangle, \boldsymbol{p}^*)$ is a competitive equilibrium for $\mathcal{E}$, and we define $\boldsymbol{z}_i^*$ by:

$$\boldsymbol{z}_i^* = \sum_{k \in L} s_{ik} y_k^*,$$

for $i = 1, \ldots, m$, then $(\langle \boldsymbol{x}_i^*, \boldsymbol{z}_i^* \rangle, \boldsymbol{p}^*)$ is a competitive equilibrium for $\mathbb{E}$.

2. If $(\langle \boldsymbol{x}_i^*, \boldsymbol{z}_i^* \rangle, \boldsymbol{p}^*)$ is a competitive equilibrium for $\mathbb{E}$, then $\boldsymbol{z}_i^*$ is of the form:

$$\boldsymbol{z}_i^* = \sum_{k \in L} s_{ik} \boldsymbol{y}_k^i,$$

for $i = 1, \ldots, m$: and if each Y_k is convex, and we define $\boldsymbol{y}_k^*$ by:

$$\boldsymbol{y}_k^* = \sum_{i \in M} s_{ik} \boldsymbol{y}_k^i,$$

for $i = 1, \ldots, m$, then $(\langle \boldsymbol{x}_i^* \rangle, \langle \boldsymbol{y}_k^* \rangle, \boldsymbol{p}^*)$ is a competitive equilibrium for $\mathcal{E}$.

11.3 The Core of a Production Economy

As mentioned earlier,in this chapter when we say $\mathbb{E} = \langle (X_i, P_i, Z_i) \rangle$ *is an economy, we will always suppose that the following condition holds:*

$$X_i \cap Z_i \neq \emptyset \quad \text{for } i = 1, \ldots, m; \tag{11.12}$$

in other words, for each $i \in M$, *we suppose that there exist* $\bar{\boldsymbol{x}}_i \in X_i$ *and* $\bar{\boldsymbol{z}}_i \in Z_i$ *such that:*

$$\bar{\boldsymbol{x}}_i = \bar{\boldsymbol{z}}_i. \tag{11.13}$$

The assumption expressed as equation (11.12) is fairly restrictive; in a modern industrialized society, individuals specialize in the expectation of being able to purchase (or trade for) necessities which they themselves do not produce. On the other hand, in much of the literature, it is supposed that $X_i = \mathbb{R}^n_+$ for each i, and that $\boldsymbol{r}_i \in \mathbb{R}^n_+$ as well; consequently, the condition in (11.12) generalizes the assumption commonly-used. In any event, the present assumption will be important in our development here; so much so that we will make use of the following definition.

11.10 Definition. In an economy $\mathbb{E} = \langle (X_i, P_i, Z_i) \rangle$ we define the (**individually**) **attainable set for** $\boldsymbol{i}$, X_i^*, by:

$$X_i^* = X_i \cap Z_i; \tag{11.14}$$

that is:

$$X_i^* = \{\boldsymbol{x}_i \in X_i \mid (\exists \boldsymbol{z}_i \in Z_i) \colon \boldsymbol{x}_i = \boldsymbol{z}_i\}. \tag{11.15}$$

The definition of X_i^* is extended to coalitions of consumers in the following.[2]

11.11 Definition. Let S be a non-empty subset of M (so that $S \in \mathbb{S}$). We will say that $\langle (\boldsymbol{x}_i, \boldsymbol{z}_i) \rangle_{i \in S}$ is **attainable for** $\boldsymbol{S}$, or **feasible for** $\boldsymbol{S}$, iff:

$$\boldsymbol{x}_i \in X_i \text{ and } \boldsymbol{z}_i \in Z_i \text{ for all } i \in S, \tag{11.16}$$

and:

$$\sum_{i \in S} \boldsymbol{x}_i = \sum_{i \in S} \boldsymbol{z}_i. \tag{11.17}$$

11.12 Definition. Let $\langle \boldsymbol{x}_i^* \rangle_{i \in M}$ be a consumption allocation for $\mathbb{E}$, and let $S \in \mathbb{S}$ be a coalition. We shall say that $\langle \boldsymbol{x}_i^* \rangle_{i \in M}$ **can be improved upon by the coalition** $\boldsymbol{S}$ (or is **blocked by** $\boldsymbol{S}$) iff there exists an allocation, $\langle (\boldsymbol{x}_i, \boldsymbol{z}_i) \rangle_{i \in S}$, which is feasible for S, and satisfies:

$$(\forall i \in S) \colon \boldsymbol{x}_i P_i \boldsymbol{x}_i^*. \tag{11.18}$$

11.13 Definition. The **core of an economy** $\mathbb{E} = \langle (X_i, P_i, Z_i) \rangle$, is defined as the set of all attainable *consumption allocations* for $\mathbb{E}$ which cannot by improved upon (or blocked) by any coalition, $S \in \mathbb{S}$. We will denote the set of all core allocations for $\mathbb{E}$ by '$\boldsymbol{C}(\mathbb{E})$.'

[2]Remember that a 'coalition' is simply a nonempty subset of M. The collection of all coalitions available in an economy is then the set of all nonempty subsets of M, which set we denote by '$\mathcal{S}$.'

Notice that $\boldsymbol{C}(\mathbb{E}) \subseteq \boldsymbol{X}^*(\mathbb{E})$. In the material to follow, we will frequently be concerned with another subset of $\boldsymbol{X}^*(\mathbb{E})$, defined as follows.

11.14 Definition. Given an economy, $\mathbb{E} = \langle (X_i, P_i, Z_i) \rangle$, and a consumer $h \in M$, we shall say that a consumption allocation $\langle \boldsymbol{x}_i^* \rangle$ is **individually rational for *h*** iff for every $\boldsymbol{x}_h \in X_h^*$:

$$\boldsymbol{x}_h^* G_h \boldsymbol{x}_h,$$

where 'G_h' denotes the negation of P_h. We denote the set of all individually rational allocations for h by $\boldsymbol{I}_h^*(\mathbb{E})$; and, finally, we define the set of **individually rational allocations for** $\mathbb{E}$, denoted by '$\boldsymbol{I}(\mathbb{E})$,' by:

$$\boldsymbol{I}(\mathbb{E}) = \bigcap_{i \in M} \boldsymbol{I}_i^*(\mathbb{E}).$$

Notice that $\boldsymbol{C}(\mathbb{E}) \subseteq \boldsymbol{I}(\mathbb{E}) \subseteq \boldsymbol{X}^*(\mathbb{E})$ [to see the first inclusion, consider the coalition $S = \{i\}$, for an arbitrary $i \in M$]. Moreover, if we denote the set of all Pareto efficient consumption allocations for $\mathbb{E}$ by '$\boldsymbol{P}(\mathbb{E})$,' then we also have:

$$\boldsymbol{C}(\mathbb{E}) \subseteq \boldsymbol{P}(\mathbb{E}) \cap \boldsymbol{I}(\mathbb{E}).$$

Of course, if $M = 2$, then the only non-empty subsets of M are $\{1\}, \{2\}$, and $\{1, 2\}$; and thus $\boldsymbol{C}(\mathbb{E}) = \boldsymbol{I}(\mathbb{E}) \cap \boldsymbol{P}(\mathbb{E})$, which should explain the Edgeworth Box diagram, Figure 11.1, below.[3]

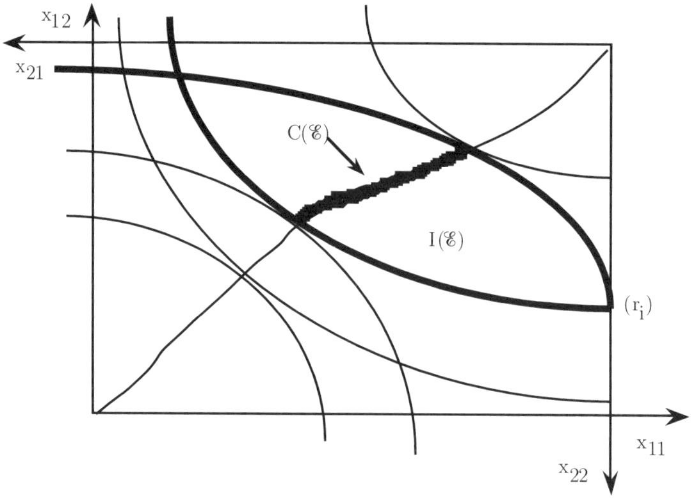

Figure 11.1: Individually Rational Allocations and the Core.

It may be helpful at times to consider a slightly different, but logically equivalent way of defining the core of an economy. We begin by defining, for each $S \in \mathbf{S}$, the

[3]In which we have an exchange economy in mind, of course. Notice that in terms of the framework being utilized here, $\mathbb{E} = \langle (X_i, P_i, Z_i) \rangle$ is an exchange economy if, and only if for each $i \in M$, $X_i = \mathbb{R}_+^n$ and there exists $\boldsymbol{r}_i \in \mathbb{R}_+^n$ such that $Z_i = \{\boldsymbol{r}_i\}$.

collection of all attainable allocations for $\mathbb{E}$ which can be improved upon by S, by '$\boldsymbol{D}(S)$, that is:

$$\boldsymbol{D}(S) = \Big\{ \langle \boldsymbol{x}_i^* \rangle_{i \in M} \in \boldsymbol{X}^*(\mathbb{E}) \mid \big(\exists \langle (\boldsymbol{x}_i, \boldsymbol{z}_i) \rangle_{i \in S}\big) \colon \sum_{i \in S} \boldsymbol{x}_i = \sum_{i \in S} \boldsymbol{z}_i$$
$$\text{and } (\forall i \in S) \colon \boldsymbol{x}_i \in X_i,\ \boldsymbol{z}_i \in Z_i \;\&\; \boldsymbol{x}_i P_i \boldsymbol{x}_i^* \Big\} \quad (11.19)$$

We can then define the core of $\mathbb{E}$ by:

$$\boldsymbol{C}(\mathbb{E}) = \boldsymbol{X}^*(\mathbb{E}) \setminus \bigcup_{S \in \mathbb{S}} \boldsymbol{D}(S). \quad (11.20)$$

By analogy with the definition of 'individually rational' allocations, we might say that the collection of core allocations, $\boldsymbol{C}(\mathbb{E})$, is exactly the set of all allocations which are 'coalition-rational' for each possible coalition, S. Because of this, we will think of the core as being primarily a welfare criterion; that is, it would appear that there is at least some interest in studying $\boldsymbol{C}(\mathbb{E})$ on the grounds that many, if not most people might accept the criterion that a 'good' allocation should be in the core.[4] However, many economists, for example W. Hildenbrand and A. P. Kirman, have felt that the core of an economy is of great interest as a solution concept; that is, they feel that the natural outcome of unfettered economic activity is that one attains an outcome in $\boldsymbol{C}(\mathbb{E})$. In fact, this is the case if the economy is competitive in the classical sense, as is shown by the following result.[5] We will not prove Theorem 11.15 here, incidentally, since it is a special case of Theorem 11.17, which we will present and prove in the next section.

11.15 Theorem. *If $\big(\langle \boldsymbol{x}_i^*, \boldsymbol{z}_i^* \rangle, \boldsymbol{p}^*\big)$ is a competitive (or Walrasian) equilibrium for $\mathbb{E}$, then $\langle \boldsymbol{x}_i^* \rangle_{i \in M}$ is a core allocation for $\mathbb{E}$; that is:*

$$\langle \boldsymbol{x}_i^* \rangle_{i \in M} \in \boldsymbol{C}(\mathbb{E}).$$

11.16 Definitions. Given an economy, $\mathbb{E} = \big\langle (X_i, P_i, Z_i) \big\rangle$, we define the set of all **Walrasian allocations for** $\mathbb{E}$, $\boldsymbol{\mathcal{W}}(\mathbb{E})$, by:

$$\boldsymbol{\mathcal{W}}(\mathbb{E}) = \big\{ \langle (\boldsymbol{x}_i^*, \boldsymbol{z}_i^*) \rangle_{i \in M} \in A(\mathbb{E}) \mid \big(\exists \boldsymbol{p}^* \in \mathbb{R}^n\big) \colon$$
$$\big(\langle \boldsymbol{x}_i^*, \boldsymbol{z}_i^* \rangle, \boldsymbol{p}^*\big) \text{ is a Walrasian equilibrium for } \mathbb{E} \big\}$$

We then define the set of **Walrasian consumption allocations for** $\mathbb{E}$, $W(\mathbb{E})$, by:

$$\boldsymbol{W}(\mathbb{E}) = \big\{ \langle \boldsymbol{x}_i^* \rangle_{i \in M} \in \boldsymbol{X}^*(\mathbb{E}) \mid (\exists \langle \boldsymbol{z}_i^* \rangle) \colon \langle (\boldsymbol{x}_i^*, \boldsymbol{z}_i^*) \rangle_{i \in M} \in \boldsymbol{\mathcal{W}}(\mathbb{E}) \big\}.$$

Thus, it follows from Theorem 11.15 that $\boldsymbol{W}(\mathbb{E}) \subseteq \boldsymbol{C}(\mathbb{E})$. Notice also that Theorem 11.15 generalizes the 'First Fundamental Theorem of Welfare Economics;' or, more specifically, the version of this result presented as Theorem 7.22. In general,

[4] Although many would also specify the further condition that this is true only after an appropriate redistribution of initial endowments takes place.

[5] We will also consider the idea of the core as a solution, or equilibrium concept in more detail in Chapter 16.

of course, the set $\boldsymbol{C}(\mathbb{E})$ is considerably larger than $\boldsymbol{W}(\mathbb{E})$. We will show, however, that with a sufficiently large number of agents, the situation changes; in some sense, as the economy 'grows large,' the core 'shrinks' to the set of competitive equilibria.

In the literature on the core, there are two basic approaches to an exact statement and proof of this last assertion; the first of which derives from a result which was first rigorously stated and proved by Debreu and Scarf [1963], although it was suggested earlier by Edgeworth, and the second of which is derived from results by Aumann [1964] and Arrow and Hahn [1971]). These two approaches yield results which can be (and have been) rather imperfectly stated in two distinct fashions, as follows.

1. As the number of agents in the economy grows, $\boldsymbol{C}(\mathbb{E})$ 'shrinks' to the set of Walrasian allocations, $\boldsymbol{W}(\mathbb{E})$.

2. If an economy is 'sufficiently large,' then any allocation in the core is a Walrasian (competitive equilibrium) allocation.

The Debreu-Scarf result is a formalization of statement 1, and in which the number of agents in the economy grows large in a very specific way; while the Aumann/Arrow-Hahn theorem more directly relates to statement 2. We will confine our discussion here to results along the lines of the Debreu-Scarf result; that is, to results along the lines of statement 1, above. For discussions of the second approach, let me recommend Anderson [1978, 1986], and Hildenbrand [1982].

11.4 The Core in Replicated Economies

Given an economy, $\mathbb{E} = \langle (X_i, P_i, Z_i) \rangle$, we consider the sequence of related economies, $\mathbb{E}_q$, defined in the following way.

$$\mathbb{E}_1 = \mathbb{E},$$

$$\dots\dots$$

$$\mathbb{E}_q = \langle (X_{hi}, P_{hi}, Z_{hi}) \rangle_{(h,i)\in Q\times M}, \text{ where } Q = \{1\dots, q\}, \text{ and :}$$
$$X_{hi} = X_i, P_{hi} = P_i, \text{ and } Z_{hi} = Z_i \text{ for } h = 1, \dots, q; i = 1, \dots, m.$$

Thus, in $\mathbb{E}_q$, the agents (consumers) have a double index; agent (h, i) is the hth agent of the ith type. If $h, h' \in Q$, and $i \in M$, then agents (h, i) and (h', i) are 'economic twins,' in the sense of having precisely the same economic characteristics. We will refer to $\mathbb{E}_q$ as the **q-fold replication of** $\mathbb{E}$. In dealing with $\mathbb{E}_q$, we will use the notation '$\langle \boldsymbol{x}_{hi} \rangle_{(h,i)\in Q\times M}$' to denote consumption allocations for $\mathbb{E}_q$; and we define $\boldsymbol{X}^*(\mathbb{E}_q)$ as the set of all consumption allocations $\langle \boldsymbol{x}_{hi} \rangle_{(h,i)\in Q\times M}$ such that there exist $\langle \boldsymbol{z}_i \rangle_{i\in M}$ satisfying:

$$\boldsymbol{x}_{hi} \in X_i \;\&\; \boldsymbol{z}_{hi} \in Z_i \quad \text{for } h = 1, \dots, q; i = 1, \dots, m, \tag{11.21}$$

and:

$$\sum_{h\in Q}\sum_{i\in M} \boldsymbol{x}_{hi} = \sum_{h\in Q}\sum_{i\in M} \boldsymbol{z}_{hi} \tag{11.22}$$

We will show that, in a sense to be explained shortly, as $q \to \infty$, $\boldsymbol{C}(\mathbb{E}_q)$ 'shrinks to $\boldsymbol{W}(\mathbb{E})$.' Our basic approach will revolve around the study of the sets $\boldsymbol{C}_q$, defined as the set of all feasible allocations, $\langle \boldsymbol{x}_i \rangle_{i \in M} \in \boldsymbol{X}^*(\mathbb{E})$ such that the allocation $\langle \boldsymbol{x}_{hi} \rangle_{(h,i) \in Q \times M}$ given by:

$$\boldsymbol{x}_{hi} = \boldsymbol{x}_i \quad \text{for } h = 1, \ldots, q; i = 1, \ldots, m; \tag{11.23}$$

is in $\boldsymbol{C}(\mathbb{E}_q)$. In other words, $\boldsymbol{C}_q$ is the projection on $\boldsymbol{X}^*(\mathbb{E})$ of the allocations $\langle \boldsymbol{x}_{hi} \rangle_{(h,i) \in Q \times M}$ from $\boldsymbol{C}(\mathbb{E}_q)$ which have the property that:

$$\boldsymbol{x}_{hi} = \boldsymbol{x}_{h'i} \quad \text{for } h, h' = 1, \ldots, q; i = 1, \ldots, m. \tag{11.24}$$

Notice, incidentally, that if $\langle \boldsymbol{x}_i \rangle_{i \in M}$ is a feasible consumption allocation for $\mathbb{E}$, and we define $\langle \boldsymbol{x}_{hi} \rangle_{(h,i) \in Q \times M}$ as in equation (11.23), then $\langle \boldsymbol{x}_{hi} \rangle_{(h,i) \in Q \times M}$ is a feasible consumption allocation for $\mathbb{E}_q$.

The following result generalizes the 'First Fundamental Theorem of Welfare Economics,' in the version presented as Theorem 7.22. Moreover, it establishes the fact that if $\boldsymbol{W}(\mathbb{E}) \neq \emptyset$, then $\boldsymbol{C}_q \neq \emptyset$, for $q = 1, 2, \ldots$.

11.17 Theorem. *For any economy, $\mathbb{E} = \langle (X_i, P_i, Z_i) \rangle$, we have:*

1. $\boldsymbol{W}(\mathbb{E}) \subseteq \boldsymbol{C}_q$, *and*
2. $\boldsymbol{C}_{q+1} \subseteq \boldsymbol{C}_q$,

for $q = 1, 2, \ldots$.

Proof.

1. Suppose $\langle \boldsymbol{x}_i^* \rangle_{i \in M} \in \boldsymbol{W}(\mathbb{E})$, let $\langle \boldsymbol{z}_i^* \rangle_{i \in M}$ and $\boldsymbol{p}^*$ be such that $(\langle \boldsymbol{x}_i^*, \boldsymbol{z}_i^* \rangle, \boldsymbol{p}^*)$ is a Walrasian equilibrium for $\mathbb{E}$, and define the consumption allocation $\langle \bar{\boldsymbol{x}}_{hi} \rangle_{(h,i) \in Q \times M}$ for $\mathbb{E}_q$ by:

$$\bar{\boldsymbol{x}}_{hi} = \boldsymbol{x}_i^* \quad \text{for } h = 1, \ldots, q; i = 1, \ldots, m. \tag{11.25}$$

If we suppose, by way of obtaining a contradiction, that $\langle \bar{\boldsymbol{x}}_{hi} \rangle_{(h,i) \in Q \times M}$ is not in $\boldsymbol{C}(\mathbb{E}_q)$, then there exists a coalition $S \subseteq Q \times M$ which can improve upon $\langle \bar{\boldsymbol{x}}_{hi} \rangle_{(h,i) \in Q \times M}$; and thus there exists $\langle \boldsymbol{x}_{hi}, \boldsymbol{z}_{hi} \rangle_{(h,i) \in Q \times M}$ such that:

$$\boldsymbol{x}_{hi} P_i \bar{\boldsymbol{x}}_{hi} \quad \text{for all } (h,i) \in S, \tag{11.26}$$

$$\boldsymbol{z}_{hi} \in Z_i \quad \text{for all } (h,i) \in S, \tag{11.27}$$

and:

$$\sum_{(h,i) \in S} \boldsymbol{x}_{hi} = \sum_{(h,i) \in S} \boldsymbol{z}_{hi}. \tag{11.28}$$

However, since $(\langle \boldsymbol{x}_i^*, \boldsymbol{z}_i^* \rangle, \boldsymbol{p}^*)$ is a competitive equilibrium for $\mathbb{E}$, it follows from (11.26) and (11.27) that:

$$\boldsymbol{p}^* \cdot \boldsymbol{x}_{hi} > \boldsymbol{p}^* \cdot \boldsymbol{z}_{hi}^* \quad \text{for all } (h,i) \in S. \tag{11.29}$$

From (11.29) we then obtain, upon adding over all $(h,i) \in S$:

$$\sum_{(h,i) \in S} \boldsymbol{p}^* \cdot \boldsymbol{x}_{hi} = \boldsymbol{p}^* \cdot \left(\sum_{(h,i) \in S} \boldsymbol{x}_{hi} \right) > \sum_{(h,i) \in S} \boldsymbol{p}^* \cdot \boldsymbol{z}_{hi} = \boldsymbol{p}^* \cdot \left(\sum_{(h,i) \in S} \boldsymbol{z}_{hi} \right);$$

which contradicts (12.28).

2. To see why condition 2 must hold, consider an allocation, $\langle \boldsymbol{x}_i^* \rangle_{i \in M}$, in $\boldsymbol{C}_{q+1}$. Then the allocation $\langle \boldsymbol{x}_{hi} \rangle_{(h,i) \in (Q+1) \times M}$ defined by:

$$\boldsymbol{x}_{hi} = \boldsymbol{x}_i^* \quad \text{for } h = 1, \ldots, q+1; i = 1, \ldots, m,$$

is such that no coalition, S, from $(Q+1) \times M$, can improve upon $\langle \boldsymbol{x}_{hi} \rangle_{(h,i) \in (Q+1) \times M}$. But then it follows that the consumption allocation $\langle \boldsymbol{x}_{hi} \rangle_{(h,i) \in Q \times M}$ defined by:

$$\boldsymbol{x}_{hi} = \boldsymbol{x}_i^* \quad \text{for } h = 1, \ldots, q; i = 1, \ldots, m,$$

must be in the core for $\mathcal{E}_q$ as well; since any coalition from $Q \times M$ which could improve upon it could also improve upon the consumption allocation $\langle \boldsymbol{x}_{hi} \rangle_{(h,i) \in (Q+1) \times M}$ in $\mathbb{E}_{q+1}$. Therefore, $\langle \boldsymbol{x}_i^* \rangle_{i \in M}$ is in $\boldsymbol{C}_q$, and it follows that:

$$\boldsymbol{C}_{q+1} \subseteq \boldsymbol{C}_q. \quad \square$$

Notice that it is an immediate consequence of Theorem 11.17 that:

$$\boldsymbol{W}(\mathbb{E}) \subseteq \bigcap\nolimits_{q=1}^{\infty} \boldsymbol{C}_q;$$

and thus that if $\boldsymbol{W}(\mathbb{E}) \neq \emptyset$, then:

$$\bigcap\nolimits_{q=1}^{\infty} \boldsymbol{C}_q \neq \emptyset.$$

It also follows from 11.17 that for all q:

$$\boldsymbol{C}_q = \bigcap\nolimits_{s=1}^{q} \boldsymbol{C}_s;$$

and thus it is natural to write:

$$\bigcap\nolimits_{q=1}^{\infty} \boldsymbol{C}_q = \lim_{q \to \infty} \boldsymbol{C}_q. \tag{11.30}$$

Debreu and Scarf [1963] showed that given any exchange economy, $\mathcal{E} = \langle (P_i, \boldsymbol{r}_i) \rangle_{i \in M}$, satisfying certain assumptions, we will have:

$$\bigcap\nolimits_{q=1}^{\infty} C_q \subseteq W(\mathcal{E});$$

which, when combined with Theorem 11.17 and equation (11.30) means that under the Debreu-Scarf conditions, we have:

$$\lim_{q \to \infty} \boldsymbol{C}_q = \boldsymbol{W}(\mathcal{E}). \tag{11.31}$$

We will prove a generalization of their result; one which applies to a private ownership economy with production. However, as was the case when we studied the 'Second Fundamental Theorem of Welfare Economics,' we will begin by introducing the idea of a 'quasi-competitive equilibrium;' this time defining said equilibrium for a private ownership economy. Since you can probably guess exactly how this definition will be stated, we will present it here in abbreviated form.

11.18 Definition. We shall say that $(\langle \boldsymbol{x}_i^*, \boldsymbol{z}_i^* \rangle, \boldsymbol{p}^*)$ is a **quasi-competitive equilibrium for the economy** $\mathbb{E} = \langle (X_i, P_i, Z_i) \rangle$, iff $(\langle \boldsymbol{x}_i^*, \boldsymbol{z}_i^* \rangle, \boldsymbol{p}^*)$ satisfies conditions 1–3 of Definition 11.8, and:

4′. for each $i \in M$, we have:
a. $\boldsymbol{p}^* \cdot \boldsymbol{x}_i^* \leq \boldsymbol{p}^* \cdot \boldsymbol{z}_i^*$, and:
b. either:

$$\boldsymbol{p}^* \cdot \boldsymbol{z}_i^* = \min \boldsymbol{p}^* \cdot X_i,$$

or:

$$(\forall \boldsymbol{x}_i \in X_i) \colon \boldsymbol{x}_i P_i \boldsymbol{x}_i^* \Rightarrow \boldsymbol{p}^* \cdot \boldsymbol{x}_i > \boldsymbol{p}^* \cdot \boldsymbol{z}_i^*.$$

We will denote the set of all consumption allocations, $\langle \boldsymbol{x}_i^* \rangle \in \boldsymbol{X}^*(\mathbb{E})$, for which there exists a production allocation $\langle \boldsymbol{z}_i^* \rangle_{i \in M}$ and a price vector $\boldsymbol{p}^*$ such that $(\langle \boldsymbol{x}_i^*, \boldsymbol{z}_i^* \rangle, \boldsymbol{p}^*)$ is a quasi-competitive equilibrium for $\mathbb{E}$ by '$\boldsymbol{W}^\dagger(\mathbb{E})$.'

In our initial result, we will establish conditions sufficient to ensure that:

$$\bigcap_{q=1}^{\infty} \boldsymbol{C}_q \subseteq \boldsymbol{W}^\dagger(\mathbb{E}).$$

In our proof, which owes a great deal to McKenzie [1988] and Nikaido [1968, Theorem 17.4, p. 291], we will need to make use of the following mathematical result, the proof of which is provided in the appendix to this chapter.

11.19 Proposition. *If $C_i \subseteq \mathbb{R}^n$ is convex and non-empty, for $i = 1, \ldots, m$, then the convex hull of $C \stackrel{def}{=} \bigcup_{i=1}^m C_i$, $co(C)$, is given by:*

$$co(C) = \{\boldsymbol{x} \in \mathbb{R}^n \mid (\exists \boldsymbol{a} \in \Delta_m \ \&\ \boldsymbol{x}_i \in C_i, \text{ for } i = 1, \ldots, m) \colon \boldsymbol{x} = \sum_{i=1}^m a_i \boldsymbol{x}_i\}. \tag{11.32}$$

The proposition just stated is one of those rather frustrating little results which appears too obvious to really need a formal proof, but for which the development of a rigorous proof is nonetheless a somewhat tricky task. *Please note, however, that the conclusion no longer holds if the C_i's are not all convex; that is, the convex hull of C is not generally given by the formula in equation (11.32) if the sets C_i are not all convex.*[6]

11.20 Theorem. *If $\mathbb{E} = \langle (X_i, P_i, Z_i) \rangle$ is an economy such that:*

1. Z_i is convex, for $i = 1, \ldots, m$;

and, for each $i \in M$:

2. X_i is convex and P_i is locally non-saturating, lower semicontinuous, and weakly convex, and:

3. $X_i \cap Z_i \neq \emptyset$,

then:

$$\bigcap_{q=1}^{\infty} \boldsymbol{C}_q \subseteq \boldsymbol{W}^\dagger(\mathbb{E}).$$

[6] We provide an example in which the formula of equation (11.32) does not hold in the appendix to this chapter.

Proof. Suppose $\langle \boldsymbol{x}_i^* \rangle_{i \in M} \in \boldsymbol{C}_q$ for all q, define $\mathbb{P}_i = P_i \boldsymbol{x}_i^* - Z_i$, for each $i \in M$,[7] and:

$$\mathbb{P} = co\Big(\bigcup\nolimits_{i=1}^{m} \mathbb{P}_i\Big);$$

that is, $\mathbb{P}$ is the convex hull of the union of the $\mathbb{P}_i$'s. The tricky part of the proof is to establish the fact that $\mathbf{0} \notin \mathbb{P}$.

Suppose, by way of establishing a contradiction, that $\mathbf{0} \in \mathbb{P}$. Then, since each $\mathbb{P}_i$ is a convex set (and non-empty, by the assumption that each P_i is locally non-saturating), it follows from Proposition 11.19 that there exist $\boldsymbol{a} \in \Delta_m$, $\boldsymbol{x}_i \in X_i$, and $\boldsymbol{z}_i \in Z_i$ for $i = 1, \ldots, m$, such that:

$$\sum\nolimits_{i=1}^{m} a_i(\boldsymbol{x}_i - \boldsymbol{z}_i) = \mathbf{0}, \tag{11.33}$$

and:

$$\boldsymbol{x}_i P_i \boldsymbol{x}_i^* \quad \text{for } i = 1, \ldots, m. \tag{11.34}$$

We will show that these two conditions allow us to construct a coalition in $\mathbb{E}_{q^*}$, for some (finite) integer, q^*, which can improve upon $\langle \boldsymbol{x}_i^* \rangle_{i \in M}$; contradicting the assumption that $\langle \boldsymbol{x}_i^* \rangle_{i \in M} \in \boldsymbol{C}_q$, for all q.

Accordingly, we begin by noting that (11.33) implies:

$$\sum\nolimits_{i=1}^{m} a_i \boldsymbol{x}_i = \sum\nolimits_{i=1}^{m} a_i \boldsymbol{z}_i. \tag{11.35}$$

We then define $I = \{i \in M \mid a_i > 0\}$, and, for each $i \in M$ and each positive integer, q, we let b_i^q be the smallest integer greater than or equal to qa_i. Now, by assumption 3, for each $i \in M$ there exist $\widehat{\boldsymbol{x}}_i \in X_i$ and $\widehat{\boldsymbol{z}}_i \in Z_i$ such that:

$$\widehat{\boldsymbol{x}}_i = \widehat{\boldsymbol{z}}_i. \tag{11.36}$$

We make use of the $\widehat{\boldsymbol{x}}_i$ to define, for each $i \in M$ and each positive integer, q:

$$\boldsymbol{x}_i^q = \Big(\frac{qa_i}{b_i^q}\Big)\boldsymbol{x}_i + \Big[1 - \Big(\frac{qa_i}{b_i^q}\Big)\Big]\widehat{\boldsymbol{x}}_i; \tag{11.37}$$

and note that, since each P_i is lower semi-continuous, and since:

$$\frac{qa_i}{b_i^q} \to 1 \ \text{ as } q \to \infty,$$

it follows from (11.34) that for each $i \in M$, there exists a positive integer, q_i such that for all $q \geq q_i$,

$$\boldsymbol{x}_i^q P_i \boldsymbol{x}_i^*. \tag{11.38}$$

But now let:

$$q^* = \max_{i \in M} q_i,$$

[7] That is:

$$\mathbb{P}_i = \big\{\boldsymbol{v} \in \mathbb{R}^n \mid (\exists \boldsymbol{x}_i \in X_i \ \& \ \boldsymbol{z}_i \in Z_i) \colon \boldsymbol{x}_i P_i \boldsymbol{x}_i^* \ \& \ \boldsymbol{v} = \boldsymbol{x}_i - \boldsymbol{z}_i\big\}.$$

let $b^* = \max\{b_1^{q^*}, \ldots, b_m^{q^*}\}$ and consider the coalition, S, in $\mathcal{E}_{b^*}$ consisting of $b_i^{q^*}$ consumers of each type $i \in M$, and the allocation $\langle \bar{\boldsymbol{x}}_{hi} \rangle_{(h,i) \in S}$ defined by:

$$\bar{\boldsymbol{x}}_{hi} = \boldsymbol{x}_i^{q^*} \quad \text{for } h = 1, \ldots, b_i^{q^*}, \text{ and each } i \in M. \tag{11.39}$$

We have $\bar{\boldsymbol{x}}_{hi} P_{hi} \boldsymbol{x}_i^*$ for each h and each $i \in I$; while by using (11.37), (11.36), and (11.35) in turn, we have:

$$\begin{aligned}
\sum\nolimits_{i \in M} \sum\nolimits_{h=1}^{b_i^{q^*}} \bar{\boldsymbol{x}}_{hi} &= \sum\nolimits_{i \in M} b_i^{q^*} \boldsymbol{x}_i^{q^*} = \sum\nolimits_{i \in M} \left[(q^* a_i)\boldsymbol{x}_i + (b_i^{q^*} - q^* a_i)\widehat{\boldsymbol{x}}_i\right] \\
&= q^* \sum\nolimits_{i \in M} a_i \boldsymbol{x}_i - q^* \sum\nolimits_{i \in M} a_i \widehat{\boldsymbol{z}}_i + \sum\nolimits_{i \in M} b_i^{q^*} \widehat{\boldsymbol{z}}_i \\
&= q^* \Big(\sum\nolimits_{i \in M} a_i \boldsymbol{z}_i \Big) - q^* \sum\nolimits_{i \in I} a_i \widehat{\boldsymbol{z}}_i + \sum\nolimits_{i \in M} b_i^{q^*} \widehat{\boldsymbol{z}}_i \\
&= + \sum\nolimits_{i \in M} b_i^{q^*} \left[\Big(\frac{q^* a_i}{b_i^{q^*}} \Big) \boldsymbol{z}_i + \widehat{\boldsymbol{z}}_i - \Big(\frac{q^* a_i}{b_i^{q^*}} \Big) \widehat{\boldsymbol{z}}_i \right].
\end{aligned}$$

Thus, since each Z_i is convex, it follows that the coalition S can improve upon $\langle \boldsymbol{x}_i^* \rangle_{i \in M}$; contradicting the assumption that $\langle \boldsymbol{x}_i^* \rangle_{i \in M} \in C_q$ for all positive integers, q. Therefore $\mathbf{0} \notin \mathbb{P}$.

Since we have now established the fact that $\mathbf{0} \notin \mathbb{P}$, it follows from the Separating Hyperplane Theorem (Theorem 6.21) that there exists a non-zero $\boldsymbol{p}^* \in \mathbb{R}^n$ satisfying:

$$(\forall \boldsymbol{v} \in \mathbb{P}) \colon \boldsymbol{p}^* \cdot \boldsymbol{v} \geq 0. \tag{11.40}$$

From the definition of $\mathbb{P}$, it then follows immediately that for each $i \in M$, we have:

$$(\forall \boldsymbol{x}_i \in X_i \ \& \ \boldsymbol{z}_i \in Z_i) \colon \boldsymbol{x}_i P_i \boldsymbol{x}_i^* \Rightarrow \boldsymbol{p}^* \cdot \boldsymbol{x}_i \geq \boldsymbol{p}^* \cdot \boldsymbol{z}_i. \tag{11.41}$$

Moreover, since P_i is locally non-saturating, it then follows easily that, for each i and each $\boldsymbol{z}_i \in Z_i$:

$$\boldsymbol{p}^* \cdot \boldsymbol{x}_i^* \geq \boldsymbol{p}^* \cdot \boldsymbol{z}_i. \tag{11.42}$$

Now, since $\langle \boldsymbol{x}_i^* \rangle_{i \in M} \in \boldsymbol{C}_q$, for each q, it follows from the definitions that there exists $\langle \boldsymbol{z}_i^* \rangle_{i \in M}$ such that:

$$\sum_{i \in M} \boldsymbol{x}_i^* = \sum_{i \in M} \boldsymbol{z}_i^*. \tag{11.43}$$

But then, letting $i \in M$ and $\boldsymbol{z}_i^\dagger \in Z_i$ be arbitrary, we see that it follows from (11.42) and (11.43) that, if we define $\langle \boldsymbol{z}_i \rangle_{i \in M} \in \prod_{i \in M} Z_i$ by:

$$\boldsymbol{z}_h = \begin{cases} \boldsymbol{z}_h^* & \text{for } h \neq i, \\ \boldsymbol{z}_i^\dagger & \text{for } h = i, \end{cases}$$

we have:

$$\sum_{h \in M} \boldsymbol{p}^* \cdot \boldsymbol{z}_h^* \geq \sum_{h \neq i} \boldsymbol{p}^* \cdot \boldsymbol{z}_h^* + \boldsymbol{p}^* \cdot \boldsymbol{z}_i^\dagger;$$

so that:

$$\boldsymbol{p}^* \cdot \boldsymbol{z}_i^* \geq \boldsymbol{p}^* \cdot \boldsymbol{z}_i^\dagger.$$

Finally, we let $i \in M$ be arbitrary once again. Then from (11.42), (11.43) we have:

$$\boldsymbol{p}^* \cdot \boldsymbol{x}_i^* = w_i(\boldsymbol{p}^*) \stackrel{\text{def}}{=} \boldsymbol{p}^* \cdot \boldsymbol{z}_i^*..$$

Furthermore, it follows from (11.41) that:

$$(\forall \boldsymbol{x}_i \in X_i)\colon \boldsymbol{x}_i P_i \boldsymbol{x}_i^* \Rightarrow \boldsymbol{p}^* \cdot \boldsymbol{x}_i \geq w_i(\boldsymbol{p}^*);$$

and thus from Proposition 7.25, it follows that either:

$$w_i(\boldsymbol{p}^*) = \min \boldsymbol{p}^* \cdot X_i,$$

or:

$$(\forall \boldsymbol{x}_i \in X_i)\colon \boldsymbol{x}_i P_i \boldsymbol{x}_i^* \Rightarrow \boldsymbol{p}^* \cdot \boldsymbol{x}_i > w_i(\boldsymbol{p}^*).$$

Therefore, $\big(\langle \boldsymbol{x}_i^*, \boldsymbol{z}_i^* \rangle, \boldsymbol{p}^*\big)$ is a quasi-competitive equilibrium for $\mathbb{E}$. □

We can strengthen the conclusion of the theorem just proved to conclude that if:

$$\langle \boldsymbol{x}_i^* \rangle \in \bigcap\nolimits_{q=1}^{\infty} \boldsymbol{C}_q,$$

and satisfies an irreducibility condition, then $\langle \boldsymbol{x}_i^* \rangle \in \boldsymbol{W}(\mathbb{E})$. In order to introduce this condition, we first define the following.

11.21 Definition. If $\mathbb{E} = \big\langle (X_i, P_i, Z_i) \big\rangle$ is an economy, and $\boldsymbol{p}^* \in \mathbb{R}^n$, we will say that $\boldsymbol{w} \in \mathbb{R}^m$ is a **feasible wealth distribution for** $\mathbb{E}$ iff for each $i \in M$, there exists $\boldsymbol{x}_i \in X_i$ such that $\boldsymbol{p}^* \cdot \boldsymbol{x}_i \leq w_i$, and:

$$\sum_{i \in M} w_i = \sup_{\boldsymbol{z} \in Z} \boldsymbol{p}^* \cdot \boldsymbol{z},$$

where we define Z by:

$$Z = \sum_{i \in M} Z_i.$$

We can then define $\big(\langle \boldsymbol{x}_i^*, \boldsymbol{z}_i^* \rangle, \boldsymbol{p}^*\big)$ to be a quasi competitive equilibrium for $\mathbb{E}$, given the wealth distribution, $\boldsymbol{w}^* \in \mathbb{R}^m$ in the obvious way: requiring that $\boldsymbol{w}^*$ be a feasible wealth distribution for $\mathbb{E}$, given $\boldsymbol{p}^*$, and requiring that Definition 11.18 be satisfied, with w_i^* replacing $\boldsymbol{p}^* \cdot \boldsymbol{z}_i^*$ in condition 4 of the definition. Our irreducibility condition is then defined as follows.

11.22 Definition. We shall say that the economy, $\mathbb{E} = \big\langle (X_i, P_i, Z_i) \big\rangle$ is **irreducible at the consumption allocation** $\langle \boldsymbol{x}_i^* \rangle \in \boldsymbol{X}^*(\mathbb{E})$ iff, given any partition of the consumers, $\{S_1, S_2\}$,[8] there exists $\langle (\boldsymbol{x}_i, \boldsymbol{z}_i) \rangle_{i \in S}$ such that:

$$(\boldsymbol{x}_i, \boldsymbol{z}_i) \in X_i \times \mathbb{R}^n \quad \text{for } i = 1, \dots, m, \tag{11.44}$$

$$\sum\nolimits_{i=1}^{m} \boldsymbol{z}_i \in Z, \tag{11.45}$$

$$\sum_{i \in S_1} (\boldsymbol{x}_i - \boldsymbol{z}_i) = \sum_{i \in S_2} (\boldsymbol{z}_i - \boldsymbol{x}_i), \tag{11.46}$$

[8] By a partition of the consumers, $\{S_1, S_2\}$, we mean $S_j \subseteq M$ & $S_j \neq \emptyset$, for $i = 1, 2$, $S_1 \cap S_2 = \emptyset$, and $S_1 \cup S_2 = M$.

and:

$$(\forall i \in S_1)\colon \boldsymbol{x}_i P_i \boldsymbol{x}_i^*. \tag{11.47}$$

We will denote the set of consumption allocations at which $\mathbb{E}$ is irreducible by '$\boldsymbol{X}^I(\mathbb{E})$.'

Those with good memories will undoubtedly already have noticed that the above definition is a straightforward modification of the condition of the same name which was presented in Chapter 7 (Definition 7.33). in the present context the vectors $\boldsymbol{z}_i$ have a more natural interpretation than was the case in Chapter 7, in that here the natural interpretation of the points $\boldsymbol{z}_i$ is to suppose that they are elements of Z_i; in which case, equation (11.45) is necessarily satisfied. However, notice that our condition does not require that $\boldsymbol{z}_i \in Z_i$. In any case, the following result is a more or less immediate application of Theorem 11.20 and the proof of Theorem 7.36. I will leave the details of the proof as an exercise.

11.23 Theorem. *If $\mathbb{E} = \langle (X_i, P_i, Z_i) \rangle$ is an economy such that:*
1. *$int(X) \cap Z \neq \emptyset$, and:*

for each $i \in M$:

2. *X_i and Z_i are convex sets,*
3. *P_i is locally non-saturating, lower semicontinuous, and weakly convex, and*
4. *$X_i \cap Z_i \neq \emptyset$,*

then:

$$X^I(\mathbb{E}) \cap \left[\bigcap_{q=1}^{\infty} \boldsymbol{C}_q\right] \subseteq \boldsymbol{W}(\mathbb{E}).$$

We can also make good use of the definition of a numéraire good in this context. For convenience, I will repeat the definition here (modified for our definition of an economy, $\mathbb{E}$).

11.24 Definitions. We will say that the jth commodity is a **numéraire good for P_i** iff for all $\boldsymbol{x} \in X_i$ and all $\theta \in \mathbb{R}_{++}$,[9] we have:

$$\boldsymbol{x} + \theta \boldsymbol{e}_j \in X_i \text{ and } (\boldsymbol{x} + \theta \boldsymbol{e}_j) P_i \boldsymbol{x}, \tag{11.48}$$

where $\boldsymbol{e}_j$ is the jth unit coordinate vector.[10] We shall say that the jth commodity is a **numéraire good for the economy**, $\mathbb{E} = \langle (X_i, P_i, Z_i) \rangle$ iff it is a numéraire good for each P_i, and for each $i \in M$ there exists $\theta_i > 0$ such that:

$$X_i \cap [Z_i - \theta_i \boldsymbol{e}_j] \neq \emptyset. \tag{11.49}$$

11.25 Theorem. *If $\mathbb{E} = \langle (X_i, P_i, Z_i) \rangle$ is an economy such that:*
1. *$int(X) \cap Z \neq \emptyset$, and:*

for each $i \in M$:

2. *X_i and Z_i are convex sets,*
3. *P_i is weakly convex and lower semicontinuous,*

and:

[9] Where $\mathbb{R}_{++} = \{x \in \mathbb{R} \mid x > 0\}$.

[10] The vector having all coordinates equal to zero except for the jth coordinate, which is equal to one.

4. for some $j' \in \{1, \ldots, n\}$, the commodity j' is a numéraire good for $\mathbb{E}$, then:

$$\bigcap_{q=1}^{\infty} \boldsymbol{C}_q = \boldsymbol{W}(\mathbb{E}).$$

Proof. Since it is an immediate implication of Theorem 11.17 that $W(\mathbb{E}) \subseteq \bigcap_{q=1}^{\infty} \boldsymbol{C}_q$, we need only prove the converse. Accordingly, let $j' \in \{1, \ldots, n\}$ be the numéraire good for $\mathbb{E}$, and note that it then follows from (11.48) and (11.49) of Definition 11.24 that, for each $i \in M$:

$$X_i \cap Z_i \neq \emptyset.$$

Consequently, since each P_i is locally non-saturating by virtue of the fact that commodity j' is a numéraire good for $\mathcal{E}$, it follows from Theorem 11.20 that if $\langle \boldsymbol{x}_i^* \rangle_{i \in M} \in \bigcap_{q=1}^{\infty} \boldsymbol{C}_q$, then there exists $\boldsymbol{p}^* \in \mathbb{R}^n \setminus \{\boldsymbol{0}\}$ and $\langle \boldsymbol{z}_i^* \rangle_{i \in M}$ such that $\big(\langle \boldsymbol{x}_i^*, \boldsymbol{z}_i^* \rangle, \boldsymbol{p}^*\big)$ is a quasi-competitive equilibrium for $\mathbb{E}$. It also follows from assumption 2 that there exists $\boldsymbol{x} \in X \stackrel{\text{def}}{=} \sum_{i \in M} X_i$, $\theta \in \mathbb{R}_{++}$, and $\boldsymbol{z} \in \sum_{i \in M} Z_i$ such that:

$$\boldsymbol{x} - \theta \boldsymbol{p}^* \in X \text{ and } \boldsymbol{x} = \boldsymbol{z}.$$

Thus, as in the proof of Theorem **??** we see that there must exist at least one $h \in M$ such that:

$$\boldsymbol{p}^* \cdot \boldsymbol{z}_h^* > \min \boldsymbol{p}^* \cdot X_i;$$

so that, by definition of a quasi-competitive equilibrium:

$$(\forall \boldsymbol{x}_h \in X_h)\colon \boldsymbol{x}_h P_h \boldsymbol{x}_h^* \Rightarrow \boldsymbol{p}^* \cdot \boldsymbol{x}_h > \boldsymbol{p}^* \cdot \boldsymbol{z}_h^*.$$

However, since commodity j' is a numéraire good for P_h, we recall that for any $\Delta x_{j'} > 0$, we have:

$$(\boldsymbol{x}_h^* + \Delta x_{j'} \boldsymbol{e}_{j'}) P_h \boldsymbol{x}_h^*,$$

where $\boldsymbol{e}_{j'}$ is the $(j')^{th}$ unit coordinate vector. It then follows that we must have $p_{j'}^* > 0$.

Now let $i \in M$ be arbitrary. Then, by definition of a numéraire good for $\mathbb{E}$, there exists $\bar{\boldsymbol{x}}_i \in X_i$, $\bar{\boldsymbol{z}}_i \in Z_i$, and $\theta_{j'} > 0$ such that:

$$\bar{\boldsymbol{x}}_i - \theta_{j'} \boldsymbol{e}_{j'} = \bar{\boldsymbol{z}}_i.$$

and, since $p_j^* > 0$, it then follows that:

$$\boldsymbol{p}^* \cdot \bar{\boldsymbol{x}}_i < \boldsymbol{p}^* \cdot \bar{\boldsymbol{z}}_{\boldsymbol{i}}. \tag{11.50}$$

Moreover, it follows from the definition of a quasi-competitive equilibrium that:

$$\boldsymbol{p}^* \cdot \bar{\boldsymbol{z}}_{\boldsymbol{i}} \leq \boldsymbol{p}^* \cdot \boldsymbol{z}_i^*. \tag{11.51}$$

From (11.50), (11.51), and the definition of a quasi-competitive equilibrium, it now follows that:

$$(\forall \boldsymbol{x}_i \in X_i)\colon \boldsymbol{x}_i P_i \boldsymbol{x}_i^* \Rightarrow \boldsymbol{p}^* \cdot \boldsymbol{x}_i > \boldsymbol{p}^* \cdot \boldsymbol{z}_i^*,$$

and we see that $\big(\langle \boldsymbol{x}_i^*, \boldsymbol{z}_i^* \rangle, \boldsymbol{p}^*\big)$ is a Walrasian equilibrium for $\mathcal{E}$. □

This last result has an interesting corollary for the case of a pure exchange economy, $\mathcal{E} = \langle (P_i, \boldsymbol{r}_i)\rangle_{i\in M}$; where, remember that if $\mathcal{E}$ is a pure exchange economy, we will always assume that:

$$X_i \subseteq \mathbb{R}^n_+ \text{ and } \boldsymbol{r}_i \in X_i \tag{11.52}$$

(and, of course, that P_i is irreflexive). I will leave the details of the proof as an exercise.

11.26 Corollary. *If $\mathcal{E} = \langle (P_i, \boldsymbol{r}_i)\rangle_{i\in M}$ is a pure exchange economy such that:*

1. *$\boldsymbol{r} = \sum_{i\in M} \boldsymbol{r}_i \gg \boldsymbol{0}$,*
2. *P_i is weakly convex and lower semi-continuous, for $i = 1, \ldots, m$; and*
3. *for some $j' \in \{1, \ldots, n\}$, the commodity j' is a numéraire good for $\mathcal{E}$,*

then:

$$\bigcap_{q=1}^{\infty} \boldsymbol{C}_q = \boldsymbol{W}(\mathcal{E}).$$

11.5 Equal Treatment

Under somewhat stronger assumptions than those we used, the only allocations in $C(\mathbb{E}_q)$ are those generated from C_q. More precisely, under familiar convexity assumptions, one can show that, if $\langle(\boldsymbol{x}_{hi}, \boldsymbol{z}_{hi})\rangle_{(h,i)\in Q\times M}$ is in $C(\mathbb{E}_q)$, then:

$$\boldsymbol{x}_{hi} = \boldsymbol{x}_{h'i} \quad \text{for } h, h' = 1, \ldots, q, \text{ and } i = 1, \ldots, m. \tag{11.53}$$

The basic fact from which this statement can be proved is the following. The proof of the result, which is an easy consequence of the definitions, I will leave as an exercise.

11.27 Proposition. *Suppose $\mathbb{E} = \big\langle (X_i, P_i, Z_i)\big\rangle$ is an economy, that $\mathbb{E}_q$ is the q-fold replication of $\mathcal{E}$, and that $\langle(\boldsymbol{x}_{hi}, \boldsymbol{z}_{hi})\rangle_{(h,i)\in Q\times M}$ is a feasible allocation for $\mathbb{E}_q$, wihere $q \geq 2$. Then, letting η be any function mapping M into Q, and letting S be the coalition formed by taking the $\eta(i)^{th}$ agent of type i, for each i; that is:*

$$S = \big\{(\eta(1), 1), (\eta(2), 2), \ldots, (\eta(m), m)\big\},$$

the allocation $\langle(\boldsymbol{x}^*_i, \boldsymbol{z}^*_i)\rangle_{i\in S}$ defined by:

$$(\boldsymbol{x}^*_{\eta(i),i}, \boldsymbol{z}^*_{\eta(i),i}) = (1/q)\sum_{h=1}^{m}(\boldsymbol{x}_{hi}, \boldsymbol{z}_{hi}) \quad \textit{for } i = 1, \ldots, m,$$

is feasible for S.

One can then use the following result, together with Proposition 11.27, to prove the 'equal treatment' result just quoted.

11.28 Proposition. *Let P be an asymmetric, negatively transitive, and convex binary relation on the convex set $X \subseteq \mathbb{R}^n$, let q be a positive integer greater than one, and let $\boldsymbol{x}, \boldsymbol{x}_1, \ldots, \boldsymbol{x}_q \in X$ and $\boldsymbol{a} \in \Delta_q$ satisfy:*

$$\neg \boldsymbol{x} P \boldsymbol{x}_h \quad \textit{for } h = 1, \ldots, q,$$

and, for some $h' \in \{1, \ldots, q\}$:

$$\boldsymbol{x}_{h'} P \boldsymbol{x} \textit{ and } a_{h'} > 0.$$

Then we have:

$$\Big(\sum_{h=1}^{q} a_h \boldsymbol{x}_h\Big) P \boldsymbol{x}.$$

11.6 Appendix

Proof of Proposition 11.14. In our proof, we will make use of the sets D_i defined by:

$$D_1 = C_1, \text{ and } D_i = C_i \setminus \big(\bigcup_{h=1}^{i-1} C_h\big) \text{ for } i = 2, \ldots, m.$$

We also choose, for each i, an arbitrary element, $\bar{\boldsymbol{x}}_i \in C_i$.

Since $C \subseteq \mathbb{R}^n$, the convex hull of C, $co(C)$, is given by:[11]

$$co(C) = \big\{\boldsymbol{x} \in \mathbb{R}^n \mid (\exists \boldsymbol{b} \in \Delta_{n+1} \ \& \ \boldsymbol{x}_j \in C, \text{ for each } j) \colon \boldsymbol{x} = \sum_{j=1}^{n+1} b_j \boldsymbol{x}_j\big\}. \tag{11.54}$$

Thus, if $\boldsymbol{x}$ is an arbitrary element of $co(C)$, there exist $\boldsymbol{b} \in \Delta_{n+1}$, and $\boldsymbol{x}_j \in C$, for $j = 1, \ldots, n+1$ such that:[12]

$$\boldsymbol{x} = \sum_{j=1}^{n+1} b_j \boldsymbol{x}_j. \tag{11.55}$$

However, each $\boldsymbol{x}_j$ is an element of one of the C_i's. Consequently, we can represent $\boldsymbol{x}$ by a different formula, as follows.

For each $i \in \{1, \ldots, m\}$, we define the set $J(i) \subseteq \{1, \ldots, n+1\}$ by:

$$J(i) = \big\{j \in \{1, \ldots, n+1\} \mid \boldsymbol{x}_j \in D_i \ \& \ b_j \neq 0\big\}. \tag{11.56}$$

Next define:

$$a_i = \begin{cases} \sum_{j \in J(i)} b_j & \text{if } J(i) \neq \emptyset, \\ 0 & \text{if } J(i) = \emptyset, \end{cases} \tag{11.57}$$

and:

$$\boldsymbol{x}_i^* = \begin{cases} \sum_{j \in J(i)} (b_j/a_j) \boldsymbol{x}_j & \text{if } J(i) \neq \emptyset, \\ \bar{\boldsymbol{x}}_i & \text{if } J(i) = \emptyset. \end{cases} \tag{11.58}$$

Since C_i is convex (and, since $D_i \subseteq C_i$, for each i), $\boldsymbol{x}_i^* \in C_i$, for each i; and from our definitions, it is obvious that:

$$\boldsymbol{x} = \sum_{i=1}^{m} a_i \boldsymbol{x}_i^*;$$

and that:

$$a_i \geq 0 \ \text{ for } i = 1, \ldots, m, \ \text{ and } \sum_{i=1}^{m} a_i = 1. \quad \square$$

In the text of this chapter, I promised to present an example showing that the conclusion of Proposition 11.14 does not necessarily hold if the sets C_i are not all convex. Here is the promised example.

11.29 Example. Define the sets C_1 and C_2 in $\mathbb{R}^3$ by:

$$C_1 = \{\boldsymbol{x} \in \mathbb{R}^3_+ \mid 0 \leq x_1 \leq 1, \ \& \ x_2 = x_3 = 0\} \cup \{\boldsymbol{x} \in \mathbb{R}^3_+ \mid x_1 = x_3 = 0 \ \& \ 0 \leq x_2 \leq 1\},$$

[11] For a proof, see, for example, Moore [1999, Theorem 5.13, p. 268].

[12] Some of the b_j's may, of course, be zero.

and:

$$C_2 = \{\boldsymbol{x} \in \mathbb{R}^3_+ \mid x_2 = 0 \leq x_1 \leq 1, \ \& \ x_3 = 1\} \cup \{\boldsymbol{x} \in \mathbb{R}^3_+ \mid x_1 = 0 \leq x_2 \leq 1 = x_3\},$$

Since the points $(1,0,0)$ and $(0,1,0)$ are both elements of C_1, and $(1,0,1)$ and $(0,1,1)$ are both contained in C_2, and since:

$$\frac{3}{8}(1,0,0) + \frac{1}{8}(0,1,0) + \frac{3}{8}(1,0,1) + \frac{1}{8}(0,1,1) = (3/4, 1/4, 1/2),$$

it follows that the vector $\boldsymbol{x}^* = (3/4, 1/4, 1/2)$ is contained in $co(C)$, where:

$$C = C_1 \cup C_2.$$

However, suppose, by way of obtaining a contradiction, that there exist $\boldsymbol{y}_i \in C_i$ $(i = 1,2)$ and $\boldsymbol{a} \in \Delta_2$ such that:

$$a_1\boldsymbol{y}_1 + a_2\boldsymbol{y}_2 = \boldsymbol{x}^* = (3/4, 1/4, 1/2).$$

Then, since $\boldsymbol{y}_1$ and $\boldsymbol{y}_2$ are in C_1 and C_2, respectively, we must have:

$$y_{13} = 0 \ \& \ y_{23} = 1.$$

consequently, we see that we must have $a_2 = 1/2$; from which it follows that we must also have $a_1 = 1/2$. We now consider two cases.

1. $y_{11} = 0$ Here we see that we must have:

$$(1/2)y_{21} = 3/4,$$

which implies $y_{21} = 3/2$; contradicting the definition of C_2.

2. $y_{11} > 0$ In this case we see that we must have $y_{12} = 0$. But then we see that we must have:

$$(1/2)y_{22} = 1/4;$$

so that $y_{22} = 1/2$. But it must then also be true that $y_{21} = 0$; which would imply that: $y_{11} = 3/2$; yielding a contradiction once again. □

Exercises.

1. Prove Proposition 11.2.

2. Suppose in a given economy, $\mathbb{E} = \langle (X_i, P_i, Z_i) \rangle$, one of the consumers, say the first, has continuously representable and strictly convex preferences. Show that if the tuple $\left(\langle \boldsymbol{x}^*_{(h,i)}, \boldsymbol{z}^*_{h,i} \rangle_{(h,i) \in Q \times M}, \boldsymbol{p}^* \right)$ is a competitive equilibrium for $\mathbb{E}_q$, then we must have:

$$\boldsymbol{x}^*_{h1} = \boldsymbol{x}^*_{11} \quad \text{for } h = 1, \ldots, q.$$

3. Consider a two-consumer, two-commodity pure exchange economy, with:

$$\boldsymbol{r}_1 = (10, 0) \ \text{ and } \ \boldsymbol{r}_2 = (0, 10),$$

and suppose the consumers' preferences can be represented by:

$$u_1(\boldsymbol{x}_1) = 10x_{11}x_{12},$$

and:

$$u_2(\boldsymbol{x}_2) = \log x_{21} + \log x_{22},$$

respectively. Is the allocation:

$$\boldsymbol{x}_1 = \boldsymbol{x}_2 = (5, 5),$$

in the core for $\mathcal{E}$? Why or why not?

4. Consider a pure exchange economy in which $m = n = 2$, and suppose that the two consumer's preferences can be represented by the utility functions:

$$u_1(\boldsymbol{x}_1) = \min\{x_{11}, x_{12}\},$$

and:

$$u_2(\boldsymbol{x}_2) = x_{21} + x_{22},$$

respectively. Suppose further that the consumers' initial endowments are given by:

$$\boldsymbol{r}_1 = (5, 5) = \boldsymbol{r}_2.$$

Find the set of core allocations for $\mathcal{E}$, $C(\mathcal{E})$.

5. Consider a pure exchange economy in which $m = n = 2$, and suppose that the two consumers' preferences can be represented by the utility functions:

$$u_1(\boldsymbol{x}_1) = \min\{x_{11}, x_{12}\},$$

and:

$$u_2(\boldsymbol{x}_2) = 2\min\{x_{21}, x_{22}\},$$

respectively. Suppose further that the consumers' initial endowments are given by:

$$\boldsymbol{r}_1 = (10, 0) \text{ and } \boldsymbol{r}_2 = (0, 10),$$

respectively. Given this information:

(a) Find $C(\mathcal{E})$.

(b) Consider replicating this economy, and the replicative cores, C_q, for $q = 1, 2, \ldots$. Find C_q, for an arbitrary q.

6. Prove Corollary 11.26

7. Prove the following proposition.

11.30 Proposition. *If $\mathcal{E} = \langle (P_i, \boldsymbol{r}_i) \rangle_{i \in M}$ is an exchange economy such that,*

1. for each $i \in M$, P_i is weakly convex, lowever semicontinuous, and strictly increasing, and

2. $\boldsymbol{r} = \sum_{i \in M} \boldsymbol{r}_i \gg \boldsymbol{0}$,

then:

$$\bigcap_{q=1}^{\infty} \boldsymbol{C}_q = \boldsymbol{W}(\mathcal{E}).$$

Chapter 12

General Equilibrium with Uncertainty

12.1 Introduction

In this chapter we will make a brief foray into the theory of general equilibrium with uncertainty. If you remember the discussion of Chapter 2, you will recall that in general equilibrium theory, a commodity is defined by (1) its physical description, (2) its location, (3) the time at which it is available, and (4) the state of the world in which it is available. Consequently, in most of this chapter we are *specializing* the theory which we have been studying; putting more structure into the model in order to account for the effects of uncertainty. Of course, when one delves more deeply into this theory, questions arise which did not appear to be relevant in our earlier studies. Moreover, if we were to pursue the subject to its current frontiers, we would find that new theoretical concepts and tools are needed to provide answers for these questions. However, in the interests of practicality, we will only attempt to provide a 'bare bones' introduction to this theory. Fortunately, in even this cursory introduction to the topic, we will find that some interesting issues and applications can be discussed. We will begin our discussion with what is known as the 'Arrow-Debreu Contingent Commodities Model.'

12.2 Arrow-Debreu Contingent Commodities

The crux of this model is that we suppose that there are two periods, $t = 0, 1$. At $t = 0$, it is supposed that we can set forth all possibilities for the state of the world at the second date, $t = 1$. We assume that there is a finite set, S, of such possible states, and we will also write $S = \#S$; denoting states by lower case 's, s',' etc. Each 'state' is a complete description of the world, and in this theory, we suppose that every agent will know which state, $s \in S$ has occurred once we reach $t = 1$.

We will suppose that there are G physically distinguishable commodities (which, in principle could also be distinguished by location), so that 'n,' the dimension of our commodity space becomes:

$$n = S \cdot G.$$

Commodity bundles then take the form:

$$\boldsymbol{x} = (x_{11}, \ldots, x_{1G}, x_{21}, \ldots, x_{sg}, \ldots, x_{SG}),$$

which is understood to be an entitlement to receive the commodity bundle:

$$\boldsymbol{x}_s = (x_{s1}, \ldots, x_{sG}),$$

if state s occurs. Thus 'x_{sg}' denotes the amount of commodity g to be received (or supplied, if $x_{sg} < 0$) if state s occurs.

In further specifying the economy, we will depart from our previous notation to denote consumer i's resource endowment, by '$\boldsymbol{\omega}_i$,' which now takes the form:

$$\boldsymbol{\omega}_i = (\omega_{i11}, \ldots, \omega_{i1G}, \omega_{i21}, \ldots, \omega_{isg}, \ldots, \omega_{iSG}); \tag{12.1}$$

that is, 'ω_{isg}' denotes consumer i's endowment of the gth commodity if state s occurs. Fortunately, we will rarely have to write out the full vector as we've done in (12.1), above. Defining

$$\boldsymbol{\omega}_{is} = (\omega_{is1}, \ldots \omega_{isG}) \quad \text{for } s = 1, \ldots, S;$$

that is, letting '$\boldsymbol{\omega}_{is}$' denote i's endowment if state s occurs, the finest detail we will usually write out is:

$$\boldsymbol{\omega}_i = (\boldsymbol{\omega}_{i1}, \ldots, \boldsymbol{\omega}_{is}, \ldots, \boldsymbol{\omega}_{iS}).$$

We suppose also that the consumer's preferences describe a weak order over X_i, denoted by '$\succsim_i$. Furthermore, we denote the ith consumer's consumption bundle, contingent upon the occurrence of state s by '$\boldsymbol{x}_{is}$;' so that we can write:

$$\boldsymbol{x}_i = (\boldsymbol{x}_{i1}, \ldots, \boldsymbol{x}_{is}, \ldots, \boldsymbol{x}_{iS}).$$

Similarly, we will let '$Y_k \subseteq \mathbb{R}^n$' denote the feasible production plans for the kth firm, and we will use the generic notation:

$$\boldsymbol{y}_k = (\boldsymbol{y}_{k1}, \ldots, \boldsymbol{y}_{ks}, \ldots, \boldsymbol{y}_{kS}), \tag{12.2}$$

to denote elements of Y_k, where '$\boldsymbol{y}_{ks}$' denotes the production vector of the firm, contingent upon the occurence of state s. We then complete the model, departing from our previous notation,[1] by letting 'θ_{ik}' denote the ith consumer's share in the kth firm's profits.

We will have to be a bit careful in dealing with individual consumption and production sets. One is tempted, for example to express the ith consumer's consumption set as:

$$X_i = \prod_{s=1}^{S} X_{is}, \tag{12.3}$$

where 'X_{is}' denotes the ith consumer's feasible consumption set if state s occurs; with similar specifications for the firms' production sets. That this will not quite do is perhaps best illustrated by considering the following production example, which is inspired by Mas-Colell, Whinston, and Green [1995, Example 19.B.2, p. 689].

[1]This change is made in order that the ith consumer's shares not be confused with the ith state of the world.

12.1 Example. Suppose there are two states, s_1 and s_2, representing good and bad weather. There are two physical commodities: seeds ($g = 1$) and crops ($g = 2$). In this case, the elements of Y_k are four-dimensional vectors. Assume that seeds must be planted before the resolution of the uncertainty about the weather and that if the weather is good, the firm's production possibilities are given by:

$$Y_{k1} = \{\boldsymbol{y}_{k1} \in \mathbb{R}^2 \mid y_{k12} \geq 0 \; \& \; 2y_{k11} + y_{k12} \leq 0\};$$

whereas in bad weather, production is given by:

$$Y_{k2} = \{\boldsymbol{y}_{k2} \in \mathbb{R}^2 \mid y_{k22} \geq 0 \; \& \; y_{k21} + 2y_{k22} \leq 0\}.$$

Recalling our assumption that the seed must be planted before the resolution of uncertainty, we see that we can represent the firm's production set as:

$$Y_k = \{\boldsymbol{y}_k \in Y_{k1} \times Y_{k2} \mid y_{k11} = y_{k21}\}. \tag{12.4}$$

Thus, for example, the production vector:

$$\boldsymbol{y}_k = (y_{k11}, y_{k12}, y_{k21}, y_{k22}) = (-2, 4, -2, 1),$$

is a feasible plan; whereas neither of the production plans:

$$\boldsymbol{y}_k = (-4, 8, -2, 1) \text{ and } \boldsymbol{y}' = (-2, 4, 0, 0),$$

is feasible. □

While the above example deals with a production set, the difficulty applies equally to consumption sets; after all, someone has to plant the seeds, and this labor must also be undertaken before the resolution of uncertainty. In order to allow for this fact, while yet being able to assume on some occasions that consumers' preferences are weakly separable over states, we will assume that for each consumer there exist sets:

$$X_{is} \subseteq \mathbb{R}^G,$$

representing the consumer's feasible consumption possibilities if state s occurs (for $s = 1, \ldots, S$), and a set $\widehat{G}_i$ (presumably a proper subset of G), such that:

$$X_i = \left\{\boldsymbol{x}_i \in \prod\nolimits_{s=1}^{S} X_{is} \mid (\forall g \in \widehat{G}_i)\colon x_{i1g} = x_{i2g} = \cdots = x_{iSg}\right\} \tag{12.5}$$

Thus, with this specification, one can make sense of the following example.

12.2 Example. Suppose that, for a given consumer, i, there exist S utility functions:

$$u_{is}\colon \mathbb{R}^G \to \mathbb{R},$$

such that:

$$\boldsymbol{x}_i \succsim_i \boldsymbol{x}'_i \iff \left[\sum_{s \in S} \pi_{is} u_{is}(\boldsymbol{x}_{is}) \geq \sum_{s \in S} \pi_{is} u_{is}(\boldsymbol{x}'_{is})\right]. \tag{12.6}$$

where 'π_{is}' denotes i's subjective (or objective) probability of the occurrence of state s. Notice that, even though we have state-dependent utility here, preferences are weakly separable on X_{is}, for each state, s. □

We will not need to assume much about the form of the firms' production sets,[2] we will simply suppose that the kth firm's technological production possibilities are given by a production set $Y_k \subseteq \mathbb{R}^n$.

We will make use of the following definition of feasible allocations for the economy.

12.3 Definition. We will say that an allocation, $(\langle \boldsymbol{x}_i^* \rangle, \langle \boldsymbol{y}_k^* \rangle) \in \mathbb{R}^{(m+\ell)n}$ is **feasible for** $\mathcal{E}$ iff:

$$\begin{aligned} \boldsymbol{x}_i^* &\in X_i \quad \text{for } i = 1, \dots, m, \\ \boldsymbol{y}_k^* &\in Y_k \quad \text{for } k = 1, \dots, \ell, \end{aligned} \tag{12.7}$$

and:

$$\sum_{i \in M} \boldsymbol{x}_i^* = \sum_{i \in M} \boldsymbol{\omega}_i + \sum_{k \in L} \boldsymbol{y}_k^*. \tag{12.8}$$

Now, at this point, you may be saying, or thinking, "Hold on! That's exactly the definition of a feasible allocation which was presented in Chapter 7!" And in saying this you are absolutely right! All we have done here so far is to present a somewhat more detailed and specialized specification of what the commodity space is. However, notice that equation (12.8) of the above definition implies that:

$$\sum_{i \in M} \boldsymbol{x}_{is}^* = \sum_{i \in M} \boldsymbol{\omega}_{is} + \sum_{k \in L} \boldsymbol{y}_{ks}^* \quad \text{for } s = 1, \dots, S; \tag{12.9}$$

so that in each state, consumption equals net supply.

To continue our interpretation of the Arrow-Debreu Contingent Commodities Model, the interpretation of the equilibrium which we are now going to discuss is that at time $t = 0$ there is a futures market for each contingent commodity. Equilibrium will require that supply equals demand for each contingent commodity.

12.4 Definition. A system of prices, $\boldsymbol{p}^* = (p_{11}^*, \dots, p_{SG}^*) \in \mathbb{R}^n$ and an allocation, $(\langle \boldsymbol{x}_i^* \rangle, \langle \boldsymbol{y}_k^* \rangle)$ will be said to be an **Arrow-Debreu equilibrium** iff:

1. $(\langle \boldsymbol{x}_i^* \rangle, \langle \boldsymbol{y}_k^* \rangle)$ is a feasible allocation,
2. for every $k \in L$, $\boldsymbol{y}_k^*$ satisfies:

$$(\forall \boldsymbol{y}_k \in Y_k) \colon \boldsymbol{p}^* \cdot \boldsymbol{y}_k^* \geq \boldsymbol{p}^* \cdot \boldsymbol{y}_k,$$

and:

3. for every $i \in M$:

$$\begin{aligned} \boldsymbol{p}^* \cdot \boldsymbol{x}_i^* &\leq \boldsymbol{p}^* \cdot \boldsymbol{\omega}_i + \sum_{k \in L} \theta_{ik} \boldsymbol{p}^* \cdot \boldsymbol{y}_k^*, \text{ and :} \\ (\forall \boldsymbol{x}_i \in X_i) \colon \boldsymbol{x}_i \succ_i \boldsymbol{x}_i^* &\Rightarrow \boldsymbol{p}^* \cdot \boldsymbol{x}_i > \boldsymbol{p}^* \cdot \boldsymbol{\omega}_i + \sum_{k \in L} \theta_{ik} \boldsymbol{p}^* \cdot \boldsymbol{y}_k^*. \end{aligned} \tag{12.10}$$

---

[2] Other than to keep in mind the fact that it is probably inappropriate to suppose that they can be written as a cartesian product of state-specific production sets.

Once again, the definition is formally identical to that presented in Chapter 7; the only difference is in the interpretation. The beauty of the situation, however, is that we can immediately deduce some important results. In particular, we can see that if $(\langle \boldsymbol{x}_i^* \rangle, \langle \boldsymbol{y}_k^* \rangle, \boldsymbol{p}^*)$ is an Arrow-Debreu equilibrium then $\big(\langle \boldsymbol{x}_i^* \rangle, \langle \boldsymbol{y}_k^* \rangle\big)$ must be Pareto efficient; at least in terms of the *ex ante* consumer preferences, $\succsim_i$.

The following example may help you to get a better 'feel' for the model and the meaning of the definition of competitive equilibrium being used here.

12.5 Example. Suppose $\mathcal{E}$ is an exchange economy with $m = 2 = S$, and $G = 1$; that is we have two consumers, one physically distinguishable commodity, and two states of the world to consider. We will also suppose that the ith consumer has a twice-differentiable Bernoullian utility function, $u_i \colon \mathbb{R}_+ \to \mathbb{R}_+$ such that, for all $x \in \mathbb{R}_+$:

$$u_i'(x) > 0 \text{ and } u_i''(x) < 0;$$

so that u_i is strictly increasing and strictly concave. If $\boldsymbol{x}_i = (x_{i1}, x_{i2})$ and $\boldsymbol{x}_i' = (x_{i1}', x_{i2}')$ are two commodity bundles in $X_i = \mathbb{R}^2_+$, consumer i will prefer $\boldsymbol{x}_i$ to $\boldsymbol{x}_i'$ if, and only if:

$$U_i(\boldsymbol{x}_i) = \pi_{i1}u_i(x_{i1}) + \pi_{i2}u_i(x_{i2}) > U_i(\boldsymbol{x}_i') = \pi_{i1}u_i(x_{i1}') + \pi_{i2}u_i(x_{i2}'),$$

where 'π_{is}' denotes i's (subjective) probability that state s will occur, for $s = 1, 2$. Supposing that these probabilities are strictly positive, and that prices for the two goods at $t = 0$ are given by $\boldsymbol{p} = (p_1, p_2) \in \mathbb{R}^2_{++}$, the ith consumer will maximize utility by setting:

$$\frac{\pi_{i1}u_i'(x_{i1})}{p_1} = \frac{\pi_{i2}u_i'(x_{i2})}{p_2}, \tag{12.11}$$

and:

$$\boldsymbol{p} \cdot \boldsymbol{x}_i = \boldsymbol{p} \cdot \boldsymbol{\omega}_i. \tag{12.12}$$

Assuming that the two consumers agree on the probabilities of the two states (so that $\pi_{1s} = \pi_{2s} \equiv \pi_s$, for $s = 1, 2$, it is easily seen that in competitive (Arrow-Debreu) equilibrium:

$$\frac{u_1'(x_{11})}{u_1'(x_{12})} = \frac{\pi_2 p_1}{\pi_1 p_2} = \frac{u_2'(x_{21})}{u_2'(x_{22})}; \tag{12.13}$$

and thus it is easy to see that the allocation will be Pareto efficient.

Now suppose that:

$$\omega_{11} + \omega_{21} = \omega_{12} + \omega_{22}; \tag{12.14}$$

that is, that the total endowment in the two states is exactly the same. Suppose further that $\pi_1 = \pi_2$; that is, that the both consumers consider the two states to be equally probable. Then (12.13) becomes:

$$\frac{u_1'(x_{11})}{u_1'(x_{12})} = \frac{p_1}{p_2} = \frac{u_2'(x_{21})}{u_2'(x_{22})}. \tag{12.15}$$

Suppose then, by way of obtaining a contradiction, that:

$$x_{11} > x_{12} \tag{12.16}$$

Then by the assumed properties of the u_i functions:

$$\frac{u_1'(x_{11})}{u_1'(x_{12})} < 1.$$

However, it then follows from (12.15) that $u_2'(x_{21})/u_2'(x_{22}) < 1$ also; in which case it follows from the assumed properties of the u_i that:

$$x_{21} > x_{22}$$

as well. But then it follows that:

$$x_{11} + x_{21} > x_{12} + x_{22};$$

which, given (12.14), contradicts the assumption that $\big(\langle \boldsymbol{x}_{is} \rangle, \boldsymbol{p}\big)$ is a competitive (Arrow-Debreu) equilibrium. A symmetric argument shows that we cannot have $x_{i1} < x_{12}$ for either $i = 1$ or $i = 2$. Therefore, we must have:

$$x_{i1} = x_{i2} \quad \text{for } i = 1, 2;$$

and from (12.15) we see that this implies that we must have $p_1 = p_2$.

Maintaining the assumption that (12.14) holds, arguments similar to those of the above paragraph establish that if both individuals believe the first state to be more probable than the second, then we must have $p_1 > p_2$ in equilibrium.

Next suppose that we have:

$$\omega_{12} = \omega_{21} = 0,$$

but that (12.14) continues to hold (so that we have **private risk**, but we do not have **social risk**). Then it follows from the reasoning above that both consumers fully insure; that is, they each sell off rights to half of their endowments in order to equalize expected consumption in the two states.

Finally, suppose that we have **social risk**; that is, suppose we have:

$$\omega_1 \stackrel{\text{def}}{=} \omega_{11} + \omega_{21} \neq \omega_2 \stackrel{\text{def}}{=} \omega_{12} + \omega_{22}, \tag{12.17}$$

but that $\pi_1 = \pi_2$. I will leave it as an exercise to show that in this situation we must have:

$$(p_1 - p_2)(\omega_1 - \omega_2) < 0. \quad \square \tag{12.18}$$

The scenario involved in the usual interpretation of the model we have been discussing is that all markets operate and are cleared in the initial period ($t = 0$), while all consumption takes place at $t = 1$. There are a couple of points which should be made with respect to this interpretation. First of all, there is a question about *ex ante* versus *ex post* efficiency. Suppose we have an Arrow-Debreu equilibrium, $(\langle \boldsymbol{x}_i^* \rangle, \langle \boldsymbol{y}_k^* \rangle, \boldsymbol{p}^*)$, but that markets are re-opened at $t = 1$, after the uncertainty is resolved, but before consumption takes place.[3] What would happen then? Strictly

[3]The markets in question here are called **spot markets**, while the markets at $t = 0$ are called **forward markets**.

speaking, we cannot say without assuming that preferences are weakly separable on X_{is} and that all consumers' *ex post* preferences are the same as their *ex ante* preferences over X_{is}. However, both of these assumptions seem to be eminently reasonable, and if both are true, then there would be no incentive for trades to take place in this situation. Why is this? Well, each consumer must be maximizing satisfaction, given the expenditure $\boldsymbol{p}_s^* \cdot \boldsymbol{x}_{is}$ at $\boldsymbol{x}_{is}^*$; for if, for some consumer i there were some $\boldsymbol{x}_{is}'$ such that:

$$\boldsymbol{p}_s^* \cdot \boldsymbol{x}_{is}' \leq \boldsymbol{p}_s^* \cdot \boldsymbol{x}_{is} \text{ and } \boldsymbol{x}_{is}' \succ_{is} \boldsymbol{x}_{is}^*,$$

the consumer would have preferred to replace the bundle $\boldsymbol{x}_{is}^*$ with $\boldsymbol{x}_{is}'$ at $t = 0$.[4] Since the allocation in state s is therefore a competitive equilibrium, given the price vector $\boldsymbol{p}_s^*$, it follows that it is also Pareto efficient. Consequently, there are no mutually beneficial trades which consumers can make among themselves after the resolution of uncertainty.

A serious objection to the interpretation of the model which we set out in the preceding paragraph is that it is clearly unrealistic to expect the existence of forward markets in each commodity. However, suppose we have an Arrow-Debreu equilibrium, $(\langle \boldsymbol{x}_i^* \rangle, \langle \boldsymbol{y}_k^* \rangle, \boldsymbol{p}^*)$. If prices in each state are correctly anticipated by all agents, and we have a futures market for only one commodity, with trading only in that one commodity at $t = 0$, then we can achieve that same consumption allocation, $\langle \boldsymbol{x}_i^* \rangle$, if re-trading is possible (at the anticipated prices) at $t = 1$. This remarkable fact was first noted by Arrow [1953]. The formal extension of this idea which we will be studying in the next section was, however, developed by Professor Roy Radner [1968, 1982].

12.3 Radner Equilibrium

For the sake of simplicity, in the remainder of this chapter we will confine our discussion to the context of a pure exchange economy, and we will retain the notation and basic assumptions regarding consumers which were introduced in the previous section; so that consumer i has a preference relation $\succsim_i$ on X_i, and has the initial endowment $\boldsymbol{\omega}_i$, as before. Once again we will deal with a two-period model; with uncertainty being resolved in the second period ($t = 1$). This time, however, we will allow no commodity trading in the first period ($t = 0$). We will, however, introduce the idea of some tradeable assets, which can be purchased (or sold short) in the first period. There are three pivotal assumptions which we will make in this context. First, we will suppose that at $t = 0$ consumers have *expectations* of the prices which will occur (and at which trading will take place) at $t = 1$, for each possible state of nature (each $s \in S$). Secondly, we suppose that all consumers expect the same vector of prices to prevail at $t = 1$ if state $s \in S$ occurs; we denote this vector by '$\boldsymbol{p}_s$,' and we denote the full vector of such prices by '$\boldsymbol{p}$;' that is:

$$\boldsymbol{p} = (\boldsymbol{p}_1, \ldots, \boldsymbol{p}_s, \ldots, \boldsymbol{p}_S).$$

[4] Notice, however, that both weak separability and the identity of ex post and ex ante preferences are needed to make this argument correct.

Thirdly, we suppose that at $t = 0$ there are K assets available, all of which pay off a conditional return in the first commodity; which we take to be a numéraire.

12.6 Definition. A unit of an **asset** is a title to receive an amount r_s of good 1 at date $t = 1$ if state s occurs. An asset is therefore characterized by its **return vector** $\boldsymbol{r} = (r_1, \ldots, r_S) \in \mathbb{R}^S$.

Thus, a checking account (in a fully-insured bank) might be characterized as having the return vector $\boldsymbol{r} = \boldsymbol{1}$, where $\boldsymbol{1} \in \mathbb{R}^S_+$ is the vector all of whose coordinates equal 1. Very useful examples are also provided by assets whose return vectors are of the form $\boldsymbol{r}_s = \boldsymbol{e}_s$, where $\boldsymbol{e}_s \in \mathbb{R}^S_+$ is the sth unit coordinate vector. In other words, such an asset provides a return of one unit of the numéraire if state s occurs, and nothing otherwise.

We suppose there is a given set of K assets (an **asset structure**), which can be traded at $t = 0$. We denote the return vector associated with the kth asset by $\boldsymbol{r}_k \in \mathbb{R}^S$. We assume that there are no initial endowments of assets, and that short sales are possible. The price of the kth asset at $t = 0$ is denoted by q_k, and a vector of quantities of the K assets, $\boldsymbol{z} \in \mathbb{R}^K$ is called a **portfolio**. Thus the expenditure required to obtain the portfolio $\boldsymbol{z}$ at $t = 0$, given the asset price vector $\boldsymbol{q}$ is equal to $\boldsymbol{q} \cdot \boldsymbol{z}$. Since we assume that consumers have no initial endowments of assets, consumer i must choose a portfolio $\boldsymbol{z}_i$ satisfying:

$$\boldsymbol{q} \cdot \boldsymbol{z}_i \leq 0.$$

Notice that we have put no restriction on the sign of $\boldsymbol{z}_i$; a negative value for z_{ik} means the consumer is selling the kth asset *short*; that is, the consumer will *owe* $r_{ks} z_{ik}$ units of the numéraire if state s should occur.

Since we are assuming that the first commodity is a numéraire, we can normalize to set $p_{s1} = 1$, for $s = 1, \ldots, S$.[5] Given an asset structure, we can then define the $S \times K$ **return matrix**, $\boldsymbol{R}$, by:

$$\boldsymbol{R} = \begin{pmatrix} r_{11} & r_{12} & \cdots & r_{1k} & \cdots & r_{1K} \\ \cdots & \cdots & \cdots & \cdots & \cdots & \cdots \\ r_{s1} & r_{s2} & \cdots & r_{sk} & \cdots & r_{sK} \\ \cdots & \cdots & \cdots & \cdots & \cdots & \cdots \\ r_{S1} & r_{S2} & \cdots & r_{Sk} & \cdots & r_{SK} \end{pmatrix}. \tag{12.19}$$

Notice that in this matrix, the rows correspond to states, while the columns correspond to assets. Thus, if $\boldsymbol{r}_k$ is the vector of returns for the kth asset, $\boldsymbol{r}_k$ becomes the kth column of the return matrix. We will be a bit sloppy in our terminology in dealing with asset structures and the return matrices which they determine in that *we will often identify the asset structure with the return matrix which it determines.* That is, we will often refer to a nonnegative $S \times K$ matrix as an asset structure.

Now, if a consumer holds the portfolio $\boldsymbol{z} \in \mathbb{R}^K$, the return from this portfolio if state s occurs is given by:

$$\boldsymbol{r}_{s\cdot} \cdot \boldsymbol{z} = \sum_{k=1}^{K} r_{sk} z_k; \tag{12.20}$$

[5]It may not be apparent at this point why it is that we can normalize in this manner, but it will be clear once we define consumer equilbrium for the model to be presented in this section. See Exercise 5, at the end of this chapter.

where '$\boldsymbol{r}_{s\cdot}$' denotes row s of $\boldsymbol{R}$.[6] Thus, we can make use of this return matrix (and given the price normalization just mentioned), to write the ith consumer's consumption budget constraint, given the price vector, $\boldsymbol{p}^* = (\boldsymbol{p}_1^*, \boldsymbol{p}_2^*, \ldots, \boldsymbol{p}_s^*, \ldots, \boldsymbol{p}_S^*)$ and the portfolio, $\boldsymbol{z}_i^* \in \mathbb{R}^K$ as:

$$\begin{pmatrix} \boldsymbol{p}_1^* \cdot (\boldsymbol{x}_{i1} - \boldsymbol{\omega}_{i1}) \\ \cdots \\ \boldsymbol{p}_s^* \cdot (\boldsymbol{x}_{is} - \boldsymbol{\omega}_{is}) \\ \cdots \\ \boldsymbol{p}_S^* \cdot (\boldsymbol{x}_{iS} - \boldsymbol{\omega}_{iS}) \end{pmatrix} \leq \begin{pmatrix} r_{11} & r_{12} & \ldots & r_{1k} & \ldots & r_{1K} \\ \ldots & \ldots & \ldots & \ldots & \ldots & \ldots \\ r_{s1} & r_{s2} & \ldots & r_{sk} & \ldots & r_{sK} \\ \ldots & \ldots & \ldots & \ldots & \ldots & \ldots \\ r_{S1} & r_{S2} & \ldots & r_{Sk} & \ldots & r_{SK} \end{pmatrix} \begin{pmatrix} z_{i1}^* \\ \cdots \\ z_{ik}^* \\ \cdots \\ z_{iK}^* \end{pmatrix} = \boldsymbol{R}\boldsymbol{z}_i^*. \quad (12.21)$$

Thus, in each state s, the consumer's expenditure on commodities, minus the value of its endowment, must be no greater than the value of the return on its portfolio. Thus, given the vector of asset prices, $\boldsymbol{q}^*$ at $t = 0$, the vector of commodity prices, $\boldsymbol{p}^* \in \mathbb{R}^n$, and the return matrix, $\boldsymbol{R}$, the ith consumer's budget set, $B_i(\boldsymbol{p}^*, \boldsymbol{q}^*, \boldsymbol{R})$, can be written as:

$$B_i(\boldsymbol{p}^*, \boldsymbol{q}^*, \boldsymbol{R}) = \big\{\boldsymbol{x}_i \in X_i \mid (\exists \boldsymbol{z}_i \in \mathbb{R}^K)\colon \boldsymbol{q}^* \cdot \boldsymbol{z}_i \leq 0, \ \&\ (\boldsymbol{x}_i, \boldsymbol{z}_i) \text{ satisfies } (12.21)\big\}.$$

12.7 Example. An important asset structure for comparisons and examples is given by the set S of assets (often called *Arrow securities*), where the sth asset has the return vector:

$$\boldsymbol{r}_s = \boldsymbol{e}_s \quad \text{for } s = 1, \ldots, S. \quad (12.22)$$

That is, the sth asset pays a return of 1 unit of the first good if state s occurs, and nothing otherwise. In this case, the return matrix is the $S \times S$ identity matrix, $\boldsymbol{I}_S$.

Thus, if we suppose that there are only two states of the world in period 1, the ith consumer's budget set, given the commodity price vector $\boldsymbol{p}^* = (\boldsymbol{p}_1^*, \boldsymbol{p}_2^*)$, and security prices $\boldsymbol{q}^* = (q_1^*, q_2^*)$, is given by:

$$B_i(\boldsymbol{p}^*, \boldsymbol{q}^*; R) = \left\{\boldsymbol{x}_i \in X_i \;\middle|\; (\exists \boldsymbol{z} \in \mathbb{R}^2)\colon \boldsymbol{q}^* \cdot \boldsymbol{z} \leq 0 \ \&\ \begin{pmatrix} \boldsymbol{p}_1^* \cdot (\boldsymbol{x}_{i1} - \boldsymbol{\omega}_{i1}) \\ \boldsymbol{p}_2^* \cdot (\boldsymbol{x}_{i2} - \boldsymbol{\omega}_{i2}) \end{pmatrix} \leq \begin{pmatrix} z_1 \\ z_2 \end{pmatrix}\right\}. \quad \square$$

In this context, we will be considering the following definition of equilibrium.

12.8 Definition. A tuple, $\big(\langle \boldsymbol{x}_i^*, \boldsymbol{z}_i^* \rangle, \langle \boldsymbol{p}_s^* \rangle, \boldsymbol{q}^*\big)$, where $\boldsymbol{q}^* \in \mathbb{R}^K$ is a vector of asset prices at $t = 0$, $\boldsymbol{p}_s^* \in \mathbb{R}^G$ is a vector of spot prices at state s $(s = 1, \ldots, S)$, $\boldsymbol{x}_i^* \in \mathbb{R}^n$ is the consumption bundle (at $t = 1$) and $\boldsymbol{z}_i^* \in \mathbb{R}^K$ the asset portfolio of the ith consumer, is a **Radner equilibrium given the asset structure $\boldsymbol{R}$** iff:

1. for each i $(i = 1, \ldots, m)$, the pair $(\boldsymbol{x}_i^*, \boldsymbol{z}_i^*)$ solves the consumer's problem:

$$\max_{w.r.t.(\boldsymbol{x}_i, \boldsymbol{z}_i)} \succsim_i \quad \text{subject to:}$$

$$\boldsymbol{q}^* \cdot \boldsymbol{z}_i \leq 0 \ \text{ and } \boldsymbol{p}_s^* \cdot \boldsymbol{x}_{is} \leq \boldsymbol{p}_s^* \cdot \boldsymbol{\omega}_{is} + \sum\nolimits_{k=1}^{K} r_{sk} z_{ik}, \ \text{for } s = 1, \ldots, S.$$

2. $\sum_{i=1}^m \boldsymbol{z}_i^* \leq \boldsymbol{0}$, and:
3. $\sum_{i=1}^m \boldsymbol{x}_{is}^* \leq \sum_{i=1}^m \boldsymbol{\omega}_{is}$ for $s = 1, \ldots, S$.

---

[6] More exactly, of course, the sum in (12.20) is the number of units of the first commodity which the consumer will receive if state s occurs. However, since we are setting $p_{s1} = 1$, for each $s \in S$, this is also the value of the return.

While the above definition is a bit difficult to state, the first condition simply says that each $\boldsymbol{x}_i^*$ is an element of $B_i(\boldsymbol{p}^*, \boldsymbol{q}^*; \boldsymbol{R})$, and that given any $\boldsymbol{x}_i \in B_i(\boldsymbol{p}^*, \boldsymbol{q}^*; \boldsymbol{R})$, we must have:

$$\boldsymbol{x}_i^* \succsim_i \boldsymbol{x}_i.$$

The second condition says that trading of assets must *balance*; that is, positive purchases of assets by some consumers must be balanced by short sales of the assets by others. Finally, the third condition simply says that aggregate consumption in each state of the world must be equal to the aggregate commodity endowment in that state.

If you re-examine the definition of the consumers' maximization problem in 12.8, it will be apparent that an equilibrium vector of asset prices can be multiplied by a positive scalar without changing the consumers' choices; that is, if $\big(\langle \boldsymbol{x}_i^*, \boldsymbol{z}_i^* \rangle, \langle \boldsymbol{p}_s^* \rangle, \boldsymbol{q}^*\big)$ is a Radner equilibrium, and θ is a positive real number, then $\big(\langle \boldsymbol{x}_i^*, \boldsymbol{z}_i^* \rangle, \boldsymbol{p}^*, \theta \boldsymbol{q}^*\big)$ is also a Radner equilibrium. Consequently, we can normalize to set the price of one of the assets equal to one (1), and we will sometimes find it convenient to do so.

12.9 Definition. We shall say that a vector of asset prices, $\boldsymbol{q} \in \mathbb{R}^K$ is **arbitrage-free**, given the return matrix, $\boldsymbol{R}$, iff there exists no portfolio, $\boldsymbol{z} \in \mathbb{R}^K$ satisfying:

$$\begin{pmatrix} -\boldsymbol{q} \\ \boldsymbol{R} \end{pmatrix} \boldsymbol{z} > \boldsymbol{0}. \tag{12.23}$$

Notice that if $\boldsymbol{z}$ satisfies (12.23), then either:

$$\boldsymbol{q} \cdot \boldsymbol{z} < 0 \text{ and } \boldsymbol{R}\boldsymbol{z} \geq \boldsymbol{0}, \tag{12.24}$$

or

$$\boldsymbol{q} \cdot \boldsymbol{z} \leq 0 \text{ and } \boldsymbol{R}\boldsymbol{z} > \boldsymbol{0}. \tag{12.25}$$

We will generally be assuming that each column of the return matrix, $\boldsymbol{R}$, is semi-positive (each asset provides a positive return in some state, and nonnegative returns in all other states), and given this, it can be shown (see Exercise 4, at the end of this chapter) that if (12.24) holds, then there exists a portfolio, $\boldsymbol{z}'$ satisfying (12.25). On the other hand, if (12.25) holds, consumers can obtain an unbounded return in at least one state, while earning a nonnegative return in all other states. Consequently, if an asset price vector $\boldsymbol{q}$ is *not* arbitrage-free, and if preferences are increasing, or if the first commodity is a numéraire good for the economy in each state, there will exist no solution for the consumers' maximization problems in our definition of a Radner equilibrium. *We will assume throughout the remainder of this chapter that preferences are weakly separable over states, and increasing in consumption goods within each state.* As a consequence of this second assumption it will follow that *if* $\big(\langle \boldsymbol{x}_i^*, \boldsymbol{z}_i^* \rangle, \langle \boldsymbol{p}_s^* \rangle, \boldsymbol{q}^*\big)$ *is a Radner equilibrium, given the asset structure,* $\boldsymbol{R}$, *then* $\boldsymbol{q}^*$ *must be arbitrage free, given* $\boldsymbol{R}$. (We will re-visit this statement shortly in a more formal fashion.)

If the asset price vector $\boldsymbol{q}$ *is* arbitrage free, given $\boldsymbol{R}$, then, as we will prove shortly, $\boldsymbol{q}$ must be in the polyhedral cone determined by the rows of $\boldsymbol{R}$; that is, there must exist $\boldsymbol{\mu} \in \mathbb{R}_+^S$ such that:

$$\boldsymbol{\mu}^\top \boldsymbol{R} = \boldsymbol{q}^\top. \tag{12.26}$$

Following Duffie [2001], we will refer to such a vector as a **state price vector for** $(\boldsymbol{R}.\boldsymbol{q})$.

12.10 Examples.

1. Let $S = 2$ and $K = 3$, and suppose:

$$\boldsymbol{R} = \begin{pmatrix} 1 & 0 & 1 \\ 0 & 1 & 1 \end{pmatrix} \text{ and } \boldsymbol{q} = \begin{pmatrix} 1 \\ 1 \\ 2 \end{pmatrix}.$$

Now the question is, is $\boldsymbol{q}$ arbitrage free, given $\boldsymbol{R}$? To answer this, suppose $\boldsymbol{z} \in \mathbb{R}^3$ is such that $\boldsymbol{q} \cdot \boldsymbol{z} \leq 0$. Then from the definitions we see that we must have:

$$z_1 \leq -z_2 - 2z_3.$$

But then:

$$\boldsymbol{Rz} = \begin{pmatrix} z_1 + z_3 \\ z_2 + z_3 \end{pmatrix} \leq \begin{pmatrix} -z_2 - 2z_3 + z_3 \\ z_2 + z_3 \end{pmatrix} = \begin{pmatrix} -z_2 - z_3 \\ z_2 + z_3 \end{pmatrix}.$$

Thus, it is clear that if $\left[\boldsymbol{Rz}\right]_1 > 0$, then $\left[\boldsymbol{Rz}\right]_2 < 0$, and conversely. Therefore, $\boldsymbol{q}$ is arbitrage-free, given $\boldsymbol{R}$.

If we define $\boldsymbol{\mu} = (1, 1)$, then we see that:

$$\boldsymbol{\mu}^\top \boldsymbol{R} = (1 \;\; 1) \begin{pmatrix} 1 & 0 & 1 \\ 0 & 1 & 1 \end{pmatrix} = (1 \;\; 1 \;\; 2) = \boldsymbol{q}^\top;$$

and thus $\boldsymbol{\mu}$ is a state-price vector for $(\boldsymbol{R}, \boldsymbol{q})$.

2, Suppose the asset structure is that defined by the set of Arrow securities; that is, suppose $\boldsymbol{R} = \boldsymbol{I}_S$. Then if $\boldsymbol{q}$ is a vector of security prices, and $\boldsymbol{\mu}$ is a state-price vector for $(\boldsymbol{R}, \boldsymbol{q})$, we must have:

$$\boldsymbol{\mu}^\top = \boldsymbol{\mu}^\top \boldsymbol{I}_S = \boldsymbol{q}^\top.$$

In other words, $\mu_s = q_s$, for $s = 1, \ldots, S$. □

While we earlier interpreted equation (12.26) as meaning that $\boldsymbol{q}$ must be in the polyhedral cone generated by the rows of $\boldsymbol{R}$; the economic interpretation of (12.26) is that, given such a state-price vector for $(\boldsymbol{R}, \boldsymbol{q})$, the price of each asset is then simply the sum over the possible states of the world of the state-price-weighted return of the asset in that state; that is (12.26) implies:

$$\sum_{s=1}^{S} \mu_s r_{sk} = q_k \quad \text{for } k = 1, \ldots, K.$$

In the following proposition we establish the fact that there always exists a state-price vector if $\boldsymbol{q}$ is arbitrage free, given $\boldsymbol{R}$.

12.11 Proposition. *Suppose $\boldsymbol{R}$ is an $S \times K$ return matrix and that $\boldsymbol{q} \in \mathbb{R}^K_+$ is an asset price vector. Then $\boldsymbol{q}$ is arbitrage-free, given $\boldsymbol{R}$ if, and only if, there exists a state-price vector for $(\boldsymbol{R}, \boldsymbol{q})$, $\boldsymbol{\mu} \in \mathbb{R}^S_{++}$, such that:*

$$\boldsymbol{\mu}^\top \boldsymbol{R} = \boldsymbol{q}^\top. \tag{12.27}$$

Proof. Define the $K \times (S+1)$ matrix $\boldsymbol{A}$ by:

$$\boldsymbol{A} = \begin{pmatrix} \boldsymbol{R}^\top & -\boldsymbol{q} \end{pmatrix},$$

and suppose first that there exists a state-price vector, $\boldsymbol{\mu}$, satisfying (12.27). Then we have:

$$\boldsymbol{A}\begin{pmatrix} \boldsymbol{\mu} \\ 1 \end{pmatrix} = \begin{pmatrix} \boldsymbol{R}^\top & -\boldsymbol{q} \end{pmatrix}\begin{pmatrix} \boldsymbol{\mu} \\ 1 \end{pmatrix} = \boldsymbol{R}^\top \boldsymbol{\mu} - \boldsymbol{q} = \boldsymbol{0};$$

and it follows from Stiemke's Theorem (Theorem 6.40) that there exists no $\boldsymbol{z} \in \mathbb{R}^K$ such that:

$$\boldsymbol{z}^\top \boldsymbol{A} = \boldsymbol{z}^\top \begin{pmatrix} \boldsymbol{R}^\top & -\boldsymbol{q} \end{pmatrix} > \boldsymbol{0}.$$

Thus we see that $\boldsymbol{q}$ is arbitrage-free, given $\boldsymbol{R}$.

On the other hand, suppose $\boldsymbol{q}$ is arbitrage-free. Then there exists no $\boldsymbol{z} \in \mathbb{R}^K$ such that:

$$\boldsymbol{z}^\top \boldsymbol{A} > \boldsymbol{0}.$$

Then using Stiemke's Theorem (Theorem 6.40) once again, it follows that there exists $\boldsymbol{x}^* \in \mathbb{R}^{S+1}$ such that $\boldsymbol{x}^* \gg \boldsymbol{0}$, and such that, writing:

$$\boldsymbol{x}^* = \begin{pmatrix} \boldsymbol{y}^* \\ x^*_{S+1} \end{pmatrix},$$

we have:

$$\boldsymbol{0} = \boldsymbol{A}\boldsymbol{x}^* = \begin{pmatrix} \boldsymbol{R}^\top & -\boldsymbol{q} \end{pmatrix}\begin{pmatrix} \boldsymbol{y}^* \\ x^*_{S+1} \end{pmatrix} = \boldsymbol{R}^\top \boldsymbol{y}^* - x^*_{S+1}\boldsymbol{q}. \tag{12.28}$$

Consequently, defining $\boldsymbol{\mu} \in \mathbb{R}^S$ by:

$$\boldsymbol{\mu} = (1/x^*_{S+1})\boldsymbol{y}^*,$$

we have $\boldsymbol{\mu} \gg \boldsymbol{0}$, and from (12.28):

$$\boldsymbol{\mu}^\top \boldsymbol{R} = \boldsymbol{q}^\top. \quad \square$$

12.12 Example. Suppose there is an asset, say the first, which is **risk-free**; that is, suppose the return vector, $\boldsymbol{r}_1 \in \mathbb{R}^S$, is given by:

$$\boldsymbol{r}_1 = \boldsymbol{1},$$

where $\boldsymbol{1}$ is the column vector in $\mathbb{R}^S$ having all of its entries equal to 1. It seems entirely natural in this case to normalize by setting the price of this asset equal to 1; and, as we noted earlier, we can do this without loss of generality, insofar as our definition of a Radner equilibrium is concerned. Notice also that if an asset price vector, $\boldsymbol{q}^*$, is arbitrage-free, then so is the asset price vector $\boldsymbol{q}'$, given by:

$$\boldsymbol{q}' = (1/q_1^*)\boldsymbol{q}^*.$$

If this normalized asset price vector is used in the proof of Proposition 12.11, then the first of equations (12.28) becomes:

$$\sum\nolimits_{s=1}^{S} r_{s1} y_s^* - x^*_{S+1} q_1 = \sum\nolimits_{s=1}^{S} y_s^* - x^*_{S+1} = 0. \tag{12.29}$$

Therefore, the state-price vector obtained in the proof satisfies:

$$\sum\nolimits_{s=1}^{S} \mu_s = 1. \tag{12.30}$$

Consequently, it is easy to show that for all k, we must have:

$$\min_s r_{sk} \leq q_k \leq \max_s r_{sk}. \quad \square$$

Suppose the first commodity is a numéraire in each state of the world. If any consumer, say the first, has preferences representable by a von Neumann-Morgenstern utility function, and is able to find a utility-maximizing choice of $(\boldsymbol{x}_i, \boldsymbol{z}_i)$, given prices $(\boldsymbol{p}, \boldsymbol{q})$, then the asset price vector, $\boldsymbol{q}$, must be arbitrage-free, given $\boldsymbol{R}$. This is the content of the following proposition, whose proof will be left as an exercise.

12.13 Proposition. *Suppose an agent, i, has the von Neumann-Morgenstern utility function:*

$$U_i(\boldsymbol{x}_i) = \sum\nolimits_{s=1}^{S} \pi_s u_{is}(\boldsymbol{x}_{Is}),$$

where π_s is the probability of state s, for $s = 1, \ldots, S$; and maximizes utility at $(\boldsymbol{x}_i^*, \boldsymbol{z}_i^*) \gg \boldsymbol{0}$, given prices $(\boldsymbol{p}^*, \boldsymbol{q}^*)$ and the return matrix, $\boldsymbol{R}$; and that:

$$\frac{\partial u_{is}(\boldsymbol{x}_{is}^*)}{\partial x_{is1}} \stackrel{def}{=} \frac{\partial u_{is}}{\partial x_{is1}}\bigg|_{\boldsymbol{x}_i^*} > 0 \quad \text{for } s = 1, \ldots, S.$$

Then there exists a positive scalar, ν_i, such that the vector $\boldsymbol{\mu}_i^*$ given by:

$$\mu_{is}^* = \frac{\pi_s}{\nu_i p_{s1}^*} \times \frac{\partial u_{is}(\boldsymbol{x}_{ia}^*)}{\partial x_{is1}} \quad \text{for } s = 1, \ldots, S,$$

is a state-price vector for $(\boldsymbol{R}, \boldsymbol{q}^*)$ (and thus $\boldsymbol{q}^*$ is arbitrage-free, given $\boldsymbol{R}$).

While the above proposition is, I believe, of some interest, it can obviously be generalized considerably. In particular, if preferences are weakly separable over states, and non-saturating within a state, then it is clear that if $(\langle \boldsymbol{x}_i^*, \boldsymbol{z}_i^* \rangle, \langle \boldsymbol{p}_s^* \rangle, \boldsymbol{q}^*)$ is a Radner equilibrium, then $\boldsymbol{q}^*$ must be arbitrage-free.

12.4 Complete Markets

An economy of the type studied here and in the previous section has very good normative properties if the return matrix is *complete*, as defined in the following.

12.14 Definition. An asset structure with the $S \times K$ return matrix $\boldsymbol{R}$ is said to be **complete** iff the rank of $\boldsymbol{R}$ is S; that is, iff there is some subset of the K assets, containing S elements, whose return vectors are linearly independent.

Obviously there needs to be at least as many assets available as there are possible states in order for this condition to hold. Perhaps the simplest example of a complete asset structure is provided by the Arrow securities of Example 12.7. However, it is all too easy to construct examples in which there are at least as many securites as states, but in which the return matrix has rank less than S.

12.15 Example. Suppose there are three states of the world at $t = 1$, and consider the return matrix:

$$\boldsymbol{R} = \begin{pmatrix} 2 & 1 & 1 \\ 1 & 0 & 1 \\ 1 & 1 & 0 \end{pmatrix}.$$

Remembering that the rows in the matrix correspond to states, while the columns correspond to assets (and thus the first asset yields a return of 2 units if state one occurs, 1 if $s = 2$, and so on, we see that there are as many assets here as there are states. However, $\boldsymbol{R}$ does not have full rank,[7] and thus the asset structure is *not* complete. On the other hand, the asset structure corresponding to the return matrix:

$$\boldsymbol{R} = \begin{pmatrix} 2 & 1 & 0 \\ 0 & 0 & 1 \\ 1 & 0 & 1 \end{pmatrix}$$

is complete, since the matrix has full rank. □

The importance of the asset structure's being complete is brought out by the following result.

12.16 Theorem. *Suppose that the asset structure, $\boldsymbol{R}$, is complete. Then we have the following.*

1. If $(\langle \boldsymbol{x}_i^* \rangle, \boldsymbol{p}^*)$ is an Arrow-Debreu equilibrium for $\mathcal{E}$, then there are asset prices $\boldsymbol{q}^* \in \mathbb{R}^K_{++}$ and portfolio plans $\boldsymbol{z}^* = (\boldsymbol{z}_1^*, \ldots, \boldsymbol{z}_m^*) \in \mathbb{R}^{mK}$ such that $(\langle \boldsymbol{x}_i^*, \boldsymbol{z}_i^* \rangle, \langle \boldsymbol{p}_s^* \rangle, \boldsymbol{q}^*)$ is a Radner equilibrium, given the asset structure $\boldsymbol{R}$

2. Conversely, if $(\langle \boldsymbol{x}_i^*, \boldsymbol{z}_i^* \rangle, \langle \boldsymbol{p}_s^* \rangle, \boldsymbol{q}^*)$ is a Radner equilibrium given $\boldsymbol{R}$, then there exist a state-price vector, $\boldsymbol{\mu} = (\mu_1, \ldots, \mu_S) \in \mathbb{R}^S_{++}$, such that, defining $\overline{\boldsymbol{p}}$ by:

$$\overline{\boldsymbol{p}}_s = \mu_s \boldsymbol{p}_s^* \quad \text{for } s = 1, \ldots, S,$$

$(\langle \boldsymbol{x}_i^* \rangle, \overline{\boldsymbol{p}})$ is an Arrow-Debreu equilibrium for $\mathcal{E}$.

Proof. Since the two definitions of equilibrium have the same feasibility requirement, we needn't bother to prove that the allocation is feasible in either part of our proof. In the argument to follow, we will denote the budget set for the ith consumer, given the Arrow-Debreu price vector, $\boldsymbol{p}$, by '$B_i^A(\boldsymbol{p})$;' while, given a vector of commodity prices, $\boldsymbol{p}$ and security prices, $\boldsymbol{q}$, we will denote the ith consumer's budget set in the Radner sense, and given the return matrix, $\boldsymbol{R}$, by '$B_i^R(\boldsymbol{p}, \boldsymbol{q}; \boldsymbol{R})$.'

1. Suppose $(\langle \overline{\boldsymbol{x}}_i \rangle, \boldsymbol{p})$ is an Arrow-Debreu equilibrium, where:

$$\boldsymbol{p} = (\boldsymbol{p}_1, \ldots, \boldsymbol{p}_S) \in \mathbb{R}^{GS}_+,$$

and define $\boldsymbol{q} \in \mathbb{R}^K$ by:

$$\boldsymbol{q} = \boldsymbol{1}^\top \boldsymbol{R}; \tag{12.31}$$

where once again '$\boldsymbol{1}$' denotes the column vector all of whose coordinates equal 1.

Now, defining:

$$\boldsymbol{y}_i = \big(\boldsymbol{p}_1 \cdot (\overline{\boldsymbol{x}}_{i1} - \boldsymbol{\omega}_{i1}), \ldots, \boldsymbol{p}_S \cdot (\overline{\boldsymbol{x}}_{iS} - \boldsymbol{\omega}_{iS})\big)^\top,$$

[7] For example, $\boldsymbol{y}\boldsymbol{R} = \boldsymbol{0}$, where $\boldsymbol{y} = (1, -1, -1)$.

we have by definition of an Arrow-Debreu equilibrium that, for each i:

$$\mathbf{1} \cdot \boldsymbol{y}_i = 0, \tag{12.32}$$

and, by feasibility:

$$\sum_{i=1}^{m} \boldsymbol{y}_i = \mathbf{0}. \tag{12.33}$$

From the assumption that the asset structure is complete, $\boldsymbol{R}$ has full rank (equal to S), and thus, for each $i = 1, \ldots, m$, we can find $\overline{\boldsymbol{z}}_i$ such that:

$$\boldsymbol{R}\overline{\boldsymbol{z}}_i = \boldsymbol{y}_i.$$

Therefore, if the ith consumer has the portfolio $\overline{\boldsymbol{z}}_i$, we have:

$$\boldsymbol{p}_s \cdot (\overline{\boldsymbol{x}}_i - \boldsymbol{\omega}_i) = \sum_{k=1}^{K} r_{sk}\overline{z}_{ik} \quad \text{for } s = 1, \ldots, S,$$

and, by the definition of $\boldsymbol{q}$ and (12.32):

$$\boldsymbol{q} \cdot \overline{\boldsymbol{z}}_i = \mathbf{1}^{\top}\boldsymbol{R}\overline{\boldsymbol{z}}_i = \mathbf{1}^{\top}\boldsymbol{y}_i = 0 \quad \text{for } i = 1, \ldots, m.$$

Thus $\overline{\boldsymbol{x}}_i \in B_i^R(\boldsymbol{p}, \boldsymbol{q}; \boldsymbol{R})$. Furthermore, it follows from (12.33) that:

$$\mathbf{0} = \sum_{i=1}^{m} \boldsymbol{y}_i = \sum_{i=1}^{m} \boldsymbol{R}\overline{\boldsymbol{z}}_i = \boldsymbol{R}\Big(\sum_{i=1}^{m} \overline{\boldsymbol{z}}_i\Big);$$

and thus, since $\boldsymbol{R}$ has full rank:

$$\sum_{i=1}^{m} \overline{\boldsymbol{z}}_i = \mathbf{0} \tag{12.34}$$

Now let i be arbitrary, and let $\boldsymbol{x}_i \in X_i$ and $\boldsymbol{z}_i \in \mathbb{R}^K$ be such that:

$$\boldsymbol{q} \cdot \boldsymbol{z}_i \leq 0,$$

and, for each $s \in S$:

$$\boldsymbol{p}_s \cdot \boldsymbol{x}_{is} \leq \boldsymbol{p}_s \cdot \boldsymbol{\omega}_{is} + \sum_{k \in K} r_{sk} z_{ik}.$$

Then we have:

$$\sum_{s \in S} \boldsymbol{p}_s \cdot (\boldsymbol{x}_{is} - \boldsymbol{\omega}_{is}) \leq \sum_{s \in S}\sum_{k \in K} r_{sk} z_{ik} = \sum_{k \in K} z_{ik} \sum_{s \in S} r_{sk} = \sum_{k \in K} q_k z_{ik} \leq 0;$$

where the last equality is by definition of $\boldsymbol{q}$. Consequently, if $\boldsymbol{x}_i \in B_i^R(\boldsymbol{p}, \boldsymbol{q}, \boldsymbol{R})$, then $\boldsymbol{x}_i \in B_i^A(\boldsymbol{p})$; and since $\overline{\boldsymbol{x}}_i \in B_i^R(\boldsymbol{p}, \boldsymbol{q}, \boldsymbol{R})$, and maximizes $\succsim_i$ over $B_i^A(\boldsymbol{p})$ it follows that $\overline{\boldsymbol{x}}_i$ maximizes $\succsim_i$ on $B_i^R(\boldsymbol{p}, \boldsymbol{q}, \boldsymbol{R})$ as well. Combining this fact with (12.34), it follows that $\big(\langle \overline{\boldsymbol{x}}_i, \overline{\boldsymbol{z}}_i \rangle, \boldsymbol{p}, \boldsymbol{q}\big)$ is a Radner equilibrium, given $\boldsymbol{R}$.

2. Now suppose $\big(\langle \boldsymbol{x}_i^*, \boldsymbol{z}_i^* \rangle, \langle \boldsymbol{p}_s^* \rangle, \boldsymbol{q}^*\big)$ is a Radner equilibrium, given the return matrix $\boldsymbol{R}$. Then $\boldsymbol{q}^*$ must be arbitrage-free, given $\boldsymbol{R}$, and it follows from Proposition 12.11 that there exists a state-price vector $\boldsymbol{\mu} \gg \mathbf{0}$ such that:

$$\boldsymbol{\mu}^{\top}\boldsymbol{R} = (\boldsymbol{q}^*)^{\top}. \tag{12.35}$$

Making use of this state-price vector, we define $\overline{\boldsymbol{p}} \in \mathbb{R}^{SG}$ by:

$$\overline{\boldsymbol{p}}_s = \mu_s \boldsymbol{p}_s^* \quad \text{for } s = 1, \ldots, S. \tag{12.36}$$

Now let $i \in \{1, \ldots, m\}$ be arbitrary, and suppose that $\boldsymbol{x}_i \in B_i^A(\overline{\boldsymbol{p}})$; so that:

$$\overline{\boldsymbol{p}} \cdot (\boldsymbol{x}_i - \boldsymbol{\omega}_i) = \sum_{s=1}^S \overline{\boldsymbol{p}}_s \cdot (\boldsymbol{x}_{is} - \boldsymbol{\omega}_{is}) = \sum_{s=1}^S \mu_s \boldsymbol{p}_s^* \cdot (\boldsymbol{x}_{is} - \boldsymbol{\omega}_{is}) \leq 0; \tag{12.37}$$

and define $\boldsymbol{y}_i \in \mathbb{R}^S$ by:

$$y_s = \boldsymbol{p}_s^* \cdot (\boldsymbol{x}_{is} - \boldsymbol{\omega}_{is}) \quad \text{for } s = 1, \ldots, S.$$

Since $\boldsymbol{R}$ is complete, there exists $\boldsymbol{z}_i \in \mathbb{R}^K$ such that:

$$\boldsymbol{R}\boldsymbol{z}_i = \boldsymbol{y}_i; \tag{12.38}$$

and we note that:

$$\begin{aligned}\boldsymbol{q}^* \cdot \boldsymbol{z}_i = (\boldsymbol{q}^*)^\top \boldsymbol{z}_i = \boldsymbol{\mu}^\top \boldsymbol{R}\boldsymbol{z}_i = \boldsymbol{\mu}^\top \boldsymbol{y}_i \\ = \sum_{s=1}^S \mu_s \boldsymbol{p}_s^* \cdot (\boldsymbol{x}_{is} - \boldsymbol{\omega}_{is}) = \sum_{s=1}^S \overline{\boldsymbol{p}}_s \cdot (\boldsymbol{x}_{is} - \boldsymbol{\omega}_{is}) \leq 0.\end{aligned}$$

Therefore $\boldsymbol{x}_i \in B_i^R(\boldsymbol{p}^*, \boldsymbol{q}^*; \boldsymbol{R})$, and it follows that:

$$B_i^A(\overline{\boldsymbol{p}}) \subseteq B_i^R(\boldsymbol{p}^*, \boldsymbol{q}^*; \boldsymbol{R}). \tag{12.39}$$

Furthermore, defining $\boldsymbol{y}_i^* \in \mathbb{R}^S$ by:

$$y_s^* = \boldsymbol{p}_s^* \cdot (\boldsymbol{x}_{is}^* - \boldsymbol{\omega}_{is}) \quad \text{for } s = 1, \ldots, S,$$

we have:

$$\overline{\boldsymbol{p}} \cdot (\boldsymbol{x}_i^* - \boldsymbol{\omega}_i) = \sum_{s=1}^S \mu_s \boldsymbol{p}_s^* \cdot (\boldsymbol{x}_{is}^* - \boldsymbol{\omega}_{is}) = \sum_{s=1}^S \mu_s y_s^* \leq \boldsymbol{\mu}^\top \boldsymbol{R}\boldsymbol{z}_i^* = \boldsymbol{q}^* \cdot \boldsymbol{z}_i^* \leq 0.$$

Therefore $\boldsymbol{x}_i^* \in B_i^A(\overline{\boldsymbol{p}})$; and, since $\boldsymbol{x}_i^*$ maximizes $\succsim_i$ over $B_i^R(\boldsymbol{p}^*, \boldsymbol{q}^*; \boldsymbol{R})$, it now follows from (12.39) that $\boldsymbol{x}_i^*$ maximizes $\succsim_i$ over $B_i^A(\overline{\boldsymbol{p}})$ as well. □

Notice that it follows from the above result that, given the assumptions of the theorem, the allocation associated with a Radner equilibrium is Pareto efficient; at least in the *ex ante* sense. As in the case of Arrow-Debreu equilibrium, it also follows that if preferences are weakly separable over states, and the *ex post* preferences, given that a state has occurred are the same as the *ex ante* preferences, then the allocation will be Pareto efficient in the *ex post* sense as well (recall the discussion at the end of Section 2). Of course, this is only half of the story in any case; with appropriate convexity assumptions, and given a complete asset structure, it also follows from Theorem 12.16 (and our work in Chapter 7) that, given a Pareto efficient allocation, there exist endowments as well as asset and (spot) commodity prices such that the allocation is achieved as a Radner equilibrium.

Clearly, the assumption that the asset structure is complete is of critical importance in establishing Theorem 12.16 and thus for the implications discussed in the above paragraph. Moreover, it is clear why this assumption is so important; it enables consumers to transfer consumption between states in any desired fashion that is consistent with their beginning wealth. We will discuss this assumption, and some of its implications further in the next section. In the meantime, let's consider a technical aspect of Theorem 12.16.

If you will recall the relationship between the state-price vector obtained in the second part of our proof and the vector of asset prices, $\boldsymbol{q}^*$ [see equation (12.36)], you will note that, since the return matrix, $\boldsymbol{R}$, is assumed to have full rank, the state-price vector is unique. Consequently, it may appear that the vector of prices which yields an Arrow-Debreu equilibrium at the allocation $\langle \boldsymbol{x}_i^* \rangle$ is also unique. It is, of course, true that equation (12.36) uniquely defines a vector of Arrow-Debreu prices, but there may be other vectors of prices which define an Arrow-Debreu equilibrium at the same allocation, as is shown in the following.

12.17 Example. We will suppose there are two consumers, two states of the world at $t = 1$ ($S = 2$), and two physically distinguishable commodities ($G = 2$). We suppose further that each consumer has the von Neumann-Morgenster utility function:

$$U_i(\boldsymbol{x}_i) = \pi_1 u(\boldsymbol{x}_{i1}) + \pi_2 u(\boldsymbol{x}_{i2}), \tag{12.40}$$

where $u(\cdot)$ is given by:

$$u(\boldsymbol{x}_{is}) = \left[\min\{x_{is1}, x_{is2}\}\right]^{1/2} \tag{12.41}$$

We will suppose that the return matrix is given by:

$$\boldsymbol{R} = \begin{pmatrix} 1 & 0 \\ 0 & 1 \end{pmatrix},$$

and that the initial endowments are given by:

$$\boldsymbol{\omega}_1 = (\boldsymbol{\omega}_{11}, \boldsymbol{\omega}_{12}) = (2, 2; 0, 0) \text{ and } \boldsymbol{\omega}_2 = (0, 0; 2, 2); \tag{12.42}$$

Now, if $\boldsymbol{p}_s$ is the vector of spot prices at $t = 1$, given that s is the state of the world, and the consumer has allocated income y_s to state s, then the indirect utility obtained is given by:

$$v_{is}(\boldsymbol{p}_s, y_s) = \left[\frac{y_s}{p_{s1} + p_{s2}}\right]^{1/2} \quad \text{for } s = 1, 2. \tag{12.43}$$

Thus, to maximize expected utility, given the probabilities π_1, π_2 and security prices $\boldsymbol{q}^* = (q_1^*, q_2^*)$, the consumer must maximize (with respect to $\boldsymbol{y}, \boldsymbol{z}$):

$$\pi_1 \left[\frac{y_1}{p_{11} + p_{12}}\right]^{1/2} + \pi_2 \left[\frac{y_2}{p_{21} + p_{22}}\right]^{1/2}, \tag{12.44}$$

subject to:

$$\boldsymbol{p}_s \cdot \boldsymbol{\omega}_{is} + z_s - y_s = 0 \text{ for } s = 1, 2, \text{ and } \boldsymbol{q}^* \cdot \boldsymbol{z} \leq 0. \tag{12.45}$$

In the special case in which $\pi_1 = \pi_2 = 1/2$ and $q_1^* = q_2^* = 1$ [I will leave you the problem of finding the solution for arbitrary positive π_1, π_2 and (q_1^*, q_2^*)], we see that $\boldsymbol{y}^*$ must satisfy:

$$\frac{y_1^*}{p_{21} + p_{22}} = \frac{y_2^*}{p_{11} + p_{12}}. \tag{12.46}$$

Thus, if we set:

$$p_{sj}^* = 1 \quad \text{for } s, j = 1, 2;$$

we see that $\big(\langle \boldsymbol{x}_i^*, \boldsymbol{z}_i^* \rangle, \langle \boldsymbol{p}_s^* \rangle, \boldsymbol{q}^*\big)$, where:

$$\boldsymbol{x}_{is}^* = (1,1), \text{for } i, s = 1, 2,\ \ \boldsymbol{z}_1^* = (-2, 2), \text{ and } \boldsymbol{z}_2^* = (2, -2);$$

is a Radner equilibrium. Furthermore, in this case the unique state-price vector for $(\boldsymbol{R}, \boldsymbol{q})$ is given by $\boldsymbol{\mu} = (1, 1)$; and thus the price vector for the Arrow-Debreu case, as defined in (12.36) of the proof of Theorem 12.16 is given by $\boldsymbol{p} = \boldsymbol{p}^*$.

However, consider the Arrow-Debreu price vector $\overline{\boldsymbol{p}} = (1, 2, 1, 2)$. You can easily verify the fact that $(\langle \boldsymbol{x}_i^* \rangle, \overline{\boldsymbol{p}})$ is an Arrow-Debreu equilibrium. □

12.5 Complete Markets and Efficiency

In the previous section, we have seen that Radner equilibria are both *efficient* and *unbiased* if the asset structure is complete. In this section we will begin by presenting an example which shows that completeness of the asset structure is not a necessary condition for the Pareto efficiency of a Radner equilibrium. We will then present an example (with an incomplete asset structure) in which a Radner equilibrium is not Pareto efficient, before moving on to discuss a bit more of the theory.

12.18 Example. Consider an economy in which there are two consumers, two physically distinguishable commodities, and three states; which we will suppose to be equally probable. We will suppose that each of the two consumers has the von Neumann-Morgenstern utility function:

$$U(\boldsymbol{x}_i) = (1/3)\big[u(\boldsymbol{x}_{i1})^{1/2} + u(\boldsymbol{x}_{i2})^{1/2} + u(\boldsymbol{x}_{i3})^{1/2}\big], \tag{12.47}$$

where:

$$u(\boldsymbol{x}_{is}) = \min\{x_{is1}, x_{is2}). \tag{12.48}$$

The consumers' initial endowment vectors are given by:

$$\boldsymbol{\omega}_1 = (4, 4; 0, 0; 4, 4) \text{ and } \boldsymbol{\omega}_2 = (0, 0; 4, 4; 4, 4). \tag{12.49}$$

We further suppose that the return matrix, $\boldsymbol{R}$, is given by:

$$\boldsymbol{R} = \begin{pmatrix} 1 & 0 \\ 0 & 1 \\ 0 & 0 \end{pmatrix}; \tag{12.50}$$

and that the vector of asset prices is given by:

$$\boldsymbol{q}^* = \begin{pmatrix} 1 \\ 1 \end{pmatrix}; \tag{12.51}$$

and we note that $\boldsymbol{q}^*$ is arbitrage-free, given $\boldsymbol{R}$ (see Exercise 10). From the form of the utility functions and the initial endowments, it is clear that we will have supply equal to demand within each of the three possible states if we have:

$$\boldsymbol{p}_s^* = (1,1)^\top \quad \text{for } s = 1,2,3.$$

Thus consumer 1 can maximize expected utility by choosing c_{11}, consumption in period 1, and z_{12}, so as to maximize (see Exercise 10):

$$\sqrt{c_{11}} + \sqrt{z_{12}},$$

subject to:

$$c_{11} = 8 - z_{12}.$$

Solving, we find $c_{11}^* = z_{12}^* = c_{12}^* = 4 = -z_{11}^*$; and, of course, $c_{13}^* = 8$. By symmetry, consumer 2 will set:

$$c_{21}^* = z_{21}^* = 4 = -z_{22}^*, \text{ and } c_{23}^* = 8.$$

If we then define $\boldsymbol{x}_{is}^*$ by:

$$x_{isj}^* = \frac{c_{is}^*}{2} \quad \text{for } i,j = 1,2, \ \& \ s = 1,2,3,$$

it is easy to show that $\big(\langle \boldsymbol{x}_i^*, \boldsymbol{z}_i^* \rangle, \langle \boldsymbol{p}_s^* \rangle, \boldsymbol{q}^*\big)$ is a Radner equilibrium, given $\boldsymbol{R}$.

Now suppose that there are futures markets for each commodity, and consider the price vector:

$$\overline{\boldsymbol{p}} = (\overline{\boldsymbol{p}}_1, \overline{\boldsymbol{p}}_2, \overline{\boldsymbol{p}}_3) = (1,1;1,1;1/\sqrt{2},1/\sqrt{2}).$$

I will leave it to you to verify the fact that $\big(\langle \boldsymbol{x}_i^* \rangle, \overline{\boldsymbol{p}}\big)$ is an Arrow-Debreu equilibrium. Consequently, it follows that $\langle \boldsymbol{x}_i^* \rangle$ is Pareto efficient. □

In the above example, both individuals would be made better off if *each* of their endowments were reduced by taking away a unit of each commodity in state 3, while adding a unit of each commodity to the aggregate endowment in both states 1 and 2. However, the structure of the model does not allow this sort of transfer; nor, of course, does the real-life situation we are trying to model. In effect, the existence of the assets in this model allows consumers to trade amounts from endowments within states, but the aggregate endowment must remain fixed within each state.

In our next example, we will consider a situation in which the initial Radner equilibrium is not Pareto efficient, but in which the addition of some additional securities allows a Radner equilibrium to be attained which *is* (*ex ante*) Pareto efficient.

12.19 Example. We will here modify the previous example by considering 4 equally probable states, keeping the consumers' (Bernoullian) utility functions the same, while letting the endowments be given by:

$$\boldsymbol{\omega}_1 = (4,4;0,0;3,3;1,1) \text{ and } \boldsymbol{\omega}_2 = (0,0;4,4;1,1;3,3),$$

and the return matrix be given by:

$$\boldsymbol{R} = \begin{pmatrix} 1 & 0 \\ 0 & 1 \\ 0 & 0 \\ 0 & 0 \end{pmatrix}.$$

I will leave you to show that $\big(\langle \boldsymbol{x}_i^*, \boldsymbol{z}_i^* \rangle, \langle \boldsymbol{p}_s^* \rangle, \boldsymbol{q}^*\big)$ is a Radner equilibrium, with:

$$\boldsymbol{x}_1^* = (2,2;2,2;3,3;1,1),\ \boldsymbol{x}_2^* = (2,2;2,2;1,1;3,3),$$
$$\boldsymbol{z}_1^* = (-4,4),\ \boldsymbol{z}_2^* = (4,-4),\ \boldsymbol{p}^* = (1,1;1,1;1,1;1,1) \text{ and } \boldsymbol{q}^* = (1,1).$$

However, suppose we introduce two new assets, having the return vectors:

$$\boldsymbol{r}_3 = \begin{pmatrix} 0 \\ 0 \\ 1 \\ 0 \end{pmatrix} \text{ and } \boldsymbol{r}_4 = \begin{pmatrix} 0 \\ 0 \\ 0 \\ 1 \end{pmatrix}.$$

You can now verify the fact that $\big(\langle \boldsymbol{x}_i', \boldsymbol{z}_i' \rangle, \boldsymbol{p}^*, \boldsymbol{q}'\big)$ is a Radner equilibrium, with:

$$\boldsymbol{x}_1' = \boldsymbol{x}_2' = (2,2;2,2;2,2;2,2),$$
$$\boldsymbol{z}_1' = (-4,\ 4, -2,\ 2),\ \boldsymbol{z}_2' = (4, -4,\ 2, -2),\ \boldsymbol{q}' = (1,1,1,1),$$

and the vector of commodity prices remains as it was, $\boldsymbol{p}^* = \boldsymbol{1}$. It is easy to see that the allocation $\langle \boldsymbol{x}_i' \rangle$ is unanimously preferred to $\langle \boldsymbol{x}_i^* \rangle$. Obviously, the original Radner equilibrium did not result in a Pareto efficient allocation. □

While the above example shows that a Radner equlibrium is not necessarily Pareto efficient, it follows from Theorem 12.16 that, if the asset structure is complete, then a Radner equilibrium necessarily yields a Pareto efficient allocation. In fact, in the sort of situation we have been considering in examples (von Neumann-Morgenstern utility functions and a concave Bernoullian utility function), any Pareto efficient allocation can (with possible redistributions within states) be attainable as an Arrow-Debreu equilibrium. But this means that, if the asset structure is complete, and given the same assumptions about consumer preferences, that any Pareto efficient allocation can be attained as a Radner equilibrium. This sets the stage for the final result of this chapter, which makes use of the following definition.

12.20 Definition. If $\boldsymbol{R}$ is an $S \times K$ return matrix, we define $r(\boldsymbol{R})$, the **range of** $\boldsymbol{R}$, by:

$$r(\boldsymbol{R}) = \{\boldsymbol{v} \in \mathbb{R}^S \mid (\exists \boldsymbol{z} \in \mathbb{R}^K) \colon \boldsymbol{v} = \boldsymbol{R}\boldsymbol{z}\}.$$

In the next result, we make use of the following notation. If $\big(\langle \boldsymbol{x}_i^*, \boldsymbol{z}_i^* \rangle, \langle \boldsymbol{p}_s^* \rangle, \boldsymbol{q}^*\big)$ is a Radner equilibrium, given the asset structure, $\boldsymbol{R}$, we define the vectors $\boldsymbol{y}_i^*$ $(i = 1, \ldots, m)$ by:

$$y_{is}^* = \boldsymbol{p}^* \cdot (\boldsymbol{x}_i^* - \boldsymbol{\omega}_{is}) \quad \text{for } s = 1, \ldots, S; \tag{12.52}$$

in other words, under our maintained assumptions:

$$\boldsymbol{y}_i^* = \boldsymbol{R}\boldsymbol{z}_i^* \quad \text{for } i = 1, \ldots, m. \tag{12.53}$$

12.21 Proposition. *Suppose* $(\langle \boldsymbol{x}_i^*, \boldsymbol{z}_i^* \rangle, \langle \boldsymbol{p}_s^* \rangle, \boldsymbol{q}^*)$ *is a Radner equilibrium, given an asset structure with the* $S \times K$ *return matrix* $\boldsymbol{R}$*, and suppose a second asset structure for* E *has the* $S \times K'$ *return matrix,* $\boldsymbol{R}'$*. If* $r(R) = r(R')$*, then* $(\langle \boldsymbol{x}_i^*, \boldsymbol{z}_i' \rangle, \boldsymbol{p}^*, \boldsymbol{q}')$ *is a Radner equilibrium for* $\mathcal{E}$*, given the asset structure* $\boldsymbol{R}'$*; where, for each* i*,* $\boldsymbol{z}_i'$ *is any vector in* $\mathbb{R}^{K'}$ *satisfying:*

$$\boldsymbol{R}'\boldsymbol{z}_i' = \boldsymbol{y}_i^*,$$

and where:

$$\boldsymbol{q}' = (\boldsymbol{R}')^\top \boldsymbol{\mu},$$

and $\boldsymbol{\mu}$ *is a state price vector associated with* $(\boldsymbol{R}, \boldsymbol{q}^*)$*.*[8]

Proof. Suppose $(\langle \boldsymbol{x}_i^*, \boldsymbol{z}_i^* \rangle, \langle \boldsymbol{p}_s^* \rangle, \boldsymbol{q}^*)$ is a Radner equilibrium, given the asset structure $\boldsymbol{R}$, and that $\boldsymbol{R}'$ is a second asset structure for E with $r(\boldsymbol{R}) = r(\boldsymbol{R}')$. Under the maintained assumptions of this chapter, it must be the case that $\boldsymbol{q}^*$ is arbitrage free, and thus there exists a state price vector associated with $(\boldsymbol{R}, \boldsymbol{q}^*)$, $\boldsymbol{\mu}$, satisfying:

$$\boldsymbol{R}^\top \boldsymbol{\mu} = \boldsymbol{q}^*. \tag{12.54}$$

We then define the asset price vector for $\boldsymbol{R}'$, $\boldsymbol{q}'$, by:

$$\boldsymbol{q}' = (\boldsymbol{R}')^\top \boldsymbol{\mu}. \tag{12.55}$$

Next, with $\boldsymbol{y}_i^*$ defined as in (12.52) and (12.53), above, we make use of the assumption that $r(\boldsymbol{R}') = r(\boldsymbol{R})$ to define assert the existence of $\boldsymbol{z}_i'$ satisfying:

$$\boldsymbol{R}'\boldsymbol{z}_i' = \boldsymbol{y}_i^* \quad \text{for } i = 1, \ldots, m-1, \tag{12.56}$$

and define $\boldsymbol{z}_m'$ by:

$$\boldsymbol{z}_m' = -\sum_{i=1}^{m-1} \boldsymbol{z}_i'. \tag{12.57}$$

and note that:

$$\boldsymbol{R}'\boldsymbol{z}_m' = \boldsymbol{R}\Big(-\sum_{i=1}^{m-1} \boldsymbol{z}_i'\Big) = -\sum_{i=1}^{m-1} \boldsymbol{R}'\boldsymbol{z}_i' = -\sum_{i=1}^{m-1} \boldsymbol{y}_i^*$$

$$= \begin{pmatrix} -\sum_{i=1}^{m-1} \boldsymbol{p}_1^* \cdot (\boldsymbol{x}_{i1}^* - \boldsymbol{\omega}_{i1}) \\ \vdots \\ -\sum_{i=1}^{m-1} \boldsymbol{p}_s^* \cdot (\boldsymbol{x}_{is}^* - \boldsymbol{\omega}_{is}) \\ \vdots \\ -\sum_{i=1}^{m-1} \boldsymbol{p}_S^* \cdot (\boldsymbol{x}_{iS}^* - \boldsymbol{\omega}_{iS}) \end{pmatrix} = \begin{pmatrix} \boldsymbol{p}_1^* \cdot (\boldsymbol{x}_{m1}^* - \boldsymbol{\omega}_{m1}) \\ \vdots \\ \boldsymbol{p}_s^* \cdot (\boldsymbol{x}_{ms}^* - \boldsymbol{\omega}_{ms}) \\ \vdots \\ \boldsymbol{p}_S^* \cdot (\boldsymbol{x}_{mS}^* - \boldsymbol{\omega}_{mS}) \end{pmatrix} = \boldsymbol{y}_m^*.$$

We wish now to prove that $(\langle \boldsymbol{x}_i^*, \boldsymbol{z}_i' \rangle, \boldsymbol{p}^*, \boldsymbol{q}')$ is a Radner equilibrium for E, given $\boldsymbol{R}'$.

Accordingly, we begin by noting that, for each i:

$$\boldsymbol{q}' \cdot \boldsymbol{z}_i' = (\boldsymbol{q}')^\top \boldsymbol{z}_i' = \boldsymbol{\mu}^\top \boldsymbol{R}\boldsymbol{z}_i' = \boldsymbol{\mu}^\top \boldsymbol{y}_i^* = \boldsymbol{\mu}^\top \boldsymbol{R}\boldsymbol{z}_i^* = (\boldsymbol{q}^*)^\top \boldsymbol{z}_i^* = 0;$$

while from (12.57), we have:

$$\sum_{i=1}^{m} \boldsymbol{z}_i' = \boldsymbol{0}.$$

[8] Both the statement and the proof of this result are derived from Proposition 19.E.2 of Mas-Colell, Whinston, and Green [1995].

Our proof that $(\langle \boldsymbol{x}_i^*, \boldsymbol{z}_i' \rangle, \boldsymbol{p}^*, \boldsymbol{q}')$ is a Radner equilibrium is completed by showing that for each i, if $\boldsymbol{x}_i \in B_i(\boldsymbol{p}^*, \boldsymbol{q}^*, \boldsymbol{R})$, then it is also in $B_i(\boldsymbol{p}^*, \boldsymbol{q}', \boldsymbol{r}')$, and conversely. Since this demonstration can pretty much follow along the same lines as the argument of the preceding paragraph, I will leave this part of the proof as an exercise. □

Notice that as a corollary of the above proof, it follows that if $\boldsymbol{R}$ is a complete asset structure, then any allocation attainable as a Radner equilibrium, given $\boldsymbol{R}$, is also attainable as such an equilibrium with the Arrow security asset structure, $\boldsymbol{I}_S$, and conversely. Moreover, if we look back at the initial steps of the above proof, letting $\boldsymbol{R}' = \boldsymbol{I}_S$, then $\boldsymbol{q}'$ becomes:

$$\boldsymbol{q}' = (\boldsymbol{R}')^\top \boldsymbol{\mu} = \boldsymbol{I}_S \boldsymbol{\mu} = \boldsymbol{\mu}.$$

Thus, given a complete asset structure, $\boldsymbol{R}$, and arbitrage-free asset price vector, $\boldsymbol{q}$, the associated state price vector, $\boldsymbol{\mu}$, can be interpreted as the asset price vector for the (equivalent) Arrow security asset structure.

Before leaving this topic, it should be mentioned that it is sometimes possible to add so-called *derivative securities* to an incomplete asset structure in such a way as to result in a complete asset structure. A favorite example is the (European) *call option.* Such a security is, essentially, a guarantee of the right to buy a basic security at a fixed price *after the resolution of uncertainty.* If the price at which the option can be exercised is denoted by q^*, and it is an option to buy the kth primiary security then the new security then has the return vector:

$$r_s^* = \max\{r_{sk} - q^*, 0\} \quad \text{for } s = 1, \ldots, S;$$

since the option will only be exercised if it is profitable to do so. Thus, let's return to Example 12.15. We there considered the asset structure:

$$\boldsymbol{R} = \begin{pmatrix} 2 & 1 & 1 \\ 1 & 0 & 1 \\ 1 & 1 & 0 \end{pmatrix},$$

which is incomplete. Suppose we now add an option to buy the first of these assets at the price of 1. Then this derived security has the return vector:

$$\boldsymbol{r}_{\cdot 0} = (1, 0, 0)^\top.$$

If we add this to the initial asset structure, we obtain:

$$\boldsymbol{R}^* = \begin{pmatrix} 1 & 2 & 1 & 1 \\ 0 & 1 & 0 & 1 \\ 0 & 1 & 1 & 0 \end{pmatrix},$$

You can easily show that the asset structure is now complete. (However, see Exercise 14, below.)

12.6 Concluding Notes

As I warned you in the introduction to this chapter, I have not attempted to provide any more than a very elementary introduction to this topic. Those interested in pursuing the topic further will find a more complete introduction provided in Chapter 19 of Mas-Colell, Whinston, and Green [1995], and I have used a notation very nearly identical to theirs; which should make it relatively easy for you to avail yourself of that source. For those interested in pursuing their study of this material still further, let me recommend Duffie [2001] for the financial aspects of this sort of model, and Magill and Quinzii [1996] for a more advanced development of the theory of general equilibrium under uncertainty than is presented here. Let me also recommend the survey by Magill and Shafer [1991].

Exercises. The first three of the following exercises, are set within the general context of Example 12.5.

1. Show that, in the context of Example 12.5, utility maximization implies that equations (12.11) and (12.12) must hold.

2. Show that, in the context of Example 12.5, if (12.14) holds, but $\pi_1 > \pi_2$, then in equilibrium we must have $p_1 > p_2$.

3. In the context of Example 12.5, show that, given (12.17), (12.18) must hold.

4. Show that if equation (12.24) holds, and each column of $\boldsymbol{R}$ is semi-positive, then there exists $\boldsymbol{z}' \in \mathbb{R}^K$ satisfying (12.25).

5. Show that if the first commodity is a numéraire good for each state of the economy, then we can normalize to set $p_{s1} = 1$, for each state s, in dealing with Radner equilibria. [Pay close attention to equation (12.21) of the text.)

6. Prove Proposition 12.13. (Hint: Notice that you can apply the classical Lagrangian method here.)

7. Suppose there are two states, and that a consumer has a Bernoulli utility function, $u_s \colon \mathbb{R}^G_+ \to \mathbb{R}$, which is concave on $\mathbb{R}^G_+$. Show that, for fixed probabilities for the two states, (π_1, π_2), the expected utility function:

$$U(\boldsymbol{x}_1, \boldsymbol{x}_2) = \pi_1 U_1(\boldsymbol{x}_1) + \pi_2 u_2(\boldsymbol{x}_2),$$

is concave on $\mathbb{R}^G_+$.

Now generalize your result by supposing that there are S states.

8. In this example, we consider a contingent-commodity model of pure exchange, with two possible states of nature, two commodities, and two consumers. Denoting commodity bundles as:

$$\boldsymbol{x}_i = (\boldsymbol{x}_{i1}, \boldsymbol{x}_{i2}) = (x_{i11}, x_{i12}, x_{i21}, x_{122}),$$

we suppose that the ith consumer has the preferences described by:

$$\boldsymbol{x}_i P_i \boldsymbol{x}_i' \iff \pi_1(\log x_{i11} + \log x_{i12}) + \pi_2(\log x_{i21} + \log x_{i22})$$
$$> \pi_1(\log x_{i11}' + \log x_{i12}') + \pi_2(\log x_{i21}' + \log x_{i22}') \quad \text{for } 1 = 1, 2.$$

where 'π_s' denotes the probability that state s will occur, for $s = 1, 2$

a. Denoting the ith consumer's initial resource endowment by '$\boldsymbol{\omega}_i$'. for $i = 1, 2$, show that for a given vector of prices, the ith consumer's demand for x_{sj} is given by:

$$x_{isj} = \frac{\pi_s \boldsymbol{p} \cdot \boldsymbol{\omega}_i}{2p_{sj}} \quad \text{for } i, s, j = 1, 2.$$

b. Suppose now that the two consumers' endowments are given by:

$$\boldsymbol{\omega}_1 = (2, 1, 1, 2) \text{ and } \boldsymbol{\omega}_2 = (1, 2, 2, 1),$$

respectively; and that both individuals have a subjective probability of 1/2 for the occurrence of both states (so that for both agents, $\pi_1 = \pi_2 = 1/2$). Show that if:

$$p_{sj} = 1 \quad \text{for } s = 1, 2, \ \& \ j = 1, 2,$$

then an Arrow-Debeu equilibrium is sattained with:

$$x_{isj} = 3/2 \quad \text{for } i = 1, 2, s = 1, 2, \ \& \ j = 1, 2.$$

c. Now suppose that only the first commodity can be traded initially, but that both consumers correstly anticipate the price vector at $t = 1$ for each of the two possible states. Find a Radner equilibrium corresponding to the Arrow-Debreu equilibrium found in part b, above.

9. Suppose there are G physically distinguishable commodities, that there are S states of nature, and that a consumer has the von Neumann-Morgenstern utility function:

$$U(\boldsymbol{x}_i) = \sum_{s=1}^{S} \left[u(\boldsymbol{x}_{is})\right]^a,$$

where:

$$0 < a < 1,$$

and $u \colon \mathbb{R}_+^G \to \mathbb{R}_+$ is positively homogeneous of degree one. Show that if $\boldsymbol{p}^*$ and $\boldsymbol{q}^*$ are the vectors of commodity and securities prices, respectively, and $\boldsymbol{R}$ is the return matrix for the economy (and we denote row s of this matrix by '$\boldsymbol{r}_{s\cdot}$', for $s = 1, \ldots, S$, then the consumer will maximize expected utility by choosing $\boldsymbol{c}_i \in \mathbb{R}_+^S$ and $\boldsymbol{z}_i \in \mathbb{R}_+^K$ so as to maximize;

$$\sum_{s=1}^{S} \pi_s \left[\frac{c_{is}}{\gamma(\boldsymbol{p}_s^*)}\right]^a$$

subject to:

$$\boldsymbol{p}^* \cdot \boldsymbol{\omega}_{is} + \boldsymbol{r}_{s\cdot} \cdot \boldsymbol{z}_i - c_{is} = 0 \quad \text{for } s = 1, \ldots, S,$$

and:

$$\boldsymbol{q}^* \cdot \boldsymbol{z}_i = 0,$$

and then setting:

$$\boldsymbol{x}_{is} = c_{is}\boldsymbol{g}(\boldsymbol{p}_s^*) \quad \text{for } s = 1, \ldots, S;$$

where $\gamma\colon \mathbb{R}_{++}^G \to \mathbb{R}_{++}$ is the cost-of-living index and $\boldsymbol{g}(\cdot)$ is the 'unit income demand function' associated with $u(\cdot)$ (see Section 4.9).

Thus, in particular, if $\boldsymbol{p}_s^* = \boldsymbol{p}_1^*$ and $\pi_s = 1/S$, for $s = 2, \ldots, S$, then the consumer can choose $\boldsymbol{c}_i$ and $\boldsymbol{z}_i$ so as to maximize:

$$\sum\nolimits_{s=1}^S c_{is}^a,$$

subject to the above constraints.

10. Verify the details of Example 12.18.

11. In the context of Example 12.18, suppose we add a third asset, with return vector:

$$\boldsymbol{r}_3 = \begin{pmatrix} 0 \\ 0 \\ 1 \end{pmatrix};$$

so that the asset structure then corresponds to Arrow securities, and the return matrix becomes the 3×3 identity matrix. Can you find a Radner equilibrium which yields the same consumption allocation, $\langle \boldsymbol{x}_i^* \rangle$, as was obtained in the original example?

12. Return to Example 12.8, and, keeping all other data the same, suppose now that the two consumers have the Bernoullian utility function $u(\boldsymbol{x}_{is}) = \sqrt{x_{is1} \cdot x_{is2}}$. Find the new Radner equilibrium.

13. Complete the details of the proof of Proposition 12.21

14. Consider the asset structure:

$$\boldsymbol{R} = \begin{pmatrix} 1 & 0 & 1 \\ 1 & 1 & 0 \\ 1 & 0 & 1 \end{pmatrix}$$

Show that this asset structure is incomplete. Can you add a (one) call option in such a way as to make the resultant asset structure complete?

Follow the same procedure and question for the asset structure:

$$\boldsymbol{R} = \begin{pmatrix} 2 & 0 & 1 & 0 \\ 1 & 1 & 0 & 0 \\ 3 & 0 & 1 & 1 \\ 1 & 1 & 0 & 0 \end{pmatrix}.$$

Chapter 13

Further Topics in General Equilibrium Theory

13.1 Introduction

In this chapter, we will consider the explicit introduction of time into the model; beginning with a finite time horizon, and then briefly considering two extensions to an infinite time horizon. In our second such extension, the 'overlapping generations model,' we not only consider an infinite number of time periods, but also an infinite number of consumers. In Section 5 we will return to the case of a finite number of time periods, but suppose that there are a continuum of consumers. We will undertake only a very brief introduction to each of these topics. This should not be construed as an implicit commentary on their relative importance; indeed, all of these topics are important, and much interesting research is currently being conducted in each of these areas. However, the introduction of time into a general equilibrium model is the primary focus of the required courses in macroeconomic theory in most graduate programs; which is my reason for not pursuing the topic at great length here. The reason for making the introduction to the 'continuum of traders' approach so brief is somewhat different; the fact of the matter is that one cannot proceed very far in the development of this approach without making use of some general measure and integration theory, topics with which most graduate students in economics are unlikely to be familiar.

13.2 Time in the Basic Model

We found in the last chapter that we could introduce uncertainty into the model simply by appropriately interpreting some of the variables in the standard general equilibrium model. In fact, we can also begin to examine the role of time in the model in much the same way. We consider a finite number of distinct time periods, $t = 1, \ldots, T$, and distinguish commodities by both physical characteristics and time of availability. If, as in the last chapter, we suppose that there are G physically distinguishable commodities, then the total number of commodities, n, is given by $n = T \times G$. We will also use a notation very similar to that developed in the previous chapter; however, we will make a change in our notation for commodity bundles, using $\boldsymbol{c}, \boldsymbol{c}'$, etc., as our generic notation for commodity bundles. Thus the commodity bundle available to the ith consumer is denoted by $\boldsymbol{c}_i \in \mathbb{R}^n$, and we write:

$$\boldsymbol{c}_i = (\boldsymbol{c}_{i1}, \ldots, \boldsymbol{c}_{it}, \ldots, \boldsymbol{c}_{iT}),$$

with '$\boldsymbol{c}_{it} \in \mathbb{R}^G$' denoting the commodity bundle available to the ith consumer in the tth period. Consistently with this notation, in this chapter we will denote the ith consumer's consumption set by 'C_i,' rather than 'X_i' (the reasons for this change will soon be apparent). It is also worth noting that in most of our work in this and the next section the assumptions of the model incorporate as a special case the situation in which, for each i there exist positive integers t_i and t_i' such that:

$$1 \leq t_i < t_i' \leq T \tag{13.1}$$

and such that C_i takes the form:

$$C_i = \mathbf{0}_{n_i} \times C_i^* \times \mathbf{0}_{n_i'}, \tag{13.2}$$

where:

$$n_i = t_i \cdot G, \; n_i' = (T - t_i') \cdot G, \tag{13.3}$$

and, defining $n_i'' = t_i' - t_i$:

$$C_i^* \subseteq \mathbb{R}^{n_i'' \cdot G}. \tag{13.4}$$

The idea here is that, for a consumer who is born in the t_i^{th} period and dies in period t_i', only its consumption in periods $t_i + 1, \ldots, t_i'$ affects its survival or preferences. In general, a set of this form can satisfy all of the assumptions which we used in Chapters 5, 7 and 8. For example, if C_i^* is bounded below, closed, or convex, then C_i is bounded below, closed, or convex, respectively. It is also common practice in the literature to assume that the ith consumer's preferences can be represented by a utility function of the form:

$$U_i(\boldsymbol{c}_i) = \sum\nolimits_{t=1}^{T} \delta_i^{t-1} u_i(\boldsymbol{c}_i), \tag{13.5}$$

where $0 < \delta_i < 1$. This also is not inconsistent with a consumption set of the form set out in equation (13.2); although such a utility function does imply that the consumer's preferences are weakly separable and stationary over time periods (see Section 2.8). We will look at a special case of this sort of model, the 'overlapping generations' model, later on in this chapter, but for now let's examine some additional general considerations.

It is clear that the model we set out in chapters 7 and 8 incorporates the situation under examination here. As was true in the previous chapter, we are simply being more specific in our interpretation of the variables. We can define competitive equilibrium for the present case, and, clearly, the First and Second Fundamental Theorems developed in Chapter 7 apply to this case with no need for any modification of assumptions. There is, however, a bit of difficulty in the interpretation of equilibrium in this case. The simplest interpretation, and the one which seems to be the most often used, is that all consumption and production plans and contracts are made in the first period. This is not very plausible as a description of how real economies operate, however. Moreover, if we allow for the possibility of consumption sets of the form specified in (13.1)–(13.4), this interpretation loses all plausibility. In any case, the 'first-period contracts with forward prices' interpretation/treatment does not allow us to investigate the effects of time at all. Consequently, economists have specialized the model in order to carry out this investigation, and we will undertake a rather cursory examination of a fairly standard model of this type.

In our treatment, we will not initially concern ourselves with individual consumers; concentrating instead on aggregate consumption, and the aggregate consumption set:

$$C = \sum_{i=1}^{m} C_i. \tag{13.6}$$

We will also suppose that all of the individual consumption sets can be written in the form:

$$C_i = \prod_{t=1}^{T} C_{it}; \tag{13.7}$$

which, of course, implies that the aggregate consumption set can also be expressed as a similar cartesian product:

$$C = \prod_{t=1}^{T} C_t \quad \text{where } C_t = \sum_{i=1}^{m} C_{it}, \text{ for } t = 1, \ldots, T. \tag{13.8}$$

Notice that consumption sets of the form specified in (13.1)–(13.4) are of this form if, and only if, the sets C_i^* can be written as cartesian products. The simplest sort of assumption which guarantees the desired form is that for each i there exists $\widehat{C}_i \subseteq \mathbb{R}^G$ such that C_i takes the form specified in (13.7) with $C_{it} = \widehat{C}_i$, for $t = 1, \ldots, T$; but our assumptions do not require this to be the case.

Turning now to the production side of the economy, the fundamental empirical fact of which the theory attempts to take account is that production takes time. One could take account of this fact by taking the duration of each time period to be sufficiently long that all production processes can be completed within the period, but this is rather evading the issue, and certainly does not allow us to analyze the effect of time lags in production. In this section we will make a very common assumption; namely that production processes can all be completed by the end of the period in which they are initiated (and thus the output from same is available at the beginning of the following period). We will also consider only the aggregate production set for the economy; supposing that production can be characterized by $T - 1$ sets $Y_t \subseteq \mathbb{R}^{2G}$ $(t = 1, \ldots, T - 1)$; where:

$$Y_t = \{(\boldsymbol{x}_t, \boldsymbol{y}_t) \in \mathbb{R}_+^{2G} \mid \boldsymbol{x}_t \text{ can produce } \boldsymbol{y}_t\}. \tag{13.9}$$

Thus, the aggregate vector of inputs for the economy is denoted by '$\boldsymbol{x}_t$,' and the output vector chosen in the tth period (which becomes available at the beginning of period $t + 1$) is denoted by '$\boldsymbol{y}_t$.' As indicated in the definition of Y_t, we will follow the convention of supposing both of these vectors to be nonnegative elements of $\mathbb{R}^G$.

The assumption that all production processes which are initiated in a period can be completed by the end of the period is considerably less stringent than it appears to be at first glance, for one can include partially-finished goods (*goods in process*) as outputs in one period and inputs in the next. Thus, consider the following example.

13.1 Example. Suppose we have two firms, the first of which has a production process which takes one period to complete, while the second process requires two periods for completion. Allowing for labor, we then suppose that there are four physically distinguishable goods in the economy. We will let the first coordinate of production vectors measure the output of the first firm, the second coordinate will measure the quantity of goods in process for the second firm, while the third coordinate will apply to the finished good production of this firm, and, finally, the fourth coordinate will measure quantities of labor. We suppose that the input-requirement function of the first firm is given by $\ell_1 = g_1(y_1)$. The labor requirement for goods in process for the second firm will be supposed to be given by:

$$\ell_2 = g_2(y_2),$$

while the production function for final goods output for the second firm is given by:

$$y_3 = \min\{y_2/a_1, \ell_3/a_2\},$$

where $a_1$and a_2 are both positive. The production set for the economy can then be defined by:

$$Y_t = \big\{(\boldsymbol{x}_t, \boldsymbol{y}_t) \in \mathbb{R}^8_+ \mid x_{t2} = a_1 \cdot y_{t3} \ \& \ x_{t4} = g_1(y_{t1}) + g_2(y_{t2}) + a_2 y_{t3} \ \& \ y_{t4} = 0\big\} \quad \square$$

If we denote the net production chosen in the t-minus-first period (and which then becomes available at the beginning of the tth period) by '$\boldsymbol{y}_{t-1}$,' the vector of inputs to be applied to production in the tth period by '$\boldsymbol{x}_t$,' and the commodity bundle available for consumption during the tth period by '$\boldsymbol{c}_t$,' feasibility requires that:

$$\boldsymbol{x}_t + \boldsymbol{c}_t = \boldsymbol{y}_{t-1} \quad \text{for } t = 1, \ldots, T. \tag{13.10}$$

As suggested by this feasibility requirement, we will be ignoring initial commodity endowments.

13.2 Example. Consider an economy with two commodities, labor and a produced good which can either be consumed or used as an input for this period's production ('capital').[1] Suppose further that technology can be characterized by the production function $f\colon \mathbb{R}^2_+ \to \mathbb{R}_+$, where we take the first coordinate to be the quantity of labor, and the second to be the quantity of the produced good ('capital') which is applied to production. We can then take Y_t to be the set:

$$Y_t = \{(x_1, x_2; y_1, y_2) \in \mathbb{R}^4_+ \mid y_1 = 0 \ \& \ y_2 = f(x_1, x_2)\}.$$

In this case, if $(\boldsymbol{x}_{t-1}, \boldsymbol{y}_{t-1}) \in Y_{t-1}$ and $(\boldsymbol{x}_t, \boldsymbol{y}_t) \in Y_t$, we must have:

$$\boldsymbol{x}_t + \boldsymbol{c}_t = \boldsymbol{y}_{t-1},$$

and thus $x_{t1} + c_{t1} = 0$. In other words, by taking $\boldsymbol{x}_t \in \mathbb{R}^G_+$, we are correspondingly following the convention that negative coordinates of consumption vectors $\boldsymbol{c}_t$ represent quantities of primary inputs (particularly labor) supplied by consumers to the production sector. $\square$

[1] It may help to think of the produced good as 'wheat,' which can either be consumed this year or planted as seed to produce next year's crop.

We will take $\boldsymbol{y}_0$, which is the initial stock of commodities, as given, and then make use of the following definition of feasibility in the present context.

13.3 Definitions. Let $\boldsymbol{y}_0 \in \mathbb{R}^G_+$ be given. We will say that the sequence (**program**) $\langle(\boldsymbol{x}_t, \boldsymbol{y}_t, \boldsymbol{c}_t)\rangle \in R^{3GT}$ is **feasible** iff:

1. $\boldsymbol{c}_t \in C_t$,
2. $(\boldsymbol{x}_t, \boldsymbol{y}_t) \in Y_t$

and

3. $\boldsymbol{x}_t + \boldsymbol{c}_t = \boldsymbol{y}_{t-1}$,

for $t = 1, \ldots, T$.

We will say that a sequence $\langle(\boldsymbol{x}_t, \boldsymbol{y}_t)\rangle_{t=1}^T$ is **a production program** iff it satisfies condition 2, above.

In the material to follow, we will refer to any finite sequence $\langle(\boldsymbol{x}_t, \boldsymbol{y}_t, \boldsymbol{c}_t)\rangle_{t=1}^T$ such that:

$$\boldsymbol{x}_t, \boldsymbol{y}_t, \boldsymbol{c}_t \in \mathbb{R}^G \quad \text{for } t = 1, \ldots, T,$$

as a **program**. In dealing with such programs, we will often find it convenient to use the notation $\boldsymbol{x}, \boldsymbol{y}$ and $\boldsymbol{c}$ to denote the whole vector of corresponding variables; that is, for example:

$$\boldsymbol{c} = (\boldsymbol{c}_1, \boldsymbol{c}_2, \ldots, \boldsymbol{c}_t, \ldots, \boldsymbol{c}_T).$$

We make use of this notation in the following definition.

13.4 Definition. Let $\langle(\boldsymbol{x}_t, \boldsymbol{y}_t, \boldsymbol{c}_t)\rangle$ and $\langle(\boldsymbol{x}'_t, \boldsymbol{y}'_t, \boldsymbol{c}'_t)\rangle$ be two programs. We will say that $\langle(\boldsymbol{x}_t, \boldsymbol{y}_t, \boldsymbol{c}_t)\rangle$ **dominates** $\langle(\boldsymbol{x}'_t, \boldsymbol{y}'_t, \boldsymbol{c}'_t)\rangle$ iff $\boldsymbol{c} > \boldsymbol{c}'$; that is, iff:

$$\boldsymbol{c}_t \geq \boldsymbol{c}'_t \ \text{ for } t = 1, \ldots, T, \ \text{ and } (\exists t^* \in \{1, \ldots, T\})\colon \boldsymbol{c}_{t^*} > \boldsymbol{c}'_{t^*}.$$

13.5 Definition. We will say that a feasible program, $\langle(\boldsymbol{x}_t, \boldsymbol{y}_t, \boldsymbol{c}_t)\rangle$, is **efficient, given $\boldsymbol{y}_T$** iff there exists no feasible program, $\langle(\boldsymbol{x}'_t, \boldsymbol{y}'_t, \boldsymbol{c}'_t)\rangle$, dominating $\langle(\boldsymbol{x}_t, \boldsymbol{y}_t, \boldsymbol{c}_t)\rangle$ and satisfying $\boldsymbol{y}'_T \geq \boldsymbol{y}_T$.

As indicated in the above definition, we only consider the resultant stream of consumption vectors in determining whether a feasible program is or is not efficient. However, notice the qualification, "given $\boldsymbol{y}_T$," in the above definition. Without this qualification, and supposing that $(\boldsymbol{x}_T, \boldsymbol{y}_T) \in Y_T$ and $\boldsymbol{y}_T > \boldsymbol{0}$ implies $\boldsymbol{x}_T > \boldsymbol{0}$, no program $\langle(\boldsymbol{x}_t, \boldsymbol{y}_t, \boldsymbol{c}_t)\rangle$ having $\boldsymbol{y}_T > \boldsymbol{0}$ could be efficient!

13.6 Definition. Let $\boldsymbol{p} = (\boldsymbol{p}_1, \ldots, \boldsymbol{p}_t, \ldots, \boldsymbol{p}_T, \boldsymbol{p}_{T+1})$ be a vector (or, if you prefer, a finite sequence) of prices. We will say that a production program, $\langle(\boldsymbol{x}_t, \boldsymbol{y}_t)\rangle_{t=1}^T$ is **competitive, given $\boldsymbol{p}$**, iff:

$$\boldsymbol{y}_{t-1} - \boldsymbol{x}_t \in C_t \quad \text{for } t = 1, \ldots, T,$$

and for all production programs, $\langle(\boldsymbol{x}'_t, \boldsymbol{y}'_t)\rangle_{t=1}^T$ satisfying:

$$\boldsymbol{y}'_{t-1} - \boldsymbol{x}'_t \in C_t \quad \text{for } t = 1, \ldots, T,$$

we have:

$$\boldsymbol{p}_{t+1} \cdot \boldsymbol{y}_t - \boldsymbol{p}_t \cdot \boldsymbol{x}_t \geq \boldsymbol{p}_{t+1} \cdot \boldsymbol{y}'_t - \boldsymbol{p}_t \cdot \boldsymbol{x}'_t \quad \text{for } t = 1, \ldots, T.$$

Notice that we make use of a sequence of $T+1$ (G-dimensional) price vectors in the above definition. In the definition, we are essentially requiring that a competitive program maximizes profits in each period, so it may look a bit strange to require that this maximization be subject to $\boldsymbol{y}_{t-1}-\boldsymbol{x}_t \in C_t$ for each t. However, this requirement can be justified by the fact that full competitive equilibrium will require that this condition is satisfied.

In connection with the definition just presented, we define:

$$\pi_t(\boldsymbol{p}_t, \boldsymbol{p}_{t+1}; \boldsymbol{y}_{t-1}) = \max\left\{\boldsymbol{p}_{t+1}\cdot\boldsymbol{y}_t - \boldsymbol{p}_t\cdot\boldsymbol{x}_t \mid (\boldsymbol{x}_t, \boldsymbol{y}_t) \in Y_t \;\&\; \boldsymbol{y}_{t-1}-\boldsymbol{x}_t \in C_t\right\} \tag{13.11}$$

We will demonstrate shortly that if $\langle(\boldsymbol{x}_t, \boldsymbol{y}_t, \boldsymbol{c}_t)\rangle$ is a program, $\boldsymbol{p}$ is a price vector, and the production program $\langle(\boldsymbol{x}_t, \boldsymbol{y}_t)\rangle$ is competitive, given $\boldsymbol{p}$, then $\langle(\boldsymbol{x}_t, \boldsymbol{y}_t, \boldsymbol{c}_t)\rangle$ is efficient, given $\boldsymbol{y}_T$.[2] However, it will be convenient to prove this by making use of the following result.

13.7 Proposition. *Suppose* $\langle(\boldsymbol{x}_t^*, \boldsymbol{y}_t^*)\rangle$ *is a production program which is competitive, given the price vector* $\boldsymbol{p}^*$*, and define* $\boldsymbol{c}^*$ *by:*

$$\boldsymbol{c}_t^* = \boldsymbol{y}_{t-1}^* - \boldsymbol{x}_t^* \quad \textit{for } t = 1, \ldots, T.$$

Then $\langle(\boldsymbol{x}_t^*, \boldsymbol{y}_t^*, \boldsymbol{c}_t^*)\rangle$ *is feasible, and given any feasible program,* $\langle(\boldsymbol{x}_t, \boldsymbol{y}_t, \boldsymbol{c}_t)\rangle$*, we must have:*

$$\sum_{t=1}^{T} \boldsymbol{p}_t^*\cdot\boldsymbol{c}_t^* + \boldsymbol{p}_{T+1}^*\cdot\boldsymbol{y}_T^* \geq \sum_{t=1}^{T} \boldsymbol{p}_t^*\cdot\boldsymbol{c}_t + \boldsymbol{p}_{T+1}^*\cdot\boldsymbol{y}_T. \tag{13.12}$$

Proof. Obviously $\langle(\boldsymbol{x}_t^*, \boldsymbol{y}_t^*, \boldsymbol{c}_t^*)\rangle$ is feasible, and since it is, we have:

$$\boldsymbol{p}_1^*\cdot\boldsymbol{c}_1^* + \boldsymbol{p}_1^*\cdot\boldsymbol{x}_1^* = \boldsymbol{p}_1^*\cdot\boldsymbol{y}_0,$$

and:

$$\boldsymbol{p}_2^*\cdot\boldsymbol{c}_2^* + \boldsymbol{p}_2^*\cdot\boldsymbol{x}_2^* = \boldsymbol{p}_2^*\cdot\boldsymbol{y}_1;$$

Thus, defining:

$$\pi_t^* = \pi_t(\boldsymbol{p}_t^*, \boldsymbol{p}_{t+1}^*; \boldsymbol{y}_{t-1}^*) \quad \text{for } t = 1, \ldots, T-1,$$

we have:

$$\boldsymbol{p}_1^*\cdot\boldsymbol{c}_1^* + \boldsymbol{p}_2^*\cdot\boldsymbol{c}_2^* = \pi_1^* - \boldsymbol{p}_2^*\cdot\boldsymbol{x}_2^* + \boldsymbol{p}_1^*\cdot\boldsymbol{y}_0.$$

Suppose now that we have, for $2 \leq t \leq T-1$:

$$\sum_{s=1}^{t} \boldsymbol{p}_s^*\cdot\boldsymbol{c}_s^* = \sum_{s=1}^{t-1} \pi_s^* - \boldsymbol{p}_t^*\cdot\boldsymbol{x}_t^* + \boldsymbol{p}_1^*\cdot\boldsymbol{y}_0.$$

Then by the feasibility of $\langle(\boldsymbol{x}_t^*, \boldsymbol{y}_t^*, \boldsymbol{c}_t^*)\rangle$ and the fact that $\langle(\boldsymbol{x}_t^*, \boldsymbol{y}_t^*)\rangle$ is competitive, given $\boldsymbol{p}^*$, we have:

$$\begin{aligned}\sum_{s=1}^{t} \boldsymbol{p}_s^*\cdot\boldsymbol{c}_s^* + \boldsymbol{p}_{t+1}^*\cdot\boldsymbol{c}_{t+1}^* &= \sum_{s=1}^{t-1} \pi_s^* - \boldsymbol{p}_t^*\cdot\boldsymbol{x}_t^* - \boldsymbol{p}_1^*\cdot\boldsymbol{y}_0 + \boldsymbol{p}_{t+1}^*\cdot\boldsymbol{y}_t^* - \boldsymbol{p}_{t+1}^*\cdot\boldsymbol{x}_{t+1}^* \\ &= \sum_{s=1}^{t} \pi_s^* - \boldsymbol{p}_{t+1}^*\cdot\boldsymbol{x}_{t+1}^* + \boldsymbol{p}_1^*\cdot\boldsymbol{y}_0.\end{aligned}$$

[2] In choosing the approach taken here, in the development of this and the next result, I am exhibiting my indebtedness to my study of lecture notes prepared by Professor Mukul Majumdar for his course in Intertemporal Economics at Cornell University.

Therefore:

$$\sum_{t=1}^{T} \boldsymbol{p}_t^* \cdot \boldsymbol{c}_t^* = \sum_{t=1}^{T-1} \pi_t^* - \boldsymbol{p}_T^* \cdot \boldsymbol{x}_T^* + \boldsymbol{p}_1^* \cdot \boldsymbol{y}_0,$$

and consequently:

$$\sum_{t=1}^{T} \boldsymbol{p}_t^* \cdot \boldsymbol{c}_t^* + \boldsymbol{p}_{T+1}^* \cdot \boldsymbol{y}_T^* = \sum_{t=1}^{T} \pi_t^* + \boldsymbol{p}_1^* \cdot \boldsymbol{y}_0, \tag{13.13}$$

A similar argument establishes:

$$\sum_{t=1}^{T} \boldsymbol{p}_t^* \cdot \boldsymbol{c}_t + \boldsymbol{p}_{T+1}^* \cdot \boldsymbol{y}_T = \sum_{t=1}^{T} (\boldsymbol{p}_{t+1}^* \cdot \boldsymbol{y}_t - \boldsymbol{p}_t^* \cdot \boldsymbol{x}_t) + \boldsymbol{p}_1^* \cdot \boldsymbol{y}_0;$$

and thus, since $(\boldsymbol{x}_t, \boldsymbol{y}_t) \in Y_t$, for $t = 1, \ldots, T$:

$$\sum_{t=1}^{T} \boldsymbol{p}_t^* \cdot \boldsymbol{c}_t + \boldsymbol{p}_{T+1}^* \cdot \boldsymbol{y}_T \leq \sum_{t=1}^{T} \pi_t^* + \boldsymbol{p}_1^* \cdot \boldsymbol{y}_0.. \tag{13.14}$$

Combining (13.13) and (13.14) yields the desired result. □

One can then prove the following; although I will leave the proof as an exercise.

13.8 Proposition. *Suppose $\langle(\boldsymbol{x}_t^*, \boldsymbol{y}_t^*, \boldsymbol{c}_t^*)\rangle$ is a feasible program, that $\mathbf{p}^*$ is a price vector satisfying:*

$$\boldsymbol{p}_t^* \gg \mathbf{0} \quad \textit{for } t = 1, \ldots, T+1,$$

and that for all feasible $\langle(\boldsymbol{x}_t, \boldsymbol{y}_t, \boldsymbol{c}_t)\rangle$, we have:

$$\sum_{t=1}^{T} \boldsymbol{p}_t^* \cdot \boldsymbol{c}_t^* + \boldsymbol{p}_{T+1}^* \cdot \boldsymbol{y}_{T+1}^* \geq \sum_{t=1}^{T} \boldsymbol{p}_t^* \cdot \boldsymbol{c}_t + \boldsymbol{p}_{T+1}^* \cdot \boldsymbol{y}_{T+1}.$$

Then $\langle(\boldsymbol{x}_t^*, \boldsymbol{y}_t^*, \boldsymbol{c}_t^*)\rangle$ is efficient, given $\boldsymbol{y}_T^*$.

Now suppose that at each time period there is a production manager whose job it is to choose a pair $(\boldsymbol{x}_t^*, \boldsymbol{y}_t^*) \in Y_t$ which maximizes profits over Y_t, given $(\boldsymbol{p}_t, \boldsymbol{p}_{t+1}; \boldsymbol{y}_{t-1})$, and subject to $\boldsymbol{y}_{t-1} - \boldsymbol{x}_t^* \in C_t$, and suppose a strictly positive price vector, $\boldsymbol{p}^*$, is given. Then the tth production manager needs to know only her/his own production set, the production (endowment), $\boldsymbol{y}_{t-1}$, available at the beginning of the period, and the pair of price vectors $(\boldsymbol{p}_t^*, \boldsymbol{p}_{t+1}^*)$ in order to carry out her/his assignment. Moreover, and most remarkably, it *almost* follows from these last two propositions that, given *any* sequence of strictly positive price vectors, $\langle \boldsymbol{p}_t^* \rangle_{t=1}^{T+1}$, the independent actions of these T production managers will result in a production program $\langle(\boldsymbol{x}_t^*, \boldsymbol{y}_t^*)\rangle$ which is efficient, given the Tth production manager's choice of $\boldsymbol{y}_T^*$!

Unfortunately, the statement of the above paragraph is not quite true, much as we would like it to be! The problem is that the choice of output, $\boldsymbol{y}_t^*$ may not enable both production and consumption to take place in period $t+1$. It should also be mentioned that if a *particular* value for $\boldsymbol{y}_T^*$ is specified in advance, that is, at $t = 1$, then the sequence of prices will have to be chosen in a way that is consistent with this desired outcome. We will return to this second point, but first consider the following example.

13.9 Example. We consider an economy with two commodities, labor and a produced good, and suppose consumption sets are given by:

$$C_t = \{\boldsymbol{c} \in \mathbb{R}^2 \mid -16t \leq c_1 \leq 0 \ \& \ c_2 \geq 2t\} \quad \text{for } t = 1, 2, \ldots, T;$$

while the production sets are given by:

$$Y_t = \{(\boldsymbol{x}, \boldsymbol{y}) \in \mathbb{R}^4_+ \mid y_1 = 0 \ \& \ y_2 = 2(x_1 x_2)^{1/4}\},$$

and $\boldsymbol{y}_0 = (0, 3)$. Basically, the idea here is that the economy is growing over time, in terms of numbers of consumers. This growth leads to difficulties if production managers simply maximize profits in a myopic fashion. In fact, I will leave it to you to verify (Exercise 2) that if $\boldsymbol{p}_1 = (1/16, 1)$, then the production manager for the first period maximizes profits at $x^*_{1,1} = 16, x^*_{1,2} = 1$, and $\boldsymbol{y}^*_1 = (0, 4)$. This choice would be fine if $C_2 = C_1$, but given the actual form of C_2, the set of pairs $(\boldsymbol{x}, \boldsymbol{y}) \in Y$ such that $\boldsymbol{y}^*_1 - \boldsymbol{x} \in C_2$ is equal to $\{\boldsymbol{0}\}$! □

As mentioned earlier, the second barrier to a decentralized development of an efficient intertemporal program is that the calculation of the price sequence which will yield a specific, predetermined value of $\boldsymbol{y}_T$ will generally require a knowledge of both C_t and Y_t for each t ($t = 1, \ldots, T$). We will discuss this issue further in the next section.

13.3 An Infinite Time Horizon

As we have suggested in the previous section, whether or not a program is efficient is conditional upon the value specified for $\boldsymbol{y}_T$; in fact, it is generally true that the values of all the variables appropriate for an efficient program are dependent upon the specified value for $\boldsymbol{y}_T$. Moreover, it is not clear what value one should take for T. But of course, the two problems are very much interrelated. If one is using the model to develop, say, a 'five-year plan,' then the appropriate value for T is, of course, clear. In this case, however, the consumption values which can be attained in each period will generally depend upon the targeted value for $\boldsymbol{y}_T$. Moreover, given a specification of $\boldsymbol{y}_T$, the values for consumption which are achieved by a feasible program will generally depend upon the value chosen for T (whether we are dealing with a 5-year, or a 10-year plan, for example). The price vector which will implement a feasible plan is also dependent upon the targeted value of $\boldsymbol{y}_T$, as well as the value chosen for T; as is demonstrated by the following example..

13.10 Example. We consider en economy in which there are two commodities; labor and a produced good. The produced good can be consumed, used as a current input in production, or used to create capital. The production function for the economy is given by:

$$y_{t,2} = k_t^{1/2} \cdot (x_{t,1} \cdot x_{t,2})^{1/4},$$

where:

$$k_t = k_0 \prod_{s=1}^{t-1} \delta_s \ \& \ \delta_s = \max\left\{1, y_{s-1,2} - x_{s,2} - c_{s,2}\right\}, \text{for } s = 1, \ldots, t-1; \ t = 1, \ldots, T;$$

where $k_0 = 4$. Thus the production sets are given by:

$$Y_t = \{(\boldsymbol{x}_t, \boldsymbol{y}_t) \in \mathbb{R}^4_+ \mid y_{t,1} = 0 \ \& \ y_{t,2} = k_t^{1/2} \cdot (x_{t,1} \cdot x_{t,2})^{1/4}\} \text{ for } t = 1, \ldots, T;$$

and we suppose C_t is given by:

$$C_t = \{\boldsymbol{c} \in \mathbb{R}^2 \mid -16 \leq c_1 \leq 0 \ \& \ c_2 \geq 0\} \quad \text{for } t = 1, \ldots, T,$$

while $T = 4$, and $\boldsymbol{y}_0 = (0, 3)$. We will concentrate most of our attention here on two programs, as set out in the following tables.

Period	$y_{t,2}$	$x_{t,1}$	$x_{t,2}$	$c_{t,1}$	$c_{t,2}$	δ_t
$t = 1$	4	1	1	-1	0	2
$t = 2$	16	4	4	-4	0	1
$t = 3$	16	4	4	-4	10	2
$t = 4$	64	16	16	-16	–	–

Table 13.1: Program 1.

I will leave it to you to demonstrate the fact that Program 1 is competitive, given prices $\boldsymbol{p}_t = (1, 1)$, for $t = 1, \ldots, 5$. Consequently, it follows from Propositions 13.7 and 13.8 that Program 1 is efficient, given $\boldsymbol{y}_4^* = (0, 64)$. Program 2 is set out in the following table.

Period	$y_{t,2}$	$x_{t,1}$	$x_{t,2}$	$c_{t,1}$	$c_{t,2}$	δ_t
$t = 1$	8	8	2	-8	1	1
$t = 2$	8	8	2	-8	6	1
$t = 3$	8	8	2	-8	6	1
$t = 4$	8	8	2	-8	–	–

Table 13.2: Program 2.

Once again I will leave it to you to demonstrate the fact that Program 2 is competitive; this time with the prices $\boldsymbol{p}_t' = (1/4, 1)$, for $t = 1, \ldots, 5$. Consequently, it follows that Program 2 is efficient, given $\boldsymbol{y}_4' = (0, 8)$. Thus these two very different programs are both competitive, and therefore are both efficient, given their respective target terminal values. (See also Exercise 3, at the end of this chaper.) □

Because of the sensitivity of competitive programs to the specification of the target terminal value of $\boldsymbol{y}_T$, and the awkwardness stemming from the fact that the Tth period must be treated differently from the other periods in the analysis, many researchers have argued that to have a satisfactory model, one must take $T = +\infty$. While this change to an infinite time horizon (technically, we are here assuming a countable number of time periods) eliminates the asymmetry associated with the finite time horizon model, and eliminates some other ambiguities encountered there as well, it also creates some new problems. For example, consider the following example.

13.11 Example. (Hurwicz and Majumdar [1988, p. 237].)) Here we consider an economy with one produced good, and suppose $f\colon \mathbb{R}_+ \to \mathbb{R}_+$ is strictly concave and satisfies:

$$(\forall x \in \mathbb{R}_{++})\colon f(x) > 0,\ f'(x) > 0 \;\&\; f''(x) < 0.$$

We assume that: $C = \mathbb{R}_+$, and that $y_0 > 0$. The program given by:

$$x_t = y_{t-1},\ y_t = f(x_t),\ \text{and } c_t = 0 \ \text{ for } t = 1, 2, \ldots,$$

is clearly *not* efficient. However, if we define the sequence of prices, $\langle p_t \rangle$ by:

$$p_1 = 1 \ \text{ and } \ p_{t+1} = p_t/f'(x_t) \text{ for } t = 2, 3, \ldots;$$

it is easy to show that the program is competitive. □

As this last example demonstrates, with an infinite time horizon a competitive program is not necessarily efficient. However, consider the following result, where we define a feasible program exactly as in Definition 13.3, except that we now define an infinite sequence $\langle(\boldsymbol{x}_t, \boldsymbol{y}_t, \boldsymbol{c}_t)\rangle$ $(t = 1, 2, \ldots)$; and similarly for a competitive program.

13.12 Proposition. *Let* $\langle(\boldsymbol{x}_t^*, \boldsymbol{y}_t^*, \boldsymbol{c}_t^*)\rangle$ *be a competitive program, given the price sequence* $\langle \boldsymbol{p}_t^* \rangle$*, where:*

$$\boldsymbol{p}_t^* \gg \boldsymbol{0} \quad \textit{for } t = 1, 2, \ldots,$$

and suppose:

$$\lim_{t\to\infty} \boldsymbol{p}_t^* \cdot \boldsymbol{y}_t^* = 0. \tag{13.15}$$

Then $\langle(\boldsymbol{x}_t^*, \boldsymbol{y}_t^*, \boldsymbol{c}_t^*)\rangle$ *is efficient.*

Proof. Suppose, by way of obtaining a contradiction, that there exists a feasible program, $\langle(\boldsymbol{x}_t, \boldsymbol{y}_t, \boldsymbol{c}_t)\rangle$, which is such that:

$$\boldsymbol{c}_t \geq \boldsymbol{c}_t^* \ \text{ for } t = 1, 2, \ldots,$$

and, for some positive integer, $\widehat{T}$:

$$c_{\widehat{T}} = c_{\widehat{T}}^* + \boldsymbol{d},$$

where $\boldsymbol{d} > \boldsymbol{0}$. Then, since $\boldsymbol{p}_{\widehat{T}}^* \gg \boldsymbol{0}$, we have for all $T \geq \widehat{T}$:

$$\sum\nolimits_{t=1}^{T} \boldsymbol{p}_t^* \cdot (\boldsymbol{c}_t - \boldsymbol{c}_t^*) \geq \boldsymbol{p}_{\widehat{T}}^* \cdot (\boldsymbol{c}_{\widehat{T}} - \boldsymbol{c}_{\widehat{T}}^*) = \boldsymbol{p}_{\widehat{T}}^* \cdot \boldsymbol{d} > 0. \tag{13.16}$$

However, if we define $\epsilon = (1/2)(\boldsymbol{p}_T^* \cdot \boldsymbol{d})$, it follows from (13.15) that there exists T^* such that for all $t \geq T^*$:

$$\boldsymbol{p}_t^* \cdot \boldsymbol{y}_t^* < \epsilon \tag{13.17}$$

But then, if we let $T \geq \max\{\widehat{T}, T^*\}$, it follows from Proposition 13.7 that:

$$\sum\nolimits_{t=1}^{T} \boldsymbol{p}_t^* \cdot (\boldsymbol{c}_t - \boldsymbol{c}_t^*) \leq \boldsymbol{p}_T^* \cdot \boldsymbol{y}_T^* - \boldsymbol{p}_T^* \cdot \boldsymbol{y}_T \leq \boldsymbol{p}_T^* \cdot \boldsymbol{y}_T^* < \epsilon;$$

which contradicts (13.16). □

How reasonable is the condition expressed by equation (13.15)? Well, if we think in terms of the possibility of a continually increasing production program, it seems very unreasonable. On the other hand, our economic universe (and, apparently, our physical universe) is actually of finite duration. Consequently, one can argue that, realistically, we only need to consider programs, $\langle(\boldsymbol{x}_t, \boldsymbol{y}_t, \boldsymbol{c}_t)\rangle$, such that there exists a finite integer, T such that:

$$\boldsymbol{c}_t = \boldsymbol{x}_t = \boldsymbol{y}_t = \boldsymbol{0} \quad \text{for } t = T+1, T+2, \ldots; \tag{13.18}$$

and such programs necessarily satisfy (13.15). However, if we confine our attention to just those programs satisfying (13.18) for some fixed value of T, we are back to the finite model of the previous section; with the attendant difficulties already discussed. On the other hand, if we require only that any feasible program satisfy (13.18) for *some* value of T (where the critical value of T may vary from program to program), then, from a mathematical point of view, we are dealing with what is called the space of finitely non-zero sequences; which is a particularly nasty space to deal with, from a mathematical point of view. However, we can always normalize the price vectors; requiring, for example, that:

$$\|\boldsymbol{p}_t\| = 1 \quad \text{for } t = 1, 2, \ldots.$$

Since the Cauchy-Schwarz inequality[3] implies:

$$\boldsymbol{p}_t \cdot \boldsymbol{y}_t \leq \|\boldsymbol{p}_t\| \cdot \|\boldsymbol{y}_t\| \quad \text{for all } t = 1, 2, \ldots,$$

condition (13.15) will hold if we simply assume that, for all feasible programs, $\langle(\boldsymbol{x}_t, \boldsymbol{y}_t, \boldsymbol{c}_t)\rangle$, we must have:

$$\lim_{t \to \infty} \|\boldsymbol{y}_t\| = 0.$$

However, the student should be warned that not all economists would agree with this assessment.

13.4 Overlapping Generations

The basic 'overlapping generations' model was introduced by Samuelson [1958],[4] and has since become the 'workhorse' of macro economics.[5] In this model, we once again consider a countably infinite number of periods ($t = 1, 2, \ldots$), but we now allow a countably infinite number of agents as well. In our treatment here, we will deal only with almost the simplest variety of such a model; an exchange economy in which each consumer is alive exactly two periods; and in which there exist only two

[3] For a statement and proof of this inequality, see, for example, Moore [1999a, p. 35].

[4] It had been developed and analyzed earlier by Allais [1947], but this work seems to have remained unknown to non-native-French-speaking economists until sometime after 1958.

[5] I am here more or less quoting from Heijdra and van der Ploeg [2002, p. 590]; although, strictly speaking, they are referring to the extension of the model (with production) which was developed by Peter Diamond.

consumers in each period. We will, however, suppose that there is a finite number, n, of commodities.

We will denote the consumer 'born' in period t by 'i_t,' and denote i_t's initial endowment by '$\boldsymbol{\omega}_{i_t}$,' where:

$$\boldsymbol{\omega}_{i_t} = (\boldsymbol{\omega}_{ty}, \boldsymbol{\omega}_{to}),$$

where the subscripts y and o are intended to suggest 'young' and 'old,' respectively, and $\boldsymbol{\omega}_{ty}$ and $\boldsymbol{\omega}_{to}$ are elements of $\mathbb{R}^n_+$.. Formally, the tth consumer's consumption set, C_t, is of the form set out in equations (13.1)–(13.4), with:

$$C_{i_t} = \{\mathbf{0}\} \times \{\mathbf{0}\} \times \ldots \{\mathbf{0}\} \times \mathbb{R}^n_+ \times \mathbb{R}^n_+ \times \{\mathbf{0}\} \times \ldots;$$

but, since we will be assuming that each consumer has a consumption set of the same general form, we will speak of the tth consumer's choosing a pair $\boldsymbol{x}_t = (\boldsymbol{x}_{ty}, \boldsymbol{x}_{to}) \in \mathbb{R}^{2n}_+$; and we will suppose that each consumer has an asymmetric preference relation on $\mathbb{R}^{2n}_+$. To make things work out symmetrically, we will also need to suppose that there is a consumer i_0 who is old in period 1; although to avoid some really cumbersome notation, we will denote this consumer's initial endowment by '$\boldsymbol{w}_0$,' and the consumption bundle available to consumer i_0 by '$\boldsymbol{x}_0$.'

We can then define competitive equilibrium in more or less the usual way:

13.13 Definition. We will say that $\big(\boldsymbol{x}^*_0, \langle \boldsymbol{x}^*_t \rangle, \langle \boldsymbol{p}^*_t \rangle\big)$ is a competitive equilibrium for $\mathcal{E} = \langle P_t, \boldsymbol{\omega}_t \rangle$ iff::

1. $\boldsymbol{x}^*_0 \in \mathbb{R}^n_+$, $\boldsymbol{x}^*_t \in \mathbb{R}^{2n}_+$ and $\boldsymbol{p}^*_t \in \mathbb{R}^n_+$ for $t = 1, 2, \ldots,$
2. $\boldsymbol{x}^*_{1y} + \boldsymbol{x}^*_0 = \boldsymbol{\omega}_{1y} + \boldsymbol{w}_0,$
3. $\boldsymbol{x}^*_{ty} + \boldsymbol{x}^*_{t-1,o} = \boldsymbol{\omega}_{ty} + \boldsymbol{\omega}_{t-1,o}$ for $t = 2, 3, \ldots,$
4. $\boldsymbol{p}^*_1 \cdot \boldsymbol{x}^*_0 \leq \boldsymbol{p}^*_1 \cdot \boldsymbol{w}_0$ & $(\forall \boldsymbol{x} \in \mathbb{R}^n_+)\colon \boldsymbol{x} P_0 \boldsymbol{y}^* \Rightarrow \boldsymbol{p}^*_1 \cdot \boldsymbol{x} > \boldsymbol{p}^*_1 \cdot \boldsymbol{w}_0$, and
5. for each t:

$$\boldsymbol{p}^*_t \cdot \boldsymbol{x}^*_{ty} + \boldsymbol{p}^*_{t+1} \cdot \boldsymbol{x}^*_{to} \leq \boldsymbol{p}^*_t \cdot \boldsymbol{\omega}_{ty} + \boldsymbol{p}^*_{t+1} \cdot \boldsymbol{\omega}_{to},$$

while:

$$\big(\forall (\boldsymbol{x}_1, \boldsymbol{x}_2) \in \mathbb{R}^{2n}_+\big)\colon (\boldsymbol{x}_1, \boldsymbol{x}_2) P_t (\boldsymbol{x}^*_{ty}, \boldsymbol{x}^*_{to}) \Rightarrow \boldsymbol{p}^*_t \cdot \boldsymbol{x}_1 + \boldsymbol{p}^*_{t+1} \cdot \boldsymbol{x}_2 > \boldsymbol{p}^*_t \cdot \boldsymbol{\omega}_{ty} + \boldsymbol{p}^*_{t+1} \cdot \boldsymbol{\omega}_{to}.$$

Since this definition appears to be simply a very natural extension of the notion of competitive equilibrium to the present context, and in fact probably looks very familiar, it is quite distressing to discover that it does not have the same normative properties that such an equilibrium has in finite economies. In fact, such a competitive equilibrium may not be Pareto efficient, as is demonstrated by the following example (which is taken from Geanakoplos' and Polemarchakis' survey [1991, p. 1927]).

13.14 Example. We suppose that there is only one commodity in each period, and that the consumers' utility functions are given by:

$$u_{i_0}(x_0) = x_0 \text{ and } u_{i_t}(\boldsymbol{x}_t) = \big(x^2_{ty} \cdot x_{to}\big)^{1/3} \text{ for } t = 1, 2, \ldots;$$

while the initial endowments are given by:

$$w_0 = 1,\ \boldsymbol{\omega}_t = (\omega_{ty}, \omega_{to}) = (5, 1),\ \text{for } t = 1, 2, \ldots.$$

It is easy to verify that if we set:

$$p_1^* = 1,\ p_2^* = 5/2, \ldots, p_{t+1}^* = (5/2) \cdot p_t^*, \ldots,$$

then $\big(\boldsymbol{x}_0^*, \langle \boldsymbol{x}_t^* \rangle, \langle \boldsymbol{p}_t^* \rangle\big)$ is a competitive equilibrium for $\mathcal{E}$, where:

$$x_0^* = 1, \boldsymbol{x}_t^* = (5,1) = \boldsymbol{\omega}_t \text{ for } t = 1, 2, \ldots.$$

However, the allocation $(x_0', \langle \boldsymbol{x}_t' \rangle)$ defined by:

$$x_0' = 2, \boldsymbol{x}_t' = (x_{ty}', x_{to}') = (4,2) \text{ for } t = 1, 2, \ldots,$$

is both feasible and unanimously preferred to $(\boldsymbol{x}_0^*, \langle \boldsymbol{x}_t^* \rangle)$. □

As if this example weren't troublesome enough, Hendricks *et. al* [1980] show that the core may be empty in even this sort of simple overlapping generations economy, while Kovenock [1984] presents an example of an economy in which there are two commodities and two consumers for $t = 1, 2, , \ldots,$ where there exists a Pareto efficient allocation which is also Walrasian, but which is not in the core.

However, there is also some good news in this context. The following result can be proved by essentially the same argument which established Theorem 7.22; the details will be left as an exercise. In the result, we define a feasible allocation as being Pareto efficient if there exists no alternative efficient allocation in which *each* consumer is better off.

13.15 Theorem. *Under the assumptions of this section, if* $\big(\boldsymbol{x}_0^*, \langle \boldsymbol{x}_t^* \rangle, \langle \boldsymbol{p}_t^* \rangle\big)$ *is a competitive equilibrium for* $\mathcal{E}$ *which is such that:*

$$\boldsymbol{p}_1^* \cdot (\boldsymbol{x}_0^* + \boldsymbol{\omega}_{1y}) + \sum\nolimits_{t=2}^{\infty} \boldsymbol{p}_t^* \cdot (\boldsymbol{\omega}_{ty} + \boldsymbol{\omega}_{t-1,o}), \tag{13.19}$$

is finite,[6] *then* $\big(\boldsymbol{x}_0^*, \langle \boldsymbol{x}_t^* \rangle\big)$ *is Pareto efficient for* $\mathcal{E}$.

If we define strong Pareto efficiency for $\mathcal{E}$ in the obvious way, then we can establish the following, although once again I will leave the proof as an exercise.

13.16 Corollary. *If, in addition to the other assumptions of Theorem 13.15, we suppose that each preference relation,* P_t *is locally non-saturating and negatively transitive, then* $\big(\boldsymbol{x}_0^*, \langle \boldsymbol{x}_t^* \rangle\big)$ *is strongly Pareto efficient for* $\mathcal{E}$.

Once again, if we view the economic universe as being fundamentally finite, then competitive equilibria are Pareto efficient (and contained in the core; see Exercise 9). However, a great many economists firmly believe that in an overlapping generations model one *must* assume an infinite time horizon (see, for example, Geanakoplos and Polemarchakis [1991, p. 1900]). In this case, the sum in equation (13.19) is an infinite series which will converge *only if*:

$$\lim_{t \to \infty} \boldsymbol{p}_t^* \cdot (\boldsymbol{\omega}_{ty} + \boldsymbol{\omega}_{t-1,o}) = 0.$$

Of course, this can be true even if $T = \infty$, but it is an awkward and unintuitive assumption if T is infinite; and it is not sufficient to guarantee that the sum in equation (13.19) is finite in any case. However, I will leave this topic here, and proceed to the consideration of models in which we have an uncountably infinite number (a continuum) of consumers.

[6] In other words, if the value of the aggregate endowment is finite, given the prices $\langle \boldsymbol{p}_t^* \rangle$.

13.5 A Continuum of Traders

There are, it would seem, two main arguments that one might make to justify the study of markets wtih a continuum of traders. In order to present the first reason, we can do no better than to quote the theorist who introduced the continuum of traders model into the literature, Robert Aumann. In his seminal article (Aumann [1964]), he argues as follows:

> The notion of *perfect competition* is fundamental in the treatment of economic equilibrium. The essential idea of this notion is that the economy under consideration has a "very large" number of participants, and that the influence of each individual participant is "negligible." ...
>
> Though writers on economic equilibrium have traditionally assumed perfect competition, they have, paradoxically, adopted a mathematical model that does not fit this assumption. Indeed, the influence of an individual participant on the economy cannot be mathematically negligible, as long as there are only finitely many participants. Thus, *a mathematical model appropriate to the intuitive notion of perfect competition must contain infinitely many participants.* We submit that the most natural model for this purpose contains a *continuum* of participants, similar to the ontinuum of points on a line or the continuum of particles in a fluid. Very succintly, the reason for this is that one can integrate over a continuum, and changing the integrand at a single point does not affect the value of the integral, that is, the actions of a single individual are negligible.

Aumann's argument is certainly eloquently stated, and has been very influential, but there is, of course, a counter-argument; individuals may behave as if they believe they have no influence on markets when they, in fact, do. For example, an individual may feel that the expected gain from haggling over price does not justify the time and trouble involved in the negotiations. There is, however, another reason for being interested in continuum of traders models. To introduce this second reason, consider the following, which is a modification of an example due to Scotchmer [2002, p. 2010].

13.17 Example. Consider an exchange economy, in which agents each have a choice of living in one of two locations. There are two tradeable commodities, and agent i attains a utility of:

$$u(\boldsymbol{x}_i) = x_{i1} + x_{i2},$$

if she/he resides in Location One (L1), and a utility of:

$$u(\boldsymbol{x}_i) = (\sqrt{2})x_{i1} + x_{i2}/2,$$

if Location Two (L2) is the chosen location.[7] We will suppose that each agent has an initial endowment of one unit of each of the two commodities.

[7] We might, for example, take L1 to be California, and L2 to be the Upper Peninsula of Michigan; while good 1 is brandy and good 2 is wine. Wine is okay as far as 'Yoopers' are concerned (and, yes, I am a Yooper), but no good for warming you up on a cold winter's day. On the other hand cold is not a problem in sunny California, so (with an appropriate choice of units of measurement), an agent residing in California may be indifferent between the two beverages.

Interestingly enough, if no trade is allowed, then each agent would prefer L1. We can verify this by noting that $\sqrt{2} + 1/2 < 2$. However, such a situation is not Pareto efficient. In fact, if an agent moves to L2, and trades $1/2 + \epsilon$ units of good 2 to an agent in L1 in exchange for $1/2 - \epsilon$ units of good one, both agents will gain as long as $0 < \epsilon < (2\sqrt{2} - 1)/2(1 + 2\sqrt{2})$. The question is, however, can a market system work in this environment; that is, can we find prices for the two commodities which result in a competitive equilibrium?

Proceeding with this question, we note first that we can obviously take the second commodity to be a numéraire, and set its price equal to one. It is also apparent that with this normalization, the price of good one, which we will denote by 'p,' must be greater than one. This being the case, each agent in L1 will wish to sell her/his endowment of good 1, and consume only good 2. On the other hand, if $p \leq \sqrt{2}$ an agent in L2 will sell her/his endowment of good 2, and consume only good 1. However, if we are to have equilibrium, then these two situations have to result in the same utility; in other words, we must have:

$$1 + p = \sqrt{2}\Big(\frac{p+1}{p}\Big),$$

that is, $p = \sqrt{2}$. On the other hand, the supply of good one on the market must equal the demand for same. Thus, denoting the number of agents choosing L1 by 'm_1, the number choosing L2 by 'm_2,' and, as usual, denoting the total number of agents by 'm;' we see that the number of units of good one offered on the market will be equal to m_1. On the other hand, with $p = \sqrt{2}$, each agent at L2 will have excess demand of $1/\sqrt{2}$ units of good 1. Thus, in order to have equilibrium, we must have:

$$m_1 = m_2/\sqrt{2} = (m - m_1)/\sqrt{2};$$

so that we must have:

$$\frac{m_1}{m} = \frac{1}{1 + \sqrt{2}}.$$

But this means that no competitive equilibrium exists in this case, for if m_1 and m are both positive integers, then m_1/m is a rational number; whereas $1/(1 + \sqrt{2})$ is not! □

As we noted in the above example, a competitive equilibrium would exist if it were possible for the *proportion* of agents choosing L1 to be equal to $1/(1 + \sqrt{2})$. While this is not possible if there are only a finite number of agents, a continuum of traders model handles this with ease, as we shall see. In fact, in a continuum of traders model this proportion could (with other choices of parameters) be any number between zero and one. Since it is very natural to think in terms of proportions of consumers or households having this or that property, or making this or that choice, this makes a continuum of traders model a very flexible and convenient tool. Moreover, those of you who have had a real analysis course know that, given any real number between zero and one, there is a sequence of rational numbers converging to that number. This means that if we obtain a result in the continuum model which involves a certain proportion of consumers making some specific choice, we can be confident that there is a finite model in which approximately the same result obtains. We will present only the rudiments of such a model here, and we will confine our attention to pure exchange economies; but, hopefully, this development will be useful in and of itself, as well as providing a basic idea of how such models work.

In our discussion to this point, we have always effectively identified the set of consumers with $M = \{1, \ldots, m\}$, the first m positive integers. In contrast, in this section we will identify the set of consumers with the set $A = [0, 1]$, the unit interval in $\mathbb{R}_+$.[8] Again, in our work to this point, we have denoted consumption allocations by $\langle \boldsymbol{x}_i \rangle, \langle \boldsymbol{x}'_i \rangle$, and so on; where the notation has been intended to suggest a finite sequence of commodity bundles (elements of $\mathbb{R}^n$), one for each consumer. However, a finite sequence with m terms is, from a formal point of view, a function from $M = \{1, \ldots, m\}$ into $\mathbb{R}^n$. Thus, it should not cause great confusion if we now denote allocations by '$\langle \boldsymbol{x}_a \rangle, \langle \boldsymbol{x}'_a \rangle$, and so on, where we think of, say, $\langle \boldsymbol{x}_a \rangle$ as being a function from A into $\mathbb{R}^n$, that is, $\boldsymbol{x} \colon A \to \mathbb{R}^n$. However, we will often denote the value of the function at a point $a \in A$ by '$\boldsymbol{x}_a$,' rather than $\boldsymbol{x}(a)$; since this is the commodity bundle available to consumer (agent) $a \in A$.

In a pure exchange economy a (competitive) consumer is fully identified by her/his characteristics $(P_i, \boldsymbol{\omega}_i)$; and, in effect, an m-consumer exchange economy is fully defined by the finite sequence, $\langle (P_i, \boldsymbol{\omega}_i) \rangle$. Similarly, in the case at hand, an economy is fully identified by $\mathcal{E} = \langle (P_a, \boldsymbol{\omega}_a) \rangle$; where, from a formal point of view, this is a function from A into 'characteristics space.' However, the meaning and significance of the notation should be clear enough without worrying about a formal definition of 'characteristics space.'

The next question is, how do we define feasible allocations for such an economy? It is pretty clear that we cannot compare the sum of individual commodity bundles with the sum of individual endowments. On the other hand, what criterion can we use to define a feasible allocation function? In order to consider this question with some precision, let's suppose for the moment that there is only one commodity, and to simplify things still further, consider an allocation $\langle \boldsymbol{x}_a \rangle$ having the property that there exist m numbers, $d_1, \ldots, d_m$, and $\langle x_i \rangle_{i=1}^m$ such that:

$$0 < d_1 < d_2 < \cdots < d_{m-1} < d_m = 1, \tag{13.20}$$

and such that, defining $d_0 = 0$, we have:

$$(\forall a, \in [d_{i-1}, d_i\, [\,) \colon x_a = x_i \quad \text{for } i = 1, \ldots, m. \tag{13.21}$$

In other words, our allocation function is constant on each sub-interval $I_i = [\, d_{i-1}, d_i\, [$. In this case, the *proportion* of consumers who will receive the quantity x_i of the commodity is given by $m_i = d_i - d_{i-1}$. Consequently, it is consistent with standard useage to define *per capita consumption* of the commodity by:

$$x = \sum\nolimits_{i=1}^m m_i x_i = \sum\nolimits_{i=1}^m (d_i - d_{i-1}) x_i. \tag{13.22}$$

[8]Denoting this set by 'A' is intended to suggest the set of *agents*.

But in fact, given the form of the function $x(\cdot)$, the expression on the right in equation (13.22) is just:[9]

$$x = \int_0^1 x(a)da \tag{13.23}$$

Now suppose that the initial endowment of the (single) commodity has a similar distribution; so that $\omega\colon A \to \mathbb{R}_+$ and there exist $2m'$ numbers, $d'_1, \ldots, d'_{m'}$, and $\omega_1, \ldots, \omega_{m'}$ such that:

$$0 < d'_1 < \cdots < \ldots d'_{m'-1} < d'_{m'} = 1, \tag{13.24}$$

and, defining $d'_0 = 0$:

$$(\forall a \in [d'_{j-1}, d'_j\,[)\colon \omega_a = \omega_j \quad \text{for } j = 1, \ldots, m'. \tag{13.25}$$

Then the *per capita* initial endowment is given by:

$$\sum\nolimits_{j=1}^{m'} (d'_j - d'_{j-1})\omega_j = \int_0^1 \omega(a)da;$$

and feasibility will require that:

$$x = \int_0^1 x(a)da = \int_0^1 \omega(a)da \tag{13.26}$$

(see Exercise 13, at the end of this chapter).

Now, in terms of fully characterizing an economy of this type, we are left with two obvious problems. First, how do we deal with more than one commodity? Secondly, how can we characterize distributions of the commodities, and the endowments, in such a way as to guarantee that the integrals we need are always well-defined?

The first of these problems is easy to handle. If $\boldsymbol{x}\colon A \to \mathbb{R}^n_+$ is a commodity distribution, it defines n coordinate functions, $x_j\colon A \to \mathbb{R}_+$, for $j = 1, \ldots, n$. We simply define the desired per capita commodity bundle by:

$$\int_0^1 \boldsymbol{x}(a)da = \begin{pmatrix} \int_0^1 x_1(a)da \\ \vdots \\ \int_0^1 x_j(a)da \\ \vdots \\ \int_0^1 x_n(a)da \end{pmatrix}; \tag{13.27}$$

although I will omit the limits of integration hereafter, as they will always be the same. The feasibility requirement:

$$\int \boldsymbol{x}(a)da = \int \boldsymbol{\omega}(a)da, \tag{13.28}$$

[9] Or, to make this expression look more familiar:

$$x = \int_0^1 x(t)dt.$$

then simply asserts that the per capita consumption of each commodity is equal to the per capita endowment of that commodity.

The second problem is much more difficult. You probably remember that if we interpret our integrals to be standard Riemann integrals, then we need the allocation and endowment functions to be continuous, except, possibly, at a finite set of points, in order for the integrals to exist.[10] This is, in principle, a very troublesome and restrictive requirement. If we were worried about one, and only one distribution function, assuming it to be continuous except at a finite number of points is fairly innocuous; for, if we stretch our imaginations a bit, we can imagine that the agents have been labeled in such a way as to ensure that this continuity occurs. However, we may find that, for example, the endowment function has to have the agents labeled in a different order in order to ensure that it is continuous. In practice, more advanced developments of models of this type deal with distribution and endowment functions which are *measurable*, which ensures that the Lebesgue integrals of these functions, which we will denote by:

$$\int_A \boldsymbol{x},$$

always exist. A very important aspect of this assumption is that the set of all measurable functions from A to $\mathbb{R}^n$ is a real linear space; and thus one can add or consider scalar multiples of such functions.[11] Moreover, and again most conveniently, it turns out that if, say, $\boldsymbol{x}$ and $\boldsymbol{x}^*$ are both measurable functions from A to $\mathbb{R}^n$, then:

$$\int_A (\boldsymbol{x} + \boldsymbol{x}^*) = \int_A \boldsymbol{x} + \int_A \boldsymbol{x}^*,$$

Since I do not expect most students to have had any previous exposure to the idea of measurable functions and Lebesgue integration, we will not pursue these ideas further here. In any case, however, most applications of continuum of traders models make use of distribution and endowment functions which are *step functions*; that is, functions of the type defined in equations (13.20) and (13.21), above. In order to deal further with this theory, however, let's define such functions in a bit more abstract fashion. First, we define the following.

13.18 Definition. If X is a nonempty interval of real numbers, we will say that a family, $\mathcal{D} = \{D_1, \ldots, D_m\}$, of sets is a **finite interval partition** of X iff:

1. for all $i \in \{1, \ldots, m\}, D_i$ is a non-empty sub-interval of X,
2. for all $i, j \in \{1, \ldots, m\}$, such that $i \neq j$, $D_i \cap D_j = \emptyset$,
3. $\bigcup_{i=1}^m D_i = X$.

We make use of this to state our (slightly) more abstract definition of a step function.

[10]More correctly, we need the set of points at which the functions are discontinuous to be a set of measure zero.

[11]Which are defined in the obvious way; for example, if $\boldsymbol{x}$ and $\boldsymbol{x}^*$ are two such functions, we define $\boldsymbol{x} + \boldsymbol{x}^*$ by:

$$(\boldsymbol{x} + \boldsymbol{x}^*)(a) = \boldsymbol{x}(a) + \boldsymbol{x}^*(a) \quad \text{for } a \in A.$$

13.19 Definition. Let A be a nonempty interval of real numbers. We will say that $\boldsymbol{x}: A \to \mathbb{R}^n$ is a **step function** iff there exist a finite interval partition of A, $\mathcal{D} = \{D_1, \ldots, D_m\}$, and a subset, $\{\boldsymbol{x}_1, \ldots, \boldsymbol{x}_m\}$ such that:

$$(\forall a \in D_i) \colon \boldsymbol{x}(a) = \boldsymbol{x}_i \quad \text{for } i = 1, \ldots, m.$$

Of course, in the remainder of this section, we will always take A to be equal to the unit interval, $A = [0, 1]$. In this context, the following should be fairly obvious, although I will leave the proof as an exercise (Exercise 10, at the end of this chapter.)

13.20 Proposition. *Suppose $\mathcal{D} = \{D_1, \ldots, D_m\}$ and $\mathcal{D}' = \{D'_1, \ldots, D'_{m'}\}$ are two finite interval partitions of A. If we then define the family of sets $\mathcal{D}^*$ by:*

$$\mathcal{D}^* = \{D \subseteq A \mid (\exists D_i \in \mathcal{D} \;\&\; D'_j \in \mathcal{D}') \colon [D = D_i \cap D'_j \;\&\; D \neq \emptyset]. \tag{13.29}$$

Then $\mathcal{D}^*$ is a finite interval partition of A.

Given two step functions, $\boldsymbol{x}$ and $\boldsymbol{x}^*$ on A, and a real number $\alpha \in \mathbb{R}$, we define $\boldsymbol{x} + \boldsymbol{x}'$ and $\alpha\boldsymbol{x}$ by:

$$(\boldsymbol{x} + \boldsymbol{x}^*)(a) = \boldsymbol{x}(a) + \boldsymbol{x}^*(a) \;\text{ for } a \in A, \tag{13.30}$$

and:

$$(\alpha\boldsymbol{x})(a) = \alpha\boldsymbol{x}(a) \;\text{ for } a \in A, \tag{13.31}$$

respectively. One can then make use of Proposition 13.20 to prove the following.

13.21 Proposition. *Let $\mathbf{S}$ be the family of all step functions (into $\mathbb{R}^n$) defined on A. Then $\mathbf{S}$ is a real linear space.*

The key to proving Proposition 13.21, and the primary reason for stating it here, is that if we define the sum of two step functions as in equation (13.30), above, we get another step function on A. Similarly, the scalar multiple of a step function on A is again a step function on A. The proof of the latter statement is obvious, and the proof of the first statement follows fairly easily from Proposition 13.20. It also is an immediate application of the elementary theory of Riemann integration that for all $\boldsymbol{x}, \boldsymbol{x}^* \in \mathbf{S}$ and all $\alpha \in \mathbb{R}$:

$$\int_0^1 (\boldsymbol{x} + \boldsymbol{x}^*)(a)da = \int_0^1 \boldsymbol{x}(a)da + \int_0^1 \boldsymbol{x}^*(a)da,$$

and:

$$\int_0^1 (\alpha\boldsymbol{x})(a)da = \alpha \cdot \int_0^1 \boldsymbol{x}(a)da.$$

Consequently, we can *almost* deal with our simplified continuum of traders model in the same way that we have dealt with economies with a finite number of traders. The difficulty is, of course, that we can only deal with allocation and initial endowment functions which treat many consumers in exactly the same way.

Consider the problem of defining a competitive equilibrium for the simplified continuum of traders economy, as we have set it out here. We can make use of the following definitions.

13.22 Definitions. We shall say that $\mathcal{E} = \langle (P_a, \boldsymbol{\omega}_a) \rangle$ is a **(simplified) continuum of traders economy** iff P_a is an asymmetric relation on $\mathbb{R}^n_+$, $\boldsymbol{\omega}_a \in \mathbb{R}^n_+$, for each $a \in A$, and the function $\boldsymbol{\omega} \colon A \to \mathbb{R}^n_+$ defined by:

$$\boldsymbol{\omega}(a) = \boldsymbol{\omega}_a \quad \text{for } a \in A,$$

is a step function. A function $\boldsymbol{x} \colon A \to \mathbb{R}^n_+$ is said to be an allocation for $\mathcal{E}$ iff if is a step function (on A), and is said to be **feasible** for $\mathcal{E}$ iff, in addition, it satisfies:

$$\int_0^1 \boldsymbol{x}(a)da = \int_0^1 \boldsymbol{\omega}(a)da.$$

We can then define a competitive equilibrium as follows.

13.23 Definition. We shall say that $(\boldsymbol{x}^*, \boldsymbol{p}^*)$ is a competitive equilibrium for a (simplified) continuum of traders economy, $\mathcal{E} = \langle (P_a, \boldsymbol{\omega}_a) \rangle$, iff:

1. $\boldsymbol{p}^* \in \mathbb{R}^n$,
2. $\boldsymbol{x}^* \colon A \to \mathbb{R}^n_+$ is a feasible allocation for $\mathcal{E}$, and
3. for each $a \in A$, $\boldsymbol{p}^* \cdot \boldsymbol{x}^*_a \leq \boldsymbol{p}^* \cdot \boldsymbol{\omega}_a$, and:

$$(\forall \boldsymbol{x} \in \mathbb{R}^n_+) \colon \boldsymbol{x} P_a \boldsymbol{x}^*_a \Rightarrow \boldsymbol{p}^* \cdot \boldsymbol{x} > \boldsymbol{p}^* \cdot \boldsymbol{\omega}_a.$$

Everything works out just fine if we apply the definitions here to the example with which we introduced this section, for consider the following.

13.24 Example. Let $\mathcal{E} = \langle (P_a, \boldsymbol{\omega}_a) \rangle$ be as in Example 13.17, except that we take the set of agents to be $A = [0, 1]$ insteacd of $M = \{1, \ldots, m\}$ (notice that the initial endowment function here is constant, and thus is a step function on A). If we define the price vector $\boldsymbol{p}^*$ by $\boldsymbol{p}^* = (\sqrt{2}, 1)$, the intervals:

$$D_1 = \Big[[0, \frac{1}{1+\sqrt{2}}\Big[\text{ and } D_2 = \Big[\frac{1}{1+\sqrt{2}}, 1\Big],$$

and the allocation function $\boldsymbol{x}^* \colon A \to \mathbb{R}^2_+$ by:

$$\boldsymbol{x}^*_a = \begin{cases} (0, 1+\sqrt{2}) & \text{for } a \in D_1 \\ (\frac{1+\sqrt{2}}{\sqrt{2}} & \text{for } a \in D_2. \end{cases}$$

it is easy to verify the fact that $(\boldsymbol{x}^*, \boldsymbol{p}^*)$ is a competitive equilibrium for $\mathcal{E}$. (Formally, we also need to specify $(u_a, \boldsymbol{\omega}_a) = (u_1, \boldsymbol{\omega})$, for $a \in D_1$, and $(u_a, \boldsymbol{\omega}_a) = (u_2, \boldsymbol{\omega})$, for $a \in D_2$, where $\boldsymbol{\omega} = (0, 0)$.) □

While the above example works out as desired, notice that our definitions imply that if $\mathcal{E} = \langle (P_a, \boldsymbol{\omega}_a) \rangle$ is a (simplified) continuum of traders economy, then there can be only a finite number of different consumer endowments. In fact, no competitive equilibrium is possible unless our continuum of consumers demand only a finite number of different commodity bundles. The simplest way to ensure that both of these things will be so is to assume that there are only a finite number of consumer *types*; that is, suppose there is a finite interval partition, $\mathcal{D} = \{D_1, \ldots, D_m\}$ and a finite sequence of consumer characteristics, $\langle(P_i, \boldsymbol{\omega}_i)\rangle_{i=1}^m$, such that the economy $\mathcal{E} = \langle(P_a, \boldsymbol{\omega}_a)\rangle$ satisfies:

$$(\forall a \in D_i)\colon (P_a, \boldsymbol{\omega}_a) = (P_i, \boldsymbol{\omega}_i) \quad \text{for } i = 1, \ldots, m.$$

If we also assume that P_i is negatively transitive (as well as asymmetric), continuous, and *strictly* convex on $\mathbb{R}^n_+$, then, given any (strictly positive) price vector, $\boldsymbol{p}^*$, each consumer of a given characteristic will demand the same commodity bundle. More precisely, if $\boldsymbol{x}_i^*$ satisfies:

$$\boldsymbol{p}^* \cdot \boldsymbol{x}_i^* \leq \boldsymbol{p}^* \cdot \boldsymbol{\omega}_i,$$

and:

$$(\forall \boldsymbol{x} \in \mathbb{R}^n_+)\colon \boldsymbol{x} P_i \boldsymbol{x}_i^* \Rightarrow \boldsymbol{p}^* \cdot \boldsymbol{x} > \boldsymbol{p}^* \cdot \boldsymbol{\omega}_i;$$

then each consumer in D_i will demand $\boldsymbol{x}_i^*$ $(i = 1, \ldots, m)$. Thus, the function $\boldsymbol{x}^*$ defined by:

$$\boldsymbol{x}^*(a) = \boldsymbol{x}_i^* \quad \text{for each } a \in D_i, \text{and for } i = 1, \ldots, m,$$

is an allocation function for $\mathcal{E}$, and if:

$$\int_0^1 \boldsymbol{x}^*(a)da = \int_0^1 \boldsymbol{\omega}(a)da, \tag{13.32}$$

then $(\boldsymbol{x}^*, \boldsymbol{p}^*)$ is a competitive equilibrium for $\mathcal{E}$.

As will no doubt have occurred to you, once we have added all of the assumptions set out in the previous paragraph, the economy will look very much like the finite exchange economy, $\mathcal{E}^* = \langle(P_i, \boldsymbol{\omega}_i)\rangle_{i=1}^m$. However, there is a very important difference. Suppose we denote the length of the interval D_i by 'μ_i,' for each i. Then the integral on the left-hand-side of equation (13.32) is given by:

$$\int_0^1 \boldsymbol{x}^*(a)da = \sum\nolimits_{i=1}^m \mu_i \boldsymbol{x}_i^*.$$

In effect, the economy $\mathcal{E}^*$ corresponds to the special case in which $\mu_i = 1/m$, for $i = 1, \ldots, m$.

13.6 Suggestions for Further Reading

As you were warned in the introduction to this chapter, we have barely scratched the surface of the areas of literature being introduced in this chapter. For those of you interested in pursuing the sort of analysis we developed in Sections 2 and 3 of this chapter, let me recommend the symposium in the *Journal of Economic Theory;* 45, 2; August, 1988. MGW offers a considerably more extensive treatment of this material (Sections 2 and 3) than was presented here, and of the 'Overlapping Generations' model as well. The most complete recent survey of the latter topic is the Geanakoplos and Polemarchakis article [1991], which I cited earlier; however, a very readable development of the topic from the point of view of macroeconomics is presented in Heijdra and van der Ploeg [2002]. In his textbook, Ellickson [1993] presents a very enthusiastic and extensive coverage of the continuum of traders model. An interesting topic related to the material in this chapter is known as 'temporary equilibrium,' and is surveyed by Grandmont [1982]. A work which I cited in the last chapter will also be of interest here: Magill and Quinzii [1996] develop general equilibrium theory with both uncertainty and an explicit representation of time, and do so in a quite readable fashion.

Exercises.

1. Prove Proposition 13.8.

2. Verify the details of Example 13.9

3. Given the assumptions of Example 13.10, show that the following program is competitive: given the prices $\boldsymbol{p}_t^* = (1,1)$ for $t = 1, \ldots, 5$. Compare this program to

Period	$y_{t,2}$	$x_{t,1}$	$x_{t,2}$	$c_{t,1}$	$c_{t,2}$	δ_t
$t=1$	4	1	1	-1	2	1
$t=2$	4	1	1	-1	3	1
$t=3$	4	1	1	-1	3	1
$t=4$	4	1	1	-1	–	–

Table 13.3: Program 3.

Program 1 of Example 13.10.

4. Show that if $T > 1$, the finite program defined in Example 13.11 for $t = 1, \ldots, T$ is efficient, given $y_T = f(x_{T-1})$.

5. Verify the details of Example 13.14

6. Show that if we define a finite economy by letting preferences and endowments be as in Example 13.14, except that we have only, say 4 periods ($T = 4$, and no consumer 'born' in the last period), then the allocation:

$$x_0^* = 1 \text{ and } \boldsymbol{x}_t^* = \boldsymbol{\omega}_t, \text{ for } t = 1, \ldots, 4,$$

is Pareto efficient. What happens if you add a Tth consumer, who lives only one period, and has the utility function $u_T(x_T) = x_T$ and $\omega_T = 5$?

7. Prove Theorem 13.15.

8. Prove Corollary 13.16.

9. Show that if the definition of the core of an economy is extended to the context of an overlapping generations economy in the natural way, then, under the assumptions of Theorem 13.15, a competitive equilibrium is in the core.

10. Prove Proposition 13.4.

11. Prove Proposiion 13.5.

12. Show that the pair $(\boldsymbol{x}^*, \boldsymbol{p}^*)$ defined in Example 13.8 is a competitive equilibrium.

13. Specialize the situation set out in equations (13.20)–(13.25) of Section 5, by supposing that:

$$d_i - d_{i-1} = a_i/b_i \quad \text{for } i = 1, \ldots, m,$$

and that:

$$d'_j - d'_{j-1} = a'_j/b'_j \quad \text{for } j = 1, \ldots, m';$$

where a_i, b_i, a'_j and b'_j are positive integers, for all i, j, suppose (13.26 holds, and let k be any integer such that for each i there exists a positive integer, q_i such that:

$$k = q_i b_i \quad \text{for } i = 1, \ldots, m;$$

while for each j there exists a positive integer r_j such that:

$$k = r_j b'_j \quad \text{for } j = 1, \ldots, m'.$$

Suppose $\mathcal{E} = \langle (P_h, \boldsymbol{\omega}_h) \rangle_{h=1}^k$ is any finite exchange economy such that $r_j a' j$ consumers have the initial endowment $\boldsymbol{\omega}_j$ $(j = 1, \ldots, m')$, and $\langle \boldsymbol{x}_h^* \rangle_{i=1}^k$ is a distribution such that $q_i a_j$ consumers receive the bundle $\boldsymbol{x}_i$, for $i = 1, \ldots, m$ [where $\langle \boldsymbol{x}_i \rangle_{i=1}^m$ is from (13.21)]. Show that the allocation is feasible; that is:

$$\sum_{h=1}^k \boldsymbol{x}_h^* = \sum_{h=1}^k \boldsymbol{\omega}_h.$$

Chapter 14

Social Choice and Voting Rules

14.1 Introduction

In this chapter we will be spending most of our time exploring the borderline between Political Science and Economics, or perhaps the intersection of the two disciplines. The origins of this study stemmed from:

1. Efforts in economics to define 'the economic good.'

2. The study of political processes: how do they work and how should they work?

and can be traced as far back as the investigations of Jean-Charles de Borda [1781], the Marquis de Condorcet [1785], C. L. Dodgson (Lewis Carroll) [1876], and E. J. Nanson [1882]. In the economics literature *per. se.* the notion of aggregating individual preferences to obtain a social preference relation apparently originated in the writings of Jeremy Bentham [1789], who originated the idea of a utilitarian social welfare function, and thought of society's utility as being literally the sum of individual (cardinal) utilities. In the next chapter, we will see that there were numerous problems with this approach.

The introduction of the idea of a Pareto improvement was a major innovation in this development, and enabled economists to begin to put normative economic analysis on a much firmer footing than had previously been possible. However, economists were also frustrated by the fact that most allocations could not be compared via the Pareto criterion. The next major innovation in welfare (normative) economics was the Compensation Principle, which appeared momentarily to allow a much broader class of cases to be compared. However, it was eventually pointed out, as will be shown in Chapter 15, that this criterion did not really allow many more allocations to be compared than did the Pareto criterion. Consequently, the publication of Bergson's classic article on social welfare functions [Bergson (Burk), 1938] was a very promising development (and one which we will study in Chapter 15).

While the idea of a social welfare function was a major innovation, and once again helped economists clarify much of their thinking on policy issues, it fell short of providing a 'universally acceptable' criterion for economic improvement; as we will see. However, it led Kenneth Arrow to ask some fundamental questions regarding the possibility of developing such a function which would be widely acceptable, or more generally, finding a social preference relation as a function of individual preference relations. More specificially, Arrow [1950, 1951b] presented three basic properties which it would seem eminently reasonable that we would require such a function to satisfy. First, he wanted it to be defined for any m-tuple of weak orders (where m is the number of consumers [agents] in the economy), and to provide a weak order of the social alternatives available as a function of these individual preferences. Secondly, he wanted this function to extend the Pareto criterion. Thirdly, he asked that the social preference between two allocations depend only upon the individual preferences regarding these same two allocations. The very startling conclusion which he derived from his analysis is that any function which satisfies all three of these criteria must be dictatorial; that is, the function must simply pick out one of the individuals' preference relations, and order the social alternatives according to this individual's preferences. This is Arrow's famous 'general possibility theorem,' and we will be studying this result in some detail in Section 4 of this chapter.

Before turning to a formal development of Arrow's result, however, we will first discuss the general idea of voting rules, and majority voting over two alternatives in particular; which is the main subject of the next section.[1]

14.2 The Basic Setting

In the remainder of this chapter we will be considering a situation in which there is a non-empty, but finite set of alternatives, X, from which a (group) choice is being contemplated. We will denote the number of elements in X by '$\#X$;' and we will generally assume that $\#X \geq 3$. We suppose that there are m agents who have preferences defined on X, and who are concerned with the social choice that is to be made from X. We will always assume that m is an integer greater than or equal to 2, and we will often denote the set $\{1, \ldots, m\}$ by 'M' (we will refer to this as the 'set of agents').

In our treatment here it will be convenient to deal with asymmetric relations; or, if you prefer, the asymmetric parts of some familiar orderings. We will let:

$\mathcal{Q}$ = the family of all asymmetric orderings of X,

$\mathcal{P}$ = the family of all (asymmetric parts of) weak orders on X,[2] and:

$\mathcal{L}$ = the family of all strict linear orders on X;

where by a strict linear order, I mean a relation which is asymmetric, transitive, and total (I will leave it as an exercise for you to show that such a relation is also negatively transitive). We will denote the m-fold cartesian products of these families by '$\mathcal{Q}^m$, $\mathcal{P}^m$,' and '$\mathcal{L}^m$,' respectively. We will generally use the generic notation, '$\boldsymbol{Q}$, $\boldsymbol{Q}'$, $\boldsymbol{P}$,' and '$\boldsymbol{P}'$,' etc., to denote elements of $\mathcal{Q}^m$, $\mathcal{P}^m$, and $\mathcal{L}^m$. Thus, when we write, for example, $\boldsymbol{Q} \in \mathcal{Q}^m$, we will mean that $\boldsymbol{Q}$ is of the form:

$$\boldsymbol{Q} = (Q_1, \ldots, Q_m),$$

[1] For an elegant, readable, and much more complete introduction to this area than I have provided here, see Suzumura [2002].

[2] That is, $\mathcal{P}$ is the family of all asymmetric and negatively transitive binary relations on X.

where:

$$Q_i \in \mathcal{Q} \quad \text{for } i = 1, \dots, m$$

(that is, each Q_i is an asymmetric ordering of A); and similarly for $\boldsymbol{Q} \in \mathcal{P}^m$, or $\boldsymbol{Q} \in \mathcal{L}^m$. We will refer to elements of $\mathcal{Q}^m$, $\mathcal{P}^m$, and $\mathcal{L}^m$ as **profiles**, or **preference profiles**. Notice that it follows from our work in Chapter 1 that:

$$\mathcal{L} \subseteq \mathcal{P} \subseteq \mathcal{Q},$$

and thus:

$$\mathcal{L}^m \subseteq \mathcal{P}^m \subseteq \mathcal{Q}^m.$$

We will seek a 'good' means by which an element of X, the set of alternatives, can be selected, given the preferences of the m agents. Formally, we can describe this search as a matter of finding a 'voting rule' with desirable properties, where we define a 'voting rule' as follows.

14.1 Definition. Let X be a nonempty set and m be a positive integer, let $\mathcal{Q}$ be the family of all asymmetric orders on X, and $\mathcal{D}$ be a nonempty subset of $\mathcal{Q}$. We will say that a function, $f\colon \mathcal{D}^m \to X$ is a **voting rule (with admissible preferences $\mathcal{D}$)**.

In our treatment we will generally suppose that m, the number of agents, is greater than one, and that $\#X$, the number of alternatives in X is at least three. However, we will begin our analysis by considering the most familiar example of a voting rule, namely majority voting,[3] and, strictly speaking, majority voting is only applicable to the situation in which X, the set of alternatives, has two elements. Consequently, let's begin our considerations here by supposing that we are interested in defining a voting rule for a group of m individuals over a set, X, with $\#X = 2$, and where we write $X = \{x, y\}$. We suppose that each of the m individuals has a weak order, P_i, on X,[4] and we will consider the formal definition of majority voting in this case.

Since there are only 3 possible weak orders over a set of two elements, $X = \{x, y\}$, we can usefully characterize the three possibilities by one of the three numbers, $1, 0, -1$, as follows. Given $P_i \in \mathcal{P}$, we replace P_i by $d(P_i) = d_i$, where:

$$d_i = d(P_i) = \begin{cases} 1 & \text{if } xP_iy, \\ 0 & \text{if } xI_iy, \\ -1 & \text{if } yP_ix. \end{cases} \tag{14.1}$$

A preference profile can then be characterized as a finite sequence, $\boldsymbol{d} = \langle d_i \rangle_{i=1}^m$ drawn from $D = \{1, 0, -1\}$ (that is, a preference profile is an element of $\boldsymbol{D} \stackrel{\text{def}}{=} D^m$); and, in this context, a voting rule can be characterized as a mapping from $\boldsymbol{D}$ to X. However, we will shift our focus a bit here initially, to consider a **social preference function**, which in this context can be defined as a mapping $\delta\colon \boldsymbol{D} \to D$. Thus, the

[3]The treatment here, particularly in the first part of this section, owes a great deal to Kelly [1988].

[4]Notice that any asymmetric order on a two element set is negatively transitive.

social preference function corresponding to simple majority rule, which we will call the **simple majority social preference**, is the mapping, $\delta^s \colon \boldsymbol{D} \to D$ defined by:

$$\delta^s(\boldsymbol{d}) = \begin{cases} 1 & \text{if } \sum_{i=1}^m d_i > 0, \\ 0 & \text{if } \sum_{i=1}^m d_i = 0, \\ -1 & \text{if } \sum_{i=1}^m d_i < 0,. \end{cases} \tag{14.2}$$

While the simple majority rule is probably the one most frequently used in practice, a variant is sometimes adopted; namely, absolute majority voting. Working with the same representation of preference profiles as was just introduced, we define:

$$\begin{aligned} M(x, y; \boldsymbol{d}) &= \#\{i \in \{1, \ldots, m\} \mid d_i = 1\}, \text{ and} \\ M(y, x; \boldsymbol{d}) &= \#\{i \in \{1, \ldots, m\} \mid d_i = -1\}. \end{aligned} \tag{14.3}$$

The **absolute majority social preference**, $\delta^a \colon \boldsymbol{D} \to D$, is then defined by

$$\delta^a(\boldsymbol{d}) = \begin{cases} 1 & \text{if } M(x, y; \boldsymbol{d}) > m/2, \\ -1 & \text{if } M(y, x; \boldsymbol{d}) > m/2, \\ 0 & \text{otherwise.} \end{cases} \tag{14.4}$$

Simple majority rule and absolute majority rule share a common defect in many social choice situations where the choice is over pairs: they may (in fact often will) fail to pick a winner. Consequently, neither actually defines a voting rule, as we have defined the term, unless $\mathcal{D} = \mathcal{L}$. In fact, the kind of situation to which they are most applicable is that in which each of the m individuals is assumed to have a *linear* order over the two alternatives, and in addition, where m, the number of individuals, is odd. Given these assumptions, each of the two voting methods will always pick a unique winner. Interestingly enough, however, simple and absolute majority rule produce identical social preference relations in this case (the proof of these statements I will leave as an exercise).

Fifty-odd years ago, K. O. May [1952] published an interesting and important characterization of simple majority rule, and it may be useful for us to begin our more formal analysis of voting rules and social preference functions by considering May's development. Retaining our characterization of preference profiles as elements of $\boldsymbol{D}$, May considered social preference functions, δ, such that $\delta \colon \boldsymbol{D} \to D$. Obviously both simple majority and absolute majority social preference functions are examples of such functions. May Introduces three properties which it appears one would like such a function to satisfy, the first of which, anonymity, is defined as follows.

14.2 Definition. We say that a function $\delta \colon \boldsymbol{D} \to D$ satisfies **anonymity** iff whenever two profiiles in $\boldsymbol{D}$, $\boldsymbol{d}$ and $\boldsymbol{d}'$, are such that $\boldsymbol{d}'$ is a permutation of $\boldsymbol{d}$, we have $\delta(\boldsymbol{d}) = \delta(\boldsymbol{d}')$.

It is easily shown that both δ^s and δ^a satisfy this condition; and that both satisfy the condition of neutrality, defined as follows.

14.3 Definition. We say that a function, $\delta \colon \boldsymbol{D} \to D$ satisfies **neutrality** iff whenever $\boldsymbol{d}, \boldsymbol{d}' \in \boldsymbol{D}$ are such that $\boldsymbol{d}' = -\boldsymbol{d}$, we have $\delta(\boldsymbol{d}') = -\delta(\boldsymbol{d})$.

In order to state May's third condition, we introduce a bit of notation which we will frequently find useful. Given $\boldsymbol{d} \in \boldsymbol{D}$ and $d_i' \in D$, we denote the profile $\boldsymbol{d}^* \in \boldsymbol{D}$ defined by:

$$d_k^* = \begin{cases} d_k & \text{for } k \neq i, \\ d_i' & \text{for } k = i, \end{cases}$$

by '$(d_i', \boldsymbol{d}_{-i})$.' We then define the following.

14.4 Definition. The social preference function $\delta\colon \boldsymbol{D} \to D$ satisfies **positive responsiveness** iff for all $\boldsymbol{d} \in \boldsymbol{D}$, all $i \in M$, and all $d_i' \in D$, we have:

$$[\delta(\boldsymbol{d}) \geq 0 \;\&\; d_i' > d_i] \Rightarrow \delta(d_i', \boldsymbol{d}_{-i}) = 1.$$

I will leave as an exercise the task of showing that, while δ^s satisfies positive responsiveness, δ^a does not. In fact, May established the following.[5]

14.5 Theorem. (May) The only social preference function, $\delta\colon \boldsymbol{D} \to D$, satisfying anonymity, neutrality, and positive responsiveness is δ^s, simple majority rule.

Thus, simple majority rule fares very well as a social preference function when X, the set of alternatives, contains only two elements. The next question is, however, how do we extend this rule to cover the case in which $\#X$, the number of disinct alternatives in X is greater than or equal to three? We will take up this issue in the next section.

14.3 Voting Rules

In this section we will be considering some voting rules which can be regarded as extensions of simple or absolute majority voting to the situation in which the number of alternatives, $\#X \geq 3$.

14.6 Example. The Condorcet Winner. The first extension of simple majority voting that we'll consider is that analyzed by one of the first people to systematically investigate this problem; the Marquis de Condorcet [1785]. We will say that $x \in B \subseteq X$ is a **Condorcet winner on B, given the profile $\boldsymbol{P} \in \mathcal{P}^m$** iff, for all $y \in B \setminus \{x\}$ x wins, or at least ties in a simple majority vote against y. In other words, x is a Condorcet winnner on B iff there is no alternative element in B which is a clear simple majority winner over x.

It will simplify our discussion of this, and other extensions of majority voting to introduce the following notation. Let's return to the representation of preferences which we used in discussing May's theorem, and given a profile $\boldsymbol{P} \in \mathcal{D}^m$, and $x, y \in X$, define $C^v(\{x,y\})$, the **majority voting choice from the pair $\{x, y\}$**, by:

$$C^v(\{x,y\}) = \begin{cases} \{x\} & \text{if } \sum_{i=1}^m d_i > 0, \\ \{y\} & \text{if } \sum_{i=1}^m d_i < 0, \text{ and} \\ \{x,y\} & \text{if } \sum_{i=1}^m d_i = 0. \end{cases} \tag{14.5}$$

[5] For a proof, see May [1952], or Kelly [1988, pp. 12–13].

An alternative, $x \in B \subseteq X$ is then a Condorcet winner on B iff, for every $y \in B \backslash \{x\}$:

$$x \in C^v(\{x, y\}). \quad \square$$

While it seems to be perfectly natural and appropriate to pick a Condorcet winner, as a way of extending simple majority voting from pairs to arbitrary finite sets of alternatives, this definition does not yield a well-defined voting rule, as the following example demonstrates.

14.7 Example. Suppose $\#X \geq 3$, with x, y, and z three distinct elements of X, let $m = 2q + 1$, where $q \geq 1$, and let $\boldsymbol{Q}$ be any preference profile for which the rankings of the three distinct alternatives, x, y, and z, is as follows:

Agent 1	**Group A**	**Group B**
x	y	z
y	z	x
z	x	y,

where we suppose that both Groups A and B have exactly q members. You can easily verify that in this case, we will have:

$$C^v(\{x, y\}) = \{x\}, C^v(\{x, z\}) = \{z\}, \text{ and } C^v(\{y, z\}) = \{y\}.$$

Thus the set $B = \{x, y, z\}$ does not contain a Condorcet winner. $\square$

The above example shows that we cannot simply define a voting rule by taking $f(\boldsymbol{P})$ to be the Condorcet winner in X; the problem being that a set X may not contain a Condorcet winner, if $\#X \geq 3$.[6] One way of overcoming this difficulty, while retaining the spirit of majority voting is to use a **staging procedure**, defined as follows. Let $\succ$ be an arbitrary strict linear order on X, which we will call the **agenda ordering**. We begin by using $\succ$ to label the alternatives in X according to $\succ$; that is, we write $X = \{x_1, \ldots, x_n\}$, where:

$$x_1 \succ x_2 \succ \cdots \succ x_j \succ x_{j+1} \succ \cdots \succ x_n.$$

We then define a voting rule as follows. We first compare x_1 with x_2. The simple majority vote winner is then compared with x_3, and so on. If there is a tie at any stage, we take the element with the smaller index to use in our next pairwise vote, or as the singleton element in $f(\boldsymbol{P})$, if the voting has progressed to a choice between x_n and the winner of the preceding stage..

It is fairly easily shown that this procedure does result in a well-defined voting rule, for each $\boldsymbol{Q} \in \mathcal{Q}^m$. On the other hand, the staging procedure has some rather severe defects. Consider the following example.

[6] Another problem, of course, is that even if a Condorcet winner does exist, it may not be unique. Consequently, one also needs to have a tie-breaking rule to obtain a voting rule from this procedure. However, this is probably a much less serious defect than the fact that there may not be a Condorcet winner at all!

14.8 Example. Suppose $X = \{x_1, x_2, x_3, x_4\}$, that $m = 3$, and consider $f(\boldsymbol{P})$ when the three agents have the preference profile set out as follows.

Agent 1	**Agent 2**	**Agent 3**
x_1, x_2	x_3	x_2
x_4	x_4	x_4
x_3	x_1	x_3
	x_2	x_1

Here the staging procedure chooses x_4. But x_2 is the unique Condorcet winner on B! □

To illustrate a second defect with the staging procedure, consider the following example.

14.9 Example. Once again we suppose $X = \{x_1, x_2, x_3, x_4\}$, and that $m = 3$; this time considering $f(\boldsymbol{P})$ when the three agents have the preference profile set out as follows.

Agent 1	**Agent 2**	**Agent 3**
x_2	x_3	x_1
x_1	x_2	x_4
x_4	x_1	x_3
x_3	x_4	x_2

Here the staging procedure yields $C(B) = \{x_4\}$, as you can readily verify. However, all three agents strictly prefer x_1 to x_4! □

The staging procedure is also very sensitive to the choice of the agenda (the strict linear ordering, $\succ$, which determines the order of the pairwise votes). To see this, if you return to Example 14.7, you can readily verify that if the agenda ordering coincides with agent one's ordering of the alternatives, then the staging procedure will choose z. On the other hand, if the agenda ordering coincides with Group A's ordering of alternatives, then the staging procedure selects x; while if Group B's order constitutes the agenda, then y wins.

A much more complete discussion of the defects of the staging procedure, as well as a development of several other attempts to consistently extend simple majority voting to a voting rule on sets containing 3 or more elements is provided in Kelly [1988], Chapters 2 and 5, 15–22 and pp 50 –6. Let's now consider a different sort of variant of majority voting; namely, **plurality voting**. Plurality voting defines a voting rule on $\mathcal{P}^m$ as follows. For $\boldsymbol{P} \in \mathcal{P}^m$, $F(\boldsymbol{P})$ consists of that element (or those elements) which is (or which are) the most preferred choice in X for the largest number of voters.[7] The following example should make the idea of plurality voting clearer, as well as illustrating a very serious problem with the procedure.

14.10 Example. Let $X = \{x, y, z\}$, and suppose $m = 15$, with the agents' orderings over X as follows:

Group 1, 6 agents	**Group 2, 5 agents**	**Group 3, 4 agents**
x	y	z
z	z	y
y	x	x

[7] Once again a tie-breaking procedure is needed to obtain a well-defined voting rule.

In this case, plurality voting yields $F(\boldsymbol{P}) = \{x\}$, but x is a Condorcet loser in X! That is, a majority prefers y to x, and a majority prefers z to x. In fact, a majority of the agents consder x to be the *worst* of the three alternatives! Moreover, there is a Condorcet winner in this case, namely z, which is not chosen by the plurality rule. □

14.11 Example. The Borda Count. You are undoubtedly already familiar with the essential idea of the Borda count, because it is very often used in situations in which one is trying to obtain some sort of aggregate ranking of some alternatives. If, for example, we have four distinct alternatives, w, x, y, and z; each agent assigns a weight of 4 to her or his first-choice alternative, 3 to the second, 2 to the third, and 1 to the last-place alternative. The weight assigned to each alternative is then added over individuals to obtain social weights, with the alternative receiving the largest total being the socially most-preferred alternative, the alternative with the second largest total being society's second-ranked alternative, and so on.[8] Thus suppose we have seven individuals, whose ranking of the alternatives is as in the following table.[9]

Agent 1	Agent 2	Agent 3	Agent 4	Agent 5	Agent 6	Agent 7
w	x	y	w	x	y	w
x	y	z	x	y	z	x
y	z	w	y	z	w	y
z	w	x	z	w	x	z

If we follow the formulas just set out, we get the social (aggregate) utility values of the rankings as follows:

$$W(w) = 18, W(x) = 19, W(y) = 20, \text{ and } W(z) = 13;$$

so that the social ranking is:

$$y \succ x \succ w \succ z.$$

However, suppose z is eliminated from consideration,[10] and the voting is over just the first three alternatives, with the first-place alternative receiving an individual weight 3, and individual's second-place alternative receiving a weight of 2, and so on. If you carry out the calculations, you will find that the social utility values are now given by:

$$W(w) = 15, W(x) = 14, \text{ and } W(y) = 13.$$

[8]It should be noted that the Borda count is often developed by assigning weights in a manner exactly opposite to that used here; that is, an agent's first choice is assigned a weight of one, second choice a weight of two, and so on. The social ordering is then given by: xPy iff x has received a *lower* total than has y. It should also be noted that the social preference relation over the set of alternatives defined by the Borda count is not necessarily antisymmetric; it may be that two alternatives are tied for the first-place ranking, and so on.

[9]This example was first developed and reported by Fishburn [1974].

[10]There are a number of reasons why it might be eliminated. If the alternatives were political candidates, it might be that z recognizes before a formal vote that he or she is likely to receive the fewest votes. If the alternatives are public projects, z might be eliminated from consideration because everyone prefers y to it, and so on.

Thus the social ordering is:

$$w \succ x \succ y;$$

which is exactly the reverse of the original social ranking! □

Since it would seem desirable to choose a Condorcet winner, if one is available, the following would seem to be a very reasonable axiom to require of a 'good' voting rule.

14.12 Definition. A **Condorcet-consistent voting rule** is a voting rule which always chooses a Condorcet winner, if one exits.

Example 14.10 shows that the plurality voting rule is not Condorcet-consistent; and it can also be shown that the Borda count method fails this test as well. However, both plurality voting and the Borda count are examples of **positional voting**, the basic idea of which is to assign weights w_n to each first place vote, w_{n-1} to each second place vote, and so on down to w_1 for the last-place alternative; where $w_1, \ldots, w_n$ satisfy:[11]

$$0 \leq w_1 \leq w_2 \leq \cdots \leq w_n \text{ and } w_n > w_1,$$

with the choice over the set being the alternative with the highest point total when we add these weights over individual agents. Thus, for a three-element set, X, plurality voting uses a weight of 1 for first-place votes, and 0 for second or third-place votes. The Borda count uses a weight of 3 for first-place votes, 2 for second-place votes, and 1 for third-place votes.[12] Now, even though neither plurality voting nor the Borda count defines a Condorcet-consistent voting rule, one might hope that some other positional voting rule might be Condorcet-consistent. Unfortunately, we are once again doomed to disappointment. Fishburn [1973, Chapter 17] has proved the following proposition.[13]

14.13 Proposition (Fishburn). *There are profiles for which the Condorcet winner is never elected by any positional voting rule.*

Proof. In order to prove this, we only need to present a profile satisfying the indicated property. Consider the following profile, in which we have seven agents and three alternatives, $X = \{x, y, z\}$:[14]

3 agents	**2 agents**	**1 agent**	**1 agent**
z	x	x	y
x	y	z	z
y	z	y	x

[11] An apparently more general definition would not require that $w_1 \geq 0$. However, since there are only a finite number of alternatives in X, we can assume nonnegativity without loss of generality.

[12] Or equvalently, 2 for first-place votes, 1 for second-place votes, and 0 for third-place votes. See Exercise 3, at the end of this chapter.

[13] The treatment here follows Moulin [1988, pp. 231–2].

[14] This example was actually presented in Fishburn [1984].

In this case, alternative z is the Condorcet winner. However, if the weights assigned to the alternatives satisfy:

$$0 \leq w_1 < w_2 < w_3,$$

the totals for x and z will satisfy:

$$W(x) = 3w_3 + 3w_2 + w_1 > W(z) = 3w_3 + 2w_2 + 2w_1.$$

While this establishes our result for the case of strictly increasing weights, we can exhibit a profile in which the same thing happens with non-decreasing weights; although in this case we need 17 agents and 3 alternatives. The profile in question is:

6 agents	**3 agents**	**4 agents**	**4 agents**
x	z	y	y
y	x	x	z
z	y	z	x

In this case, x is the Condorcet winner. However, if the weights satisfy:

$$0 \leq w_1 \leq w_2 \leq w_3 \text{ and } w_3 > w_1 \geq 0,$$

then we have:

$$\begin{aligned} W(y) - W(x) &= 8w_3 + 6w_2 + 3w_1 - (6w_3 + 7w_2 + 4w_1) \\ &= 2w_3 - w_2 - w_1 \geq w_3 - w_1 > 0. \end{aligned}$$

□

While the result just proved shows that no positional voting rule is Condorcet-consistent, there do exist voting rules which are. The two most well-known are the 'Copeland rule' and the 'Simpson rule.' These are defined in the exercises at the end of this chapter, where you are asked to prove that the rules are indeed Condorcet-consistent.

It turns out that even though the Borda count method has the kind of defect illustrated in Example 14.11, Saari [1996] is able to make a very strong case for its being superior to other positional voting methods. The reason that the Borda count has some desirable properties which are generally not posessed by other positional voting methods in that the Borda count method posesses an internal consistency which the others generally (unless they are equivalent to the Borda count method) lack. The internal consistency which Saari has in mind relates to the fact that the Borda count (and other positional voting schemes as well) actually defines a social preference relation, and/or a social choice function; concepts which we will take up in the next section.

14.4 Arrow's General Possibility Theorem

In the previous section we saw that the Borda count, in fact, any positional voting rule, can be used to obtain a social preference ranking. Such a social preference ranking can, of course, be used to determine a social choice function, or correspondence. In principle, such a rule could be extremely useful in a situation where a society is to be faced with a succession of choices from a known alternative set. In this section we will consider social preference functions and social choice functions, beginning with social preference functions.

The concept of a social preference function was introduced into the economics (and political science) literature by Kenneth Arrow [1951].[15] The basic problem introduced by Arrow was that of somehow arriving at a social ordering of alternatives which took account, in a 'reasonable fashion,' of individual preferences over those states. Obviously this can be viewed as a problem of arriving at some sort of ordering as a function of the individual agents' orderings of the alternative set. In our discussion, we will consider this to be the problem of defining a 'reasonable' social preference function, or an Arrovian social preference function, defined as follows, where we use the notation introduced in Section 2.

14.14 Definition. If $\mathfrak{D}$ is a non-empty subset of $\mathcal{Q}^m$, we shall say that a function, $f\colon \mathfrak{D} \to \mathcal{Q}$ is a **social preference function**. In the special case in which $f\colon \mathcal{P}^m \to \mathcal{P}$ (that is, where $\mathfrak{D} = \mathcal{P}^m$, and f maps into $\mathcal{P}$ as well), we shall say that f is an **Arrovian social preference function**.

It is easy to define examples of such functions; the difficulty, as we will see, is to define such functions which also satisfy apparently 'reasonable' properties. In the meantime, consider the following; the first two of which are the simplest possible examples of social preference functions..

14.15 Examples.

1. Let $P^* \in \mathcal{Q}$ be fixed, and define $f^*\colon \mathcal{Q}^m \to \mathcal{Q}$ by:

$$f^*(\boldsymbol{Q}) = P^* \quad \text{for each } \boldsymbol{Q} = (Q_1, \ldots, Q_m) \in \mathcal{Q}^m;$$

in other words, f^* is a constant function. Arrow [1951] called this type of social preference function an *imposed* social preference function. In the extreme, this is the case in which society is ruled by convention, or a 'sacred code.'

2. Let $j \in \{1, \ldots, m\}$; and define $f_j\colon \mathcal{Q}^m \to \mathcal{Q}$ by:

$$f_j(Q_1, \ldots, Q_j, \ldots, Q_m) = Q_j;$$

in other words, let f_j be the jth projection function. Notice that if we take the domain of this function to be $\mathcal{P}^m$, rather than $\mathcal{Q}^m$, then this will be an Arrovian social preference function, as we have just defined the term. However, this is the undesirable (unless you are agent j) situation in which agent j is a dictator.

3. **The Unanimity ordering: A social preference function**. Given a preference profile, $\boldsymbol{Q} = (Q_1, \ldots, Q_m) \in \mathcal{Q}^m$, define $P = F(\boldsymbol{Q})$ on X by:

$$xPy \iff [xQ_iy \ \text{ for } i = 1, \ldots, m]. \tag{14.6}$$

[15] One should also mention, however, the contribution of Duncan Black [1958], which was almost contemporaneous with Arrow's; and the work of Julian Blau (for example, [1972]), which helped to clarify much of this area early on.

4. **The Pareto ordering: An (almost) Arrovian social preference function.** We define a function $f\colon \mathcal{P}^m \to \mathcal{Q}$ in the following way: given a profile $\boldsymbol{P} \in \mathcal{P}^m$, we define $Q = f(\boldsymbol{P})$ by:

$$xQy \iff [\neg yP_kx \text{ for } k = 1, \ldots, m \;\&\; (\exists i \in \{1, \ldots, m\})\colon xP_iy]. \tag{14.7}$$

Notice that this second variation is not quite an Arrovian social preference function, because, as you can easily verify, $Q = f(\boldsymbol{P})$ is not generally negatively transitive, although it is an asymmetric order.[16] □

Basically, Arrow's General Possibility Theorem states that no Arrovian social preference function can satisfy all of three very reasonable (and independent) conditions, if $\#X \geq 3$. In this section, we will state and discuss the three conditions, then state, discuss, and prove a version of Arrow's General Possibility Theorem. We will then go on to discuss the formal idea of a social choice function. .

In our statements of the properties introduced by Arrow, we will suppose only that f is a social preference function; so that $f\colon \mathcal{D} \to \mathcal{Q}$, where $\boldsymbol{\mathcal{D}}$ is simply taken to be a non-empty subset of $\mathcal{Q}^m$.

Property 1. The (Weak) Pareto Principle (WPP). For each profile, $\boldsymbol{Q} \in \mathcal{D}$, and each $x, y \in X, Q = f(\boldsymbol{Q})$ extends the unanimity ordering; that is:

$$[xQ_iy \text{ for } i = 1, \ldots, m] \Rightarrow xQy.$$

Property 2. Independence of Irrelevant Alternatives [IIA]. For each $\boldsymbol{Q}, \boldsymbol{Q}' \in \mathcal{D}$, and any $x, y \in X$, we have that if:

$$Q_{i\{x,y\}} = Q'_{i\{x,y\}} \quad \text{for } i = 1, \ldots, m,$$

then, writing $Q = f(\boldsymbol{Q})$ and $Q' = f(\boldsymbol{Q}')$, we must have:

$$Q_{\{x,y\}} = Q'_{\{x,y\}};$$

where, for a binary relation, G, and $\{x, y\} \subseteq X$, we denote the restriction of G to $\{x, y\}$ by '$G_{\{x,y\}}$.' In order to define our third property, we will need a definition.

14.16 Definition. If $f\colon \mathcal{D} \to \mathcal{Q}$, we shall say that $i \in \{1, \ldots, m\}$ is a **dictator for** $\boldsymbol{f}$ iff, for all $\boldsymbol{Q} \in \mathcal{D}$, and for all $x, y \in X$, we have, writing $P = f(\boldsymbol{Q})$:

$$xQ_iy \Rightarrow xPy.$$

Property 3. Absence of a dictator. No individual, $i \in \{1, \ldots, m\}$, is a dictator for f.

You should have no trouble in proving that our Example 14.15.1 satisfies Properties 2 and 3 of the above list; and that Example 14.15.2 satisfies Properties 1 and 2. As you have probably already noticed (but I will leave the proof as an exercise), the Borda Count function (Example 14.11) satisfies Properties 1 and 3, but not 2. These three examples collectively demonstrate that the three conditions are independent of one another, and that none (in fact, no pair) of the three conditions is self-contradictory. It is also an easy exercise to prove the following.

[16]Both of these properties were established in Section 5 of Chapter 5.

14.17 Proposition. *The Pareto ordering, Variation 1, as defined in Example 14.15.3, is a social preference function which satisfies all three of the above properties (and with $\mathbf{\mathcal{D}} = \mathcal{Q}^m$).*

The problem with the Pareto ordering, of course, is that for most alternatives, x and y, we will have neither xPy, nor yPx; in other words, most of the alternatives in X will be non-comparable. Because of this difficulty, investigators in the area spent a great deal of time trying to find (in effect) an Arrovian social preference function. Unfortunately, as we will demonstrate, if $\#X \geq 3$, there is no Arrovian social preference function which satisfies all of Properties 1–3. We will prove this by establishing several supporting results, which, I believe, are of some interest in their own right.

The basic strategy of our proof of Arrow's Theorem is adapted from Blau [1972] (see also Blau [1957], and Arrow [1963, 98–100]). In it we will often use the following notation. Suppose E is a non-empty subset of agents, that is suppose:

$$E \subseteq \{1, \ldots, m\} \text{ and } E \neq \emptyset. \tag{14.8}$$

and denote the complement of E by 'E^c,' that is:

$$E^c = \{1, \ldots, m\} \setminus E. \tag{14.9}$$

If $x, y \in X$, for example, and $\boldsymbol{Q} \in \mathcal{Q}^m$, we will write:

$$\begin{array}{cc} E & E^c \\ x & y \\ y & x, \end{array} \tag{14.10}$$

as shorthand for the statements:

$$(\forall i \in E)\colon xQ_iy \text{ and } (\forall j \in E^c)\colon yQ_jx.$$

Secondly, if we have a social preference function $f\colon \mathcal{D} \to \mathcal{Q}$, where $\mathcal{L}^m \subseteq \mathbf{\mathcal{D}}$ and $\boldsymbol{Q} \in \mathbf{\mathcal{D}}$, we will always denote $f(\boldsymbol{Q})$ by 'P' (even though, remember, we are dealing with social preference functions, and not Arrovian social preference functions, in our first several results). Given a social preference function, $f\colon \mathcal{D} \to \mathcal{Q}$, and a non-empty subset of agents, E, we define the relation D_E on X by:

$$xD_Ey \iff (\exists Q \in \mathbf{\mathcal{D}})\colon \boldsymbol{Q} \text{ satisfies (14.10) and } xPy. \tag{14.11}$$

(In the special case in which $E^c = \emptyset$, we will say that (14.10) holds if the left-hand column is true for E. Notice that in this case, xD_Ey iff there exists $\boldsymbol{Q} \in \mathcal{D}$ such that x is Pareto superior to y, given $\boldsymbol{Q}$. Notice also that, since E is non-empty, and, for each $\boldsymbol{Q} \in \mathcal{D}$ and each $i \in E$, Q_i is irreflexive, it follows that D_E is irreflexive. In both of the next two results, we will suppose that we are given a social preference function, f, satisfying Property 2 (IIA). Notice that, in this case, if we have xPy for one $\boldsymbol{Q} \in \mathcal{D}$ satisfying (14.10), then we will have $xP'y$ for $P' = f(\boldsymbol{Q}')$ and $\boldsymbol{Q}'$ *any* profile from $\mathcal{D}$ which satisfies (14.10). We can formalize this a bit for later reference, as follows. We begin with a definition.

14.18 Definition. Let f be a social preference function, E be a non-empty subset of agents, and x and y be a pair of distinct alternatives in X. If, for each $\boldsymbol{Q} \in \mathcal{D}$ satisfying (14.10), above, we have xPy, where $P = f(\boldsymbol{Q})$, then we shall say that **E is decisive for $\{x, y\}$ (given f)**.

The fact which we have just noted can then be stated as follows. The proof is immediate.

14.19 Proposition. *Suppose $f : \mathcal{D} \to \mathcal{Q}$ satisfies IIA, let E be a non-empty subset of agents, and x and y be a pair of distinct alternatives in X. Then xD_Ey if, and only if, E is decisive for $\{x, y\}$.*

We can now make use of this result, and these last definitions, to obtain a sharper characterization of social preference functions, as follows. (From here on, however, we will suppose that $\mathfrak{D}$ takes the form $\mathfrak{D} = \mathcal{D}^m$, for some $\mathcal{D}$ satisfying $\mathcal{L} \subseteq \mathcal{D} \subseteq \mathcal{P}$.)

14.20 Theorem. *Suppose $f : \mathcal{D}^m \to \mathcal{Q}$ satisfies WPP and IIA, where $\mathcal{D}$ contains $\mathcal{L}$, that $\#X \geq 3$, and let E be a non-empty subset of agents. Then either E is decisive for all pairs of distinct $x, y \in X$, or E is decisive for no such pair.*

In order to prove this result, we first prove the following lemma.

14.21 Lemma. *Suppose $f : \mathcal{D}^m \to \mathcal{Q}$ satisfies WPP and IIA, where $\mathcal{D}$ contains $\mathcal{L}$, and that $\#X \geq 3$, let E be a non-empty subset of agents, and let a, b, c and d be elements of X. Then if aD_Eb, and $c \neq a$ [respectively, $d \neq b$], then aD_Ec [respectively, dD_Eb].*

Proof. Suppose first that $c \neq a$. If $b = c$, then it follows at once that aD_Ec. On the other hand, if $b \neq c$, and using the assumption that $\mathcal{L}^m \subseteq \mathcal{D}$, consider a profile $\boldsymbol{Q} \in \mathcal{D}$ for which we have:

E	E^c
a	b
b	c
c	a.

Then, writing $P = f(\boldsymbol{Q})$, we have, applying aD_Eb and WPP in turn:

$$aPb \text{ and } bPc.$$

Therefore, since P is transitive, it follows that:

$$aPc;$$

and, since we have aQ_ic for all $i \in E$, and cQ_ja for all $j \in E^c$, it then follows that aD_Ec.

Suppose now that $d \neq b$. If $a = d$, it follows trivially that dD_Eb. Otherwise, consider a profile $\boldsymbol{Q}' \in \mathcal{D}$ in which

E	E^c
d	b
a	d
b	a;

and define $P' = f(\boldsymbol{Q}')$. Applying WPP and the assumption that aD_Eb in turn, we have:

$$dP'a \text{ and } aP'b.$$

Therefore, since P' is transitive, $dP'b$; and it then follows that dD_Eb. □

Proof of Theorem 14.20. Suppose E is decisive for some pair of elements, $a, b \in X$. Then we must have aD_Eb. Suppose now that $x, y \in X$ are such that $x \neq y$. We wish to prove that xD_Ey, but to do so we consider several cases.

1. Suppose $b \neq x$. Then from the lemma (14.21), we see that xD_Eb. But then, since $x \neq y$, it also follows from the lemma (letting x take the place of a and y the place of c) that xD_Ey.
2. Suppose $b = x$. Since we must here have aD_Ex, and $y \neq x$, it then follows from the lemma that yD_Ex.
3. Suppose $a = x$. Then, since $y \neq x$, it follows from the lemma that xD_Ey.
4. Suppose $a \neq x$ Then by the lemma, we must have aD_Ex, and then, since $x \neq y$, we have, by making use of the respective statement of the lemma, we have yD_Ex.

We have now shown that if E is decisive for one pair of distinct alternatives, then for any $x, y \in X$ such that $x \neq y$ we must have either xD_Ey or yD_Ex. It then follows from IIA that E is decisive for the pair $\{x, y\}$. Therefore we see that if E is decisive for some pair of distinct alternatives, it is decisive for any pair of distinct alternatives in X. □

14.22 Definition. Let $f: \mathcal{D} \to \mathcal{Q}$ be a social preference function. We shall say that f is **neutral (with respect to alternatives)** if, given any non-empty subset of agents, E, and any $x, y \in X$, we have that if E is decisive for $\{x, y\}$, then it is decisive for any pair of distinct points from X.

It is certainly not clear that this neutrality condition is a desirable property for a social preference function to satisfy; there may be some 'local' issues which are part of the overall choice set. X. but which should be decided by a subset of the agents who would not be decisive for other choices.[17] For example, a decision as to whether or not to build a city park should presumably be determined only by the preferences of those living in the area, and be independent of agents' preferences who live well outside the area. However, we have shown in Theorem 14.20 that if f satisfies the hypotheses set out there, then f is neutral with respect to alternatives. We can state this formally as follows (the result follows immediately from Theorem 14.20).

14.23 Corollary. *If $\#X \geq 3$, if $f: \mathcal{D}^m \to \mathcal{Q}$ is a social preference function satisfying WPP and IIA, and if $\mathcal{L} \subseteq \mathcal{D}$, then f is neutral.*

Now, define the set M by:

$$M = \{1, \ldots, m\};$$

[17] On this issue, see Sen [1970] and Sen [1986, pp. 1155-6].

that is, we will think of the set M as being the set of all agents; and suppose that a social preference function, f, and alternative set, X, satisfy the hypotheses of Theorem 14.20. By that result we have that each non-empty set of agents, E, is either decisive for all distinct pairs of alternatives, x and y, or it is decisive for no such pair. [It is also worth noting, incidentally, that M is necessarily decisive for all distinct pairs, by the Weak Pareto Principle.] Consequently, we can partition the collection of all subsets of M into two subcollections: the set $\mathcal{W}$ (for 'winners'), consisting of all subsets of M which are decisive for all distinct pairs, and the set $\mathcal{N}$ (for 'non-winners') consisting of all subsets of M which are decisive for no distinct pair of alternatives. We can then establish the following result.

14.24 Proposition. *Suppose that $\#X \geq 3$, that $f\colon \mathcal{D}^m \to \mathcal{P}$ is a social preference function satisfying IIA and WPP, and that $\mathcal{L} \subseteq \mathcal{D}$. Then if $\mathcal{N}$ is defined as above, the union of any finite collection of pairwise disjoint sets from $\mathcal{N}$ is again an element of $\mathcal{N}$.*

Proof. We will prove this for the case of two sets, $E, F \in \mathcal{N}$; the general case follows by an easy induction argument. Denote $\{1, \ldots, m\}\backslash(E \cup F)$ by 'H,' let x and y be distinct alternatives from X; and suppose, by way of obtaining a contradiction, that $E \cup F$ is decisive for $\{x, y\}$. Accordingly, let (using the assumption that $\#X \geq 3$) z be an alternative from X which is distinct from both x and y, and let $\boldsymbol{Q} \in \mathcal{D}$ be such that:

$$\begin{array}{ccc} E & F & H \\ x & z & y \\ y & x & z \\ z & y & x. \end{array} \tag{14.12}$$

Then, letting $P = f(\boldsymbol{Q})$, it follows from our assumption that $E \cup F$ is decisive for $\{x, y\}$, that

$$xPy. \tag{14.13}$$

However, since z is distinct from both x and y, it follows from (14.13) and the negative transitivity of P that either

$$xPz, \tag{14.14}$$

or

$$zPy. \tag{14.15}$$

But neither (14.14) nor (14.15) can hold! For example, if (14.14) holds, then notice that it follows from (14.12) that:

$$xD_Ez;$$

and it then follows from Proposition 14.20 that E is decisive for $\{x, z\}$, contradicting the assumption that $E \in \mathcal{N}$. Similarly, if (14.15) holds, it would follow that F is decisive for $\{z, y\}$; contradicting the assumption that $F \in \mathcal{N}$. Since either possibility [(14.14) or (14.15)] involves us in a contradiction, it follows that (14.13) cannot hold, given (14.12); and thus that $E \cup F$ cannot be decisive for $\{x, y\}$. $\square$

In our final two results, we will suppose that $\mathcal{D}$ satisfies the following property.

14.25 Definition. We shall say that a set $\mathcal{D} \subseteq \mathcal{Q}^m$ satisfies the **Arrow condition** iff:

1. $\mathcal{L} \subseteq \mathcal{D}$, and
2. given any $Q \in \mathcal{D}$, and any $\{x, y\} \subseteq X$, there exists $Q' \in \mathcal{D}$ and $z \in X$ such that:

$$Q_{\{x,y\}} = Q'_{\{x,y\}}, zQ'x, \text{ and } zQ'y.$$

While the condition just stated is admittedly a bit strange-looking, notice that the family of all linear orders on X satisfies the Arrow Condition. That is, if we set $\mathcal{D} = \mathcal{L}$, then $\mathcal{D}$ satisfies the Arrow Condition. At rather another extreme, if we set $\mathcal{D} = \mathcal{P}$, or if $\mathcal{D} = \mathcal{Q}$, then the Arrow Condition is satisfied. To give an example of a case in which the condition is not satisfied, suppose $X = \{x, y, z\}$, where all three points are distinct; and that $\mathcal{D}$ contains all 6 linear orderings of X, together with the relation in which all three elements are indifferent to one another. Then $\mathcal{D}$ does not satisfy the Arrow Condition, since it does not satisfy condition 2 of the definition.

14.26 Proposition. *Suppose that $\#X \geq 3$, that $f : \mathcal{D}^m \to \mathcal{Q}$ is a social preference function satisfying IIA and WPP, and that $\mathcal{D}$ satisfies the Arrow condition. Then if $E \in \mathcal{W}$; $x, y \in X$, and $\boldsymbol{Q} \in \mathcal{D}$ are such that:*

$$(\forall i \in E) \colon xQ_iy, \tag{14.16}$$

then, defining $P = f(\boldsymbol{Q})$, we have xPy.

Proof. Let $\boldsymbol{Q} \in \mathcal{D}$ be a profile satisfying (14.16). By the fact that $\mathcal{D}$ satisfies the Arrow condition, there exists $\boldsymbol{Q}' \in \mathcal{D}$ and $z \in X$ such that:

$$Q_{i\{x,y\}} = Q'_{i\{x,y\}} \quad \text{for } i = 1, \ldots, m; \tag{14.17}$$

while (using part 1 of the Arrow condition):

$$(\forall i \in E) \colon xQ'_iz \text{ and } zQ'_iy,$$

and (using part 2 of the Arrow condition):

$$(\forall j \in E^c) \colon zQ'_jx \text{ and } zQ'_jy.$$

We can indicate this in our shorthand notation as:

$$\begin{array}{cc} E & E^c \\ x & z \\ z & \\ y & \{x, y\}. \end{array}$$

Then writing $P' = f(\boldsymbol{Q}')$, and using the fact that $E \in \mathcal{W}$, and WPP in turn,we see that:

$$xP'z \text{ and } zP'y.$$

Therefore, since P' is transitive:

$$xP'y. \tag{14.18}$$

Letting $P = f(\boldsymbol{Q})$, and using the fact that f satisfies IIA, it then follows from (14.17) that xPy as well. □

Notice the distinction involved in the conclusion of Proposition 14.26. If $E \in \mathcal{W}$, and x and y are two distinct alternatives from X, then E is decisive for $\{x, y\}$; meaning that whenever every agent in E prefers x to y, and every agent in E^c prefers y to x, then xPy in terms of the social preference, P. In Proposition 14.26, however, we have extended this idea to show that it must be true that whenever all the agents in E prefer x to y, then society will prefer x to y whatever the preferences of the agents in E^c. This is only 'common sense,' but in saying this what we mean is that it seems 'right,' in a normative sense, that if society should prefer x to y whenever the agents in E feel this way and the agents in E^c have the opposite preferences, then social preferences should also be this way when the agents in E have the same ranking for x vis-a-vis y, but the agents in E^c are not necessarily so unalterably opposed. Thus, this is a 'reasonable' property to require of a social preference function; but, since we have not directly assumed that f will satisfy it, we had to prove that the other conditions we required f to satisfy do imply this property.

We can now prove the following, which is a slight generalization of Arrow's classic 'General Possibility Theorem.'

14.27 Theorem. (**Arrow**) If $\#X \geq 3$, and $\mathcal{D}$ satisfies the Arrow condition, then any social preference function $f\colon \mathcal{D}^m \to \mathcal{P}$ which satisfies IIA and WPP must be dictatorial.

Proof. Suppose, by way of obtaining a contradiction, that f satisfies IIA, WPP, and, in addition, that no individual is a dictator for f. Then, using the notation of Propositions 14.24 and 14.26, we see from Proposition 14.26 and the definition of $\mathcal{N}$ that:

$$i \in \mathcal{N}, \quad \text{for } i = 1, \ldots, m;$$

and by Proposition 14.24 it then follows that

$$\bigcup_{i=1}^{m} \{i\} = M \equiv \{1, \ldots, m\} \in \mathcal{N}.$$

But this is impossible, since by WPP we must have $M \in \mathcal{W}$. □

We then obtain as an immediate corollary, the following result; which is the original statement of the 'General Possibility Theorem.'

14.28 Theorem. Arrow's 'General Possibility Theorem.' If $\#X \geq 3$, then any social preference function $f\colon \mathcal{P}^m \to \mathcal{P}$ which satisfies IIA and WPP must be dictatorial.

Social preference functions and social choice functions are often lumped together into one category, but in principle there are significant differences between the two. We can illustrate the differences and similarities between the two ideas by making use of much of the notation and some of the definitions developed in Chapter 3.

Let $\mathcal{B}$ be a family of non-empty subsets of X (to be held fixed in this discussion), and let '$\mathcal{C}$' denote the family of all choice correspondences, C on $\langle X, \mathcal{B}\rangle$. That is, $\mathcal{C}$ is the family of all correspondences $C\colon \mathcal{B} \mapsto X$ satisfying, for all $B \in \mathcal{B}$:

$$C(B) \neq \emptyset \text{ and } C(B) \subseteq B. \tag{14.19}$$

14.29 Definitions. If $\mathcal{D}$ is a non-empty subset of $\mathcal{Q}^m$, we shall say that a function $g\colon \mathcal{D} \to \mathcal{C}$ is a **social choice rule**. Given such a function, and a profile $\boldsymbol{Q} \in \mathcal{D}$, we will refer to the correspondence, $C = g(\boldsymbol{Q})$ as the **social choice correspondence determined by $(g, \boldsymbol{Q})$**.

While the above definition is a bit ambiguous on this score, it would appear that there would be no good reason for trying to develop a theory of social choice functions for any context other than that in which $\mathcal{B}$ consists of all non-empty subsets of X. Of course, we can weaken this stipulation by requiring only that $\mathcal{B}$ consist of all subsets of X containing two or more distinct elements of X; however different two social choice correspondences may be on non-singleton sets, their values must coincide on all singleton sets.

From our work in Chapter 3, we know that if $f\colon \mathcal{D} \to \boldsymbol{Q}$ is a social preference function, then it will always determine a social choice rule in the following way.

14.30 Definition. Let $\mathcal{D}$ be a non-empty subset of $\mathcal{Q}^m$, and $f\colon \mathcal{D} \to \mathcal{Q}$ be a social preference function. We define the social choice rule corresponding to f, C^f on $\mathcal{B}$ by:

$$C^f(B; \boldsymbol{Q}) = \{x \in B \mid (\forall y \in X)\colon yQx \Rightarrow y \notin B\},$$

where $Q = f(\boldsymbol{Q})$.

On the other hand, we also know from our work in Chapter 3 that there must be social choice rules which determine social choice correspondences which cannot be derived from any social preference function. However, one would like the social choice correspondence to display some consistency.[18] For example, it would certainly appear to be reasonable to require that, if an alternative, x, is chosen when a second alternative, y, is also available, then whenever we have $y \in C(B)$ and $x \in B$ as well, then we should also have $x \in C(B)$. But, in terms of the concepts introduced in Chapter 3, what we are saying here is that the social choice correspondence, C, should satisfy Richter's V-Axiom. Furthermore, it then follows at once from Theorem 3.11 that if C satisfies this condition, then it can be derived from a social preference function. In fact, a very complete theory of social choice rules can be built up by straightforward applications of the results from Chapter 3. For example, if $g\colon \mathcal{P}^m \to \mathcal{C}$ is derived from an Arrovian social preference function, then, for each $\boldsymbol{P} \in \mathcal{P}^m$, $C(\cdot)$ must satisfy Richter's Congruence Axiom. One can make use of the results of Chapter 3 to derive a a number of additional results involving social choice rules, but I will leave this as a 'project for the interested reader.'

[18] In this connection, see Plott [1973].

14.5 Appendix. A More Sophisticated Borda Count

In order to more generally define the Borda count relation, we begin by recalling the following notation. For $Q_i \in \mathcal{Q}$, we define $Q_i \colon X \mapsto X$ by:

$$Q_i(x) = \{y \in X \mid yQ_ix\} \quad \text{for } x \in X.$$

We then define the function $u_i \colon X \to \mathbb{R}_+$ by:

$$u_i(x) = N - \#Q_i(x) \quad \text{for } x \in X. \tag{14.20}$$

where N is the total number of distinct elements in X; that is:

$$N = \#X.$$

Next, we define the function $n \colon X \to \mathbb{R}$ by:

$$n(x) = \#\{y \in X \mid u_i(y) = u_i(x)\}, \tag{14.21}$$

and we then define $\bar{u}_i \colon X \to \mathbb{R}_{++}$ by:

$$\bar{u}_i(x) = u_i(x) - \frac{n(x) - 1}{2}. \tag{14.22}$$

In effect, the functions $\bar{u}_i$ are individual utility functions quasi-representing (I will explain this term shortly) the asymmetric orders Q_i. We use these individual 'utility functions' to define a social welfare function, $W \colon X \to \mathbb{R}_+$ by:

$$W(x) = \sum\nolimits_{i=1}^{m} \bar{u}_i(x). \tag{14.23}$$

Finally, we define $\succ_B = \boldsymbol{\beta}(\boldsymbol{Q})$ by:

$$x \succ_B y \iff W(x) > W(y). \tag{14.24}$$

Interestingly enough, we have actually defined a social preference function here whose domain is $\mathcal{Q}^m$ and whose range is $\mathcal{P}$; that is,

$$\boldsymbol{\beta} \colon \mathcal{Q}^m \to \mathcal{P}.$$

Of greater interest for the matter at hand, however, is the fact that the restriction of $\boldsymbol{\beta}$ to $\mathcal{P}^m$ is an Arrovian social preference function. Unfortunately, as we will see shortly, it does not satisfy all of the remaining conditions considered by Arrow. Moreover this function is particularly sensitive to certain kinds of strategies which might be employed by the individual agents, as we saw in Section 6.

It is worth noting a couple of things about the development which we used here. First of all, notice that the function u_i defined in the above material is a well-defined function for any $Q_i \in \mathcal{Q}$; *whether or not Q_i satisfies negative transitivity.* In fact, it is easy to show that u_i satisfies the following condition: for any $x, y \in X$:

$$xQ_iy \Rightarrow u_i(x) > u_i(y). \tag{14.25}$$

This is what I meant earlier by the comment that the function u_i 'quasi-represents Q_i.' In fact, if Q_i is negatively transitive, then it can be shown that u_i actually represents Q_i. Of course, if Q_i does not satisfy negative transitivity, then u_i cannot represent Q_i; and thus it cannot satisfy

$$u_i(x) > u_i(y) \Rightarrow xQ_iy, \tag{14.26}$$

as well as (14.25) [why is this?]

It is also worthwhile to take a moment to consider why it is that one might wish to take the extra step to go from the nice simply-defined functions u_i to use the more complicated functions in defining the social welfare function $W(\cdot)$, and thus the social preference relation, $\succ_B$. The reason amounts to this: each individual's preferences are used to determine a set of weights, $W(x)$, to be assigned to the elements $x \in X$. In effect, individual i gets to vote for the desirability of alternative x, and can add the amount $\bar{u}_i(x)$ to the social evaluation of x. The extra step (using $\bar{u}_i$ in place of u_i) ensures that each agent gets to cast the same total number of votes. We can show this, for the case in which each Q_i satisfies negative transitivity (and thus is a weak order), as follows. Let:

$$u_i(x) \equiv u^\dagger, n(x) = p \geq 1,$$

and consider the utility assigned to the next best alternative which is not indifferent to x.[19] If y is such an alternative, then clearly the p alternatives tied with x in the ranking are preferred to y, but not to x, while every other alternative preferred to x is also preferred to y. Thus:

$$u_i(y) \equiv u^* = N - [p + \#\{z \in X \mid zQ_ix\}] = [N - \#\{z \in X \mid zQ_ix\}] - p = u^\dagger - p. \tag{14.27}$$

If the alternatives tied with x in agent i's ranking were ordered linearly, then one of these alternatives would have the utility (under the u_i function) of $u^* + 1$, the next $u^* + 2$, and so on up to $u^* + p$. Remembering the formula for the sum of the first n positive integers,[20] we see that the total utility weights assigned to the elements tied with x (that is, in the equivalence class for $x, [x]$) is given by:

$$u^* + 1 + u^* + 2 + \cdots + u^* + p = pu^* + p(p+1)/2. \tag{14.28}$$

By using the function $\bar{u}_i$ defined in (14.22), above, we give each of these elements the weight $u^\dagger - (p-1)/2$; and thus the total of the weights assigned to these elements is

$$p\Big[u^\dagger - \frac{(p-1)}{2}\Big] = pu^\dagger - \frac{p^2 - p}{2}. \tag{14.29}$$

Substituting from (14.27) into (14.29), we see that this sum is:

$$pu^\dagger - \frac{p^2 - p}{2} = p(u^* + p) - \frac{p^2 - p}{2} = pu^* + \frac{p(p+1)}{2},$$

[19] Technically, the utility assigned the next equivalence class down from $[x]$.

[20] Which is equal to $n(n+1)/2$.

which agrees with (14.28). Thus what the use of the function accomplishes, is to give everyone the same total weights to be allocated over the alternatives.[21]

Exercises.

1. Suppose $\#X = 2$, that the number of agents, m is odd ($m = 2q + 1$, for some integer $q \geq 1$) and that $\mathcal{D} = \mathcal{L}^m$. Show that in this case the simple majority voting and absolute majority voting rules produce identical results.

2. Suppose $\#X = 2$, and that the domain over which we define a voting rule, $\mathcal{D}$, is equal to $\mathcal{P}^m$. Show that the simple majority voting rule satisfies May's positive responsiveness condition, but that the absolute majority voting rule does not.

3. Prove that the following Borda count methods are equivalent, where $\#X = n$, and we have $m \geq 2$ agents, each of whom has a preference relation on X which is a linear order.

a. Each agent assigns a weight n to her/his most preferred alternative, $n - 1$ to her/his second-place alternative, and so on.

b. Each agent assigns a weight of $n - 1$ to her/his most preferred alternative, $n - 2$ to her/his second-pland alternative, and so on.

4. Suppose $X = \{w, x, y, z\}$, that $m = 3$, and consider the set $B = X$ when the three agents have the preference profiles set out as follows.

Agent 1	**Agent 2**	**Agent 3**
w	x	x
x	y	y
y	z	w
z	w	z

(a) Is there a Condorcet winner in this case? If there is, what is it? (b) What is the plurality winner? (c) Find the Borda count ranking of the four alternatives.

5. Show that if $\#X = 2$, and m is an odd number greater than or equal to 3, then there exists an Arrovian social preference function satisfying IIA and WPP which is non-dictatorial.

6. Suppose Y is a non-empty subset of X, and that Q is an asymmetric order (respectively a weak order) on Y. Show that there exists an asymmetric order on X (respectively, a weak order on X), Q^*, such that Q is the restriction of Q^* to Y. [**Hint:** See Exercise 7, at the end of Chapter 1.]

7. In this exercise, we return to the notation utilized in our discussion of May's

[21]Technically, what we have shown (or what you can show with very little additional work), is that the $\bar{u}_i$ function assigns the same total weights as if the alternatives were ordered linearly. Consequently, it follows that the total is the same for any weak order.

theorem; that is, for $\{x, y\} \subseteq X$, we define $d_i(x, y)$ by:

$$d_i(x,y) = \begin{cases} 1 & \text{if } xP_iy, \\ 0 & \text{if } xI_iy, \text{ and} \\ -1 & \text{if } y_iPx; \end{cases}$$

and we consider the domain $\mathcal{P}^m$, that is the set of all profiles of weak orders on X.

Given a preference profile, $\boldsymbol{P}$, and for a given $x \in X$, we define, for each $y \in X \setminus x$, $f(y; x)$ by:

$$f(y;x) = \begin{cases} 1 & \text{if } \sum_{i=1}^m d_i(x,y) \geq 0, \\ 0 & \text{otherwise.} \end{cases}$$

We then define the **Copeland score**, $C(x)$, of an alternative $x \in X$ by:

$$C(x) = \sum_{y \in X \setminus \{x\}} f(y;x).$$

The Copeland winner is then the alternative which has the highest Copeland score (although a tie-breaking procedure is needed here). Show that this voting method is Condorcet-consistent, and satisfies May's anonymity and neutrality conditions.

8. Given the context of the previous exercise, and for a given preference profile, $\boldsymbol{P}$, we define, for each $x \in X$, and $y \in X \setminus x$, the number $N(x, y)$ by:

$$N(x,y) = \#\{i \mid xG_iy\},$$

where G_i is the negation of P_i. We then define the social utility of x, $U(x)$, by:

$$U(x) = \min_{y \in X \setminus \{x\}} N(x,y).$$

The alternative chosen is then one which has the highest 'social utility' (once again, however, we must have a tie-breaking procedure).

Show that this voting method (the **Simpson Rule**) is Condorcet-consistent, and satisfies May's anonymity and neutrality conditions.[22]

9. Returning to the context of Exercise 7, suppose we define a voting rule in the following way: for a given profile, $\boldsymbol{P} \in \mathcal{P}^m$, define, for each $x \in X$:

$$D(x) = \sum_{y \in X \setminus \{y\}} \sum_{i=1}^{m} d_i(x,y);$$

and let $F(\boldsymbol{P})$ be the alternative having the highest value for $D(x)$ (with an appropriate tie-breaking rule). Is this voting rule Condorcet-consisten?

10. Show that the Borda count *does not* satisfy IIA. Notice that an appropriate example to show this must involve a fixed choice set.

[22] Both the Copeland and the Simpson rules satisfy a 'Pareto optimality' rule as well. For an excellent, and much more thorough discussion, see Moulin [1988, pp. 233–40].

Chapter 15

Some Tools of Applied Welfare Analysis

15.1 Introduction

In this chapter, we will examine a number of tools which are, or can be used in applied welfare economics. We will begin by continuing our consideration of social preference functions with an investigation of the so-called 'Bergson-Samuelson Social Welfare Function.' This function was introduced into the English-language economics literature by Abram Bergson [1938], while Samuelson [1947] emphasized its importance and illustrated new usages for such a function. Interestingly enough, Pareto had introduced the idea earlier, although he did not analyze and develop the idea as extensively as did Bergson (who developed the idea independently in any case); consequently, we will give credit to all three economists. We define a 'Pareto-Bergson-Samuelson (PBS) Social Welfare Function' as a real-valued function whose domain is the space of allocations in an economy and which can be written as a composition, $W = F \circ \boldsymbol{u}$, where $\boldsymbol{u}$ is a vector of individual utility functions, with u_i defined over individual i's commodity bundle $(i = 1, \ldots, m)$, and F is an increasing real-valued function on $\mathbb{R}^m$ (which we will call the *aggregator function*); that is, a function whose domain is utility space.

Perhaps no tool of theoretical normative economics has been used as much or criticized as severely as has the 'Pareto-Bergson-Samuelson Social Welfare Function.' On the one hand, such functions are commonly-used in applied normative analysis, and are frequently used in theoretical policy analyses as well. On the other hand, we have been trained to think of preferences as being only (at most) ordinally measurable, whereas the concept seems to require that individual utilities be not only 'cardinally measurable,' but 'interpersonally comparable.' The obvious question then arises, is there a way out of this dilemma? We will explore this issue in Sections 2–4.

In Section 5 we continue our exploration of tools of applied welfare analysis by examining the 'compensation principle:' the principal tool of what was once known as 'The New Welfare Economics.' In Section 6, we extend the ideas of Sections 2–4 to define indirect social preferences and indirect social welfare functions. We then make use of these notions to examine some ideas about the measurement of 'real national income' in Section 7. We then continue our exploration of the applications of indirect social preferences and welfare functions by considering consumers' surplus in Section 8.

15.2 The Framework

In the remainder of this chapter, we will, as usual, suppose that there are m consumers and n commodities in the economy. We will also assume throughout (except where otherwise explictly stated) that each consumer has a (weak) preference relation G_i, which will always be assumed to be a continuous, strictly convex, and increasing weak order; and, as usual we will use 'I_i' and 'P_i' to denote the symmetric and asymmetric parts of G_i, respectively. More formally, we will always be assuming that each G_i is an element of a family, $\mathcal{G}^c$, defined as follows.

15.1 Definitions. We denote by '$\mathcal{G}^c$' the family of all weak orders on $\mathbb{R}^n_+$ which are also:

a. continuous,
b. increasing,[1] and
c. strictly convex.[2]

We then let '$\boldsymbol{\mathcal{G}}^c$' denote the collection of all m-tuples of elements of $\mathcal{G}^c$; that is:

$$\boldsymbol{\mathcal{G}}^c = (\mathcal{G}^c)^m,$$

and we will refer to m-tuples, $\boldsymbol{G} = (G_1, \ldots, G_m) \in \boldsymbol{\mathcal{G}}^c$ as **preference profiles**.

While we will always assume that each consumer's preference relation is an element of $\mathcal{G}^c$, we will often assume that each relation is homothetic as well.

15.2 Definition. We denote by '$\mathcal{G}^h$' the subset of $\mathcal{G}^c$ consisting of all elements of $\mathcal{G}^c$ which are also homothetic; that is, $\mathcal{G}^h$ is the family of all weak orders on $\mathbb{R}^n_+$ which are continuous, increasing, strictly convex, and homothetic. We use '$\boldsymbol{\mathcal{G}}^h$' to denote the collection of all m-tuples of elements of $\mathcal{G}^h$.

Since each consumer's consumption set is equal to $\mathbb{R}^n_+$, an allocation $\langle \boldsymbol{x}_i \rangle$ will be a finite sequence (of m terms), with:

$$\boldsymbol{x}_i \in \mathbb{R}^n_+ \quad \text{for } i = 1, \ldots, m.$$

Thus we could consider allocations to be elements of $\mathbb{R}^{mn}_+$. However, to avoid possible confusion, we will denote the allocation space by '$\mathcal{X}$;' that is:

$$\mathcal{X} = \big\{ \langle \boldsymbol{x}_i \rangle \mid \boldsymbol{x}_i \in \mathbb{R}^n_+, \text{ for } i = 1, \ldots, m \big\}.$$

[1]That is, if $\boldsymbol{x}, \boldsymbol{x}' \in \mathbb{R}^n_+$ are such that $\boldsymbol{x} \gg \boldsymbol{x}'$, then $\boldsymbol{x} P \boldsymbol{x}'$.

[2]That is: if $\boldsymbol{x}, \boldsymbol{x}^* \in \mathbb{R}^n_+$ are such that $\boldsymbol{x} G_i \boldsymbol{x}^*$ and $\boldsymbol{x} \neq \boldsymbol{x}^*$, and if $0 < \theta < 1$, then:

$$\big[\theta \boldsymbol{x} + (1-\theta)\boldsymbol{x}^*\big] P_i \boldsymbol{x}^*.$$

Strict convexity will not really be needed in the vast majority of our work in this chapter, but making use of it greatly simplifies many of our definitions and proofs.

In this chapter, we will use the generic notation 'G' to denote the weak Pareto ordering of allocations; that is, if $\boldsymbol{G} = (G_1, \ldots, G_m)$ is an element of $\mathcal{G}^c$, we define G on $\mathfrak{X}$ by:

$$\langle \boldsymbol{x}_i \rangle G \langle \boldsymbol{x}'_i \rangle \iff \boldsymbol{x}_i G_i \boldsymbol{x}'_i, \text{ for } i = 1, \ldots, m.$$

Where needed, we will use the generic notation 'P' to denote the asymmetric part of G (the strict Pareto ordering).

15.3 Measurement Functions

In this chapter, we will look at utility functions in a bit different way than is usual. Let $\mathcal{U}$ be defined by:

$$\mathcal{U} = \{f \mid f\colon \mathbb{R}^n_+ \to \mathbb{R}_+\}.$$

We then define the following:

15.3 Definition. We will say that a function $\mu\colon \mathcal{G}^c \to \mathcal{U}$ is a **utility measurement function** (for $\mathcal{G}^c$) iff for each $G \in \mathcal{G}^c$, $f = \mu(G)$ satisfies:

$$(\forall \boldsymbol{x}, \boldsymbol{x}' \in \mathbb{R}^n_+)\colon f(\boldsymbol{x}) \geq f(\boldsymbol{x}') \iff \boldsymbol{x} G \boldsymbol{x}'. \tag{15.1}$$

While the above definition may seem a bit strange to you, the next example may help to clear things up a bit. In fact, the method of utility measurement, or class of utility measurement functions of which we will make use, is based upon the classic representation theorem of Herman Wold (Theorem 4.21, of chapter 4), and is set out in the following example.

15.4 Example. We will define a function $\varphi\colon \mathcal{G}^c \times \mathbb{R}^n_{++} \to \mathcal{U}$: given $G \in \mathcal{G}^c$, and $\boldsymbol{x}^* \in \mathbb{R}^n_{++}$,[3] we define $u = \varphi(G, \boldsymbol{x}^*)$ as follows. For $\boldsymbol{x} \in \mathbb{R}^n_+$, there exists a unique value of θ satisfying:

$$\boldsymbol{x} \, I \, \theta \boldsymbol{x}^*, \tag{15.2}$$

where I is the indifference relation for G, and we let $u(\boldsymbol{x}) = \theta$. In other words, $u(\boldsymbol{x}) = \varphi(G, \boldsymbol{x}^*)(\boldsymbol{x})$ is that unique real number satisfying:

$$\boldsymbol{x} \, I \, [u(\boldsymbol{x})\boldsymbol{x}^*], \tag{15.3}$$

In terms of this notation, we showed in chapter 4 (Theorem 4.21) that $u = \varphi(G, \boldsymbol{x}^*)$ is a continuous function satisfying (15.1), above; that is, it represents G on $\mathbb{R}^n_+$. Thus, for a fixed $\boldsymbol{x}^* \in \mathbb{R}^n_{++}$, the function $\mu(\cdot) = \varphi(\cdot; \boldsymbol{x}^*)$ is a utility measurement function for $\mathcal{G}^c$. □

It is important to notice that, given an element, $\boldsymbol{x}^* \in \mathbb{R}^n_{++}$ (**a unit of measure**), the function $\mu(\cdot) = \varphi(\cdot, \boldsymbol{x}^*)$ is a utility measurement function for $\mathcal{G}^c$; that is, for each $G \in \mathcal{G}^c$, $u = \mu(G)$ is a continuous utility function which represents G. In fact,

[3] Where '$\mathbb{R}^n_{++}$' denotes the set of strictly positive elements of $\mathbb{R}^n$; that is:

$$\mathbb{R}^n_{++} = \{\boldsymbol{x} \in \mathbb{R}^n \mid \boldsymbol{x} \gg \boldsymbol{0}\}.$$

this is the only type of utility measurement function which we will consider in this chapter. *When we say that* $\mu\colon \mathcal{G}^c \to \mathcal{U}$ *is a measurement function for* $\mathcal{G}^c$, *we will mean that* μ *is defined as in Example 15.4; that is, there exists* $\boldsymbol{x}^* \in \mathbb{R}^n_{++}$ *such that* $\mu(\cdot) = \varphi(\cdot, \boldsymbol{x}^*)$, and even though this is the only type of measurement function we'll consider, we will often refer to such a function as a **Wold measurement function**. Continuing with this idea, if we speak of μ^* and $\mu^\dagger$ as being two different measurement functions for $\mathcal{G}$, we will mean that, while both are Wold measurement functions, there exist $\boldsymbol{x}^*, \boldsymbol{x}^\dagger \in \mathbb{R}^n_{++}$ such that $\boldsymbol{x}^* \neq \boldsymbol{x}^\dagger$ and for each $G \in \mathcal{G}$:

$$\mu^*(G) = \varphi(G, \boldsymbol{x}^*) \ \&\ \mu^\dagger(G) = \varphi(G, \boldsymbol{x}^\dagger),$$

where φ is defined in Example 15.4. That is, μ^* and $\mu^\dagger$ will be obtained by the same process, but may use different 'units of measure' ($\boldsymbol{x}^*$ versus $\boldsymbol{x}^\dagger$ in this example).

The utility measurement function just defined has especially interesting properties in the homothetic case. Recall that in chapter 4, we proved the following (Theorem 4.36).

15.5 Proposition. *If* $\boldsymbol{x}^* \in \mathbb{R}^n_{++}$ *then, given any* $G \in \mathcal{G}^h$, *the function* $u^* = \varphi(P, \boldsymbol{x}^*)$ *defined in Example 15.4 [satisfies (15.1), above, and] is concave, continuous, increasing, and positively homogeneous of degree one.*

Our next result shows that, effectively, we lose no generality in confining our attention to measurement functions of the Wold type when dealing with homothetic preference relations.

15.6 Proposition. *Let* $G \in \mathcal{G}^h$, *and let* $u\colon \mathbb{R}^n_+ \to \mathbb{R}_+$ *be any function representing* G *which is positively homogeneous of degree one. Then there exists* $\boldsymbol{x}^* \in \mathbb{R}^n_{++}$ *such that* $u = \varphi(G; \boldsymbol{x}^*)$, *where* $\varphi\colon \mathcal{G}^c \times \mathbb{R}^n_{++} \to \mathcal{U}$ *is defined as in Example 15.4.*

Proof. Since G is increasing and $u(\cdot)$ is positively homogeneous of degree one, there exists $\boldsymbol{x}^* \in \mathbb{R}^n_{++}$ such that $u(\boldsymbol{x}^*) = 1$, and we let $u^* = \varphi(G; \boldsymbol{x}^*)$; where $\varphi(\cdot)$ is from Example 15.4. We then note that, for an arbitrary $\boldsymbol{x} \in \mathbb{R}^n_+$, we have, since $\boldsymbol{x}I[u^*(\boldsymbol{x})\boldsymbol{x}^*]$ and $u(\cdot)$ represents G:

$$u(\boldsymbol{x}) = u[u^*(\boldsymbol{x})\boldsymbol{x}^*].$$

However, since $u(\cdot)$ is positively homogeneous of degree one, we then have:

$$u(\boldsymbol{x}) = u^*(\boldsymbol{x})u(\boldsymbol{x}^*) = u^*(\boldsymbol{x}),$$

and our result follows. □

Our last result of this section re-states a fact which we had already established in chapter 4. Given its importance in our endeavors of this chapter, it has seemed worthwhile to reprise both its statement and proof.

15.7 Proposition. *If* $G \in \mathcal{G}^h$, *and* $u\colon \mathbb{R}^n_+ \to \mathbb{R}_+$ *and* $u^*\colon \mathbb{R}^n_+ \to \mathbb{R}_+$ *are any two functions representing* G *which are also positively homogeneous of degree one, then there exists* $a \in \mathbb{R}_{++}$ *such that for all* $\boldsymbol{x} \in \mathbb{R}^n_+$, *we have* $u(\boldsymbol{x}) = au^*(\boldsymbol{x})$.

Proof. Let u and u^* satisfy the stated hypotheses. Making use of Proposition 15.6, we let $\boldsymbol{x}^* \in \mathbb{R}^n_{++}$ be such that $u^* = \varphi(G; \boldsymbol{x}^*)$, and define $a = u(\boldsymbol{x}^*)$. As in the proof of Proposition 15.6, we note that for arbitrary $\boldsymbol{x} \in \mathbb{R}^n_+$, we must have $\boldsymbol{x} I[u^*(\boldsymbol{x})\boldsymbol{x}^*]$; and thus, since u is positively homogeneous of degree one and represents G, we then conclude that:

$$u(\boldsymbol{x}) = u[u^*(\boldsymbol{x})\boldsymbol{x}^*] = u^*(\boldsymbol{x})u(\boldsymbol{x}^*) = au^*(\boldsymbol{x}). \quad \square$$

15.4 Social Preference Functions

As in Chapter 14, we will refer to a function which maps preference profiles into asymmetric orders on the allocation space as a social preference function. However, this time we will always take the domain of such a function to be a subset of $\mathcal{G}^c$. Formally, by a **social preference function** we will mean a function $\omega \colon \mathcal{G} \to \mathcal{Q}$, where $\mathcal{G} \subseteq \mathcal{G}^c$, and '$\mathcal{Q}$' denotes the family of asymmetric orders on the allocation space $\boldsymbol{X} = \mathbb{R}^{mn}$. Moreover, in this chapter, our principal concern will be with a special case of such functions, defined as follows.

15.8 Definition. We will say that a social preference function, $\omega \colon \mathcal{G}^c \to \mathcal{Q}$, is a **Pareto-Bergson-Samuelson (PBS) Social Preference Function** iff there exists $\mu \colon \mathcal{G}^c \to \mathcal{U}$ and an increasing function $F \colon \mathbb{R}^m_+ \to \mathbb{R}$ such that for all $\boldsymbol{G} \in \mathcal{G}^c$, if we define $Q = \omega(\boldsymbol{G})$ and $u_i = \mu(G_i)$ for $i = 1, \ldots, m$; we have:

$$(\forall \langle \boldsymbol{x}_i \rangle, \langle \boldsymbol{x}_i^* \rangle \in \mathcal{X}) \colon \langle \boldsymbol{x}_i \rangle Q \langle \boldsymbol{x}_i^* \rangle \iff F\Big[\boldsymbol{u}\big(\langle \boldsymbol{x}_i \rangle\big)\Big] > F\big[\boldsymbol{u}\big(\langle \boldsymbol{x}_i^* \rangle\big)\big],$$

where we define:

$$\boldsymbol{u}\big(\langle \boldsymbol{x}_i \rangle\big) = (u_1(\boldsymbol{x}_1), \ldots, u_m(\boldsymbol{x}_m)) \text{ and } \boldsymbol{u}\big(\langle \boldsymbol{x}_i^* \rangle\big) = (u_1(\boldsymbol{x}_1^*), \ldots, u_m(\boldsymbol{x}_m^*)).$$

In dealing with PBS Social Preference Functions, we will refer to the function μ as the **measurement function**, F, as the **aggregator function**, and the composite function, $W = F \circ \boldsymbol{u}$ as a **Pareto-Bergson-Samuelson (PBS) Social Welfare Function**. Since each such social preference function has the property that it can be represented by the composition of a measurement function, μ, and an aggregator function, F, we shall speak of such a social preference function as being **determined by** a pair (μ, F), where μ is a measurement function, and F is an aggregator function.[4] Notice that if ω is a PBS social preference function, then, for any preference profile, $\boldsymbol{G}$, in the domain of ω, the allocation ordering, $Q = \omega(\boldsymbol{G})$, extends the Pareto ordering, G, determined by $\boldsymbol{G}$.

15.9 Examples. Consider the aggregator function $F \colon \mathbb{R}^m_+ \to \mathbb{R}_+$ defined by:

$$F(\boldsymbol{u}) = \sum\nolimits_{i=1}^{m} u_i. \tag{15.4}$$

[4]Of course, such a function can generally be determined by many such (effectively equivalent) pairs.

The function F defines a Bergson-Samuelson Social Welfare Function when paired with any Wold measurement function;[5] as does the aggregator function F^* defined by:

$$F^*(\boldsymbol{u}) = \prod\nolimits_{i=1}^{m} (u_i)^{a_i}, \tag{15.5}$$

where $a_i \in \mathbb{R}_{++}$, for $i = 1, \ldots, m$, and $\sum_{i=1}^{m} a_i = 1$. We shall refer to the first of these two examples as the **utilitarian aggregator function**, and any function of the type in equation (15.5) as a **Cobb-Douglas-Eisenberg (CDE) aggregator function**.[6] A final example of interest is the **Rawlsian aggregator function**, defined by:

$$F(\boldsymbol{u}) = \min\{u_1, \ldots, u_m\}. \quad \square \tag{15.6}$$

In connection with the examples just presented, it should be noted that, using the method of Example 15.4, one obtains a significantly different measurement function for each different value of $\boldsymbol{x}^* \in \mathbb{R}^n_{++}$. This in turn means that, for a given aggregator function, F, one may obtain very different social preference functions if one combines one measurement function, μ^*, defined from $\boldsymbol{x}^*$, than one does from $\mu^\dagger$, say, defined from a second unit of measure, $\boldsymbol{x}^\dagger \in \mathbb{R}^n_{++}$; as is demonstrated by the following example.

15.10 Example. Consider the two-commodity, two-consumer economy in which the preferences of the consumers can be represented by the Cobb-Douglas utility functions:

$$u_1(\boldsymbol{x}_1) = A_1(x_{11})^2 \cdot x_{12} \text{ and } u_2(\boldsymbol{x}_2) = A_2 x_{21} \cdot (x_{22})^2, \tag{15.7}$$

respectively, and where $A_i > 0$ is a positive constant, for $i = 1, 2$. We begin by considering the measurement function $\tilde{\mu} = \varphi(\cdot, \tilde{\boldsymbol{x}})$, where:

$$\tilde{\boldsymbol{x}} = (1, 1).$$

For an arbitrary $\boldsymbol{x} \in \mathbb{R}^2_+$, $\tilde{u}_1(\boldsymbol{x}_1)$ can be found by solving the equation:

$$A_1(x_{11})^2 \cdot x_{12} = A_1(\tilde{u}_1(\boldsymbol{x}_1) \cdot 1)^2 \cdot (\tilde{u}_1(\boldsymbol{x}_1) \cdot 1) = A_1[\tilde{u}_1(\boldsymbol{x}_1)]^3;$$

so that:

$$\tilde{u}_1(\boldsymbol{x}_1) = (x_{11})^{2/3}(x_{12})^{1/3}. \tag{15.8}$$

Similarly, with this measurement function, consumer 2's utility function is given by:

$$\tilde{u}_2(\boldsymbol{x}_2) = (x_{21})^{1/3}(x_{22})^{2/3}. \tag{15.9}$$

If we now define the allocations $\boldsymbol{x}$ and $\bar{\boldsymbol{x}}$ by:

$$\boldsymbol{x} = \big((27, 1), (8, 8)\big) \text{ and } \bar{\boldsymbol{x}} = \big((8, 8), (1, 27)\big),$$

---

[5] Of course, one could equally well pair this aggregator function with a measurement function which is not of the Wold type, but we will not be considering such a possibility in this chapter.

[6] The two-person and symmetric version of this function was introduced by John Nash in his analysis of the bargaining problem (1950). For this reason, Moulin refers to the symmetric version of this function (all $a_i = 1/m$) as the 'Nash CUF' (See Moulin [1988]). As to the inclusion of Eisenberg's name in my labeling of the function, see Eisenberg [1961].

and make use of the utility functions defined in (15.8) and (15.9), the corresponding vectors of utilities are given by:

$$\tilde{\boldsymbol{u}}(\boldsymbol{x}) = (9, 8) \text{ and } \tilde{\boldsymbol{u}}(\bar{\boldsymbol{x}}) = (8, 9),$$

respectively. Thus, if a decision-maker has the PBS social preference function defined by the pair $(F, \tilde{\mu})$, where the aggregator function, F, is the utilitarian aggregator given by:

$$F(\boldsymbol{u}) = \sum_{i=1}^{m} u_i, \tag{15.10}$$

said decision-maker will be indifferent between the two allocations. This same social indifference will occur if the decision-maker has the PBS social preference function defined by the pair $(F^R, \tilde{\mu})$, with F^R being the Rawlsian function:

$$F^R(\boldsymbol{u}) = \min_i u_i. \tag{15.11}$$

The situation changes, however, if we take our unit of measure to be the bundle $\boldsymbol{x}^*$ given by:

$$\boldsymbol{x}^* = (64, 1),$$

while continuing to use the measurement function $[\varphi(\cdot, P)]$ defined in Example 15.4. In this case we can obtain $u_1(\boldsymbol{x}_1)$, for an arbitrary bundle $\boldsymbol{x}_1 \in \mathbb{R}^2_+$, by solving the equation:

$$(x_{11})^{2/3}(x_{12})^{1/3} = \left[u_1^*(\boldsymbol{x}_1) \cdot 64\right]^{2/3} \cdot \left[u_1^*(\boldsymbol{x}_1) \cdot 1\right]^{1/3} = 16u_1^*(\boldsymbol{x}_1);$$

so that:[7]

$$u_1^*(\boldsymbol{x}_1) = (1/16)(x_{11})^{2/3}(x_{12})^{1/3} = (1/16)\tilde{u}_1(\boldsymbol{x}_1). \tag{15.12}$$

Similarly, we find $u_2^*(\boldsymbol{x}_2)$ by solving:

$$(x_{21})^{1/3} \cdot (x_{22})^{2/3} = \left(u_2^*(\boldsymbol{x}_2) \cdot 64\right)^{1/3} \cdot \left(u_2^*(\boldsymbol{x}_2) \cdot 1\right)^{2/3} = 4u_2^*(\boldsymbol{x}_2).$$

Therefore, $u_2^*(\boldsymbol{x}_2) = (1/4)\tilde{u}_2(\boldsymbol{x}_2)$. Our vectors of utilities at the two allocations now become:

$$\boldsymbol{u}^*(\boldsymbol{x}) = (9/16, 2) \text{ and } \boldsymbol{u}^*(\bar{\boldsymbol{x}}) = (1/2, 9/4);$$

so that our utilitarian now prefers $\bar{\boldsymbol{x}}$ to $\boldsymbol{x}$, while our Rawlsian decision-maker now prefers $\boldsymbol{x}$ to $\bar{\boldsymbol{x}}$.

On the other hand, suppose we take our unit of measure equal to $\widehat{\boldsymbol{x}}$, where:

$$\widehat{\boldsymbol{x}} = (1, 64),$$

while continuing to use the measurement function defined in Example 15.4. Here the same basic reasoning as before establishes that $\widehat{u}_i = \varphi(P_i, \widehat{\boldsymbol{x}})$ is given by:

$$\widehat{u}_i(\boldsymbol{x}_1) = (1/4)\tilde{u}_i(\boldsymbol{x}_1) \text{ and } \widehat{u}_2(\boldsymbol{x}_2) = (1/16)\tilde{u}_2(\boldsymbol{x}_2).. \tag{15.13}$$

[7] We knew from Proposition 15.7 that the new function was going to be a scalar multiple of the old one.

Thus our vectors of utilities at the allocations of interest become:

$$\widehat{\boldsymbol{u}}(\boldsymbol{x}) = (9/4, 1/2) \text{ and } \widehat{\boldsymbol{u}}(\bar{\boldsymbol{x}}) = (2, 9/16),$$

respectively. Thus our utilitarian decision-maker now prefers $\boldsymbol{x}$ to $\bar{\boldsymbol{x}}$, while our Rawlsian now prefers $\bar{\boldsymbol{x}}$ to $\boldsymbol{x}$; reversing their previous preferences. □

In connection with the preceding example it is worth noting, first of all, how and why the utilitarian ordering can be manipulated as per the above example. In the example, consumer one basically likes commodity one better than commodity two; while consumer two has the opposite preferences. Suppose, then, that you are consumer two and that, while you are bound by sacred oath, or in some other fashion, to tell the truth in response to questions about your individual preferences, but you know that the aggregator function to be used is the utilitarian aggregator function, and you are to be allowed to choose the unit of measure for the measurement function. You can then gain by choosing a unit of measure with the proportions of the two commodities more to consumer one's liking than to your own. It will then tend to take a larger multiple of the unit of measure to yield a commodity bundle indifferent to an arbitrary bundle in your case than it will for consumer one. Consequently, your utility numbers will tend to be higher than consumer one's, and the utilitarian aggregator will tend to favor you. Of course, if you know that the aggregator to be used is the Rawlsian one, then you want to choose a bundle as the unit of measure which has proportions which you like. Your utility numbers will then tend to be smaller than consumer one's, and the Rawlsian aggregator will then tend to favor you. In fact, I suspect that any of us who have siblings have implicitly tried to apply these principles in bargaining with parents; sometimes trying 'I should get that rather than his getting it, because it is my favorite and he likes other things better' (thus exploring the possibility that the family social welfare aggregator is utilitarian), at other times the alternative, 'I should get that rather than his getting it, because he has (or has had) lots of such things compared to my paltry few' (the Rawlsian approach).

Now, while Example 15.10 and the above discussion emphasize the dependence of the social ordering upon the unit of measure used, in the utilitarian and the Rawlsian cases, this is not to say that a decision-maker might not have and use a social ordering of one of these types. The salient point is that the social ordering in these cases is determined jointly by the aggregator and the unit of measurement used in the utility-measurement process; and both are critical in defining the resulting social preference ordering. *Interestingly, however, it follows from Propositions 15.6 and 15.7 that if we confine our attention to $\mathcal{G}^h$, then a PBS function of the CDE form induces a social preference ordering which is independent of the measurement function, μ, with which it is paired.* The proof of this fact is quite simple, and will be left as an exercise. A more difficult question to answer, however, is whether this is the *only* PBS function which has this property. Unfortunately, however, it can be shown that this is the only one.[8]

The example just considered demonstrates the fact that PBS social preference

[8]This can be proved by a slight modification of Moulin's proof of his Theorem 2.3 [1988, p. 37]

functions do not satisfy Arrow's Independence of Irrelevant Alternatives condition.[9] As we have already noted, however, any PBS function satisfies the Weak Pareto Principle. The non-dictatorship condition is a bit tricky, for if we only require the aggregator function to be increasing, then a PBS function may be dictatorial; after all, the function $F\colon \mathbb{R}^m_+ \to \mathbb{R}_+$ defined by:

$$F(\boldsymbol{u}) = u_1,$$

is an increasing function on $\mathbb{R}^m_+$. In order to avoid this possibility, we could require the aggregator function to be strictly increasing, but this would eliminate the CDE aggregator function defined in (15.5), above, which is not strictly increasing on the boundary of $\mathbb{R}^m_+$. A reasonable way to eliminate this difficulty would be to require aggregator functions to be increasing, and to be strictly increasing on $\mathbb{R}^m_{++}$; on the other hand, this requirement would eliminate the Rawlsian aggregator function from consideration. As it turns out, however, all of the formal theory of PBS functions which we will be considering requires only that the aggregator function be increasing; and consequently we make use of this condition in our definition of PBS social preference functions, rather than the stronger condition that F is strictly increasing. It is important to notice, however, that this means that some of what we are calling PBS social preference functions may have some highly undesirable properties.

Before concluding this section, let's consider the Independence of Irrelevant Alternatives issue a bit further. Suppose that you are, in fact, a benign dictator/social planner for an economy, and that you want to make economic decisions in a manner which takes into account the preferences of the individual agents comprising the economy. How can you then determine whether one allocation should be chosen over another (that is, is better, in terms of your social preference) without knowing the full preference relation of every consumer in the economy? It was the profound insight of Bergson (1938) that, whatever the form of your social preferences in other respects, if you believe that one allocation is better that a second whenever every consumer in the economy considers it to be so [that is, if your social preference relation extends the (unanimity) Pareto order), then whatever allocation you consider best will necessarily be Pareto efficient, and if two allocations are such that every consumer is better off in one than in the other, then you (in fact, anyone ordering allocations by a PBS social preference function) would prefer the former allocation to the latter. As we all know, however, being able to compare allocations only when one Pareto dominates another is extremely limiting, and indeed when appeal is made to a PBS social welfare function in economic policy analyses, the motivation for its use is usually precisely in order to be able to compare allocations on a broader basis than is possible using only Pareto dominance. But, as a practical matter, wouldn't such extended comparisons require that you know each consumer's full preference relation? or, more stringently still, that you develop a utility function to represent each consumer's preferences? On the face of it, it may appear that a PBS social welfare function does require the development/estimation of a utility function for each

[9]While the CDE aggregator induces a social preference relation which is independent of which *Wold* measurement function with which it is paired, given that preferences are in $\mathcal{G}^h$, other measurement functions (in particular, if they could yield non homogeneous utility functions) may yield a different social preference relation when paired with such an aggregator.

consumer in order to allow comparisons to be made on anything other than Pareto dominance; and it was to avoid such astronomical informational requirements that Arrow introduced the requirement that a social preference function should satisfy independence of irrelevant alternatives. However, while PBS social welfare functions do not satisfy IIA, they also do not require full knowledge of each consumer's utility function in order to compare two allocations. In fact, suppose the social preference function is determined by the pair (μ, F), where $\mu = \varphi(\cdot, \boldsymbol{x}^*)$, and that it is desired to compare the allocations $\langle \boldsymbol{x}_i^1 \rangle$ and $\langle \boldsymbol{x}_i^2 \rangle$. In order to determine which of these allocations is to be preferred, it is only necessary to determine $\boldsymbol{u}(\boldsymbol{x}^t)$, for $t = 1, 2$, and this can be done by comparisons of the $\boldsymbol{x}_i^t$ with $\boldsymbol{x}^*$. This is a vastly weaker informational/estimation requirement than determining the form of each $u_i(\cdot)$! Of course, the process of finding the value of θ_t $[= u(\boldsymbol{x}^t)]$ is a great deal more complex an operation than I am making it sound, but it is at least theoretically possible.

15.5 The Compensation Principle

The Compensation Principle has a long history in economics, having been proposed in slightly different form independently by Hicks [1939] and Kaldor [1939]. The basic idea was that if we were to contemplate adopting a policy which would result in a change from situation A to a second situation, B, the change could be viewed as desirable if those who gained from the change from A to B could compensate those who lost by the change, and still be better off. While this statement may appear at first glance to be rather unambiguous, it was soon pointed out that this was not the case. In the first place, if we were regarding the compensation as being monetary, then a large change would likely change prices, so that monetary compensation which at first seemed adequate might fail to be so after the price changes. Consequently, it was soon decided that the compensation should be in real terms, at least in theory; although to analyze this idea, we will need to add a little notation to that set out in Section 2, as follows.

Suppose the allocation which initially prevails in the economy is $\langle \boldsymbol{x}_i^1 \rangle$, and that the adoption of a policy measure under consideration would result in the new consumption allocation, $\langle \boldsymbol{x}_i^2 \rangle$. Define:

$$\boldsymbol{y}^2 = \sum_{i=1}^{m} \boldsymbol{x}_i^2,$$

and, for $\boldsymbol{y} \in \mathbb{R}_+^n$, define:

$$A(\boldsymbol{y}) = \Big\{ \langle \boldsymbol{x}_i \rangle \in \mathbb{R}_+^n \mid \sum_{i=1}^{m} \boldsymbol{x}_i = \boldsymbol{y} \Big\}.$$

Then the policy would be said to result in an improvement if there exists $\langle \boldsymbol{x}_i^3 \rangle \in A(\boldsymbol{y}^2)$ such that $\langle \boldsymbol{x}_i^3 \rangle P \langle \boldsymbol{x}_i^1 \rangle$; where '$P$' denotes the strict Pareto ordering. The basic motivation behind the introduction of this condition as a criterion for inprovement is that, while economists cannot, as economists, recommend one consumption allocation over another (this being a 'political question'), the satisfaction of the criterion would mean that the government could, if it so desires, redistribute the gains from the change in such a way as to result in an allocation Pareto superior to the original allocation.

Probably something already strikes you as being odd about such a criterion, and in any event it was soon discovered that this criterion is intransitive; in fact it is not acyclic. If you think about it carefully, one of the things you may notice is that if $\langle \boldsymbol{x}_i^* \rangle$ and $\langle \boldsymbol{x}_i' \rangle$ are two allocations such that:

$$\sum_{i=1}^{m} \boldsymbol{x}_i^* = \sum_{i=1}^{m} \boldsymbol{x}_i',$$

then the two are equivalent, insofar as this criterion is concerned. From a formal point of view, for a given aggregate commodity bundle, $\boldsymbol{y} \in \mathbb{R}_+^n$, every allocation in $A(\boldsymbol{y})$ is in the same equivalence class, insofar as the ordering of allocations which is induced by this criterion is concerned. Another way of putting this is that the criterion should be viewed as an ordering of aggregate bundles, not of allocations; a point which was made in a similar fashion by Samuelson [1950], and emphasized by Chipman and Moore [1971]. In fact, in the latter paper it was noted that the criterion could easily be extended and formalized as follows. First let's extend the attainable allocations notion to sets, in the obvious way:

$$A(Y) = \left\{ \langle \boldsymbol{x}_i \rangle \in \mathbb{R}_+^{mn} \mid \sum_{i=1}^{m} \boldsymbol{x}_i \in Y \right\}$$

(in general, we would interpret Y as being the aggregate production set). We then define $\succcurlyeq$, which we will refer to as the 'Kaldor-Hicks-Samuelson (KHS) ordering' on the subsets of $\mathbb{R}_+^n$ by:

$$Y \succcurlyeq Y' \iff \big(\forall \langle \boldsymbol{x}_i' \rangle \in A(Y')\big)\big(\exists \langle \boldsymbol{x}_i \rangle \in A(Y)\big) \colon \langle \boldsymbol{x}_i \rangle G \langle \boldsymbol{x}_i' \rangle,$$

where 'G' denotes the weak Pareto ordering:

$$\langle \boldsymbol{x}_i \rangle G \langle \boldsymbol{x}_i' \rangle \iff \boldsymbol{x}_i G_i \boldsymbol{x}_i' \text{ for } i = 1, \ldots, m.$$

Following the terminology of Chipman and Moore [1971], we will refer to subsets, Y, of $\mathbb{R}_+^n$ as '**situations**.'

The idea here is that an economic policy change, other than one which is purely re-distributive, will usually result in a different potential aggregate supply set. For example, if a country adopts a free trade policy, as opposed to autarky, the set of potentially available aggregate supply vectors will now consist of those reflecting the net results of trading with other countries. The KHS ordering, in principle, would then allow a comparison between the situation attainable before the change and the possibilities attainable after the change. As it is formulated here, the KHS criterion does in fact correct some of the problems connected with the earlier Kaldor-Hicks formulations. For example, you can easily prove the following.

15.11 Proposition. *Suppose G_i is reflexive and transitive, for $i = 1, \ldots, m$. Then the KHS ordering, $\succcurlyeq$, will be reflexive and transitive, and its asymmetric part will be (asymmetric and) transitive. Furthermore, for each $Y_1, Y_2 \subseteq \mathbb{R}_+^n$, we have:*

$$Y_1 \subseteq Y_2 \Rightarrow Y_2 \succcurlyeq Y_1.$$

While the above proposition establishes the fact that the KHS ordering has some nice properties, it is important to notice that it is not generally total. In fact, strengthening the assumptions of the proposition to require that each G_i be total, as well as reflexive and transitive, does not correct this shortcoming. The following is a simple example demonstrating the fact that this is the case.

15.12 Example. Suppose $m = 2$, and that the two consumers' preference relations can be represented by the utility functions:

$$u_1(\boldsymbol{x}_1) = \min\{x_{11}, x_{12}/2\} \text{ and } u_2(\boldsymbol{x}_{21}/2, x_{22}\};$$

and let $Y_t = \boldsymbol{y}_t$, for $t = 1, 2$, where:

$$\boldsymbol{y}_1 = (2, 4) \text{ and } \boldsymbol{y}_2 = (4, 2).$$

In this case it is easy to show (see Exercise 2, at the end of this chapter) that we have neither $Y_1 \succcurlyeq Y_2$ nor $Y_2 \succcurlyeq Y_1$. □

While the above example rules out the possibility of the KHS ordering's always being total, there remains the chance that one could compare situations on some basis other than the rather trivially obvious set inclusion criterion mentioned in Proposition 15.11. In fact, one can prove the following (see Chipman and Moore [1971], Theorem 3, p. 9).

15.13 Theorem. *Suppose there exists an increasing, continuous, concave and homogeneous of degree one function, $g \colon \mathbb{R}^n_+ \to \mathbb{R}_+$ which is such that the ith consumer's preferences can be represented by the utility function:*

$$u_i = g(\boldsymbol{x}_i) \quad \textit{for } i = 1, \ldots, m.$$

Then we have the following, for all $Y_1, Y_2 \subseteq \mathbb{R}^n_+$:

$$Y_1 \succcurlyeq Y_2 \iff (\forall \boldsymbol{y} \in Y_2)(\exists \boldsymbol{y}' \in Y_1) \colon g(\boldsymbol{y}') \geq g(\boldsymbol{y}).$$

If we are willing to confine our attention to situations which are compact and nonempty subsets of $\mathbb{R}^n_+$, then we can simplify the statement of the conclusion to:

$$Y_1 \succcurlyeq Y_2 \iff \max_{\boldsymbol{y} \in Y_1} g(\boldsymbol{y}) \geq \max_{\boldsymbol{y} \in Y_2} g(\boldsymbol{y}).^{10}$$

Obviously, the assumptions of Theorem 15.13 allow a comparison of situations on a much broader basis than mere set inclusion. The price of this expansion of comparability is high, however; in the result we not only make quite strong assumptions about individual preferences, we also require all m individuals to have the same preferences!

In fact, however, the difficulty with Theorem 15.13 goes beyond the strong assumptions regarding individual preferences; in order to make use of the criterion, one needs to know the function $g(\cdot)$, which is tantamount to requiring that we know

[10] Notice that this means that, under the assumptions of the theorem, the relation $\succcurlyeq$ is total on the family of compact and nonempty subsets of $\mathbb{R}^n_+$.

the preference relation of each and every consumer in the economy! Obviously this is a hopelessly impractical requirement. In general we are at most likely to be willing to assume only that each preference relation, G_i, satisfies some specific *qualitative* properties; for example, that each G_i is a continuous weak order. This is equivalent to requiring that the preference profile, $\boldsymbol{G} = (G_1, \ldots, G_m)$ be such that each G_i is an element of some well-defined family of preference relations, $\mathcal{G}$. Let's formalize this idea a bit as follows.

First of all, as in Section 2 of this chapter let '$\mathcal{G}^c$' denote the family of continuous, increasing, and strictly convex weak orders on $\mathbb{R}^n_+$, and let '$\boldsymbol{\mathcal{G}}^c$' denote the m-fold cartesian product of $\mathcal{G}^c$; that is, let:

$$\boldsymbol{\mathcal{G}}^c = \prod_{i=1}^{m} \mathcal{G}_i, \quad \text{where } \mathcal{G}_i = \mathcal{G}^c, \text{ for } i = 1, \ldots, m;$$

For the remainder of this discussion, let's denote the family of situations (that is, the family of nonempty subsets of $\mathbb{R}^n_+$) which are **weakly disposable** by '$\mathcal{Y}$;' that is, we let $\mathcal{Y}$ be the family of all nonempty subsets of $\mathbb{R}^n_+$ satisfying the condition:

$$(\forall \boldsymbol{y}, \boldsymbol{y}' \in \mathbb{R}^n)\colon [\boldsymbol{y} \in Y \;\&\; \boldsymbol{0} \leq \boldsymbol{y}' \leq \boldsymbol{y}] \Rightarrow \boldsymbol{y}' \in Y. \tag{15.14}$$

The reason we will want to deal only with sets satisfying the weak disposability condition stems from the fact that we can trivially extend the last part of the conclusion of Proposition 15.11 to note that if each G_i is non-decreasing (as we will usually be assuming to be the case), and if Y_1 and Y_2 are such that there exist $\boldsymbol{y}_t \in \mathbb{R}^n_+$ such that:

$$Y_t = \{\boldsymbol{y}_t\} \quad \text{for } t = 1, 2,$$

where $\boldsymbol{y}_2 \geq \boldsymbol{y}_1$, then $Y_2 \succcurlyeq Y_1$. We will not include such singleton sets in $\mathcal{Y}$ (unless $\boldsymbol{y} = \boldsymbol{0}$), but we will include all sets of the form:

$$Y_t = \{\boldsymbol{y} \in \mathbb{R}^n_+ \mid \boldsymbol{y} \leq \boldsymbol{y}_t\},$$

for $\boldsymbol{y}_t \in \mathbb{R}^n_+$. This enables us to incorporate the trivial extension of Proposition 15.11 just discussed within the statement:

$$(\forall Y_1, Y_2 \in \mathcal{Y})\colon Y_2 \supseteq Y_1 \Rightarrow Y_2 \succcurlyeq Y_1;$$

or, more succinctly, by the statement: 'the KHS order *extends* $\supseteq$ on $\mathcal{Y}$.' Now consider the following definition.

15.14 Definition. If $\mathcal{G} \subseteq \boldsymbol{\mathcal{G}}^c$, we define the **extended KHS relation for** $\mathcal{G}$, $\succcurlyeq_{\mathcal{G}}$, on $\mathcal{Y}$ by:

$$Y_1 \succcurlyeq_{\mathcal{G}} Y_2 \iff (\forall \boldsymbol{G} \in \mathcal{G})\colon Y_1 \succcurlyeq Y_2. \tag{15.15}$$

It is then very easy to show that for any $\mathcal{G} \subseteq \boldsymbol{\mathcal{G}}^c$, $\succcurlyeq_{\mathcal{G}}$ will *extend* $\supseteq$ on $\mathcal{Y}$; that is, given any $\mathcal{G} \subseteq \boldsymbol{\mathcal{G}}^c$, and for all $Y_1, Y_2 \in \mathcal{Y}$:

$$Y_1 \supseteq Y_2 \Rightarrow Y_1 \succcurlyeq_{\mathcal{G}} Y_2. \tag{15.16}$$

The question is, can we find an admissible preference space such that $\succcurlyeq_{\mathcal{G}}$ *significantly* extends $\supseteq$? Unfortunately, a result established in Chipman and Moore [1971, Theorem 4, p. 13] dashes our hopes here.

15.15 Theorem. *Let $\mathcal{G}^\dagger \subseteq \mathcal{G}^c$ be the set of all preference profiles, $\boldsymbol{G}$ such that:*

$$G_1 = G_2 = \cdots = G_m,$$

and G_i is homothetic and strictly increasing (in addition to being a continuous and strictly convex weak order). Then for all $Y_1, Y_2 \in \mathcal{Y}$, we have:

$$Y_1 \succcurlyeq_{\mathcal{G}^\dagger} Y_2 \Rightarrow \overline{Y}_1 \supseteq Y_2;$$

where '$\overline{Y}_1$' denotes the closure of Y_1.

In particular, then, if we confine our attention to the subset of situations, $\mathcal{Y}^c$, consisting of only those elements of $\mathcal{Y}$ which are closed, then the relation $\succcurlyeq_{\mathcal{G}^\dagger}$ coincides with $\supseteq$ on $\mathcal{Y}^c$.

While the above theorem doesn't definitively rule out the possiblity of finding a subset of $\mathcal{Y}$ for which $\succcurlyeq_{\mathcal{G}}$ significantly extends $\supseteq$, it certainly suggests that a search for such a subset is extremely likely to be a fruitless endeavor. After all, we know that the assumption that preferences are homothetic generally yields much stronger aggregative conclusions than we can obtain without this hypothesis; and, in addition, Theorem 15.13 establishes the fact that if each consumer's preference relation is the same homothetic, continuous, increasing, and strictly convex weak order that characterizes every other consumer's preferences, then the KHS relation is a very significant extension of $\supseteq$. More to the point, however, the theorem states that unless we are willing to *rule out* the case of identical homothetic preferences, then the extended KHS relation will not significantly extend $\supseteq$; for notice that if $\mathcal{G}$ is a subset of $\mathcal{G}^c$ which contains $\mathcal{G}^\dagger$, then for all $Y_1, Y_2 \in \mathcal{Y}$:

$$Y_1 \succcurlyeq_{\mathcal{G}} Y_2 \Rightarrow Y_1 \succcurlyeq_{\mathcal{G}^\dagger} Y_2.$$

On that note, let's turn our attention to indirect social preferences.

15.6 Indirect Preferences: Individual and Social

In many economic contexts, both theoretical and applied, one can fruitfully make use of the concept of an indirect social preference relation. In fact, in the next section we will be considering measures of 'real national income,' a topic which depends crucially on such indirect preferences; and in Section 8 of this chapter we will be making use of this idea in deriving some results concerning consumers' surplus. In our analysis here, and generally in making use of the idea of indirect social preferences, we will be concentrating our attention upon the problem of comparing situations which are competitive equilibria from the standpoint of the consumers in the economy. We define this sort of equilibrium as follows.

15.16 Definition. Let $\boldsymbol{G} \in \mathcal{G}^c$. We will say that a tuple $(\langle \boldsymbol{x}_i^* \rangle, \boldsymbol{p}^*) \in \mathcal{X} \times \mathbb{R}^n_{++}$ is a **consumers' competitive equilibrium for $\boldsymbol{G}$** iff, for each i $(i = 1, \ldots, m)$, the following condition holds:

$$(\forall \boldsymbol{x}_i \in \mathbb{R}^n_+)\colon \boldsymbol{p}^* \cdot \boldsymbol{x}_i^* \geq \boldsymbol{p}^* \cdot \boldsymbol{x}_i \Rightarrow \boldsymbol{x}_i^* G_i \boldsymbol{x}_i.$$

Our development of the idea of social indirect preferererences is based upon the concept of individual indirect preferences, which we studied extensively in Section 4.7. For convenient reference, however, we will present a summary of this material here.

Given any weak order, $G_i \in \mathcal{G}^c$, G_i induces an **indirect preference relation**, G_i^*, on:

$$\Omega \stackrel{\text{def}}{=} \mathbb{R}^n_{++} \times \mathbb{R}_+$$

by:

$$(\boldsymbol{p}, w) G_i^* (\boldsymbol{p}', w') \iff \boldsymbol{h}_i(\boldsymbol{p}, w) G_i \boldsymbol{h}_i(\boldsymbol{p}', w');$$

where '$\boldsymbol{h}_i(\cdot)$' denotes the ith consumer's demand function (**the demand function determined by**) $\boldsymbol{G}_i$.[11] We say that a function $v_i \colon \Omega \to \mathbb{R}$ is an **indirect utility function corresponding to $\boldsymbol{G}_i$** iff v_i represents G_i^* on Ω. In the present context, one most conveniently obtains an indirect utility function in the following way: if u_i is a utility function representing G_i, and if $\boldsymbol{h}_i(\cdot)$ is the demand function determined by G_i, then the composite function $v_i \colon \Omega \to \mathbb{R}_+$ defined by:

$$v_i(\boldsymbol{p}, w) = u_i[\boldsymbol{h}_i(\boldsymbol{p}, w)] \quad \text{for } (\boldsymbol{p}, w) \in \Omega,$$

is an indirect utility function representing G_i^* on Ω.

Recall also that, if $G \in \mathcal{G}^h$, then there exists an indirect utility function for G which takes the particularly simple and useful form:

$$v_i^*(\boldsymbol{p}, w_i) = \frac{w_i}{\gamma_i^*(\boldsymbol{p})};$$

where $\gamma_i^*(\cdot)$ is a **cost-of-living function** for G_i; that is, For $G_i \in \mathcal{G}^c$, the function γ_i^* is defined as:

$$\gamma_i^*(\boldsymbol{p}) = \frac{1}{u_i^*[\boldsymbol{h}_i(\boldsymbol{p}, 1)]} \quad \text{for } \boldsymbol{p} \in \mathbb{R}^n_{++},$$

where u_i^* is any positively homogeneous of degree one function representing G_i.[12]

Now, given any preference profile $\boldsymbol{G} \in \mathcal{G}^c$, any social preference function ω defined on $\mathcal{G}^c$ induces an **indirect social preference relation**, Q^* on:

$$\boldsymbol{\Omega} \stackrel{\text{def}}{=} \mathbb{R}^n_{++} \times \mathbb{R}^m_+.$$

defined by:

$$(\boldsymbol{p}'', \boldsymbol{w}'') Q^* (\boldsymbol{p}', \boldsymbol{w}') \iff \boldsymbol{H}(\boldsymbol{p}'', \boldsymbol{w}'') Q \boldsymbol{H}(\boldsymbol{p}', \boldsymbol{w}');$$

where $Q = \omega(\boldsymbol{G})$ is the social preference relation determined by $(\omega, \boldsymbol{G})$, and we define $\boldsymbol{H} \colon \boldsymbol{\Omega} \to \mathbb{R}^{mn}_+$ by:

$$\boldsymbol{H}(\boldsymbol{p}, \boldsymbol{w}) = \big(\boldsymbol{h}_1(\boldsymbol{p}, w_1), \ldots, \boldsymbol{h}_m(\boldsymbol{p}, w_m)\big).$$

We will then refer to Q^* as the **indirect social preference relation induced by $\boldsymbol{\omega}(\boldsymbol{G})$**.

---

[11] The ith consumer's demand function for the jth commodity will then be denoted by '$h_{ij}(\cdot)$.'

[12] See Section 4.9 for details.

So, what is going on here is very much the same kind of procedure by which we define indirect (individual) preferences from direct preferences. Two vectors of prices and (m-tuples of) incomes determine allocations via the demand functions of the individual consumers. The first $(\boldsymbol{p}, \boldsymbol{w})$-pair is (indirectly) preferred to the second if the first allocation is socially preferred to the second allocation.

In much the same way that we define an indirect utility function to represent (individual) indirect preferences, we can then define an indirect social welfare function. In particular, if ω is of the PBS form, we can define an indirect social welfare function to represent Q^*, as follows. If ω is determined by (μ, F), we define $\boldsymbol{v}\colon \boldsymbol{\Omega} \to \mathbb{R}^m_+$ by:

$$\boldsymbol{v}(\boldsymbol{p}, \boldsymbol{w}) = \big(u_1[h_1(\boldsymbol{p}, w_1)], \ldots, u_m[h_m(\boldsymbol{p}, w_m)]\big) = \big(v_1(\boldsymbol{p}, w_1), \ldots, v_m(\boldsymbol{p}, w_m)\big),$$

where $u_i = \mu(G_i)$ and $v_i = u_i \circ \boldsymbol{h}_i$, for $i = 1, \ldots, m$. It is then easily seen that $F \circ \boldsymbol{v}$ represents the indirect allocation ordering, Q^*; that is, for any $\boldsymbol{G} \in \mathcal{G}^c$, and all $(\boldsymbol{p}, \boldsymbol{w}), (\boldsymbol{p}', \boldsymbol{w}') \in \boldsymbol{\Omega}$:

$$(\boldsymbol{p}, \boldsymbol{w})Q^*(\boldsymbol{p}', \boldsymbol{w}') \iff F\big[\boldsymbol{v}(\boldsymbol{p}, \boldsymbol{w})\big] > F\big[\boldsymbol{v}(\boldsymbol{p}', \boldsymbol{w}')\big].$$

We shall refer to the composite function, $F \circ \boldsymbol{v}$, as the **indirect social welfare function for ω.**[13]

15.7 Measures of Real National Income

In this section we will investigate the problem of measuring real national income for an economy; exploring the implications for this problem of the ideas we have been presenting in this chapter.[14]

If $\boldsymbol{G} \in \mathcal{G}^c$, each element of $\boldsymbol{\Omega}$ defines a unique (consumers') competitive equilibrium, and conversely. In this section we exploit this fact in that, when we say that $(\boldsymbol{p}, \boldsymbol{w})$ is an element of $\boldsymbol{\Omega}$, we will always suppose that consumer i is choosing the bundle:

$$\boldsymbol{x}_i = \boldsymbol{h}_i(\boldsymbol{p}, w_i) \quad \text{for } i = 1, \ldots, m.$$

We will make use of the following notation: for $(\boldsymbol{p}, \boldsymbol{w}) \in \boldsymbol{\Omega}$ [respectively, $(\boldsymbol{p}', \boldsymbol{w}') \in \boldsymbol{\Omega}$, etc.], we will use the notation 'w' and '$\overline{w}$' [respectively, 'w'' and '$\overline{w}'$,' etc.] to denote total and average income; that is:

$$w = \sum_{i=1}^m w_i \text{ and } \overline{w} = (1/m)\sum_{i=1}^m w_i = (1/m)w.$$

We will assume throughout this section that we are concerned with maximizing a social preference function, ω, of the PBS form; where ω is determined by a pair (μ, F), and F is increasing, quasi-concave, and positively homogeneous of degree

[13] There are, of course, other functions which represent Q^*, but this function is uniquely determined by the pair (μ, F).

[14] The relationship between improvements in 'real national income,' as conventionally measured, and the KHS criterion is examined extensively in Chipman and Moore [1973], [1976b]. The fact that the results obtained in those studies were essentially negative provided the motivation for the rather unconventional approach to be developed in this section.

one. We will also begin with the assumption that the admissible preference space is the set $\mathcal{G}^* \subseteq \mathcal{G}^c$ given by:

$$\mathcal{G}^* = \{\boldsymbol{G} \in \mathcal{G}^h \mid G_1 = G_2 = \cdots = G_m\}; \tag{15.17}$$

in other words, we will suppose that each preference profile consists of identical homothetic preferences.

Now, given $\boldsymbol{G} \in \mathcal{G}^*$, let $\gamma(\cdot)$ be the cost-of-living function for G_1 determined by μ. Then the indirect social welfare function for ω, denote it by 'V,' can be written:

$$V(\boldsymbol{p}, \boldsymbol{w}; \boldsymbol{G}) = F\Big[\frac{w_1}{\gamma(\boldsymbol{p})}, \ldots, \frac{w_m}{\gamma(\boldsymbol{p})}\Big] = \frac{w}{\gamma(\boldsymbol{p})} F\Big(\frac{w_1}{w}, \ldots, \frac{w_m}{w}\Big). \tag{15.18}$$

Thus, V factors into the product of a measure of 'real national income,' $w/\gamma(\boldsymbol{p})$, and a function of the distribution of income, $f \colon \Delta_m \to \mathbb{R}_+$, where f is simply the restriction of F to Δ_m.

The vector, $\boldsymbol{d} \in \Delta_m$ defined by:

$$\boldsymbol{d} = (1/w)\boldsymbol{w} = \Big(\frac{w_1}{w}, \ldots, \frac{w_m}{w}\Big),$$

can be thought of as the *income distribution vector* associated with $\boldsymbol{w}$; and, since we are supposing that the aggregator function, F, is positively homogeneous of degree one, it is of great interest to see at what value of $\boldsymbol{d} \in \Delta_m$ the aggregator function is maximized. Define $\mathcal{F}$ as the set of all functions $F \colon \mathbb{R}^m_+ \to \mathbb{R}_+$ which are increasing, quasi-concave, and positively homogeneous of degree one; and then define $\delta \colon \mathcal{F} \mapsto \Delta_m$ by:

$$\delta(F) = \{\boldsymbol{d} \in \Delta_m \mid (\forall \boldsymbol{a} \in \Delta_m) \colon F(\boldsymbol{d}) \geq F(\boldsymbol{a})\}. \tag{15.19}$$

If F is *strictly* quasi-concave, then $\delta(F)$ will be single-valued, and, as we have done in similar contexts, we can think of δ as being a function. However, it will not be necessary to do this in our present discussion.

Given that F is positively homogeneous of degree one it is easy to show that if $\boldsymbol{d} \in \delta(F)$, then, given any value of $w \in \mathbb{R}_+$, F is maximized, subject to $\sum_{i=1}^m w_i = w$, at $\boldsymbol{w} = (wd_1, \ldots, wd_m) = w\boldsymbol{d}$. We can therefore define a measure of the *efficiency* of the income distribution, which we will denote by '$E(\boldsymbol{w}; F)$' by:

$$E(\boldsymbol{w}; F) = \frac{F(\boldsymbol{w})}{F(w\boldsymbol{d})} = \frac{F(w_1, \ldots, w_m)}{F(wd_1, \ldots, wd_m)}. \tag{15.20}$$

Since any $F \in \mathcal{F}$ is quasi-concave and positively homogeneous of degree one, it is easy to show that, for all $\boldsymbol{w} \in \mathbb{R}^m_+ \setminus \{\boldsymbol{0}\}$, we must have:

$$0 \leq E(\boldsymbol{w}; F) \leq 1;$$

with:

$$E(\boldsymbol{w}; F) = 1 \iff (1/w)\boldsymbol{w} \in \delta(F).$$

From the standpoint of making use of $F \circ \boldsymbol{v}$ as a social preference function, it is clear that one wants to keep $E(\boldsymbol{w}; F)$ as close to one as possible; in fact, the larger the value of $E(\boldsymbol{w}; F)$, the better. We formalize our definition of this index in the following.

15.17 Definition. Given $F \in \mathcal{F}$, and $\boldsymbol{d} \in \delta(F)$, we define $E(\boldsymbol{w}; F)$, for $\boldsymbol{w} \in \mathbb{R}^m_+ \setminus \{\boldsymbol{0}\}$, the **distribution (efficiency) index**, by:

$$E(\boldsymbol{w}; F) = \frac{F(\boldsymbol{w})}{F(w\boldsymbol{d})} = \frac{F(w_1/w, \ldots, w_m/w)}{F(\boldsymbol{d})}. \tag{15.21}$$

Before proceeding further, let's take a look at some examples.

15.18 Example. Suppose F takes the CES form:

$$F(\boldsymbol{w}) = \Big(\sum\nolimits_{i=1}^m (d_i)^{1-a} \cdot (w_i)^a\Big)^{1/a}, \tag{15.22}$$

where:

$$d_i > 0 \text{ for } i = 1, \ldots, m, \ \sum_{i=1}^m d_i = 1, \text{ and } a \leq 1. \tag{15.23}$$

It is an easy exercise to prove (see Exercise 3, at the end of this chapter) that F is maximized, subject to $\sum_{i=1}^m w_i = 1$, when:

$$\boldsymbol{w} = \boldsymbol{d};$$

that is, in this case, we can take $\delta(F) = \boldsymbol{d} = (d_1, \ldots, d_m)$. The distribution index, $E(\boldsymbol{w}; F)$ is then given by:

$$E(\boldsymbol{w}; F) = \frac{F(\boldsymbol{w})}{F(w\boldsymbol{d})} = \frac{F(w_1/w, \ldots, w_m/w)}{F(\boldsymbol{d})} = \frac{\big[\sum_{i=1}^m (d_i)^{1-a} \cdot (w_i)^a\big]^{1/a}}{w\big(\sum_{i=1}^m d_i\big)^{1/a}}. \tag{15.24}$$

Of course, in this particular case, $F(\boldsymbol{d}) = 1$ for $a \neq 0$. so that the index reduces to:

$$E(\boldsymbol{w} : F) = F(w_1/w, \ldots, w_m/w) = \Big[\sum\nolimits_{i=1}^m (d_i)^{1-a} \cdot \big(\frac{w_i}{w}\big)^a\Big]^{1/a}.$$

On the other hand, when $a = 0$ (the CDE case), the index is given by:

$$E(\boldsymbol{w} : F) = \frac{\prod_{i=1}^m (w_i)^{d_i}}{\prod_{i=1}^m (d_i w)^{d_i}}. \quad \square \tag{15.25}$$

Now, in the special case in which $(1/m)\boldsymbol{1} \in \delta(F)$, we will say that $\boldsymbol{F}$ is **egalitarian**;[15] for in this case, for a given aggregate income, w, F will always be maximized when individual incomes are all the same. Notice also that in this case, the distribution index becomes:

$$E(\boldsymbol{w}; F) = \frac{F(\boldsymbol{w})}{F(\overline{w}, \ldots, \overline{w})}; \tag{15.26}$$

Consequently, in this egalitarian case, the **inequality index** generally used in the literature as a measure of income inequality (see Exercise 5, at the end of this chapter), $J(\boldsymbol{w}; F)$ is given by:

$$J(\boldsymbol{w}; F) = 1 - E(\boldsymbol{w}; F). \tag{15.27}$$

[15] Recall that we use '$\boldsymbol{1}$' to denote the vector each of whose coordinates equals 1.

Under these assumptions, it is usual in this literature to argue that the *lower* the value of this index (which corresponds to the higher the value of $E(\boldsymbol{w}; F)$), the better.[16]

At first glance, it is difficult to think of a good reason why one shouldn't favor an egalitarian aggregator; after all, if you were to be one of the consumers, but were not sure which number you would be labeled by (what value of i would be yours), it would be safer for you to argue in favor of an equally-weighted (egalitarian) aggregator function than otherwise. In other words, if you were in favor of using an aggregator function of the form indicated in (15.22), wouldn't you want to set the weights all equal; that is, let:

$$d_i = 1/m \quad \text{for } i = 1, \ldots, m?$$

Upon reflection, however, it isn't so clear that this is reasonable. After all, would it really be reasonable for someone who is able, but unwilling to work to have the same income as someone who has a difficult and dangerous occupation? or, perhaps more to the point, would anyone elect to pursue a difficult and dangerous (but necessary) occupation if she or he could earn the same income without working at all? In fact, a move toward a more equal income distribution in an economy may mean that the labor market is not functioning effectively; with a corresponding decline in economic efficiency. In any event, we will not confine our attention to egalitarian aggregator functions in the discussion to follow; on the other hand, we will not rule out this case either.

Despite the fact that the cost of living function conveniently cancels out in the derivation of the distribution index under the assumptions being utilized here, there are problems involved with the use of this index when prices have changed. To be more precise, if one observes two price-income pairs, $(\boldsymbol{p}^1, \boldsymbol{w}^1)$ and $(\boldsymbol{p}^2, \boldsymbol{w}^2)$, social welfare may have decreased in the move from situation 1 to situation 2 under the present assumptions even if individual preferences are unchanged and:

$$E(\boldsymbol{w}^2; F) > E(\boldsymbol{w}^1; F).$$

On the other hand, we do have the following, fairly obvious, proposition; the proof of which I will leave as an exercise.

15.19 Proposition. *Suppose the social preference function, ω, is determined by a pair (μ, F), where F is positively homogeneous of degree one, and that $\boldsymbol{G} \in \mathcal{G}^*$, V is an indirect social welfare function for ω, that γ is a cost of living index for each G_i, and let $(\boldsymbol{p}^1, \boldsymbol{w}^1), (\boldsymbol{p}^2, \boldsymbol{w}^2) \in \boldsymbol{\Omega}$ be two (consumer) competitive equilibrium situations. Then we have the following.*

1. *If:*

$$\frac{w^2}{\gamma(\boldsymbol{p}^2)} \geq \frac{w^1}{\gamma(\boldsymbol{p}^1)}, \tag{15.28}$$

and:

$$E(\boldsymbol{w}^2; F) \geq E(\boldsymbol{w}^1; F), \tag{15.29}$$

[16] For more on the measurement of inequality of income, and the properties of the inequality index, see Dutta [2002], Foster and Sen [1997], Moulin [1988, pp. 51-52], or Myles [1995, Chapter 3].

then:

$$V(\boldsymbol{p}^2, \boldsymbol{w}^2) \geq V(\boldsymbol{p}^1, \boldsymbol{w}^1). \tag{15.30}$$

2. Conversely, if (15.30) holds, and $w^2/\gamma(\boldsymbol{p}^2) \leq w^1/\gamma(\boldsymbol{p}^1)$*, then (15.29) holds as well.*

The above proposition shows that, under the assumptions of that result, the comparison of values of the distribution index, $E(\boldsymbol{w}; F)$, in two different equilibria has an unambiguous interpretation in terms of the defining social welfare function in the case where prices are unchanged. That is, if prices are unchanged, and $E(\boldsymbol{w}^2; F) > E(\boldsymbol{w}^1; F)$, we can be sure that $V(\boldsymbol{p}^2, \boldsymbol{w}^2; \boldsymbol{G}) > V(\boldsymbol{p}^1, \boldsymbol{w}^1; \boldsymbol{G})$ as well. In order to begin to extend this analysis, consider the product $y(\boldsymbol{p}, \boldsymbol{w}; F)$, defined by:

$$y(\boldsymbol{p}, \boldsymbol{w}; \boldsymbol{F}) = \frac{w}{\gamma(\boldsymbol{p})} \times E(\boldsymbol{w}; F). \tag{15.31}$$

It is easily seen from (15.18) that this product is an increasing transformation (in fact, a positive scalar multiple) of the indirect social welfare function, $V(\boldsymbol{p}, \boldsymbol{w})$. Consequently, a comparison of values of $y(\cdot; F)$ at different (consumer) competitive equilibria has unambiguous welfare implications. Moreover, notice that the function $y(\cdot; F)$ defined in (15.31) has an interesting and natural interpretation: we can think of it as **income distribution-adjusted real national income**.

The above discussion, and Proposition 15.19 shows that, under the assumptions of the proposition, the income distribution-adjusted real national income function has an unambiguous meaning in terms of the underlying social welfare function. However, suppose we take the admissible preference space to be all of $\mathfrak{G}^h$, rather than just $\mathfrak{G}^*$. In this case, we do not get the convenient factorization which we have presented in (15.18); in fact the best we can do is something like:

$$V(\boldsymbol{p}, \boldsymbol{w}) = F\Big[\frac{w_1}{\gamma_1(\boldsymbol{p})}, \ldots, \frac{w_m}{\gamma_m(\boldsymbol{p})}\Big] = w \cdot F\Big(\frac{w_1/\gamma_1(\boldsymbol{p})}{w}, \ldots, \frac{w_m/\gamma_m(\boldsymbol{p})}{w}\Big); \tag{15.32}$$

or, perhaps:

$$V(\boldsymbol{p}, \boldsymbol{w}) = \left[\frac{w}{\boldsymbol{\gamma}(\boldsymbol{p})}\right] \cdot F\left[\frac{w_1/\gamma_1(\boldsymbol{p})}{w/\boldsymbol{\gamma}(\boldsymbol{p})}, \ldots, \frac{w_m/\gamma_m(\boldsymbol{p})}{w/\boldsymbol{\gamma}(\boldsymbol{p})}\right], \tag{15.33}$$

where $\boldsymbol{\gamma}(\boldsymbol{p})$ is any sort of average of the cost of living functions, for example:

$$\boldsymbol{\gamma}(\boldsymbol{p}) = (1/m)\sum_{i=1}^{m} \gamma_i(\boldsymbol{p}) \text{ or } \gamma^*(\boldsymbol{p}) = \Big[\prod_{i=1}^{m} \gamma_i(\boldsymbol{p})\Big]^{1/m}. \tag{15.34}$$

However we express the equation, the basic problem remains; once we allow individual preferences to differ, our distribution index, which is defined as a function of nominal income, may have no dependable relationship with the value of the indirect social preference function; *even when comparing equilibria with unchanged prices.* Consider the following example:

15.20 Example. Consider an economy with two consumers, whose utility functions are given by:

$$u_1(\boldsymbol{x}_1) = \min\{x_{11}/2, x_{12}\} \text{ and } u_2(\boldsymbol{x}_2) = \min\{x_{21}, x_{22}/4\};$$

and suppose the social welfare function is given by:

$$W(\langle \boldsymbol{x}_i \rangle) = \left([u_1(\boldsymbol{x}_1)]^{1/2} + [u_2(\boldsymbol{x}_2)]^{1/2}\right)^2, \tag{15.35}$$

(we suppose that the given utility functions are the appropriate functions to be paired with the aggregator function in order to define W). Now suppose the vector of prices is given by $\boldsymbol{p} = (1, 2)$, and consider the income figures $\boldsymbol{w}^1 = (144, 36)$ and $\boldsymbol{w}^2 = (36, 144)$. Then (see Exercise 6, at the end of this chapter):

$$V(\boldsymbol{p}, \boldsymbol{w}^1) = 64 \text{ while } V(\boldsymbol{p}, \boldsymbol{w}^2) = 49.$$

However, the distribution index is given by:

$$E(\boldsymbol{w}^1; F) = 9/10 = E(\boldsymbol{w}^2; F).$$

It should also be noted that, since the aggregator function is egalitarian, the inequality index, $J(\boldsymbol{w}; F)$, is equal to $1/10$ in both equilibria. □

This last example shows that even if the aggregator function is egalitarian, and both prices and aggregate incomes are unchanged in two situations, a change in the distribution of income (which leaves both the distribution index and inequality index unchanged) may nonetheless result in a decline in welfare. However, there is at least one case in which things work out much better. Consider the generic CDE case with the aggregator function:

$$F(\boldsymbol{u}) = \prod\nolimits_{i=1}^{m} u_i^{a_i}, \tag{15.36}$$

where:

$$\sum\nolimits_{i=1}^{m} a_i = 1 \text{ and } a_i > 0, \quad \text{for } i = 1, \ldots, m. \tag{15.37}$$

In this case, the corresponding indirect social welfare function becomes:

$$V(\boldsymbol{p}, \boldsymbol{w}) = \prod\nolimits_{i=1}^{m} \left(\frac{w_i}{\gamma_i(\boldsymbol{p})}\right)^{a_i} = \left[\frac{w}{\prod_{i=1}^{m} \gamma_i(\boldsymbol{p})^{a_i}}\right] \cdot \left[\prod\nolimits_{i=1}^{m} \left(\frac{w_i}{w}\right)^{a_i}\right]; \tag{15.38}$$

so that:

$$y(\boldsymbol{p}, \boldsymbol{w}; F) = \frac{w}{\prod_{i=1}^{m} \gamma_i(\boldsymbol{p})^{a_i}} \times E(\boldsymbol{w}; F); \tag{15.39}$$

is once again a scalar multiple of $V(\boldsymbol{p}, \boldsymbol{w})$. Consequently, if two consumers' competitive equilibria, $(\boldsymbol{p}^1, \boldsymbol{w}^1)$ and $(\boldsymbol{p}^2, \boldsymbol{w}^2)$ are such that:

$$\boldsymbol{p}^1 = \boldsymbol{p}^2,$$

our decision-maker will prefer the equilibrium having the higher value of $y(\boldsymbol{p}, \boldsymbol{w}; F)$. Of course, it should also be noted that in this situation, that is, with $\boldsymbol{p}^1 = \boldsymbol{p}^2$, we will have:

$$V(\boldsymbol{p}^2, \boldsymbol{w}^2) \geq V(\boldsymbol{p}^1, \boldsymbol{w}^1) \iff \prod_{i=1}^{m} (w_i^2)^{a_i} \geq \prod_{i=1}^{m} (w_i^1)^{a_i}.$$

However, it nonetheless seems worthwhile to consider the index number idea a bit further.

Define $\gamma^*(\boldsymbol{p})$ by:

$$\gamma^*(\boldsymbol{p}) = \prod\nolimits_{i=1}^{m} \gamma_i(\boldsymbol{p})^{a_i}, \tag{15.40}$$

then we can write:

$$y(\boldsymbol{p}, w; F) = \frac{w}{\gamma^*(\boldsymbol{p})} \times E(\boldsymbol{w}; F). \tag{15.41}$$

The γ^* function could reasonably be estimated by taking a manageable sample of consumers, and estimating cost of living functions for the sample. Of course the function γ^* is determined by the parameters $a_1, \ldots, a_m$ as well; however, one might hope that the function values might not be too sensitive to the values of the parameters chosen, since γ^* is, essentially, a geometric mean. Indeed, in the special case in which $\gamma_1(\boldsymbol{p}) \equiv \gamma_2(\boldsymbol{p}) \equiv \cdots \equiv \gamma_m(\boldsymbol{p})$, the function is independent of the values of the parameters a_i. Consequently, it might be possible to construct such an aggregate index which would be meaningful whatever one's feelings as to what were the appropriate values of the a_i parameters. However, this is probably not the place for further speculation along these lines, and it is time we turned our attention to consumers' surplus.

15.8 Consumers' Surplus

As we saw in Section 10 of Chapter 4, the analysis of consumer's surplus for a single consumer gets complicated very quickly if one attempts to do the analysis carefully. Obviously even more complications arise when one is trying to develop a justifiable measure of (aggregate) consumers' surplus, and in fact in most applied work it appears that investigators implicitly assume the existence of a 'representative consumer.'[17] However, it turns out that we can make some progress in the development of such a measure without making such a restrictive assumption.

In our analysis we will make use of the general framework and assumptions which we have been using throughout the rest of this chapter. We now define:

$$\boldsymbol{\Omega} = \mathbb{R}^n_{++} \times \mathbb{R}^m_{+},$$

and expand upon our definition of an acceptable (integral) measure of welfare change, as presented in Section 10 of Chapter 4, as follows.[18]

We will take as given a preference profile $\boldsymbol{G} \in \mathcal{G}^c$ and a PBS social welfare function, W, determined by a pair (μ, F), and we will denote the indirect social preference relation determined by W, given $\boldsymbol{G}$, by 'Q^*.' In the following, we will let $\boldsymbol{\Omega}^*$ be a nonempty open subset of $\boldsymbol{\Omega}$, and let $\mathcal{P}(\boldsymbol{\Omega}^*)$ be the set of all polygonal paths, $\omega\colon [0,1] \to \boldsymbol{\Omega}^*$, connecting points of $\boldsymbol{\Omega}^*$, and lying entirely within the set (see Section 4.10 for a discussion of polygonal paths, as well as an explanation of the idea of line integrals, which will be used in the following definition).

[17] That is, the method used is typically only theoretically correct in the situation where consumers as a whole behave as if they were maximizing a single utility function. See Section 3 of Chapter 5 for a discussion of this topic.

[18] In our treatment of integral measures of change in Q^*, we are generally following the development presented in Chipman and Moore [1994]; although in a somewhat simplified form.

15.21 Definition. We will say that a function $\boldsymbol{f}\colon \boldsymbol{\Omega}^* \to \mathbb{R}_+^{n+m}$ furnishes an **acceptable indicator of change in Q^* on $\boldsymbol{\Omega}^*$** iff:

1. for all $\omega, \omega^* \in \mathcal{P}(\boldsymbol{\Omega}^*)$ satisfying $\omega(0) = \omega^*(0)$ and $\omega(1) = \omega^*(1)$, we have:

$$\int_0^1 \boldsymbol{f}[\omega(t)] \cdot d\omega(t) = \int_0^1 \boldsymbol{f}[\omega^*(t)] \cdot d\omega^*(t), \tag{15.42}$$

(independence of path) and:

2. for all $(\boldsymbol{p}^0, \boldsymbol{w}^0), (\boldsymbol{p}^1, \boldsymbol{w}^1) \in \boldsymbol{\Omega}^*$ and for all $\omega \in \mathcal{P}(\boldsymbol{\Omega}^*)$ satisfying:

$$\omega(0) = (\boldsymbol{p}^0, \boldsymbol{w}^0) \text{ and } \omega(1) = (\boldsymbol{p}^1, \boldsymbol{w}^1), \tag{15.43}$$

we have:

$$\int_0^1 \boldsymbol{f}[\omega(t)] \cdot d\omega(t) \geq 0 \iff (\boldsymbol{p}^1, \boldsymbol{w}^1) Q^* (\boldsymbol{p}^0, \boldsymbol{w}^0). \tag{15.44}$$

The basic rationale for the above definition is a straightforward extension of the definition for the single-consumer case which was presented in Chapter 4. We have simply allowed for m consumers, rather than just one, in the specification of the range of the polygonal paths; and we have substituted an indirect social preference relation for a single consumer's indirect preference relation. I will refer you to Section 10 of Chapter 4 for a justification of the use of line integrals in this definition, as well as for the independence of path requirement.

Very much as was the case in Chapter 4, standard results on line integrals tell us that if $\boldsymbol{f}$ satisfies Condition 1 of the above definition, then there exists a twice-differentiable *potential function*, $V\colon \boldsymbol{\Omega}^* \to \mathbb{R}$ such that, for all $(\boldsymbol{p}^0, \boldsymbol{w}^0), (\boldsymbol{p}^1, \boldsymbol{w}^1) \in \boldsymbol{\Omega}^*$ and all $\omega \in \mathcal{P}(\boldsymbol{\Omega}^*)$ satisfying (15.43), we will have:

$$\int_0^1 \boldsymbol{f}[\omega(t)] \cdot d\omega(t) = V(\boldsymbol{p}^1, \boldsymbol{w}^1) - V(\boldsymbol{p}^0, \boldsymbol{w}^0); \tag{15.45}$$

and, for all $(\boldsymbol{p}, \boldsymbol{w}) \in \boldsymbol{\Omega}^*$, V and $\boldsymbol{f}$ will satisfy:

$$\left.\frac{\partial V}{\partial p_j}\right|_{(\boldsymbol{p},\boldsymbol{w})} = f_j(\boldsymbol{p}, \boldsymbol{w}) \quad \text{for } j = 1, \ldots, n, \tag{15.46}$$

$$\left.\frac{\partial V}{\partial w_i}\right|_{(\boldsymbol{p},\boldsymbol{w})} = f_{n+i}(\boldsymbol{p}, \boldsymbol{w}) \quad \text{for } i = 1, \ldots, m, \tag{15.47}$$

and, for example (remember that V is twice-differentiable):

$$\left.\frac{\partial f_j}{\partial w_i}\right|_{(\boldsymbol{p},\boldsymbol{w})} = \left.\frac{\partial f_{n+i}}{\partial p_j}\right|_{(\boldsymbol{p},\boldsymbol{w})} \quad \text{for } j = 1, \ldots, n; i = 1, \ldots, m. \tag{15.48}$$

It then follows from (15.45), that if $\boldsymbol{f}$ furnishes an acceptable indicator of welfare change in Q^* on $\boldsymbol{\Omega}^*$, then the corresponding potential function, V, must be an indirect social welfare function representing Q^*. Now that we know this to be the case, the following generalization of the 'Antonelli-Allen-Roy' equation (Theorem 4.28)[19] shows us what form the function $\boldsymbol{f}$ must take.

[19] A slightly more general result is established in Chipman and Moore [1990, Theorem 4, p. 484].

15.22 Theorem. *Let $\boldsymbol{\Omega}^*$ be a nonempty open subset of $\boldsymbol{\Omega}$, let Q^* be an indirect social preference relation on $\boldsymbol{\Omega}^*$ which extends the weak Pareto ordering, and let $V\colon \boldsymbol{\Omega}^* \to \mathbb{R}$ be a differentiable function representing Q^* on $\boldsymbol{\Omega}^*$. Then, given any $(\boldsymbol{p}^0, \boldsymbol{w}^0) \in \boldsymbol{\Omega}^*$, V and $\boldsymbol{H}$ satisfy:*

$$\frac{\partial V}{\partial p_j}\bigg|_{(\boldsymbol{p}^0,\boldsymbol{w}^0)} = -\sum\nolimits_{i=1}^{m} \left[\frac{\partial V}{\partial w_i}\bigg|_{(\boldsymbol{p}^0,\boldsymbol{w}^0)}\right] h_{ij}(\boldsymbol{p}^0, w_i^0) \quad \textit{for } j = 1, \ldots, n. \tag{15.49}$$

Proof. From the definitions of individual demand and indirect preferences, it follows that, for each i ($i = 1, \ldots, m$);

$$\big(\forall(\boldsymbol{p}, w_i) \in \Omega\big)\colon \boldsymbol{p}\cdot \boldsymbol{h}_i(\boldsymbol{p}^0, w_i^0) \leq w_i \Rightarrow (\boldsymbol{p}, w_i) G_i^*(\boldsymbol{p}^0, w_i^0), \tag{15.50}$$

where G_i^* is the ith consumer's indirect preference relation. Therefore, since V represents Q^* on $\boldsymbol{\Omega}^*$, and Q^* extends the weak Pareto ordering, it follows that V is minimized at $(\boldsymbol{p}^0, \boldsymbol{w}^0)$, subject to:

$$\boldsymbol{p}\cdot \boldsymbol{h}_i(\boldsymbol{p}^0, w_i^0) \leq w_i \quad \text{for } i = 1, \ldots, m.$$

Consequently, it follows from the classical theory of constrained minimization that there exist multipliers $\lambda_i \in \mathbb{R}$, for $i = 1, \ldots, m$, such that:

$$\frac{\partial V}{\partial p_j}\bigg|_{(\boldsymbol{p}^0,\boldsymbol{w}^0)} - \sum\nolimits_{i=1}^{m} \lambda_i h_{ij}(\boldsymbol{p}^0, w_i^0) = 0 \quad \text{for } j = 1, \ldots, n; \tag{15.51}$$

and:

$$\frac{\partial V}{\partial w_i}\bigg|_{(\boldsymbol{p}^0,\boldsymbol{w}^0)} + \lambda_i = 0 \quad \text{for } i = 1, \ldots, m. \tag{15.52}$$

We then obtain (15.49) by substituting (15.52) into (15.51). □

From the above result and our earlier discussion we can now see a great deal about what form a function which furnishes an acceptable (integral) indicator of change in Q^* on $\boldsymbol{\Omega}^*$ must take. We must have:

$$f_j(\boldsymbol{p}, \boldsymbol{w}) = -\sum\nolimits_{i=1}^{m} f_{n+i}(\boldsymbol{p}, \boldsymbol{w}) h_{ij}(\boldsymbol{p}, w_i), \tag{15.53}$$

and:

$$f_j(\boldsymbol{p}, \boldsymbol{w}) = \frac{\partial V}{\partial w_i}\bigg|_{(\boldsymbol{p},\boldsymbol{w})} \quad \text{for } j = n+1, \ldots, n+m; \tag{15.54}$$

where V is the potential function associated with $\boldsymbol{f}$.

Turning things around, suppose Q^* is the indirect social preference relation determined by a PBS social welfare function which is, in turn, determined by the aggregator-measure function pair (μ, F). Suppose further that F is twice-differentiable, and that each utility function $u_i = \mu(G_i)$ is twice-differentiable as well. Then we know that the function $\boldsymbol{f}\colon \boldsymbol{\Omega} \to \mathbb{R}^{n+m}$ defined by:

$$f_j(\boldsymbol{p}, \boldsymbol{w}) = -\sum\nolimits_{i=1}^{m} \left(\frac{\partial F}{\partial u_i}\frac{\partial v_i}{\partial w_i}\bigg|_{(\boldsymbol{p},\boldsymbol{w})}\right) h_{ij}(\boldsymbol{p}, w_i) \quad \text{for } j = 1, \ldots, n; \tag{15.55}$$

$$f_{n+i}(\boldsymbol{p}, \boldsymbol{w}) = \frac{\partial F}{\partial u_i}\frac{\partial v_i}{\partial w_i}\bigg|_{(\boldsymbol{p},\boldsymbol{w})} \quad \text{for } i = 1, \ldots, m. \tag{15.56}$$

furnishes an acceptable measure of change in Q^* on $\mathbf{\Omega}^*$.

The problem with the formulas in (15.55) and (15.56), from an applied standpoint is that there are some really thorny estimation problems that must be dealt with in order to evaluate the integrals of interest. In the first place, each individual consumer's demand function must be estimated; a quite impractical chore in and of itself; but in fact, if each such function is estimated, then, generally speaking, one could then obtain each consumer's indirect utility function. Given these functions, and a knowledge of the appropriate aggregator function, the issue of whether a given $(\boldsymbol{p}, \boldsymbol{w})$-pair does or does not dominate a second such pair can be determined directly and exactly without resort to any line integral! Once again, however, things are a bit better in some special cases; one of which we set out in the following example.

15.23 Example. Suppose the PBS social welfare function of interest is of the CDE form, and that each u_i is positively homogeneous of degree one (as well as twice-differentiable). In this case, the indirect social preference function can be expressed in the form:

$$V(\boldsymbol{p}, \boldsymbol{w}) = \prod_{i=1}^{m} \left(\frac{w_i}{\gamma_i(\boldsymbol{p})}\right)^{a_i}, \tag{15.57}$$

where $\boldsymbol{a} \in \mathbb{R}^m_+$ is such that:

$$\sum_{i=1}^{m} a_i = 1 \text{ and } a_i > 0, \text{ for } i = 1, \ldots, m; \tag{15.58}$$

and $\gamma_i(\cdot)$ is the ith consumer's cost-of-living function. It will be convenient in this development, however, to take the log of the function in (15.57), to obtain the equivalent indirect social preference function given by:

$$v(\boldsymbol{p}, \boldsymbol{w}) = \sum_{i=1}^{m} a_i\big[\log w_i - \log \gamma_i(\boldsymbol{p})\big]. \tag{15.59}$$

In this case it follows from equations (15.55) and (15.56) that the function $\boldsymbol{f}\colon \mathbf{\Omega} \to \mathbb{R}^{n+m}$ defined by:

$$f_j(\boldsymbol{p}, \boldsymbol{w}) = -\sum_{i=1}^{m} (a_i/w_i)h_{ij}(\boldsymbol{p}, w_i) \quad \text{for } j = 1, \ldots, m; \tag{15.60}$$

$$f_{n+i}(\boldsymbol{p}, \boldsymbol{w}) = a_i/w_i \quad \text{for } i = 1, \ldots, m; \tag{15.61}$$

furnishes an acceptable (integral) measure of change in Q^* on $\mathbf{\Omega}$. We can actually simplify this function a bit further. Recall that, since each consumer's preferences are homothetic, $\boldsymbol{h}_i(\cdot)$ can be expressed in the form:

$$\boldsymbol{h}_i(\boldsymbol{p}, w_i) = \boldsymbol{g}_i(\boldsymbol{p})w_i \quad \text{for } i = 1, \ldots, m.$$

Consequently, we can simplify (15.60) to:

$$f_j(\boldsymbol{p}, \boldsymbol{w}) = -\sum_{i=1}^{m} a_i \cdot g_{ij}(\boldsymbol{p}) = -\sum_{i=1}^{m} h_{ij}(\boldsymbol{p}, a_i) \quad \text{for } i = 1, \ldots, m. \tag{15.62}$$

Now, in order to estimate $f_j(\cdot)$, one would presumably need to estimate the function:

$$\boldsymbol{h}(\boldsymbol{p}, \boldsymbol{w}) \stackrel{\text{def}}{=} \sum_{i=1}^{m} \boldsymbol{h}_i(\boldsymbol{p}, w_i). \tag{15.63}$$

However, notice that, while to estimate this function one needs observations on individual consumer incomes, one only needs observations on aggregate commodity demands. Of course, in practice, one would need to aggregate consumers into income classes, or occupation, or in some other meaningful fashion, as well as aggregating over commodities in some standard fashion. It is worth noting, however, that if one obtains an estimate of the function defined in (15.63), then one can define a function $\boldsymbol{f}$ to furnish an acceptable indicator of change in Q^*, for *any* indirect social preference function of the CDE form. That is, any $\boldsymbol{a} \in \mathbb{R}^m_+$ satisfying (15.58) determines a PBS social welfare function of the CDE form, and one can then define a function which furnishes an acceptable indicator of change for the associated indirect social preference by:

$$f_j(\boldsymbol{p}, \boldsymbol{w}) = -h_j(\boldsymbol{p}, \boldsymbol{a}) = -\sum\nolimits_{i=1}^{m} g_{ij}(\boldsymbol{p}) a_i \quad \text{for } j = 1, \ldots, m, \tag{15.64}$$

$$f_{n+i}(\boldsymbol{p}, \boldsymbol{w}) = a_i / w_i \quad \text{for } i = 1, \ldots, m. \quad \square \tag{15.65}$$

Let's now turn our attention to the development of a measure of consumers' surplus which is of a different form; namely adding values of compensating variation. Recall that in Section 10 of Chapter 4, we defined the **compensating variation criterion for welfare improvement** for a single consumer by:

$$W_i^C\left[(\boldsymbol{p}^1, w_i^1), (\boldsymbol{p}^2, w_i^2)\right] = w_i^2 - \mu_i(\boldsymbol{p}^2; \boldsymbol{p}^1, w_i^1). \tag{15.66}$$

This definition is then easily extended to obtain the **aggregate compensating variation obtained in moving from** $(\boldsymbol{p}^1, \boldsymbol{w}^1)$ **to** $(\boldsymbol{p}^2, \boldsymbol{w}^2)$ as:

$$W^C\left[(\boldsymbol{p}^1, \boldsymbol{w}^1), (\boldsymbol{p}^2, \boldsymbol{w}^2)\right] = \sum\nolimits_{i=1}^{m} \left[w_i^2 - \mu_i(\boldsymbol{p}^2; \boldsymbol{p}^1, w_i^1)\right]. \tag{15.67}$$

It is intuitively appealing to say that if a project or policy would result in a change from a first price-wealth pair, $(\boldsymbol{p}^1, \boldsymbol{w}^1)$, to a second, $(\boldsymbol{p}^2, \boldsymbol{w}^2)$, and it is estimated (accurately, we will assume) that $W^C\left[(\boldsymbol{p}^1, \boldsymbol{w}^1), (\boldsymbol{p}^2, \boldsymbol{w}^2)\right]$ is positive, then the change should be made. However, it is doubtful whether anyone would advocate this as a welfare criterion in and of itself; in fact, a quite persuasive critique of its use as a welfare measure is contained in Blackorby and Donaldson [1990], and I will refer you to their article for a detailed critique of the use of aggregate compensating variation in applied welfare analysis. In the remainder of this section, we will turn our attention to a closely related concept; aggregate 'willingness to pay.' In our discussion we will make use of some new concepts and notation, as follows.

It is reasonable to suppose that a consumer's preferences depend upon not only her or his privately-purchased consumption goods but also on publicly-provided commodities (for example, parks, bridges, roads, and so on). Consequently, in the remainder of this section, we will suppose the ith consumer's consumption set takes the form $C_i = \mathbb{R}^n_+ \times Y_i$, where Y_i is something which we will call the ith consumer's 'public goods space,' and we will use the generic notation '$(\boldsymbol{x}_i, \boldsymbol{y}_i)$' to denote elements of C_i. We will study public goods in some detail in the next chapter, but for now we will simply suppose that elements $y_i \in Y_i$ are things which contribute to consumer i's well-being, but which are not privately purchased. The vector of private goods chosen by consumer i, given a price vector for private goods, $\boldsymbol{p}$, and an income (or wealth), w_i, may depend not only on these variables, but also upon the public goods available to her or him. Consequently, we should in principle consider the ith consumer's demand function h_i to be defined on $\Omega \times Y_i$, and write '$\boldsymbol{h}_i(\boldsymbol{p}, w_i, y_i)$' in place of '$\boldsymbol{h}_i(\boldsymbol{p}, w_i)$' to denote values of i's demand function. Allowing for this complication in the analysis to follow will not complicate things for us at all, for we will be making use of indirect preferences, which are now defined on $\Omega \times Y_i$ for the ith consumer by:

$$(\boldsymbol{p}^2, w_i^2, \boldsymbol{y}_i^2)G_i^*(\boldsymbol{p}^1, w_i^1, \boldsymbol{y}_i^1) \iff \big(h_i(\boldsymbol{p}^2, w_i^2, \boldsymbol{y}^2), \boldsymbol{y}_i^2\big)G_i\big(h_i(\boldsymbol{p}^1, w_i^1, \boldsymbol{y}^1), \boldsymbol{y}_i^1\big).$$

We will make use of the notational framework just introduced in the example which follows. It is a fairly extended discussion of the notion of 'aggregate willingness to pay,' which is a concept closely related to aggregate compensating variation.

15.24 Example. Suppose a project or a policy change is being considered for the economy, which we will characterize as involving a change from $(\boldsymbol{p}^1, \boldsymbol{w}^1, \boldsymbol{y}^1)$ to $(\boldsymbol{p}^2, \boldsymbol{w}^2, \boldsymbol{y}^2)$; where:

$$y_i^t \in Y_i \quad \text{for } i =, 1, \ldots, m; t = 1, 2.$$

The change may, in fact involve the building of some public project, and be such as to cause no change (or a negligible change) in the prices of marketed commodities; in which case we would suppose $\boldsymbol{y}^2 \neq \boldsymbol{y}^1$, while $\boldsymbol{p}^1 = \boldsymbol{p}^2$ (although we might nonetheless have $\boldsymbol{w}^1 \neq \boldsymbol{w}^2$). As an opposite case, we could be dealing with a policy change which would result in $\boldsymbol{y}^1 = \boldsymbol{y}^2$, while changing some prices and income. As it turns out, in our treatment we needn't distinguish between such cases; the change might be a public project or a policy change, or involve changes in both. However, we will refer to the change being contemplated as being the construction of some public project.

We will suppose that, in order to assess the desireability of undertaking the project, a survey is taken in which each consumer is questioned about her or his willingness to pay for this project. We will denote by 'c_i' the ith consumer's stated willingness to pay for the project; which may be a negative number if the consumer does not wish the project to be undertaken; and we will suppose that c_i, is such that if w_i^2 is the income consumer i expects to earn after the change, in the absence of any tax assessed to pay for the project, that we have:

$$(\boldsymbol{p}^2, w_i^2 - c_i, \boldsymbol{y}_i^2)G_i^*(\boldsymbol{p}^1, w_i^1, \boldsymbol{y}_i^1), \tag{15.68}$$

for each i. In other words, we are supposing that consumer i states an amount c_i which would leave her or him at least as well off after the change as before even if she or he were to have the amount c_i subtracted from her or his income after the change. Before proceeding further, we should note that if we ignore the presence of public goods, then the condition in (15.68) will hold if, and only if:

$$w_i^2 - c_i = \mu_i(\boldsymbol{p}^2; \boldsymbol{p}^2, w_i^2 - c_i) \geq \mu_i(\boldsymbol{p}^2; \boldsymbol{p}^1, w^1);$$

so that:

$$c_i \leq W_i^C\big[(\boldsymbol{p}^1, w_i^1), (\boldsymbol{p}^2, w_i^2)\big] \quad \text{for } i = 1, \ldots, m. \tag{15.69}$$

Returning to the general case, suppose that we have a project for which:

$$\sum_{i=1}^{m} c_i > C, \tag{15.70}$$

where C is the cost of carrying out the project; and let $\delta > 0$ be any number such that:

$$0 < \delta \leq \sum_{i=1}^{m} c_i - C. \tag{15.71}$$

Imagine now that, as the project is completed, we first increase every consumer's income in the amount δ/m, and then assess a 'tax', t_i, on each consumer to pay for the project (and the increase in incomes), where t_i is defined as follows. We begin by defining the set I_1 by:

$$I_1 = \{i \in \{1, \ldots, m\} \mid c_i \leq 0\}.$$

Thus I_1 is the set of consumers who perceive themselves as receiving no benefit, or as being damaged by the project. We then define $I_2 = \{1, \ldots, m\} \setminus I_1$:

$$\lambda = -\sum_{i \in I_1} c_i \text{ and } \gamma = \sum_{i \in I_2} c_i, \tag{15.72}$$

and $a \in \mathbb{R}_+$ by:

$$a = \frac{C + \delta + \lambda}{\gamma}. \tag{15.73}$$

Finally, we define $\boldsymbol{t} = (t_1, \ldots, t_m)$ by:

$$t_i = \begin{cases} c_i & \text{for } i \in I_1, \text{ and} \\ a \cdot c_i & \text{for } i \in I_2. \end{cases} \tag{15.74}$$

Now, the ith consumer's net income after the change and after the adjustments are made is given by:

$$\widehat{w}_i = w_i^2 + \delta/m - t_i. \tag{15.75}$$

For $i \in I_1$, it then follows at once from the definition of t_i and (15.68) that:

$$(\boldsymbol{p}^2, \widehat{w}_i, \boldsymbol{y}_i^2) P_i^* (\boldsymbol{p}^1, w_i^1, \boldsymbol{y}_i^1);$$

since for each such consumer we obviously have $\widehat{w}_i > w_i^2 - c_i$. Furthermore, we note that it follows from (15.71) that

$$\gamma - \lambda - C - \delta = \sum_{i=1}^{m} c_i - C - \delta \geq 0,$$

and therefore we have:

$$0 < a \leq 1. \tag{15.76}$$

Consequently, for $i \in I_2$, we see that:

$$t_i = a \cdot c_i \leq c_i,$$

and thus:

$$\widehat{w}_i = w_i^2 + \delta/m - t_i > w_i^2 - c_i;$$

and thus it follows readily from (15.68) that each individual $i \in I_2$ is also better off than before the change. Finally, we have:

$$\sum_{i=1}^{m} t_i = \sum_{i \in I_1} t_i + \sum_{i \in I_2} t_i = -\lambda + \sum_{i \in I_2} a \cdot c_i = -\lambda + a \cdot \gamma = -\lambda + C + \delta + \lambda = C + \delta$$

Therefore, the tax bill covers both the cost of the project and the income subsidy. □

In the above example we have shown that if aggregate willingness to pay for a project is greater than the cost of the project, then there may be a strongly Pareto superior improvement if the project is undertaken, *and if there is an appropriate allocation of the cost and appropriate compensation is paid.* However, it is obvious that the fact that aggregate willingness to pay for a project exceeds its cost does not guarantee that carrying it out will result in Pareto improvement in the absence of such cost allocation and compensation. It should also be noted that if consumers are aware of the plan to allocate costs and compensation after the adoption of the project or policy, then they have a very strong incentive to misrepresent their 'willingness-to-pay' amounts, c_i. In fact, there is every incentive for consumers to claim that they will suffer large damage if the policy is adopted; given the scheme we have set out in the above example. We will return to a discussion of this issue in Chapter 18

Exercises.

1. Prove Proposition 15.11

2. Verify the assertion of Example 15.12

3. Verify the calculations in Example 15.18.

4. Show that if we confine our admissible preference profiles to $\mathcal{G}^h$, then a PBS social welfare function defined from an aggregator function of the CDE form induces the same social preference relation on $\mathcal{X}$ regardless of the measurement function with which it is paired.

5. Moulin [1988, pp. 51–2] defines, for a given $\boldsymbol{w} \in \mathbb{R}^m_+$, the *equally distributed equivalent income*, $e(\boldsymbol{w}; F)$, as that income, which given any $\boldsymbol{p} \in \mathbb{R}^n_{++}$, solves the equation:

$$F\Big[\frac{e(\boldsymbol{w}; F)}{\gamma(\boldsymbol{p})}, \ldots, \frac{e(\boldsymbol{w}; F)}{\gamma(\boldsymbol{p})}\Big] = F\Big[\frac{w_1}{\gamma(\boldsymbol{p})}, \ldots, \frac{w_m}{\gamma(\boldsymbol{p})}\Big]. \tag{15.77}$$

Show that, given the assumptions of Section 6, this yields:

$$e(\boldsymbol{w}; F) = \frac{F(\boldsymbol{w})}{F(\boldsymbol{1})}. \tag{15.78}$$

Moulin then defines the **inequality index**, $J(\boldsymbol{w}; F)$, by:

$$J(\boldsymbol{w}; F) = 1 - \frac{e(\boldsymbol{w}; F)}{\overline{w}}. \tag{15.79}$$

Show that, if F is egalitarian, then this agrees with the income distribution index defined in Section 6.

6. Verify the figures given in Example 15.20

7. Returning to the sort of situation contemplated in Example 15.24, show that if we ignore the public goods aspect of that example, then it is true that if aggregate compensating variation is greater than the cost of undertaking a policy change, there is a way of allocating the cost of the change which guarantees that the net effect will be strongly Pareto-improving. [Hint: consider equation (15.69).]

Chapter 16

Public Goods

16.1 Introduction

In this chapter we will examine a bit of the economic theory of public goods. We will begin our study by conducting a brief analysis of a simple general equilibrium model which can be used in the analysis of both public goods and externalities (and thus will be used in Chapter 17, as well as the present chapter). In Section 3 we discuss the basic definition of public goods, as well as the distinction between public and private goods. Section 4 deals with the simple general equilibrium model which will be the primary tool used in our analysis of public goods allocation. This model is then used in Section 5 to present the basic theory of Lindahl and Ratio equilibrium. In Section 6 we develop a much more general model of public goods production, and show that in this context, the Lindahl equilibrium is both 'non-wasteful' and 'unbiased;' in other words, we show that Lindahl equilibria in this model are Pareto efficient; and, with appropriate assumptions, given any Pareto efficient allocation, there exist Lindahl prices such that the allocation is (theoretically) attainable as a Lindahl equilibrium.

16.2 A Simple Model

In this section we consider an m-agent, $(1+n)$-commodity economy, E. We will use the generic notation, '$(w_i, \boldsymbol{y}_i)$,' where $w_i \in \mathbb{R}$ and $\boldsymbol{y}_i \in \mathbb{R}^n$ to denote the commodity bundle available to the ith agent, for $i = 1, \ldots, m$. We will suppose that the ith consumer's 'consumption' set, Z_i, takes the form:

$$Z_i = W_i \times Y_i,$$

where Y_i is a nonempty subset of $\mathbb{R}^n$, and W_i takes the form:

$$W_i = \{w_i \in \mathbb{R} \mid w_i \geq \widehat{w}_i\},$$

for some (fixed) $\widehat{w}_i \in \mathbb{R}$. We then suppose that the ith agent's payoff, or utility function, u_i, is of the 'quasi-linear' form:

$$u_i(w_i, \boldsymbol{y}_i) = w_i + v_i(\boldsymbol{y}_i);$$

and we take as given a 'cost function;'

$$c\colon \boldsymbol{Y} \to \mathbb{R}_+$$

[in some applications we may take $c(\cdot)$ to be identically zero], where $\boldsymbol{Y}$ is a nonempty subset of $\prod_{i=1}^m Y_i$. Finally, we suppose that the attainable set for E, $A(E)$, takes the form:

$$A(E) = \Big\{\big\langle(w_i, \boldsymbol{y}_i)\big\rangle \in \boldsymbol{Z} \mid \langle \boldsymbol{y}_i\rangle \in \boldsymbol{Y} \ \&\ \sum\nolimits_{i=1}^m w_i + c(\boldsymbol{y}) \leq \bar{w}\Big\},$$

where:

$$\boldsymbol{Z} = \prod\nolimits_{i=1}^m Z_i,$$

$\bar{w} > 0$ is the total endowment of the 'commodity,' w, and we assume that:

$$\bar{w} > \sum\nolimits_{i=1}^m \widehat{w}_i.$$

We will use the generic notation:

$$\boldsymbol{y} = (\boldsymbol{y}_1, \ldots, \boldsymbol{y}_m),$$

to denote elements of $\prod_{i=1}^m Y_i$, we define:

$$\boldsymbol{W} = \prod\nolimits_{i=1}^m W_i,$$

and we use the generic notation:

$$\boldsymbol{w} = (w_1, \ldots, w_m),$$

to denote elements of $\boldsymbol{W}$.

The model, and the variables therein, can be interpreted in many different ways, and this is one of the strengths of the model. On the other hand, the absence of a specific interpretation may also make it a bit more difficult to understand what is going on in the discussion to follow. Consequently, I think it is helpful as we begin our analysis to look at one specific interpretation and application for which the model can be utilized. Let's suppose that there are n public goods and one private good in an economy, and in this application we can take:

$$Y_i = \mathbb{R}^n_+ \quad \text{for } i = 1, \ldots, m.$$

In a pure public goods model, which we will consider here, each consumer consumes the same amount of the public goods; and to express this in the present framework, we will take $\boldsymbol{Y}$ to be the set defined by:

$$\boldsymbol{Y} = \{\boldsymbol{y} \in \prod_{i=1}^m Y_i \mid \boldsymbol{y}_1 = \boldsymbol{y}_2 = \cdots = \boldsymbol{y}_m\}.$$

The production possibilities for public goods are then summarized by the cost function, $c(\cdot)$, which can then be viewed as expressing the quanitity of the private good which must be used as an input to produce the public goods vector $\boldsymbol{y}$ (which in this context can be viewed as an element of $\mathbb{R}^n_+$). Finally, in this particular interpretation, we would probably want to take $\widehat{w}_i = 0$, for each i.

Now, returning to the generic model, if we assume that $\boldsymbol{Y}$ is compact,[1] that $v_i \colon Y_i \to \mathbb{R}$ is continuous, for $i = 1, \ldots, m$ and that c is continuous as well, then there exists $\boldsymbol{y}^* \in \boldsymbol{Y}$ satisfying:[2]

$$(\forall \boldsymbol{y} \in \boldsymbol{Y})\colon \sum_{i=1}^m v_i(\boldsymbol{y}_i^*) - c(\boldsymbol{y}^*) \geq \sum_{i=1}^m v_i(\boldsymbol{y}_i) - c(\boldsymbol{y}). \tag{16.1}$$

This fact provides the motivation for the following.

16.1 Proposition. *If $\boldsymbol{y}^* \in \boldsymbol{Y}$ satisfies (16.1), and if $c(\boldsymbol{y}^*) \leq \bar{w}$, then for all $\langle w_i^* \rangle \in \boldsymbol{W}$ satisfying:*

$$\sum_{i=1}^m w_i^* = \bar{w} - c(\boldsymbol{y}^*), \tag{16.2}$$

the allocation $\langle (w_i^*, \boldsymbol{y}_i^*) \rangle$ is strongly Pareto efficient for E.

Proof. Obviously such an allocation is feasible for E. To complete our proof, suppose, by way of obtaining a contradiction, that there exists $\langle (w_i, \boldsymbol{y}_i) \rangle \in A(E)$ such that:

$$u_i(w_i, \boldsymbol{y}_i) \geq u_i(w_i^*, \boldsymbol{y}_i^*) \quad \text{for } i = 1, \ldots, m, \tag{16.3}$$

and, for some $j \in \{1, \ldots, m\}$:

$$u_j(w_j, \boldsymbol{y}_j) > u_j(w_j^*, \boldsymbol{y}_j^*). \tag{16.4}$$

Adding (16.3) and (16.4) over i, we obtain:

$$\sum_{i=1}^m u_i(w_i, \boldsymbol{y}_i) > \sum_{i=1}^m u_i(w_i^*, \boldsymbol{y}_i^*);$$

which, making use of the quasi-linear form of the utility functions and equation (16.2), implies:

$$\sum_{i=1}^m v_i(\boldsymbol{y}_i) + \sum_{i=1}^m w_i > \sum_{i=1}^m v_i(\boldsymbol{y}_i^*) + \bar{w} - c(\boldsymbol{y}^*). \tag{16.5}$$

However, since $\langle (w_i, \boldsymbol{y}_i) \rangle$ is feasible for E, we have:

$$\bar{w} - \sum_{i=1}^m w_i - c(\boldsymbol{y}) \geq 0,$$

and thus from (16.5), we obtain:

$$\sum_{i=1}^m v_i(\boldsymbol{y}_i) + \sum_{i=1}^m w_i + \bar{w} - \sum_{i=1}^m w_i - c(\boldsymbol{y}) > \sum_{i=1}^m v_i(\boldsymbol{y}_i^*) + \bar{w} - c(\boldsymbol{y}^*),$$

so that:

$$\sum_{i=1}^m v_i(\boldsymbol{y}_i) - c(\boldsymbol{y}) > \sum_{i=1}^m v_i(\boldsymbol{y}_i^*) - c(\boldsymbol{y}^*);$$

contrary to our assumption. □

We can also prove a partial converse of the above result, as follows.[3]

[1] As will be the case, for example, if each Y_i is compact, and $\boldsymbol{Y} = \prod_{i=1}^m Y_i$.

[2] Under the present assumptions the tuple $\boldsymbol{y}^*$ is not necessarily unique, but this makes no difference whatsoever in our analysis.

[3] Notice that we did not use any continuity assumptions in Proposition 16.1; nor did we need to assume that $\boldsymbol{Y}$ was compact.

16.2 Proposition. *Suppose that $\boldsymbol{Y}$ is a non-empty convex subset of $\prod_{i=1}^m Y_i$, that v_i is concave and continuous, for $i = 1, \ldots, m$, and that $c(\cdot)$ is convex. Then if $\langle(w_i^*, \boldsymbol{y}_i^*)\rangle$ is a Pareto efficient allocation for E which satisfies:*

$$w_i^* > \widehat{w}_i \quad \text{for } i = 1, \ldots, m, \tag{16.6}$$

it must be the case that $\langle(w_i^*, \boldsymbol{y}_i^*)\rangle$ also satisfies:

$$\sum_{i=1}^m w_i^* + c(\boldsymbol{y}^*) = \bar{w} \tag{16.7}$$

and:

$$(\forall \langle \boldsymbol{y}_i \rangle \in \boldsymbol{Y})\colon \sum_{i=1}^m v_i(\boldsymbol{y}_i^*) - c(\boldsymbol{y}^*) \geq \sum_{i=1}^m v_i(\boldsymbol{y}_i) - c(\boldsymbol{y}). \tag{16.8}$$

Proof. Obviously (16.7) must hold, so that we need only to prove that (16.8) is satisfied. Accordingly, suppose, by way of obtaining a contradiction, that there exists $\langle \widehat{\boldsymbol{y}}_i \rangle \in \boldsymbol{Y}$ satisfying:

$$\sum_{i=1}^m v_i(\widehat{\boldsymbol{y}}_i) - c(\widehat{\boldsymbol{y}}) > \sum_{i=1}^m v_i(\boldsymbol{y}_i^*) - c(\boldsymbol{y}^*). \tag{16.9}$$

Now, from (16.9), the concavity of the v_i functions, and the convexity of c, we see that for all $\theta \in \,]0, 1]$, we have:

$$\begin{aligned}\sum_{i=1}^m v_i\big[\theta\widehat{\boldsymbol{y}}_i + (1-\theta)\boldsymbol{y}_i^*\big] &- c\big[\theta\widehat{\boldsymbol{y}} + (1-\theta)\boldsymbol{y}^*\big] \\ &\geq \theta\Big(\sum_{i=1}^m v_i(\widehat{\boldsymbol{y}}_i) - c(\widehat{\boldsymbol{y}})\Big) + (1-\theta)\Big(\sum_{i=1}^m v_i(\boldsymbol{y}_i^*) - c(\boldsymbol{y}^*)\Big) \\ &> \sum_{i=1}^m v_i(\boldsymbol{y}_i^*) - c(\boldsymbol{y}^*).\end{aligned}$$

Moreover, from the continuity of each v_i and (16.6), we see that there exists $\theta^\dagger \in \,]0, 1]$ small enough so that, defining:

$$\boldsymbol{y}_i^\dagger = \theta^\dagger\widehat{\boldsymbol{y}}_i + (1-\theta^\dagger)\boldsymbol{y}_i^* \quad \text{for } i = 1, \ldots, m,$$

we must have:

$$w_i^* - \widehat{w}_i - \big[v_i(\boldsymbol{y}_i^\dagger) - v_i(\boldsymbol{y}_i^*)\big] > 0 \quad \text{for } i = 1, \ldots, m. \tag{16.10}$$

Next, define $g > 0$ by:

$$g = \Big(\sum_{i=1}^m v_i(\boldsymbol{y}_i^\dagger) - c(\boldsymbol{y}^\dagger)\Big) - \Big(\sum_{i=1}^m v_i(\boldsymbol{y}_i^*) - c(\boldsymbol{y}^*)\Big),$$

the (adjustments to w_i^*) terms t_i by:

$$t_i = v_i(\boldsymbol{y}_i^*) - v_i(\boldsymbol{y}_i^\dagger) + g/m$$

and the 'wealth' terms $w_i^\dagger$ by:

$$w_i^\dagger = w_i^* + t_i,$$

for $i = 1, \ldots, m$. It then follows from (16.10) that, for each i:

$$w_i^\dagger - \widehat{w}_i = w_i^* + t_i - \widehat{w}_i = w_i^* + \left[v_i(\boldsymbol{y}_i^*) - v_i(\boldsymbol{y}_i^\dagger) + g/m\right] - \widehat{w}_i$$
$$> w_i^* + \left[v_i(\boldsymbol{y}_i^*) - v_i(\boldsymbol{y}_i^\dagger)\right] - \widehat{w}_i > 0;$$

while from the definitions, we have:

$$\begin{aligned}\sum_{i=1}^m w_i^\dagger + c(\boldsymbol{y}^\dagger) &= \sum_{i=1}^m w_i^* + \sum_{i=1}^m t_i + c(\boldsymbol{y}^\dagger)\\ &= \sum_{i=1}^m w_i^* + \sum_{i=1}^m \left[v_i(\boldsymbol{y}_i^*) - v_i(\boldsymbol{y}_i^\dagger) + g/m\right] + c(\boldsymbol{y}^\dagger)\\ &= \sum_{i=1}^m w_i^* + \sum_{i=1}^m v_i(\boldsymbol{y}_i^*) - \sum_{i=1}^m v_i(\boldsymbol{y}^\dagger) + g + c(\boldsymbol{y}^\dagger)\\ &= \sum_{i=1}^m w_i^* + c(\boldsymbol{y}^*).\end{aligned}$$

Thus we see that the allocation $\langle(w_i^\dagger, \boldsymbol{y}_i^\dagger)\rangle$ is feasible for E. However, we also have, for each i:

$$\begin{aligned}v_i(\boldsymbol{y}_i^\dagger) + w_i^\dagger &= v_i(\boldsymbol{y}_i^\dagger) + w_i^* + \left[v_i(\boldsymbol{y}_i^*) - v_i(\boldsymbol{y}_i^\dagger) + g/m\right]\\ &= v_i(\boldsymbol{y}_i^*) + w_i^* + g/m > v_i(\boldsymbol{y}_i^*) + w_i^*.\end{aligned}$$

and thus we have obtained a contradiction to the assumption that $\langle(w_i^*, \boldsymbol{y}_i^*)\rangle$ is Pareto efficient for E. □

16.3 Public Goods

The basic distinction between public, private, semi-private, and so on, goods hinges around two issues: is the good **rivalrous**? Is the good **excludable**? A good is non-rivalrous if its consumption by one consumer does not prevent another's consumption. For example, scenery, or the enjoyment thereof, is a non-rivalrous good. Another example is national defense. A good is non-excludable if property rights or convention do not allow the owner of the good to exclude individuals from consuming it. For example, TV broadcasts (over the airways) are a non-excludable good. Thus we have the categorization in Table 16.1, below.

	Fully Rivalrous	**Partially Rivalrous**	**Non-Rivalrous**
Fully Excludable	Private Goods	Club Goods (Golf courses)	Merit Goods (Opera)
Partially Excludable		[Specialty Stores (?)]	(Satellite TV)
Non-excludable		(Public Parks)	Pure Public Goods (Pollution Control)

Table 16.1: Categorization of Commodities.

Up to this point we have been considering private goods in this book; however, in the next sections we will be discussing pure public goods. In the meantime, let me note that, while I think the table is self-explanatory, a couple of the categories could probably use a word or two of explanation. First of all, while opera is often used as an example of a 'merit good' (fully excludable and non-rivalrous), one's enjoyment of an opera is likely to depend upon where one is sitting during the performance. Consequently, we cannot simply suppose that everyone attending an opera is consuming (the same quantity of) the same good; which is the assumption typically made regarding non-rivalrous goods. The 'specialty shop' category (partially rivalrous and partially excludable) is not a category, nor an example, which is standard in the literature on public goods. The idea here is that specialty shops, for example, camera shops, supply two goods: a physical commodity and a service (expertise). Unfortunately for such shops, it is fairly easy for individuals to 'free ride' on the expertise, and then buy the physical commodity at a discount store. Finally, the example 'Satellite TV' might better be included in the merit good category, particularly if one is thinking of the newer DSS type satellites. However, one used to be able to purchase decoders which enabled one to receive programs on large satellite dishes without paying a fee for the broadcast.[4]

To repeat the statement made earlier, in the next three sections we will be considering only pure public goods; those which are non-excludable and non-rivalrous.

16.4 A Simple Public Goods Model

In this section we will develop the basics of a standard public goods model. We will suppose that there are fixed quantities of a private good available, and that this private good can either be made available for consumption by one of the m consumers, or can be used to produce one or all of n (public) goods. We suppose that the technology of public goods production can be represented by the cost function $c\colon \mathbb{R}^n_+ \to \mathbb{R}_+$; and, since we can normalize to set the price of the private good equal to one, for a given $\boldsymbol{y} \in \mathbb{R}^n_+$, the value of $c(\boldsymbol{y})$ is, effectively, the amount of the private good which must be used as an input in order to produce:

$$\boldsymbol{y} = (y_1, \dots, y_n).$$

We will use the generic notation $(x_i, \boldsymbol{y}) \in \mathbb{R}^{1+n}_+$ to denote the consumption bundle available to the ith consumer. Notice that there is no subscript attached to the public goods; we are here considering pure public goods, so that the vector of the public goods ($\boldsymbol{y}$) available is the same for every consumer. In this development we will suppose that the ith consumer's preferences can be represented by a continuously differentiable utility function, and, for the sake of simplicity, that for all $(x_i, \boldsymbol{y}) \in \mathbb{R}^{1+n}_+$:

$$\frac{\partial u_i}{\partial x_i} > 0 \text{ and } \frac{\partial u_i}{\partial y_j} \geq 0, \tag{16.11}$$

and that $u_i(\cdot)$ is strictly quasi-concave. Finally, we will denote the total amount of the private good available in the economy by 'w,' and we will assume that $w > 0$. Thus we make use of the following definition of a feasible allocation.

[4]These decoders were and are, apparently, illegal to use; however, they could be purchased legally. Go figure!

16.3 Definition. We shall say that an allocation $\boldsymbol{a}^* = (\boldsymbol{x}^*, \boldsymbol{y}^*) \in \mathbb{R}_+^{m+n}$ is **feasible** (or **attainable**) iff:

$$\sum\nolimits_{i=1}^{m} x_i^* + c(\boldsymbol{y}^*) \leq w.$$

We begin by considering necessary conditions for Pareto efficiency. We can develop these by considering the problem:

$$\max_{w.r.t.\boldsymbol{x},\boldsymbol{y}} u_1(x_1, \boldsymbol{y}),$$

subject to:

$$\begin{aligned} &(\lambda_i) \qquad && u_i(x_i, \boldsymbol{y}) - \overline{u}_i = 0 \quad \text{for } i = 2, \ldots, m, \\ &(\mu) && w - \sum\nolimits_{i=1}^{m} x_i - c(\boldsymbol{y}) = 0. \end{aligned} \tag{16.12}$$

Forming the appropriate Lagrangian (with the multipliers indicated in the above equations), and taking first derivatives, we obtain the first order conditions:

$$\frac{\partial u_1}{\partial x_1} - \mu = 0, \tag{16.13}$$

$$\frac{\partial u_1}{\partial y_j} + \sum\nolimits_{i=2}^{m} \lambda_i \frac{\partial u_i}{\partial y_j} - \mu \frac{\partial c}{\partial y_j} = 0, \quad \text{for } j = 1, \ldots, n; \tag{16.14}$$

$$\lambda_i \frac{\partial u_i}{\partial x_i} - \mu = 0 \quad \text{for } i = 2, \ldots, m. \tag{16.15}$$

Now, from (16.15), we have:

$$\lambda_i = \mu \Big/ \frac{\partial u_i}{\partial x_i} \quad \text{for } i = 2, \ldots, m; \tag{16.16}$$

while from (16.13), we have:

$$\frac{\partial u_1}{\partial x_1} = \mu. \tag{16.17}$$

Substituting (16.16) into (16.14, we have:

$$\frac{\partial u_1}{\partial y_j} + \sum\nolimits_{i=2}^{m} \mu \Big(\frac{\partial u_i}{\partial y_j} \Big/ \frac{\partial u_i}{\partial x_i} \Big) = \mu \frac{\partial c}{\partial y_j};$$

so that, upon making use of (16.17), we obtain the **Samuelson Conditions**:

$$\sum\nolimits_{i=1}^{m} \Big(\frac{\partial u_i}{\partial y_j} \Big/ \frac{\partial u_i}{\partial x_i} \Big) = \frac{\partial c}{\partial y_j}, \quad \text{for } j = 1, \ldots, n; \tag{16.18}$$

While we will consider a more general public goods model in Section 6 of this chapter, it is also useful to consider several special cases as well. Notice that our formulation allows for joint production of the public goods. Interesting special cases of the assumed production technology involve separable cost functions, which we will consider in connection with Lindahl and Ratio Equilibria:

$$c(\boldsymbol{y}) = \sum\nolimits_{j=1}^{n} c_j(y_j); \tag{16.19}$$

or, still more specialized:

$$c(\boldsymbol{y}) = \sum_{j=1}^{n} \gamma_j \cdot y_j, \tag{16.20}$$

where γ_j is a positive constant, for $j = 1, \ldots, n$.

Turning to the other side of the market, it is very common in the literature to make use of the assumption that each utility function takes the quasi-linear form:

$$u_i(x_i, \boldsymbol{y}) = x_i + v_i(\boldsymbol{y}). \tag{16.21}$$

In this case, the present model becomes a special case of the 'simple model' presented in Section 2 of this chapter. In fact, it follows from Proposition 16.1 that if $\boldsymbol{y}^* \in \mathbb{R}_+^n$ is such that $c(\boldsymbol{y}^*) \leq w$ and satisfies:

$$(\forall \boldsymbol{y} \in \mathbb{R}_+^n) \colon \sum_{i=1}^{m} v_i(\boldsymbol{y}^*) - c(\boldsymbol{y}^*) \geq \sum_{i=1}^{m} v_i(\boldsymbol{y}) - c(\boldsymbol{y}), \tag{16.22}$$

then, given any $\boldsymbol{x}^* \in \mathbb{R}_+^m$ satisfying:

$$\sum_{i=1}^{m} x_i^* = w - c(\boldsymbol{y}^*),$$

the allocation $\big(\langle x_i^* \rangle, \boldsymbol{y}^*\big)$ is Pareto efficient for E. If each $v_i(\cdot)$ and $c(\cdot)$ are differentiable, then a necessary condition for (16.22) to hold is that (16.18) holds; which brings us back to the Samuelson conditions by a somewhat different route. From these considerations, we can also deduce the following result, although I will leave the details of the proof as an exercise (see Exercise 4, at the end of this chapter).

16.4 Proposition. *Given the assumptions of this section, and that the consumers' utility functions are of the form indicated in (16.21), above, we have the following. If:*

1. *$v_i(\cdot)$ is concave, for $i = 1, \ldots, m$,*
2. *the cost function, $c(\cdot)$ is convex, and*
3. *at least one of the v_i functions is strictly concave, or c is strictly convex,*

then the Pareto efficient vector of public goods (if one exists) is unique.

Incidentally, if $c(\cdot)$ is continuous and:

$$Y \stackrel{\text{def}}{=} \{\boldsymbol{y} \in \mathbb{R}_+^n \mid c(\boldsymbol{y}) \leq w\}$$

is bounded, while the v_i functions appearing in (16.21) are all continuous, then it follows from Proposition 16.1 that a Pareto efficient allocation will exist for the economy under the maintained assumptions of this section.

Now let's turn to the special case of this model most often appearing in the literature; that in which we have just one public good, and where a production function $f \colon \mathbb{R}_+ \to \mathbb{R}_+$ delineates the technology for producing the public good from the private good. In this case, the Samuelson conditions take the form (see Exercise 6, at the end of this chapter):

$$\sum_{i=1}^{m} \left(\frac{\partial u_i}{\partial y} \Big/ \frac{\partial u_i}{\partial x_i}\right) = \frac{1}{f'(z^*)}, \tag{16.23}$$

where $f(z^*)$ is the Pareto efficient quantity of the public good, and z^* is the amount of the private good needed to produce it.

Now suppose we compare this with what might be achieved in the way of public goods production if individuals decided independently (and selfishly) how much to contribute to public goods production. As before we will normalize to set p, the price of the private good, equal to one. Consequently if we denote the contribution made by individuals other than agent i by 'z_{-i},' individual i's maximization problem can be expressed as:

$$\max_{w.r.t.\ z} u_i\big[w_i - z_i, f(z_i + z_{-i})\big]. \tag{16.24}$$

Taking first-order conditions, we have:

$$-\frac{\partial u_i}{\partial x_i} + \frac{\partial u_i}{\partial y} f'(z_i + z_{-i}) = 0;$$

so that:

$$\frac{\partial u_i}{\partial y} \Big/ \frac{\partial u_i}{\partial x_i} = \frac{1}{f'(z_i + z_{-i})}. \tag{16.25}$$

Comparing (16.25) and (16.23), it is apparent that this 'voluntary contributions equilibrium' will not achieve Pareto efficiency. The possible extent of the difference is illustrated in the following simple example.

16.5 Example. Suppose that each of the m consumers has the utility function:

$$u_i(x_i, y) = x_i + \alpha \log y$$

where $\alpha > m$, and that the production function for the public good is given by $y = z$. In this case, equation (16.25) becomes:

$$\frac{\alpha}{z_i + z_{-i}} = 1,$$

which implies $z_i = \alpha/m$ and $y = \alpha$. On the other hand, (16.23) becomes:

$$\frac{m\alpha}{y} = 1,$$

or $y = m\alpha$. Thus, denoting the Pareto efficient quantity of the public good by 'y^*,' and the voluntary contributions quantity by '$\overline{y}$,' we have $y^* = m \cdot \overline{y}$.

It is also interesting to consider what agent i's contribution would be in this case if she or he expected to have $z_{-i} = 0$. With such an expectation, equation (16.25) yields $z_i = \alpha$, and perhaps rather surprisingly, if each agent were to make such a contribution the Pareto efficient allocation would be achieved. In fact, however, if agent i expects the other agents to make a positive contribution to the production of the public good (so that $z_{-i} > 0$), she or he will (if purely a utility-maximizer) 'free ride' on the contributions of others; resulting in a sub-optimal level of contributions.[5]

□

[5] The seminal, and still fairly definitive work on voluntary contributions equilibria is Bergstrom, Blume, and Varian [1986].

While the above example illustrates the incentive individual agents have to 'free ride' in making public goods contributions, public goods experiments have indicated that individuals tend to contribute more than is suggested by the above analysis.[6] However, mechanisms which would insure that individuals would make the Pareto efficient level of contributions purely by following their own (utility-maximizing) self interest are obviously of fundamental importance. We will discuss the search for such mechanisms, which we will label 'incentive-compatible' and Pareto efficient, in Chapter 18. In the meantime, in the next section we will discuss two equilibrium concepts, which while not themselves necessarily incentive-compatible, result in Pareto efficient allocations, if individuals behave in a specified (rather myopic) manner. Consequently, these mechanisms are quite worthy of study for their own sake, and have been used as a part of the framework of larger mechanisms which are, in some sense, incentive-compatible. We will consider some of these latter mechanisms in Chapter 18.

16.5 Lindahl and Ratio Equilibria

A very clever theoretical invention for the allocation of production and consumption in a public goods economy is the **Lindahl equilibrium**, which was, in fact, invented by the distinguished Swedish economist Erik Lindahl [1919]. We will develop this idea within the context of the model presented in the previous section, but the definitions and results obtained here can be extended to the case in which there is production of both private and public goods (and we will present such a model in Section 6). In the present development, we will change the model set out in the previous section only very slightly. We will, at least initially, suppose only that individuals have asymmetric strict preference relations, P_i. Further, we will specify initial endowments of the private good for the consumers, which we will denote by 'w_i'; and we suppose that:

$$w_i \geq 0 \quad \text{for } i = 1, \ldots, m.$$

Finally, we will suppose throughout that:

$$c(\mathbf{0}) = 0.$$

A feasible allocation for this economy is then defined by the obvious modification of Definition 16.3, of the previous section.

In a Lindahl equilibrium, each consumer pays an individual price per unit, q_{ij}, for the jth public good; and we will assume throughout our treatment that the consumers' preferences are strictly increasing in the private good, so that we can normalize to set its price equal to one. We will denote the vector of prices for the public goods which is levied upon consumer i by '$\boldsymbol{q}_i$,' and we will make use of the following definition.

16.6 Definition. We will say that an m-tuple $\boldsymbol{s}$ is a **distribution of shares for $\boldsymbol{E}$** iff:

$$\boldsymbol{s} \in \mathbb{R}^m_+ \text{ and } \sum_{i=1}^{m} s_i = 1.$$

[6]For an excellent survey of experimental results dealing with public goods, see Ledyard [1995].

Of course, s_i is the ith consumer's share of ownership in the firm producing the public goods; and if $\boldsymbol{q}_i$ is the consumer's personal vector of prices for the public goods, and π is the firm's profits, then in the context of a Lindahl equilibrium, the ith consumer's budget constraint is given by:

$$x_i + \boldsymbol{q}_i \cdot \boldsymbol{y} \leq w_i + s_i \pi.$$

16.7 Definition. We shall say that $(\boldsymbol{x}^*, \boldsymbol{y}^*, \langle \boldsymbol{q}_i^* \rangle)$ is a **Lindahl equilibrium for $\boldsymbol{E}$**, given the distribution of shares $\boldsymbol{s}^*$ iff:

1. $(\boldsymbol{x}^*, \boldsymbol{y}^*)$ is feasible for E,
2. $\boldsymbol{q}_i^* \in \mathbb{R}_+^n$, for $i = 1, \ldots, m$,
3. $\boldsymbol{y}^* \in \mathbb{R}_+^n$ maximizes the firm's profits, given $\boldsymbol{q}^*$; where we define:

$$\boldsymbol{q}^* = \sum_{i=1}^m \boldsymbol{q}_i^*;$$

that is, for all $\boldsymbol{y} \in \mathbb{R}_+^n$:

$$\pi^* \stackrel{\text{def}}{=} \boldsymbol{q}^* \cdot \boldsymbol{y}^* - c(\boldsymbol{y}^*) \geq \boldsymbol{q}^* \cdot \boldsymbol{y} - c(\boldsymbol{y}).$$

4. for each i,

$$x_i^* + \boldsymbol{q}_i^* \cdot \boldsymbol{y}^* \leq w_i + s_i \pi^*,$$

and, for all $(x_i, \boldsymbol{y}) \in \mathbb{R}_+^{1+n}$, we have:

$$(x_i, \boldsymbol{y}) P_i (x_i^*, \boldsymbol{y}^*) \Rightarrow x_i + \boldsymbol{q}_i^* \cdot \boldsymbol{y} > w_i + s_i \pi^*.$$

If one now assumes that the individual consumer preferences are representable by continuously differentiable utility functions, and that the firm's production function is continuously differentiable, one can show that the Samuelson conditions are satisfied at a Lindahl equilibrium.[7] If we then, for example, assume additionally that the consumers' utility functions are all concave and the firm's cost function is convex, we can establish that whenever the Samuelson conditions are satisfied at a feasible allocation, then it is Pareto efficient. From this it follows that, given all these assumptions, any Lindahl equilibrium must be Pareto efficient. As it happens, however, we can prove a much more general result much more simply, as follows.

16.8 Theorem. *Under the assumptions of this section, if $(\boldsymbol{x}^*, \boldsymbol{y}^*, \langle \boldsymbol{q}_i^* \rangle)$ is a Lindahl equilibrium for E, given the distribution of shares $\boldsymbol{s}^*$, then $\boldsymbol{a}^* = (\boldsymbol{x}^*, \boldsymbol{y}^*)$ is Pareto efficient.*

Proof. Suppose, by way of obtaining a contradiction, that there exists a feasible allocation, $\boldsymbol{a} = (\boldsymbol{x}, \boldsymbol{y})$ such that:

$$(x_i, \boldsymbol{y}) P_i (x_i^*, \boldsymbol{y}^*) \quad \text{for } i = 1, \ldots, m.$$

Then by the definition of a Lindahl equilibrium, we have:

$$x_i + \boldsymbol{q}_i^* \cdot \boldsymbol{y} > w_i + s_i \pi^* \quad \text{for } i = 1, \ldots, m. \tag{16.26}$$

[7] I will leave the proof of this fact as an exercise.

Adding both sides of (16.26) over i, we have:

$$\sum_{i=1}^{m} x_i + \boldsymbol{y} \cdot \Big(\sum_{i=1}^{m} \boldsymbol{q}_i^*\Big) > \sum_{i=1}^{m} w_i + \pi^*\Big(\sum_{i=1}^{m} s_i\Big);$$

so that, making use of our definitions:

$$\sum_{i=1}^{m} x_i + \boldsymbol{q}^* \cdot \boldsymbol{y} > \sum_{i=1}^{m} w_i + \boldsymbol{q}^* \cdot \boldsymbol{y}^* - c(\boldsymbol{y}^*). \tag{16.27}$$

However, from the assumption that $\boldsymbol{a}$ is feasible, we have:

$$\sum_{i=1}^{m} x_i + c(\boldsymbol{y}) \leq \sum_{i=1}^{m} w_i.$$

Making use of this last inequality and (16.27), we then obtain:

$$\boldsymbol{q}^* \cdot \boldsymbol{y} - c(\boldsymbol{y}) > q^* \cdot \boldsymbol{y}^* - c(\boldsymbol{y}^*);$$

which contradicts the assumption that $\boldsymbol{y}^*$ maximizes the firm's profits. □

If we strengthen our assumptions to something like those used in the previous section, then we can obtain the stronger conclusion of the following; the proof of which will be left as an exercise.

16.9 Theorem. *If each preference relation can be represented by a continuous utility function which is strictly increasing in the private good, and if $(\boldsymbol{x}^*, \boldsymbol{y}^*, \langle \boldsymbol{q}_i^* \rangle)$ is a Lindahl equilibrium for E, given the distribution of ownership, $(\boldsymbol{w}^*, \boldsymbol{s}^*)$, then $\boldsymbol{a}^* = (\boldsymbol{x}^*, \boldsymbol{y}^*)$ is strongly Pareto efficient.*

While the concept of a Lindahl equilibrium is a very clever invention, and has some very nice properties, there are a number of problems with the mechanism. First, of course, it is not clear that any practical market mechanism exists which would drive the public goods prices toward their equilibrium levels (of course, our theory is also a bit weak on that score in the case of a private goods economy). It has sometimes been claimed that a public agency/planner could calculate the equilibrium prices, and then simply announce these prices to the consumers; however, it is not clear how the planner could obtain all of this information (among other things, the consumers will have an incentive to misrepresent their willingness to pay for the public good, as we will see). We will consider these two difficulties in Chapter 18; at the moment, let's consider three additional difficulties. First, there may be positive profits earned by the producer; in which case the existence of the Lindahl equilibrium may depend upon how these profits are distributed. Secondly, if one defines the core of the economy in the manner which seems to be standard in the context of this model, then a Lindahl equilibrium may not be in the core. Thirdly, if the cost function is concave (increasing returns), it may be that there is no profit-maximizing level of public goods production, and thus no Lindahl equilibrium will exist. As to the first difficulty, consider the following example.

16.10 Example. Suppose $m = 2$ and $n = 1$; with $c(y)$ given by:

$$c(y) = (1/2)(y)^2;$$

while the consumers' preferences are represented by the utility functions:

$$u_1(x_1, y) = x_1 + 4\log y \text{ and } u_2(x_2, y) = x_2 + 12\log y,$$

respectively. Suppose further that $w_1 = 2$, while $w_2 = 6$.

Making use of the Samuelson conditions, it is easily shown that the unique Pareto efficient quantity of the public good is given by:

$$y^* = 4;$$

and notice that $c(y^*) = 16/2 = 8$, which is equal to $w_1 + w_2$. If a Lindahl equilibrium exists in this case, it follows from the first-order conditions for utility maximization that we must have:

$$4/y^* = 1 = q_1 \text{ and } 12/y^* = 3 = q_2.$$

However, we then have, in particular:

$$q_1 y^* = 4,$$

and thus:

$$q_1 y^* - w_1 = 4 - 2 = 2.$$

Now, while the producer maximizes profits at $y^* = 4$, to earn profits of:

$$(q_1 + q_2)y^* - (1/2)(y^*)^2 = 16 - (1/2)16 = 8;$$

we see that, if s_1, consumer one's share of the firm's profits, is less than 1/4, Ms. 1 cannot achieve the desired Lindahl utility maximization outcome. However, if $s_1 > 1/4$, then $s_2 < 3/4$; and then the desired utility maximization bundle for Mr. 2, $(0, y^*) = (0, 4)$ is not in his budget set. In fact, the desired Lindahl equilibrium is unattainable unless the distribution of ownership in the firm is given by:

$$s_1 = 1/4 \text{ and } s_2 = 3/4. \quad \square$$

The above example shows that the informational requirements of the Lindahl equilibrium are truly formidable. If a planner is to implement this equilibrium he or she needs to somehow determine true willingness to pay in order to determine the Pareto efficient quantity of the public good and the personalized price for the public good in the one public good example just considered. In addition, however, the two consumers need to have the proper shares of ownership in the firm producing the public good in order to achieve the equilibrium even if the planner has somehow determined the proper personalized prices for the public good as well as determining its Pareto efficient level of production.

In order to see the difficulty which arises in relation to the core, we need the following definition.[8]

---

[8]This is essentially the definition originally introduced by Duncan Foley [1970]; and is the one generally used in public goods models.

16.11 Definition. We will say that a coalition $S \subseteq M$ can **block an allocation** $(\boldsymbol{x}^*, \boldsymbol{y}^*) \in R_+^{m+n}$ iff there exists $x_i \in \mathbb{R}_+$ for each $i \in S$ and $\boldsymbol{y} \in \mathbb{R}_+^n$ such that:

$$(\forall i \in S)\colon (x_i, \boldsymbol{y}) P_i (x_i^*, \boldsymbol{y}^*) \tag{16.28}$$

and:

$$c(\boldsymbol{y}) \leq \sum\nolimits_{i \in S} (w_i - x_i). \tag{16.29}$$

A feasible state, $(\boldsymbol{x}^*, \boldsymbol{y}^*) \in \mathbb{R}_+^{m+n}$ will be said to be in the **core**, or to be a **core allocation** iff it cannot be blocked by any coalition $S \subseteq M$.

It should be noted before we go any farther that this may not be an appropriate definition of the core in this context. We will discuss this issue further shortly; in the meantime, consider the following example.

16.12 Example. Once again we suppose that $m = 2$ and $n = 1$, but we suppose now that the consumers have the utility functions:

$$u_1(x_1, y) = x_1 + 8\sqrt{y} \text{ and } u_2(x_2, y) = x_2 + 4\sqrt{y},$$

respectively; while $w_1 = 10$, $w_2 = 0$, and $c(y) = (y)^{3/2}$.

From the Samuelson conditions, we find the unique Pareto efficient quantity of the public good to be $y^* = 4$, and proceeding as before, we find the Lindahl prices for the two individuals to be given by:

$$q_1 = 2 \text{ and } q_2 = 1,$$

respectively. Moreover, given the price $q = q_1 + q_2 = 3$, the producer maximizes profits at $y^* = 4$, to earn a profit of $\pi^* = 4$.

Now, with the indicated personalized prices, the second consumer (Mr. 2) has the budget constraint:

$$x_2 + y \leq s_2 \pi^* = s_2 \cdot (4).$$

Since x_2 must be nonnegative, this means that in order to maximize utility at $y^* = 4$, we must have:

$$4s_2 \geq 4;$$

so that we must have $s_2 = 1$. However, this means that Ms. 1 has the consumption (x_1^*, y^*) at the Lindahl equilibrium, where:

$$x_1^* = w_1 - q_1 y^* = 10 - (2)4 = 2;$$

with a utility of $u_1(x_1^*, y^*) = 2 + 8\sqrt{4} = 18$. On the other hand, were Ms. 1 to defect, she maximizes:

$$w_2 - c(y) + 8\sqrt{y} = 10 - (y)^{3/2} + 8\sqrt{y},$$

at $y' = 8/3$, with private good consumption $x_1' = 10 - (8/3)^{3/2}$, yielding utility:

$$u_1(x_1', y') = 10 - (8/3)^{3/2} + 8\sqrt{8/3} = 10 + [(8/3)^{1/2}](8 - 8/3) = 10 + [(8/3)^{1/2}](16/3).$$

Comparing the two utility values, we have:

$$u_1(x_1', y') - u_1(x_1^*, y^*) = [(8/3)^{1/2}](16/3) - 8 = (16/3)\big[(8/3)^{1/2} - 3/2\big].$$

But then, since $4\sqrt{2} > 3\sqrt{3}$, we see that Ms. 1 gains utility by defecting from the Lindahl equilibrium and paying the full cost of producing a level of $y = 8/3$. □

As mentioned earlier, it is not at all clear that the definition of the core given in Definition 16.11 is appropriate in the context of this model; although it is equally unclear what the appropriate definition should be. The basic problem with the definition presented earlier is that it is not apparent what production possibilities should be viewed as being attainable by a coalition which is a proper subset of M. In particular, in the example just considered, we supposed that Ms. 1 could produce any amount of the public good she desired, as long as she paid the full cost of this production. There are at least two difficulties in making this supposition. First, the manner in which the model is formulated suggests that the production facility for producing the public good is publicly (that is, jointly) owned. If this is so, it is not clear why it is that Ms. 1 can make use of this production capability for purely her own gain. Secondly, since we are dealing with a public good, Ms. 1's choice of $y' = 8/3$ leaves her better off than she was in the Lindahl equilibrium situation, but it leaves Mr. 2 worse off; in effect, the y' solution is simply the outcome when Ms. 1 is the dictator.[9]

Foley's original definition of the core for public goods economies (Foley [1970]), is essentially equivalent to that presented here (Definition 16.11), but *he assumed that the production technology for producing the public goods was a convex cone!* Since this means that the production technology satisfies the additivity condition, it makes sense to imagine that a coalition can produce whatever they desire, as long as they pay the full cost of doing so; because simultaneously the group left out of the first coalition can produce whatever they desire as well, given only that they pay the full cost of producing their choice. Moreover, given these production conditions, the producer's profit at a Lindahl equilibrium will always be zero; which eliminates the difficulty arising in the two examples we have just presented. In fact, for what it is worth, we can easily prove the following; where we say that a preference relation, P_i is **increasing in the public goods component** iff, for all $(x_i, \boldsymbol{y}) \in \mathbb{R}_+^{1+n}$ and all $\boldsymbol{y}' \in \mathbb{R}_+^n$, we have:

$$\boldsymbol{y}' \geq \boldsymbol{y} \Rightarrow (x_i, \boldsymbol{y}')G_i(x_i, \boldsymbol{y}) \;\;\&\;\; \boldsymbol{y}' \gg \boldsymbol{y} \Rightarrow (x_i, \boldsymbol{y}')P_i(x_i, \boldsymbol{y}),$$

where G_i is the negation of P_i.

16.13 Proposition. *Suppose:*

1. each preference relation, P_i, is asymmetric and increasing in the public goods component, and

2. the cost function, c, is linear.

Then if $\big(\boldsymbol{x}^*, \boldsymbol{y}^*, \langle \boldsymbol{q}_i^* \rangle\big)$ is a Lindahl equilibrium, given the distribution of shares $\boldsymbol{s}$, the allocation $(\boldsymbol{x}^*, \boldsymbol{y}^*)$ is a core allocation for E.

Proof. Suppose for some coalition, $S \subseteq M$, there exist $x_i' \in \mathbb{R}_+$, for each $i \in S$, and $\boldsymbol{y}' \in \mathbb{R}_+^n$ such that:

$$(\forall i \in S)\colon (x_i', \boldsymbol{y}')P_i(x_i^*, \boldsymbol{y}^*).$$

[9] Of course, in the example as we developed it, it would certainly not be surprising if this dictatorial outcome were forthcoming. After all, some of the private good has to be given up in order to obtain a positive amount of the public good, and Ms. 1 initially holds all of the private good in the economy. Notice also that this dictatorial outcome is actually individually rational for Mr. 2, since his only alternative operating on his own, even given full access to the production technology, is the bundle $(0, 0)$.

Then for each $i \in S$, we must have:

$$x_i' + \boldsymbol{q}_i^* \cdot \boldsymbol{y}' > w_i + s_i \pi^*; \tag{16.30}$$

although, since the cost function is linear, and the producer must be maximizing profits at $\boldsymbol{y}^*$, we must have $\pi^* = 0$. Taking this into account, and adding the inequalities in (16.30) over i, we obtain:

$$\sum_{i \in S} x_i' + \Big(\sum_{i \in S} \boldsymbol{q}_i^*\Big) \cdot \boldsymbol{y}' > \sum_{i \in S} w_i. \tag{16.31}$$

However, since each P_i is increasing in the public goods component, and $(\boldsymbol{x}^*, \boldsymbol{y}^*, \langle \boldsymbol{q}_i^* \rangle)$ is a Lindahl equilibrium, we see that we must have $\boldsymbol{q}_i^* \geq \boldsymbol{0}$, for each i. Therefore:

$$\Big(\sum_{i \in S} \boldsymbol{q}_i^*\Big) \cdot \boldsymbol{y}' - c(\boldsymbol{y}') \leq \boldsymbol{q}^* \cdot \boldsymbol{y}' - c(\boldsymbol{y}'); \tag{16.32}$$

and, since the producer has a maximum profit, given $\boldsymbol{q}^*$, it follows that this last difference in (16.32) must be nonpositive. Therefore, we see that:

$$\Big(\sum_{i \in S} \boldsymbol{q}_i^*\Big) \cdot \boldsymbol{y}' \leq c(\boldsymbol{y}');$$

and combining this with (16.31), we see that:

$$\sum_{i \in S} x_i' + c(\boldsymbol{y}') > \sum_{i \in S} w_i,$$

and it follows that the allocation $\langle (x_i', \boldsymbol{y}') \rangle$ is not feasible for S. □

Thus, one of the difficulties we have mentioned regarding Lindahl equilibria disappears if the cost function is linear, although the price we pay for this gain is fairly high. However, the next equilibrium concept we're going to examine has some advantages over the Lindahl equilibrium concept.

Kanecko [1977] introduced an equilibrium concept for public goods economies which, while formally equivalent to Lindahl equilibria in some contexts, has some real advantages over the latter notion in other contexts. Kanecko's definition has been extended and refined in Diamantaras and Wilkie [1994], and Tian [1994] (see also Corchon and Wilkie [1996], Tian [2000], and van den Nouweland and Wooders [2002]), while a similar concept was introduced in Mas-Colell and Silvestre [1989]. The definition to be presented here is from this last reference. We begin with the following.

16.14 Definition. A **cost share system** is a family of m functions, $g_i \colon \mathbb{R}_+^n \to \mathbb{R}$ such that:

$$(\forall \boldsymbol{y} \in \mathbb{R}_+^n) \colon \textstyle\sum_{i=1}^m g_i(\boldsymbol{y}) = c(\boldsymbol{y}).$$

Interesting examples of cost share systems include that introduced by Kanecko [1977], for the case in which the cost function takes the form:

$$c(\boldsymbol{y}) = \sum_{j=1}^{n} c_j(y_j). \tag{16.33}$$

This is, of course, the case in which the n public goods can be produced independently, and with no external effects on one another. In this situation, Kanecko defines a cost share system by a nonnegative $m \times n$ matrix, $[r_{ij}]$ satisfying:

$$\sum_{i=1}^{m} r_{ij} = 1 \quad \text{for } j = 1, \ldots, n;$$

the cost share system then being given by:

$$g_i(\boldsymbol{y}) = \sum_{j=1}^{n} r_{ij} c_j(y_j) \quad \text{for } i = 1, \ldots, m. \tag{16.34}$$

Mas-Colell and Silvestre actually devote most of their attention to **linear cost share systems**, which take the form:

$$g_i(\boldsymbol{y}) = \boldsymbol{a}_i \cdot \boldsymbol{y} + b_i c(\boldsymbol{y}) \quad \text{for } i = 1, \ldots, m; \tag{16.35}$$

where $\boldsymbol{a}_i \in \mathbb{R}^n, b_i \in \mathbb{R}_+$ for each i, and:

$$\sum_{i=1}^{m} \boldsymbol{a}_i = \boldsymbol{0} \ \& \ \sum_{i=1}^{m} b_i = 1.$$

16.15 Definition. A feasible state $(\boldsymbol{x}^*, \boldsymbol{y}^*)$ is a **cost share equilibrium**, given the cost share system, $\boldsymbol{g} = (g_1, \ldots, g_m)$ iff, for each $i \in M$:

$$x_i^* = w_i - g_i(\boldsymbol{y}^*) \text{ and } \left(\forall\, (x_i, \boldsymbol{y}) \in \mathbb{R}_+^{1+n}\right)\colon (x_i, \boldsymbol{y}) P_i (x_i^*, \boldsymbol{y}^*) \Rightarrow x_i + g_i(\boldsymbol{y}) > w_i.$$

As you may have guessed already, in the case where the cost function satisfies (16.33) above, we can define a **ratio equilibrium** (Kanecko [1977]) as a cost share equilibrium in which the cost share system takes the form indicated in (16.34), above. (See also Exercise 6, at the end of this chapter.)

If each g_i function is nonnegative-valued, then such an equilibrium is always in the core, as is noted in the following.

16.16 Proposition. *If $(\boldsymbol{x}^*, \boldsymbol{y}^*)$ is a cost share equilibrium, given $\boldsymbol{g}$, where $\boldsymbol{g}$ is such that $g_i\colon \mathbb{R}_+^n \to \mathbb{R}_+$, for $i = 1, \ldots, m$, then $(\boldsymbol{x}^*, \boldsymbol{y}^*)$ is in the core for E.*

Proof. Suppose $S \subseteq M$, $\boldsymbol{y} \in \mathbb{R}_+^n$ and x_i $(i \in S)$ are such that:

$$(\forall i \in S)\colon x_i \in \mathbb{R}_+ \ \& \ (x_i, \boldsymbol{y}) P_i (x_i^*, \boldsymbol{y}^*).$$

Then, by definition of a cost share equilibrium, we must have:

$$(\forall i \in S)\colon x_i + g_i(\boldsymbol{y}) > w_i. \tag{16.36}$$

Adding the inequalities in (16.36), we obtain:

$$\sum_{i \in S} x_i + \sum_{i \in S} g_i(\boldsymbol{y}) > \sum_{i \in S} w_i.$$

However, since each g_i is nonnegative-valued, we have:

$$\sum_{i \in S} g_i(\boldsymbol{y}) \leq \sum\nolimits_{i=1}^{m} g_i(\boldsymbol{y}) = c(\boldsymbol{y});$$

and it follows that:

$$\sum_{i \in S} x_i + c(\boldsymbol{y}) > \sum_{i \in S} w_i,$$

and thus we see that $(\boldsymbol{x}, \boldsymbol{y})$ is not feasible for S. □

An argument very similar to that used in the above proof shows that if $(\boldsymbol{x}^*, \boldsymbol{y}^*)$ is a cost share equilibrium, given $\boldsymbol{g}$, then $(\boldsymbol{x}^*, \boldsymbol{y}^*)$ is Pareto efficient, even if $\boldsymbol{g}$ is not nonnegative-valued. I will leave the details of this proof as an exercise (albeit a rather trivial one).

While the cost share (or ratio-) equilibrium approach does not solve all of the problems associated with Lindahl equilibria, it does at least alleviate some of them. For instance, suppose we re-visit Example 16.10, as follows.

16.17 Example. Recall that in this example (16.10) the unique Pareto efficient public goods output was $y^* = 4$. If we set:

$$r_1 = 1/4, r_2 = 3/4, \text{ and } g_i(y) = r_i c(y) = r_i (y)^2/2 \text{ for } i = 1, 2,$$

then you can easily verify that $(\boldsymbol{x}^*, y^*)$ is a cost share equilibrium, given $\boldsymbol{g}$, where:

$$x_1^* = x_2^* = 0. \quad \square$$

Thus the ratio equilibrium on the above example does not require any redistribution of wealth; whereas attaining a Lindahl equilibrium in the same situation required a precise calculation of the distribution of shares. A more significant advantage of the cost share/ ratio equilibrium idea is illustrated by the following example.

16.18 Example. Suppose $n = 1$, that m is a finite integer greater than one, and that the ith consumer's preferences can be represented by the utility function:

$$u_i(x_i, y) = x_i + \gamma_i \log y \quad \text{for } i = 1, \ldots, m; \tag{16.37}$$

and suppose $w_i \geq 2\gamma_i > 0$, for $i = 1, \ldots, m$. Finally, suppose the cost function for producing the public good is given by:

$$c(y) = 2\sqrt{y}.$$

In this case, if the producer is presented with a positive price for the public good, no profit-maximizing output exists. Consequently, no Lindahl equilibrium exists in this case.

However, we have:

$$\frac{d}{dy}\Big[\sum\nolimits_{i=1}^{m} \gamma_i \log y - 2\sqrt{y}\Big] = (1/y)\sum\nolimits_{i=1}^{m} \gamma_i - (y)^{-1/2};$$

which is positive if, and only if:

$$y < \left[\sum_{i=1}^{m} \gamma_i\right]^2.$$

Consequently, it follows from Proposition 16.1 that the unique Pareto efficient quantity of the public good is given by:

$$y^* = \left[\sum_{i=1}^{m} \gamma_i\right]^2.$$

If we define:

$$r_i = \frac{\gamma_i}{\sum_{h=1}^{m} \gamma_h}$$

and the cost share system:

$$g_i(y) = r_i c(y) = 2r_i\sqrt{y} \quad \text{for } i = 1, \ldots, m,$$

similar considerations show that the ith consumer maximizes:

$$u_i\big[w_i - r_i c(y), y\big] = w_i - 2r_i\sqrt{y} + \gamma_i \log y,$$

at y^*. Consequently (I will leave you to verify the details), $(\boldsymbol{x}^*, y^*)$ is a cost share equlibrium, with:

$$x_i^* = w_i - 2\gamma_i \quad \text{for } i = 1, \ldots, m. \quad \square$$

While the above example does not show that a cost share equilibrium always exists, even in the face of increasing returns in the production of the public good, it does show that one *may* exist; and, since we know that a Lindahl equilibrium will never exist in such a case, there are circumstances in which the cost share equilibrium approach has a real advantage over the Lindahl equilibrium approach. We will consider some incentive and informational aspects of these mechanisms in Chapter 18. In the meantime, we devote the next section to the analysis of Lindahl equilibria in the context of a much more general model than the one we have been using here.

16.6 The 'Fundamental Theorems' for Lindahl Equilibria

In this secton we will develop the basic theory of Lindahl equilibria in the context of a more general equilibrium model; one which allows for multiple private and public goods, and for production of private, as well as public goods. While the model to be developed here is quite general in many ways, and is sufficiently rich in structure as to allow one to develop the basic theory of Lindahl equilibria at a quite general level, it is still fairly simple and tractable. Moreover, in the terminology introduced originally by Hurwicz, and discussed in Chapters 5 and 7, we will be able to show that Lindahl equilibria are both 'non-wasteful' and 'unbiased' in the context of this model.

16.6.1 The 'First Fundamental Theorem'

We suppose that there are:

I consumers (indexed by 'i'),
L private goods,
M public goods.

The ith consumption set, C_i, is taken to be a subset of $\mathbb{R}^{L+M}$ satisfying:

$$\big(\forall(\boldsymbol{x}_i, \boldsymbol{y}) \in C_i\big) : \boldsymbol{y} \geq \boldsymbol{0}, \tag{16.38}$$

and:

$$\big(\forall(\boldsymbol{x}_i, \boldsymbol{y}), (\boldsymbol{x}'_i, \boldsymbol{y}') \in \mathbb{R}^{L+M}\big) : [(\boldsymbol{x}_i, \boldsymbol{y}) \in C_i \;\&\; (\boldsymbol{x}'_i, \boldsymbol{y}') \geq (\boldsymbol{x}_i, \boldsymbol{y})] \Rightarrow (\boldsymbol{x}'_i, \boldsymbol{y}') \in C_i, \tag{16.39}$$

and in much of this discussion, we will simply suppose that the ith consumer's (strict) preferences can be represented as an asymmetric binary relation, P_i, on C_i. As usual, we define the 'no-worse-than' relation, G_i, as the negation of P_i; that is:

$$(\boldsymbol{x}_i, \boldsymbol{y}) G_i (\boldsymbol{x}'_i, \boldsymbol{y}') \iff \neg[(\boldsymbol{x}'_i, \boldsymbol{y}') P_i (\boldsymbol{x}_i, \boldsymbol{y})].$$

The aggregate production set for the economy is a non-empty subset of $\mathbb{R}^L \times \mathbb{R}^M_+$:

$$T = \{(\boldsymbol{z}, \boldsymbol{y}) \in \mathbb{R}^L \times \mathbb{R}^M_+ \mid (\boldsymbol{z}, \boldsymbol{y}) \text{ is a feasible aggregate net production vector}\}.$$

As usual, we will suppose that net demand of the production sector for (private goods) inputs is indicated by a negative coordinate of $\boldsymbol{z}$, while positive net output is indicated by a positive coordinate. Thus if the production sector is presented with a pair of price vectors, $(\boldsymbol{p}, \boldsymbol{q})$, for private and public goods, respectively, and chooses a pair $(\boldsymbol{z}, \boldsymbol{y}) \in T$, then aggregate profits will be given by:

$$\pi = \boldsymbol{p} \cdot \boldsymbol{z} + \boldsymbol{q} \cdot \boldsymbol{y} = (\boldsymbol{p}, \boldsymbol{q}) \cdot (\boldsymbol{z}, \boldsymbol{y}).$$

We will use the generic notation:

$(\boldsymbol{x}_i, \boldsymbol{y}) \in C_i$ to denote the ith consumer's consumption vector.
$\boldsymbol{a} = (\langle \boldsymbol{x}_i \rangle, \boldsymbol{z}, \boldsymbol{y}) \in \mathbb{R}^{I \cdot L + L + M}$ to denote allocations.

16.19 Definition. We will say that an allocation $\boldsymbol{a} = (\langle \boldsymbol{x}_i \rangle, \boldsymbol{z}, \boldsymbol{y})$ is **feasible** (or **attainable**) iff:

1. $\sum_{i=1}^{I} \boldsymbol{x}_i = \boldsymbol{z}$,
2. $(\boldsymbol{z}, \boldsymbol{y}) \in T$, and
3. $(\boldsymbol{x}_i, \boldsymbol{y}) \in C_i$, for $i = 1, \ldots, I$.

You will probably be able to define Pareto efficiency for this economy without my help, but I will state the formal definition nonetheless.

16.20 Definition. We shall say that an allocation, $\boldsymbol{a} = (\langle \boldsymbol{x}_i \rangle, \boldsymbol{z}, \boldsymbol{y})$ is **Pareto efficient** iff (a) it is feasible, and (b) there exists no feasible $\boldsymbol{a}^* = (\langle \boldsymbol{x}^*_i \rangle, \boldsymbol{z}^*, \boldsymbol{y}^*)$, satisfying:

$$(\boldsymbol{x}^*_i, \boldsymbol{y}^*) P_i (\boldsymbol{x}_i, \boldsymbol{y}) \quad \text{for } i = 1, \ldots, I.$$

In dealing with Lindahl equilibria, we will let '$\boldsymbol{q}_i$' denote the ith agent's personalized price vector for the public goods, the vector:

$$\boldsymbol{q} = \sum\nolimits_{i=1}^{I} \boldsymbol{q}_i,$$

will be the producers' vector of selling prices for the public goods, and '$\boldsymbol{p}$' will denote the vector of prices of the private goods. We will be assuming competitive behavior in the production sector, and we will let 'Π' denote the set of price vectors $(\boldsymbol{p}, \boldsymbol{q})$ in $\mathbb{R}^{L+M}$ for which a profit-maximizing production vector $(\boldsymbol{z}, \boldsymbol{y})$ exists in T. Given $(\boldsymbol{p}, \boldsymbol{q}) \in \Pi$, we define:

$$\pi(\boldsymbol{p}, \boldsymbol{q}) = \max_{(\boldsymbol{z},\boldsymbol{y}) \in T} (\boldsymbol{p}, \boldsymbol{q}) \cdot (\boldsymbol{z}, \boldsymbol{y}),$$

and we let:

$$\sigma(\boldsymbol{p}, \boldsymbol{q}) = \{(\boldsymbol{z}, \boldsymbol{y}) \in T \mid (\boldsymbol{p}, \boldsymbol{q}) \cdot (\boldsymbol{z}, \boldsymbol{y}) = \pi(\boldsymbol{p}, \boldsymbol{q})\}.$$

16.21 Definition. We will say that a vector $\boldsymbol{w} = (w_1, \ldots, w_I) \in \mathbb{R}^I$ is a **wealth assignment for E, given the prices** $(\boldsymbol{p}, \boldsymbol{q}) \in \Pi$ iff

$$\sum\nolimits_{i=1}^{I} w_i = \pi(\boldsymbol{p}, \boldsymbol{q}), \tag{16.40}$$

and, for each i, there exists $(\boldsymbol{x}_i, \boldsymbol{y}_i) \in C_i$ such that:

$$\boldsymbol{p} \cdot \boldsymbol{x}_i + \boldsymbol{q}_i \cdot \boldsymbol{y}_i \leq w_i.$$

We then define a Lindahl equilibrium as follows.

16.22 Definition. We shall say $(\langle \boldsymbol{x}_i^* \rangle, \boldsymbol{z}^*, \boldsymbol{y}^*; \boldsymbol{p}^*, \langle \boldsymbol{q}_i^* \rangle)$ is a **Lindahl equilibrium, given the wealth assignment** $\boldsymbol{w}^* = (w_1^*, \ldots, w_I^*)$ iff:

1. $(\boldsymbol{p}^*, \langle \boldsymbol{q}_i^* \rangle) \in \mathbb{R}_+^{L+I \cdot M}$ and is nonnull,
2. $\boldsymbol{a}^* = (\langle \boldsymbol{x}_i^* \rangle, \boldsymbol{z}^*, \boldsymbol{y}^*)$ is feasible,
3. defining $\boldsymbol{q}^* = \sum_{i=1}^{I} \boldsymbol{q}_i^*$, we have:

$$(\forall (\boldsymbol{z}, \boldsymbol{y}) \in T) \colon \boldsymbol{p}^* \cdot \boldsymbol{z} + \boldsymbol{q}^* \cdot \boldsymbol{y} \leq \boldsymbol{p}^* \cdot \boldsymbol{z}^* + \boldsymbol{q}^* \cdot \boldsymbol{y}^* \stackrel{\text{def}}{=} \pi(\boldsymbol{p}^*, \boldsymbol{q}^*),$$

4. for each i:

$$\boldsymbol{p}^* \cdot \boldsymbol{x}_i^* + \boldsymbol{q}_i^* \cdot \boldsymbol{y}^* \leq w_i^*;$$

and, for all $(\boldsymbol{x}_i, \boldsymbol{y}) \in C_i$:

$$(\boldsymbol{x}_i, \boldsymbol{y}) P_i (\boldsymbol{x}_i^*, \boldsymbol{y}^*) \Rightarrow \boldsymbol{p}^* \cdot \boldsymbol{x}_i + \boldsymbol{q}_i^* \cdot \boldsymbol{y} > w_i^*.$$

In investigating the properties of a Lindahl equilibrium in this setting, and with the above definition, we gain a great deal of flexibility and lose very little generality, as we will demonstrate in the final subsection of this chapter.

We can now state a version of the 'First Fundamental Theorem' for Lindahl equilibria, as follows.

16.23 Theorem. *If $(\langle \boldsymbol{x}_i^* \rangle, \boldsymbol{z}^*, \boldsymbol{y}^*; \boldsymbol{p}^*, \langle \boldsymbol{q}_i^* \rangle)$ is a Lindahl equilibrium, given the wealth assignment $\boldsymbol{w}^* = (w_1^*, \ldots, w_I^*)$, then $\boldsymbol{a}^* = (\langle \boldsymbol{x}_i^* \rangle, \boldsymbol{z}^*, \boldsymbol{y}^*)$ is Pareto efficient.*

Proof. Suppose, by way of obtaining a contradiction, that there exists a feasible allocation, $\boldsymbol{a} = (\langle \boldsymbol{x}_i \rangle, \boldsymbol{z}, \boldsymbol{y})$, such that:

$$(\boldsymbol{x}_i, \boldsymbol{y}) P_i (\boldsymbol{x}_i^*, \boldsymbol{y}^*) \quad \text{for } i = 1, \ldots, I. \tag{16.41}$$

By (16.41) and the definition of a Lindahl equilibrium, we have:

$$\boldsymbol{p}^* \cdot \boldsymbol{x}_i + \boldsymbol{q}_i^* \cdot \boldsymbol{y} > w_i^* \quad \text{for } i = 1, \ldots, I. \tag{16.42}$$

Adding both sides of (16.42) over i, we have:

$$\sum_{i=1}^{I} \boldsymbol{p}^* \cdot \boldsymbol{x}_i + \sum_{i=1}^{I} \boldsymbol{q}_i^* \cdot \boldsymbol{y} > \sum_{i=1}^{I} w_i^* = \pi(\boldsymbol{p}^*, \boldsymbol{q}^*). \tag{16.43}$$

However, from the fact that $\boldsymbol{a}$ is feasible, we have:

$$\sum_{i=1}^{I} \boldsymbol{x}_i = \boldsymbol{z}; \tag{16.44}$$

and, substituting (16.44) into (16.43), we have:

$$\boldsymbol{p}^* \cdot \boldsymbol{z} + \boldsymbol{q}^* \cdot \boldsymbol{y} > \pi(\boldsymbol{p}^*, \boldsymbol{q}^*) = \boldsymbol{p}^* \cdot \boldsymbol{z}^* + \boldsymbol{q}^* \cdot \boldsymbol{y}^*,$$

and thus:

$$\boldsymbol{p}^* \cdot \boldsymbol{z} + \boldsymbol{q}^* \cdot \boldsymbol{y} > \boldsymbol{p}^* \cdot \boldsymbol{z}^* + \boldsymbol{q}^* \cdot \boldsymbol{y}^*;$$

which contradicts the assumption that $(\boldsymbol{z}^*, \boldsymbol{y}^*)$ maximizes profits over T. □

16.6.2 The 'Second Fundamental Theorem'

In this subsection we will show that the Lindahl mechanism is 'unbiased,' in the same sense as is the competitive mechanism with private goods. The following condition will prove to be useful in our further investigations.

16.24 Definition. We will say that $E = \big(\langle C_i, P_i \rangle, T\big)$ is a **productive public goods economy** iff there exist $(\overline{\boldsymbol{z}}, \overline{\boldsymbol{y}}) \in T$, and $(\overline{\boldsymbol{x}}_i, \overline{\boldsymbol{y}}_i) \in C_i$ for each i such that:

$$\overline{\boldsymbol{y}}_i = \overline{\boldsymbol{y}}, \text{ for } i = 1, \ldots, I, \text{ and } \overline{\boldsymbol{z}} \gg \sum_{i=1}^{I} \overline{\boldsymbol{x}}_i. \tag{16.45}$$

Roughly speaking, the economy E is a productive public goods economy if some (possibly zero) level of public goods can be produced with an aggregate input/private goods production which provides aggregate levels of goods and services which are greater than the collective minimums required by the consumers in the economy. Next we define a numéraire good for a public goods economy.

16.25 Definition. We will say that the jth (*private*) good is a **numéraire good for $\boldsymbol{P}_i$** iff, for all $(\boldsymbol{x}_i, \boldsymbol{y}_i) \in C_i$ and all $\theta \in \mathbb{R}_{++}$, we have:

$$(\boldsymbol{x}_i + \theta \boldsymbol{e}_j, \boldsymbol{y}) \in C_i \ \text{ and } \ (\boldsymbol{x}_i + \theta \boldsymbol{e}_j, \boldsymbol{y}) P_i (\boldsymbol{x}_i, \boldsymbol{y}_i),$$

where $\boldsymbol{e}_j \in \mathbb{R}_+^L$ is the jth unit coordinate vector. We will say that the jth private good is a **numéraire good for the economy $\boldsymbol{E}$ at an attainable allocation**, $\big(\langle \boldsymbol{x}_i^* \rangle; \boldsymbol{z}^*, \boldsymbol{y}^*\big)$ iff it is a numéraire good for each i, and for each i there exists $\theta_i > 0$ such that:

$$(\boldsymbol{x}_i^* - \theta_i \boldsymbol{e}_j, \boldsymbol{y}^*) \in C_i.$$

We will also make use of the following two conditions, both of which are probably sufficiently natural as to not really need a formal definition.

16.26 Definition. We say P_i is **non-decreasing** on C_i iff for all $(\boldsymbol{x}_i, \boldsymbol{y}), (\boldsymbol{x}'_i, \boldsymbol{y}') \in C_i$:[10]

$$(\boldsymbol{x}_i, \boldsymbol{y}) \geq (\boldsymbol{x}'_i, \boldsymbol{y}') \Rightarrow (\boldsymbol{x}_i, \boldsymbol{y}) G_i (\boldsymbol{x}'_i, \boldsymbol{y}').$$

16.27 Definition. We will say that P_i is weakly convex iff, for all $(\boldsymbol{x}_i, \boldsymbol{y}) \in C_i$, the set of all $(\boldsymbol{x}'_i, \boldsymbol{y}')$ such that $(\boldsymbol{x}'_i, \boldsymbol{y}') P_i (\boldsymbol{x}_i, \boldsymbol{y})$ is a convex set.

The variant of the 'Second Fundamental Theorem' which we will prove here is the following.

16.28 Theorem. *Suppose E is a productive public goods economy satisfying:*

1. P_i is weakly convex, lower semi-continuous, and non-decreasing on C_i, for $i = 1, \ldots, I$, and

2. T is convex;

and that $\boldsymbol{a}^* = (\langle \boldsymbol{x}_i^* \rangle, \boldsymbol{z}^*, \boldsymbol{y}^*)$ is an allocation which is Pareto efficient for E and satisfies:

3. there exists $(\boldsymbol{z}^\dagger, \boldsymbol{y}^\dagger) \in T$ such that $\boldsymbol{y}^\dagger \gg \boldsymbol{y}^*$, and:

4. for some $j \in \{1, \ldots, L\}$, the jth private good is a numéraire good for E at $\boldsymbol{a}^*$.

Then there exists $(\boldsymbol{p}^*, \boldsymbol{q}_1^*, \ldots, \boldsymbol{q}_I^*) \in \mathbb{R}^{L+I \cdot M}$, and a wealth assignment $\boldsymbol{w}^*$, such that the tuple:

$$(\langle \boldsymbol{x}_i^* \rangle, \boldsymbol{z}^*, \boldsymbol{y}^*; \boldsymbol{p}^*, \langle \boldsymbol{q}_i^* \rangle)$$

is a Lindahl equilibrium with the wealth assignment $\boldsymbol{w}^*$.

Proof.[11] Suppose $\boldsymbol{a}^*$ is Pareto efficient, and define the sets A and B by:

$$A = \Big\{(\boldsymbol{z}, \boldsymbol{y}_1, \ldots, \boldsymbol{y}_I) \in \mathbb{R}^L \times \mathbb{R}_+^{I \cdot M} \mid \boldsymbol{y}_1 = \cdots = \boldsymbol{y}_I \ \& \ (\boldsymbol{z}, \boldsymbol{y}_1) \in T\Big\} \tag{16.46}$$

and:

$$B = \Big\{(\boldsymbol{z}, \boldsymbol{y}_1, \ldots, \boldsymbol{y}_I) \in \mathbb{R}^L \times \mathbb{R}_+^{I \cdot M} \mid \big(\exists (\boldsymbol{x}_1, \ldots, \boldsymbol{x}_I) \in \mathbb{R}_+^{I \cdot L}\big): (\boldsymbol{x}_i, \boldsymbol{y}_i) P_i (\boldsymbol{x}_i^*, \boldsymbol{y}^*) \text{ for } i = 1, \ldots, I \text{ and } \sum\nolimits_{i=1}^{I} \boldsymbol{x}_i = \boldsymbol{z}\Big\} \tag{16.47}$$

It follows readily from the convexity of T that the set A is convex, and it follows easily from the assumption that each P_i is weakly convex that B is convex as well. Moreover, you should have little difficulty in seeing that:

$$A \cap B = \emptyset.$$

Therefore, there exist a vector $(\boldsymbol{p}^*, \boldsymbol{q}_1^*, \ldots, \boldsymbol{q}_I^*) \in \mathbb{R}^{L+I \cdot M}$ and a real number, α, such that:

$$(\boldsymbol{p}^*, \boldsymbol{q}_1^*, \ldots, \boldsymbol{q}_I^*) \neq \boldsymbol{0}, \tag{16.48}$$

[10] Recall that we are assuming that C_i satisfies (16.39) throughout this section.

[11] This proof owes a great deal to the construction invented by Duncan Foley [1970].

and:

$$\sup_{(\boldsymbol{z},\boldsymbol{y}_1,\ldots,\boldsymbol{y}_I)\in A} \boldsymbol{p}^* \cdot \boldsymbol{z} + \sum_{i=1}^{I} \boldsymbol{q}_i^* \cdot \boldsymbol{y}_i \stackrel{\text{def}}{=} \alpha \leq \inf_{(\boldsymbol{z},\boldsymbol{y}_1,\ldots,\boldsymbol{y}_I)\in B} \boldsymbol{p}^* \cdot \boldsymbol{z} + \sum_{i=1}^{I} \boldsymbol{q}_i^* \cdot \boldsymbol{y}_i. \quad (16.49)$$

For $(\boldsymbol{z}, \boldsymbol{y}) \in T$, we have:

$$(\boldsymbol{z}, \boldsymbol{y}, \ldots, \boldsymbol{y}) \in A.$$

Thus, from (16.49), we see that:

$$\big(\forall(\boldsymbol{z}, \boldsymbol{y}) \in T\big) : \boldsymbol{p}^* \cdot \boldsymbol{z} + \sum_{i=1}^{I} \boldsymbol{q}_i^* \cdot \boldsymbol{y} = \boldsymbol{p}^* \cdot \boldsymbol{z} + \boldsymbol{q}^* \cdot \boldsymbol{y} \leq \alpha; \quad (16.50)$$

where we have defined $\boldsymbol{q}^*$ by:

$$\boldsymbol{q}^* = \sum_{i=1}^{I} \boldsymbol{q}_i^*.$$

On the other hand, since the jth private good is a numéraire for E at $\boldsymbol{a}^*$, we see that for each $\epsilon > 0$, we have:

$$\big(\boldsymbol{x}_i^* + (\epsilon/I \cdot p_j^*)\boldsymbol{e}_j, \boldsymbol{y}^*\big) P_i(\boldsymbol{x}_i^*, \boldsymbol{y}^*) \quad \text{for } i = 1, \ldots, I.$$

Therefore, defining:

$$\widehat{\boldsymbol{x}} = \sum_{i=1}^{I} \big(\boldsymbol{x}_i^* + (\epsilon/I \cdot p_j^*)\boldsymbol{e}_j\big) = \sum_{i=1}^{I} \boldsymbol{x}_i^* + (\epsilon/p_j^*)\boldsymbol{e}_j = \boldsymbol{z}^* + (\epsilon/p_j^*)\boldsymbol{e}_j,$$

we see that $(\widehat{\boldsymbol{x}}, \boldsymbol{y}^*, \ldots, \boldsymbol{y}^*) \in B$, and therefore, by (16.49), we have:

$$\alpha \leq \boldsymbol{p}^* \cdot \widehat{\boldsymbol{x}} + \boldsymbol{q}^* \cdot \boldsymbol{y}^* = \boldsymbol{p}^* \cdot \boldsymbol{z}^* + \boldsymbol{q}^* \cdot \boldsymbol{y}^* + \epsilon.$$

Since $\epsilon > 0$ was arbitrary, it now follows, using (16.50), that:

$$\boldsymbol{p}^* \cdot \boldsymbol{z}^* + \boldsymbol{q}^* \cdot \boldsymbol{y}^* = \alpha = \pi(\boldsymbol{p}^*, \boldsymbol{q}^*). \quad (16.51)$$

From the assumption that P_i is non-decreasing for each i, it is apparent that the set B satisfies the following condition: given any $(\boldsymbol{z}, \boldsymbol{y}_1, \ldots, \boldsymbol{y}_I) \in B$ and $(\boldsymbol{z}', \boldsymbol{y}_1', \ldots, \boldsymbol{y}_I') \in \mathbb{R}^{L+I\cdot M}$, we have:

$$(\boldsymbol{z}', \boldsymbol{y}_1', \ldots, \boldsymbol{y}_I') \geq (\boldsymbol{z}, \boldsymbol{y}_1, \ldots, \boldsymbol{y}_I) \Rightarrow (\boldsymbol{z}', \boldsymbol{y}_1', \ldots, \boldsymbol{y}_I') \in B.$$

Consequently, it follows easily from (16.46) and (16.47) that:

$$(\boldsymbol{p}^*, \boldsymbol{q}_1^*, \ldots, \boldsymbol{q}_I^*) \in \mathbb{R}_+^{L+I\cdot M} \setminus \{\boldsymbol{0}\}. \quad (16.52)$$

Now suppose, by way of obtaining a contradiction, that:

$$\boldsymbol{p}^* = \boldsymbol{0}. \quad (16.53)$$

Then it follows from (16.52) that one of the $\boldsymbol{q}_i^*$ is semi-positive, and thus that:

$$\boldsymbol{q}^* = \sum_{i=1}^{I} \boldsymbol{q}_i^* > \boldsymbol{0}. \quad (16.54)$$

However, by hypothesis 3, there exists $(\boldsymbol{z}^\dagger, \boldsymbol{y}^\dagger) \in T$ such that:

$$\boldsymbol{y}^\dagger \gg \boldsymbol{y}^*,$$

and by (16.53) and (16.54), it then follows that:

$$\boldsymbol{q}^* \cdot \boldsymbol{y}^\dagger + \boldsymbol{p}^* \cdot \boldsymbol{z}^\dagger = \boldsymbol{q}^* \cdot \boldsymbol{y}^\dagger > \boldsymbol{q}^* \cdot \boldsymbol{y}^* = \boldsymbol{q}^* \cdot \boldsymbol{y}^* + \boldsymbol{p}^* \cdot \boldsymbol{z}^*;$$

which contradicts (16.51). We conclude, therefore, that (16.53) cannot hold; and thus that:

$$\boldsymbol{p}^* > \boldsymbol{0}. \tag{16.55}$$

Next, define the wealth assignment vector $\boldsymbol{w}^* = (w_1^*, \ldots, w_i^*)$ by:

$$w_i^* = \boldsymbol{p}^* \cdot \boldsymbol{x}_i^* + \boldsymbol{q}_i^* \cdot \boldsymbol{y}^* \quad \text{for } i = 1, \ldots, I; \tag{16.56}$$

and notice that it then follows from (16.51) and the fact that $\boldsymbol{a}^* = (\langle \boldsymbol{x}_i^* \rangle, \boldsymbol{z}^*, \boldsymbol{y}^*)$ is feasible for E that:

$$\sum_{i=1}^{I} w_i^* = \boldsymbol{p}^* \cdot \sum_{i=1}^{I} \boldsymbol{x}_i^* + \Big(\sum_{i=1}^{I} \boldsymbol{q}_i^* \Big) \cdot \boldsymbol{y}^* = \boldsymbol{p}^* \cdot \boldsymbol{z}^* + \boldsymbol{q}^* \cdot \boldsymbol{y}^* = \pi(\boldsymbol{p}^*, \boldsymbol{q}^*);$$

so we see that $\boldsymbol{w}^*$ is a wealth assignment for E, given $(\boldsymbol{p}^*, \boldsymbol{q}_i^*)$.

To complete our proof[12] that $(\langle \boldsymbol{x}_i^* \rangle, \boldsymbol{z}^*, \boldsymbol{y}^*; \boldsymbol{p}^*, \langle \boldsymbol{q}_i^* \rangle)$ is a Lindahl equilibrium, we must show that each consumer is maximizing satisfaction at $(\boldsymbol{x}_i^*, \boldsymbol{y}^*)$, given $(\boldsymbol{p}^*, \boldsymbol{q}^*)$. Accordingly, let $h \in \{1, \ldots, I\}$ be arbitrary, and suppose $(\widehat{\boldsymbol{x}}_h, \widehat{\boldsymbol{y}}_h) \in C_h$ is such that:

$$(\widehat{\boldsymbol{x}}_h, \widehat{\boldsymbol{y}}_h) P_h (\boldsymbol{x}_h^*, \boldsymbol{y}^*),$$

and let $\epsilon > 0$ be arbitrary. Making use of the fact that the jth private good is a numéraire for E, we then define:

$$\widehat{\boldsymbol{x}}_i = \boldsymbol{x}_i^* + \Big(\frac{\epsilon}{p_j^*(I-1)} \Big) \boldsymbol{e}_i \text{ and } \widehat{\boldsymbol{y}}_i \quad \text{for all } i \neq h.$$

Defining:

$$\widehat{\boldsymbol{x}} = \sum\nolimits_{i=1}^{I} \widehat{\boldsymbol{x}}_i$$

we see that $(\widehat{\boldsymbol{x}}, \widehat{\boldsymbol{y}}_1, \ldots, \widehat{\boldsymbol{y}}_M) \in B$, and therefore:

$$\alpha \leq \boldsymbol{p}^* \cdot \widehat{\boldsymbol{x}} + \sum_{i=1}^{I} \boldsymbol{q}_i^* \cdot \widehat{\boldsymbol{y}}_i = \sum_{i \neq h} \boldsymbol{p}^* \cdot \boldsymbol{x}_i^* + \sum_{i \neq h} \boldsymbol{q}_i^* \cdot \boldsymbol{y}^* + \boldsymbol{p}^* \cdot \widehat{\boldsymbol{x}}_h + \boldsymbol{q}_i^* \cdot \widehat{\boldsymbol{y}}_h + \epsilon;$$

and thus it follows from (16.51), the fact that $\boldsymbol{a}^*$ is feasible, and the definition of $\boldsymbol{w}^*$, that:

$$\boldsymbol{p}^* \cdot \widehat{\boldsymbol{x}}_h + \boldsymbol{q}_i^* \cdot \widehat{\boldsymbol{y}}_h + \epsilon \geq \boldsymbol{p}^* \cdot \boldsymbol{x}_h^* + \boldsymbol{q}_h^* \cdot \boldsymbol{y}^* = w_h^*.$$

[12]From this point onward we will assume that $I \geq 2$. The argument for the case in which $I = 1$ will be left as an exercise.

Since h and $\epsilon > 0$ were arbitrary, we can now conclude that, for each i and each $(\boldsymbol{x}_i, \boldsymbol{y}_i) \in C_i$, we have:

$$(\boldsymbol{x}_i, \boldsymbol{y}_i) P_i (\boldsymbol{x}_i^*, \boldsymbol{y}^*) \Rightarrow \boldsymbol{p}^* \cdot \boldsymbol{x}_i + \boldsymbol{q}_i^* \cdot \boldsymbol{y}_i \geq w_i^*. \tag{16.57}$$

Now, since E is a productive public goods economy, there exist $(\overline{\boldsymbol{z}}, \overline{\boldsymbol{y}}) \in T$, and $(\overline{\boldsymbol{x}}_i, \overline{\boldsymbol{y}}_i) \in C_i$ for each i such that:

$$\overline{\boldsymbol{y}}_i = \overline{\boldsymbol{y}}, \text{ for } i = 1, \ldots, I, \text{ and } \overline{\boldsymbol{z}} \gg \sum\nolimits_{i=1}^{I} \overline{\boldsymbol{x}}_i. \tag{16.58}$$

Consequently, it follows from (16.55) that:

$$\sum_{i=1}^{I} (\boldsymbol{p}^* \cdot \overline{\boldsymbol{x}}_i + \boldsymbol{q}_i^* \cdot \overline{\boldsymbol{y}}_i) = \boldsymbol{p}^* \cdot \Big(\sum_{i=1}^{I} \overline{\boldsymbol{x}}_i \Big) + \boldsymbol{q}^* \cdot \overline{\boldsymbol{y}} < \boldsymbol{p}^* \cdot \overline{\boldsymbol{z}} + \boldsymbol{q}^* \cdot \overline{\boldsymbol{y}} \leq \pi(\boldsymbol{p}^*, \boldsymbol{q}^*) = \sum_{i=1}^{I} w_i^*.$$

Therefore, for some consumer, h, we must have:

$$w_h^* > \boldsymbol{p}^* \cdot \overline{\boldsymbol{x}}_h + \boldsymbol{q}_h^* \cdot \overline{\boldsymbol{y}} \geq \min_{(\boldsymbol{x}_h, \boldsymbol{y}_h) \in C_h} \boldsymbol{p}^* \cdot \boldsymbol{x}_h + \boldsymbol{q}_h^* \cdot \boldsymbol{y}_h.$$

Since P_h is lower semi-continuous, it follows from (16.57) and Proposition 7.25 that h is maximizing preferences at $(\boldsymbol{x}_i^*, \boldsymbol{y}^*)$, subject to $\big((\boldsymbol{p}^*, \boldsymbol{q}_i^*), w_i^*\big)$. But then, since the jth commodity is a numéraire for P_h, it follows that we must have $p_j^* > 0$. Consequently, since the jth private good is a numéraire for E at $\boldsymbol{a}^*$, it now follows that for each i:

$$w_i^* > \min_{(\boldsymbol{x}_i, \boldsymbol{y}_i) \in C_i} \boldsymbol{p}^* \cdot \boldsymbol{x}_i + \boldsymbol{q}_i^* \cdot \boldsymbol{y};$$

and, making use of Proposition 7.25 once again, we see that for each i and all $(\boldsymbol{x}_i, \boldsymbol{y}_i) \in C_i$, we have:

$$(\boldsymbol{x}_i, \boldsymbol{y}_i) P_i (\boldsymbol{x}_i^*, \boldsymbol{y}^*) \Rightarrow \boldsymbol{p}^* \cdot \boldsymbol{x}_i + \boldsymbol{q}_i^* \cdot \boldsymbol{y}_i > w_i^*.$$

Thus we can now conclude that $(\langle \boldsymbol{x}_i^* \rangle, \boldsymbol{z}^*, \boldsymbol{y}^*; \boldsymbol{p}^*, \langle \boldsymbol{q}_i^* \rangle)$ is a Lindahl equilibrium for E, given the wealth distribution, $\boldsymbol{w}^*$. □

16.6.3 The 'Metatheorem'

I should begin by admitting that it is more than a bit ludicrous to call the result to be set out here a 'metatheorem.' My excuse for using such a label is that I want to emphasize the fact that the model and results presented in the previous subsections have immediate implications for general equilibrium models which initially appear to be quite different. We will consider the 'private-ownership economy,'

$$\mathcal{E} = \big(\langle C_i, P_i \rangle, \langle T_j \rangle, \langle \boldsymbol{r}_i \rangle, [s_{ij}]\big)$$

in which:

$$T_j \subseteq \mathbb{R}^L \times \mathbb{R}_+^M;$$

where a pair $(\boldsymbol{z}_j, \boldsymbol{y}_j) \in T_j$ represents net private goods production $\boldsymbol{z}_j \in \mathbb{R}^L$, and public goods production $\boldsymbol{y}_j \in \mathbb{R}^M_+$. Given:

$$(\boldsymbol{z}_j, \boldsymbol{y}_j) \in T_j \quad \text{for } j = 1, \ldots, J,$$

the net amounts made available to the consumption sector are given by:

$$(\boldsymbol{z}, \boldsymbol{y}) = \sum_{j=1}^{J} (\boldsymbol{z}_j, \boldsymbol{y}_j).$$

We then define feasible allocations as follows.

16.29 Definition. We shall say that an allocation, $\boldsymbol{a} = (\langle \boldsymbol{x}_i \rangle, \boldsymbol{z}, \boldsymbol{y})$ is **feasible** (or **attainable**) for $\mathcal{E}$ iff:

1. $(\boldsymbol{z}_j, \boldsymbol{y}_j) \in T_j$, for $j = 1, \ldots, J$,
2. $(\boldsymbol{x}_i, \boldsymbol{y}) \in C_i$ for $i = 1, \ldots, I$, where

$$\boldsymbol{y} = \sum_{j=1}^{J} \boldsymbol{y}_j,$$

and:

3. $\sum_{i=1}^{I} \boldsymbol{x}_i = \sum_{i=1}^{I} \boldsymbol{r}_i + \sum_{j=1}^{J} \boldsymbol{z}_j$.

In keeping with our treatment of this as a private ownership economy, we will suppose that:

$$\sum_{i=1}^{I} s_{ij} = 1 \quad \text{for } j = 1, \ldots, J.$$

Given a price vector $(\boldsymbol{p}, \boldsymbol{q}) \in \Pi_j$, we will, as usual let:

$$\pi_j(\boldsymbol{p}, \boldsymbol{q}) = \max_{(\boldsymbol{z}_j, \boldsymbol{y}_j) \in T_j} (\boldsymbol{p}, \boldsymbol{q}) \cdot (\boldsymbol{z}_j, \boldsymbol{y}_j),$$

for $j = 1, \ldots, J$. We will also allow for a tax/transfer scheme, $\boldsymbol{\tau} = (\tau_1, \ldots, \tau_I)$ such that:

$$\sum_{i=1}^{I} \tau_i = 0.$$

In principle, we will think of τ_i as a transfer payment (which may, of course, be negative).

16.30 Definition. We shall say $(\langle \boldsymbol{x}_i^* \rangle, \langle \boldsymbol{z}_j^*, \boldsymbol{y}_j^* \rangle; \boldsymbol{p}^*, \langle \boldsymbol{q}_i^* \rangle)$ is a **Lindahl equilibrium for $\mathcal{E}$**, given the tax/transfer scheme $\boldsymbol{\tau}^* = (\tau_1^*, \ldots, \tau_I^*)$ iff:

1. $(\boldsymbol{p}^*, \boldsymbol{q}_1^*, \ldots, \boldsymbol{q}_I^*) \in \mathbb{R}_+^{L+I \cdot M}$ and is nonnull,
2. $\boldsymbol{a}^* = (\langle \boldsymbol{x}_i^* \rangle, \boldsymbol{z}^*, \boldsymbol{y}^*)$ is feasible for $\mathcal{E}$,
3. $\sum_{i=1}^{I} \tau_i^* = 0$,
4. defining $\boldsymbol{q}^* = \sum_{i=1}^{I} \boldsymbol{q}_i^*$, we have:

$$(\forall (\boldsymbol{z}_j, \boldsymbol{y}_j) \in T_j)\colon \boldsymbol{p}^* \cdot \boldsymbol{z}_j + \boldsymbol{q}^* \cdot \boldsymbol{y}_j \leq \pi_j(\boldsymbol{p}^*, \boldsymbol{q}^*) = \boldsymbol{p}^* \cdot \boldsymbol{z}_j^* + \boldsymbol{q}^* \cdot \boldsymbol{y}_j^*,$$

5. defining $\boldsymbol{y}^* = \sum_{j=1}^{J} \boldsymbol{y}_j^*$, we have for each i:

$$\boldsymbol{p}^* \cdot \boldsymbol{x}_i^* + \boldsymbol{q}_i^* \cdot \boldsymbol{y}^* \leq w_i^* \stackrel{\text{def}}{=} \boldsymbol{p}^* \cdot \boldsymbol{r}_i + \sum_{j=1}^{J} s_{ij} \pi_j(\boldsymbol{p}^*, \boldsymbol{q}^*) + \tau_i^*;$$

and, for all $(\boldsymbol{x}_i, \boldsymbol{y}) \in C_i$:

$$(\boldsymbol{x}_i, \boldsymbol{y}) P_i (\boldsymbol{x}_i^*, \boldsymbol{y}^*) \Rightarrow \boldsymbol{p}^* \cdot \boldsymbol{x}_i + \boldsymbol{q}_i^* \cdot \boldsymbol{y} > w_i^*.$$

Now, given a private ownership economy, $\mathcal{E} = (\langle C_i, P_i\rangle, \langle T_j\rangle, \langle \boldsymbol{r}_i\rangle, [s_{ij}])$, we define the **aggregate economy corresponding to $\mathcal{E}$**, $E = (\langle C_i, P_i\rangle, T)$ by letting:

$$T = \sum_{i=1}^{I} \boldsymbol{r}_i + \sum_{j=1}^{J} T_j$$

(preferences and consumption sets are the same in $\mathcal{E}$ and E). We then have the following result; the proof of which I will leave as an exercise.

16.31 Proposition. *Let $\mathcal{E} = (\langle C_i, P_i\rangle, \langle T_j\rangle, \langle \boldsymbol{r}_i\rangle, [s_{ij}])$ be a private ownership economy, and $E = (\langle C_i, P_i\rangle, T)$ be the aggregate economy corresponding to $\mathcal{E}$. Then we have the following.*

1. If $(\langle \boldsymbol{x}_i^*\rangle, \langle \boldsymbol{z}_j^*, \boldsymbol{y}_j^*\rangle; \boldsymbol{p}^*, \langle \boldsymbol{q}_i^*\rangle)$ is a Lindahl equilibrium for $\mathcal{E}$, given the tax/transfer vector $\boldsymbol{\tau}^* = (\tau_1^*, \ldots, \tau_I^*)$, then $(\langle \boldsymbol{x}_i^*\rangle, \boldsymbol{z}^*, \boldsymbol{y}^*; \boldsymbol{p}^*, \langle \boldsymbol{q}_i^*\rangle)$ is a Lindahl equilibrium for E, given the wealth assignment $\boldsymbol{w}^*$ defined by:

$$w_i^* = \boldsymbol{p}^* \cdot \boldsymbol{r}_i + \sum_{j=1}^{J} s_{ij}\pi_j(\boldsymbol{p}^*, \boldsymbol{q}^*) + \tau_i^* \quad \textit{for } i = 1, \ldots, I.$$

2. Conversely, if $(\langle \boldsymbol{x}_i^*\rangle, \boldsymbol{z}^*, \boldsymbol{y}^*; \boldsymbol{p}^*, \langle \boldsymbol{q}_i^*\rangle)$ is a Lindahl equilibrium for E, given the wealth assignment $\boldsymbol{w}^* = (w_1^*, \ldots, w_I^*)$, and $(\boldsymbol{z}_j^*, \boldsymbol{y}_j^*) \in T_j$ $(j = 1, \ldots, J)$ are such that:

$$\sum_{j=1}^{J} (\boldsymbol{z}_j^*, \boldsymbol{y}_j^*) = (\boldsymbol{z}^*, \boldsymbol{y}^*),$$

then $(\langle \boldsymbol{x}_i^*\rangle, \langle \boldsymbol{z}_j^*, \boldsymbol{y}_j^*\rangle; \boldsymbol{p}^*, \langle \boldsymbol{q}_i^*\rangle)$ is a Lindahl equilibrium for $\mathcal{E}$, given the tax/transfer vector $\boldsymbol{\tau}^*$ defined by:

$$\tau_i^* = w_i^* - \boldsymbol{p}^* \cdot \boldsymbol{r}_i - \sum_{j=1}^{J} s_{ij}\pi_j(\boldsymbol{p}^*, \boldsymbol{q}^*) \quad \textit{for } i = 1, \ldots, I.$$

Exercises.

1. Consider an economy with two consumers, one private good, and one public good, suppose that $X_i = \mathbb{R}_+^2$, and that the ith consumer's preferences can be represented by the utility functions:

$$u_1(x_1, y) = x_1 + 6\sqrt{y} \text{ and } u_2(x_2, y) = x_2 + 10\sqrt{y},$$

respectively; where 'x_i' and 'y' denote the respective quantities of the private and public goods. Suppose further that the consumers have the initial endowments:

$$\boldsymbol{\omega}_1 = (40, 0) \text{ and } \boldsymbol{\omega}_2 = (60, 0);$$

and that one unit of the private good can always be used to produce one unit of the public good. On the basis of this information:

a. Find the (unique) Pareto efficient production of the public good.

b. Find the Lindahl prices which will suport the allocation you have found in part (a), or prove that none exist.

2. Consider an economy with m consumers, one private good, and one public good. Suppose that the ith consumer's preferences can be represented by the utility function:

$$u_i(x_i, y) = x_i + 2\beta_i\sqrt{y} \quad \text{for } i = 1, \ldots, m,$$

where 'x_i' and 'y' denote the quantities of the private and public goods, respectively:

$$\beta_i > 0 \quad \text{for } i = 1, \ldots, m, \quad \text{and} \quad \sum_{i=1}^{m} \beta_i = 40.$$

Suppose further that the production function for the production of the public good is given by:

$$y = f(z) = z/4,$$

where 'z' denotes the quantity of the private good devoted to public goods production. Given this information, find the Pareto efficient quantity of the public good.

3. Consider an economy with m consumers, one private good, and one public good. Suppose that the ith consumer's preferences can be represented by the utility function:

$$u_i(x_i, y) = x_i + 2\beta_i\sqrt{y} \quad \text{for } i = 1, \ldots, m,$$

where 'x_i' and 'y' denote the quantities of the private and public goods, respectively:

$$\beta_i > 0 \quad \text{for } i = 1, \ldots, m, \quad \text{and} \quad \sum_{i=1}^{m} \beta_i = 81.$$

Suppose further that the input-requirement finction for the production of the public good is given by:production function for the production of the public good is given by:

$$g(y) = 6y^{3/2},$$

(only the public good is produced); and that the ith consumer's initial endowment is given by:

$$\boldsymbol{r}_i = (r_{i1}, 0),$$

where:

$$r_{i1} \geq 3 \quad \text{for } i = 1, \ldots, m.$$

Given this information:

a. Find the Pareto efficient quantity of the public good.

b. Find a (the) Lindahl equilibrium for this economy. Is this equilibrium unique in this case?

4. Prove Proposition 16.4.

5. Prove the Samuelson condition for the one-public good case, as set out in (16.23) of Section 4 of this chapter.

6. In the context of the model set out in Section 17.5, suppose the cost function takes the form given in equation (16.33) of the text; with:

$$c_j(y_j) = \gamma_j \cdot y_j \quad \text{for } j = 1, \ldots, n;$$

where γ_j is a positive constant for $j = 1, \ldots, n$, and suppose $(\boldsymbol{x}^*, \boldsymbol{y}^*)$ is a feasible state for E.

a. Show that if $(\boldsymbol{x}^*, \boldsymbol{y}^*)$ is a ratio equilibrium [that is, is a cost share equilibrium for a cost share system of the form (16.34)], and we define:

$$q_{ij}^* = r_{ij}\gamma_j \quad \text{for } i = 1, \ldots, m \text{ and } j = 1, \ldots, n,$$

then $\big((\boldsymbol{x}^*, \boldsymbol{y}^*), \langle \boldsymbol{q}_i^* \rangle\big)$ is a Lindahl equilibrium for E.

b. Show that if $\big((\boldsymbol{x}^*, \boldsymbol{y}^*), \langle \boldsymbol{q}_i^* \rangle\big)$ is a Lindahl equilibrium for E, and we define $[r_{ij}]$ by:

$$r_{ij} = q_{ij}^*/\gamma_j \quad \text{for } i = 1, \ldots, m \text{ and } j = 1, \ldots, n,$$

then $(\boldsymbol{x}^*, \boldsymbol{y}^*)$ is a ratio equilibrium for E.

7. Prove Proposition 16.31

8. Determine the conditions which must hold in the model of Section 4 in order to be able to apply Theorem 16.28 in concluding that to each Pareto efficient outcome there exist prices and a wealth distribution which make the allocation a Lindahl equilibrium.

9. Prove Theorem 16.9

Chapter 17

Externalities

17.1 Introduction

When studying the First and Second Fundamental Theorems of Welfare Economics, we noted the fact that both results depended upon 'an absence of externalities;' something which we have not as yet defined. However, before attempting a formal definition of economic externalities, or external effects, let's consider some examples.

1. If you build a fence separating our lots in a suburban neighborhood, then the utility I gain from owning power grass trimmers increases.
2. If I live next door to a coal-burning power plant, the production choices made by the power plant affects my willingness to pay for a clothes dryer. In fact the power plant's choices are likely to have a profound effect on my general well-being.
3. If I am a bee-keeper, my farmer-neighbor's decision to devote more of his land to pasture and/or hay production may result in an unexpected increase in my honey production.

The common thread in the above examples revolves around the fact that one economic agent's choices directly affects the well-being of one or more other agents. Moreover, in the normal functioning of the price system, the agent inflicting or conferring the externality takes only the direct gain or loss to her- or himself into account in choosing the level of the externality-causing activity. A more formal definition is the following, which is used by Baumol and Oates [1988].

17.1 Definition. (Baumol and Oates, p. 17.) An **externality** is present whenever some individual (say A's) utility or production relationships include real (that is, nonmonetary) variables, whose values are chosen by others (persons, corporations, governments) without particular attention to the effects on A's welfare.

In this chapter we will look at some of the theory of externalities; in the context of a general equilibrium model, insofar as is feasible. In Section 2 we undertake the development of the fundamentals of the analysis of externalities in the context of a simple model; which is then extended somewhat in Section 3. In Section 4 we will discuss the famous 'Coase Theorem,' which advocates the assignment of appropriate property rights as a solution to the problems caused by external effects. Finally, in Section 5, we briefly consider the optimality properties of a Lindahl-type equilibrium in the context of a simple general equilibrium-type model. In the context of the model developed in Section 5, we find that this equilibrium is 'non-wasteful;' that is, it results in a Pareto efficient allocation.

17.2 Externalities: A First Look

In beginning our analysis of externalities, or external effects, let's reverse our usual procedure by beginning with a fairly general model, which we will then specialize a bit for most of our analysis. We will suppose initially that the externality is produced by a consumer, say the first (consumer number 1), and is inflicted upon some or all of the remaining consumers. We will suppose there are $M \geq 2$ consumers, and, as usual, denote consumer i's bundle of marketed commodities by '$\boldsymbol{x}_i$,' use the generic notation '$\boldsymbol{p}$' to denote the vector of commodity prices for these commodities, and denote the initial value of the ith consumer's income (wealth) by 'w_i.' We then denote the level of the externality-generating activity by 'z,' and suppose the ith consumer's preferences are represented by the utility function $u_i(\boldsymbol{x}_i, z)$. In order to proceed with our analysis, we develop a variant of indirect utility, as follows. Given $(\boldsymbol{p}, m_i; z)$, let '$h(\boldsymbol{p}, m_i; z)$' denote the value of $\boldsymbol{x}_i$ which solves the problem:

$$\max_{w.r.t.\ \boldsymbol{x}_i} u_i(\boldsymbol{x}_i, z) \quad \text{subject to: } \boldsymbol{p} \cdot \boldsymbol{x}_i \leq m_i, \tag{17.1}$$

given the level of z. We then define $v_i \colon \mathbb{R}_+^{n+1} \times \mathbb{R} \to \mathbb{R}$ by:

$$v_i(\boldsymbol{p}, m_i; z) = u_i\big[h(\boldsymbol{p}, m_i; z); z\big]. \tag{17.2}$$

However, we can, and will simplify matters somewhat by supposing that the level of the externality has no effect upon prices; so that we can consider this indirect utility function to be determined by the pair (m_i, z); in other words, for fixed $\boldsymbol{p}$, say $\boldsymbol{p} = \boldsymbol{p}^*$, we define $\varphi \colon \mathbb{R}_+ \times \mathbb{R} \to \mathbb{R}$ by:

$$\varphi_i(m_i, z) = v_i(\boldsymbol{p}^*, m_i; z). \tag{17.3}$$

Of course, the initial values of m_i will be equal to w_i in our analysis.[1]

We will assume initially that there exists $z^0 > 0$ such that, for all $z < z^0$:

$$\frac{\partial}{\partial z}\big[\varphi_1(w_1, z)\big] > 0 \text{ and } \left.\frac{\partial \varphi_1}{\partial z}\right|_{(w_1, z^0)} = 0; \tag{17.4}$$

and, in this initial specification, the only thing we will suppose about the other consumers is that for each $z \geq 0$, there exists $j \in \{2, \ldots, M\}$ such that for all $m_j \geq 0$:

$$\frac{\partial}{\partial z}\big[\varphi_2(m_j, z)\big] \neq 0. \tag{17.5}$$

[1]The definition of the function φ_i is borrowed from Mas-Colell , Whinston, and Green [1995, p. 352].

Before proceeding further with the analysis, however, it should be pointed out that, even if, for example, there exists some consumer, j for whom, say:

$$\frac{\partial}{\partial z}\left[\varphi_2(m_j, z)\right] < 0 \quad \text{for all } (m_i, z) \gg \mathbf{0};$$

consumer j's choice of $\boldsymbol{x}_j$ may be independent of z, for all $(\boldsymbol{p}, m_i)$; in other words, j's demand function may depend only on the pair $(\boldsymbol{p}, m_i)$, and be completely independent of z. I, for example, consider myself to be worse off because of the multitude of cars with amazingly powerful stereo systems which their owners use to play rap at ridiculous volumes. On the other hand, this has not as yet induced me to buy ear plugs to wear while walking outside. This is perhaps not the best of examples, but the point is that a consumer's utility may be affected by another consumer's activity even though his or her consumption choices do not change as a result of the activity.

To proceed with our analysis, suppose each φ_i is strictly increasing in m_i, in fact that:

$$\frac{\partial}{\partial m_i}\left[\varphi_i(m_i, z)\right] > 0 \quad \text{for all } (m_i, z) \in \mathbb{R}^2_+; \tag{17.6}$$

recall that we are supposing that consumer one conducts an externality-generating activity, and that Ms. 1's indirect utility function satisfies (17.4). Pareto efficiency requires that the allocation $(\langle m_i \rangle, z)$ solves the problem:

$$\max_{w.r.t.\ (\langle m_i \rangle, z)} \varphi_1(m_1, z) \tag{17.7}$$

subject to:

$$\begin{aligned} \varphi_i(m_i, z) &\geq \overline{u}_i \quad \text{for } i = 2, \ldots, M, \\ \overline{w} &\geq \sum\nolimits_{i=1}^{M} m_i, \end{aligned} \tag{17.8}$$

where:

$$\overline{w} = \sum\nolimits_{i=1}^{M} w_i$$

(and w_i is the ith consumer's initial income [wealth]). Of course, under the present assumptions, each of the constraints in (17.8) must be an equality at the solution. Consequently, we can make use of the classical Lagrangian method i analyzing this problem. Denoting the multipliers associated with the first $M-1$ constraints in (17.8) by λ_i $(i = 2, \ldots, M)$, and that for the last equality by μ, our first-order conditions become:

$$\frac{\partial \varphi_1}{\partial m_i} - \mu = 0 \tag{17.9}$$

$$\frac{\partial \varphi_1}{\partial z} + \sum\nolimits_{i=2}^{M} \lambda_i \frac{\partial \varphi_i}{\partial z} = 0 \tag{17.10}$$

$$\lambda_i \frac{\partial \varphi_i}{\partial m_i} - \mu = 0 \quad \text{for } i = 2, \ldots, M. \tag{17.11}$$

From (17.11) we obtain:

$$\lambda_i = \mu \Big/ \left(\frac{\partial \varphi_i}{\partial m_i}\right) \quad \text{for } i = 2, \ldots, M. \tag{17.12}$$

Substituting into (17.10), and making use of (17.9), we then obtain:

$$\frac{\partial \varphi_1}{\partial z} \Big/ \frac{\partial \varphi_1}{\partial m_1} = -\sum_{i=2}^{M} \left(\frac{\partial \varphi_i}{\partial z} \Big/ \frac{\partial \varphi_i}{\partial m_i} \right). \tag{17.13}$$

The ratio:

$$\frac{\partial \varphi_i}{\partial z} \Big/ \frac{\partial \varphi_i}{\partial m_i},$$

evaluated at a point (m_i, z), is the slope, with respect to the z-axis, of the indifference curve through (m_i, z), at that point. Consequently, it can be interpreted as the the marginal willingness-to-pay for z, in the case of Ms. 1; or the monetary value of the affect upon agent i, for $i \neq 1$. Given that φ_1 satisfies (17.4), and assuming utility-maximizing behavior on the part of Ms. 1, as well as an absence of regulation, Ms. 1 will set the value of z equal to z^0. This sets the left-hand-side of equation (17.13) equal to zero; whereas, in the presence of external effects, the right-hand-side of (17.13) will generally not be equal to zero, when evaluated at $(\langle w_i \rangle, z^0)$. Correspondingly, the unrestricted choice of level of z by Ms. 1 will generally not result in a Pareto efficient situation.

A number of different proposals have been made as to how to achieve Pareto efficiency in this sort of situation, and we will consider some of these shortly. In the meantime, let's discuss something which you have probably already noticed; namely, that equation (17.13) looks an awful lot like the Samuelson conditions! In fact, the formal theory of externalities is not very different from that of pure public goods; indeed, it is possible to develop the theory of pure public goods as a special case of an externality. I have chosen not to do so in this text largely because the tools which seem most appropriate for the analysis of externalities are different from those most suited to the analysis of public goods production and allocation.

Of the proposals which have been introduced to correct the undesirable effects of economic externalities, we will discuss three: Pigouvian taxes or subsidies, bargaining solutions, and regulation. We will begin our discussion of these proposals in a simplified model; namely one in which we consider only two consumers.

Suppose now that $M = 2$ (or that the consumers other than Mr. 2 are not affected by the externality), that φ_1 satisfies (17.4), and that for all (m_2, z):

$$\frac{\partial}{\partial z}\left[\varphi_2(m_2, z)\right] < 0.$$

As noted previously, if agent one is unrestricted in her choice of level of z, she will set $z = z^0$. We can depict the initial situation in a variant of the Edgeworth Box diagram, as in Figure 17.1, on the next page. In the diagram, we measure the quantity of m_1 (which we will think of as the income available for expenditure on [marketed] commodities) along the horizontal axis in the conventional direction; that is, m_1 increases as we move to the right. On the other hand, Mr. 2's level of m_2 increases as we move left. The quantity of z is measured along the vertical axis for both consumers. Thus, Ms. 1's utility increases as we move to the northeast in our diagram, while Mr. 2's utility is increasing as we move to the southwest. We will indicate a distribution of wealth, or income (or 'expenditure on marketed commodities') by, for example, 'm_i,' or 'w_i;' with the distance from O_i to m_i being

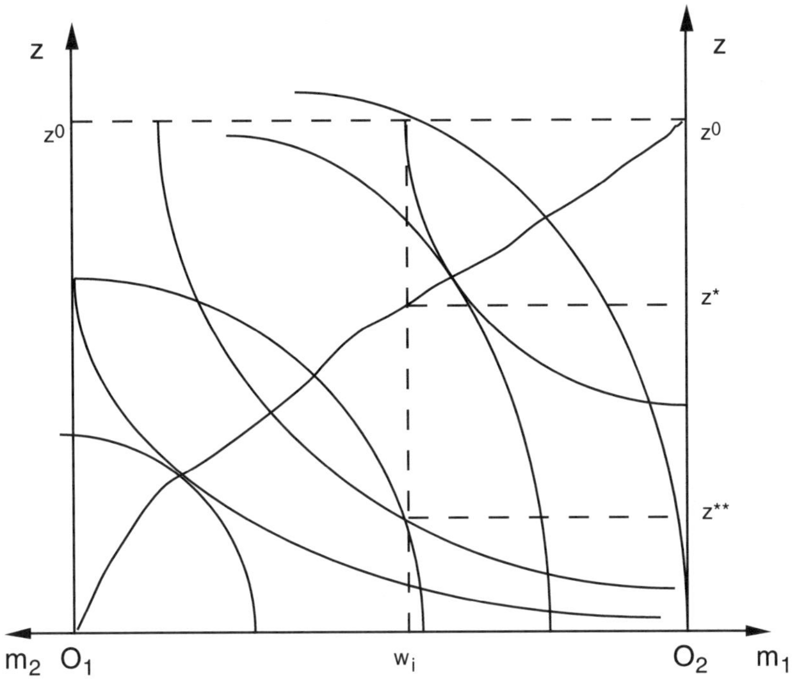

Figure 17.1: Unconstrained Equilibrium at (w_i, z^0).

the value of m_i. Thus, for example, in the diagram, the distance from O_1 to w_i (moving left to right)is w_1, Ms. 1's initial wealth; while the distance from O_2 to w_i (moving from right to left). equals w_2. The initial equilibrium, then, is at (w_i, z^0) in our diagram, at the point where Ms. 1's 'marginal willingness-to-pay for z equals zero.

Now, it is easily seen that Pareto efficient allocations in our diagram occur at tangency points of the two consumers' indifference curves (since in the context of Figure 17.1, the slope of Mr. 2's indifference curve with respect to the z-axis is the negative of the conventional slope), verifying equation (17.13), and we have indicated the locus of efficient allocations in Figure 17.1 as the curve connecting $(O_1, 0)$ and (O_2, z^0). Moreover, it is apparent that, in the situation depicted in Figure 17.1, the initial situation will not be Pareto efficient.

With the aid of the diagram we can obtain some insight into the good and the bad of government regulation in a situation like that under study here. If, for example, goveernment regulation sets a legal maximum of z^{**} on the production of z, at but allows Ms. 1 the choice of its quantity, subject to this regulation, then the resultant allocation will be (w_i, z^{**}) in our diagram. In general, of course, this allocation will not be Pareto efficient (although it should be noted that any regulation which reduces the quantity of z below z^0 will be beneficial to Mr. 2). It is only if the regulator chooses the quantity z^* in this situation that Pareto efficiency will be achieved, and obviously determining exactly what this optimal maximum is requires the regulator to obtain a great deal of information about the agents' preferences which will be extremely difficult to obtain.

In the preceding discussion we have supposed that, in the absence of regulation, Ms. 1 has been assigned full property rights to z; in other words, that Ms. 1 can choose whatever level of z she chooses. Let's continue to make that assumption, but suppose now that she is willing to make an 'all-or-nothing' bargain with Mr. 2 to reduce the level of z. If, in fact, Ms. 1 knows Mr. 2's preferences, and is a utility maximizer, she will extract a payment, T, from 2 in an amount which maximizes $\varphi_1(m_1, z)$ subject to $\varphi_2(w_2 - T, z) \geq \varphi_2(w_2, z^0)$.[2] Thus, we would expect to arrive at the point (m_i^*, z^*) in Figure 17.2, below, where:

$$m_1^* - w_1 = T^* = w_2 - m_2^*.$$

Notice that, under the assumptions being employed here, the bargain will achieve

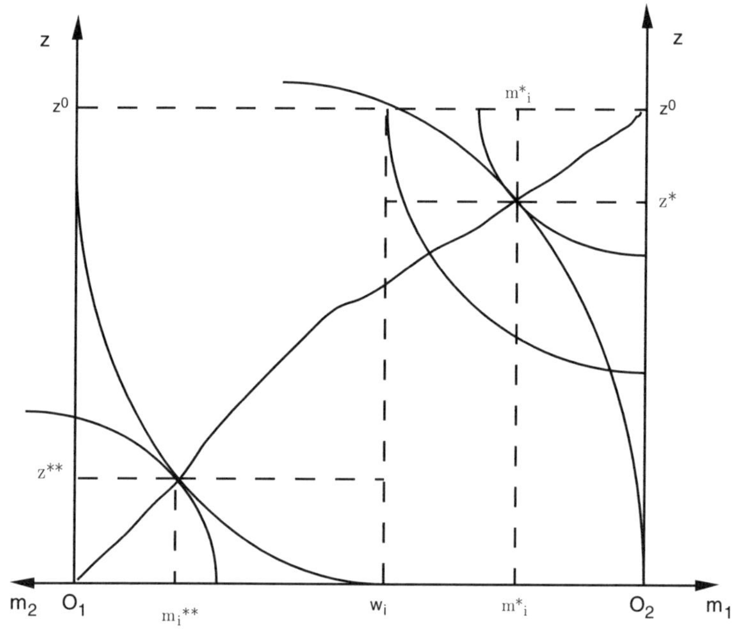

Figure 17.2: The Bargaining Solutions.

a Pareto efficient solution. However, suppose now that Mr. 2 is given the property right, and has the right to demand that $z = 0$, that Mr. 2 is willing to enter into an 'all-or-nothing' bargain with Ms. 1, is a utility-maximizer, *and knows Ms. 1's preferences.* Now the bargain struck will be at (m_i^{**}, z^{**}) in our diagram, where

[2]Presumably, we should use a constraint of the form $\varphi_2(w_2 - T, z) \geq \varphi_2(w_2, z^0) + \epsilon$, for some $\epsilon > 0$; but, as is usually done in the related literature, we will ignore this complication, and suppose that Mr. 2 will accept the bargain as long as he doesn't lose utility.

$w_1 - m_1^{**} = T^{**} = m_2^{**} - w_2$. Once again the resultant allocation will be Pareto efficient, but the relative well-being of the two consumers, the level of the externality, and the payments made are all likely to be quite different.

There is a case where the externality level is independent of the assignment of property rights, however. If the consumers' indirect utility functions take the form:

$$\varphi_i(m_i, z) = m_i + \phi_i(z) \quad \text{for } i = 1, 2, \tag{17.14}$$

where ϕ_i is concave, for $i = 1, 2$, then it follows from Proposition 16.2 that to obtain a Pareto efficient allocation, we must maximize $\phi_1(z) + \phi_2(z)$. Assuming that both functions are differentiable, this implies that at the optimal level of z, call it z^*, (and assuming that both m_1 and m_2 are positive) we must have $\phi_1'(z^*) = -\phi_2(z^*)$. In fact, in this case, it follows from Propositions 16.1 and 16.2 that if the functions ϕ_i are differentiable and concave, then every interior Pareto efficient allocation[3] will involve the same quantity of the externality-producing activity (z). In particular, since we have shown that the all-or-nothing bargaining solutions result in Pareto efficient outcomes, it follows that in this quasi-linear utility case both solutions result in the same level of the externality. Another way of looking at this is to note that in this situation, equation (17.13) becomes (since $\partial\varphi_i/\partial m_i = 1$, for each i):

$$\partial\varphi_1/\partial z = -\partial\varphi_2/\partial z;$$

or, given the form of φ_i, $\phi_1'(z) = -\phi_2'(z)$. Consequently, in the context of our 'Edgeworth Box-like' diagram, the interior of the locus of Pareto efficient points will be a horizontal line; as in Figure 17.3, below.

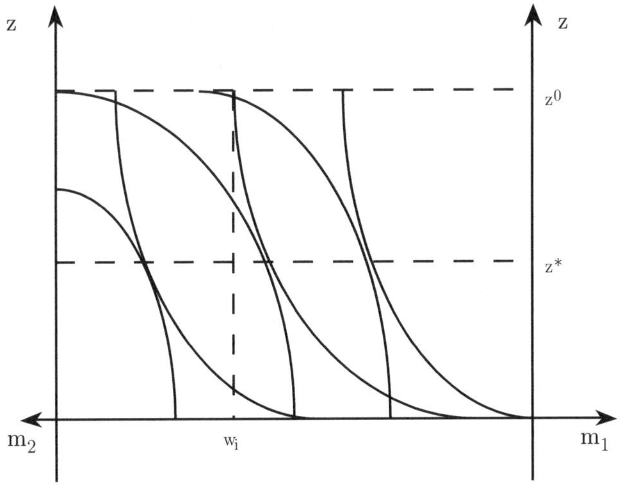

Figure 17.3: Pareto Efficiency with Quasi-Linear Utility.

[3]By 'interior Pareto efficient allocations,' I mean those Pareto efficient allocations at which $m_i > 0$, for $i = 1, 2$.

Continuing with our discussion of policy tools to correct for external effects, let's now return to our original assumption of M consumers, with the first producing the externality, and consider the idea of a **Pigouvian tax**. Suppose a tax of t per unit production of z is imposed on Ms. 1, where:

$$t = -\sum\nolimits_{i=2}^{M} \Big(\frac{\partial \varphi_i}{\partial z} \Big/ \frac{\partial \varphi_i}{\partial m_i} \Big). \tag{17.15}$$

If Ms. 1 is a utility maximizer and retains the right to choose the level of z, she will maximize $\varphi_1(w_1 - tz, z)$; which, as you can readily verify, will result in the satisfaction of the Samuelson condition [equation (17.13)]. Or will it? We have been very sloppy in this analysis; having ignored a number of complications which might be pertinent to the analysis. In the first place, the Pigouvian tax we have specified in equation (17.15) is not actually well-defined, in that we have not indicated the values of (m_i, z) at which the derivatives are to be evaluated. This problem is particularly serious in that the relevant partials are functions of both variables; and while it may seem obvious that there will be a Pareto efficient allocation with $z = z^*$ and in which:

$$m_1 = w_1 - tz^* \text{ and } m_i = w_i \text{ for } i = 2, \ldots, M,$$

it is certainly not clear that this will always be the case. In fact, if we view our model as incorporating the entire economy, there definitely will not be a Pareto efficient allocation of the indicated form; the tax collected from Ms. 1 will have to be distributed among the other consumers in some way in order that Pareto efficiency can be achieved.[4] Finally, the Samuelson condition is *necessary* for Pareto efficiency (given differentiability) at an interior allocation, and the assumptions which we have made so far are not sufficient to ensure that an allocation at which the Samuelson condition is satisfied will be Pareto efficient.

In common with most of the literature on this topic, in the remainder of our discussion we will ignore most of the complications mentioned in the above paragraph; a posture which is made possible by the simple device of assuming that all of our consumers have quasi-linear utility functions. Specifically, we will suppose that each φ_i satisfies (17.14), that Ms. 1's utility function satisfies:

$$(\forall z \in [0, z^0]) \colon \phi_1'(z) > 0, \phi_1''(z) \leq 0, \quad \text{and } \phi'(z^0) = 0; \tag{17.16}$$

and, defining:

$$\Phi(z) = \sum\nolimits_{i=2}^{M} \phi_i(z) \quad \text{for } z \in \mathbb{R}_+, \tag{17.17}$$

we will assume that for all $z \in]\,0, z^0]$:

$$\Phi'(z) < 0 \text{ and } \Phi''(z) \leq 0. \tag{17.18}$$

With these assumptions, things become quite straightforward; it follows from Propositions 16.1 and 16.2 that we have Pareto efficiency if, and only if:

$$\phi_1'(z) = -\Phi'(z).$$

[4]Or spent by government on a project desired by consumers. However, this is an aspect of the situation that we will not try to cover further here.

In fact, it also follows that if either ϕ_1 or Φ is strictly concave, then there will exist a unique value of z, z^* satisfying (17.18). Under these assumptions, the ambiguities regarding the optimal Pigouvian tax disappear; if the regulator sets:

$$t = -\Phi(z^*), \tag{17.19}$$

then Ms. 1's utility-maximizing choice of z achieves Pareto efficiency with any distribution of the corresponding tax receipts among the other $M-1$ consumers. However, notice that Ms. 1 must be excluded from this distribution of tax receipts; if Ms. 1 knows she will receive a share of the tax receipts, she will generally choose a level of z greater than z^*.

I will leave it as an exercise to show that if Ms. 1 receives a subsidy of t per unit of reduction of z below z^0, where t is given by (17.19), then the Pareto efficient level of z will be chosen by Ms. 1. (See Exercise 2 at the end of this chapter.) In this case, it follows from Propositions 16.1 and 16.2 that the subsidy can be paid for by taxes (for example, a tax of $[1/(M-1)]t$) levied on the other $M-1$ consumers. We will not pursue these ideas further here; instead let's consider what happens if the externality is inflicted by a firm.

17.3 Extending the Model

Let's begin our discussion by considering the case in which a firm generates the externalities. How does our analysis change in this case? To take a simple case, suppose we have a firm producing a product x, which we will suppose for simplicity is sold on a perfectly competitive market at a price which will be fixed in our analysis at $p > 0$. Suppose further that our firm uses k inputs, $y_1, \ldots, y_k$ in its production process, and that all of these inputs are purchased on competitive markets at fixed prices $q_1, \ldots, q_k$. Finally, suppose that the firm's production results in an external effect whose quantity can be measured by z. In this case, it is convenient to suppose that the firm has effectively two differentiable production functions, f and g, which produce the respective outputs, x and z; that is our variables are related by:

$$x = f(\boldsymbol{y}) \text{ and } z = g(\boldsymbol{y}).$$

Now, in the absence of regulation, a profit-maximizing competitive firm will ignore the g function, and choose $\boldsymbol{y}^* \in \mathbb{R}^k_+$ satisfying:

$$p\Big(\frac{\partial f}{\partial y_j}\Big) = q_j \quad \text{for } j = 1, \ldots, k. \tag{17.20}$$

However, if we suppose the consumers' indirect utility can be represented by functions φ_i of the form indicated in (17.3) and which satisfy (17.6); the firm's choice of inputs will have an impact upon the consumers which the firm does not take into account in profit-maximization. In fact, if we allow for the fact that the firm may have to be paid a subsidy or pay a tax in order to achieve Pareto efficiency, our Pareto problem can be expressed as follows;

$$\max_{w.r.t.\,\boldsymbol{m},\boldsymbol{y}} pf(\boldsymbol{y}) - \boldsymbol{q}\cdot\boldsymbol{y} + m_0, \tag{17.21}$$

subject to:

$$\begin{aligned} &(\lambda_i) \qquad \varphi_i\big[m_i, g(\boldsymbol{y})\big] \geq \overline{u}_i \quad \text{for } i = 1, \ldots, M, \\ &(\mu) \qquad w - \sum\nolimits_{i=0}^{M} m_i \geq 0. \end{aligned} \tag{17.22}$$

Given the assumptions which we have made, the constraints must be satisfied as equalities; so that we obtain the following necessary conditions for a solution at a point $(\boldsymbol{m}^*, \boldsymbol{y}^*)$:

$$1 - \mu = 0, \tag{17.23}$$

$$\lambda_i \frac{\partial \varphi_i}{\partial m_i} - \mu = 0 \quad \text{for } i = 1, \ldots, M, \tag{17.24}$$

$$p\frac{\partial f}{\partial y_j} - q_j + \sum_{i=1}^{M} \lambda_i \frac{\partial \varphi_i}{\partial z}\frac{\partial g}{\partial y_j} = 0 \text{ for } j = 1, \ldots, k. \tag{17.25}$$

From (17.23) and (17.24) we obtain:

$$\lambda_i = 1 \Big/ \frac{\partial \varphi_i}{\partial m_i} \quad \text{for } i = 1, \ldots, M;$$

and substituting into (17.25), we obtain:

$$p\frac{\partial f}{\partial y_j} = q_j - \frac{\partial g}{\partial y_j}\sum_{i=1}^{M}\Big(\frac{\partial \varphi}{\partial z}\Big/\frac{\partial \varphi_i}{\partial m_i}\Big) \quad \text{for } j = 1, \ldots, k. \tag{17.26}$$

Thus it can be seen that if a Pigouvian tax in the amount:

$$t_j = -\frac{\partial g}{\partial y_j}\sum_{i=1}^{M}\Big(\frac{\partial \varphi_i}{\partial z}\Big/\frac{\partial \varphi_i}{\partial m_i}\Big)$$

per unit of y_j is levied on the firm (for $j = 1, \ldots, k$), then, subject to the sorts of qualifications introduced in our initial discussion of Pigouvian taxes in the previous section, Pareto efficiency may be achieved. However, we now have the additional ambiguity in that we have left open the question of who receives the profits and the factor payments. Once again the conclusions become much less ambiguous, and the analysis becomes much more straightforward if each consumer's (indirect) utility function is of the quasi-linear form considered in the latter part of the previous section. Moreover, we can do a bit better than this, as follows.

We begin by defining the function $u_0 \colon \mathbb{R} \times \mathbb{R}^k_+ \to \mathbb{R}$ by:

$$u_0(m_0, \boldsymbol{y}) = m_0 + v_0(\boldsymbol{y}), \tag{17.27}$$

where:

$$v_0(\boldsymbol{y}) = pf(\boldsymbol{y}) - \boldsymbol{q} \cdot \boldsymbol{y}. \tag{17.28}$$

Next, define $u_i \colon \mathbb{R}^{k+1}_+ \to \mathbb{R}$ by:

$$u_i(m_i, \boldsymbol{y}) = m_i + v_i(\boldsymbol{y}), \tag{17.29}$$

where:

$$v_i(\boldsymbol{y}) = \phi_i[g(\boldsymbol{y})] \quad \text{for } i = 1, \ldots, M. \tag{17.30}$$

If we assume that the production function and each of the v_i functions ($i = 1, \ldots, M$) is concave (we will consider sufficient conditions for the latter shortly), and if the production function, f, or at least one of the v_i functions is strictly concave, then it follows from Propositions 16.1 and 16.2 that there will exist a unique Pareto efficient vector of inputs, $\boldsymbol{y}^* \in \mathbb{R}^k_+$; and which maximizes the sum of the v_i functions. Thus, if all of the v_i functions are differentiable and $\boldsymbol{y}^* \gg \boldsymbol{0}$, Pareto efficiency will be achieved if, and only if:

$$\sum\nolimits_{i=0}^{M} \frac{\partial v_i}{\partial y_j} = p\Big(\frac{\partial f}{\partial y_j}\Big) - q_j + \Big(\frac{\partial g}{\partial y_j}\Big) \sum\nolimits_{i=1}^{M} \phi'\big[g(\boldsymbol{y}^*)\big] = 0 \quad \text{for } j = 1, \ldots, k. \tag{17.31}$$

Thus in this case, if we suppose that the firm knows the form of the function g, then the Pigouvian tax, τ^*, per unit of z produced, where τ^* is given by::

$$\tau^* = -\sum\nolimits_{i=1}^{M} \phi'\big[g(\boldsymbol{y}^*)\big], \tag{17.32}$$

will result in the firm's making the correct (Pareto efficient) choice of $\boldsymbol{y}$. I will leave you to verify the truth of this assertion (see Exercise 3, at the end of this chapter). Notice also that, under these assumptions, we needn't worry about who receives the profits or the factor payments, as long as they're all paid out to someone (once again this follows from Propositions 16.1 and 16.2). Moreover, In this case, the firm will choose the most efficient way of reducing the production of the externality (taking into account the factor's marginal productivity in producting the desired output), despite the fact that, in principle, the regulating agency need not know the form of g at all. We will pursue this idea a bit further shortly; in the meantime, let's return to the problem of finding sufficient conditions for the concavity of the v_i functions, for $i = 1, \ldots, M$. In this connection, we have the following. I will leave the proof as an exercise (see Exercise 4 at the end of this chapter).

17.2 Proposition. *If ϕ_i is concave and decreasing, and g is convex, then the composite function:*

$$v_i(\boldsymbol{y}) = \phi_i\big[g(\boldsymbol{y})\big],$$

is concave.

It should be noted that some of the agents, other than the 0th may be firms; all that is needed to incorporate this case is that the effect of z on the firms' profits can be described by a function ϕ_i which is decreasing and concave. Furthermore, the ϕ_i functions need not all be strictly decreasing, in order that the analysis of this section goes through; that is, it need not be the case that the externality is negative for each agent i ($i = 1, \ldots, M$). However, v_i will not be concave if ϕ_i is strictly increasing, given that g is convex. In other words, our analysis breaks down if z provides a positive externality for some of the consumers, while being strictly negative for others. In fact, the problem here is not simply a matter of not having the assumptions needed for our derivations, as we can see by considering the extreme case. Suppose each ϕ_i is increasing, and that some are strictly increasing;

so that we have a *positive* externality. In this case, if g is convex, then the Pareto efficient level of $\boldsymbol{y}$ may well not exist, in that the sum of the consumers' utilities may increase indefinitely as $g(\boldsymbol{y})$ increases. On the other hand, if g is concave, and the ϕ_i functions are increasing and concave, our analysis will go through as before, except that the optimal 'tax' levied on z will now be negative (that is, will be a subsidy). I will leave the details of these extensions as exercises.

Let's take a look now at a situation in which there are multiple firms producing an externality; for example an air pollutant. We could allow for an arbitrary finite number of polluting firms in our analysis, but the essential points can be made, and with a great simplification in notation, if we have only two firms generating the externality. So, we suppose now that there are M agents in the economy, each having an 'indirect utility function' of the quasi-linear form. Moreover, while the analysis is applicable to firms generating any sort of negative externality, for simplicity in exposition we will refer to the externality being generated as 'air pollution.' Thus, we suppose that the first two agents are the polluting firms, with:

$$u_i(m_i, \boldsymbol{y}_i) = m_i + v_i(\boldsymbol{y}_i) = m_i + p_i f_i(\boldsymbol{y}_i) - \boldsymbol{q} \cdot \boldsymbol{y}_i \quad \text{for } i = 1, 2,$$

while:[5]

$$u_i(m_i, \boldsymbol{y}_i, \boldsymbol{y}_2) = m_i + v_i(\boldsymbol{y}_1, \boldsymbol{y}_2) = m_i + \phi_i[g_1(\boldsymbol{y}_i) + g_2(\boldsymbol{y}_2)] \quad \text{for } i = 3, \ldots, M.$$

In the formulation here we are supposing that the effect of the firms' pollution is additive. We will re-examine this assumption later.

Now, under the present assumptions, it follows from Propositions 16.1 and 16.2 that Pareto efficiency is achieved in this case if, and only if the sum of the individual 'utility functions' is maximized. Thus, if all these functions are differentiable, we have Pareto efficiency if, and only if, the following conditions hold (where we denote the Pareto efficient quantity of z by 'z^*'):

$$p_i\Big(\frac{\partial f_i}{\partial y_{ij}}\Big) - q_j + \sum\nolimits_{h=3}^{M} \phi_h'(z^*)\frac{\partial g_i}{\partial y_{ij}} = 0. \tag{17.33}$$

Consequently, if we define the tax, t^*, by:

$$t^* = \sum\nolimits_{h=3}^{M} \phi_h'(z^*), \tag{17.34}$$

it is easy to show (see Exercise 5, at the end of this chapter) that the Pareto efficient outcome is achieved if the two firms are charged a tax of t^* per unit of z produced.

While the tax (or emissions charge) solution achieves a Pareto efficient outcome under the present assumptions (and this result can be generalized, as we will demonstrate in Section 5), there are several problems associated with this policy. First of all, while, in principle, the regulating agency needn't know the form of the g_i functions, it does need to know the precise monetary evaluation of the damages incurred by the other agents at the Pareto efficient outcome in order to know the level at which the tax should be set. Secondly, the amount of tax actually paid by

[5]Once again some of these agents may also be firms.

the firms may be quite high; in fact it could be high enough to cause some firms to shut down, which is one of the reasons that this scheme may be politically infeasible. A policy measure which at least partially solves both of these problems is the issuance of tradeable or marketable emission permits. The idea is this: the regulatory agency determines the total quantity of emissions to be permitted. In principle, one would like this quantity to be the Pareto efficient quantity, but the calculation of this quantity may require more information than the agency is able to obtain. In any event, having determined the allowable total level of total emissions (the value of $z = z_1 + z_2$ in our example), the agency issues permits which allow a given amount of z to be generated per period. For simplicity we will suppose that each permit allows a production of one unit of z per period; so that, if the allowable level of emissions is, say $\overline{z} > 0$, the agency will issue a total of $\overline{z}$ permits. These permits are sometimes auctioned off to the firms, but doing this involves the same sort of political difficulties as the tax, and consequently, they are more often issued to the firms at no charge. This results in some political problems revolving around the issue of how these permits are distributed among the polluting firms, but as we will see, the final outcome in terms of the values of the z_i's is independent of the initial distribution of permits. That this is so comes about because the regulatory agency sets up a market for trading the permits; enabling a firm which needs a larger amount of the permits to buy the additional permits from firms which are able to reduce their emissions below the number of issued permits.

So, in terms of the model we are using here, Firm i will maximize:

$$p_i f(\boldsymbol{y}_i) - \boldsymbol{q} \cdot \boldsymbol{y}_i - \tau\big[g_i(\boldsymbol{y}_i) - z_i\big], \tag{17.35}$$

where:

z_i is the number of permits originally issued to firm i,

τ is the market price of the permits,

and:

$g_i(\boldsymbol{y}_i)$ is the measure of the emission produced by Firm i (and thus is the number of permits needed by Firm i). We will denote by '$\overline{z}$' the total number of permits issued, so that:

$$z_1 + z_2 = \overline{z};$$

and we will denote Firm i's profit-maximizing choice of $\boldsymbol{y}_i$ by '$\boldsymbol{y}_i^*$,' for $i = 1, 2$. I will then leave it to you to show that $(\boldsymbol{y}_1^*, \boldsymbol{y}_2^*)$ solves the problem:

$$\max_{w.r.t.(\boldsymbol{y}_1, \boldsymbol{y}_2)} \sum\nolimits_{i=1}^{M} u_i(m_i, \boldsymbol{y}_1, \boldsymbol{y}_2) \quad \text{subject to: } g_1(\boldsymbol{y}_1) + g_2(\boldsymbol{y}_2) = \overline{z} \tag{17.36}$$

(see Problem 6, at the end of this chapter). It then follows from Proposition 16.1 that the resulting outcome is second-best Pareto efficient; that is, no choice of $(\boldsymbol{y}_1, \boldsymbol{y}_2)$ which satisfies $g_1(\boldsymbol{y}_1) + g_2(\boldsymbol{y}_2) = \overline{z}$ strictly Pareto dominates the profit-maximizing outcome $(\boldsymbol{y}_1^*, \boldsymbol{y}_2^*)$.

Now suppose we complicate things bit by supposing that the effect of the pollution is not the same for the two firms; that is, suppose that the total effect of the emissions discharge from the standpoint of the last $M - 2$ agents is given by:

$$z = g_1(\boldsymbol{y}_1) + a g_2(\boldsymbol{y}_2),$$

where $a < 1$. So, the idea here is that if we were to measure Firm 2's emission discharge at the source, the amount would be given by $g_2(\mathbf{y}_2)$, but the effect of the pollution upon the rest of the population is smaller than the emissions by Firm 1.[6] The problem that this creates for the regulatory agency is that the permit scheme just set out here must now be modified in that the firms cannot in this instance trade permits on a one-for-one basis. However, I will not pursue this discussion further here. For those interested in further reading on this topic, let me suggest Baumol and Oates (1988), Chapters 11 and 12; Helfand, Berck, and Maull (2003); and Stavins (2003).

17.4 The 'Coase Theorem'

Several different assertions are variously represented as 'the Coase Theorem.' However, the following three statements pretty much cover the versions which have been stated.

1. The existence of a negative externality does not necessarily mean that the activity should be banned.

2. In general, the assignment of property rights in connection with an externality will:

a. result in the same level of the externalities being generated–if an equilibrium exists (waiving the question of what the equilibrium concept is).

b. result in a Pareto efficient allocation's being attained if the parties are allowed to bargain freely, and if there are no transactions costs.[7]

Thus, for example Hurwicz [1999, p. 239] states the 'Coase Theorem' as, "...in the absence of transaction costs, institutional factors such as liability rules will not affect the level of the externality."

In the previous section we have already seen that statement 1, above, is probably true, but that statement 2.a is not necessarily true. The bargaining solutions which we looked at in that section did support statement 2.b; however, the following two informal examples cast a somewhat different light on the whole issue.

17.3 Example. (Aivazian & Callen [1981]) This example involves 3 firms: Factories A and B, which exert a negative externality (via air pollution) on a laundry, Firm C. In the status quo, the firm's profits per day are as follows:

$$V(A) = \$3,000/\text{day}, \ V(B) = \$8,000/\text{day}, \ \text{and} \ V(C) = \$24,000/\text{day}.$$

We suppose that if firms A and B merge, their combined profits will be given by:

$$V(A, B) = 15,000/\text{day}.$$

[6] For example, it may be that Firm 1 is situated in a heavily populated area, while Firm 2 is quite isolated.

[7] This phrase '...bargain freely, and if there are no transactions costs...' appears frequently in the literature, but is something of an oxymoron. It is not clear how any reasonable definition of transactions costs could fail to include the necessity for bargaining as such a cost.

We suppose further that if firms A and C were to merge, that the combined firm would simply shut A's operation down. This would result in a loss of profits from A's production of \$3,000/day, but we will suppose that C's profits would increase by \$7,000/day, so that the merged firm's profits are given by:

$$V(A, C) = \$31,000/\text{day}.$$

Similarly, if firms B and C were to merge, firm B's operation would be shut down, but we will suppose that C's profits would increase by \$12,000/day; so that the merged firm's profits would be:

$$V(B, C) = \$36,000/\text{day}.$$

Finally, it is supposed that if the three firms all merge, then both Firm A and firm B will be shut down, and while firm C's profits will then be supposed to increase, it is reasonable to suppose that, probably because of decreasing returns, the increase in profits will be less than the sum of the two separate increases. Suppose, therefore, that the net profits of the three-firm coalition will be:

$$V(A, B, C) = \$40,000/\text{day}.$$

It is easily checked that total profit is maximized at the grand coalition, which is, therefore, the Pareto efficient solution. However, the issue to be explored here is whether or not this solution can be attained if '...the parties are allowed to bargain freely, and in the absence of transaction costs.' Presumably, the solution concept which is applicable here is the core; so that what we wish to investigate is whether or not the 'grand coalition' solution is in the core. If it is, then letting π_A, π_B, and π_C denote the firms' payoffs, we must have:

$$\pi_A + \pi_B + \pi_C = \$40,000/\text{day},$$

and

$$\begin{aligned} \pi_A + \pi_B &\geq V(A, B) \\ \pi_A + \pi_C &\geq V(A, C) \\ \pi_B + \pi_C &\geq V(B, C); \end{aligned} \tag{17.37}$$

and, of course:

$$\pi_A \geq V(A), \pi_B \geq V(B), \text{ and } \pi_C \geq V(C).$$

However, if we add up the inequalities in (17.37). we see that we must have:

$$2(\pi_A + \pi_B + \pi_C) \geq V(A, B) + V(A, C) + V(B, C);$$

or:

$$\pi_A + \pi_B + \pi_C \geq (1/2)\big[V(A, B) + V(A, C) + V(B, C)\big] = (1/2) \times (82,000) = 41,000.$$

Since:

$$\pi_A + \pi_B + \pi_C = 40,000,$$

it follows that the core is empty in this case. □

17.4 Example. [8] A farmer can choose to plant within, say, a 10-foot margin of a railroad track, or leave it fallow. If he plants this extra area of his field, he earns an additional return of 10 (hundred, thousand, whatever). However, if he does plant this margin, there is a high probability (which we will take to be equal to one) that the marginal area will catch fire (from sparks from the train's passage), and if it does catch there is also a high probability (which we will also take to be equal to one) that the fire will spread to and destroy the main part of the field. If this happens, the loss to the farmer is 10 (loss of the margin) + 50 (loss in the main field). The gross return to the raliroad of running one train per day is 36.

So, we have the situation shown in Table 18.1, if the railroad is liable for damages. It is then readily seen that the equilibrium in this situation is at (Plant, 0 trains); which is *not* socially optimal, since total profits are highest at the (Don't plant, 1 train) outcome. On the other hand if the railroad is *not* liable for damages, we have the payoffs shown in Table 18.2.

Farmer	**Railroad**	
	No train	1 train
Plant	(10, 0)	(10, −24)
Don't Plant	(0, 0)	(0, 36)

Table 17.1: Payoffs if the Railroad is liable.

Farmer	**Railroad**	
	No train	1 train
Plant	(10, 0)	(−50, 36)
Don't Plant	(0, 0)	(0, 36)

Table 17.2: Payoffs if the Railroad is not liable.

Thus, in this second situation the unique Nash equilibrium is at (Don't plant, 1 train); which *is* the socially optimal outcome.

Now, while the Nash equilibria seem quite natural and obvious, they are different in the two situations. The change in property rights results in a socially optimal solution if the railroad is not liable, whereas the Nash equilibrium in the first situation was definitely sub-optimal. On the other hand, in the first situation it seems quite likely that if bargaining is allowed, a deal might be struck in which the railroad paid the farmer 10 (or 10.1) not to plant the margin, while the railroad runs one train per day. This results in the payoff pair $(10, 26)$ [or $(10.1, 25.9)$] which is also Pareto efficient. Moreover, notice that when one allows for bargaining in this example, the Coase conjecture is supported. The actions, though not the payoffs, of the two parties are exactly the same [(don't plant, 1 train)] under the two different assignments of property rights, if bargaining is allowed.

[8] This example is a modified version of an example originally discussed by Pigou [1932], and revisited in Coase [1960]. In its present form, it has been borrowed from Marcus Berliant (private correspondence).

However, suppose we change the example slightly, to have five farmers rather than one; with each farmer having the same payoff possibilities as did the farmer in the original example. We will now suppose that the railroad runs by each of the five farms, and that the railroad has exactly the same profit possibilities as before. In this situation, the socially optimal solution is for each farmer to plant, and the railroad to run no trains, and this will be the Nash equilibrium solution if the railroad is liable for damages.[9] On the other hand, if the railroad is not liable, the Nash equilibrium is attained when the railroad runs one train per day, and no farmer plants his or her margin; a situation which is definitely not Pareto efficient. It is, of course, true that a bargaining solution might now be attained in which a Pareto efficient outcome results; namely, if the farmers pay the railroad 36 to not run the train, and each farmer then plants his or her margin. However, notice that, while a coalition of any four of the farmers can profitably pay the requisite 36 to the railroad in return for not running the train, the fifth farmer can 'freeride' on the agreement to earn an extra profit of 10, rather than the other farmers' share of the surplus of 4 attained by the agreement. Thus, it would certainly appear that the Pareto efficient solution is much less likely to result under the second assignment of property rights (that in which the railroad is not liable for damages) than it is if the railroad is liable. □

17.5 Example. The example presented here is a slight modification of an example presented in Donald Campbell's excellent text (Campbell [1987], pp. 6–7). Campbell deals with the choice of a group of n individuals deciding whether to choose hard or soft coal to heat their homes. This example was inspired by the Clean Air Act passed in England in 1956, and more or less represents the choices of a household in London prior to the passage of this act. We suppose that n individuals live in a city of area $A > 0$, and that an individual choosing a source of heat for her or his home incurs costs of two sorts: first, there is the cost of the coal itself; and secondly, if the individual chooses soft coal for heat there is an additional cost due to the greater air pollution. In fact, we will suppose that, given constant choices of other consumption goods, and a given level of heat in the home, each individual, i's utility is given by:

$$u_i(m_i, \rho) = m_i - \rho,$$

where m_i is income available for expenditure on other goods, and ρ is a measure of air pollution. Insofar as the measure of air pollution is concerned, we suppose that an individual's heating with soft coal injects a quantity of k particulates into the atmosphere, but that this is spread over the area of measure $A > 0$, so that a measure of increased air pollution is $\gamma = k/A$. On the other hand, if n_s individuals choose to heat with soft coal, then the total quantity of air polution is given by $\gamma \cdot n_s$. We suppose also that the out-of-pocket cost of providing the desired amount of heat is £100 for hard coal (H), and £90 for soft coal (S). Thus, if there are n_s individuals already burning soft coal, and an additional individual, i, chooses to burn soft coal as well, then i's utility is given by:

$$u_i^S = w_i - 90 - \gamma(1 + n_s);$$

[9] Notice that in this case, planting the margin is a dominant strategy for each farmer, while the payoff to the railroad if each farmer plants, and the railroad runs one train, is -214.

while if i chooses hard coal instead, then:

$$u_i^H = w_i - 100 - \gamma \cdot n_s.$$

Therefore, for any value of n_s:

$$u_i^S - u_i^H = 10 - \gamma.$$

Consequently, we see that if $\gamma < 10$ (that is, if $k/A < 10$), then soft coal is the dominant choice for each individual, regardless of the value of n_s. However, if each individual chooses their dominant strategy, then each achieves a utility of:

$$\overline{u}_i^S = w_i - 90 - \gamma \cdot n;$$

whereas if the burning of soft coal is banned, so that each individual is forced to burn hard coal, then each achieves a utility of:

$$\overline{u}_i^H = w_i - 100.$$

Therefore, all are better off, given this legislation, if:

$$\gamma \cdot n > 10.$$

So, burning soft coal is a dominant strategy in the absence of regulation, but regulation makes everyone better off if:

$$10/n < \gamma < 10. \quad \square$$

So, in the first of the three examples presented here we had a situation in which the core of the implied game was empty; suggesting that bargaining would not lead to an optimal solution. In the second, we saw how the assignment of property rights could change the probable outcomes in the absence of bargaining, but that bargaining might nonetheless result in an optimal solution. Finally, we examined a case in which everyone was made better off by the outright banning of an activity. Clearly, one cannot assume that one blanket solution should be applied whenever and wherever negative externalities are encountered.

17.5 Lindahl and Externalities

In this chapter we have emphasized the formal similarity between the economic theories of externalities and public goods. In essence, they are both examples of the failure of normal markets to provide a Pareto efficient outcome; but there is more to this similarity in that they can be regarded as 'commodities' which lack the excludability feature which allows markets to develop and to function well. One may well ask, then, whether one might not be able to formulate a slight modification of the Lindahl equilibrium concept to apply to externalities. Well, in fact one can, at least to some extent. In particular, we can develop a version of the First Fundamental Theorem for an economy with externalities, as follows.

We consider an exchange economy with m consumers, n (private) commodities, and as in Section 2, we will eventually be supposing that the first consumer (Ms. 1) generates an externality which is regarded as detrimental by the other consumers. However, we will initially suppose only that each consumer has a preference relation, P_i, on $\mathbb{R}_+^{n+1}$ which is asymmetric. In our modified Lindahl equilibrium, we will proceed as if each consumer is independently choosing the quantity of z; so that, as you are no doubt anticipating, we will say that a tuple $\langle \boldsymbol{x}_i, z_i \rangle$ is an **allocation for *E*** iff $(\boldsymbol{x}_i, z_i) \in \mathbb{R}_+^{n+1}$, for $i = 1, \ldots, m$, and

$$z_1 = z_2 = \ldots, z_m;$$

and that it is a **feasible allocation for *E*** if, in addition:

$$\sum\nolimits_{i=1}^{m} \boldsymbol{x}_i = \boldsymbol{r},$$

where $r \in \mathbb{R}_+^n$ is the aggregate resource endowment for E. To complete our preliminaries, we will say that $\boldsymbol{w} \in \mathbb{R}_+^m$ is a **feasible wealth distribution for *E* given $\boldsymbol{p} \in \mathbb{R}_+^n$** iff:

$$\sum\nolimits_{i=1}^{m} w_i = \boldsymbol{p} \cdot \boldsymbol{r}.$$

17.6 Definition. We will say that $(\langle \boldsymbol{x}_i^*, z_i^* \rangle, \boldsymbol{p}^*, \boldsymbol{t}^*)$ is a **Lindahl (externalities) equilibrium for *E*, given the wealth distribution $\boldsymbol{w} \in \mathbb{R}_+^n$** iff:

1. $\langle \boldsymbol{x}_i, z_i \rangle$ is feasible for E,
2. $\boldsymbol{w}$ is a feasible wealth distribution for E, given $\boldsymbol{p}^*$,
3. $\boldsymbol{t}^* \in \mathbb{R}_+^n$ and satisfies:

$$\sum\nolimits_{i=1}^{m} t_i^* = 0, \tag{17.38}$$

4. for each $i \in \{1, \ldots, m\}$, we have:

$$\boldsymbol{p}^* \cdot \boldsymbol{x}_i^* \leq w_i + t_i^* z_i \tag{17.39}$$

and:

$$\left(\forall (\boldsymbol{x}_i, z_i) \in \mathbb{R}_+^{n+1}\right) : (x_i, z_i) P_i (x_i^*, z_i^*) \Rightarrow \boldsymbol{p}^* \cdot \boldsymbol{x}_i > w_i + t_i^* z_i. \tag{17.40}$$

As usual, we will say that a feasible allocation $\langle \boldsymbol{x}_i^*, z^* \rangle$ is **Pareto efficient for *E*** iff there exists no feasible allocation, $\langle x_i, z \rangle$, such that:

$$(\boldsymbol{x}_i, z) P_i (\boldsymbol{x}_i^*, z^*) \quad \text{for } i = 1, \ldots, m.$$

We then have the following.

17.7 Proposition. *If $(\langle x_i^*, z^* \rangle, \boldsymbol{p}^*, \boldsymbol{t}^*)$ is a Lindahl equilibrium for E, given the wealth distribution $\boldsymbol{w} \in \mathbb{R}_+^n$, then $\langle x_i^*, z^* \rangle$ is Pareto efficient for E.*

Proof. Suppose $\langle \boldsymbol{x}_i, z \rangle$ is an allocation such that:

$$(\boldsymbol{x}_i, z) P_i (\boldsymbol{x}_i^*, z^*) \quad \text{for } i = 1, \ldots, m. \tag{17.41}$$

Then, since $(\langle x_i^*, z^*\rangle, \boldsymbol{p}^*, \boldsymbol{t}^*)$ is a Lindahl equilibrium for E, given the wealth distribution $\boldsymbol{w}$, we must have:

$$\boldsymbol{p}^* \cdot \boldsymbol{x}_i - t_i z > w_i \quad \text{for } i = 1, \ldots, m. \tag{17.42}$$

Adding the inequalities in (17.42, we have:

$$\sum\nolimits_{i=1}^{m} \boldsymbol{p}^* \cdot \boldsymbol{x}_i - \sum\nolimits_{i=1}^{m} t_i z > \sum\nolimits_{i=1}^{m} w_i. \tag{17.43}$$

However, since $\boldsymbol{w}$ is a feasible wealth distribution for E, given $\boldsymbol{p}^*$, we have:

$$\sum\nolimits_{i=1}^{m} w_i = \boldsymbol{p}^* \cdot \boldsymbol{r};$$

and making use of this and (17.38), we have from (17.42) that:

$$\boldsymbol{p}^* \cdot \left(\sum\nolimits_{i=1}^{m} \boldsymbol{x}_i\right) > \boldsymbol{p}^* \cdot \boldsymbol{r};$$

and we see that $\langle \boldsymbol{x}_i, z\rangle$ is not feasible for E. It follows that $\langle \boldsymbol{x}_i^*, z^*\rangle$ is Pareto efficient for E. □

While we have not assumed anything about the form of any externalities in the above result, it is of interest to consider the form of the equilibrium in the situation analyzed in Section 2. Specifically, suppose one consumer, say the first, inflicts an externality on the remaining consumers, that each P_i is increasing in $\boldsymbol{x}_i$, given z; while, given $\boldsymbol{x}_i$:

$$P_1 \text{ is strictly increasing in } z, \text{ for } 0 \leq z < z^0; \tag{17.44}$$

and:

$$P_i \text{ is strictly decreasing in } z, \text{ for } 0 \leq z < z^0, \quad \text{for } i = 2, \ldots, m. \tag{17.45}$$

In this case, it is easily seen that if $(\langle x_i^*, z^*\rangle, \boldsymbol{p}^*, \boldsymbol{t}^*)$ is a Lindahl equilibrium for E, given the wealth distribution $\boldsymbol{w}$, then we must have:

$$t_1 < 0 \text{ and } t_i > 0 \text{ for } i = 2, \ldots, m.$$

Moreover, all the problems associated with achieving a Lindahl equilibrium for public goods are present here, and some are even more severe, if anything. First of all, notice that in the special case in which $n = 1$ and P_i can be represented by a differentiable utility function of the form:

$$u_i(x_i, z_i) = x_i + \phi_i(z_i) \quad \text{for } i = 1, \ldots, m;$$

a Lindahl equilibrium requires that, for each i:

$$t_i^* = -\phi_i'(z^*).$$

Thus the individual tax/subsidy quantities t_i^* must reflect the exact benefit, or the exact monetary representation of the marginal damage to the consumer at the

Pareto efficient quantity of z. However, if one were to try to elicit a schedule of these damages or benefits from the individual consumers (that is, to obtain an estimate of ϕ_i', for each i), then given the assumptions of this paragraph, Ms. 1 has an incentive to understate the benefit of the externality to her, while the other consumers have an incentive to overstate the damage done to them.

We will discuss these incentive problems further in the next chapter. In the meantime, let me mention that a similar Lindahl-type equilibrium can be defined for the model and situation analyzed in Section 3, but I will leave this as (a fairly extensive) exercise for 'the interested reader.'

17.6 Postscript

In this chapter we have barely scratched the surface in terms of an analysis of the effects and possible solutions for economic externalities. Two problems which are normally considered as involving types of externalities, but which we have not touched upon at all here are the management of a common property resource and coordination failures. For a very readable introduction to these two areas, let me recommend Leach [2004], Chapters 8 and 9. For a further discussion of the Lindahl mechanism approach and the problem of missing markets in connection with externalities, see Starrett[2003].

Exercises.

1. Suppose in an economy there are two firms, both selling their outputs in competitive markets. Firm 1 produces x with the cost function:

$$c(x) = (1/2)x^2;$$

while firm 2 produces y, and has the cost function:

$$c(y, x) = (1/20)y^2 + x^2,$$

where 'x' denotes the output of firm 1. Suppose also that the market prices for the two commodities are given by:

$$p_x = 20 \text{ and } p_y = 10.$$

On the basis of this information, answer the following questions.

a. Find the competitive outputs for firms 1 and 2.

b. Find the socially optimal production of x and y, given the data at your disposal.

c. Is there a Pigovian tax/subsidy scheme which will result in the firms' choosing the socially optimal output quantitites as their profit-maximizing choices? If there is, what is it?

2. Show that, given the assumptions set out in (17.16) and (17.18) of the text, if Ms. 1 is paid a subsidy of t per unit reduction of z from z^0, where t is given by (17.19), then she will maximize utility by choosing the Pareto efficient level of z.

3. Verify the fact that, in the context of the model developed at the end of Section 3, and assuming that the firm knows the form of the function g, then requiring the firm to pay a tax of τ^* per unit production of z [where τ^* is from (17.32)], will result in the firm's producing the Pareto efficient levels of x and z.

4. Prove Proposition 17.2

5. Show that in the two-firm emissions case considered in Section 3 that the tax t^* defined in (17.34) will result in the Pareto efficient outcome.

6. Complete the details of the analysis of the marketable emissions case considered in Section 3 [for the case in which the total effect of the emissions is given by $z = g_1(\boldsymbol{y}_1) + g_2(\boldsymbol{y}_2)$].

Chapter 18

Incentives and Implementation Theory

18.1 Introduction

In this chapter we will undertake an introductory survey of a portion of a very complicated, but exciting and important area of economics; the study of implementation theory. Much economic research has been focused on the topic of 'mechanisms,' or 'allocation mechanisms,' a line of research initiated by L. Hurwicz [1960]. In general, such mechanisms proceed by eliciting agents' preferences as to possible quantities of, say, a public good. Given this information concerning preferences, the function of the mechanism is to arrive at an allocation for the economy, including a production level for the public good. We then say that the mechanism *implements* a given social choice correspondence if the allocation determined by the mechanism is always contained in the correspondence. For example, we would say that our mechanism implements the Pareto correspondence, if the allocation chosen is always (for any admissible preferences on the part of the agents) Pareto efficient (Hurwicz termed such mechanisms '**non-wasteful**'). There are, in fact, a number of mechanisms of this sort which are 'non-wasteful,' *assuming that agents are truthful in announcing their preferences!* Unfortunately, it may well be the case that agents have an incentive to announce preferences which are different from their true preferences; that is, an agent may expect to do better by announcing false preferences (which we term 'acting strategically'), than by truthful responses. This is a major difficulty in trying to design a 'good' allocation mechanism, because a mechanism which has very good properties when agents respond truthfully may have very bad properties if agents respond strategically.

One case in which we can be reasonably sure that agents will respond truthfully is that in which, given the mechanism, truth-telling is a 'dominant strategy' for each agent. Unfortunately, Gibbard [1973] and Satterthwaite [1975] showed that if a general type of mechanism makes truth-telling a dominant strategy, then the mechanism must be dictatorial. The fundamental result here is the famous 'Gibbard-Satterthwaite Theorem,' which we will study intensively in Sections 2 and 3.

While the above paragraph may make it seem that this chapter is filled with nothing but negative results, let me assure you that we will be developing some related positive results as well. One of the critical conditions in both the Gibbard-Satterthwaite Theorem and Arrow's General Possibility Theorem is that the domain of the 'outcome function' in the first case, or the 'social preference function' in the second, includes at least every possible n-tuple of linear orders, where n is the number of agents involved. In many economic contexts, however, it is very natural to assume that the agents have preferences of a more specific type; and in some of these contexts we will be able to develop some mechanisms for which truth-telling is a dominant strategy, and which are not dictatorial.

18.2 Game Forms and Mechanisms

The concept of a game form was introduced by Gibbard [1973] and can be viewed as a generalization of the concept of a voting rule. With a voting rule, the individual agents can be viewed as announcing their preference relations.[1] In contrast, a game form admits of wider possibilities for the messages to be sent by the individual agents, and is formally defined as follows.

18.1 Definition. An ***n*-agent game form** is defined by a set of n strategy spaces, $S_1, \ldots, S_n$, a set of alternatives, X, and a function $g\colon \prod_{i=1}^n S_i \to X$ (called the **outcome function**). We will use the notation '$\Gamma = \langle \boldsymbol{S}, X, g\rangle$' to denote a game form, where:

$$\boldsymbol{S} = \prod\nolimits_{i=1}^{n} S_i.$$

For example, a voting rule, $f\colon \mathcal{D}^n \to X$ is a game form in which:

$$S_i = \mathcal{D} \quad \text{for } i = 1, \ldots, n.$$

In terms of formal structure, a game form is equivalent to a simplified definition of a 'mechanism,' as was originially introduced by Hurwicz [1960]. We define this as follows.

18.2 Definition. An ***n*-agent mechanism** is defined by n **message spaces**, M_i, a set of alternatives, X, and an **outcome function**, $g\colon \prod_{i=1}^n M_i \to X$. We will use the generic notation '$\mathcal{M} = \langle \boldsymbol{M}, X, g\rangle$,' to denote an (abstract) mechanism, where:

$$\boldsymbol{M} = \prod\nolimits_{i=1}^{n} M_i.$$

Of course, a mechanism is simply a game form in which the strategy spaces are the individual message spaces. However, the terminology is more evocative of economic 'mechanisms,' and may sometimes help us to focus on some pragmatic issues which arise in trying to implement some of our theories involving game forms. On the other hand, the game form definition tends to focus our attention on some strategic and incentive issues which we might otherwise overlook. Two of the examples which we discussed in Section 3 of Chapter 14 may help to illustrate what I have in mind here.

[1] Or of something about their preferences. We will return to this point later.

Plurality voting is quite naturally viewed as the mechanism in which $M_i = X$ for each i, and where $g(\boldsymbol{m})$, the value of the outcome function at:

$$\boldsymbol{m} = (m_1, \ldots, m_n),$$

is that message (alternative) sent by the largest number of agents (although once again we need a tie-breaking rule). Stating things in this fashion may help bring two important aspects of this voting rule into sharper focus: it is instructionally and computationally simple. That is, it is easy to instruct the agents as to what they are to do, and it is easy for the agency or individual implementing the rule to determine the winner.

The simplicity of the plurality voting rule which is brought out in sharp relief by the mechanism language quite emphatically differentiates it from the Borda count voting rule. From the mechanism standpoint, the messages sent by the individual agents are their complete rankings of the alternative set; a much more complicated message than for the plurality voting mechanism. Moreover, the calculations which must be made by the implementing agency are more complicated as well.

On the other hand, thinking of the Borda count method as a game form brings out a very pertinent consideration, which is important in evaluating not only the Borda count, but most other voting rules as well. It is easy to show that the Borda count voting rule is strongly Pareto efficient; no matter what tie-breaking rule is used, it cannot be the case that another outcome is at least as good for each agent, and strictly preferred by at least one agent, to the Borda choice. However, this statement assumes that everyone votes truthfully; that is, announces her or his true preference ranking. But, looking at this voting rule as a game form leads to a natural, and very pertinent question; namely, is it a rational strategy for each agent to announce her or his true preferences? Consider the following example.

18.3 Example. Suppose X consists of three distinct elements, $X = \{x, y, z\}$, that $n = 3$, and a choice is to be made by the Borda count rule, with alphabetical order of the alternatives to be used as the tie-breaking rule; if w_1 and w_2 receive the same Borda score, and w_2 *follows* w_1 in alphabetical order, then w_2 is the social choice. Suppose now that Ms. 1 orders the three alternatives as xP_1y & yP_1z, and that she believes that the other two agents have the preferences indicated in the following:

agent 2	***agent 3***
$z\ y$	z
x	y
	$x.$

If these are the preferences actually indicated by agents 2 and 3 and agent 1 announces her true preferences, then the Borda scores for the three alternatives are:[2]

$$W(x) = 5,\ W(y) = 2 + 2 + 2 = 6,\ \text{and } W(z) = 1 + 2 + 3 = 6,$$

[2]See the Appendix to Chapter 14 for the weights to be used when indifference is allowed for. The essential fact is that if there are three alternatives, and two of them are tied for first place in someone's ranking, then they each get a weight of 2.

so that, given the tie-break rule, the alternative chosen will be z.

However, suppose Ms. 1 announces that her preferences are yP_1x & xP_1z. Then the Borda scores would be given by:

$$W(x) = 4,\ W(y) = 7 \text{ and } W(z) = 6;$$

so that y, which Ms. 1 prefers to z (perhaps quite strongly), is the clear winner! On the other hand, suppose that (unkown to Ms. 1) both agents 2 and 3 are indifferent between x and y, and prefer either to z. Then if Ms. 1 announces the false preferences, that is, votes strategically rather than truthfully, the Borda scores (assuming the other two vote truthfully) are:

$$W(x) = 6,\ W(y) = 7, \text{ and } W(z) = 3;$$

so that the alternative chosen is y. However, Ms. 1 prefers x to y, and the other two agents consider x at least as good as y! □

Now, you may now be thinking something like, "Well, if we start assuming crazy things about peoples' beliefs, then almost anything can happen!" Well, this is pretty much true, but it isn't the point: the point is that truth-telling is not a dominant strategy in the case of the Borda count voting rule; that is, agents may perceive a possible gain to be made by voting *strategically*. Further, if agents vote strategically, rather than truthfully, then a mechanism may fail to satisfy some of the good properties which it seemed to possess before we raised this issue. Moreover, lest you think that this kind of difficulty is somehow unique to the Borda count rule, consider the following example, which illustrates an even more insidious problem which can arise in connection with plurality voting.

18.4 Example. Suppose X and the number of agents are the same as in the previous example, and that the same tie-break rule is to be used as was presented in the previous example. We also suppose that Ms. 1's preferences are the same as before, but that she *correctly* believes that the other two agents have the following preferences:

***agent* 2**	***agent* 3**
$x\ z$	y
y	x
	z.

Suppose, however, that the person/agency implementing the rule requires that if an agent is indifferent between two (best) alternatives, then she/he should simply pick one of the two, and send that as her/his message. Suppose further that Ms. 1 does *not* know what tie-break rule Mr. 2 will use in picking one of the alternatives $\{x, z\}$ to call his top choice. If, in fact, Mr. 2 uses the same tie-breaking rule as is used in the mechanism itself, then each of the three alternatives will receive one vote, and the social choice will be z, which is strictly Pareto dominated by x. If Ms. 1 is sure that this is the tie-break rule that Mr. 2 will use, then her best response is to vote for y, which is then the social alternative chosen whatever Mr. 2 does (assuming that agent 3 votes truthfully). However, x is the Condorcet winner in this case, and will be the social choice if Mr. 2 chooses x as the message to send and Ms. 1 votes truthfully. Once again the mechanism fails to have properties that it appears to possess if individuals have an incentive to vote strategically. □

If one is trying to design a mechanism to achieve, for example, a Pareto efficient production level of a public good (a topic we will examine in some detail later in this chapter), one needs to examine the question of how individuals are likely to behave if one attempts to implement the mechanism. As we have just seen, even in the case of simple voting rules, some of the desirable properties which they appear to have if agents vote truthfully may disappear if agents have an incentive to vote strategically. The next question, however, is when *don't* individual agents have an incentive to vote strategically, rather than to give truthful responses or votes? In the case of a voting rule, it would appear that the only sure means for (more or less) guaranteeing that individuals will vote truthfully is if truthful voting is in her or his best interests; that is, if truthful voting yields an outcome which, for each agent is at least as good as that resulting from a non-truthful response, whatever the actions of the other agents. Before pursuing this issue further, however, let's consider the relationship of this issue to general game forms.

Suppose we are considering a game form, $\Gamma = \langle \boldsymbol{S}, X, g \rangle$, which is supposed to achieve a desirable level of public goods production, say. The two paramount issues which arise in our attempt to carry out this evaluation are first, what exactly do we mean by a desirable outcome? Secondly, what strategies will the agents choose if the game form/mechanism is actually implemented? Obviously the second question is the one which needs to be addressed first, because we cannot evaluate the desirability of the outcome until we can make a prediction as to what it will be. In this connection, it would appear that we are likely to feel most comfortable about such predictions if each of the individual agents has a dominant strategy; for in such a case, we can feel reasonably confident that this is the strategy that will be chosen. However, before proceeding further with our discussion, we had better provide formal definitions of some of the terms which we have been using.

We need first to introduce some notation. Let $\Gamma = \langle \boldsymbol{S}, X, g \rangle$ be an n-agent game form, let $\boldsymbol{s} = (s_1, \ldots, s_n) \in \boldsymbol{S}$ be a vector of strategies chosen by the n agents, and let $i \in N \overset{\text{def}}{=} \{1, \ldots, n\}$. We will denote the strategy vector in which i's strategy, s_i is replaced by some $s_i' \in S_i$, while every other agent's strategy remains unchanged, by '$(s_i', \boldsymbol{s}_{-i})$.' Furthermore, we will use the generic notation, '$\boldsymbol{s}_{-i}$, $\boldsymbol{s}'_{-i}$, and so on, to denote an $n-1$ tuple of strategy choices by every agent except i, and define:

$$\boldsymbol{S}_{-i} = \prod_{j \neq i} S_j.$$

In chapter 14 we always assumed that individual preferences were drawn from the family of asymmetric orders on the outcome (or choice) set. In effect, we always dealt with strict preferences rather than the 'at-least-as-good-as' relations for the agents. In this chapter, however, we will find it more convenient to deal with reflexive relations as our basic concepts; in fact, not only is this more convenient for us, but it is the usual pattern in the literature we are going to be discussing here. In effect, we are simply going to make use of the negations of the families of preferences we made use of in Chapter 14. More specifically, given a nonempty outcome set, X, we will let:

$\mathcal{L}$ = the family of all linear orders on X; that is the family of all antisymmetric weak orders on X,

$\mathcal{G}$ = the family of all weak orders on X, and

$\mathcal{Q}$ = the family of all quasi-orders on X; that is, the family of all binary relations which are total, reflexive, and whose asymmetric part is transitive.

This last type of relation has not been used much in this book, but it was introduced in Definition 2.18 of Chapter 2; where it was noted that a total and reflexive binary relation is a quasi-order if, and only if its asymmetric part is an asymmetric order. I will leave it as an easy exercise to show that if $\succsim$ is a linear order by the above definition, then its asymmetric part, $\succ$ is total, asymmetric, and transitive (and thus negatively transitive as well). Consequently, we have the inclusions:

$$\mathcal{L} \subseteq \mathcal{G} \subseteq \mathcal{Q}.$$

18.5 Definitions. Given a game form, $\Gamma = \langle \boldsymbol{S}, X, g \rangle$, we will say that $s_i^* \in S_i$ is a **weakly dominant strategy** for agent i with preference relation $G_i \in \mathcal{Q}$ (and given Γ) iff, for every $\boldsymbol{s} \in \boldsymbol{S}$, we have:

$$g(s_i^*, \boldsymbol{s}_{-i})\, G_i\, g(\boldsymbol{s}). \tag{18.1}$$

If, in addition, we have that for each $s_i' \in S_i \setminus \{s_i^*\}$, there exists $\boldsymbol{s}_{-i} \in \boldsymbol{S}_{-i}$ such that:

$$g(s_i^*, \boldsymbol{s}_{-i}) P_i g(s_i', \boldsymbol{s}_{-i}), \tag{18.2}$$

where P_i is the asymmetric part of G_i, then s_i^* is said to be a **dominant strategy** for agent i, given G_i.

The strategy s_i^* would be said to be **strictly** (or **strongly**) **dominant** for i, given G_i, iff for each $s_i' \in S_i \setminus \{s_i^*\}$, and each $\boldsymbol{s}_{-i} \in \boldsymbol{S}_{-i}$ we have:

$$g(s_i^*, \boldsymbol{s}_{-i}) P_i g(s_i', \boldsymbol{s}_{-i}),$$

but we will rarely, if ever, be able to find a game form in which this condition is satisfied, for each i. Notice that if a dominant strategy exists for i, given G_i and Γ [which qualification we will abbreviate as '... given (G_i, Γ)'], then it is unique. On the other hand, agent i may have many weakly dominant strategies, given (G_i, Γ). Nonetheless, we will only make use of weakly dominant strategies in the following definition.

18.6 Definition. Let X be a set of alternatives, $N = \{1, \ldots, n\}$ be a set of agents, and $\mathcal{D}$ be a subset of $\mathcal{Q}$. We will say that an n-agent game form, $\Gamma = \langle \boldsymbol{S}, X, g \rangle$ is **straightforward on $\mathcal{D}$** iff, given any $\boldsymbol{G} \in \mathcal{D}^n$ each agent $i \in N$ has a weakly dominant strategy, given (G_i, Γ).

One might prefer to make use of a strengthened form of the above definition in which it is required that a dominant strategy exists for i; since we can be reasonably confident that an agent will choose a dominant strategy if one is available (and is recognized as such), the uniqueness of dominant strategies means that a game form satisfying this stronger condition allows us to make a fairly confident prediction of what the outcome will be, given any profile $\boldsymbol{G} \in \mathcal{D}^n$. On the other hand, a game form satisfying this stronger condition is obviously straightforward, and the distressing conclusion of the 'Gibbard-Satterthwaite Theorem' is that that any game form which is nontrivial and straightforward is also dictatorial. We will study this result in detail in the next section.

18.3 The Gibbard-Satterthwaite Theorem

In this section we will retain the notation and definitions introduced in the previous section; for example, letting $\mathcal{G}$ be the set of all weak orders on X, and so on; where X itself, the set of alternatives, is assumed to be finite, and to contain at least three distinct elements. We will use the generic notation 'G, G', G^*,' and so forth to denote elements of $\mathcal{Q}$; with 'P, P'' and 'P^*,' and so on, denoting their respective asymmetric parts. We suppose, as before, that we are dealing with a situation in which n agents have preference relations defined on X, and that $n \geq 2$. We will denote the set of agents by 'N,' that is, we define:

$$N = \{1, \ldots, n\}.$$

Given a game form, $\Gamma = \langle \boldsymbol{S}, X, g \rangle$, we define X_g, the **outcome set for $\boldsymbol{\Gamma}$**, by:

$$X_g = \{x \in X \mid (\exists \boldsymbol{s} \in \boldsymbol{S}) \colon x = g(\boldsymbol{s})\}. \tag{18.3}$$

The set of assumptions and notation set out in the preceding paragraph will be maintained throughout this section without further explicit mention; and in this context, we define the following.

18.7 Definition. An agent i is a **dictator for a game form** $\Gamma = \langle \boldsymbol{S}, X, g \rangle$ iff, for each $x \in X_g$, there exists $s_i \in S_i$ such that for all $\boldsymbol{s}_{-i} \in \boldsymbol{S}_{-i}$, $x = g(s_i, \boldsymbol{s}_{-i})$. A game form Γ is **dictatorial** if there is a dictator for Γ.

The principal result with which we will be concerned in this section is the following; a slightly generalized version of Gibbard's original statement of what has become known as the 'Gibbard-Satterthwaite Theorem.'[3]

18.8 Theorem. (**Gibbard-Satterthwaite**) If $\Gamma = \langle \boldsymbol{S}, X, g \rangle$ is a game form which has at least three possible outcomes and is straightforward on $\mathcal{D}$, where:

$$\mathcal{L} \subseteq \mathcal{D} \subseteq \mathcal{Q},$$

then Γ is dictatorial.

In the remainder of this section, we will be concerned with the construction of a proof of this theorem. However, most of our argument will be framed in terms of voting rules, rather than game forms. It was Gibbard's rather profound insight

[3]This result was developed independently by Gibbard [1973] and Satterthwaite [1975].

that a straightforward game form is formally equivalent to a voting rule in which no agent has a positive incentive to vote strategically, rather than honestly. This is the 'revelation principle,' and in order to provide a formal proof of this, we begin by defining some notation and concepts regarding voting rules.

As in Chapter 14 (except that here we will use 'n,' rather than 'm,' to denote the number of agents) we will deal with voting rules, $f\colon \mathcal{D}^n \to X$, where $\mathcal{D} \subseteq \mathcal{Q}$, and we denote the range of f (the **outcome set**) by 'X_f,' that is:

$$X_f = \{x \in X \mid (\exists \boldsymbol{G} \in \mathcal{D}^n)\colon f(\boldsymbol{G}) = x\}.$$

We will refer to n-tuples:

$$\boldsymbol{G} = (G_1, \ldots, G_n) \in \boldsymbol{\mathcal{Q}}^n,$$

as **preference profiles**, and given such a profile, $\boldsymbol{G}$, and $G_i' \in \mathcal{Q}$, we denote the profile obtained from $\boldsymbol{G}$ by replacing G_i with G_i', all other preference relations remaining the same, by '$(G_i', \boldsymbol{G}_{-i})$.' We make use of this notation in the following definition.

18.9 Definitions. We will say that a voting rule is **manipulable** (**at** $\boldsymbol{G} \in \mathcal{D}^n$) iff there exists $i \in N$, and $G_i^* \in \mathcal{D}$, such that:

$$f(G_i^*, \boldsymbol{G}_{-i}) P_i f(\boldsymbol{G}).$$

If f is not manipulable, it will be said to be **strategy proof** (abbreviated '**SP**').

Notice that if f is strategy-proof, then for every profile $\boldsymbol{G} \in \mathcal{D}^n$, every $i \in N$, and every $G_i^* \in \mathcal{D}$, we will have:

$$f(\boldsymbol{G}) G_i f(G_i^*, \boldsymbol{G}_{-i}).$$

In other words, every agent always does at least as well by reporting her/his true preferences as she or he would by pretending to have a different preference relation.

A particularly undesirable property for a voting rule to satisfy is the following.

18.10 Definition. We will say that a voting rule is **dictatorial**, if there exists $i \in N$ such that for every $\boldsymbol{G} \in \mathcal{D}^n$, and for any $x \in X_f$, $f(\boldsymbol{G}) G_i x$.

Of course, if X_f only contains one element, then each agent is a dictator, according to the above definition; and any reasonable voting rule will be dictatorial if $n = 1$. However, our interest here is centered around voting schemes which have at least three possible outcomes, and where $n \geq 2$.

Now let's return to the **Revelation Principle**; where Gibbard's original observation proceeds as follows. Let $\Gamma = \langle \boldsymbol{S}, X, g \rangle$ be a straightforward game form on $\mathcal{D}$. Then given any $G_i \in \mathcal{D}$, the set of (weakly) dominant strategies for (G_i, Γ), which we will denote by '$D(G_i, \Gamma)$,' is non-empty. If we let σ_i be a function satisfying:

$$(\forall G_i \in \mathcal{D})\colon \sigma_i(G_i) \in D(G_i, \Gamma),$$

the composition of g and $\boldsymbol{\sigma}$, $f(\boldsymbol{G}) = g[\boldsymbol{\sigma}(\boldsymbol{G})]$, where we define:

$$\boldsymbol{\sigma}(\boldsymbol{G}) = \big(\sigma_1(G_1),\ \sigma_2(G_2), \ldots, \sigma_n(G_n)\big),$$

then defines a voting rule. Moreover, since $\sigma_i(G_i)$ is a weakly dominant strategy for i, given G_i, it follows that, given any $\boldsymbol{G} \in \mathcal{D}^n$, and any $G_i' \in \mathcal{D}$, we must have:

$$f(\boldsymbol{G})\, G_i\, f(G_i', \boldsymbol{G}_{-i}); \tag{18.4}$$

that is, f is strategy proof (SP). Consequently, the following theorem is often referred to as the Gibbard-Satterthwaite Theorem.

18.11 Theorem. (Gibbard [1973], Satterthwaite [1975]) *Suppose $\mathcal{D}$ satisfies:*

$$\mathcal{L} \subseteq \mathcal{D} \subseteq \mathcal{Q}, \tag{18.5}$$

and that $f\colon \mathcal{D}^n \to X$ is a strategy-proof voting rule. If $n \geq 2$, and X_f contains at least three elements, then f is dictatorial.

Before beginning our proof of the main theorem, let's consider a supporting lemma, the proof of which I'll leave as an easy exercise.[4]

18.12 Lemma. *Let $\succsim$ be an arbitrary linear order on X, and let $\{Y, Z\}$ be a partition of X.*[5] *If $\succsim^*$ is a linear order on Y, and we define the relation Q on X by:*

$$aQb \iff \begin{cases} a \succsim^* b & \text{if } a, b \in Y, \\ a \succsim b & \text{if } a, b \in Z, \text{ or} \\ a \in Y \;\&\; b \in Z, \end{cases}$$

then Q is a linear order on X.

The key thing about the ordering defined in the above lemma is that everything in the set Y strictly dominates (is preferred to, if Q is a preference relation) everything in Z.

Now, back to our proof of the Gibbard-Satterthwaite Theorem. Our proof of the theorem is an adaptation of Gibbard's original [1973] proof, although we will borrow a bit from Barbera and Peleg [1990] as well (see also Barbera [2001, pp. 625–6]). In outline, we proceed as follows. We will make use of the Revelation Principle to go from a straightforward game form, $\Gamma = \langle \boldsymbol{S}, X, g\rangle$ to the derived strategy-proof voting rule, $f(\boldsymbol{G}) = g[\boldsymbol{\sigma}(\boldsymbol{G})]$. Consequently, we begin by examining some general properties of strategy-proof voting rules. We then show that a strategy-proof (**SP**) voting rule can be used to define a social preference function which satisfies Arrow's Independence of Irrelevant Alternatives (IIA) and Weak Pareto Principle (WPP) conditions. It then follows that this derived social preference function is dictatorial. We conclude our proof by showing that this in turn implies that the voting rule is dictatorial; and that if it is derived from a straightforward game form, then that game form is dictatorial as well.

Accordingly, suppose that $f\colon \mathcal{D}^n \to X$ is a strategy-proof voting rule, that (18.5) holds, and that X_f contains at least three elements; in fact, we can, and will suppose throughout that $X = X_f$. We will also define, for each $x \in X$, the family,

---

[4]See Exercise 15, at the end of Chapter 1.

[5]So that both Y and Z are non-empty, $Y \cap Z = \emptyset$, and $X = Y \cup Z$.

$\mathcal{L}_x$, consisting of all linear orders of X for which x is the maximal element of X;[6] that is:

$$\mathcal{L}_x = \{ \succsim \in \mathcal{L} \mid (\forall y \in X \setminus \{x\}) \colon x \succ y \}. \tag{18.6}$$

(Note: we will generally use the generic notation '$\succsim, \succsim^*$,' and so on, to denote a linear order, with '$\succ, \succ^*$,' and so on, denoting the respective asymmetric parts.)

A property which appears to be quite desirable in a voting rule is the following.

18.13 Definition. A voting rule satisfies the **positive association property** (**PAP**) iff, for every $\boldsymbol{G} \in \mathcal{D}^n$, every $i \in N$, and every $G_i^* \in \mathcal{D}$, we have: if $f(\boldsymbol{G}) = x$ and G_i^* and G_i are such that for all $y \in X_f$ we have:

$$[xG_iy \text{ and } y \neq x] \Rightarrow xP_i^*y,$$

then $x = f(G_i^*, \boldsymbol{G}_{-i})$.

Basically, the positive association property (PAP) says that if a voting rule chooses an alternative, x, for some profile of preferences, and one agent's preferences are changed in such a way that x is now preferred to every alternative agent i considered no worse than x previously, then x should remain the social choice. This is certainly a property that one would like to be satisfied by a voting rule, and our first result establishes the fact that it must indeed be satisfied by any strategy-proof voting rule; in fact, *it must be satisfied by a strategy-proof voting rule even if its domain does not contain* $\mathcal{L}^n$.

18.14 Lemma. *If $f \colon \mathcal{D}^n \to X$ is a strategy-proof voting rule, where $\mathcal{D}$ is a nonempty subset of $\mathcal{Q}$, then f must satisfy PAP.*

Proof. Suppose $\boldsymbol{G} \in \mathcal{D}^n, i \in N$ and $G_i^* \in \mathcal{D}$ are such that $f(\boldsymbol{G}) = x$ and for all y in X_f:

$$[xG_iy \;\&\; y \neq x] \Rightarrow xP_i^*y. \tag{18.7}$$

We then suppose, by way of obtaining a contradiction, that:

$$f(G_i^*, \boldsymbol{G}_{-i}) = z \neq x.$$

Since $z \neq x$, we have by (18.7) that if zG_i^*x, then zP_ix. But then:

$$fG_i^*, \boldsymbol{G}_{-i})P_if(\boldsymbol{G}),$$

contradicting SP. On the other hand, if xP_i^*z, then:

$$f(\boldsymbol{G})P_i^*f(G_i^*, \boldsymbol{G}_{-i}),$$

which again contradicts SP. □

Our next result establishes a second desirable property satisfied by SP voting rules.

[6]Notice that, not only do such linear orders always exist, but there are in fact always $(q-1)!$ of them, where $q = \#X$.

18.15 Lemma. *Given any $x \in X_f$, if $G_i \in \mathcal{L}_x$, for $i = 1, \ldots, n$, then $f(\boldsymbol{G}) = x$.*

Proof. Given $x \in X_f$, let $\boldsymbol{G}^* \in \mathcal{D}^n$ be such that $x = f(\boldsymbol{G}^*)$, let $G_i \in \mathcal{L}_x$ for each i, and suppose, by way of obtaining a contradiction, that:

$$x \neq f(\boldsymbol{G}).$$

Given this, we see that if we define:

$$\begin{aligned} z_0 &= f(\boldsymbol{G}^*) = x, \\ z_i &= f(G_1, \ldots, G_i, G_{i+1}^*, \ldots, G_n^*) \quad \text{for } i = 1, \ldots, n, \end{aligned}$$

we have $z_0 = x$ and $z_n \neq x$. Therefore, there exists $j \in N$ such that:

$$z_{j-1} = x \text{ and } z_j \neq x.$$

But then we see that:

$$f(G_1, \ldots, G_{j-1}, G_j^*, G_{j+1}^*, \ldots, G_n^*) P_j f(G_1, \ldots, G_j, G_{j+1}^*, \ldots, G_n^*);$$

contradicting the assumption that f is strategy-proof. □

Notice that it follows from this last result that if we denote the restriction of f to $\mathcal{L}^n$ by 'f^*,' then $X_{f^*} = X_f$; that is, any alternative which might be chosen for some $\boldsymbol{G} \in \mathcal{D}^n$ will also be chosen for some (not necessarily distinct) $\boldsymbol{G}' \in \mathcal{L}^n$. As mentioned earlier, Gibbard noted that a SP voting rule can be used to define a social preference function. We will modify his approach slightly, in that, while he uses a voting rule, $f \colon \mathcal{G}^n \to X$ to define an Arrovian social preference function, we will use the somewhat more general assumptions regarding f which were set out above, and then use its restriction to $\mathcal{L}^n$ to define a social preference function. We begin as follows. Let $\succsim$ be an arbitrary linear order of X, which we will hold fixed throughout the remainder of our proof. Given $\boldsymbol{G} \in \mathcal{L}^n$ and $Z \subseteq X$, we define the preference profile $\boldsymbol{G} * Z$ as follows. For each $i \in N$, we define $G_i^* \stackrel{\text{def}}{=} G_i * Z$ by:

$$xG_i^*y \iff \begin{cases} xG_iy & \text{if } x, y \in Z, \text{ or} \\ x \in Z \;\&\; y \in X \setminus Z & \text{or} \\ x \succsim y & \text{if } x, y \in X \setminus Z. \end{cases} \tag{18.8}$$

Notice that it follows at once from Lemma 18.12 that each $G_i^* = G_i * Z$ is a linear order of X. We can then easily prove the following (recall that we are denoting the restriction of f to $\mathcal{L}^n$ by 'f^*').

18.16 Lemma. *If x, y and z are distinct elements of X, $\boldsymbol{G} \in \mathcal{L}^n$, and:*

$$x = f^*\big(\boldsymbol{G} * \{x, y, z\}\big),$$

then:

$$x = f^*(\boldsymbol{G} * \{x, y\}) \text{ and } x = f^*(\boldsymbol{G} * \{x, z\}),$$

as well.

Proof. Define $\boldsymbol{G}^* = \boldsymbol{G} * \{x, y, z\}, \boldsymbol{G}' = \boldsymbol{G} * \{x, y\}$, and $\boldsymbol{G}'' = \boldsymbol{G} * \{x, z\}$. Notice that it then follows from the definitions that for each $i \in N$, and each $w \in X$:

$$xG_i^* w \ \& \ x \neq w \Rightarrow xP_i' w.$$

Consequently, it follows from Lemma 18.14 that $x = f^*(\boldsymbol{G}') = f^*(\boldsymbol{G} * \{x, y\})$. Similar considerations establish that $x = f^*(\boldsymbol{G} * \{x, z\})$. □

A property of the operation just defined which lends added importance to the above lemma, and is critical to our further development of Gibbard's construction is the following:

18.17 Lemma. *If Z is a nonempty subset of X, and $\boldsymbol{G} \in \mathcal{L}^n$, then:*

$$f^*(\boldsymbol{G} * Z) \in Z.$$

Proof. Suppose, by way of obtaining a contradiction, that for some $\boldsymbol{G} \in \mathcal{L}^n$ and some nonempty subset, Z, of X, we have $f\big(\boldsymbol{G} * Z\big) \notin Z$; and define/denote $G_i^* = G_i * Z$, for $i = 1, \ldots, n$. Let x be any element of Z, and let $\widehat{\boldsymbol{G}} \in \mathcal{L}^n$ be a profile such that:

$$\widehat{G}_i \in \mathcal{L}_x \quad \text{for } i = 1, \ldots, n;$$

and note that it follows from Lemma 18.15 that $f(\widehat{\boldsymbol{G}}) = x$. Next we define the sequence $z_0, \ldots, z_n$ by:

$$\begin{aligned} z_0 &= f(\widehat{\boldsymbol{G}}) = x, \\ z_i &= f(G_1^*, \ldots, G_i^*, \widehat{G}_{i+1}, \ldots, \widehat{G}_n) \quad \text{for } i = 1, \ldots, n. \end{aligned}$$

We can then see that there must exist $i \in N$ such that:

$$f(G_1^*, \ldots, G_{i-1}^*, \widehat{G}_i, \ldots, \widehat{G}_n) \in Z,$$

and:

$$f(G_1^*, \ldots, G_i^*, \widehat{G}_{i+1}, \ldots, \widehat{G}_n) \notin Z.$$

But then it follows from the definition of P_i^* that:

$$f(G_1^*, \ldots, G_{i-1}^*, \widehat{G}_i, \ldots, \widehat{G}_n) P_i^* f(G_1^*, \ldots, G_i^*, \widehat{G}_{i+1}, \ldots, \widehat{G}_n);$$

contradicting the assumption that f is SP. □

We then define the social preference function, $F\colon \mathcal{L}^n \to \mathcal{L}$ as follows. Let $\boldsymbol{G} \in \mathcal{L}^n$, and let $x, y \in X$. Denoting $F(\boldsymbol{G})$ by '$\succsim$,' we then define:

$$x \succsim y \iff x = f^*(\boldsymbol{G} * \{x, y\}). \tag{18.9}$$

We will show that this defines a social preference function via a series of lemmas; the proof of the first of which is more or less immediate; and the proof of the second of which I will leave as an exercise.

18.18 Lemma. *The function F satisfies IIA.*

18.19 Lemma. *If $\boldsymbol{G} \in \mathcal{L}^n$, and $\succsim = F(\boldsymbol{G})$, then $\succsim$ is a linear order of X.*

18.20 Lemma. *The function F satisfies the Weak Pareto Principle (WPP).*

Proof. Suppose $x, y \in X$ are such that: xP_iy, for $i = 1, \ldots, n$. Then we note that it follows from the definition of $G_i * \{x, y\}$ that:

$$G_i * \{x, y\} \in \mathcal{L}_x \quad \text{for } i = 1, \ldots, n;$$

and it then follows from Lemma 18.15 that $x = f^*(\boldsymbol{G} * \{x, y\})$. Therefore $x \succ y$. □

It follows from Lemmas 18.18–18.20 and Arrow's Theorem (Theorem 14.27) that F is dictatorial.[7] *The following result then completes our proof of Theorem 18.11.*

18.21 Lemma. *If $f \colon \mathcal{D}^n \to X$ is a SP voting rule, where $\mathcal{D}$ satisfies (18.5), and $i \in N$ is a dictator for the derived social preference function, $F \colon \mathcal{L}^n \to \mathcal{L}$, then i is also a dictator for f.*

Proof. Suppose agent k is the dictator for F; and, by way of obtaining a contradiction, that k is not a dictator for f. Then there exists $\boldsymbol{G} \in \mathcal{D}^n$ and $x, y \in X_f$ such that:

$$f(\boldsymbol{G}) = x \text{ and } yP_kx.$$

However, let $\widehat{G}_i \in \mathcal{L}_x$ for each $i \neq k$, let $\widehat{G}_k = G_k$; and consider the preference profile $\boldsymbol{G}^* \in \mathcal{L}^n$ defined by $\boldsymbol{G}^* = \widehat{\boldsymbol{G}} * \{x, y\}$. Clearly we have, for all $i \in N$:

$$(\forall z \in X_f) \colon xP_iz \Rightarrow xP_i^*z;$$

and consequently it follows from PAP (Lemma 18.14) that $x = f(\boldsymbol{G}^*)$. However, since yP_kx and k is a dictator for F, we must have:

$$f(\boldsymbol{G}^*) = f^*(\widehat{\boldsymbol{G}} * \{x, y\}) = y. \quad \square$$

Proof of Theorem 18.8. Suppose $\Gamma = \langle \boldsymbol{S}, X, g \rangle$ is a game form which is straightforward on $\mathcal{D}$, where $\mathcal{L} \subseteq \mathcal{D} \subseteq \mathcal{Q}$, and that $\#X_g \geq 3$. Letting $\boldsymbol{\sigma} \colon \mathcal{D}^n \to \boldsymbol{\mathcal{S}}$ be a function mapping preference profiles $\boldsymbol{G} \in \mathcal{D}^n$ into weakly dominant strategies, we then define the voting rule $f \colon \mathcal{D}^n \to X$ by $f(\boldsymbol{G}) = g[\boldsymbol{\sigma}(\boldsymbol{G})]$. A critical fact regarding this voting rule is set out in the following, the proof of which will be left as an exercise (see Exercise 4, at the end of this chapter).

18.22 Lemma. *Given the above assumptions and definitions, the voting rule f satisfies $X_f = X_g$.*

So, the above result asserts that any alternative which is attainable by use of the game form is attainable by use of the voting rule. Consequently, if $\#X_g \geq 3$, then $\#X_f \geq 3$. It then follows from Theorem 18.11 and the fact that f is strategy-proof that there exists a dictator for f. It remains only to show that the dictator for f is also a dictator for the game form.

[7]Notice that $F \colon \mathcal{L}^n \to \mathcal{L}$, and $\mathcal{L}$ satisfies the Arrow condition.

Suppose for the sake of convenience that agent one is a dictator for f, and let x be arbitrary element of $X_g = X_f$. We then let $G_1 \in \mathcal{L}_x$, define $s_1^* = \sigma_1(G_1)$, and suppose, by way of obtaining a contradiction, that there exists an $(n-1)$-tuple $\boldsymbol{s}_{-1} \in \boldsymbol{\mathcal{S}}_{-1}$ such that:

$$g(s_1^*, \boldsymbol{s}_{-1}) \stackrel{\text{def}}{=} y \neq x.$$

Let $G' \in \mathcal{L}_y$, and let $G_i = G'$, for $i = 2, \ldots, n$. If we then let $s_i' = \sigma_i(G_i)$ for $i = 2, \ldots, n$, it follows from the fact that agent one is a dictator for f that $x = g(s_1^*, \boldsymbol{s}'_{-1})$. But now if we define the sequence $z_1, \ldots, z_n$ by:

$$g(s_1^*, \boldsymbol{s}'_{-1}) = x \stackrel{\text{def}}{=} z_1,$$

and:

$$z_j = g(s_1^*, s_2, \ldots, s_j, s'_{j+1}, \ldots, s'_n) \quad \text{for } j = 2, \ldots, n;$$

we see that there exists $j \in \{2, \ldots, n\}$ such that $z_j = y$ and $z_{j-1} \neq y$. However, this means that:

$$g(s_1^*, s_2, \ldots, s_{j-1}, s_j, s'_{j+1}, \ldots, s'_n) P_j g(s_1^*, s_2, \ldots, s_{j-1}, s'_j, s'_{j+1}, \ldots, s'_n);$$

contradicting the fact that s'_j was defined as a dominant strategy for agent j. It now follows that agent one is a dictator for $\Gamma = \langle \boldsymbol{S}, X, g \rangle$. □

Both the Gibbard-Satterthwaite Theorem and the Revelation Principle have profound implications for implementation theory, a topic which we will begin to explore in the next section.

18.4 Implementation Theory

We will use a framework and notation in our development of implementation theory which, in the main, is exactly as set out in the previous sections of this chapter. We deal with a finite group of agents, $N = \{1, \ldots, n\}$, and a set of outcomes, X (which we will usually assume to be finite). We suppose that individual agents have (weak) preferences over outcomes, G_i, with the asymmetric and symmetric parts denoted by 'P_i' and 'I_i,' respectively. The set of admissible profiles of preferences is denoted by '$\boldsymbol{\mathcal{D}}$,' which we will usually assume takes the form $\boldsymbol{\mathcal{D}} = \mathcal{D}^n$, for some set of quasi-orders, $\mathcal{G}$. The notation:

$$\boldsymbol{G} = (G_1, \ldots, G_n) \in \boldsymbol{\mathcal{D}},$$

will be used to denote preference profiles; with '$(\overline{G}_i, \boldsymbol{G}_{-i})$' denoting the profile $\boldsymbol{G}$ with G_i replaced by $\overline{G}_i$.

We continue to use the term '**game form**' to denote a triple $\Gamma = \langle \boldsymbol{S}, X, g \rangle$, where $\boldsymbol{S}$ and g are the **strategy space** and **outcome function**, respectively. However, we will often substitute 'mechanism' for 'game form,' and make use of a triple, $\mathcal{M} = \langle \boldsymbol{M}, X, g \rangle$ in place of $\Gamma = \langle \boldsymbol{S}, X, g \rangle$. As we discussed earlier, game forms and mechanisms, as we have defined them in this chapter, are formally identical. However, there will be occasions in which the strategies chosen by agents are naturally described as messages; and where this is the case, it is much more natural to use the mechanism terminology and notation rather than that for game forms.

The general problem with which implementation theory deals is this: We suppose that there is a **social choice correspondence**, $F\colon \mathfrak{D} \mapsto X$, which expresses the desired social outcome of a planner or planners (or society as a whole). We assume, however, that the planner does not have detailed information about the preferences of individual agents, and cannot, therefore, simply derive a desired outcome as a straightforward mathematical derivation. Instead, our planner must choose or design a mechanism in such a way that the individual choices of the agents in the context of this mechanism brings about the desired outcome. Thus we seek a game form, or mechanism, which *implements* the social choice correspondence; a term we will define shortly.

If we are given a game form, we cannot analyze its effectiveness in bringing about the desired social choice unless we make a prediction as to how individuals will behave in the context of the game form, or mechanism. Thus we proceed as follows.

18.23 Definitions. Given a game form, $\Gamma = \langle \boldsymbol{S}, X, g\rangle$, we will say that a correspondence, $\boldsymbol{\sigma}\colon \mathfrak{D} \mapsto \boldsymbol{S}$, is a **solution concept (with the admissible preference space, $\mathfrak{D}$)**. The corresponding **equilibrium outcome correspondence**, $O_{\boldsymbol{\sigma}}(\boldsymbol{G};\Gamma)$, is then defined by:

$$O_{\boldsymbol{\sigma}}(\boldsymbol{G};\Gamma) = \Big\{x \in X \mid \big(\exists \boldsymbol{s} \in \boldsymbol{\sigma}(\boldsymbol{G})\big) \colon g(\boldsymbol{s}) = x\Big\}.$$

To this point, our favorite example of a solution concept is the mapping from preferences to dominant strategies. However, we can and will consider other examples; for instance, the Nash equilibrium correspondence, or the mapping to subgame-perfect equilibria. We make use of this last definition to formulate our basic requirement for 'good performance' by a game form/mechanism, as follows.

18.24 Definition. A social choice correspondence, $F\colon \mathfrak{D} \mapsto X$, where $\mathfrak{D} \subseteq \mathfrak{Q}^m$, is **implemented** (respectively, **fully implemented**) **by the game form** $\Gamma = \langle \boldsymbol{S}, X, g\rangle$ **via the solution** $\boldsymbol{\sigma}$, iff:

$$(\forall \boldsymbol{G} \in \mathfrak{D})\colon O_{\boldsymbol{\sigma}}(\boldsymbol{G};\Gamma) \subseteq F(\boldsymbol{G}) \text{ [respectively, } O_{\boldsymbol{\sigma}}(\boldsymbol{G};\Gamma) = F(\boldsymbol{G})]. \tag{18.10}$$

The correspondence F is then said to be **implementable** (respectively, **fully implementable**) **via the solution** $\boldsymbol{\sigma}$ iff there exists a game form Γ which implements it (respectively, fully implements it) via the solution, $\boldsymbol{\sigma}$. (This definition is easily transformed to an effectively equivalent definition for mechanisms; and we will often make use of the equivalent definition without further comment.)

We will often supplement the language of the above definition somewhat; for example, if the solution concept via which a game form/mechanism, $\Gamma = \langle \boldsymbol{S}, X, g\rangle$, implements a social choice correspondence, F, is the mapping taking preferences into dominant strategies, then we will say that $\Gamma = \langle \boldsymbol{S}, X, g\rangle$ **implements F in dominant strategies**, and that **F is implementable in dominant strategies**.

We will say that a mechanism, $\mathcal{M} = \langle \boldsymbol{M}, X, g\rangle$, which has the property that, for each $i \in N$, and each $G_i \in \mathcal{D}$, there exists a domininant strategy/message, $m_i^* \in M_i$, for $(G_i, \mathcal{M})$, is a **dominant strategy mechanism.** One of the things which should be noted regarding such mechanisms is that the uniqueness of a dominant strategy for each preference relation means that $\boldsymbol{\sigma}(\boldsymbol{G})$ is a function, in the case of a dominant strategy mechanism. Thus the equilibrium outcome correspondence for such a mechanism is single-valued; and consequently a dominant strategy mechanism cannot fully implement any social choice correspondence which is not single-valued.

We will often refer to the set $\mathfrak{D}$ in Definition 18.24 as the **admissible preference space**. Our definition implicitly takes implementability as being conditional upon the admissible preference space specified, in that the admissible preference space is part of the specification of the social choice correspondence; namely, its domain. It should be apparent that the more general (that is, the larger) is this set, the more difficult it will be find a game form/mechanism which implements a social choice correspondence mapping $\mathfrak{D}$ into X.

A special type of game form/mechanism of particular interest is defined in the following, in which we use the mechanism language.

18.25 Definition. We will say that a mechanism, $\mathcal{M} = \langle \boldsymbol{M}, X, g\rangle$ is a **direct revelation mechanism** iff the message space,

$$M = \prod\nolimits_{i=1}^{n} M_i,$$

is such that $M_i \subseteq \mathcal{Q}$, for $i = 1, \ldots, n$.

We will say that a direct mechanism is **incentive compatible** iff the outcome function, g, is strategy-proof, as defined in Definition 18.9. It follows from Gibbard's Revelation Principle that if a social choice correspondence is implementable in dominant strategies, then it is implementable by an incentive-compatible direct revelation mechanism. Of course, it is also a discouraging consequence of the Gibbard-Satterthwaite Theorem that if $\#X \geq 3$, $n \geq 2$, and $\mathcal{L} \subseteq \mathcal{D} \subseteq \mathcal{Q}$, then no social choice correspondence, $F\colon \mathcal{D}^n \mapsto X$ is implementable in weakly dominant strategies, unless it allows dictatorship. Happily, however, if we restrict the admissible preferences in some not-altogether-unreasonable ways, then there do exist social choice correspondences which are implementable in dominant strategies. We consider the two most prominent examples of such restrictions in the next two sections.

18.5 Single-Peaked Preferences and Dominant Strategies

18.5.1 Single-Peaked Preferences

A situation in which one can develop meaningful dominant strategy mechanisms, is that in which all of the agents have 'single-peaked' preferences; defined as follows.

18.26 Definitions. Let X be a non-empty set, let $\succsim$ be a linear order on X, and let G_i be a quasi order on X. We will say that G_i is **single-peaked on $\boldsymbol{X}$ (with**

respect to $\succsim$) iff G_i achieves a maximum at some point $m(G_i) \in X$, and, for all $x, y \in X$, we have:

$$[m(G_i) \succsim y \succ x] \Rightarrow yP_ix,$$

and:

$$[x \succ y \succsim m(G_i)] \Rightarrow yP_ix.$$

In essence, preferences are single-peaked if there is a most-preferred element in X, and the individual orders other alternatives in the set according to how far away (as measured by the linear order, $\succsim$) the elements are from this most-preferred alternative. For example, consider a group of citizens, all of whom live along a straight road, voting on the location of, say, a public park along this road. In this situation, we might find that the voters all would like to have the park in whichever suitable location is closest to them, ordering the other alternatives in terms of how far away they were from this most desired location.

In the next subsection, we will find that, given quite standard assumptions, individual preference relations will be single-peaked in a familiar public goods problem. In the meantime, let's consider some examples of single-peaked preferences.

18.27 Examples.

1. Let $X = [0, 2]$, and define the function $f\colon X \to \mathbb{R}_+$ by:

$$f(x) = \begin{cases} x/2 & \text{for } 0 \leq x < 1, \text{ and} \\ 2 - x/2 & \text{for } 1 \leq x \leq 2. \end{cases}$$

I will leave it to you to show that if G is defined by:

$$xGy \iff f(x) \geq f(y),$$

then G is a linear order on X which is single-peaked with respect to $\geq$. (See exercise 7, at the end of this chapter.)

2. Let $\succsim$ be a linear order on a nonempty set, X, and let x^* be an arbitrary element of X. We then define the relation, P, on X via its lower contour set correspondence, as follows:

$$xP = \begin{cases} \{y \in X \mid x \succ y\} & \text{if } x^* \succsim x, \\ \{y \in X \mid y \succ x\} & \text{if } x \succsim x^*. \end{cases} \tag{18.11}$$

Letting Q be the negation of P, you can then show (Exercise 8 at chapter's end) that Q is a quasi-order which is single-peaked with respect to $\succsim$. □

The second of the above examples is, in a sense, a 'canonical' example of a single-peaked quasi-order. We can modify it to obtain a very large number (an infinite number, if the set X contains an infinite number of elements) of other quasi-orders, all of which have the same maximal element, x^*, by specifying the relationship between elements of $L \stackrel{\text{def}}{=} \{x \in X \mid x^* \succsim x\}$ and $R \stackrel{\text{def}}{=} \{x \in X \mid x \succsim x^*\}$. For example, we might expand the xP sets for $x \in R$ by letting:

$$xP = \{y \in X \mid y \succ x\} \cup L;$$

or, for another example, if X is a subset of $\mathbb{R}$, we can define Q on X by:

$$xQy \iff |x^* - y| \geq |x^* - x|.$$

In this section we will be considering social choice functions of the form $F\colon \mathcal{D}^n \to X$, where $\mathcal{D} \subseteq \mathcal{Q}$. Two properties of interest with respect to such functions are the following.

18.28 Definitions. If $F\colon \mathcal{D}^n \to X$ is a social choice function, we will say that F is:
1. **anonymous** iff interchanging any two agents (and their preferences) does not change the outcome.
2. **efficient** iff it satisfies the Weak Pareto Principle.

Moulin [1980] has proved the following result. In its statement, the elements $a_j \in X$ are called '**phantom voters**.'

18.29 Theorem. (Moulin [1980, 1984]) Let '$\mathcal{P}$' denote the family of all quasi-orders on a nonempty set, X, which are single-peaked with respect to a (fixed) linear order on X; let $a_1, \ldots, a_{n-1}$ be elements of X, and define $F\colon \mathcal{P}^n \to X$ by:

$$F(\boldsymbol{G}) = \text{med}\{m(G_1), \ldots, m(G_n), a_1, \ldots, a_{n-1}\}. \tag{18.12}$$

Then F is anonymous, efficient, and strategy-proof; in fact, it is coalitional strategy-proof.

Proof. Since $n + (n-1) = 2n - 1$ is necessarily an odd integer, the function in (18.12) is well-defined. Moreover, it is obviously anonymous. To prove that F is efficient, define the sequence $\langle y_j \rangle \subseteq X$ by:

$$y_1 = \min_{w.r.t. \succsim} \{m(G_1), \ldots, m(G_n), a_1, \ldots, a_{n-1}\}, \tag{18.13}$$

that is, y_1 is an element of the set which satisfies:

$$y_1 \precsim m(G_i) \text{ for } i = 1, \ldots, n, \text{ and } y_1 \precsim a_j, \text{ for } j = 1, \ldots, n-1;^8 \tag{18.14}$$

and, for $j = 2, \ldots, 2n-1$:

$$y_j = \min_{w.r.t. \succsim} \big\{\{m(G_1), \ldots, m(G_n), a_1, \ldots, a_{n-1}\} \setminus \{y_1, \ldots, y_{j-1}\}\big\}. \tag{18.15}$$

Then we note that $F(\boldsymbol{G}) = y_n$; and, since there are only $n - 1$ 'phantom voters,' it follows that:

$$\min_i m(G_i) \precsim F(\boldsymbol{G}) \precsim \max_i m(G_i);$$

and thus $F(\boldsymbol{G})$ is efficient; its coalitional strategy-proofness follows from our next result. □

Before turning to our next result, there are several aspects of Theorem 18.29 which deserve some discussion. The first such item, probably, is what do we mean by the median with respect to an abstract linear order? An example may help to clarify things here.

---

[8]The elements of the set may not be distinct, in which case several elements of the set may satisfy (18.14). If this is the case, any tie-break rule can be used to pick the element to be labeled 'y_1.' Of course, in this event, y_2 will be equal to y_1.

18.30 Example. Let $X = \{a, b, c, d\}$, and suppose there are five agents ($n = 5$), whose preferences are as follows.

agent 1	agent 2	agent 3	agent 4	agent 5
c	b	d	b	a
b	a	c	c	b
a	c	b	a	c
d	d	a	d	d

I will leave it up to you to verify the fact that the five agents' preferences are all single-peaked with respect to the linear order $\succsim$ defined by $a \succ b \succ c \succ d$. Obviously, then the values of $m(G_i)$ are given by:

$$m(G_1) = c, \ m(G_2) = b, \ m(G_3) = d, m(G_4) = b, \text{ and } m(G_5) = a.$$

Consequently, if we define the sequence $\langle y_j \rangle$ as in (18.13)–(18.15) in the proof of Theorem 18.29, we have:

$$y_1 = d, \ y_2 = c, \ y_3 = y_4 = b, \text{ and } y_5 = a.$$

Therefore;

$$F(\boldsymbol{G}) = \text{med}_{w.r.t.\succsim}\{c, b, d, b, a\} = y_3 = b. \quad \square$$

Your next question regarding Theorem 18.29 probably is, "Why the phantom voters?" Well, first of all, notice that, whether n is even or odd, the number $n + (n-1) = 2n - 1$ is necessarily odd, and therefore the function in (18.12) is always well-defined. Another example may also be of use here.

18.31 Example. Let $X = [0, 3] \subseteq \mathbb{R}_+$, let $n = 4$, and suppose the agents' preferences are represented by the utility functions:

$$u_i(x) = -|i/2 - x| \quad \text{for } i = 1, 2, 3, 4.$$

Then all four preference relations are single-peaked with respect to $\geq$, the usual weak inequality for the real numbers, and denoting $m(G_i)$ by 'm_i,' we have:

$$m_1 = 1/2, m_2 = 1, m_3 = 3/2, \text{ and } m_4 = 2.$$

Now, suppose we modify our median definition in the way it is often done when taking the median of an even number of elements; that is, define:

$$F^*(1/2, 1, 3/2, 2) = \frac{1 + 3/2}{2} = 5/4.$$

The problem with this is, it is no longer strategy-proof; if everyone else responds truthfully, agent 3, for example, can gain by claiming $m(G_3) = 2$, which would yield a social choice of 3/2. If we add 3 'phantom voters' here, with $a_1 = a_2 = 0$ and $a_3 = 3$, our strategy-proof median function will choose the second largest of the actual agents' $m(G_i)$ values. In our example, we would have $F(\boldsymbol{G}) = 1$. What happens if we take $a_1 = 0$, and $a_2 = a_3 = 3$? $\square$

While the above example and discussion suggests some of the reasons for incorporating 'phantom voters' into a generalized median scheme, the most important reason for introducing the notion is that Moulin [1980, 1984] has proved that a social choice function $F\colon \mathcal{P}^n \to X$ is anonymous, efficient, and strategy-proof *only if* it is a generalized median function of the form defined in (18.12). He and Barbera have developed necessary and sufficient conditions for such a function to be anonymous and strategy-proof (not necessarily efficient), and to simply be strategy-proof. However, we will not attempt a proof of such necessary conditions here. For a complete discussion of these results, see the survey by Barbera [2001].

In our next result, it will be convenient to make use of a 'mechanism formulation' of the social choice function being considered in this section, as follows. We consider the mechanism, $\mathcal{M} = \langle \boldsymbol{M}, X, g\rangle$, where:

$$\boldsymbol{M} = X^n \quad \text{and} \quad g(\boldsymbol{m}) = \text{med}_{w.r.t.\succsim}\{m_1, \ldots, m_n\}, \tag{18.16}$$

and where $n = 2q - 1$, with q an integer greater than one (we will not explicitly incorporate 'phantom voters' in the material to follow). As in the preceding result, we will let '$\mathcal{P}$' denote the family of all quasi-orders on X which are single-peaked with respect to a given linear ordering of $X, \succsim$. Given a nonempty subset, S, of N, and an m-tuple of messages, $\boldsymbol{m} \in \boldsymbol{M}$, we will denote the vector of messages $\boldsymbol{m}'$, in which $m'_i = m^*_i$, for each $i \in S$, while $m'_i = m_i$, for each $i \in N \setminus S$ by $\boldsymbol{m}' = (\boldsymbol{m}^*_S, \boldsymbol{m}_{-S})$.

18.32 Proposition. *Given the mechanism defined in the above paragraph, let $S \subseteq N$ be a nonempty subset of agents, let $Q_i \in \mathcal{P}$, for each $i \in S$, and define:*

$$m^*_i = m(Q_i) \quad \textit{for each } i \in S.$$

Then there exists no $m \in \boldsymbol{M}$ such that:

$$g(\boldsymbol{m})P_i g(\boldsymbol{m}^*_S, \boldsymbol{m}_{-S}) \quad \textit{for each } i \in S.$$

Moreover, given any $i \in N$, and any $Q_i \in \mathcal{P}$, $m_i = m(Q_i)$ is a dominant strategy for i.

Proof. In order to prove the first part of our result, let S be a nonempty subset of N, let $Q_i \in \mathcal{P}$ and define $m^*_i = m(Q_i)$, for each $i \in S$. We then suppose, by way of obtaining a contradiction, that there exists $\boldsymbol{m} \in \boldsymbol{M}$ such that for each $i \in S$:

$$g(\boldsymbol{m})P_i g(\boldsymbol{m}^*_S, \boldsymbol{m}_{-S}). \tag{18.17}$$

Then we must have $g(\boldsymbol{m}) \neq g(\boldsymbol{m}^*_S, \boldsymbol{m}_{-S})$. Suppose, then, that:

$$g(\boldsymbol{m}) \succ g(\boldsymbol{m}^*_S, \boldsymbol{m}_{-S}), \tag{18.18}$$

where $\succ$ is the asymmetric part of the linear order, $\succsim$. Then from (18.17) we see that we must have, for each $i \in S$:

$$m(Q_i) \succ g(\boldsymbol{m}^*_S, \boldsymbol{m}_{-S}). \tag{18.19}$$

Now, defining $(\boldsymbol{m}_S^*, \boldsymbol{m}_{-S}) = \boldsymbol{m}'$, we see from the definition of g that there exists $h \in N$ such that:

$$g(\boldsymbol{m}') \equiv g(\boldsymbol{m}_S^*, \boldsymbol{m}_{-S}) = m_h' \text{ and } \#\{i \in N \mid m_h' \succsim m_i'\} \geq q. \tag{18.20}$$

However, from (18.19) we see that $h \notin S$; and, since for all $i \in S, m_i' \succ m_h'$, it follows that:

$$\{i \in N \mid m_h' \succsim m_i'\} \subseteq \{i \in N \mid m_h' \succsim m_i\}.$$

It follows, therefore, that:

$$\#\{i \in N \mid m_h' \succsim m_i\} \geq q,$$

and thus that:

$$m_h' = g(\boldsymbol{m}_S^*, \boldsymbol{m}_{-S}) \succsim g(\boldsymbol{m}) = \text{med}\{m_i, \ldots, m_n\};$$

contradicting (18.18).

To prove the second part of our conclusion, we note first that it follows from the preceding argument that for any $i \in N$ and any $Q_i \in \mathcal{P}$, $m_i = m(Q_i)$ is a weakly dominant strategy for i. To complete our proof, and I will leave this as an exercise, you need to show that for any $m_i' \in X$, there exists $\boldsymbol{m}_{-i} \in \boldsymbol{M}_{-i}$ such that:

$$g[m(Q_i), \boldsymbol{m}_{-i}]P_i g(m_i', \boldsymbol{m}_{-i}). \quad \square$$

Notice that the first part of this last proof applies equally well to the 'phantom voter' case, and thus completes the proof of Theorem 18.29. Truth-telling on the part of the real agents can also be shown to be dominant, as opposed to just weakly dominant strategies in the 'phantom voter' case as well, but the proof is considerably more complicated, and it did not seem worthwhile to develop it here. Let's instead turn our attention to an economic application, the basic idea of which was introduced in Bowen [1943].

18.5.2 The Bowen Model

Let's return to the sort of two-commodity, n-consumer public goods model which we discussed in Chapter 16. Here, however, we will suppose that the public good is produced by a firm which has the cost function $c(y)$, where the function c is continuous and convex. We will also suppose that the ith consumer's preferences can be represented by a utility function, u_i, which is continuous, increasing, and strictly quasi-concave. We will further suppose that a choice is to be made of the quantity of the public good to be provided, and that the ith consumer knows that she/he will be required to pay the fraction a_i, where:

$$0 \leq a_i \leq 1, \text{ for each } i, \text{ and } \sum_{i \in N} a_i = 1,$$

of the total cost.

Let's consider the ith consumer's choice as to her/his most preferred quantity of the public good. The consumer's budget set will consist of all combinations of

$(x_i, y) \in \mathbb{R}^2_+$ satisfying (we have normalized by setting the price of the private good equal to one):

$$x_i + a_i c(y) \le w_i.$$

However, since preferences are increasing, the optimal choice will be on the budget line (which may be a curve in this case); and, since utility functions are continuous and strictly quasi concave, there will be a unique bundle, (x_i^*, y^*), which maximizes utility on this budget line. Moreover, on this budget line/curve, we have:

$$x_i = w_i - a_i c(y),$$

and thus, on the budget line we can consider utility to be a function of y alone; that is, the function v_i defined by:

$$v_i(y) = u_i\big[w_i - a_i c(y), y\big],$$

represents the consumer's preferences on this budget line.

Suppose now that (x_i^t, y^t), $(t = 1, 2)$ are two bundles on this budget line (curve) satisfying:

$$y^2 > y^1 > y^*,$$

and define $\theta \in]0,1[$ by:

$$\theta = \frac{y^1 - y^*}{y^2 - y^*}.$$

Then we see that:

$$y^1 = \theta y^2 + (1-\theta) y^*, \tag{18.21}$$

and thus, defining:

$$x_i^\dagger = \theta x_i^2 + (1-\theta) x_i^*,$$

it follows from the strict quasi-concavity of u_i that:

$$u_i(x_i^\dagger, y^1) > u_i(x_i^2, y^2). \tag{18.22}$$

Moreover, we have from (18.21) and the convexity of $c(\cdot)$:

$$c(y^1) \le \theta c(y^2) + (1-\theta) c(y^*), \tag{18.23}$$

and therefore:

$$\begin{aligned} x_i^\dagger &= \theta x_i^2 + (1-\theta) x_i^* = \theta[w_i - a_i c(y^2)] + (1-\theta)[w_i - a_i c(y^*)] \\ &= \theta w_i + (1-\theta) w_i - a_i[\theta c(y^2) + (1-\theta) c(y^*)] \\ &\le w_i - a_i c(y^1) = x_i^1. \end{aligned} \tag{18.24}$$

Making use of (18.22), (18.24) and the fact that preference is increasing, we see that:

$$v(y^*) = u_i(x_i^*, y^*) > u_i(x_i^1, y^1) = v(y^1) > v(y^2) = u_i(x_i^2, y^2).$$

A similar argument shows that if:

$$y_2 < y^1 < y^*,$$

then:

$$v(y^2) < v(y^1) < v(y^*).$$

Consequently, it follows that preferences are single-peaked (with respect to $\geq$) over the potential choices of y.

Having shown that individual preferences are single-peaked over the relevant region, let's now consider the following mechanism for the collective choice of a quantity of the public good. We suppose that the policy-maker (government) announces a tax share representing each individual's contribution to the cost of the public good. For example, a_i, consumer i's tax share might be given by:

$$a_i = \frac{w_i}{\sum_{j \in N} w_j}.$$

Secondly, the cost function, $c(y)$, representing the cost of the public good in terms of the amounts of the private good which must be provided as an input, is also announced (or in any case, assumed to be public information, that is, commmon knowledge). The individuals are then asked to determine their most-preferred quantity of the public good to be produced, and send their personal 'vote,' y^i to the policy-maker. The public choice is to be made by setting production of $y = y^m$, where 'y^m denotes the median quantity, that is, the median of the y^i values, *and this decision rule is known in advance by all of the agents (consumers).*

So the mechanism to be considered here is (g, M), where:

$$M = \prod_{i=1}^{n} M_i,$$

where $M_i = \mathbb{R}_+$, for each i; and the outcome function, $g \colon M \to \mathbb{R}_+$ is defined by:

$$g(\boldsymbol{m}) = g(m_1, \ldots, m_n) = \text{med}\{m_1, \ldots, m_n\}.$$

From the analysis of the previous subsection, it follows that the dominant strategy for each agent is to truthfully announce her/his most preferred quantity of the public good. The resultant choice of the public good quantity will satisfy a 'second-best' efficiency, in that any change in the quantity (given the stated tax shares) would make someone (in fact, in general, $q + 1$ someones) worse off. However, this mechanism will not, in general result in a fully Pareto efficient allocation. (See Exercises 9 and 10, at the end of this chapter).

The allocation obtained with this mechanism is also often called the **majority voting solution**; for suppose each agent is asked to submit for consideration the quantity of the public good which they would most like to see produced, and then these proposed quantities are voted upon. It is easy to see that the median quantity would be the Condorcet winner in pairwise-majority voting.

18.6 Quasi-Linearity and Dominant Strategies

In this section we will deal with a situation in which we have n agents, who are to make a collective choice from a nonempty set of alternatives, A. We will suppose that in the context of this choice, we can treat the situation rather as if we were dealing with a two-commodity economy; a public good, a, and a private good, x, which might be considered to be money or income available for expenditure on a set of private goods. We will further suppose that the ith agent's preferences can be represented by a utility function of the quasi-linear form:

$$u_i(a, x_i) = v_i(a) + x_i, \tag{18.25}$$

where x_i is the quantity of the private good (or income available for expenditure on private goods). It will be convenient, and presumably harmless, to assume that $0 \in A$; that is, that one of the choices available is the status quo. This being the case, we can normalize to set $v_i(0) = 0$, for each $i \in N$. Given this normalization, consider how much of the private good consumer (agent) i would just be willing to give up in order to have $a \in A$ be adopted. Denoting this quantity by 'c,' we see that if the agent's income is x_i, then c must satisfy:

$$u_i(a, x_i - c) = u_i(0, x_i);$$

that is:

$$v_i(a) + x_i - c = x_i;$$

and therefore $v_i(a) = c$. Consequently, we can interpret $v_i(a)$ as agent i's willingness to pay for $a \in A$ (and notice that this quantity is independent of x_i). Since preferences are quasi-linear, we need only know the willingness-to-pay function, $v_i \colon A \to \mathbb{R}$, in order to fully characterize i's utility function and preferences. Thus we will specify the admissible preferences for the ith agent as:

$$\mathcal{V} = \{v \colon A \to \mathbb{R} \mid v(0) = 0\}. \tag{18.26}$$

We define $\boldsymbol{\mathcal{V}} = \mathcal{V}^n$, and denote elements of $\boldsymbol{\mathcal{V}}$ by '$\boldsymbol{v}$,' '$\boldsymbol{v}^*$, and so on; where, for example:

$$\boldsymbol{v} = (v_1, \ldots, v_i, \ldots, v_n),$$

with v_i being the ith agent's willingness-to-pay function, for $i = 1, \ldots, n$. We will suppose that it is desired to attain an allocation/situation $(a, x_1, \ldots, x_n) \in A \times \mathbb{R}^n$ such that $(a, x_1, \ldots, x_n) \in F(\boldsymbol{v})$, where:

$$F \colon \boldsymbol{\mathcal{V}} \mapsto A \times \mathbb{R}^n. \tag{18.27}$$

We will usually suppose that F is given by:

$$F(\boldsymbol{v}) = \left\{a^* \in A \mid (\forall a \in A) \colon \sum\nolimits_{i=1}^n v_i(a^*) \geq \sum\nolimits_{i=1}^n v_i(a)\right\} \times \mathbb{R}^n; \tag{18.28}$$

but we will give some consideration to some other social choice correspondences as well.

In the course of our discussions, we will be illustrating and applying our results to two basic examples, the framework of the first of which we will set out now.

18.33 Example. Suppose a group of agents is to make a choice from among a finite list of projects. In this scenario we will not concern ourselves with costs; supposing that either the projects are costless, or that the cost of the projects has already been allocated, with the alternatives all attainable with the budget available. To make the example more concrete, we will particularize things by supposing that we have a group of five agents who are attempting to make a decision about the location of a new park; and suppose that their evaluation (willingness-to-pay) functions are as follows.

Location	Agent 1	Agent 2	Agent 3	Agent 4	Agent 5
a	8	4	5	1	3
b	5	5	2	2	6
c	4	2	1	10	3

We will put ourselves in the role of an arbitrator trying to find a 'good' solution to this problem; our fundamental difficulty being that we will suppose that the information in the above table is unknown to us. □

We want to find a mechanism which implements the desired social choice correspondence in dominant strategies; and the Revelation Principle suggests that this means that we can confine our attention to Strategy-Proof voting rules.[9] Consequently, nearly all of our study will be devoted to mechanisms of the following type. (There is, incidentally, no well-established name for this general mechanism in the literature; I have named it a 'transfer mechanism' rather arbitrarily.)

18.34 Definitions. We will say that a mechanism, $\mathcal{M} = \langle \boldsymbol{M}, X, g \rangle$, is a **transfer mechanism** iff:

1. $M_i = \mathcal{V}$, for $i = 1, \ldots, n$, where $\mathcal{V}$ is defined in (18.26),
2. $X = A \times \mathbb{R}^n$, and
3. the outcome function, g, takes the form:

$$g(\boldsymbol{v}) = \big(d(\boldsymbol{v}), t_1(\boldsymbol{v}), \ldots, t_n(\boldsymbol{v})\big), sst \tag{18.29}$$

where:

$$d(\boldsymbol{v}) = \text{argmax}_{a \in A} \sum_{i=1}^{n} v_i(a), \tag{18.30}$$

and $t_i \colon \boldsymbol{\mathcal{V}} \to \mathbb{R}$, for $i = 1, \ldots, n$.

We say that a transfer mechanism is **feasible** [respectively, **budget-balanced**] iff, for all $\boldsymbol{v} \in \boldsymbol{\mathcal{V}}$:

$$\sum_{i=1}^{n} t_i(\boldsymbol{v}) \leq 0$$

[respectively, $\sum_{i=1}^{n} t_i(\boldsymbol{v}) = 0$.]

There is, of course, a bit of a problem revolving around the question of when or whether the maximum appearing in the above definition exists. In order to insure that this is the case, *until further notice, we will assume that the set A is finite.*

We will use the generic notation '$\boldsymbol{v}, \boldsymbol{v}', \boldsymbol{v}^*$,' and so on, to denote elements of $\boldsymbol{M} = \boldsymbol{\mathcal{V}}$; where, for example;

$$\boldsymbol{v} = (v_1, \ldots, v_n),$$

with $v_i \colon A \to \mathbb{R}$, and satisfying $v_i(0) = 0$. For a given agent, i, the vector of willingness-to-pay functions for the other $n-1$ agents will be denoted by '$\boldsymbol{v}_{-i}$;' we

[9]However, see Section 9 of this chapter.

will denote $\boldsymbol{M}_{-i} = \mathcal{V}^{n-1}$ by '$\boldsymbol{\mathcal{V}}_{-i}$,' or simply by '$\mathcal{V}^{n-1}$,' and we use the notation '$(v_i^*, \boldsymbol{v}_{-i})$' to denote the n-tuple of reported messages in which v_i^* replaces v_i in $\boldsymbol{v}$. Thus, a transfer mechanism is **strategy-proof** iff for all $v_i^* \in \mathcal{V}$ and all $\boldsymbol{v} \in \boldsymbol{\mathcal{V}}$, we have:

$$v_i^*[d(v_i^*, \boldsymbol{v}_{-i})] + t_i(v_i^*, \boldsymbol{v}_{-i}) \geq v_i^*[d(\boldsymbol{v})] + t_i(\boldsymbol{v}). \tag{18.31}$$

The following is the special case of the transfer mechanism which was introduced by Clarke [1971] (and independently by Groves and Loeb [1975]; see also Vickrey [1961]).

18.35 Definition. We will say that a transfer mechanism is a **pivot** (or **pivotal**) **mechanism** iff the functions t_i take the form:

$$t_i(\boldsymbol{v}) = \sum_{j \neq i} v_j[d(\boldsymbol{v})] - \max_{a \in A} \sum_{j \neq i} v_j(a). \tag{18.32}$$

In the proof of the following result, and in much of our discussion of this section, it will be convenient to make use of the following notation: for $\boldsymbol{v}_{-i} \in \boldsymbol{\mathcal{V}}_{-i}$, we define the function $v_{N \setminus i} \colon A \to \mathbb{R}$ by:

$$v_{N \setminus i}(a) = \sum_{j \neq i} v_j(a). \tag{18.33}$$

18.36 Theorem. (Clarke [1971], Groves and Loeb [1975]) *Given the assumptions set out in (18.25) and (18.26), above, and if A is finite, then the pivot mechanism is feasible and strategy-proof.*

Proof. In order to prove feasibility, let $\boldsymbol{v} \in \boldsymbol{\mathcal{V}}$, and define:

$$a^* = \operatorname{argmax}_{a \in A} \sum_{i=1}^n v_i(a) = d(\boldsymbol{v}).$$

Then for $i \in N$, we have:

$$t_i(\boldsymbol{v}) = \sum_{j \neq i} v_j(a^*) - \max_{a \in A} \sum_{j \neq i} v_j(a),$$

which is obviously non-positive; and thus:

$$\sum_{i=1}^n t_i(\boldsymbol{v}) \leq 0,$$

as well.

To prove strategy-proofness, let $i \in N$ be arbitrary, let v_i^* be i's true willingness-to-pay function, let $\boldsymbol{v} \in \boldsymbol{\mathcal{V}}$ be arbitrary, and define:

$$a^* = d(v_i^*, \boldsymbol{v}_{-i}) \text{ and } a' = d(\boldsymbol{v}).$$

Then we have:

$$\begin{aligned} v_i^*(a^*) + t_i(v_i^*, \boldsymbol{v}_{-i}) &= v_i^*(a^*) + v_{N \setminus i}(a^*) - \max_{a \in A} v_{N \setminus i}(a) \\ &\geq v_i^*(a') + v_{N \setminus i}(a') - \max_{a \in A} v_{N \setminus i}(a); \end{aligned}$$

where the inequality follows from the definition:

$$a^* = d(v_i^*, \boldsymbol{v}_{-i}) = \operatorname{argmax}_{a \in A}\left[v_i^*(a) + v_{N \setminus i}(a)\right]. \quad \square$$

18.37 Example. Returning to Example 18.33, we find the following figures for the implementation of the pivotal mechanism. We have $d(\boldsymbol{v}) = a$, and:

Agent	$t_i(\boldsymbol{v})$	$\text{argmax}_{y \in A} v_{N \setminus i}(y)$
1	$13 - 16 = -3$	c
2	$17 - 18 = -1$	c
3	$16 - 19 = -3$	c
4	$20 - 20 = 0$	a
5	$18 - 18 = 0$	a

From the above figures, we see that agents 1–3 are all pivotal; that is, without any one of them a different location would be chosen. Agents 4 and 5 are non-pivotal; without them the decision as to location would be the same as it is with them, and consequently, their 'tax,' $t_i(\boldsymbol{v})$, is zero.

Interestingly, agent 2 is pivotal despite the fact that the social choice, a, is agent 2's second choice alternative. On the other hand, an agent cannot be pivotal if $d(\boldsymbol{v})$ is her or his last-place alternative. To see this, notice that if, say $a = d(\boldsymbol{v})$, and agent i is pivotal, then there must be another social choice, call it b, such that:

$$v_{N \setminus i}(b) > v_{N \setminus i}(a). \tag{18.34}$$

However, since $a = d(\boldsymbol{v})$, we must have:

$$v_{N \setminus i}(a) + v_i(a) \geq v_{N \setminus i}(b) + v_i(b);$$

which, together with (18.34), implies $v_i(a) > v_i(b)$. □

A pivotal mechanism is a special case of the following.

18.38 Definition. We will say that a transfer mechanism is a **Vickrey-Clarke-Groves (VCG) mechanism** iff there exists an n-tuple of functions, $\langle h_i \rangle$, where $h_i \colon \boldsymbol{V}_{-i} \to \mathbb{R}$, for $i = 1, \ldots, n$, such that:

$$t_i(\boldsymbol{v}) = \sum_{h \neq i} v_h[d(\boldsymbol{v})] - h_i(\boldsymbol{v}_{-i}) \quad \text{for } i = 1, \ldots, n. \tag{18.35}$$

In the following result, we will suppose that for each $i \in N$,

$$v_{N \setminus i}(a) = \sum_{j \neq i} v_j(a),$$

is bounded above on A. The simplest condition sufficient to guarantee that this holds is that A is finite. The first part of the following result was proved for special cases in Vickrey [1961] and in Clarke [1971], and for the general case in Groves [1970]. The second part (the 'moreover statement') was established in Groves [1976]. (See also Green and Laffont [1979])

18.39 Theorem. (Vickrey-Clarke-Groves) *Given the assumptions set out in equations (18.25) and (18.26), above, a VCG mechanism is strategy-proof. Moreover, given these assumptions, any strategy-proof transfer mechanism must be a VCG mechanism.*

Proof. One can prove that a VCG mechanism is strategy-proof by almost exactly the same argument that we used in establishing this property for the pivotal mechanism (Theorem 18.36); details will be left as an exercise. To prove the converse, suppose $\mathcal{M}$ is a transfer mechanism which is strategy proof. To prove that $\mathcal{M}$ must be a VCG mechanism, we begin by defining the functions h_i by:

$$h_i(\boldsymbol{v}) = t_i(\boldsymbol{v}) - v_{N\setminus i}[d(\boldsymbol{v})]. \tag{18.36}$$

Now, suppose $v_i^*, \overline{\boldsymbol{v}}$ are such that:

$$d(v_i^*, \overline{\boldsymbol{v}}_{-i}) = d(\overline{\boldsymbol{v}}) \stackrel{\text{def}}{=} \overline{a}. \tag{18.37}$$

Then, since truth-telling is a dominant strategy, we must have:

$$\overline{v}_i(\overline{a}) + t_i(\overline{\boldsymbol{v}}) \geq \overline{v}_i(\overline{a}) + t_i(v_i^*, \overline{\boldsymbol{v}}_{-i}), \tag{18.38}$$

and:

$$v_i^*(\overline{a}) + t_i(v_i^*, \overline{\boldsymbol{v}}_{-i}) \geq v_i^*(\overline{a}) + t_i(\overline{\boldsymbol{v}}). \tag{18.39}$$

From these two inequalities we then conclude that $t_i(\overline{\boldsymbol{v}}) = t_i(v_i^*, \overline{\boldsymbol{v}}_{-i})$.

Next, suppose, by way of obtaining a contradiction, that there exist $v_i^* \in \mathcal{V}$ and $\boldsymbol{v} \in \boldsymbol{\mathcal{V}}$ such that $h_i(v_i^*, \boldsymbol{v}_{-i}) \neq h_i(\boldsymbol{v})$, and notice that we can then assume, without loss of generality, that:

$$h_i(v_i^*, \boldsymbol{v}_{-i}) < h_i(\boldsymbol{v}). \tag{18.40}$$

From the argument of the last paragraph, we see that we must have:

$$\overline{a} \stackrel{\text{def}}{=} d(\boldsymbol{v}) \neq a^* \stackrel{\text{def}}{=} d(v_i^*, \boldsymbol{v}_{-i}). \tag{18.41}$$

Now, define:

$$\delta = (1/2)\big[h_i(\boldsymbol{v}) - h_i(v_i^*, \boldsymbol{v}_{-i})\big],$$

and consider the willingness-to-pay function $v_i^\dagger \in M_i$ defined by:

$$v_i^\dagger(a) = \begin{cases} -\sum_{j\neq i} v_j(\overline{a}) & \text{for } a = \overline{a}, \\ -\sum_{j\neq i} v_j(a^*) + \delta & \text{for } a = a^*, \\ 0 & \text{for } a = 0, \\ -\sup_{a\in A} v_{N\setminus i}(a) & \text{for } a \notin \{0, \overline{a}, a^*\}. \end{cases} \tag{18.42}$$

Then, for all $a \in A \setminus \{0, a^*, \overline{a}\}$, we have:

$$v_i^\dagger(a) + v_{N\setminus i}(a) = -\sup_{a'\in A} v_{N\setminus i}(a') + v_{N\setminus i}(a) \leq 0;$$

and thus, since we also have $v_j(0) = 0$ for all $i \neq j$:

$$d(v_i^\dagger, \boldsymbol{v}_{-i}) = a^*. \tag{18.43}$$

Consequently, if truth-telling is a dominant strategy, we must have:

$$v_i^\dagger(a^*) + t_i(v_i^\dagger, \boldsymbol{v}_{-i}) \geq v_i^\dagger(\overline{a}) + t_i(\boldsymbol{v});$$

or, upon substituting from the definition of $v_i^\dagger$ and (18.36):

$$h_i(v_i^\dagger, \boldsymbol{v}_{-i}) + \delta \geq h_i(\boldsymbol{v}). \tag{18.44}$$

However, since $d(v_i^\dagger, \boldsymbol{v}_{-i}) = d(v_i^*, \boldsymbol{v}_{-i})$, we have $h_i(v_i^\dagger, \boldsymbol{v}_{-i}) = h_i(v_i^*, \boldsymbol{v}_{-i})$, and using the definition of δ, we see that (18.44) implies $h_i(v_i^*, \boldsymbol{v}_{-i}) \geq h_i(\boldsymbol{v})$; contradicting (18.40) □

It is an almost immediate consequence of our next result that truth-telling is a *dominant* strategy in a VCG mechanism. *In this result, we will temporarily drop the assumption that* $(\forall v \in \mathcal{V})\colon v(0) = 0$.

18.40 Proposition. (Groves [1974]) *Let* $\mathcal{M} = \langle \boldsymbol{M}, X, g\rangle$ *be a VCG mechanism, let* $i \in N$*, and let* $v_i^*, \widehat{v}_i \in \mathcal{V}$*. Then either there exists a nonzero constant* α *such that:*

$$(\forall a \in A)\colon |v_i^*(a) - \widehat{v}_i(a)| = \alpha,$$

or there exists $\boldsymbol{v}_{-i} \in \mathcal{V}^{n-1}$ *such that:*

$$v_i^*[d(v_i^*, \boldsymbol{v}_{-i})] + v_{N\setminus i}[d(v_i^*, \boldsymbol{v}_{-i})] - h(\boldsymbol{v}_{-i}) > v_i^*[d(\widehat{v}_i, \boldsymbol{v}_{-i})] + v_{N\setminus i}[d(\widehat{v}_i, \boldsymbol{v}_{-i})] - h(\boldsymbol{v}_{-i}).$$

Proof. Suppose $|v_i^* - \widehat{v}_i|$ is *not* constant. Then we can suppose, without loss of generality, that there exist $a^*, \widehat{a} \in A$ such that:

$$v_i^*(a^*) - \widehat{v}_i(a^*) > |\, v_i^*(\widehat{a}) - \widehat{v}_i(\widehat{a})|. \tag{18.45}$$

Define:

$$\alpha = |\, v_i^*(\widehat{a}) - \widehat{v}_i(\widehat{a})| \text{ and } \beta = v_i^*(a^*) - \widehat{v}_i(a^*) - \alpha.$$

We then choose $\boldsymbol{v}_{-i} \in \mathcal{V}^{n-1}$ to satisfy:

$$\begin{aligned} v_{N\setminus i}(\widehat{a}) &= -\,\widehat{v}_i(\widehat{a}) - \alpha, \\ v_{N\setminus i}(a^*) &= -\,\widehat{v}_i(a^*) - \alpha - \beta/2, \end{aligned}$$

and, for all $a \in A \setminus \{a', a^*\}$:

$$v_{N\setminus i}(a) = -\max\{\max_{a'\in A} v_i^*(a'), \max_{a'\in A} \widehat{v}_i(a')\} - \alpha - \beta.$$

Then we have:

$$\begin{aligned} v_i^*(a^*) + v_{N\setminus i}(a^*) &= \alpha + \beta - \alpha - \beta/2 = \beta/2, \\ v_i^*(\widehat{a}) + v_{N\setminus i}(\widehat{a}) &= v_i^*(\widehat{a}) - \widehat{v}_i(\widehat{a}) - \alpha \leq 0; \end{aligned} \tag{18.46}$$

while, for $a \in A \setminus \{\widehat{a}, a^*\}$:

$$v_i^*(a) + v_{N\setminus i}(a) \leq v_i^*(a) - \max_{a'\in A} v_i^*(a') - \alpha - \beta \leq -\alpha - \beta.$$

Therefore, we see that $d(v_i^*, \boldsymbol{v}_{-i}) = a^*$.

Similarly, we see that $d(\widehat{v}_i, \boldsymbol{v}_{-i}) = \widehat{a}$. Furthermore, it follows from (18.46) that:

$$\begin{aligned} v_i^*[d(v_i^*, \boldsymbol{v}_{-i})] + v_{N\setminus i}[d(v_i^*, \boldsymbol{v}_{-i})] - h(\boldsymbol{v}_{-i}) &= \beta/2 \\ &> v_i^*[d(\widehat{v}_i, \boldsymbol{v}_{-i})] + v_{N\setminus i}[d(\widehat{v}_i, \boldsymbol{v}_{-i})] v_{N\setminus i} - h(\boldsymbol{v}_{-i}) \end{aligned}$$ □

Now, since we have defined $\mathcal{V}$ as the set of all $v\colon A \to \mathbb{R}$ satisfying $v(0) = 0$, it follows that if $\widehat{v}_i, v_i^* \in \mathcal{V}$ are such that:

$$(\forall a \in A)\colon |\widehat{v}_i(a) - v_i^*(a)| = \alpha,$$

then we must have $\alpha = 0$, and thus $\widehat{v}_i = v_i^*$. Consequently, it follows from Theorem 18.39 that in a VCG mechanism, truth-telling is a *dominant* strategy.

The following example is adapted from Moulin [1988, pp. 205–6].

18.41 Example. In this example we consider a situation in which we have an indivisible public good; the issue to be analyzed is whether or not the good should be built or produced. Thus we take $A = \{0, 1\}$, where 0 corresponds to 'do not build,' and 1 represents 'build/produce.' We suppose that completing the project will cost an amount $c > 0$, the cost of which must be shared among the n agents. Since the willingness-to-pay functions, v_i are completely specified by the value $v_i(1)$, we will simplify our notational framework by supposing that each agent is asked to report $b_i = v_i(1)$.

Consider the following mechanism, where $\boldsymbol{a} \in \Delta_n$ is a proposed vector of cost shares:

1. If $\sum_i b_i < c$, the good is *not* produced, and agent i pays the amount τ_i defined by:

$$\tau_i = \begin{cases} 0 & \text{if } \sum_{j \neq i} b_j < (1 - a_i)c, \\ \sum_{j \neq i} b_j - (1 - a_i)c & \text{if } \sum_{j \neq i} b_j \geq (1 - a_i)c. \end{cases}$$

2. If $\sum_i b_i \geq c$, the good *is* produced, and agent i pays the amount τ_i defined by:

$$\tau_i = \begin{cases} a_i c & \text{if } \sum_{j \neq i} b_j \geq (1 - a_i)c, \\ c - \sum_{j \neq i} b_j & \text{if } \sum_{j \neq i} b_j < (1 - a_i)c. \end{cases}$$

One can prove directly that this mechanism is feasible and strategy-proof. However, by equivalently reformulating the mechanism, we can show that this follows immediately from the results of this section, as follows.

We begin by defining $B = \mathbb{R}$, and u_i, for each i and $b_i \in \mathbb{R}$, by:

$$u_i(a) = \begin{cases} 0 & \text{if } a = 0, \\ b_i - a_i c & \text{if } a = 1. \end{cases}$$

We then define $d\colon \mathbb{R}^n \to A$ by $d(\boldsymbol{b}) = \operatorname{argmax}_{a \in A} \sum u_i$, and note that:

$$d(\boldsymbol{b}) = \begin{cases} 0 & \text{if } \sum_{i=1}^n b_i < c \\ 1 & \text{if } \sum_{i=1}^n b_i \geq c. \end{cases}$$

We then note that we can equivalently define the amounts τ_i as the values of the function t_i given by:

$$t_i(\boldsymbol{b}) = \sum_{j \neq i} u_j[d(\boldsymbol{b})] - \max_{a \in A} \sum_{j \neq i} u_j(a),$$

(I will leave the verification of this as an exercise). Thus, the mechanism is equivalently expressed as a pivotal mechanism, which we know to be feasible and strategy-proof. □

Notice that we are taking the cost shares as given in the above argument; the cost shares are not determined by the mechanism. It should also be noted that the mechanism is not budget-balanced; in general the agents will be contributing more than the cost of the project. This is a common problem with VCG mechanisms; in fact, Green and Laffont have proved the following result, which we will state here without proof.[10]

18.42 Theorem. *There exists no Groves mechanism satisfying:*

$$(\forall \boldsymbol{v} \in \mathcal{V}^n) \colon \sum_{i=1}^{n} t_i(\boldsymbol{v}) = 0.$$

This result leads us to an interesting point: in the kind of model presented in Example 18.41, we know from Proposition 16.1 that if a VCG mechanism is budget-balanced, then the outcome is always Pareto efficient. Of course, if the mechanism is not budget balanced, then the outcome need not be Pareto efficient. There is another problem here, however, in that even though utility functions are quasi-linear, there may be alternatives which are Pareto efficient, but where the utilitarian social welfare function used in the VCG mechanism is not maximized. For instance, consider the following example.

18.43 Example. Let's return to the general scenario considered in Example 18.33; a group of agents attempting to arrive at a decision regarding the location of a park, except that we now assume that the agents have the following preferences.

Location	**Agent 1**	**Agent 2**	**Agent 3**	**Agent 4**	**Agent 5**
$\boldsymbol{a}$	7	3	6	5	4
$\boldsymbol{b}$	3	2	5	3	3
$\boldsymbol{c}$	1	20	2	1	2

In this case (I will leave you to verify the details), the pivotal mechanism will choose location c, despite the fact that all but one of the agents agree that location a is best. In fact, all but one of the agents considers location c to be the worst of the three alternatives. Of course, the pivotal mechanism levies a heavy tax on agent 2, namely $t_2 = 16$, so that agent 2 ends up with a utility of 4. But this means that agent 1, who would pay up to 6 units of the private good to move the choice to location a could pay agent 2 two units of the private good to give up location c in favor of location a, a change which would make all five agents better off. Consequently, the outcome of the pivotal mechanism is decidedly not Pareto efficient in this case.

It may be of some interest to note that, if in place of the function d used in the VCG mechanism, we were to use the function d^* defined by:

$$d^*(\boldsymbol{v}) = \operatorname{argmax}_{x \in A} \prod_{i=1}^{n} v_i(x),$$

the mechanism would then choose location a in our example. We could then combine this with the transfer function:

$$t_i(\boldsymbol{v}) = \sum_{j \neq i} \log v_j[d^*(\boldsymbol{v})] - \max_{x \in A} \sum_{j \neq i} \log v_j(x),$$

[10] For a proof, see Green and Laffont [1979, Theorem 5.3, p. 90], or Moulin [1988, p. 210].

to create a strategy-proof and 'feasible' mechanism. Unfortunately, since the log function is unbounded below as $a \to 0$, it could result in a 'pivotal' agent's being made very badly off indeed! □

While we have discussed only two basic types of applications for the VCG mechanism in this section, many more can and have been developed. See, for example, Groves [1982].

18.7 Implementation in Nash Equilibria

In this section, we will briefly consider the issue of implementation in Nash equilibria. In our treatment it will be convenient to make use of mechanism, rather than game form language. Accordingly, let $\mathcal{M} = \langle \boldsymbol{M}, X, g\rangle$ be a mechanism, and let $\boldsymbol{G} \in \mathcal{Q}^n$. We then define the set of Nash equilibria for $\boldsymbol{G}$, given $\mathcal{M}$ by;

$$NE(\boldsymbol{G}; \mathcal{M}) = \{\boldsymbol{m} \in \boldsymbol{M} \mid (\forall i \in N)(\forall m_i' \in M_i)\colon g(\boldsymbol{m})G_i g(m_i', \boldsymbol{m}_{-i})\} \tag{18.47}$$

In terms of Definition 18.23, $NE(\cdot)$ is a solution concept, and the resultant equilibrium outcome correspondence, which we will denote by 'O_N,' is given by:

$$O_N(\boldsymbol{G}; \mathcal{M}) = \{g(\boldsymbol{m}) \mid \boldsymbol{m} \in NE(\boldsymbol{G}; \mathcal{M})\}. \tag{18.48}$$

We can then define Nash Implementation as follows.

18.44 Definition. Let $\mathcal{M} = \langle \boldsymbol{M}, X, g\rangle$ be a mechanism and $F\colon \mathcal{D}^n \mapsto X$ be a social choice correspondence. We will say that **$\mathcal{M}$ implements F in Nash equilibria**, and **F** is said to be **Nash-implementable** iff:

$$(\forall \boldsymbol{G} \in \mathcal{D}^n)\colon O_N(\boldsymbol{G}; \mathcal{M}) \subseteq F(\boldsymbol{G}). \tag{18.49}$$

The mechanism will be said to **fully implement F in Nash equilibria** iff:

$$(\forall \boldsymbol{G} \in \mathcal{D}^n)\colon O_N(\boldsymbol{G}; \mathcal{M}) = F(\boldsymbol{G}). \tag{18.50}$$

Maskin introduced the notion of a montonic social choice correspondence into economic literature in the context of one of the most oft-cited working papers of all time. While the working paper dated from the 70's, it was eventually revised, extended and published in 1999.

18.45 Definition. A social choice correspondence, $F\colon \mathcal{D}^n \mapsto X$ is **monotonic** iff, for any $\boldsymbol{G}, \boldsymbol{G}' \in \mathcal{D}^n$, and any $x \in X$, we have the following:

if $x \in F(\boldsymbol{G})$ and $x \notin F(\boldsymbol{G}')$, then there must exist an agent, $i \in N$, and an alternative $y \in X$ such that:

$$xG_iy \text{ and } yP_i'x.$$

One can prove that this condition is a necessary condition for a social choice correspondence to be fully Nash-implementable, as follows.

18.46 Theorem. (Maskin [1999]) If a social choice correspondence, $F\colon \mathcal{D}^n \mapsto X$ is fully Nash implementable, then it is monotonic.

Proof. Suppose F is fully Nash-implementable by $\mathcal{M} = \langle \boldsymbol{M}, X, g \rangle$, and let $x \in X$ and $\boldsymbol{G}, \boldsymbol{G}' \in \mathcal{D}^n$ be such that:

$$x \in F(\boldsymbol{G}) \text{ and } x \notin F(\boldsymbol{G}').$$

Then, since F is fully Nash implementable by $\mathcal{M}$, there must exist $\boldsymbol{m} \in \boldsymbol{M}$ such that $x = g(\boldsymbol{m})$, and, since $x \notin F(\boldsymbol{G}')$, it must be that $\boldsymbol{m} \notin NE(\boldsymbol{G}'; \mathcal{M})$. Therefore, there must exist an agent, $i \in N$, and $m_i^* \in M_i$ such that:

$$g(m_i^*, \boldsymbol{m}_{-i}) P_i' g(\boldsymbol{m}).$$

Letting $y = g(m_i^*, \boldsymbol{m}_{-i})$, we see that $y P_i' x$. However, since $\boldsymbol{m}$ is a Nash equilibrium for $\boldsymbol{G}$ [that is, $\boldsymbol{m} \in NE(\boldsymbol{G}; \mathcal{M})$]:

$$g(\boldsymbol{m}) G_i g(m_i^*, \boldsymbol{m}_{-i});$$

in other words, $x G_i y$. Therefore, F is monotonic. □

If a monotonic social choice correspondence also satisfies a condition called 'no veto power,' then it can be shown to be fully Nash-implementable. This condition is presented in our next definition.

18.47 Definition. A social choice correspondence, $F \colon \mathcal{D}^n \mapsto X$ satisfies **no veto power** iff, for all $x \in X$, all $\boldsymbol{G} \in \mathcal{D}^n$, and all $i \in N$, we have that:
if, for all $y \in X$, and all $j \in N \setminus \{i\}$, we have $x G_j y$, then $x \in F(\boldsymbol{G})$.

So, the no veto power condition says that if all but at most one agent believe a given alternative is at least as good as any other, then that alternative should be in the social choice correspondence for $\boldsymbol{G}$. Clearly the condition is overly-strong for the case in which $n = 2$. Moreover, it is a questionable requirement for a social choice correspondence whose domain contains profiles of quasi-orders which are not weak orders. However, Maskin has proved[11] the following, rather surprising and important result.

18.48 Theorem. (Maskin [1999]) Suppose $n \geq 3$, and $F \colon \mathcal{D}^n \mapsto X$, where $\mathcal{D} \subseteq \mathcal{G}$. If F satisfies monotonicity and no veto power, then it is fully Nash-implementable.

I will not provide a proof of this result here; for a proof, see the Maskin article cited, Repullo [1987], or Williams [1986]. *John* Moore and R. Repullo [1990] have provided a complete characterization of Nash implementability. You will probably have noted that, while the Maskin result quoted above shows that the monotonicity and no veto power conditions are together *sufficient* for full Nash implementation, only monotonicity has been shown to be necessary. Moore and Repullo have developed conditions which are both necessary and sufficient, and, even more remarkably, have developed necessary and sufficient conditions for full Nash implementability in the case where $n = 2$. These same two authors have also developed a very interesting and significant characterization of sub-game perfect implementation [1988]. Other references and suggestions for further reading are provided in the last section of this chapter, but for now let's turn to a mechanism which implement Pareto efficient allocations in Nash equilibrium for economies with public goods .

[11] Maskin conjectured, and provided a partial proof of this result in the working paper mentioned earlier. However, the first published proof of the result appears to have been in Williams [1986].

18.8 Nash Implementation with Public Goods

While the material to be devloped here could be done more generally, we will confine our discussion to a simple mechanism which will make use of essentially the same 2-goods model we considered in Section 16.5. That is, we suppose there are n consumers, one public good and one private, and, for the sake of convenience, we will suppose that the ith consumer's preferences can be represented by the utility function:

$$u_i(x_i, y),$$

where 'x_i' denotes the quantity of the private good available for agent i's consumption, and 'y' denotes the quantity of the public good (and thus, an admissible preference profile can be represented as a vector of utility functions, $\boldsymbol{u}$). We suppose that there is an aggregate endowment, $w \in \mathbb{R}_{++}$, of the private good (and, of course, none of the public good); and we will suppose that the technology for the production of the public good is delineated by a cost function (input-requirement function), $c\colon \mathbb{R}_+ \to \mathbb{R}_+$, where $c(0) = 0$, and where, for $y \in \mathbb{R}_+$, $c(y)$ is the amount of the private good which must be used to produce a unit of the public good. Thus, the production set for the public good is given by:

$$Y = \{(z, y) \in \mathbb{R}^2_+ \mid z = c(y)\}. \tag{18.51}$$

From a formal point of view, in our discussion here we will take the consumption sets $X_i = \mathbb{R}^2_+$, for $i = 1, \ldots, n$, the aggregate private goods endowment, $w \in \mathbb{R}_{++}$, and the cost function, $c\colon \mathbb{R}_+ \to \mathbb{R}_+$ as fixed (the **economic environment**). Defining $\mathcal{U}$ as the family of all continuous non-decreasing functions on $\mathbb{R}^2_+$ which are strictly increasing in the first argument, we will consider economies $E(\boldsymbol{u})$ of the form:

$$E(\boldsymbol{u}) = \big(\langle u_i \rangle, Y, w\big),$$

where $\boldsymbol{u} \in \mathcal{U}^n$, Y is defined in (18.51) and:

$$A(E) = \Big\{(\boldsymbol{x}, y) \in R^{n+1}_+ \mid \sum\nolimits_{i=1}^n x_i + c(y) \leq w\Big\}$$

In other words, for each profile, $\boldsymbol{u}$ in the admissible space, $\mathcal{U}^n$, we obtain a well-defined (two-commodity) economy by combining the n consumers whose preferences are represented by the functions u_i with the (fixed) cost function and aggregate private goods endowment.

The mechanism, $\mathcal{M} = \langle \boldsymbol{M}, X, g \rangle$, which we are going to present here is probably best thought of not as a single mechanism, but rather a family of mechanisms, the individual members of which are determined by a vector $\boldsymbol{a} \in \Delta_n$ which we will call a **cost share vector**, and a **wealth assignment vector**, $\boldsymbol{w} \in \mathbb{R}^n_+$. (It is basically a simplified version of mechanisms developed by Tian and Li [1994] and Corchon and Wilkie [1996].) Thus $\mathcal{M}(\boldsymbol{a}, \boldsymbol{w}) = \langle \boldsymbol{M}, X, g \rangle$ is given by:

$$M = \prod\nolimits_{i=1}^n M_i, \quad \text{with } M_i = \mathbb{R}, \text{ for } i = 1, \ldots, n, \tag{18.52}$$

$$X = A(E) \tag{18.53}$$

$$g(\boldsymbol{m}) = \big(x_1(\boldsymbol{m}), \ldots, x_n(\boldsymbol{m}), y(\boldsymbol{m})\big), \tag{18.54}$$

where:

$$y(\boldsymbol{m}) = (1/n)\Big(\sum_{i=1}^{n} m_i\Big), \text{ and :} \tag{18.55}$$

$$x_i(\boldsymbol{m}) = w_i - a_i c\big[y(\boldsymbol{m})\big], \text{ for } i = 1, \ldots, n; \tag{18.56}$$

and where $\boldsymbol{a} \in \Delta_n$, that is:

$$a_i \geq 0 \text{ and } \sum_{i=1}^{n} a_i = 1,$$

and $\boldsymbol{w} \in \mathbb{R}_+^n$ satisfies:

$$\sum_{i=1}^{n} w_i = w.$$

Given $\boldsymbol{m} \in \boldsymbol{M}$, consumer i's consumption of the private good will be given by (18.56), and the amount of the private good available for the production of the public good is given by:

$$z = \sum_{i=1}^{n} a_i c[y(\boldsymbol{m})] = c[y(\boldsymbol{m})]. \tag{18.57}$$

We will show that every *Nash equilibrium*, given $\boldsymbol{u} \in \mathcal{U}^n$, of a mechanism of this form is a ratio equilibrium for $E(\boldsymbol{u})$.

Suppose $\boldsymbol{m}^*$ is a Nash equilibrium for $\mathcal{M}(\boldsymbol{a}, \boldsymbol{w})$, given $\boldsymbol{u}$. Then defining:

$$y^* = y(\boldsymbol{m}^*), z^* = c(y^*), \text{ and } x_i^* = w_i - a_i c(y^*) \quad \text{for } i = 1, \ldots, n;$$

we see, since $\boldsymbol{m}^*$ is a Nash equilibrium, given $\boldsymbol{u}$, that if $i \in N$, and $m_i \in M_i = \mathbb{R}$, then:

$$u_i(x_i^*, y^*) \geq u_i\big(w_i - a_i c[y(m_i, \boldsymbol{m}_{-i})], y(m_i, \boldsymbol{m}_{-i})\big). \tag{18.58}$$

However, given any desired nonnegative amount, y, of the public good, agent i can cause production of the good to be set equal to y by sending the message:

$$m_i = ny - \sum_{j \neq i} m_j^*.$$

Consequently, it follows from (18.58) that, for all $y \in \mathbb{R}_+$, we have:

$$u_i(x_i^*, y^*) \geq u_i\big[w_i - a_i c(y), y\big];$$

and we see that (x_i^*, y^*) maximizes u_i subject to $x_i + a_i c(y) \leq w_i$. Furthermore, it follows from (18.57) that $(z^*, y^*) \in Y$, and from (18.58) $x_i^* \geq 0$; while from the definition of x_i^* we see that:

$$\sum_{i=1}^{n} x_i^* + c(y^*) = \sum_{i=1}^{n} [w_i - a_i c(y^*)] + c(y^*) = \sum_{i=1}^{n} w_i = w.$$

Consequently, the allocation $(\boldsymbol{x}^*, y^*)$ is feasible; and therefore, $(\boldsymbol{x}^*, y^*)$ is a cost share (ratio) equilibrium, given the cost share vector $\boldsymbol{a}$ and wealth assignment vector $\boldsymbol{w}$. Since we know that cost share equilibria are Pareto efficient, it follows that $(\boldsymbol{x}^*, y^*)$ is Pareto efficient.

We can almost prove the converse; for suppose $c(\cdot)$ is linear, $c(y) = \beta y$,, where $\beta > 0$, and let $\widehat{\mathcal{U}} \subseteq \mathcal{U}$ be the family of all utility functions in $\mathcal{U}$ such that u is quasi-concave. Then if $\boldsymbol{u} \in \widehat{\mathcal{U}}^n$, and $(\boldsymbol{x}^*, y^*)$ is a Pareto efficient allocation in $E(\boldsymbol{u})$ such that $y^* > 0$, it follows from Theorem 16.28 that there exist Lindahl prices $\boldsymbol{q}^* = (q_1^*, \ldots, q_n^*)$ and a wealth assignment vector, $\boldsymbol{w}$, such that $(\boldsymbol{x}^*, y^*, \boldsymbol{q}^*)$ is a Lindahl equilibrium for $E(\boldsymbol{u})$, given $\boldsymbol{w}$. However, since profits are maximized over Y at y^*, and Y is linear, we must have:

$$q^* y^* - \beta y^* = 0;$$

and thus (since $y^* > 0$) we must have $\boldsymbol{q}^* = \beta$, where:

$$q^* = \sum\nolimits_{i=1}^{n} q_i^*.$$

Consequently, if we define the cost share vector $\boldsymbol{a}$ by:

$$a_i = q_i^*/q^* = q_i^*/\beta \quad \text{for } i = 1, \ldots, n;$$

and $\boldsymbol{m}^*$ by:

$$m_i^* = y^* \quad \text{for } i = 1, \ldots, n,$$

it is easily shown that $\boldsymbol{m}^*$ is a Nash equilibrium for $\mathcal{M}(\boldsymbol{a}, \boldsymbol{w})$, given $\boldsymbol{u}$. (Notice, incidentally, that the proof of Theorem 16.28 sets $w_i = x_i^* + q_i^* y^*$, and thus $x_i^* + a_i c(y^*) = x_i^* + (q_i^*/\beta)\beta y^* = w_i$, for each i.)

It is interesting to contrast the results obtained here with the necessary condition for full Nash implementation (monotonicity) which was discussed in the previous section. Retaining the assumptions regarding the economic environment which were made in the previous paragraph; essentially that $c(y) = \beta y$ and that the admissible preference space is $\widehat{\mathcal{U}}$, consider the social choice correspondence $\mathbb{P}\colon \widehat{\mathcal{U}}^n \mapsto A(E)$ defined by:

$$\mathbb{P}(\boldsymbol{u}) = \big\{(\boldsymbol{x}, y) \in A(E) \mid (\boldsymbol{x}, y) \text{ is Pareto efficient for } E(\boldsymbol{u}) \;\&\; y > 0\big\}.$$

The correspondence $\mathbb{P}$ does *not* necessarily satisfy the monotonicity condition, as the following example demonstrates.

18.49 Example. Suppose $n = 2$, that $c(y) = y$, $w = 13/2$, and that $u_i, u_i^* \in \widehat{\mathcal{U}}$ are given by:

$$u_1(x_1, y) = x_1 + 4\sqrt{y} \text{ and } u_1^*(x_1, y) = x_1 + 3\sqrt{y},$$

while:

$$u_2(x_2, y) = x_2 + \sqrt{y} = u_2^*(x_2, y).$$

Then we see (I will leave you to verify this) that the allocation:

$$(\boldsymbol{x}^*, y^*) = (2, 1/2, 4) \in \mathbb{P}(\boldsymbol{u}^*), \text{ while } (2, 1/2, 4) \notin \mathbb{P}(\boldsymbol{u}); \tag{18.59}$$

in particular, the allocation $\boldsymbol{x}' = (1/8, 1/8)$ and $y' = 25/4$ is feasible and such that:

$$u_1(x_1', y') = 1/8 + 4\sqrt{25/4} = 1/8 + 10 > 2 + 4\sqrt{4} = 10 = u_1(x_1^*, y^*),$$

while:

$$u_2(x_2', y') = 1/8 + \sqrt{25/4} = 21/8 > u_2(x_2^*, y^*) = 1/2 + \sqrt{4} = 5/2.$$

Now, in order that $\mathbb{P}$ satisfy the monotonicity condition, there must exist $i \in \{1, 2\}$ and $(\boldsymbol{x}'', y'') \in A(E)$ such that:

$$u_i(x_i'', y'') > u_i(x_i^*, y^*), \text{ but } u_i^*(x_i^*, y^*) \geq u_i^*(x_i'', y''). \tag{18.60}$$

However, since $u_2 = u_2^*$, it is obvious that no allocation $(\boldsymbol{x}'', y'') \in A(E)$ can satisfy (18.60) for $i = 2$. On the other hand, the function $u_i[w - c(y), y] = 26/4 - y + 4\sqrt{y}$ is maximized at $y^* = 4$; so that if (x_1'', y'') satisfies the first inequality in (18.60), it must take the form:

$$(x_1'', y'') = (x_1'', 4),$$

where $x_1'' > x_1^* = 2$. However, we then have $u_1^*(x_1'', y'') = x_1'' + 3\sqrt{4} = x_1'' + 6 > 8 = u_1(x_1^*, y^*)$. Consequently, there exists no allocation $(\boldsymbol{x}'', y'') \in A(E)$ satisfying (18.60), and it follows that $\mathbb{P}$ does not satisfy the monotonicity condition. However, the mechanism discussed in this section fully implements $\mathbb{P}$! or does it? □

The answer to the question posed at the end of the example is that the *family* of mechanisms $\mathcal{M}(\boldsymbol{a}, \boldsymbol{w})$ fully implements $\mathbb{P}$ in Nash equilibria. That is, given any $\boldsymbol{a} \in \Delta_n$, and any wealth assignment vector for E, $\boldsymbol{w}$, if $\boldsymbol{m}^* \in \boldsymbol{M}$ is a Nash equilibrium for $\boldsymbol{u}$, given $\mathcal{M}(\boldsymbol{a}, \boldsymbol{w})$, then $g(\boldsymbol{m}^*) \in \mathbb{P}(\boldsymbol{u})$.[12] Conversely, if $(\boldsymbol{x}^*, y^*) \in \mathbb{P}(\boldsymbol{u})$, where $\boldsymbol{u} \in \widehat{\mathcal{U}}$, then there exists $\boldsymbol{a} \in \Delta_n$ and a wealth assignment vector for E, $\boldsymbol{w}$, such that $\boldsymbol{m}^* = (y^*, y^*, \ldots, y^*)$ is a Nash equilibrium for $\boldsymbol{u}$, given $\mathcal{M}(\boldsymbol{a}, \boldsymbol{w})$, and $g(\boldsymbol{m}^*) = (\boldsymbol{x}^*, y^*)$. This is probably what we should expect; it is probably unrealistic to hope to find a single well-defined mechanism which fully implements a (multi-valued) social choice correspondence.

Let me also emphasize a point touched upon in the last paragraph; namely, for a given $\boldsymbol{u} \in \mathcal{U}$ and a pair $(\boldsymbol{a}, \boldsymbol{w})$, there may be no Nash equilibrium for the mechanism $\mathcal{M}(\boldsymbol{a}, \boldsymbol{w})$. If there is a Nash equilibrium, it will be Pareto efficient, but only for the pairs $(\boldsymbol{a}, \boldsymbol{w})$ satisfying (assuming that each u_i is differentiable):

$$\frac{\partial u_i}{\partial y} \Big/ \frac{\partial u_i}{\partial x_i} = a_i c'(y^*) \text{ and } w_i - a_i c(y^*) \geq 0,$$

for y^* a Pareto efficient level of public goods production (the same level for each agent i), will a Nash equilibrium exist.

18.9 The Revelation Principle Reconsidered

It has become routine for authors attempting to develop a mechanism to solve a particular allocation problem to make a statement to the effect that "...making use of the revelation principle, we can, without loss of generality, confine our attention to direct revelation mechanisms." As you may remember, however, the Revelation

[12] I am using a bit of 'poetic license' here: in order to guarantee this membership, we would need to specify conditions on $\boldsymbol{u}$ guaranteeing that $y(\boldsymbol{m}^*) > 0$.

Principle, as formulated by Gibbard, states that, given a straightforward game form we can define a direct revelation mechanism from it which is SP. However, if one is considering a solution concept other than dominant strategies, it is not so clear that one can define an 'equivalent direct revelation mechanism' to implement a given social choice correspondence. In an interesting and important article, Repullo [1985] carefully analyzes the scope of the Revelation Principle. He begins by presenting an example of a mechanism, $\mathcal{M} = \langle \boldsymbol{M}, X, g\rangle$, for which the number of agents, n, is two, $\#X = 4$, and where each of the agents has two possible preference relations, and has access to three different strategies. He then presents a social choice correspondence (actually, a function), $F\colon \mathcal{D}^n \to X$, which is implemented in *dominant* strategies by $\mathcal{M}$. Defining a direct revelation mechanism, $\mathcal{M}^*$, from $\mathcal{M}$ in the way we did in the proof of the Gibbard-Satterthwaite Theorem, he finds that truth-telling is a (weakly) dominant strategy in the direct mechanism, but, from the point of view of the agents there are equivalent (non-truthful) strategies; which can result in outcomes which are *not* consistent with the given social choice correspondence.

What is the fundamental problem here? We can illustrate the basic difficulty with a skeletal example involving 2 agents, two possible preference relations for each (denoted by 'G_i^j' for $i, j = 1, 2$, where the subscript refers to the agent), and three strategies for each agent. Suppose further that for each i, j, s_i^j is a dominant strategy for G_i^j, but that, for example, for G_1^1, we have:

$$g(s_1^1, s_2^j) I_1^1 g(s_1^2, s_2^j),\ \text{for } j = 1, 2;\ \text{while } g(s_1^1, s_2^3) P_1 g(s_1^j, s_2^3),\ \text{for } j = 2, 3.$$

If we then define the mapping σ by $\sigma_i(G_i^j) = s_i^j$, and the outcome function h by:

$$h(G_1^j, G_2^k) = g[\sigma_1(G_1^j), \sigma_2(G_2^k)] = g(s_1^j, s_2^k),$$

truth-telling is only a weakly dominant strategy. For example, the strategy which makes s_1^1 dominant, as opposed to weakly dominant, for G_1^1 is s_2^3, which is never a dominant strategy for agent 2, and is thus irrelevant, insofar as the direct mechanism is concerned. But this means that agent i may always claim to have the preferences G_i^1 (there is no loss to the agent in doing so). However, it may well be the case that, say $h(G_1^1, G_2^1) = g(s_1^1, s_2^1)$ is *not* an element of $F(G_1^2, G_2^2)$; and thus the direct mechanism may not implement the given social choice correspondence in dominant strategies.

The following is a slightly modified version of a result which is quoted by Repullo, and which he credits to Dasgupta, Hammond, and Maskin [1979]. It provides what I believe to be the correct statement of the revelation principle, as it applies to implementation.

18.50 Theorem. *If* $\mathcal{M} = \langle \boldsymbol{M}, X, g\rangle$ *is a mechanism which implements (respectively, fully implements) the social choice correspondence,* $F\colon \mathcal{D}^n \mapsto X$ *in weakly dominant strategies, and we define the direct mechanism,* $\mathcal{M}^* = \langle \mathcal{D}^n, X, h\rangle$ *by:*

$$h(\boldsymbol{G}) = g[\boldsymbol{\sigma}(\boldsymbol{G})],$$

where $\boldsymbol{\sigma}\colon \mathcal{D}^n \to \boldsymbol{M}$ *satisfies:*

$$\sigma_i(G_i) \in D(G_i, \mathcal{M}) \quad \textit{for } i = 1, \ldots, n,$$

and:

$$D(G_i, \mathcal{M}) = \{s_i \in S_i \mid s_i \text{ is weakly dominant for } (G_i, \mathcal{M})\},$$

then $\mathcal{M}^*$ *truthfully implements (respectively, truthfully fully implements)* F *in weakly dominant strategies.*

The wording of the conclusion in this last theorem may be a bit confusing; what is meant by the basic statement are two things. First, for each $i \in N$, and each $G_i \in \mathcal{D}$, G_i is a weakly dominant strategy for i, given $(G_i, \mathcal{M}^*)$. Secondly, for each $\boldsymbol{G} \in \mathcal{D}^n$, $h(\boldsymbol{G}) \in F(\boldsymbol{G})$. If $\mathcal{M}$ implements F in *dominant* strategies, then the correspondence $D(\cdot)$ appearing in the above result is single-valued. However, it folllows from the Repullo example that we nonetheless cannot strengthen the conclusion of Theorem 18.50 to assert that $\mathcal{M}^*$ truthfully implements F in *dominant* strategies.

18.10 Notes and Suggestions for Further Reading

As mentioned in the introduction to this chapter, it was L. Hurwicz , who in two classic articles ([1960], [1972]) set the stage for the study of mechanism design, and raised the issue of incentive compatibility. Groves [1970, 1973], Clark [1971], and Vickrey [1961] were the first to introduce dominant strategy mechanisms. Groves and Ledyard [1977] were the first to develop a mechanism whose Nash equilibria resulted in Pareto efficient production levels of public goods. Other authors who should be mentioned in this connection are M. Walker, who, in his 1981 article developed an elegantly simple mechanism whose Nash equilibria are Lindahl equilibria; and Hurwicz [1979a, 1979b] develops very general results regarding implementation in Nash equilibria.

Some other articles, in addition to those already cited in this chapter, which I have thought to be of particular interest, and which deal with issues touched upon (or perhaps skirted) in this chapter are Barbera, Sonnenschein, and Zhou [1991] Jackson and Moulin [1992], , Jackson [1992], Kalai and Ledyard [1998], and Ledyard and Palfrey [2002].

For anyone beginning a serious study of this area, the surveys by Barbera [2001], Jackson [2001], Maskin and Sjöström [2002],[13] and *John* Moore [1992] are must reading; as are the surveys by Groves and Ledyard [1987] and Hurwicz [1986]. While I hope my indebtedness to these surveys is not too noticeable, it should certainly be acknowledged.

Exercises

1. Prove Lemma 18.12.

2. Complete the proof of Lemma 18.19. (Hint: to prove that $\succ$ is transitive, see if you can make use of Lemma 18.16.)

[13] Actually, the whole book, or at least parts 2 and 3 of the book in which this last article is contained (Arrow, Sen, and Suzumura [2002]) is pretty much must reading.

3. Prove the following: If $f\colon \mathcal{D}^m \to X$ is a strategy-proof voting rule, where $\mathcal{L} \subseteq \mathcal{D}$ and $\#X_f \geq 3$, and $y \in X_f$ and $\boldsymbol{P} \in \mathcal{D}^m$ are such that there exists $x \in r_f$ with:

$$xP_iy \quad \text{for } i = 1, \ldots, m,$$

then $f(\boldsymbol{P}) \neq y$. (**Hint.** Consider the ordering P^* defined by:

$$\begin{matrix} x \\ y \\ Z, \end{matrix}$$

where $Z = X \setminus \{x, y\}$.)

4. Prove Lemma 18.22. (Hint: it can be proved by constructions similar to those used in Lemma 18.15 and the proof that a dictator for the voting rule derived from a straightforward game form is also a dictator for the game form itself.)

5. Prove the following (see Gibbard [1973]).

Lemma. Let G be a weak order, let Q be a linear order on a nonempty set, X; and define the relation $\succ$ on X by:

$$x \succ y \iff \begin{cases} xPy & \text{or:} \\ xIy \;\&\; xQy, \end{cases} \tag{18.61}$$

where P and I are the asymmetric and symmetric parts of G, respectively. Then $\succ$ is a linear order on X.

6. Consider an economy with n consumers, one private good, and one public good; suppose that $X_i = \mathbb{R}^2_+$, and that the ith consumer's preferences can be represented by the utility function:

$$u_i(x_i, y) = x_i + \alpha_i \log y \quad \text{for } i = 1, \ldots, n,$$

where 'x_i' and 'y' denote the quantities of the private and public goods, respectively, and:

$$\alpha_i > 0 \quad \text{for } i = 1, \ldots, n.$$

Suppose further that the consumers have the initial endowments, $\boldsymbol{r}_i = (r_{i1}, 0),$, where r_{i1} is the initial endowment of the private good, and that the input-requirement function for the production of y is given by:

$$g(y) = \beta y,$$

where:

$$0 < \beta < \sum\nolimits_{i=1}^{n} \alpha_i.$$

Given this information:

a. Find the Pareto efficient quantity of the public good.
b. Find the Lindahl equilibrium for this economy.

c. Suppose now that the tax shares for the allocation of the cost of public goods production are given by:

$$t_1 = \frac{r_{k1}}{\sum_{i=1}^{m} r_{i1}} \quad \text{for } k = 1, \ldots, n.$$

What is the majority voting solution for the public goods equilibrium in this example?

7. Prove that the relation defined in Example 18.27.1 is a linear order which is single-peaked with respect to $\geq$.

8. Prove that the relation defined in Example 18.27.2 is a quasi-order which is single-peaked with respecto to $\geq$.

9. Suppose, in the context of the Bowen model of section 5, the consumers' utility functions take the quasi-linear form:

$$u_i(x_i, y) = x_i + \phi_i(y) \quad \text{for } i = 1, \ldots, m;$$

and that there exists a quantity of the public good, y^*, such that:

$$\phi_i'(y^*) \geq 0, \text{ for } i = 1, \ldots, n, \text{ and } \sum_{i=1}^{n} \phi_i'(y^*)/c'(y^*) = 1.$$

Show that if we then define the ith consumer's cost share, a_i, by:

$$a_i = \phi_i'(y^*)/c(y^*),$$

then the median voter solution is Pareto efficient.

10. Provide a sufficient condition for the median voter solution to public goods allocation to result in a Pareto efficient outcome. (Do not simply re-state the conditions of Exercise 9.)

11. Verify the claims made in Example 18.43.

12. Show that the preferences in the profile in Example 18.33 are single-peaked with respect to a common linear order.

Chapter 19

Appendix. Solutions for Selected Exercises

19.1 Chapter 1

Problem 4. This problem may be sufficiently easy as to need no explanation, but we will look at a sample of the reasoning. One needs to show that G is reflexive, total, and transitive. We'll consider the transitivity proof. If xGy and yGz, then, by definition of G:

$$f(x) \geq f(y) \text{ and } f(y) \geq f(z).$$

But then, by the transitivity of the usual inequality for the real numbers:

$$f(x) \geq f(z);$$

and thus, by defintion of G, xGz.

Problem 5. Once again the properties you are being asked to prove here may be so obvious that it's not apparent what it is that needs to be proved. However, it my be worthwhile to go through the proofs of antisymmetry and transitivity.

Accordingly, suppose that $\boldsymbol{x}, \boldsymbol{y} \in \mathbb{R}^n$ are such that:

$$\boldsymbol{x} \geq \boldsymbol{y} \text{ and } \boldsymbol{y} \geq \boldsymbol{x}.$$

Then, by definition of the weak inequality for $\mathbb{R}^n$, we see that for each i:

$$x_i \geq y_i \text{ and } y_i \geq x_i.$$

But then, since the weak inequality for the real numbers is antisymmetric, it follows that $x_i = y_i$. Since this equality holds for each i, we then conclude that $\boldsymbol{x} = \boldsymbol{y}$.

As to transitivity, suppose $\boldsymbol{x}, \boldsymbol{y}$ and $\boldsymbol{z}$ are elements of $\mathbb{R}^n$ such that:

$$\boldsymbol{x} \geq \boldsymbol{y} \text{ and } \boldsymbol{y} \geq \boldsymbol{z}.$$

Then, for each i, we have, by definition of the weak inequality for $\mathbb{R}^n$:

$$x_i \geq y_i \text{ and } y_i \geq z_i.$$

Therefore, since the weak inequality on $\mathbb{R}$ is transitive, we see that:

$$x_i \geq z_i;$$

and since this inequality holds for each i, we conclude that $\boldsymbol{x} \geq \boldsymbol{z}$.

Problem 8. It is very easy to show that the relation P must be irreflexive and asymmetric, in order that it satisfy the given inequality. Transitivity is a bit tougher, however; but it turns out that the given condition does not imply that P is transitive. To see this, consider the following example. Let $X = \{w, x, y, z\}$, and the relation P given by wPx and xPy (with no other comparisons). Then P is not transitive; however, consider the function defined in the following table:

element	**function value**
w	2
x	1
y	0
z	2

It is easy to show that this example and function satisfy the condition:

$$aPb \Rightarrow f(a) > f(b),$$

for $a, b \in X$, despite the fact that P is not transitive. Since P is not transitive, it is also not negatively transitive. (However, see Exercise 5, at the end of Chapter 3.)

Problem 14. In this example, P is asymmetric, but it is *not* transitive.

To see that P is asymmetric, notice that, if $\boldsymbol{x}P\boldsymbol{y}$, then whether or not either or both of $\boldsymbol{x}$ and $\boldsymbol{y}$ are elements of E, we must necessarily have:

$$\min\{x_1, x_2\} > \min\{y_1, y_2\};$$

and if this is the case, then from the definition of P, we cannot have $\boldsymbol{y}P\boldsymbol{x}$.

As to transitivity, consider for example the three points defined as follows:

$$\boldsymbol{x} = (5, 4), \boldsymbol{y} = (3, 3), \text{ and } \boldsymbol{z} = (1, 2).$$

Then, since $\boldsymbol{y} \in E$ and:

$$\min\{x_1, x_2\} = \min\{5, 4\} = 4 > 3 = \min\{y_1, y_2\},$$

it follows that $\boldsymbol{x}P\boldsymbol{y}$. Furthermore, once again using the fact that $\boldsymbol{y} \in E$, together with the fact that:

$$\min\{y_1, y_2\} = 3 > 1 = \min\{z_1, z_2\};$$

we see that, even though $\boldsymbol{z} \notin E$, we have $\boldsymbol{y}P\boldsymbol{z}$. However, we do not have $\boldsymbol{x}P\boldsymbol{z}$, since $\boldsymbol{x} \notin E$, and only points in E are preferred to $\boldsymbol{z}$. That is, since $\boldsymbol{z} \notin E$, $P\boldsymbol{z}$ is given by:

$$P\boldsymbol{z} = \{\boldsymbol{x}' \in E \mid \min\{x_1', x_2'\} > 1\};$$

and, since $\boldsymbol{x} \notin E$, we see that $\boldsymbol{x} \notin P\boldsymbol{z}$. We conclude, therefore, that P is *not* transitive.

19.2 Chapter 2

Problem 4. Since, for example:

$$u(\boldsymbol{y}) \leq u(\boldsymbol{y}) + \alpha\|\boldsymbol{x} - \boldsymbol{y}\| + \beta,$$

it is obvious that P is asymmetric.

To prove that P is transitive, suppose $\boldsymbol{x}, \boldsymbol{y}$, and $\boldsymbol{z}$ are such that $\boldsymbol{x}P\boldsymbol{y}$ and $\boldsymbol{y}P\boldsymbol{z}$. Then, by definition of P, we must have:

$$u(\boldsymbol{x}) > u(\boldsymbol{y}) + \alpha\|\boldsymbol{x} - \boldsymbol{y}\| + \beta, \tag{19.1}$$

and:

$$u(\boldsymbol{y}) > u(\boldsymbol{z}) + \alpha\|\boldsymbol{y} - \boldsymbol{z}\| + \beta. \tag{19.2}$$

However, by the triangle inequality, and the fact that $\alpha \geq 0$:

$$\alpha\|\boldsymbol{x} - \boldsymbol{z}\| \leq \alpha\|\boldsymbol{x} - \boldsymbol{y}\| + \alpha\|\boldsymbol{y} - \boldsymbol{z}\|.$$

Consequently, using (19.2) and (19.1) in turn:

$$\begin{aligned} u(\boldsymbol{z}) + \alpha\|\boldsymbol{x} - \boldsymbol{z}\| + \beta &\leq u(\boldsymbol{z}) + \alpha\|\boldsymbol{x} - \boldsymbol{y}\| + \alpha\|\boldsymbol{y} - \boldsymbol{z}\| + \beta \\ &< u(\boldsymbol{y}) + \alpha\|\boldsymbol{x} - \boldsymbol{y}\| \leq u(\boldsymbol{y}) + \alpha\|\boldsymbol{x} - \boldsymbol{y}\| + \beta < u(\boldsymbol{x}). \end{aligned}$$

Consequently, P is transitive.

19.3 Chapter 3

Problem 1. If we use the choice correspondence to derive the V relation, we see that this relation is as follows.

	$\boldsymbol{a}$	$\boldsymbol{b}$	$\mathbf{c}$	$\boldsymbol{d}$
$\boldsymbol{a}$	aVa	aVb	$\dots$	$\dots$
$\boldsymbol{b}$	bVa	bVb	bVc	$\dots$
$\mathbf{c}$	cVa	$\dots$	cVc	$\dots$
$\boldsymbol{d}$	$\dots$	dVb	$\dots$	$dVd.$

i. From the above relation, we can easily answer the first two questions, because it is easy to show that the relation V rationalizes h. For example, for B_1, a and b satisfy:

$$(\forall x \in B_1)\colon aVx \& bVx;$$

and there are no other elements in B_1. Thus we should find that $h(B_1) = \{a, b\}$, and indeed this is the case. For B_2, we have:

$$(\forall x \in B_2)\colon bVx,$$

but we have $\neg cVb$; and thus we should have $h(B_2) = \{b\}$, and this is the case.

The answer to the second question is then an immediate consequence, for notice that the V relation is reflexive. Therefore it follows from the definition that h is reflexive-reational in this case.

ii. To determine whether h is total rational we need to either show that no total relation can rationalize h, or we need to exhibit a total binary relation which does rationalize h. Since *any* binary relation which rationalizes h must extend V, the V relation itself provides a good starting point for trying to accomplish either of these goals. The first question which we should ask ourselves is: "Is V total?" If the answer to this question is "yes," then, since we have already shown that V rationalizes h, we are done; that is, it follows immediately that h is total-rational. Upon checking th V relation, however, we see that it is not toal; since, for example, $\neg aVd$ and $\neg dVa$. If we check out the rows of the matrix in which V is displayed in turn, we find only one other violation of the totality, or completeness condition: we have neither cVd nor dVc. Since any total relation, G, which rationalizes h must extend V, and in addition must have:

$$\text{either } aGd \text{ or } dGa \;\; \textit{and} \;\; \text{either } cGd \text{ or } dGc;$$

in principle, we need to check 9 possible binary relations and see if one of them rationalizes h. In practice, however, we can shortcut this process by checking each of the cells which needs to be filled in to see whether we get a contradiction by filling it in, and if so look at the mirror-image cell, and so on. I'll leave the details to you, and simply note here that you can show that the following relation *does* rationalize h, and thus h is total-rational.

	a	**b**	**c**	**d**
a	aGa	aGb	...	aGd
b	bGa	bGb	bGc	...
c	cGa	...	cGc	cGd
d	...	dGb	...	dGd.

iii. It is actually a bit easier to answer the next question: "Is h tansitive-rational?" because we know that this will be the case if, and only if, h satisfies the Congruence axiom. To check out whether h does indeed satisfy the Congruence Axiom, the first step is to use the V relation to construct the W relation. Doing this, we obtain the following.

	a	**b**	**c**	**d**
a	aWa	aWb	aWc	...
b	bWa	bWb	bWc	...
c	cWa	cWb	cWc	...
d	dWa	dWb	dWc	dWd.

[You should be sure to carry out this construction for yourself. Probably the easiest way to do this is to look at each blank cell of the V matrix in turn, and to see whether it would be filled in in the transitive closure of V. For example, we do not have aVc, but we do have aVb and bVc. Thus, by defnition, aWc. On the other hand, if we look down the d-column in the V matrix, we see that nothing dominates d under the V relation, except d itself. Thus we will not be able to trace out a 'V-chain' from a to d, and consequently we don't want to fill in the fourth cell in the first row.]

Having constructed the W relation, we now see that the Congruence Axiom is violated; because, for example, we have:

$$c \in B_2, \ b \in h(B_2), \ \& \ cWb, \ \text{but } c \notin h(B_2).$$

Since this shows that h does *not* satisfy the Congruence axiom, it follows immediately that h is *not* transitive-rational (Theorem 3.17) The answer to the next question is then immediatel, for in order that h be regular-rational, it must be transitive - rational.

Problem 2. Having arrived at this question, I should begin by admitting that the discussion of the solution given for Problem 1, above is *not* generally the most efficient way to proceed. In general, the most efficient way to proceed is to go through the questions asked in reverse order; since if h is regula-rational, it is also transitive-rational, total-rational, reflexive-rational, and rational. Thus the efficient way to proceed is to (a) construct the V relation, (b) use the V relation to construct the W relation and then (c) check the Congruence Axiom. If we can show that h satisfies the Congruence Axiom, then it follows immediately from Theorem 3.21 that we are effectively finished.

Accordingly, here you can establish that the V and W relations are respectively given by the following.

	a	**b**	**c**	**d**
a	aVa	aVb	...	aVd
b	...	bVb	...	bVd
c	cVa	...	cVc	cVd
d	...	...	...	...

	a	**b**	**c**	**d**
a	aWa	aWb	...	aWd
b	...	bWb	...	bWd
c	cWa	cWb	cWc	cWd
d	...	...	...	...

To check the Congruence Axiom, we look at the budget sets in turn:

1. B_1: the only point in B_1 satisfying xWa is a itself, which is in $h(B_1)$
2. B_2: the only point in B_2 satisfying xWb is b itself, which is in $h(B_2)$
3. B_3: the only point in B_3 satisfying xWc is c itself, which is in $h(B_3)$
4. B_4: the only point in B_4 satisfying xWa is a itself, which is in $h(B_4)$

Therefore, h satisfies the Congruence Axiom, and is thus regular-rational.

Problem 3. The V relation is as follows iin this case:

	$\boldsymbol{a}$	$\boldsymbol{b}$	$\boldsymbol{c}$	$\boldsymbol{d}$
$\boldsymbol{a}$	aVa	aVb	aVc	$\ldots$
$\boldsymbol{b}$	$\ldots$	bVb	bVc	$\ldots$
$\boldsymbol{c}$	cVa	cVb	cVc	$\ldots$
$\boldsymbol{d}$	dVa	dVb	dVc	dVd;

which yields the W relation:

	$\boldsymbol{a}$	$\boldsymbol{b}$	$\boldsymbol{c}$	$\boldsymbol{d}$
$\boldsymbol{a}$	aWa	aWb	aWc	$\ldots$
$\boldsymbol{b}$	bWa	bWb	bWc	$\ldots$
$\boldsymbol{c}$	cWa	cWb	cWc	$\ldots$
$\boldsymbol{d}$	dWa	dWb	dWc	dWd.

But then if we consider the budget set B_1, we have:

$$bWa \;\&\; a \in h(B_1), \quad \text{but} \quad b \notin h(B_1).$$

Therefore, h does *not* satisfy the Congruence Axiom; and thus it is *not* regular-rational.

However, consider the relation, P, defined as follows:

	$\boldsymbol{a}$	$\boldsymbol{b}$	$\boldsymbol{c}$	$\boldsymbol{d}$
$\boldsymbol{a}$	$\ldots$	aPb	$\ldots$	$\ldots$
$\boldsymbol{b}$	$\ldots$	$\ldots$	$\ldots$	$\ldots$
$\boldsymbol{c}$	$\ldots$	$\ldots$	$\ldots$	$\ldots$
$\boldsymbol{d}$	dPa	dPb	dPc	$\ldots$.

This relation is asymmetric and transitive; however, it is not negatively transitive. Nonetheless, it does motivate h. Since it then follows that h is asymmetric-transitive motivated, it now also follows that it is total-reflexive-rational (Thoerem 3.41).

Problem 4. You should have no trouble in showing that the (competitive) demand correspondence generated by the lexicographic order is given by:

$$\boldsymbol{h}(\boldsymbol{p}, w) = \{(w/p_1, 0)\} \quad \text{for } (\boldsymbol{p}, w) \in \mathbb{R}^2_{++} \times \mathbb{R}_+.$$

This demand correspondence is, oddly enough, representative-rational, since it is also the demand correspondence gnerated by the utility function:

$$u(\boldsymbol{x}) = x_1.$$

19.4 Chapter 4

Problem 1. It is easy to verify the fact that u is increasing, and thus that G is increasing. Notice, however, that if some $x_j = 0$, then an increase in the kth coordinate of $\boldsymbol{x}$ (for $k \neq j$) does not change the function value. Therefore $u(\cdot)$ (and thus G) is *not* strictly increasing on all of $\mathbb{R}^n_+$; although it is strictly increasing at all strictly positive values of $\boldsymbol{x}$.

It is easy to show (this is a Cobb-Douglas function, after all) that the demand function generated by G is:

$$h_j(\mathbf{p}, w) = (1/n)w/p_j \quad \text{for } j = 1, \ldots, n.$$

The function is homothetic, since it can be written in the form:

$$u(\boldsymbol{x}) = F[\sigma(\boldsymbol{x})],$$

where F is given by:

$$F(y) = y^n,$$

and:

$$\sigma(\boldsymbol{x}) = \prod_{j=1}^{n} x_j^{1/n};$$

and σ is positively homogeneous of degree one, while F is strictly increasing.

4. Given that $f \colon \mathbb{R}^n_+ \to \mathbb{R}_+$ is positively homogeneous of degree $\theta > 0$, define the functions g and F on $\mathbb{R}^n_+$ and $\mathbb{R}_+$, respectively, by:

$$g(\boldsymbol{x}) = \left[f(\boldsymbol{x})\right]^{1/\theta} \text{ and } F(y) = y^\theta.$$

Then, since f is positively homogeneous of degree θ, we have, for $\lambda > 0$ and $\boldsymbol{x} \in \mathbb{R}^n_+$:

$$g(\lambda\boldsymbol{x}) = \left[f(\lambda\boldsymbol{x})\right]^{1/\theta} = \left[\lambda^\theta f(\boldsymbol{x})\right]^{1/\theta} = \lambda g(\boldsymbol{x}).$$

Therefore, g is positively homogeneous of degree one; and, since F is obviously increasing, and for each $\boldsymbol{x} \in \mathbb{R}^n_+$, we have:

$$f(\boldsymbol{x}) = F[g(\boldsymbol{x})],$$

it follows that f is a homothetic function.

19.5 Chapter 5

Problem 1. With $w_i = (1/2)W$, for $i = 1, 2$, the individual demand functions for the first commodity are given by:

$$h_{11}[\mathbf{p}, (1/2)W] = (3/4)[(1/2)W]/p_1 = (3/8)W/p_1,$$

and:

$$h_{21}[\mathbf{p}, (1/2)W] = (1/4)[(1/2)W]/p_1 = (1/8)W/p_1,$$

respectively. Therefore, aggregate demand for the first commodity is given by:

$$h_1(\mathbf{p}, W) \equiv h_{11}[\mathbf{p}, (1/2)W] + h_{21}[\mathbf{p}, (1/2)W] = (1/2)W/p_1.$$

Simalrly, aggregate demand for the second commodity is given by:

$$h_2(\mathbf{p}, W) = (1/2)W/p_2.$$

It is easy to show that the aggregate demand function is that generated by a single consumer whose preferences are representable by the Cobb-Douglas utility function:

$$u(\boldsymbol{x}) = (x_1)^{1/2} \cdot (x_2)^{1/2}.$$

Problem 2.

a. The ith consumer will maximize utility at the point where:

$$x_{i2} = (a_{i2}/a_{i1})x_{i1}. \tag{19.3}$$

Substituting from (19.3) into the budget constraint:

$$p_1 x_{i1} + p_2\Big(\frac{a_{i2}}{a_{i1}}\Big)x_{i1} = \Big(\frac{p_1 a_{i1} + p_2 a_{i2}}{a_{i1}}\Big)x_{i1} = \boldsymbol{p} \cdot \boldsymbol{r}_i,$$

so that:

$$x_{i1} = \frac{a_{i1}\boldsymbol{p} \cdot \boldsymbol{r}_i}{p_1 a_{i1} + p_2 a_{i2}} \quad \text{for } i = 1, 2. \tag{19.4}$$

Substituting equation (19.3) into (19.4, we also obtain:

$$x_{i2} = \frac{a_{i2}\boldsymbol{p} \cdot \boldsymbol{r}_i}{p_1 a_{i1} + p_2 a_{i2}} \quad \text{for } i = 1, 2. \tag{19.5}$$

b. Since the preferences of both individuals are increasing, we can take commodity 2 as the numéraire, and set $p_2 = 1$. Using Walras' Law, it suffices to solve for equilibrium in the market for the first commodity, where (given the values specified) equilibrium requires that:

$$\frac{5p_1}{p_1 + 1} + \frac{10}{p_1 + 4} = r_{11} = 5. \tag{19.6}$$

Solving, we obtain:

$$p_1 = 2;$$

which, as is easily seen satisfies the conditions for a competitive equilibrium.

c. With the change, equation (19.6) becomes:

$$\frac{5p_1}{p_1 + 1} + \frac{10}{p_1 + 2} = r_{11} = 5. \tag{19.7}$$

If we solve this for p_1, we obtain $p_1 = 0$. Oddly enough, this is a competitive equilibrium; or more precisely, the triple $\big(\langle \boldsymbol{x}_i^* \rangle, \boldsymbol{p}^*\big)$ is a competitive equilibrium for $\mathcal{E}$, where:

$$x_{11}^* = x_{12}^* = 0, x_{21}^* = 5, x_{22}^* = 10, p_1^* = 0, p_2^* = 1.$$

Problem 3.

a. Here (using Lagrangian techniques), the demand functions for the first commodity are easily found to be:

$$x_{i1} = \frac{a_i \boldsymbol{p} \cdot \boldsymbol{r}_i}{p_1} \quad \text{for } i = 1, 2.$$

b. With the given values, equilibrium in the first market requires (setting $p_2 = 1$):

$$\frac{a_1 p_1 r_{11}}{p_1} + \frac{a_2 r_{22}}{p_1} = r_{11}.$$

If we solve this for p_1, we obtain:

$$p_1 = \frac{a_2 r_{22}}{(1 - a_1) r_{11}}.$$

It is easily verified that this value for p_1 (and with $p_2 = 1$) yields a competitive equilibrium.

c. From the above solution, we see that:

$$\frac{\partial p_1}{\partial r_{22}} = \frac{a_2}{(1 - a_1) r_{11}}.$$

Notice that it is positive, and is numerically larger the larger is a_2; that is:

$$\frac{\partial^2 p_1}{\partial a_2 \partial r_{22}} = \frac{1}{(1 - a_1) r_{11}} > 0.$$

I'll leave any further verbal elaboration on this answer to you.

Problem 4.

a. It *is* homothetic; for if:

$$\min\left\{\frac{x_{11}}{2}, x_{12}\right\} > \min\left\{\frac{\bar{x}_{11}}{2}, \bar{x}_{12}\right\},$$

and $\theta > 0$, then:

$$\min\left\{\frac{\theta x_{11}}{2}, \theta x_{12}\right\} = \theta \min\left\{\frac{x_{11}}{2}, x_{12}\right\} > \theta \min\left\{\frac{\bar{x}_{11}}{2}, \bar{x}_{12}\right\} = \min\left\{\frac{\theta \bar{x}_{11}}{2}, \theta \bar{x}_{12}\right\}.$$

b. It is convex (and thus weakly convex), but not strictly convex.

c. The first consumer will set $x_{12} = x_{11}/2$, and thus:

$$p_1 x_{11} + x_{11}/2 = 3p_1;$$

so that:

$$x_{11} = \frac{6p_1}{2p_1 + 1} \text{ and } x_{12} = \frac{3p_1}{2p_1 + 1}.$$

Similarly, the second consumer will set $x_{22} = 2x_{21}$, and from this and the budget constraint, we obtain:

$$x_{21} = \frac{3}{p_1 + 2} \text{ and } x_{22} = \frac{6}{p_1 + 2}.$$

To find equilibrium, we equate the aggregate demand and aggregate supply of the first comodity, obtaining:

$$x_{11} + x_{21} = \frac{6p_1}{2p_1 + 1} + \frac{3}{p_1 + 2} = 3,$$

and solve to obtain $p_1 = 1$. It follows from Walras' Law that $(\langle \boldsymbol{x}_i^* \rangle, \boldsymbol{p}^*)$ is a competitive equilibrium, with:

$$x_{11}^* = 2, x_{12}^* = 1, x_{21}^* = 1, x_{22}^* = 2, \ \& \ \boldsymbol{p}^* = (1,1).$$

Problem 5. Individual one has:

$$w_1 = (3/5) \cdot 1 = 3/5.$$

Obviously, with the given prices, any bundle which is semi-greater than $(1,0)$ will cost more than w. The only bundle which is preferred to $(1,0)$ and which is not semi-greater is the bundle $(0,2)$, and we have:

$$\boldsymbol{p}^* \cdot (0,2) = 4/5 > w_1.$$

Therefore, the bundle $\boldsymbol{x}_1^* = (1,0)$ maximizes preferences for consumer one, given the price vector $\boldsymbol{p}^* = (3/5, 2/5)$. Similarly, the only bundle which is both preferred to $\boldsymbol{x}_2^* \equiv (1,1)$ and not semi-greater than $\boldsymbol{x}_2^*$ is $\boldsymbol{x} = (2,0)$; but:

$$\boldsymbol{p}^* \cdot (2,0) = 6/5 > 1 = w_2.$$

Therefore, $(\boldsymbol{x}_1^*, \boldsymbol{x}_2^*, \boldsymbol{p}^*)$ *is* a competitive equilibrium. However, this equilibrium is *not* strongly Pareto efficient, since the allocation:

$$\widehat{\boldsymbol{x}}_1 = (0,1), \widehat{\boldsymbol{x}}_2 = (2,0),$$

is attainable and strictly Pareto dominates $(\boldsymbol{x}_1^*, \boldsymbol{x}_2^*)$. □

19.6 Chapter 6

Problem 1. To establish the first part, suppose $(\boldsymbol{v}, \boldsymbol{x}) \in \mathbb{R}^{m+q}$ is technogically feasible, and let θ be a nonnegative real number. Then, $\theta \boldsymbol{x} \in \mathbb{R}_+^q$, and since:

$$\boldsymbol{v} + \boldsymbol{A}\boldsymbol{x} \leq \boldsymbol{0},$$

it follows that:

$$\theta \boldsymbol{v} + \boldsymbol{A}(\theta \boldsymbol{x}) = \theta\big(\boldsymbol{v} + \boldsymbol{A}\boldsymbol{x}\big) \leq \boldsymbol{0},$$

as well.

A similar argument establishes that T is convex and additive.

iii. In order to characterize efficiency, we begin by noting that if $(\boldsymbol{v}, \boldsymbol{x})$ and $(\boldsymbol{v}^*, \boldsymbol{x}^*)$ satisfy:

$$(\boldsymbol{v}, \boldsymbol{x}) > (\boldsymbol{v}^*, \boldsymbol{x}^*), \tag{19.8}$$

we must have:

$$\boldsymbol{v} + \boldsymbol{A}\boldsymbol{x} > \boldsymbol{v}^* + \boldsymbol{A}\boldsymbol{x}^*. \tag{19.9}$$

To prove this, notice that if (19.8) holds, then either $\boldsymbol{v} > \boldsymbol{v}^*$ and $\boldsymbol{x} \geq \boldsymbol{x}^*$, or $\boldsymbol{v} \geq \boldsymbol{v}^*$ and $\boldsymbol{x} > \boldsymbol{x}^*$. But then we see that in the first case:

$$\boldsymbol{0} \geq \boldsymbol{v} + \boldsymbol{A}\boldsymbol{x} > \boldsymbol{v}^* + \boldsymbol{A}\boldsymbol{x} \geq \boldsymbol{v}^* + \boldsymbol{A}\boldsymbol{x}^*.$$

In the second case, since for each $j \in \{1, \ldots, q\}$ there exists $i \in \{1, \ldots, m\}$ such that $a_{ij} > 0$, we have:

$$\boldsymbol{A}(\boldsymbol{x} - \boldsymbol{x}^*) = \sum_{j=1}^{q} \boldsymbol{a}_{\cdot j}(x_j - x_j^*) > \boldsymbol{0}.$$

Therefore, in this case:

$$\boldsymbol{v} + \boldsymbol{A}\boldsymbol{x} > \boldsymbol{v} + \boldsymbol{A}\boldsymbol{x}^* \geq \boldsymbol{v}^* + \boldsymbol{A}\boldsymbol{v}^*.$$

Now suppose that $(\boldsymbol{v}^*, \boldsymbol{x}^*) \in T$ is *not* efficient. Then there exists $(\boldsymbol{v}, \boldsymbol{x}) \in T$ such that inequality (19.8), above, holds. But then we have by the argument of the above paragraph that:

$$\boldsymbol{0} \geq \boldsymbol{v} + \boldsymbol{A}\boldsymbol{x} > \boldsymbol{v}^* + \boldsymbol{A}\boldsymbol{x}^*.$$

Consequently, we see that if $(\boldsymbol{v}^*, \boldsymbol{x}^*) \in T$ *is* efficient, then we must have:

$$\boldsymbol{v}^* + \boldsymbol{A}\boldsymbol{x}^* = \boldsymbol{0}. \tag{19.10}$$

Conversely, if (19.10) holds, reversing the steps of the above argument yields a simple proof by contradiction.

iv. From the characterization of efficiency in part (iii) and part (i), we see that there will exist a profit-maximizing nonzero output, $(\boldsymbol{v}^*, \boldsymbol{x}^*) \in T$ only if:

$$\boldsymbol{p}^* \cdot \boldsymbol{x} + \boldsymbol{w}^* \cdot \boldsymbol{v}^* = 0;$$

and $\boldsymbol{v}^* + \boldsymbol{A}\boldsymbol{x}^* = \boldsymbol{0}$; so that:

$$\boldsymbol{p}^* \cdot \boldsymbol{x}^* - (\boldsymbol{w}^*)'\boldsymbol{A}\boldsymbol{x}^* = \big[(\boldsymbol{p}^*)' - (\boldsymbol{w}^*)'\boldsymbol{A}\big]\boldsymbol{x}^* = 0.$$

Thus we see that we must have:

$$\boldsymbol{p}^* = \boldsymbol{A}'\boldsymbol{w}^*.$$

Strictly speaking, the argument just given is woefully incomplete, but would be acceptable as an exam answer. The more complete answer would start from the fact that, since this technology satisfies constant returns to scale, we see that if $(\boldsymbol{p}^*, \boldsymbol{w}^*) \in \mathbb{R}_+^{q+m}$ is such that a profit maximizing choice exists, then it must be the case that, for all *efficient* (and feasible) pairs, $(\boldsymbol{v}, \boldsymbol{x}) \in \mathbb{R}_+^{m+q}$, we have:

$$\big[(\boldsymbol{p}^*)' - (\boldsymbol{w}^*)'\boldsymbol{A}\big]\boldsymbol{x} = 0.$$

Setting $\boldsymbol{x} = \boldsymbol{e}_j$, for each $j = 1, \ldots, q$ in turn,[1] we see that:

$$p_j^* = \boldsymbol{w}^* \cdot \boldsymbol{a}_{\cdot j},$$

for $j = 1, \ldots, q$.

The remaining part of this problem can be established by arguments quite similar to those used in Chapter 6 to establish these same properties for general linear production sets.

[1]Notice that the pair $(-\boldsymbol{A}\boldsymbol{e}_j, \boldsymbol{e}_j)$ is feasible, for $j = 1, \ldots, q$.

19.7 Chapter 7

Problem 1. We know that if we can find a price vector, $\boldsymbol{p}^*$, such that $(\boldsymbol{x}^*, \boldsymbol{y}^*, \boldsymbol{p}^*)$ is a competitive equilibrium, then it will follow that the allocation $(\boldsymbol{x}^*, \boldsymbol{y}^*)$ is Pareto efficient. Since $\boldsymbol{y}^* \neq \boldsymbol{0}$, we know that such a price vector (if one exists) must be a scalar multiple of:

$$\boldsymbol{p}^* = (1, 1).$$

With this price vector, we see that the consumer's wealth is equal to $w^* \equiv 2$. Solving for the consumer's demand (in essentially the same way as was done in Problem 2), We find that:

$$\boldsymbol{x} = (-2, 4) \stackrel{\text{def}}{=} \boldsymbol{x}^*;$$

and that:

$$\boldsymbol{x}^* = (-2, 4) = \boldsymbol{r} + \boldsymbol{y}^* = (0, 2) + (-2, 2).$$

Since $\boldsymbol{y}^* \in Y$, and $\boldsymbol{p}^* \cdot \boldsymbol{y}^* = 0$, it follows that $(\boldsymbol{x}^*, \boldsymbol{y}^*, \boldsymbol{p}^*)$ is a competitive equilibrium, and therefore, $(\boldsymbol{x}^*, \boldsymbol{y}^*)$ is Pareto efficient.

b. The allocation $(\boldsymbol{x}', \boldsymbol{y}')$ is actually feasible (you should check this); but is strictly less preferred than the feasible allocation $(\boldsymbol{x}^*, \boldsymbol{y}^*)$. Since the latter allocation is feasible, the former cannot be Pareto efficient. □

Problem 2. Since the two consumers have Cobb-Douglas utility functions, it is easy to establish that all of the assumptions of Theorem 7.27 of Chapter 7 are satisfied here. Consequently, if $(\boldsymbol{x}_i^*)$ is Pareto efficient, then there will exist $\boldsymbol{p}^* \in \mathbb{R}^2_+$ and a wealth distribution, $\boldsymbol{w}^* = (w_1^*, w_2^*)$ such that $((\boldsymbol{x}_i^*), \boldsymbol{p}^*)$ is a quasi-competitive equilibrium with the wealth distribution $\boldsymbol{w}^*$. Moreover, it can be shown (we'll discuss this in class), that at any such quasi-Walrasian equilibrium, we must have $\boldsymbol{p}^* \gg \boldsymbol{0}$.

Now, given the form of the utility functions, the consumers' demand functions will be given by:

$$x_{ij}(\boldsymbol{p}) = \frac{w_i}{2p_j} \quad \text{for } i = 1, 2; j = 1, 2.$$

Since we must have $\boldsymbol{p}^* \gg \boldsymbol{0}$, we can normalize to set $p_1^* = 1$, and equilibrium then requires the satisfaction of the following two equations:

$$x_{11}(\boldsymbol{p}^*) + x_{21}(\boldsymbol{p}^*) = \frac{w_1}{2} + \frac{w_2}{2} = w/2; \tag{19.11}$$

and:

$$x_{12}(\boldsymbol{p}^*) + x_{22}(\boldsymbol{p}^*) = \frac{w_1}{2p_2^*} + \frac{w_2}{2p_2^*} = w/2p_2^*, \tag{19.12}$$

where we define:

$$w^* = w_1 + w_2. \tag{19.13}$$

Solving (19.11) and (19.12), we see that equlibrium requires:

$$\boldsymbol{p}^* = (1, 1) \text{ and } w^* = \boldsymbol{p}^* \cdot (1, 1) = 2.$$

If, for an arbitrary values of w_1 and w_2 which satisfy (19.13), we define:

$$\theta = \frac{w_1}{w_1 + w_2} = \frac{w_1}{w^*},$$

we necessarily have $w_1 = \theta w^*$ and $w_2 = (1-\theta)w^*$; and it follows from equations (19.11)–(19.13) that the commodity bundles demanded by the two consumers are given by:

$$\boldsymbol{x}_1 = \Big(\frac{\theta w^*}{2}, \frac{\theta w^*}{2}\Big) = (\theta, \theta);$$

while:

$$\boldsymbol{x}_2 = \Big(\frac{(1-\theta)w^*}{2}, \frac{(1-\theta)w^*}{2}\Big) = (1-\theta, 1-\theta). \quad \square$$

Problem 3. We can actually answer part (a) most easily by first working out parts (b) and (c).

(b.) Clearly the function $\boldsymbol{w}$ satisfies $w_i(\boldsymbol{p}) \geq 0$ for and $\boldsymbol{p} \in \mathbb{R}^2_{++}$ and $i = 1, 2$. Moreover, for any $\boldsymbol{p} \in \mathbb{R}^2_{++}$, we have:

$$w_1(\boldsymbol{p}) + w_2(\boldsymbol{p}) = 10\theta(p_1 + p_2) + 10(1-\theta)(p_1 + p_2) = 10(p_1 + p_2) = \boldsymbol{p} \cdot \boldsymbol{r}.$$

Therefore $\boldsymbol{w}(\cdot)$ is a feasible wealth-assignment function for E.

(c.) If $\big((\boldsymbol{x}_i^*), \boldsymbol{p}^*\big)$ is a Walrasian equilibrium , we must have $\boldsymbol{p}^* \gg \boldsymbol{0}$, since u_1 is strictly increasing. But then we see that we must also have $x_{21}^* = x_{22}^*$. Since we have:

$$w_2(\boldsymbol{p}^*) = 10(1-\theta)(p_1^* + p_2^*),$$

we can solve for $\boldsymbol{x}_2^*$ by:

$$p_1^* x_{21}^* + p_2^* x_{21}^* = 10(1-\theta)(p_1^* + p_2^*);$$

and thus:

$$x_{21}^* = 10(1-\theta) = x_{12}^*.$$

Since $(\boldsymbol{x}_i^*)$ is feasible, we must then have:

$$x_{11}^* = 10 - 10(1-\theta) = 10\theta,$$

and similarly, $x_{12}^* = 10\theta$.

Turning now to part (a), it is easy to see that $(\boldsymbol{x}_i^*)$ is Pareto efficient if, and only if, there exists $\theta \in [0, 1]$ such that:

$$\boldsymbol{x}_1^* = \theta(10, 10) = \theta \boldsymbol{r} \;\text{ and }\; \boldsymbol{x}_2^* = (1-\theta)(10, 10) = (1-\theta)\boldsymbol{r}.$$

I will leave the details of the argument to you.

19.8 Chapter 8

Problem 1. a. If $\boldsymbol{x}_1 \in X_1$, then $\boldsymbol{x}_1 \geq (-2, 2)$, and thus:

$$\boldsymbol{p} \cdot \boldsymbol{x}_1 \geq \boldsymbol{p} \cdot (-2, 2);$$

so that we must have:

$$w_1 \geq \boldsymbol{p} \cdot (-2, 2) = 2(p_2 - p_1).$$

b. In this case, the consumer will wish to maximize x_{i1} subject to:

$$p_1 x_{i1} + p_2 x_{i2} \leq w_i \ \ \& \ \ -2 \leq x_{i1} \ \& \ 2 \leq \ x_{i2}.$$

Obviously at the solution, we will have:

$$x_{i2} = 2;$$

so that the maximization problem reduces to:

$$\max x_{i1} \text{ subject to: } p_1 x_{i1} \leq w_i - 2p_2 \ \& \ -2 \leq x_{i1} \leq 0.$$

It is then easy to see that the consumer's demand functions are here given by:

$$x_{i1}(\boldsymbol{p}, w_i) = \frac{w_i - 2p_2}{p_1}$$

and:

$$x_{i2}(\boldsymbol{p}, w_i) = 2,$$

for $2(p_2 - p_1) \leq w_i$,

c. Consider the problem:

$$\max x_{i2} \text{ subject to: } w_i - p_1 x_1 - p_2 x_2 \geq 0 \ \& \ -2 \leq x_1.$$

Clearly this is solved by setting $x_{i1} = -2$, and thus:

$$\boldsymbol{x}_i(\boldsymbol{p}, w_i) = \Big(-2, \frac{w_i + 2p_1}{p_2} \Big).$$

d. Since feasibility of aggregate consumption requires $x_2 \geq 4$, and the aggregate resource endowment of the second commodity is only 2 units, it is clear that, if a competitive equilibrium exists, it must involve positive production. Consequently, it follows from the form of the production set that any equilibrium price vector must be a positive scalar multiple of the the vector $\boldsymbol{p}^* = (1, 1)$. With this price vector and the given resource endowments, we will have:

$$w_i = \boldsymbol{p}^* \cdot \boldsymbol{r}_i = 1, \quad \text{for } i = 1, 2;$$

and thus, given the form of the demand functions found in parts (b) and (c), aggregate demand will be given by:

$$\boldsymbol{x}_1 + \boldsymbol{x}_2 = (-1, 2) + (-2, 3) = (-3, 5).$$

Consequently,

$$\boldsymbol{x}_1 + \boldsymbol{x}_2 - \boldsymbol{r} = (-3, 5) - (0, 2) = (-3, 3) \stackrel{\text{def}}{=} \boldsymbol{y}^*.$$

Since $\boldsymbol{y}^* \in Y$, and obviously maximizes profits in Y, given $\boldsymbol{p}^*$, it follows that $((\boldsymbol{x}_i^*), \boldsymbol{y}^*, \boldsymbol{p}^*)$ is a competitive equilibrium for $\mathcal{E}$. □

Problem 2.

a. Will need $w \geq 2(p_2 - p_1)$, as before.

b. The consumer will wish to set:

$$x_2 = 4 + x_1,$$

and thus:

$$p_1 x_1 + p_2(4 + x_1) = (p_1 + p_2)x_1 + 4p_2 = w,$$

so that:

$$x_1 = \frac{w - 4p_2}{p_1 + p_2},$$

and:

$$x_2 = 4 + \frac{w - 4p_2}{p_1 + p_2} = \frac{w + 4p_1}{p_1 + p_2}.$$

d. If an equilibrium exists, production must be positive, and the form of the production set then indicates that if production is efficient, so that $y_1 = -2y_2$, we must have:

$$p_1(-2y_2) + p_2 y_2 = (-2p_1 + p_2)y_2 \equiv 0,$$

and thus:

$$p_2 = 2p_1;$$

so that we can set:

$$p_1 = 1 \;\&\; p_2 = 2.$$

We will then have:

$$w = p_2 \cdot 1 = 2;$$

and therefore:

$$\boldsymbol{x} = \left(\frac{2-8}{3}, \frac{2+4}{3}\right) = (-2, 2).$$

If $\boldsymbol{y}$ is such that:

$$\boldsymbol{x} = \boldsymbol{r} + \boldsymbol{y},$$

we must then have:

$$\boldsymbol{y} = (-2, 2) - (0, 1) = (-2, 1).$$

Since the required value of $\boldsymbol{y}$ is a profit-maximizing element of Y, it follows that we have found a competitive equilibrium. □

Problem 3. This problem is very similar to the previous two problems. This time the competitive equilibrium occurs with:

$$\boldsymbol{x}^* = (-2, 8), \quad \boldsymbol{y}^* = (-2, 6), \text{ and } \boldsymbol{p}^* = (3, 1).$$

(Of course, any positive scalar multiple of the price vector given will also suffice.) I will leave the details to you.

Problem 4. (a) The simplest way to find the consumer's demand function is probably to substitute:

$$x_2 = \frac{w - p_1 x_1}{p_2},$$

into the utility function and maximize with respect to x_1. Doing this, we maximize the function:

$$v(x_1) = (x_1)^2 \cdot \Big(\frac{w - p_1 x_1}{p_2}\Big).$$

Setting the first derivative of $v(\cdot)$ equal to zero and solving, we obtain:

$$x_1 = \frac{2w}{3p_1} \text{ and } x_2 = \frac{w - p_1 x_1}{p_2} = \frac{w}{3p_2}.$$

(b) With a Cobb-Douglas utility function, it is obvious that any equilibrium will have to involve positive production. Given the form of Y, this implies that the price vector has to be a scalar multiple of $\boldsymbol{p}^* = (1,1)$. Substituting into the demand function (noting also that with $p_1^* = 1$, we will have $w = 24$), we obtain:

$$\boldsymbol{x}^* = (x_1^*, x_2^*) = (16, 8).$$

If this is to be matched by supply, we must have:

$$\boldsymbol{y}^* = \boldsymbol{x}^* - \boldsymbol{r} = (16,8) - (24,0) = (-8,8),$$

and this value for $\boldsymbol{y}^*$ is not only in the production set, we also have:

$$\boldsymbol{p}^* \cdot \boldsymbol{y}^* = 0,$$

and thus $\boldsymbol{y}^*$ maximizes profits on Y, given $\boldsymbol{p}^*$. Consequently, we have found a competitive equilibrium.

c. We have $24 - 16 = 8$.

5. It is convenient to first find the firm's supply function, and to do this, we begin by noting that (since the two consumers both have Cobb-Douglas utility functions, labor/leisure can be taken to be a numéraire good; so that we can normalize to set the wage equal to one (1). The firm's profit function can, therefore, be written as:

$$py - (y^2/4),$$

and with the usual simple calculus, we find that the firm's supply and profit functions are given by:

$$y = 2p \text{ and } \pi(p) = p^2,$$

respectively. The consumers then have the incomes:

$$w_i = 24 + (1/2)p^2,$$

for $i = 1, 2$; and thus aggregate demand for the produced good is given by:

$$y_1(p) + y_2(p) = (3/4)\Big(\frac{24 + p^2/2}{p}\Big) + (1/4)\Big(\frac{24 + p^2/2}{p}\Big) = \frac{48 + p^2}{2p}.$$

Setting this equal to supply, we have:

$$\frac{48 + p^2}{2p} = 2p;$$

from which we obtain $p = 4$.

By Walras' Law, the market for labor/leisure should also be equilibrated at this price; and, if you check, you will find that with $p = 4$, the consumers' aggregate demand for leisure is equal to:

$$(48 + p^2)/2 = 32;$$

meaning that the aggregate labor offer is equal to $32 - 48$, or $z = -16$. Since $2\sqrt{-(-16)} = 8 = 2p$, this confirms the fact that we have found the competitive equilibrium for the economy.

6. a. The key to the first part of this question is to consider the demand of consumer 1. By drawing diagrams (this is the easiest way), it is easy to see that if $p_1 > p_2$, then consumer one's demand will be of the form:

$$\boldsymbol{x}_1^* = (0, x_{12}^*),$$

with $x_{12}^* > 4$; which can't be an equilibrium. On the other hand, if $p_1 < p_2$, the first consumer's demand will be of the form:

$$\boldsymbol{x}_1^* = (x_{11}^*, 0),$$

with $x_{11}^* > 16$; and, since the second consumer (who has the same preferences) will demand:

$$\boldsymbol{x}_2^* = \boldsymbol{r}_2 = (16, 0),$$

this can't be an equilibrium either. However, with $p_1 = p_2$, say $\boldsymbol{p}^* = (1, 1)$, any point on consumer i's indiference curve through $\boldsymbol{r}_i$ will maximize utility, given $\boldsymbol{p}^*$. Consequently, with $\boldsymbol{p} = \boldsymbol{p}^*$, any allocation $\langle \boldsymbol{x}_i^* \rangle$ satisfying:

$$\boldsymbol{x}_1^* = (20 - x_{12}^*, x_{12}^*) \ \& \ \boldsymbol{x}_2^* = (12 + x_{12}^*, 4 - x_{12}^*) \ \& \ 0 \leq x_{12}^* \leq 4,$$

is such that $(\langle \boldsymbol{x}_i^* \rangle, \boldsymbol{p}^*)$ is a competitive equilibrium.

b. Since the production set is linear, profit-maximizing output can be non-zero only if the price vector, $\boldsymbol{p}$ is normal to the set; that is, iff it is a scalar multiple of $\widehat{\boldsymbol{p}} = (1, 2)$. However, given such a price vector, the firm's maximum profits will be zero, and the ith consumer's wealth will be given by:

$$\widehat{\boldsymbol{p}} \cdot \boldsymbol{r}_i;$$

and it is then easy to show that the consumers' demands will be given by:

$$\widehat{\boldsymbol{x}}_1 = (24, 0) \ \text{ and } \ \widehat{\boldsymbol{x}}_2 = (16, 0).$$

Net aggregate non-produced demand will then be given by:

$$\widehat{\boldsymbol{x}}_1 + \widehat{\boldsymbol{x}}_2 - \boldsymbol{r}_1 - \boldsymbol{r}_2 = (40, 0) - (32, 4) = (8, -4);$$

which is not a feasible production vector ($y_1 > 0$).

From the reasoning of the above paragraph, we can see that no Walrasian equilibrium exists in which production is positive. However, if we set $\boldsymbol{p}^* = (1, 1)$, then it is easily seen that the firm's maximum profit is achieved at $\boldsymbol{y}^* = \boldsymbol{0}$; and thus from our work in part (a), it follows that an equilbrium there is an equilbrium here with $\boldsymbol{y}^* = \boldsymbol{0}$.

19.9 Chapter 9

2. Since the production set is linear, we see that the producer prices, $\boldsymbol{p}$, must be given by:

$$\boldsymbol{p} = (1, 2, 1),$$

given that we normalize to set $w = 1$. The consumer price vector then becomes:

$$\boldsymbol{q} = (1, 2 + t_1, 1 + t_2);$$

yielding demands of:

$$x_0 = 8, x_1 = \frac{8}{2 + t_1}, \text{ and } x_2 = \frac{8}{1 + t_2}, \tag{19.14}$$

respectively. Consequently, we can write the consumer's indirect utility function as:

$$v(\boldsymbol{t}) = \frac{512}{(2 + t_1)(1 + t_2)};$$

and to find the optimal tax, we maximize this function subject to the constraint:

$$t_1 x_1 + t_2 x_2 = \frac{8t_1}{2 + t_1} + \frac{8t_2}{1 + t_2} = \boldsymbol{p} \cdot \boldsymbol{x}^g = 2 \cdot 3 + 2 = 8.$$

Forming the Lagrangian function in the usual way, taking first-order derivatives, and solving, we then obtain:

$$t_1 = 2 \text{ and } t_2 = 1. \tag{19.15}$$

Checking on this solution, we note that, with these taxes, the consumer's demand vector is given by:

$$\boldsymbol{x} = (8, 2, 4),$$

and the consumer's labor offer is -16. Adding government demand to the consumer's demand yields:

$$y_1 = 5 \text{ and } y_2 = 6.$$

Since:

$$-16 + 2 \cdot 5 + 6 = 0,$$

we then see that $\boldsymbol{y} \in Y$, and that $\boldsymbol{y}$ maximizes profits on Y, verifying our solution.

19.10 Chapter 10

19.11 Chapter 11

2. Since each consumer of type 1 has the same initial endowment, they each have the same budget set. Consequently, since each has the same preferences, we must have:

$$\boldsymbol{x}_{h1}^* I_i \boldsymbol{x}_{11}^*.$$

If $\boldsymbol{x}^*_{h1} \neq \boldsymbol{x}^*_{11}$, for some h, then, defining:

$$\overline{\boldsymbol{x}}_1 = (1/q)\sum_{h=1}^{q} \boldsymbol{x}^*_{h1},$$

we have $\overline{\boldsymbol{x}}_1 P_1 \boldsymbol{x}^*_{h1}$, for $h = 1, \ldots, q$ (since P_1 is strictly convex), and:

$$\boldsymbol{p}^* \cdot \overline{\boldsymbol{x}}_1 = w_1;$$

contradicting the assumption that $\boldsymbol{x}^*_{h1}$ is the bundle demanded by consumer $(h1)$.

3. Here you should easily be able to show that $(\langle \boldsymbol{x}^*_i \rangle_{i \in M}, \boldsymbol{p}^*)$ is a Walrasian equilibrium for $\mathcal{E}$, with:

$$\boldsymbol{x}^*_i = (5, 5) \quad \text{for } i = 1, 2,$$

and $\boldsymbol{p}^* = (1, 1)$. Since $\langle \boldsymbol{x}^*_i \rangle$ is a Walrasian allocation, it is in the core.

4. We begin by noting that if $\boldsymbol{x} \in \mathbb{R}^2_+$ is such that $u_1(\boldsymbol{x}_1) > u_1(\boldsymbol{r}_1)$, then we must have $\boldsymbol{x}_1 > (5, 5) = \boldsymbol{r}_1$. But then we would have:

$$u_2(10 - x_{11}, 10 - x_{12}) < u_2(\boldsymbol{r}_2).$$

Consequently, the only Pareto efficient allocation which is individually rational is $\langle \boldsymbol{x}_i \rangle = \langle \boldsymbol{r}_i \rangle$; that is, the core of this economy is just the initial endowment position.

5. We begin by noting that if $\boldsymbol{p} \in \Delta_2$, then the demands of the two consumers (I will leave it to you to verify this) are given by:

$$x_{11} = \frac{10p_1}{p_1 + p_2} \;\&\; x_{12} = \frac{10p_1}{p_1 + p_2},$$

and:

$$x_{21} = \frac{10p_2}{p_1 + p_2} \;\&\; x_{22} = \frac{10p_2}{p_1 + p_2},$$

respectively. Since, for example:

$$\frac{10p_1}{p_1 + p_2} + \frac{10p_2}{p_1 + p_2} = 10,$$

it follows that any allocation, $(\boldsymbol{x}_1, \boldsymbol{x}_2)$ satisfying:

$$x_{i1} = x_{i2} \quad \text{for } i = 1, 2, \tag{19.16}$$

and:

$$x_{11} + x_{21} = 10, \tag{19.17}$$

is a competitive (Walrasian) allocation; and thus, by Theorem 11.15, is in $\mathcal{C}(\mathcal{E})$.

Conversely, suppose $(\boldsymbol{x}^*_1, \boldsymbol{x}^*_2)$ is an attainable allocation for $\mathcal{E}$, but that:

$$x^*_{11} > x^*_{12}.$$

Then, since:

$$x^*_{11} + x^*_{21} = 10 = x^*_{12} + x^*_{22}, \tag{19.18}$$

we must have:

$$x_{11}^* - x_{12}^* = x_{22}^* - x_{21}^*. \tag{19.19}$$

Consider, then, the allocation $(\widehat{\boldsymbol{x}}_1, \widehat{\boldsymbol{x}}_2)$ defined by:

$$\widehat{x}_{11} = x_{11}^* - \frac{x_{11}^* - x_{12}^*}{2} \ \& \ \widehat{x}_{12} = x_{12}^* + \frac{x_{11}^* - x_{12}^*}{2},$$
$$\widehat{x}_{21} = x_{21}^* + \frac{x_{11}^* - x_{12}^*}{2} \ \& \ \widehat{x}_{22} = x_{22}^* - \frac{x_{11}^* - x_{12}^*}{2}.$$

You can readily check to verify the fact that:

$$u_i(\widehat{\boldsymbol{x}}_i) > u_i(\boldsymbol{x}_i^*) \quad \text{for } i = 1, 2,$$

and:

$$\widehat{\boldsymbol{x}}_{1j} + \widehat{\boldsymbol{x}}_{2j} = 10 \quad \text{for } j = 1, 2;$$

from which it follows that $(\boldsymbol{x}_1^*, \boldsymbol{x}_2^*)$ cannot be Pareto efficient. A symmetric argument establishes the fact that $(\boldsymbol{x}_1^*, \boldsymbol{x}_2^*)$ cannot be Pareto efficient if $x_{11}^* < x_{12}^*$, and thus it follows that the Pareto efficient set for $\mathcal{E}$ is exactly the set of allocations satisfying equations (19.16)–(19.18); that is, it coincides with $W(\mathcal{E})$. Since each of the allocations satisfying (19.16)–(19.18) is also individually rational, it now follows that $C(\mathcal{E})$ is exactly the set of allocations satisfying (19.16)–(19.18).

b. Since, in general, $W(\mathcal{E}) \subseteq C_q$ and $C_q \subseteq C(\mathcal{E})$ for $q = 1, 2, \ldots$, it now follows [since in this case $C(\mathcal{E}) = W(\mathcal{E})$] that:

$$C_q = C(\mathcal{E}) = W(\mathcal{E}) \quad \text{for } q = 1, 2, \ldots.$$

19.12 Chapter 12

7. In this case, the commodity space becomes $\mathbb{R}^{2G}$, and we will use the generic notation '$(\boldsymbol{x}_1, \boldsymbol{x}_2)$' to denote points the space, where:

$$\boldsymbol{x}_s \in \mathbb{R}^G \quad \text{for } s = 1, 2.$$

Now, if $(\boldsymbol{x}_1, \boldsymbol{x}_2)$ and $(\boldsymbol{x}_1^*, \boldsymbol{x}_2^*)$ are elements of $\mathbb{R}^{2G}$, and $\theta \in [0, 1]$, we will have:

$$\theta(\boldsymbol{x}_1, \boldsymbol{x}_2) + (1-\theta)(\boldsymbol{x}_1^*, \boldsymbol{x}_2^*) = (\theta\boldsymbol{x}_1 + (1-\theta)\boldsymbol{x}_1^*, \theta\boldsymbol{x}_2 + (1-\theta)\boldsymbol{x}_2^*);$$

and therefore:

$$\begin{aligned}
U\Big[\theta(\boldsymbol{x}_1, \boldsymbol{x}_2) &+ (1-\theta)(\boldsymbol{x}_1^*, \boldsymbol{x}_2^*)\Big] \\
&= \pi_1 u\big[\theta\boldsymbol{x}_1 + (1-\theta)\boldsymbol{x}_1^*\big] + \pi_2 u\big[\theta\boldsymbol{x}_2 + (1-\theta)\boldsymbol{x}_2^*)\big] \\
&\geq \pi_1\big[\theta u(\boldsymbol{x}_1) + (1-\theta)u(\boldsymbol{x}_1^*)\big] + \pi_2\big[\theta u(\boldsymbol{x}_2) + (1-\theta)u(\boldsymbol{x}_2^*)\big] \\
&= \theta\big[\pi_1 u(\boldsymbol{x}_1) + \pi_2 u(\boldsymbol{x}_2)\big] + (1-\theta)\big[\pi_1 u(\boldsymbol{x}_1^*) + \pi_2 u(\boldsymbol{x}_2^*)\big] \\
&= \theta U(\boldsymbol{x}_1, \boldsymbol{x}_2) + (1-\theta)U(\boldsymbol{x}_1^*, \boldsymbol{x}_2^*).
\end{aligned}$$

The generalization is straightforward.

19.13 Chapter 13

19.14 Chapter 14

5. Simple majority rule satisfies the stated conditions.

7. To prove Condorcet-consistency, notice that the highest possible Copeland score is $n-1$, if n is the number of elements in X. Furthermore, an alternative will have a Copeland score equal to $n-1$ if, and only if, it is a Condorcet winner. Anonymity and neutrality are even more obvious, and I'll leave the formal proofs of these properties to you.

19.15 Chapter 15

19.16 Chapter 16

1. a. Here the Samuelson conditions require:

$$\frac{3}{\sqrt{y}}+\frac{5}{\sqrt{y}}=1;$$

so that the Pareto efficient quantity of the public good is $y=64$.

b. We can obtain the required Lindahl prices by evaluating the individuals' marginal-willingness-to-pay at the optimal quantity of the public good. Doing this, we obtain:

$$\frac{3}{\sqrt{y}}=3/8=q_1,$$

and similarly:

$$q_2=\frac{5}{\sqrt{y}}=5/8.$$

You can easily verify that these prices support a Lindahl equilibrium at the optimal value of y.

2. Here the Samuelon conditions are given by:

$$\sum_{i=1}^{m}\frac{\partial u_i}{\partial y}\Big/\frac{\partial u_i}{\partial x_i}=\sum_{i=1}^{m}\frac{\partial u_i}{\partial y}=\frac{1}{\sqrt{y}}\sum_{i=1}^{m}\beta_i=\frac{40}{\sqrt{y}}=\frac{\partial z}{\partial y},$$

while the input requirement function is given by $z=4y$. Therefore, Pareto efficiency requires:

$$\frac{40}{\sqrt{y}}=4,$$

or $y=100$.

4. a. Here the Samuelson conditions are given by:

$$\sum_{i=1}^{m}\frac{\partial u_i/\partial y}{\partial u_i/\partial x_i}=\sum_{i=1}^{m}\frac{\beta_i}{\sqrt{y}}=g'(y)=9\sqrt{y}.$$

Solving, we obtain $y = 9$.

b. If we set a price of p_i per unit of consumer i's demand for the public good, i's maximization problem becomes (normalizing, with $p_2 = 1$):

$$\max_{w.r.t. x_i, y} x_i + 2\beta_i\sqrt{y},$$

subject to:

$$r_{i2} - x_i - p_i y = 0.$$

Solving, we see that i's demand for the public good is given by:

$$y = (\beta_i/p_i)^2.$$

Thus, if we are to obtain a demand of $y = 9$, we must set:

$$p_i = \beta_i/3.$$

However, we must also check to see whether we have marginal cost of production equal to the sum of these prices; that is, we need:

$$g'(9) = 9\sqrt{9} = 27,$$

to be equal to:

$$\sum_{i=1}^{m} p_i = \sum_{i=1}^{m} (\beta_i/3) = 81/3 = 27,$$

as required.

Since the Pareto efficient allocation is unique, and the Lindahl prices supporting this equilibrium are unique, it follows that, yes, there is a unique Lindahl equilibrium in this case.

19.17 Chapter 17

2. a. The competitive outputs for the two firms are:

$$x = 20 \text{ and } y = 100.$$

b. We can find the socially optimal output in one of two ways.

Method 1. If we consider the problem of maximizing the sum of profits of the two firms:

$$\begin{aligned}\max_{w.r.t. x,y} \pi_1(x) + \pi_2(x,y) &= p_1 x - c_1(x) + p_2 y - c_2(x,y) \\ &= 20x - (1/2)x^2 + 10y - (1/20)y^2 - x^2,\end{aligned}$$

we obtain, as the solution:

$$x = 20/3 \text{ and } y = 100.$$

Method 2. Here we solve for the two profit (benefit) functions as functions of the externality (x). Doing so yields:

$$\phi_1(x) = p_1 x - c_1(x) = 20x - (1/2)x^2,$$

and:

$$\phi_2(x) = 5(p_2)^2 - x^2 = 500 - x^2.$$

Differentiating the two benefit functions, and setting:

$$\phi_1'(x) = -\phi_2'(x),$$

we obtain the same solution as before.

19.18 Chapter 18

6. a. Here the Samuelson condition is:

$$\sum_{i=1}^{m} \frac{\partial u_i/\partial y}{\partial u_i/\partial x_i} = \sum_{i=1}^{m} \alpha_i/y = \beta,$$

which yields the solution:

$$y = \frac{\sum_{i=1}^{m} \alpha_i}{\beta}.$$

b. To obtain a Lindahl equilibrium, we consider the individual maximization problem:

$$\max_{w.r.t. x_i, y} \; x_i + \alpha_i \log y,$$

subject to:

$$r_{i2} - x_i - p_i y = 0.$$

Solving yields:

$$p_i = \alpha_i/y.$$

Setting

$$y = (1/\beta) \sum_{i=1}^{m} \alpha_i,$$

we obtain:

$$p_i = \frac{\alpha_i \beta}{\sum_{k=1}^{m} \alpha_k};$$

and, since the sum of the p_i's obviously equals β, it follows that we have obtained the Lindahl equilibrium.

c. Given that consumer i is paying the tax:

$$t_i = \left(\beta r_{i2} / \sum_{k=1}^{m} r_{k2}\right),$$

his/her indirect utility is given by:

$$v_i(y; t_i) = r_{i2} - \left(\frac{r_{i2}\beta}{r}\right) y + \alpha_i \log y,$$

where we have defined:

$$r = \sum_{k=1}^{m} r_{k2}.$$

It follows that:

$$\frac{\partial v_i}{\partial y} = -\frac{r_{i2}\beta}{r} + \frac{\alpha_i}{y},$$

is nonnegative if, and only if:

$$y < \frac{\alpha_i r}{\beta \cdot r_{i2}}.$$

Now suppose that:

$$\frac{\alpha_1}{r_{12}} < \frac{\alpha_2}{r_{22}} < \cdots < \frac{\alpha_m}{r_{m2}},$$

and that $n = 2q + 1$, for some positive integer, q. Then if we imagine votes to be taken on increasing x, with x increased as long as a majority favor an increase, the quantity of x produced will be equal to:

$$y = \frac{\alpha_k r}{\beta r_{k2}},$$

with:

$$k = q + 1.$$

References

Afriat, Sidney N. [1967]: The Construction of a Utility Function from Expenditure Data,' *International Economic Review;* 8, 67–77.

Afriat, Sidney N. [1973]: On a System of Inequalities in Demand Analysis: An Extension of the Classical Method,' *International Economic Review;* 14, 460–72.

Aivazian, Varouj A., and J. L. Callen [1981]: 'The Coase Theorem and the Empty Core,' *Journal of Law and Economics;* 24; 174–81.

Allen, R. G. D. [1933]: 'The Marginal Utility of Money and its Application, *Economica;* 13, 186–209.

Anderson, Robert M. [1978]: 'An Elementary Core Equivalence Theorem,' *Econometrica;* 46, 1483–87.

Anderson, Robert M. [1986]: 'Notions of Core Convergence,' *in:* Hildenbrand, Werner, and A. Mas-Colell, eds.: *Contributions to Mathematical Economics in Honor of Gerard Debreu.* North-Holland (Chapter 2), 25–46.

Antonelli, Giovanni B. [1886]: *Sulla teoria matematica della Economia politica.* Pisa. (English translation, 'On the Mathematical Theory of Political Economy,' *in:* Chipman *et al* [1971] pp. 333–64.

Armstrong, W. E. [1939]: 'The Determinateness of the Utility Function,' *Economic Journal;* 49, 453–67.

Arrow, Kenneth J. [1950]: 'A Difficulty in the Concept of Social Welfare,' *Journal of Political Economy;* 58, 328–46.

Arrow, Kenneth J. [1951a]: 'An Extension of the Basic Theorems of Classical Welfare Economics,' *in:* Neyman, Jerzy, ed.: *Proceedings of the Second Berkeley Symposium on Mathematical Statistics and Probability.* University of California Press, pp. 507–32. [Reprinted in Newman, Peter, ed.: *Readings in Mathematical Economics, Vol. I.* The Johns Hopkins Press, 1968, pp. 365–90.]

Arrow, Kenneth J. [1951b]: *Social Choice and Individual Values.* J. Wiley.

Arrow, Kenneth J. [1964]: 'The Role of Securities in the Optimal Allocation of Risk Bearing,' *Review of Economic Studies;* 31, 91–6. (Translation of: 'Le rôle de valeur doursièrs pour la répartition de la meilleure des risques,' *Econométrie*, Paris [1953]).

Arrow, Kenneth J. [1959]: 'Rational Choice Functions and Orderings,' *Economica;* 102, 121–7.

Arrow, Kenneth J. [1963]: *Social Choice and Individual Values, 2nd edition.* J. Wiley.

Arrow, Kenneth J., H. D. Block, and L. Hurwicz [1959]: 'On the Stability of Competitive Equilibrium II,' *Econometrica;* 27, 82–109.

Arrow, Kenneth J., and G. Debreu [1954]: "Existence of an Equilibrium for a Competitive Economy," *Econometrica;* 22, 265–90.

Arrow, Kenneth J., and F. H. Hahn [1971]: *General Competitive Analysis.* Holden-Day.

Arrow, Kenneth J., and L. Hurwicz [1958]: 'On the Stability of Competitive Equilibrium I,' *Econometrica;* 26, 522-52.

Arrow, Kenneth J., A. K. Sen, and K. Suzumura, eds [2002]: *Handbook of Social Choice and Welfare, Vol. 1.* North-Holland/Elsevier.

Auerbach, Alan J., and J. R. Hines, Jr. [2002]: 'Taxation and Economic Efficiency,' *in:* Auerbach, Alan J., and M. Feldstein, eds.: *Handbook of Public Economics, Vol 3;* North-Holland, pp. 1347–1421.

Aumann, Robert J. [1964]: 'Markets with a Continuum of Traders,' *Econometrica;* 32, 39–50.

Ballinger, T. Parker, and N. T. Wilcox [1997]: 'Decisions, Error, and Heterogeneity,' *Economic Journal;* 107, 1090–105.

Barbera, Salvador [2001]: 'An Introduction to Strategy-Proof Social Choice Functions,' *Social Choice and Welfare;* 18, 619–53.

Barbera, Salvador, and B. Peleg [1990]: 'Strategy-Proof Voting Schemes with Continuous Preferences,' *Social Choice and Welfare;* 7, 31–8.

Barbera, Salvador, H. Sonnenschein, and L. Zhou [1991]: 'Voting by Committees,' *Econometrica;* 59, 595–609.

Barone, Enrico [1908]: 'Il ministerio della produzione nello stato colletivista,' *Giornale degli Economisti;* 2, 37; 267–93, 391–414. English translation: 'The Ministry of Production in the Collectivist State,' *in:* Hayek, F. A., ed.: *Collectivist Economic Planning.* Routledge & Kegan Paul Ltd., 1935, pp. 245–90. [this is also reprinted in the volume edited by Peter Newman which is cited below.]

Bass, Frank M., E. A. Pessemier, and D. R. Lehmann [1972]: 'An Experimental Study of the Relationships between Attitudes, Brand Preference, and Choice,' *Behavioral Science;* 17, 532–41.

Bassett, Lowell, J. Maybee, and J. Quirk [1968]: 'Qualitative Economics and the Scope of the Correspondence Principle,' *Econometrica;* 36, 544–63.

Baumol, William J. and W. E. Oates [1988]: *The Theory of Environmental Policy,* 2nd ed. Cambridge University Press.

Bell, David E., H. Raiffa, and A. Tversky, eds. [1988a]: *Decision Making: Descriptive, Normative, and Prescriptive Interactions.* Cambridge University Press.

Bell, David E., H. Raiffa, and A. Tversky, eds. [1988b]: 'Descriptive, Normative, and Prescriptive Interactions in Decision-Making,' pp. 9–32 *in:* Bell, Raiffa, and Tversky [1988a], above.

Bentham, Jeremy [1789]: *An Introduction to the Principles of Morals and Legislation.* Payne, London. Republished in 1907 by Clarendon Press, Oxford.

Berge, Claude [1963]: *Topological Spaces.* MacMillan, New York.

Bergson (Burk), Abram [1938]: 'A Reformulation of Certain Aspects of Welfare Economics,' *Quarterly Journal of Economics;* 52, 310–34.

Bergstrom, Theodore [1976]: 'How to Discard 'Free Disposability'—At No Cost,' *Journal of Mathematical Economics;* 3, 131–4.

Bergstrom, Theodore, L. Blume, and H. Varian [1986]: 'On the Private Provision of Public Goods,' *Journal of Public Economics;* 5, 131–8.

Black, Duncan [1958]: *The Theory of Committees and Elections.* Cambridge University Press.

Blackorby, Charles, and P. Davidson [1991]: 'Implicit Separability: Characterization and Implications for Consumer Demands,' *Journal of Economic Theory;* 55, 364–99.

Blackorby, Charles, and D. Donaldson [1990]: 'The Case Against the Use of the Sum of Compensating Variations in Cost-Benefit Analysis,' *Canadian Journal of Economics;* 23, 3; 471—-93.

Blackorby, Charles, D. Primont, and R. Russell [1975]: *Duality, Separability, and Functional Structure: Theory and Economic Applications.* North-Holland.

Blackorby, Charles, and R. R. Russell [1994]: 'The Conjunction of Direct and Indirect Separability,' *Journal of Economic Theory;* 62, 480–98.

Blackorby, Charles, and R. R. Russell [1997]: 'Two-Stage Budgeting: An Extension of Gorman's Theorem,' *Economic Theory;* 9, 185–93.

Blau, Julian H. [1957]: 'The Existence of Social Welfare Functions,' *Econometrica;* 25, 302-13.

Blau, Julian H. [1972]: 'A Direct Proof of Arrow's Theorem,' *Econometrica;* 40, 61-67.

de Borda, Jean-Charles [1781]: 'Memoire sur les élections au scrutin," *Mémoires de l'Acadëmie Royale des Sciences;*, pp. 657–65.

Borglin, P. A., and H. Keiding [1976]: "Existence of Equilibrium Actions and Equilibrium: A Note on the 'New' Existence theorems," *Journal of Mathematical Economics;* 3, 313–16.

Bowen, Howard [1943]: 'The Interpretation of Voting in the Allocation of Economic Resources,' *Quarterly Journal of Economics;* 58, 27–48.

Campbell, Donald E. [1987]: *Resource Allocation Mechanisms.* Cambridge University Press.

Chipman, John S., L. Hurwicz, M. K. Richter, and H. R. Sonnenschein, eds. [1971]: *Preferences, Utility, and Demand.* Harcourt Brace Jovanovich, Inc.

Chipman, John S., and J. C. Moore [1971]: 'The Compensation Principle in Welfare Economics,' *in:* Zarley, Arvid, ed.: *Papers in Quantitative Economics, 2* (pp. 1–78) University Press of Kansas.

Chipman, John S., and J. C. Moore [1973]: 'Aggregate Demand, Real National Income, and the Compensation Principle,' *International Economic Review;* 14, 1; 153–81.

Chipman, John S., and J. C. Moore [1976a]: 'The Scope of Consumer's Surplus,' *in:* Tang, A. M., F. M. Westfield, and J. S. Worley, eds. *Evolution, Welfare and Time in Economics* (D. C. Heath and Co.), pp. 69–123.

Chipman, John S., and J. C. Moore [1976b]: 'Why an Increase in GNP Need Not Imply an Improvement in Potential Welfare,' *Kyklos;* 29, 66–85.

Chipman, John S., and J. C. Moore [1977]: 'Continuity and Uniqueness in Revewaled Preference,' *Journal of Mathematical Economics;* 4, 139–62.

Chipman, John S., and J. C. Moore [1979]: 'On Social Welfare Functions and the Aggregation of Preferences,' *Journal of Economic Theory;* 21, 111–39.

Chipman, John S., and J. C. Moore [1980]: 'Compensating Variation, Consumer's Surplus, and Welfare,' *American Economic Review;* 70, 933–49.

Chipman, John S., and J. C. Moore [1990]: 'Acceptable Indicators of Welfare Change, Consumer's Surplus Analysis and the Gorman Polar Form,'*in:* Chipman, John S., D. McFadden, and M. K. Richter, eds: *Preferences Uncertainty, and Optimality: Essays in Honor of Leonid Hurwicz,* Westview Press), pp. 68–120.

Chipman, John S., and J. C. Moore [1994]: 'The Measurement of Aggregate Welfare,' *in:* Eichhorn, Wolfgang, ed.: *Models and Measurement of Welfare and Inequality,* Springer-Verlag, pp. 552–92.

Clarke, Edward H. [1971]: 'Multipart Pricing of Public Goods,' *Public Choice;* 11, 17–33.

Coase, Ronald [1960]: 'The Problem of Social Cost,' *The Journal of Law and Economics;* 3, 1–44.

Condorcet, Marquis de [1785]: *Essai sur l'application de l'analyse à la probabilité des decisions rendues à la pluralité des voix.* Paris.

Coombs, Clyde H., R. M. Dawes, and A. Tversky [1970]: *Mathematical Psychology: An Elementary Introduction.* Prentice-Hall.

Corchon, Luis, and S. Wilkie [1996]: 'Double Implementation of the Ratio Correspondence by a Market Mechanism,' *Ecoomic Design;* 2, 325–37.

Dasgupta, Partha, P. Hammond, and E Maskin [1979]: 'The Implementation of Social Choice Rules: Some General Results on Incentive Compatibility,' *Review of Economic Studies;* 46, 185–216.

Debreu, Gerard [1954]: 'Valuation Equilibrium and Pareto Optimum,' *Proceedings of the National Academy of Sciences of the U. S. A.;* 40, 588–92.

Debreu, Gerard [1959] *Theory of Value.* John Wiley & Sons (re-published by Yale University Press.)

Debreu, Gerard [1974]: 'Excess Demand Functions,' *Journal of Mathematical Economics;* 1, 15–23.

Debreu, Gerard [1982]: "Existence of Competitive Equilibrium," *in:* Arrow, Kenneth J., and M. D. Intriligator, eds.: *Handbook of Mathematical Economics,* Volume II. North-Holland.

Debreu, Gerard, and H. Scarf [1963]: 'A Limit Theorem on the Core of an Economy,' *International Economic Review;* 4 (September), 235–46.

Diamantaras, Dimitrios, and S. Wilkie [1992]: 'A Generalization of Kaneko's Ratio Equilibrium for Economies with Private and Public Goods,' *Journal of Economic Theory;* 62, 499–512.

Diamond, Peter [1965]: 'National Debt in a Neoclassical Growth Model,' *American Economic Review;* 55, 1126–50.

Diamond, Peter [2003]: *Taxation, Incomplete Markets, and Social Security.* The MIT Press.

Diamond, Peter, and J. Mirrlees [1971]: 'Optimal Taxation and Public Production, I & II,' *American Economic Review;* 61; 8–27, 261–78.

Diewert, W. Erwin [1973]: 'Afriat and Revealed Preference Theory,' *Review of Economic Studies;* 40, 419–26.

Dodgson, C. L. [1876]: *A Method of Taking Votes on More than Two Issues.* Clarendon Press, Oxford.

Domencich, Thomas A., and Daniel McFadden [1975]: *Urban Travel Demand: A Behavioral Analysis.* North Holland/American Elsevier.

Duffie, Darrell [2001]: *Dynamic Asset Pricing Theory, 3rd ed.,* Princeton University Press.

Dutta, Bhaskar [2002]: 'Inequality, Poverty and Welfare,' Chapter 12, pp. 595–633, in: Arrow, Sen, and Suzumura [2002].

Eisenberg, E. [1961]: 'Aggregation of Utility Functions,' *Management Science;* 7, 337-50.

Ellickson, Bryan [1993]: *Competitive Equilibrium: Theory and Applications,* Cambridge University Press.

Fishburn, Peter C. [1970]: *Utility Theory for Decision Making.* New York: John Wiley & Sons.

Fishburn, Peter C. [1973]: *The Theory of Social Choice,* Princeton University Press.

Fishburn, Peter C. [1974]: 'Paradoxes of Voting,' *American Political Science Review;* 68, 537–46.

Fishburn, Peter C. [1984]: 'Discrete Mathematics in Voting and Group Choice,' *SIAM Journal of Algebraic and Discrete Methods;* 5, 263–75.

Foley, Duncan K. [1967]: 'Resource Allocation and the Public Sector,' *Yale Economic Essays;* 7,1; 45–98.

Foley, Duncan K. [1970]: 'Lindahl's Solution and the Core of an Economy with Public Goods,' *Econometrica;* 38; 66–72.

Fontaine, P., M Garbely, and M. Gilli [1991]: 'Qualitative Solvability in Economic Models,' *Computational Economics;* 4. 285–301.

Fostel, A., H. D. Scarf, and J. J. Todd [2004]: 'Two New Proofs of Afriat's Theorem,' *Economic Theory;* 24, 211–19.

Foster, James, and A. K. Sen [1997] 'On Economic Inequality after a Quarter Century, *in:* Sen, A. k¿, ed: *On Economic Inequality, 2nd ed* (Clarendon Press, Oxford), pp. 107–219.

Gale, David [1960]: 'A Note on Revealed Preference,' *Economica;* 27, 348–54.

Gale, David, and A. Mas-Collell [1975]: "An Equilibrium Existence Theorem for a General Model without Ordered Preferences," *Journal of Mathematical Economics;* 2, 9–15.

Geanakoplos, John D., and H. M. Polemarchakis [1991]: 'Overlapping Generations,' *in:* Hildenbrand, Werner, and H. Sonnenschein, eds: *Handbook of Mathematical Economics, Vol. IV* (Chapter 35, pp. 1899–1960). North-Holland.

Gibbard, Allan [1973]: 'Manipulation of Voting Schemes: A General Result,' *Econometrica;* 41, 587–601.

Goldman, Steven M. [1979]: 'Intertemporally Inconsistent Preferences and the Rate of Consumption,' *Econometrica;* 47, 621–26.

Goldman, Steven M. [1980]: 'Consistent Plans,' *Review of Economic Studies;* 47, 533-7.

Gorman, W. M. [1964]: 'More Scope for Qualitative Economics,' *Review of Economic Studies;* 31, 65–8.

Grandmont, Jean-Michel [1982]: 'Temporary General Equilibrium Theory,' *in:* Kenneth J. Arrow and M. D. Intriligator, eds: *Handbook of Mathematical Economics,*

Vol. II (pp. 879–922). North-Holland.

Green, Jerry R., and J-J. Laffont [1979]: *Incentives in Public Decision Making.* North Holland.

Groves, Theodore L. [1970]: *The Allocation of Resources Under Uncertainty.* Ph.D. dissertation, University of California, Berkeley.

Groves, Theodore [1973]: 'Incentives in Teams,' *Econometrica;* 41, 617–31.

Groves, Theodore [1976]: 'Information, Incentives, and the Internalization of Production Externalities,' *in:* Lin, S. A., edl: *Theory and Measurement of Economic Externalities.* Academic Press.

Groves, Theodore [1982]: 'On Theories of Incentive Compatible Choice with Compensation,' *in:* Hildenbrand, Werner, ed.: *Advances in Economic Theory* (pp. 1–29), Cambridge University Press.

Groves, Theodore, and J. O. Ledyard [1977]: 'Optimal Allocation of Public Goods: A Solution to the "Free Rider" Problem,' *Econometrica;* 45, 783–809.

Groves, Theodore, and J. O. Ledyard [1987]: 'Incentive Compatibility Since 1972,' *:* Theodore L. Groves, R. Radner, and S Reiter, eds.: *Information, Incentives, and Economic Mechanisms* (University of Minnesota Press, 1987), pp. 48–113.

Groves, Theodore, and M. Loeb [1975]: 'Incentives and Public Inputs,' *Journal of Public Economics;* 4, 211–26.

Hahn, Frank [1982]: 'Stability,' *in:* Kenneth J. Arrow and M. D. Intriligator, eds: *Handbook of Mathematical Economics, Vol. II* (pp. 745–93). North-Holland.

Heijdra, Ben, and F. van der Ploeg [2002]: *The Foundations of Modern Macroeconomics.* Oxford University Press.

Helfand, Gloria E., P. Berck, and T. Maull [2003]: 'The Theory of Pollution Policy,' *in*: Mäler, Karl-Göran, and J. R. Vincent [2003], pp. 249–303.

Hellwig, Martin F. [1986]: 'The Optimal Linear Income Tax Revisited,' *Journal of Public Economics;* 31, 163–70.

Hendricks, Ken, K. Judd, and D Kovenock [1980]: 'A Note on the Core of the Overlapping Generations Model,' Economics Letters; 6, 95–7.

Herrnstein, Richard J., and D. Prelec [1992]: 'Melioration,' *in:* Loewenstein, George, and J. Elster, eds.: *Choice over Time.* New York: Russell Sage Foundation (pp. 235–63).

Hey, John D., and C. Orme [1994]: 'Investigating Generalizations of Expected Utility Theory Using Experimental Data,' *Econometrica;* 62, 1291–326.

Hicks, John R. [1939]: 'The Foundations of Welfare Economics,' *Economic Journal;* 49, 696–712.

Hicks, John R. [1942]: 'Consumers' Surplus and Index Numbers,' *Review of Economic Studies;* 9, 126–37.

Hicks, John R. [1956]: *A Revision of Demand Theory.* Oxford, at the Clarendon Press.

Hildenbrand, Werner [1982]: 'Core of an Economy,' *in:* Arrow, Kenneth J., and M. Intriligator, eds.: *Handbook of Mathematical Economics, Vol II,* North-Holland (Chapter 18, pp. 831–77).

Hildenbrand, Werner, and A. P. Kirman [1988]: *Equilibrium Analysis.* North-Holland.

Houthakker, Hendrik S. [1950]: 'Revealed Preference and the Utility Function,' *Economica, N.S.;* 17, 159–174.

Houthakker, Hendrik S. [1965]: 'On the Logic of Preference and Choice,' Chapter 11, pp. 193–207, *in:* Tymieniecka, Ann-Teresa, and C. Parsons, eds.: *Contributions to Logic and Methodology in Honor of J. M., Bochenski.* North-Holland.

Hurwicz, Leonid [1960]: 'Optimality and Informational Efficiency in Resource Allocation Processes,' *in:* Arrow, Kenneth J., S. Karlin, and P. Suppes, eds.: *Mathematical Methods in the Social Sciences, 1959.* Stanford University Press (pp. 27–46).

Hurwicz, Leonid [1972]: 'On Informationally Decentralized Systems,' *in:* C. B. McGuire and R. Radner, eds.: *Decision and Organization: A Volume in Honor of Jacob Marschak* (pp. 297–336). North-Holland.

Hurwicz, Leonid [1979a]: 'Outcome Functions Yielding Walrasian and Lindahl Allocations at Nash Equilbrium Points,' *Review of Economic Studies;* 46, 217–25.

Hurwicz, Leonid [1979b]: 'On Allocations Attainable Through Nash Equilibria,' *Journal of Economic Theory;* 21, 140–65.

Hurwicz, Leonid [1986]: 'Incentive Aspects of Decentralization,' *:* Kenneth J. Arrow and M. D. Intriligator, eds: *Handbook of Mathematical Economics, v. III* (pp. 1441–82) North-Holland).

Hurwicz, Leonid [1999]: 'Revisiting Externalities,' *Journal of Public Economic Theory;* 1, 225–45.

Hurwicz, Leonid, and M. Majumdar [1988]: 'Optimal Intertemporal Allocation Mechanisms and Decentralization of Decisions,' *Journal of Economic Theory;* 45, 228– 61.

Hurwicz, Leonid, and H. Uzawa [1971]: 'On the Integrability of Demand Functions,' *in:* Chipman, *et al* [1971], pp. 114–48.

Jackson, Matthew O. [1992]: 'Implementation in Undominated Strategies: A Look at Bounded Mechanisms,' *Review of Economic Studies;* 59, 757–75.

Jackson, Matthew O. [2001]: 'A Crash Course in Implementation Theory,' *Social Choice and Welfare;* 18, 655–708.

Jackson, Matthew O., and H. Moulin [1992]: 'Implementing a Public Project and Distributing its Cost,' *Journal of Economic Theory;* 57, 125–40.

Kalai, Ehud, and J. O. Ledyard [1998]: 'Repeated Implementation,' *Journal of Economic Theory;* 83, 308–17.

Kaldor, Nicholas [1939]: 'Welfare Propositions in Economics and Interpersonal Comparisons of Utility,' *Economic Journal;* 49, 549–52.

Kaneko, Mamoru [1977]: 'The Ratio Equilibrium and a Voting Game in a Public Goods Eonomy,' *Journal of Economic Theory;* 16, 123–36.

Katzner, Donald W. [1970]: *Static Demand Theory.* London: The MacMillan Co.

Kelly, Jerry S. [1988]: *Social Choice Theory: An Introduction.* Springer-Verlag.

Kihlstrom, Richard, A. Mas-Colell, and H. Sonnenschein [1976]: 'The Demand Theory of the Weak Axiom of Revealed Preference,' *Econometrica;* 44, 5; 971–78.

Kim, Taesung [1987]: 'Intransitive Indifference and Revealed Preference,' *Econometrica;* 55, 1; 163–7.

Kim, Taesung, and M. K. Richter [1986]: 'Nontransitive-Nontotal Consumer Theory,' *Journal of Economic Theory;* 38, 324–63.

Koopmans, Tjalling C. [1957]: *Three Essays on the State of Economic Science.* McGraw-Hill. [Re-published by Yale University Press.]

Koopmans, Tjalling C. [1972a]: 'Representation of Preference Orderings with Independent Componenets of Consumption,' *in:* McGuire, C. E., and R. Radner [1972]; pp. 54–78.

Koopmans, Tjalling C. [1972b]: 'Representation of Preference Orderings over Time,' *in:* McGuire, C. E., and R. Radner [1972]; pp.79–100.

Koopmans, Tjalling C., and A. F. Bausch [1959]: 'Selected Topics in Economics Involving Mathematical Reasoning,' *SIAM Review;* 1. 79–128.

Kovenock, Dan [1984]: 'A Second Note on the Core of the Overlapping Generations Model,' *Economics Letters;* 14, 101–6.

Kreps, David M. [1990]: *A Course in Microeconomic Theory.* Princeton University Press.

Laffont, Jean-Jacques, and D. Martimort [2002]: *The Theory of Incentives.* Princeton University Press.

Lancaster, Kelvin J. [1962]: 'The Scope of Qualitative Economics, *Review of Economic Studies;* 29, 99–123.

Lang, K., J. C. Moore, and A B. Whinston [1995]: 'Computational Systems for Qualitative Economics,' *Computational Economics;* 8, 1–26.

LeBreton, Michel, and A. Sen [1999]: 'Separable Preferences, Strategy-Proofness, and Decomposability,' *Econometrica;* 67, 605–28.

Ledyard, John O. [1995]: 'Public Goods: A Survey of Experimental Research,' *in:* Kagel, John H., and A. E. Roth, eds.: *The Handbook of Experimental Economics* (pp. 111–94) Princeton University Press.

Ledyard, John O., and T. R. Palfrey [2002]: 'The Approximation of Efficient Public Good Mechanisms by Simple Voting Schemes,' *Journal of Public Economics;* 83, 153–71.

Leach, John [2004]: *A Course in Public Economics.* Cambridge University Press.

Lindahl, Erik [1919]: *Die Gerechtigkeit der Besteuerung.* Translated (in part) as 'Just Taxation: A Positive Solution,' *in:* R. A. Musgrave and A. T. Peacock, eds. [1967]: *Classics in the Theory of Public Finance* (pp. 168–76). London, Macmillan.

Loewenstein, George, and D. Prelec [1992]: 'Anomalies in Intertemporal Choice: Evidence and an Interpretation,' *Quarterly Journal of Economics;* 52, 573–97.

Loomes, Graham, C. Starmer, and R. Sugden [1991]: 'Observing Violations of Transitivity by Experimental Methods,' *Econometrica;* 59, 425–39.

Luce, R. Duncan [1956]: 'Semiorders and a Theory of Utility Discrimination,' *Econometrica;* 24, 178–91.

Luce, R. Duncan, and P. Suppes [1965]: 'Preference, Utility, and Subjective Probability,' Chapter 19, pp. 149–410, *in:* Luce, R. Duncan, R. R. Bush, and E. Galanter, eds.: *Handbook of Mathematical Psychology,* Vol. III. New York: John Wiley & Sons.

Machina, Mark J. [1987]: 'Choice Under Uncertainty: Problems Solved and Unsolved,' *Journal of Economic Perspectives;* 1, 121–54.

Magill, Michael, and M. Quinzii [1996]: *Theory of Incomplete Markets, Vol 1*, MIT Press.

Magill, Michael, and W. Shafer [1991]: 'Incomplete Markets,' *in:* W. Hildenbrand and H. Sonnenschein, eds: *Handbook of Mathematical Economics, Vol. IV* (pp. 1523–1614). North-Holland.

Mak, King-Tim [1984]: 'Notes on Separable Preferences,' *Journal of Economic Theory;* 33, 2; 309–21.

Mak, King-Tim [1986]: 'On Separability: Functional Structure,' *Journal of Economic Theory;* 40, 250–82.

Mäler, Karl-Göran, and J. R. Vincent [2003]: *Handbook of Environmental Economics, V. 1.* North-Holland/ Elsevier.

Mangasarian, Olvi L. [1969]: *Nonlinear Programming.* McGraw-Hill.

Mas-Colell, Andreu [1977]: 'The Recoverability of Consumers' Preferences from Market Demand Behavior,' *Econometrica;* 45, 6; 1409–30.

Mas-Colell, Andreu [1978]: 'On Revealed Preference Analysis,' *Review of Economic Studies;* 45, 121–31.

Mas-Colell, Andreu, and J. Silvestre [1989]: 'Cost Share Equilibria: A Lindahlian Approach,' *Journal of Economic Theory;* 47, 239–56.

Mas-Colell, Andreu, M. Whinston, and J. Green [1995]: *Microeconomic Theory.* Oxford University Press.

Maskin, Eric [1999]: 'Nash Equilibrium and Welfare Optimality,' *Review of Economic Studies;* 66, 23–38.

Maskin, Eric, and T. Sjöström [2002]: 'Implementation Theory,' *in:* K. J. Arrow, A. K. Sen, and K. Suzumura, eds: *Handbook of Social Choice and Welfare, Vol. 1* (pp. 237–288). North-Holland.

Matzkin, Rosa L., and M. K. Richter [1991]: 'Testing Strictly Concave Rationality,' *Journal of Economic Theory;* 53, 287–303.

May, Kenneth O. [1952]: 'A Set of Independent Necessary and Sufficient Conditions for Simple Majority Decision,' *Econometrica;* 20, 680-4.

May, Kenneth O. [1953]: 'A Note on the Complete Independence of the Conditions for Simple Majority Decision,' *Econometrica;* 21, 172-3.

May, Kenneth O. [1954]: 'Intransitivity, Utility, and the Aggregation of Preference Patterns,' *Econometrica;* 22, 1–13.

McFadden, Daniel [2001]: 'Economic Choices,' *American Economic Review;* 91, 3; 351–78.

McGuire, C. B., and Roy Radner, eds. [1972]: *Decision and Organization.* North-Holland.

McKenzie, Lionel W. [1954]: 'On Equilibrium in Graham's Model of World Trade and other Competitive Systems,' *Econometrica;* 22, 147–61.

McKenzie, Lionel W. [1959]: 'On the Existence of General Equilibrium for a Competitive Market,' *Econometrica;* 27, 54–71.

McKenzie, Lionel W. [1961]: 'On the Existence of General Equilibrium: Some Corrections,' *Econometrica;* 29, 247–48.

McKenzie, Lionel W. [1981]: 'The Classical Theorem on Existence of Competitive Equilibrium,' *Econometrica;* 49, 819–41.

McKenzie, Lionel W. [1988]: 'A Limit Theorem on the Core,' *Economics Letters;* 27, 7–9.

McKenzie, Lionel W. [2002]: *Classical General Equilibrium Theory.* Cambridge, MA, and London: The MIT Press.

Milgrom, Paul, and J. Roberts [1994]: 'Comparing Equilibria,' *American Economic Review;* 84, 441–59.

Mirrlees, James A. [1971]: 'An Exploration in the Theory of Optimum Income Taxation,' *Review of Economic Studies;* 38, 175–208.

Mirrless, James A. [1986]: 'The Theory of Optimal Taxation,' *in:* Arrow, Kenneth J., and M. D. Intriligator, eds: *Handbook of Mathematical Economics, Vol III,* North-Holland, Chapter 24, pp. 1197–1249.

Moore, James C. [1970]: 'On Pareto Optima and Competitive Equilibria: Part I, Relationships Among Equilibria and Optima,' *Krannert Graduate School of Industrial Administration, Purdue University: Institute Paper No. 268;* April, 1970.

Moore, James C. [1973]: 'Pareto Optimal Allocations as Competitive Equilibria,' *Krannert Graduate School of Industrial Administration, Purdue University: Institute Paper No. 386;*, January, 1973.

Moore, James C. [1975]: 'The Existence of 'Compensated Equilibrium' and the Structure of the Pareto Efficiency Frontier,' *International Economic Review;* 16, 267–300.

Moore, James C. [1999]: *Mathematical Methods for Economic Theory, Vol. 1.* Springer-Verlag.

Moore James C. [2002] 'Real national income and some principles of aggregation,' Manuscript, Purdue University, October, 2002.

Moore, John [1992]: 'Implementation, Contracts, and Renegotiation in Environments with Complete Information,' *in:* Laffont, Jean-Jacques, ed.: *Advances in Economic Theory: Sixth World Congress, v. 1* (Cambridge University press), Chapter 5, pp. 182–282.

Moore, John, and R. Repullo [1988]: 'Subgame Perfect Implementation,' *Econometrica;* 56, 1191–1220.

Moore, John, and R. Repullo [1990]: 'Nash Implementation: A Full Characterization,' *Econometrica;* 58, 1083–99.

Moulin, Hervé [1980]: 'On Strategy-Proofness and Single Peakedness,' *Public Choice;* 35, 437–55.

Moulin, Hervé [1984]: 'Generalized Condorcet Winners for Single-Peaked and Single Plateau Preferences,' *Social Choice and Welfare;* 1, 127–47.

Moulin, Hervé [1988]: *Axioms of Cooperative Decision Making.* Cambridge University Press.

Myles, Gareth [1995]: *Public Economics.* Cambridge University Press.

Nanson, E. J. [1882]: 'Methods of Election,' *Transactions and Proceedings of the Royal Society of Victoria;* 19, 197-240.

Nash, John [1950]: 'The bargaining problem,' *Econometrica;* 18, 155–62.

Negishi, Takashi [1962]: 'The Stability of a Competitive Economy: A Survey Article,' *Econometrica;* 30, 635–69.

Nikaido, Hukukane [1968]: *Convex Structures and Economic Theory.* Academic Press.

Pareto, Vilfredo [1894]: 'Il massimo di utilità dato dalla libera concorrenza,' *Giornale degli Economisti;* 2. 9; 48–66.

Phelps, Edmund S., and R. A. Pollak [1968]: 'On Second-Best National Saving and Game-Equilibrium Growth.' *Review of Economic Studies;* 35, 185–99.

Pigou, A. C. [1932]: *The Economics of Welfare,* 4th ed. London, Macmillan.

Plott, Charles R. [1973]: 'Path Independence, Rationality and Social Choice,' *Econometrica;* 41, 1075–91.

Quirk, James, and R. Saposnik [1968]: *Introduction to General Equilibrium Theory and Welfare Economics.* McGraw-Hill.

Rabin, Matthew [1998]: 'Psychology and Economics,' *Journal of Economics Literature;* 36, 11–46.

Rabin, Matthew [2002]: 'A Perspective on Psychology and Economics,' *European Economic Review;* 46, 657–85.

Ramsey, Frank [1927]: 'A Contribution to the Theory of Taxation,' *Economic Journal;* 37, 47–61.

Rawls, J. [1971]: *A Theory of Justice.* Cambridge, MA, Belknap.

Repullo, Rafael [1985]: 'Implementation in Dominant Strategies under Complete and Incomplete Information,' *Review of Economic Studies;* 52, 223–9.

Repullo, Rafael [1988]: 'A Simple Proof of Maskin's Theorem on Nash Implementation,' *Social Choice and Welfare;* 4, 39–41.

Richter, Marcel K. [1966]: 'Revealed Preference Theory,' *Econometrica;* 34, 3; 635- 645.

Richter, Marcel K. [1971]: 'Rational Choice,' Chapter 2, pp. 29–58, in Chipman, Hurwicz, and Sonnenschein [1971].

Roy, René [1942]: *De l'utilité: Contribution à la théorie des Choix.* Paris: Hermann & Cie.

Saari, Donald G. [1996]: 'Election Relations and a Partial Ordering for Positional Voting,' *in:* Schofield, Norman, ed.: *Collective Decision-Making: Social Choice and Political Economiy.* Kluwer Academic Publishers (Chapter 5, pp. 93–110.)

Samuelson, Paul A. [1938]: 'A Note on the Pure Theory of Consumer's Behavior,' *Economica, N. S.;* 5, 61–71, 353–4.

Samuelson, Paul A. [1942]: 'Constancy of the Marginal Utility of Income,' *in:* Lange, Oscar, F. McIntyre, and T. O. Yntema, eds: *Studies in Mathematical Economics and Econometrics* (University of Chicago Press), pp. 75–91.

Samuelson, Paul A. [1947]: *Foundations of Economic Analysis.* Harvard University Press.

Samuelson, Paul A. [1948]: 'Consumption Theory in Terms of Revealed Preference,' *Economica, N. S.;* 15, 243–53.

Samuelson, Paul A. [1950]: 'Evaluation of Real National Income,' *Oxford Economic Papers;* N. S., 2, 1–29.

Samuelson, Paul A. [1958]: 'An Exact Consumption-Loan Model of Interest with or without the Social Contrivance of Money,' *Journal of Political Economy;* 66, 467–82.

Satterthwaite, Mark A. [1975]: 'Strategy-Proofness and Arrow's Conditions: Existence and Correspondence Theorems for Voting Procedures and Social Welfare Functions,' *Journal of Economic Theory;* 10, 187–217.

Scarf, Herbert [1960]: 'Some Examples of Global Instability of the Competitive Equilibrium,' *International Economic Review;* 1, 157–72.

Scotchmer, Suzanne [2002]: 'Local Public Goods and Clubs,' *in:* A. J. Auerbach and M. Feldstein, eds.: *Handbook of Public Economics, Vol. 4* (pp. 1997–2941). North-Holland.

Sen, Amartya K. [1986]: 'Social Choice Theory,' Chapter 23, pp. 1073 - 1181, *in:* Arrow, Kenneth J., and M. D. Intriligator, eds.: *Handbook of Mathematical Economics, vol. III.* North-Holland.

Shafer, Wayne J. [1974]: 'The Nontransitive Consumer,' *Econometrica;* 42, 5; 913–19.

Shafer, Wayne J. [1976]: "Equilibrium in Economies without Ordered Preferences or Free Disposal," *Journal of Mathematical Economics;* 3, 135–37.

Shafer, Wayne J., and H. F. Sonnenschein [1975]: "Equilibrium in Abstract Economies without Ordered Preferences," *Journal of Mathematical Economics;* 2, 345–48.

Shafer, Wayne, and H. Sonnenschein [1982]: 'Market Demand and Excess Demand Functions,' *in:* Arrow, Kenneth J., and M. D. Intriligator, eds.: *Handbook of Mathematical Economics, II* (Chapter 14, pp. 671–93). North-Holland/American Elsevier.

Sonnenschein, Hugo F. [1973]: 'Do Walras' Identity and Continuity Characterize the Class of Community Excess Demand Functions?' *Journal of Economic Theory;* 6, 345–54.

Sonnenschein, Hugo F. [1974]: 'Market Excess Demand Functions,' *Econometrica;* 40, 549–63.

Starmer, Chris [1996]: 'Explaining Risky Choices Without Assuming Preferences,' *Social Choice and Welfare;* 13, 201–13.

Starmer, Chris [2000]: 'Developments in Non-Expected Utility Theory: The Hunt for a Descriptive Theory of Choice under Risk,' *Journal of Economic Literature;* 38, 332–82.

Starrett, David A. [2003]: 'Property Rights, Public Goods and the Environment,' *in:* Mäler, Karl-Göran, and J. R. Vincent [2003], pp. 97–125.

Stavins, Robert N. [2003]: 'Experience with Market-Based Environmental Policy Instruments,' *in:* Mäler, Karl-Göran, and J. R. Vincent [2003], pp. 355–435.

Stiglitz, Joseph E [1987]: 'Pareto Efficient and Optimal Taxation and the New New Welfare Economics,' *in:* Auerbach, Alan J., and M. Feldstein, eds: *Handbook of Public Economics, Vol II;* North-Holland, Chapter 15, pp. 991–1042.

Strawczynski, Michel [1998]: 'Social Insurance and the Optimum Piecewise Linear Income Tax,' *Journal of Public Economics;* 69, 371–88.

Strotz, Robert H. [1955]: 'Myopia and Inconsistency in Dynamic Utility Maximization,' *Review of Economic Studies;* 23, 165–80.

Suzumura, Kotaro [2002]: 'Introduction,' *in:* Arrow, Kenneth J., A. K. Sen, and K. Suzumura, eds.: *Handbook of Social Choice and Welfare, Volume 1,* pp. 1–32, North Holland/Elsevier.

Takayama, Akira [1985]: *Mathematical Economics,* 2nd Ed. Cambridge University Press.

Tian, Guoqiang [2000]: 'Double Implementation of Linear Cost Share Equilibrium Allocations,' *Mathematical Social Sciences;* 40, 175–90.

Tian, Guoqiang, and Q. Li [1994]: 'An Implementable State-Ownership System with General Variable Returns,' *Journal of Economic Theory;* 64, 286–97.

Train, Kenneth [1986]: *Qualitative Choice Analysis: Theory, Econometrics, and an Application to Automobile Demand.* The M. I. T. Press.

Tversky, Amos [1969]: 'Intransitivity of Preferences,' *Psychological Review;* 75, 31–48.

Tversky, Amos, and D. Kahneman [1988]: 'Rational Choice and the Framing of Decisions,' pp. 167–92, *in:* Bell, Raiffa, and Tversky [1988a].

Tversky, Amos, P. Slovic, and D. Kahneman [1990]: 'The Causes of Preference Reversal,' *American Economic Review;* 80, 204–17.

Tversky, Amos, and R. H. Thaler [1990]: 'Anomalies: Preference Reversal,' *Journal of Economic Perspectives;* 4, 201–11.

van den Nouweland, Anne, S. Tijs, and M. H. Wooders [2002]: 'Axiomatization of Ratio Equilibria in Public Good Economies,' *Social Choice and Welfare;* 19, 627–36.

Varian, Hal R. [1982]: 'The Non-Parametric Approach to Demand Analysis,' *Econometrica;* 50, 945–74.

Vickrey, William [1961]: 'Counterspeculation, Auctions, and Competitive Sealed Tenders,' *Journal of Fionance;*, 16, 8–37.

Walker, Mark [1981]: 'A Simple Incentive Compatible Scheme for Attaining Lindahl Allocations,' *Econometrica;* 49, 65–71.

Wald, Abraham [1936]: Über einige Gleichungssysteme der Mathematischen Ökonomie," *Zeitschrift für Nationalökonomie;* 7, 637–70. Translated as: 'On Some Systems of Equations of Mathematical Economics,' *Econometrica;* 19 [1951], 368–403.

Williams, Steven R. [1986]: 'Realization and Nash Implementation: Two Aspects of Mechanism Design,' *Econometrica;* 54, 139–52.

Wold, Herman [1943]: 'A Synthesis of Pure Demand Analysis, Parts I and II,' *Skandinavisk Aktuarietidskrift;* 26, 85–118, & 220–63.

Wold, Herman [1944]: 'A Synthesis of Pure Demand Analysis, Part III,' *Skandinavisk Aktuarietidskrift;* 27, 69–120.

Yannelis, Nicholas C. [1987]: 'Equilibra in Noncooperative Models of Competition,' *Journal of Economic Theory;* 41, 96–111.

Author Index

Afriat, S. 75, 76
Aivazian, V. 480
Allen, R. 101
Anderson, R. 320
Antonelli, G. 101
Armstrong, W. 50
Arrow, K. 63, 150, 212, 293, 302, 305, 306, 309, 320, 336, 339, 384, 393, 395, 400, 416, 527
Auerbach, A. 259
Aumann, R. 320, 372

Ballinger, T. 46
Barbera, S. 497, 508, 527
Barone, E. 150
Bass, F. 29
Bassett, L. 294
Baumol, W. 480
Bentham, J. 383
Berliant, M. 482
Berck, P. 480
Bergson, A. 415, 407
Bergstrom, T. 238, 445
Black, D. 393
Blackorby, C. 33, 37, 432
Blau, J. 393, 395
Block, H. 306
Blume, L. 445
de Borda, J-C. 383

Callen, J. 480
Campbell, D. 483
Chipman, J. 63, 101, 111, 115, 118, 121, 141, 417, 418, 419, 422, 428, 429
Clarke, E. 514
Coase, R. 482
Condorcet, M. 383, 387
Coombs, C. 46
Corchon, L. 452, 522

Dasgupta, P. 526
Davidson, P. 33
Dawes, R. 46
Debreu, G.15, 17, 27, 96, 102, 141, 174, 194, 212, 227, 232, 320, 322, 336
Diamantaras, D. 452
Diamond, P. 369, 257
Diewert, E. 75
Dodgson, C. 383
Donaldson, D. 432
Duffie, D. 343, 355
Dutta, B. 425

Eisenberg, E. 412
Ellickson, B. 380
Fishburn, P. 390, 391
Foley, D. 449, 451, 459
Fontaine, P. 295,
Fostel, A. 75
Foster, J. 425

Gale, D. 71, 201, 227
Garbely, M. 295
Geanakoplos, J. 370, 370
Gibbard, A. 489, 490, 495, 497
Gilli, M. 295
Goldman, S. 44, 44
Gorman, W. 294,
Grandmont, J-M. 380
Green, J. 27, 283, 286, 293, 334, 353, 380, 468, 515, 519
Groves, T. 514, 517, 520, 527

Hahn, F. 302, 309, 320
Hammond, P. 526
Heijdra, B. 380
Helfand, G. 480
Hellwig, M. 268
Herrnstein, R. 44
Hey, J. 46

Hicks, J. 119, 286, 416
Hildenbrand, W 319, 320
Hines, J. 259
Houthakker, H. 71
Hurwicz, L. 150, 206, 293, 302, 305, 306, 368, 455, 480, 489, 490, 527

Kahneman, D. 27
Kalai, E. 527
Kaldor, N. 416
Kanecko, M. 452, 453
Katzner, C. 33
Kelly, J. 385, 387, 389
Kihlstrom, R. 71
Kim, T. 77
Kirman, A. 319
Koopmans, T. 44, 196, 212

Laffont, J-J. 515, 519
Lancaster, K. 294
Lang, R. 295,
Leach, J. 487
Le Breton, M. 37
Ledyard, J. 446, 527, 527
Lehmann, D. 29
Leontief, W. 166
Li, Q. 522
Lindahl, E. 446
Loeb, M. 514
Loewenstein, G. 44, 44
Loomes, G. 27, 53
Luce, R. 50

Machina, M. 27
Magill, M. 355, 380
Majumdar, M. 364, 368
Mak, K-T. 33
Mangasarian, O. 186
Mas-Colell, A. 27, 71, 201, 227, 283, 286, 293, 334, 353, 380, 452, 453, 468
Maskin, E. 521, 526
Matzkin, R. 76
Maull, T. 480
May, K. 51, 386, 387
Maybee, J. 294
McFadden, D. 46 49
McKenzie, L. 284, 285, 302, 323
Milgrom, P. 298
Mirrlees, J. 259, 259
Moore, James 63, 97, 101, 111, 115, 118, 121, 141, 295, 418, 419, 422, 428, 429
Moore, John 521, 527
Moulin, H. 391, 405, 412, 414, 425, 435, 506, 508, 518, 519, 527
Myles, G. 257, 425

Nanson, E. 383
Nash, J. 412
Negishi, T. 302, 309
Nikaido, N. 170, 178, 184, 186, 323

Oates, W. 480
Orme, C. 46

Pareto, V. 407
Peleg, B. 497
Pessemier, E. 29
Phelps, E. 44
Pigou, A¿ 482
Plott, C. 401
Polemarchakis, H. 370, 380
Pollak, R. 44
Prelec, D. 44, 44
Primont, D. 33

Quinzii, M. 355, 380
Quirk, J. 294, 295, 298, 302, 309

Rabin, M. 43, 44
Radner, R. 339
Ramsey, F. 258
Repullo, R. 521, 526
Richter, M. 59, 65, 67, 68, 69, 71, 72, 73, 76, 77, 230
Roberts, J. 298
Roy, R. 101
Russell, R. 33, 37

Saari, D. 392
Samuelson, P. 48, 70, 174, 282, 294, 407, 417, 443
Saposnik, R. 295, 298, 302, 309
Satterthwaite, M. 489, 489
Scarf, H. 75, 306, 320, 322
Scotchmer, S. 372
Sen, A. 37, 397, 425, 527
Shafer, W. 141, 238, 355
Silvestre, J. 453

Sonnenschein, H. 71, 141, 527
Starmer, C. 27, 46, 53
Starrett, D. 487
Stavins, R. 480
Stiglitz, J. 259
Strawczynski, M. 268
Strotz, R. 44
Sugden, R. 27, 53
Suzumura, K. 384, 527

Takayama, A. 302
Thaler, R. 27
Tian, G. 452, 522
Todd, J. 75
Train, K. 49
Tversky, A. 27, 46, 52

van den Nouweland, A. 452
van der Ploeg, F. 380
Varian, H. 75, 445
Vickrey, W. 527

Walker, M. 527
Whinston, A 295,
Whinston, M. 27, 283, 286, 293, 334, 353, 380, 468
Wilcox, N. 46
Wilkie, S. 452, 522
Williams, S. 521
Wold, H. 96, 409
Wooders, M. 452

Zhou, L. 527

Subject Index

absolute majority voting rule, 386
additive separability, 40
Afriat's Theorem, 75
agenda ordering, 388
agent monotonicity, 264
aggregate production set, 178
Antonelli-Allen-Roy Condition, 101, 430
Arrovian social preference function, 393
Arrow Condition, 399
Arrow Securities, 341
Arrow-Debreu model, 333
assets
  complete asset structure, 345
  in the Radner model, 340
  portfolio, 340
  return vector, 340
  risk-free, 344
  structure, 340
asymmetric order, 55
asymmetric relations
  uppersemicontinuous, 55
  lower semicontinuous, 55
  continuous, 55
  strongly continuous, 55

binary preference probability, 45
binary relations
  acyclic, 78
  antisymmetric, 2
  asymmetric, 2
  asymmetric part of, 5
  cyclic, 78
  homothetic, 104
  irreflexive, 2
  negation of, 7
  reflexive, 2
  representable, 12
  symmetric, 2
  symmetric part of, 5
  total, 2
  transitive, 2
blocking coalition, 317
Borda count, 390
budget balance condition, 91
budget correspondence, 88
budget space, 60

choice correspondence, 60
  competitive, 68
  irrational, 60
  rational, 60
  regular rational, 60
  reflexive rational, 60
  representable, 68
  transitive rational, 60
Coase Theorem, 482
Cobb-Douglas-Eisenberg (CDE) aggregator function, 412
compensating variation, 119
Compensation Principle, The, 416
competitive (or Walrasian) equilibrium,
  for a continuum of traders, 378
  for an exchange economy, 316
  for a production economy, 193
  given a level of government activity, 252
conditional preference relation, 34
Condorcet-consistent voting rule, 391
Condorcet winner, 383
congruence axiom, 65
consumers' competitive equilibrium, 420
consumer's surplus, 111
consumers' surplus, 428
continuum of traders model, 378
contour sets, 11
convex cone, 160

Copeland voting rule, 405
core of a competitive economy, 317
core of a public goods economy, 450
cost-of-living index, 109
cost share equilibrium, 453

demand correspondence
  aggregate (exchange economy), 138
  generated by P, 88
dictator
  for a social preference function, 394
  for a game form, 495
  dictatorial voting rule, 496
direct revelation mechanism, 504
disposability, 159
distribution of ownership (for an economy), 194
dominant strategy, 494
dual cone, 177

efficient production, 171
egalitarian social welfare function, 424
equivalence relation, 6
equivalent variation, 124
excess demand correspondence
  for an individual consumer, 138
  aggregate (exchange economy), 138
  aggregate (production economy), 203
excludability, 441
extends a binary relation, 62
externality, definition of, 467

feasible allocations
  for an exchange economy, 132
  for a production economy, 193
  given a level of government, 252
feasible program, 363
feasible wealth assignment function, 201
First Fundamental Theorem of Welfare Economics ('Non-Wastefulness')
  for exchange economies, 150, 150
  for production economies, 208, 209
  for public goods economies, 447, 448, 457
  with externalities, 485
forward market, 338
function
  concave, 94
  convex, 94
  homothetic, 105
  quasi-concave, 94
  quasi-convex, 94
  semi-concave, 95
  strictly quasi-concave, 95
  strictly quasi-convex, 95

Gale/Mas-Colell Existence Theorem, 239
game form, definition of, 490
General Possibility Theorem (Arrow), 400
GARP (generalized Axiom of Revealed Preference), 75
Gibbard-Satterthwaite Theorem, 495, 497
gross substitutes, 292

homothetic relation, 104
  homothetic indirect preferences, 109

implementation, 503
income compensation function, 103
income distribution-adjusted real nat'l. income, 426
income distribution condition, 289
income distribution function, 288
income distribution index, 424
Independence of Irrelevant Alternatives (IIA), 394
indirect preference relation, 98
indirect social preference function, 422
indirect social preference relation, 421
indirect utility function, 99
individually rational allocations
input requirement correspondence, 161
irreducibility, 219

Kaldor-Hicks-Samuelson (KHS) ordering
  basic definition, 417
  extended KHS ordering, 419

Law of Demand, 287
lexicographic order, 14
Lindahl equilibrium
  for the simple public goods model, 447
  for the general model, 457
  for the private ownership economy, 463
  with externalities, 485
linear production set, 161
lump-sum transfers, 195

May's Theorem, 387
(utility) measurement function, 411
mechanism, general definition, 490
Minkowski-Farkas Lemma, 186
monotonic social choice corresp., 520
motivates a choice correspondence, 77

Nash implementation, 520
negatively transitive binary relation, 7
no veto power, 521
numéraire good, 92
  for a public goods economy, 458

option, (European) call, 354

Pareto-Bergson-Samuelson (PBS) social welfare function, 411
Pareto efficiency
  Pareto efficient allocations (exchange economy), 144
  strongly Pareto efficient (exchange economy), 144
  Pareto efficient allocations, 206
  strongly Pareto efficient, 206
Pigouvian tax, 482, 477
pivot (or pivotal) mechanism, 514
plurality voting rule, 389
polyhedral cone, 187
Positive Association Principle, 498
preference profile, 496
preference relations
  convex, 94
  increasing, 91
  locally non-saturating, 91
  non-decreasing, 91
  non-saturating, 91
  single-peaked, 504
  strictly convex, 94
  strictly increasing, 91
  weakly convex, 94, 212
price adjustment mechanism, 303
private ownership economy, 194
production possibility corresp., 160
profit function, 173
proper linear technology, 166
public goods (basic definition), 441

quasi-competitive equilibrium, 207
quasi order, 56

Radner equilibrium, 341
ratio equilibrium, 453
Rawlsian aggregator function, 412
relatively closed sets, 16
returns to scale, 156
revealed preferred, 62
Revelation Principle, 496, 526
risk
  private, 338
  social, 338
rivalrous goods, 441

Samuelson condition, 443
SARP (Strong Axiom of Revealed Preference), 71
Second Fundamental Theorem of Welfare Economics ('Unbiasedness')
  for private goods production, 213, 217, 220, 222
  for public goods economies, 459
semi-order, 56
separating hyperplane theorem, 172
sign-preserving function, 301
simple majority voting, 386
Simpson voting rule, 405
social choice rule, 401
social preference function, 411
Stability
  global, 303
  local, 304
  Marshallian, 299
  system, 304
  Walrasian, 299
staging procedure, 388
state price vector, 343
Stiemke's Theorem, 184
straightforward game form, 494
spot market, 338
strict Pareto ordering, 142

transfer mechanism, 513
transitive closure, 66
Tucker's Theorem, 185

unanimity (strong Pareto) ordering, 142
unit simplex, 94

utility function, 412
  represents a binary relation, 12
  represents an asymmetric relation, 13
utilitarian aggregator function, 412
utility measurement function, 409

V axiom, 62
vector inequality (for $\mathbb{R}^n$), 3
Vickrey-Clarke-Groves (VCG) mechanism, 515
voting rule, definition of, 385
  manipulable, 496
  strategy-proof, 496

Walras' Law
  weak form, 203
  strong form, 204
  original form, 204
WARP (Weak Axiom of Revealed Preference), 71, 283
Weak* Axiom (WA*), for aggregate excess demand, 283
weak order, 4
  continuous, 16
  lower semi-continuous, 16
  upper semi-continuous, 16
weak Pareto ordering, 142
Weak Pareto Principle (WPP), 394
weak separability, 35
wealth assignment (for an economy), 193
Wold measurement function, 410
Wold representation theorem, 96